Fundamental Electromagnetic Theory and Applications

Fundamental Electromagnetic Theory and Applications

Ronold W. P. King

Harvard University
Cambridge, Massachusetts

Sheila Prasad

Northeastern University
Boston, Massachusetts

Prentice-Hall, Inc., Englewood Cliffs, N.J. 07632

Library of Congress Cataloging in Publication Data

KING, RONOLD W. P., (date)
Fundamental electromagnetic theory and applications.

Bibliography
Includes indexes.
1. Electromagnetic theory.   I. Prasad, Sheila.
II. Title.
QC670.K48 1986   530.1′41   85-9360
ISBN 0-13-336959-5

Portions of this book have been previously published under the title
of *Fundamental Electromagnetic Theory.*

Editorial/production and supervision
and interior design: *Theresa A. Soler*
Cover design: *20/20 Services, Inc.*
Manufacturing buyer: *Rhett Conklin*

Printed in the United States of America

10 9 8 7 6 5 4 3 2 1

ISBN 0-13-336959-5 01

Prentice-Hall International (UK) Limited, *London*
Prentice-Hall of Australia Pty. Limited, *Sydney*
Prentice-Hall Canada Inc., *Toronto*
Prentice-Hall Hispanoamericana, S.A., *Mexico*
Prentice-Hall of India Private Limited, *New Delhi*
Prentice-Hall of Japan, Inc., *Tokyo*
Prentice-Hall of Southeast Asia Pte. Ltd., *Singapore*
Editora Prentice-Hall do Brasil, Ltda., *Rio de Janeiro*
Whitehall Books Limited, *Wellington, New Zealand*

"I have therefore thought that a treatise would be useful which should have for its principal object to take up the whole subject in a methodical manner, and which should also indicate how each part of the subject is brought within the reach of methods of verification by actual measurement."

James Clerk Maxwell

Contents

VECTOR AND SCALAR POTENTIAL FUNCTIONS   *149*

ELECTROMAGNETIC FORCE AND ENERGY   192

ELECTROMAGNETIC WAVES IN UNBOUNDED REGIONS   220

6

GENERAL THEOREMS OF ELECTROMAGNETIC THEORY AND THEIR APPLICATIONS

244

7

SCATTERING AND DIFFRACTION OF PLANE WAVES

263

10 TRANSMISSION-LINE THEORY   *395*

13 WAVES ALONG DIELECTRIC RODS: OPTICAL FIBER TRANSMISSION

14 WAVES AND ANTENNAS NEAR AND ACROSS THE BOUNDARY BETWEEN ELECTRICALLY DIFFERENT HALF-SPACES

Appendix

I

**DIFFERENTIAL OPERATORS; VECTOR FORMULAS AND
IDENTITIES**   *553*

Appendix

II

TABLES OF FUNCTIONS $f(h)$ AND $g(h)$ *566*

Appendix

III

TABLES OF BESSEL FUNCTIONS   *575*

Appendix

IV

MATERIAL CONSTANTS   *580*

Appendix

V

DIMENSIONS AND UNITS   *584*

Appendix

Preface

Electromagnetism is one of the cornerstones of classical physics. For the student it is usually the first introduction to a fully developed field theory. The concept of a vector field (the electromagnetic field) with its sources (positive and negative electric charges) and their mutual interactions is new and difficult. The fact that such interactions require a finite time (i.e., propagate with a finite velocity) is a philosophically and mathematically complicating aspect.

Electromagnetism is also one of the cornerstones of modern technology—indeed, of modern life. Electromagnetic waves through space or along wires, optical fibers and waveguides provide our principal means of communication; electric currents activate our lights, motors, computers, television and radio transmitters and receivers. Antennas on roofs, on automobiles and ships, space shuttles and satellites expand the range of our eyes and ears.

How should the physics and mathematics of electromagnetic field theory be presented to the modern student who would rather program a computer than acquire the mental and technical expertise needed to understand the fascinating complexities of the physical world? The explicit introduction of computer-based methods into a course on electromagnetic theory is not the answer. Numerical methods and computer software are invaluable tools for the modern scientist and engineer which he must certainly learn to use. But this does not mean that skill in the use of techniques like the method of moments is an acceptable alternative to physical understanding or mathematical analysis. Accordingly, this introduction to electromagentism seeks to develop an understanding of electromagnetic phenomena, their mathematical representation, and their manifold applications without the explicit added distractions and complications of computer-related methods.

In the first section of his distinguished book, *Electrodynamics*, (volume III in his *Lectures on Theoretical Physics*) Arnold Sommerfeld describes the surprising new insights which he gained as a student from the great work of Heinrich Hertz, "On the Basic Equations of Electrodynamics for Bodies at Rest." In particular, he points out that in Hertz's presentation—in contrast with the "old-style lectures" that begin with Coulomb's law and electrostatics—the equations of Maxwell are introduced initially and axiomatically as the basis of electromagnetism. He emphasizes that from them "the entirety of electromagnetic phenomena can be deduced logically and systematically. Coulomb's law, that used to provide the initial foundation, now appears as a necessary consequence of the comprehensive theory."

Fundamental Electromagnetic Theory and Applications departs from the traditional "old style" to follow the lead of Hertz in the spirit of Maxwell. It provides a moderately comprehensive and logically coordinated introduction to electromagnetism based directly on Maxwell's equations. The basic theory in the first six chapters borrows heavily from the senior author's earlier work, *Fundamental Electromagnetic Theory*, but with appropriate rearrangement and a completely modernized symbolism and standardized notation. Since the book begins with first principles, it does not depend directly on previous work in electricity and magnetism, but does presuppose an intermediate level of physical and mathematical maturity.

Chapter 1 serves a double purpose. Although directed primarily toward defining the density functions which subsequently appear in Maxwell's equations and formulating the principle of conservation of electricity, it also introduces the vector operators in terms of fundamental physical concepts, rather than merely as a mathematical symbolism. Instead of summarizing vector analysis in a separate introductory chapter or in the appendix, it is made an integral part of the logical formulation of electromagnetic principles. In this manner, it becomes associated with tangible pictures that are basic to the very subject of study. The outline of classical electromagnetism begins in Chapter 2 with the definition of the electromagnetic vectors and continues in the succeeding three chapters with the introduction of potential and energy functions. Chapter 6 is concerned with the formulation of general theorems and their applications. Beginning with Chapter 7, the theory is applied to the scattering and diffraction of plane waves, linear antennas and arrays, the foundations of electric circuit theory and the loop antenna, transmission-line theory, the insulated antenna, the theories of metal and dielectric waveguides, and to waves and antennas near and across the boundary between electrically different half-spaces.

Because investigators concerned with outward traveling waves are accustomed to the time dependence $\exp(-i\omega t)$ while those dealing with electric circuits use $\exp(j\omega t)$, both forms appear in this book. The choice is made dependent on the particular subject being discussed. The conversion from the one to the other involves the simple relation $i = -j$ except in Hankel functions, where $H_n^{(1)}(x)$ goes over into $H_n^{(2)}(x)$ when i is changed to $-j$. Note that in the exponential the letter j is used consistently and exclusively with the positive sign, the letter i with the negative sign.

Boldface type is used for space vectors whether real or complex. Real and complex scalars are in lightface type. Readers are reminded that the so-called rotating "vectors" used to represent periodic phenomena are actually complex numbers. The "rotating vector" is a rotating pointer in the complex plane. A complex vector in boldface is a shorthand for a real space vector in boldface (often a unit vector) multiplied by a complex number in lightface.

All of the material in this book except that which appears in the last chapter has been tested in the classroom.

Acknowledgments

The first draft of the manuscript was typed in part by Barbara Cottam and Ravindran. The final draft was typed by Pat Kent and Karen Kent. Margaret Owens corrected the entire manuscript and typed the last chapter. The authors wish to acknowledge the help and encouragement given by Bernard Goodwin, Senior Editor, Prentice-Hall, Inc., throughout the preparation of the manuscript.

Ronold W. P. King
Sheila Prasad

1

Mathematical Description of Matter: Density Functions and Vector Operators

Electromagnetic engineering is concerned with solutions of mathematical equations which can be interpreted physically and applied to practical problems in electrical communication, geophysical exploration, biomedical science, and other fields. It deals with an elaborate symbolism that describes the physical models in terms of which all macroscopic electrical phenomena are explained. It requires a mathematical shorthand to express as concisely as possible the interrelations between symbols in accordance with the fundamental postulates of physical science. This chapter is concerned with the vector-analytical description of the electrical properties of matter.

STATIONARY STATES: THE STATIC STATE

1.1 ELECTRIC CHARGE

In the systematic formulation of an electrodynamical model to serve as the mathematical foundation of electromagnetic engineering, it is convenient to consider electric charge to be a basic concept that is not derivable from or expressible in terms of other concepts. This view is consistent with atomic theory, which constructs its models of matter in terms of the four concepts of space (length L), time (T), mass (M), and electric charge (Q). Its picture of matter involves vast numbers of molecules characterized by mass and random velocity in space. The molecules are combinations of atoms; the atoms are complex structures consisting of electrons and protons (neutrons, positrons). The electron is not divisible, is associated with a mass $m_e = 9.1 \times 10^{-31}$ kg, and is invariably characterized by a definite quantity of negative electric charge $q_e = -e = -1.6 \times 10^{-19}$ coulomb (C). For most engineering purposes, a simplified model of the nucleus is adequate: a mass M (e.g., mass of proton $m_p = 1836.3 m_e \gg m_e$, mass of neutron $m_n \gg m_e$) with a positive electric charge $q_i = Ze$, where Z is the number of electrons outside the nucleus. The atom as a whole is neutral. In constructing the electrodynamical model, it is adequate to treat electrons and positive nuclei as charges and masses associated with points. It is not necessary to specify shape or volume.

Two fundamental properties of electric charge are postulated. The first is its indestructibility. A basic characteristic of electric charge is that it can be neither created nor destroyed. If a charge disappears from one point, it must reappear at another. This postulate is called the *principle of conservation of electricity*. The second postulate is that of *mutual interaction at a distance*. This assumes attraction and repulsion between charges both as a characteristic of the charges themselves and of their relative motions. No attempt is made at this point to formulate these two postulates with precision. They will be involved quantitatively in describing the physical models on which the mathematical structure that constitutes the real theory of electromagnetism is based. An essential part of this theory is the formulation of the mathematical counterparts of these two qualitative principles assumed in the physical model.

The fundamental problem of engineering electrodynamics is to incorporate millions of electrons and positive nuclei in a mathematical model from which quantitative observations associated, for example, with antennas and transmission circuits may be predicted in terms of their theoretical analogues. It is formally possible to set up this problem in terms of the individual position and motion of the fabulous numbers of electric charges contained in the atomic model of matter. But this formulation is mathematically so complicated that it is of no practical value for deriving experimentally verifiable quantities. Clearly, a fundamental prerequisite of any model is a reasonable simplicity and transparency in its mathematical structure, even at the expense of consistency in the picture and of a degree of accuracy in its final predictions. In the present case, as in others involving a very large number of similar units, great simplification results from a study of average rather than of individual behavior. This permits an approximate, overall description in terms of continuous functions suitably defined throughout the body and so constructed that they assume at every point values characteristic of the average properties of a small region near the point. The number of such functions required and their complexity depend on the number of classes of units present and their relative behavior, as well as on the degree of approximation desired. For example, in representing the mass of a solid body in terms of its molecular model, a single, continuous function giving the average density D at every point is adequate to secure a good approximation. In this case, there is only one class of units, since each molecule is characterized alike by mass. The corresponding electrical problem is considerably more complicated owing to the presence of two distinct classes of units, the positive nuclei and the electrons, and the possibility of widely dissimilar behavior because of their different structures and functions in the atom. To secure a reasonably good approximation in a large variety of problems, it is necessary to construct a mathematical model that includes several continuous functions called *densities*. These take account of the magnitude, the distribution, and the relative velocities of the charges. In certain problems requiring a higher degree of approximation, additional functions and a more complicated model may be constructed. These will not be described.

To define the continuous functions required for the mathematical model, it is convenient to examine in detail separate atomic models with special structures. These are so chosen that a general case may be obtained by superposition or combination, although usually not without overlapping. These special models are considered conveniently in two important groups or states called, respectively, the *stationary* and the *nonstationary states*. The former is subdivided into the static state and the steady state.

1.2 THE STATIC STATE AND THE ATOMIC MODEL

The *static state* is more correctly called the *statistically stationary state* since it does not involve charges at rest. On the contrary, all charges are assumed to be in motion in a most general and irregular way. A motion in which no regularity exists is called *random*. Each charge moves in its own unique way unlike that of any other. For any volume as a whole there must be no preferred directions, no common axes, nothing that would in any way permanently relate the average motion of one charge to that of another. Any effect due to the motion of one charge in a given direction or around a particular orbit is undone by the motion of other charges in the opposite direction somewhere in its neighborhood, or by the combined effect of the individual motions of many charges. All effects due to the individual random motions of the charges cancel over a time average taken over a period that is long compared with the time of atomic or molecular events (such as rotations or collisions) but that may be very short from the point of view of an experimental observation. Such a time-average picture of the behavior of a large number of entities moving at random is called *statistical*. From the statistical point of view, a volume containing millions of charges moving at random is indistinguishable from the same volume containing the same charges with each fixed at an average rest position. Hence the charges may be spoken of as statistically at rest, and to each may be assigned a statistical rest position. The average overall electrical properties of the statistically static model are the same as those of the dynamic one with random motion, and one may be substituted for the other whenever convenient. Any regularity in orientation or motion of the charges may be pictured in the dynamic model as superimposed on the random motion; in the static model it is represented by a relative orientation or motion of the statistical rest positions of the charges.

It is possible to incorporate within the picture of random distribution a rather complicated model skillfully devised to represent the inner structure of the elements. It assigns a definite and different number of electrons to the atom of each element and divides the electrons so assigned into groups according to a comprehensive scheme. This defines a set of *energy levels* or *shells*, of which each is limited to a specified maximum number of electrons. Each positive nucleus has associated with it a certain number of electrons in an arrangement characteristic of a particular element. In the case of a molecule consisting of several closely bound atoms, the electronic distribution peculiar to each atom is complicated by an overlapping or sharing of shells. All the electrons in an atom or in a molecule are freely movable to the extent that each individual electron can exchange places with any other and in this way associate itself successively with different shells or nuclei belonging to different atoms or molecules throughout a body or region. However, the number of electrons belonging to each shell, nucleus, or group of nuclei in a molecule is a constant time-average characteristic of each element or compound.

In each atom two classes of electrons may be distinguished solely with reference to the type of shell to which they belong. One class is associated with complete or full shells, the other with shells that are only partly filled. In most atoms, there is only one incomplete shell, the outermost one. But in some cases (e.g., copper and silver) the difference between the energy levels of two shells is so small, as determined by the mathematical scheme used to define these, that the outermost one may already contain some electrons while the next one still has empty spaces. The electrons in partly filled shells are called *valence electrons*. It is assumed that closed shells in atoms and closely linked configurations in molecules are characterized by strong intra-atomic and intramolecular constitutive forces that act to maintain them over a time average. Such forces are not presumed to act on valence electrons of atoms unless they are a part of a molecular configuration. Electrons that are subject to strong constitutive forces are called *closed-shell electrons* or *bound charges*; valence electrons are called *free charges*. The term "bound charge" does not mean that random exchanges may not take place freely. It does mean that a definite and characteristic number of electrons is bound to each atom or molecule not only under random conditions but also under the action of strong external forces tending to disturb a random distribution. The term "free charge" refers to a charge that may leave the atom or molecule with which it is statistically associated without having another charge take its place. A motion of free electrons may occur from one section of a body to another, leaving an excess of positive charge behind and bringing with it an excess of negative charge. But such a transfer of free charge can take place only under the action of external influences that disturb the normal random conditions.

The number of free electrons characteristic of an atomic configuration is a fundamentally significant property. Atomic or molecular models in which there are very few or none will be called *closed-shell* or *bound-charge models*. Other names are *nonconductors, dielectrics,* or *insulators*. Models in which there is an abundance of free electrons will be called *free-charge models* or *conductors*. Intermediate cases are *semiconductors* or *imperfect dielectrics*. Some of their macroscopic properties may be deduced by a suitable combination of the properties of the two extremes. The free- and bound-charge models are necessarily statistically identical as long as purely random conditions prevail. But they behave quite differently under the action of external forces which seek to establish preferred directions in the motions or relative positions of the charges or in the orientation of the atoms or molecules.

1.3 VOLUME DENSITY OF CHARGE

The electrical properties of a region (or body) in which a random (statistical) distribution of charge prevails may be described approximately by *a continuous function which assigns to every point in the region a number characteristic of the average condition of total charge in the neighborhood of the point.* Any function that assigns a *scalar* to every point in a region in which it is defined is called a *scalar point function*. For a volume V, the physical meaning of the volume density of charge $\rho(\mathbf{r})$ is contained in

$$\int_V \rho(\mathbf{r})\, dV = \sum_{j=1}^{N} e_j \qquad (1.3\text{--}1a)$$

Figure 1.3–1 shows the element of volume dV.

The *average volume density of charge* is

$$\overline{\rho(\mathbf{r})} = \frac{\displaystyle\sum_{j=1}^{N} e_j}{V} \qquad (1.3\text{--}1b)$$

However, this formula tells nothing about the distribution of charge in V. If the volume V containing many millions of electric charges is divided into small elements ΔV and the density function is defined to be the limit approached by the ratio of the total charge in the element to the volume of the element as the latter is allowed

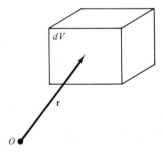

Figure 1.3–1 Location of volume element dV.

to approach zero, a microscopic definition of the volume density of charge is obtained:

$$\rho(\mathbf{r}) = \lim_{\Delta V \to 0} \frac{\sum\limits_{j=1}^{N} e_j}{\Delta V}$$

$$= \sum_{j=1}^{N} e_j \delta(|\mathbf{r} - \mathbf{r}_j|) \tag{1.3-1c}$$

Figure 1.3–2 shows the location of the charge e_j. In (1.3–1c), $\delta(|\mathbf{r} - \mathbf{r}_j|)$ is the *Dirac delta function*, which has the following properties:

$$\delta(|\mathbf{r} - \mathbf{r}_j|) = \begin{cases} 0, & \mathbf{r} \neq \mathbf{r}_j \\ \infty, & \mathbf{r} = \mathbf{r}_j \end{cases}$$

$$\int_V \delta(|\mathbf{r} - \mathbf{r}_j|) \, dV = \begin{cases} 1, & \text{when } V \text{ contains } \mathbf{r}_j \\ 0, & \text{when } V \text{ does not contain } \mathbf{r}_j \end{cases}$$

$$\int_V \rho(\mathbf{r}) \, dV = \int_V \sum_{j=1}^{N} e_j \delta(|\mathbf{r} - \mathbf{r}_j|) \, dV$$

$$= \sum_{j=1}^{N} \int_V e_j \delta(|\mathbf{r} - \mathbf{r}_j|) \, dV$$

$$= \sum_{j=1}^{N} e_j$$

which is the same as (1.3–1a).

Since every charge is endowed with mass, a corresponding scalar point function to describe the distribution of mass is

$$D(\mathbf{r}) = \lim_{\Delta V \to 0} \frac{\sum\limits_{j=1}^{N} m_j}{\Delta V} \tag{1.3-1d}$$

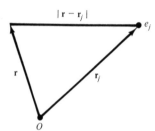

Figure 1.3–2 Location of charge e_j.

In (1.3–1c) e_j is the magnitude of any one of the N charges in the volume element. In (1.3–1d) m_j is the mass associated with the same charge. Unfortunately, functions constructed in this way are discontinuous because the charges and masses are associated with discrete points. They have values different from zero only at points characterized by charge and mass. For example, at a point locating an electron, $\rho = -\infty$, $D = \infty$ and at a point locating a positive nucleus, $\rho = \infty$, $D = \infty$. At all intermediate points both ρ and D vanish. This is precisely the representation in terms of individual charges which has already been rejected because of its great mathematical complexity.

To obtain functions that will represent the average distribution of charge and of mass in a small element of volume, it is necessary to subdivide a larger region into small elements. But these may not approach zero as a limit. Several questions arise: How large must each element of volume be in order that the discontinuous effect of individual, more or less widely separated charges and masses may be avoided? How small must each element be so that significant variations in charge and mass, which may characterize the region as a whole, are not obscured? Is it, in fact, possible to select volume elements that are at the same time sufficiently large and sufficiently small? The answers to the first two questions are easily given. Each volume element must be large enough to contain enough charges so that statistical conditions prevail. But it must also be very small compared with physically measurable magnitudes. The answer to the last question, whether these two restrictions on the size of volume elements can be fulfilled simultaneously, depends on the inner structure of the postulated atomic model, in particular on the average distance between charges. Fortunately, atomic theory requires this distance to be so small that it is possible to construct volume cells that are large enough to contain many millions of charges, and that are yet extremely minute compared with laboratory magnitudes. In fact, the mean distance between charges in the model is assumed to be so short that a volume cell which is only as thick as this mean distance may still be made large enough to be statistically regular as a whole without approaching directly measurable magnitudes in length or breadth.

Let the region or body for which the continuous functions are to be constructed be subdivided into volume cells of which $\Delta\tau_i$ is a typical one (Fig. 1.3–3). $\Delta\tau_i = d_i^3$ such that

$$\text{(mean free path) } d_c \ll d_i \ll L \text{ (laboratory dimension)} \qquad (1.3\text{–}2)$$

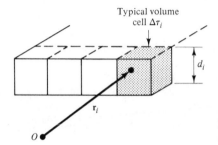

Typical volume
cell $\Delta\tau_i$

d_i

$\mathbf{r}_i$

O

Figure 1.3–3 Body divided into volume cells.

where the mean free path is the mean distance between charges. For solids and liquids,

$$d_c \sim 4 \times 10^{-10} \text{ m}; \qquad d_i \sim 4 \times 10^{-6} \text{ m}; \qquad L \sim 4 \times 10^{-2} \text{ m}$$

The inequality (1.3–2) is satisfied and $N_i = (d_i/d_c)^3 \sim 10^{12}$ particles in $\Delta\tau_i$. This satisfies the condition for statistical regularity. For gases and plasmas, d_c is usually quite large and the definition of a useful density may be impossible.

Let there be N charges—positive, negative, or both—in a typical cell. The total charge e_i in the volume $\Delta\tau_i$ is the algebraic sum of the individual charges. It is

$$e_i = \sum_{j=1}^{N} e_j \qquad (1.3\text{–}3a)$$

The total mass is

$$m_i = \sum_{j=1}^{N} m_j \qquad (1.3\text{–}3b)$$

Here e_j is the charge and m_j the mass of an electron or of a positive nucleus. Appropriate magnitudes and algebraic signs are to be used in each case. Evidently, the total charge in the volume element vanishes if it contains either no charges or equal amounts of positive and negative charge. On the other hand, the total mass vanishes only if $\Delta\tau_i$ contains no charges and hence no units of mass.

Two scalars are defined at the center of each volume cell (located by the vector $\mathbf{r}_i$) according to the formulas

$$\rho(\mathbf{r}_i) = \frac{e_i}{\Delta\tau_i} \qquad \text{C/m}^3 \qquad (1.3\text{–}4a)$$

$$D(\mathbf{r}_i) = \frac{m_i}{\Delta\tau_i} \qquad \text{kg/m}^3 \qquad (1.3\text{–}4b)$$

and giving, respectively, the average charge and mass per unit volume in the cell $\Delta\tau_i$. Corresponding scalars are defined at the center of every volume element throughout the entire body. Using all these scalars as a reference frame, two continuous scalar point functions are constructed that, by definition, assume, respectively, the values $\rho(\mathbf{r}_i)$ and $D(\mathbf{r}_i)$ at the center of each cell, while smoothly and continuously connecting them at all intermediate points. This is shown in Fig. 1.3–4. The continuous functions constructed in this way are said to be *interpolated* from the discrete values $\rho(\mathbf{r}_i)$ and $D(\mathbf{r}_i)$. They are, respectively, the *volume density of charge* (denoted by ρ) and the *volume density of mass* (denoted by D). The former has the dimensions of charge divided by volume, the latter of mass divided by volume.

$$\rho(\mathbf{r}) \sim \frac{Q}{L^3} \qquad \text{C/m}^3 \qquad (1.3\text{–}5a)$$

$$D(\mathbf{r}) \sim \frac{M}{L^3} \qquad \text{kg/m}^3 \qquad (1.3\text{–}5b)$$

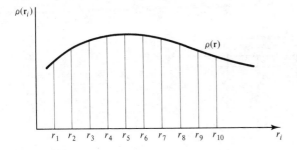

Figure 1.3–4 Volume density of charge.

From the very manner in which they are constructed, it is clear that such interpolated functions can adequately represent the average densities of charge and of mass in a body only if this possesses considerable uniformity of structure. The mean distribution of charges may vary only slowly from volume cell to volume cell. In fact, an adequate representation of the properties of a body or region in terms of volume densities of charge and mass as defined above requires that these functions be so slowly varying through the body as to be sensibly constant over distances that are large compared with the dimensions of a volume cell. Such a condition of very slow variation must be assumed to prevail wherever the density functions are used.

1.4 SURFACE DENSITY OF CHARGE

In constructing the volume density of charge ρ by interpolation from the scalars $\rho(\mathbf{r}_i)$ defined at the centers of the volume cells $\Delta\tau_i$, it was assumed that these cells were all sensibly alike throughout the region in which ρ was defined. In the interior of a region this assumption merely implies a reasonably slow variation in characteristic structure in passing from cell to cell. On the surface of a region, or on the boundary between two dissimilar regions, however, this uniformity of structure does not exist because there the volume cells are not completely surrounded by other similar ones. At a surface or boundary, all cells are asymmetrically placed because one side is necessarily exposed to surroundings that are entirely different from those experienced by the other sides. It must be expected, therefore, that at a surface or boundary, the electrical properties of a region cannot in general be represented correctly by the same continuous function used to describe the interior. In particular, the volume density of charge ρ cannot be required to be as slowly varying as is demanded for the interior and at the same time to represent correctly the rapid change that may occur near the surface. This difficulty may be overcome by treating separately a layer of very thin surface cells of which $\Delta\tau_s$ is a typical one (Fig. 1.4–1). The thickness of the surface layer is taken to be d_c, the mean free path, and the dimensions parallel to the surface are d_i, where d_c, d_i, and L are related as in (1.3–2). It is stated in Sec. 1.3 that on the basis of the atomic theory these tangential dimensions may be so large that each surface cell contains enough

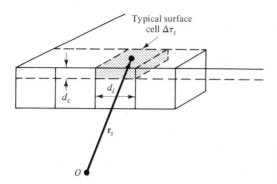

Typical surface
cell $\Delta\tau_s$

d_c

d_i

r_s

O

Figure 1.4–1 Body divided into volume and surface cells.

charges to be statistically regular and yet so small that they are of a different and much smaller order of magnitude compared with physical dimensions. It is, therefore, possible to choose the tangential dimensions of the surface cells to be of the same order of magnitude as the dimensions d_i of the interior cell and still have both surface and volume cells satisfy the limiting conditions for statistical regularity on the one hand, and physical smallness on the other. In this case the number of charges is $N_s = d_c d_i^2/d_c^3 \sim 10^8$ particles in $\Delta\tau_s$, which still satisfies the requirement for $N \geq 10^5$. Because the surface cells are like the volume cells in the tangential plane, it is reasonable to suppose that for such properties as depend only on conditions in this plane, and not on those along a perpendicular to a boundary, each surface cell will behave just like a correspondingly thin slice of an adjacent volume cell. It follows that a surface layer of cells need be used only when asymmetry along the *normal* to the surface leads to significant and different effects from those determined from conditions of charge that are characteristic of the interior. Since the volume of each surface cell is negligible compared with that of the adjacent interior cell, *surface cells may be disregarded in describing tangential effects.*

The significance of thin surface layers of charge and their relation to the volume density of charge as already defined are illustrated in the following example. Suppose that a closed region containing positive and negative charges is at first under the action of no external influence. The static condition of complete statistical equilibrium prevails, and every volume and surface element is electrically neutral because it contains on the average an equal amount of positive and negative charge. This is true for both the free-charge and the bound-charge models since a random distribution prevails in each. If an asymmetrical external force, for example, due to a positively charged body placed near its left end, acts on the region, electrons in it experience an unbalanced force of attraction toward the left, whereas positive nuclei are repelled toward the right. In the free-charge model, mobile free electrons drift toward the left to make the left surface negatively charged and leave the right-hand surface positive. The drift extends through the entire region and does not stop until the average force acting on each charge due to the external charges and the two oppositely charged surface layers is again zero. A new condition of statistical equilibrium is established in this way with a greater electron density at the surface on the left and a correspondingly diminished electron density at the surface on the right. In the interior, each volume element remains neutral.

In the bound-charge model, conditions are different. All electrons are in closed shells, and although individually free to move and exchange places with one another they are under the influence of strong intra-atomic forces that seek to maintain a distribution which is random from the point of view of a volume cell, but which on the average provides each nucleus with a characteristic number of electrons properly distributed in shells. Hence, even under the action of quite strong external forces, there can be no mean drift of electrons one way or the other. Over a time average the same number must always move away from an attracting external influence as toward it. Consequently, the statistical rest positions of the electrons can at most be shifted slightly to produce a distorted outer shell of electrons in each atom or molecule. If an external force is applied outside the left surface of a region to attract electrons as in the preceding case, each atom or molecule exhibits a slight distortion and orientation in the form of a shift of the statistical rest positions of the electrons relative to those of the positive nuclei. The mean rest position of the total negative charge associated with each atom or molecule is in this way displaced so that it no longer coincides exactly with that of the positive center. The net effect is a small shift of the electrons associated with each atom toward the left. This does not constitute a general drift in that direction throughout the charged region as in the free-charge model, but only an infinitesimal distortion and orientation within each atom. The final effect, however, is again to make the left-hand surface of the entire region negative, the right-hand surface positive while the interior remains neutral.

From these simple illustrations it is clear that similar distributions of charge may exist in both models. In particular, charges may be so distributed that a volume density of charge would have to have the value zero throughout the interior, rise continuously but very steeply to a positive value at one surface and descend to an equal negative value on the opposite surface. In general, and under identical external circumstances, the distribution, although similar in the two models, would not be the same because the conditions determining the final equilibrium are quite different. In the free-charge model, equilibrium is reached when enough negative charge has accumulated on the left-hand surface (by effectively taking it from the right) to neutralize in the interior of the region the action of the external positive charge. In the bound-charge model, equilibrium is reached when each atom has been distorted and oriented in such a way that stronger internal forces are brought into play to balance the external influence. The amount of distortion, which determines the density of charge at the surface, depends on the nature of the intra-atomic forces of constitution.

If such an equilibrium distribution of free or bound charge is to be described in terms of the volume density of charge ρ as defined above, difficulties are encountered. For a region that has been divided into similar volume cells at the surface and the interior, the following observations may be made about the discrete densities of charge ρ_i defined for the individual cells. In the free-charge model, ρ_i

vanishes in each interior cell because it contains on the average equal magnitudes of positive and negative charge; ρ_i has a small negative value in the layer of cells along the left surface; a small positive value in the layer along the right surface. The values of ρ_i in the $\Delta\tau_i$ cells along the surfaces are very small because the surface layer constitutes only a minute part of the volume of these cells. The process of averaging over cells that are extremely thick compared with the surface layer of charges (the thickness of such a layer is of molecular magnitudes) obscures the actually high concentration of charges in such a thin layer. The ρ_i's are average values defined at the centers of cubical volume cells. When ρ is interpolated, no account is taken of concentrations of charge in layers that are thin compared with the dimensions of the volume cells. In brief, a subdivision into $\Delta\tau_i$ cells of dimension d_i is too coarse near the surface or boundary where a thin layer of charge exists.

The same conclusion is reached for the bound-charge model. Since each volume cell is assumed to contain a large number of neutral atoms, a mere distortion of each atom by small shifts of the negative charges relative to the positive nucleus does not change the total charge in each volume element either in the interior or along the surface if boundaries are drawn so that they do not cut through atoms or molecules. Hence the volume density of charge is zero throughout the interior and along the surface.

In neither the free-charge nor the bound-charge model is a representation in terms of the volume density of charge ρ adequate to take account of surface conditions of charge. The reason in both cases is that a function interpolated from values averaged over thick volume cells cannot be sensitive to a thin surface distribution of charge that contributes little to the volume of these cells. What is needed is a separate treatment of the surface. This is easily accomplished by constructing a layer of extremely thin cells along each boundary, while the remainder of the volume is divided up into volume cells as before. The volume density of charge ρ, as already defined, continues to characterize the condition of charge in the interior, while a new function, η, the *surface density of charge*, is defined to describe the condition of charge along the surface.

The surface density of charge is defined in a manner analogous to that used for the volume density of charge. Let $\Delta\tau_s = d_c d_i^2$ be one of the thin surface cells each of thickness d_c that satisfies the inequality (1.3–2). A scalar $\eta(\mathbf{r}_s)$ is defined at the center of each surface cell (Fig. 1.4–1) by the relation

$$\eta(\mathbf{r}_s) = \frac{d_c \sum_{j=1}^{N} e_j}{\Delta\tau_s} = \frac{d_c e_s}{\Delta\tau_s} \tag{1.4–1}$$

A continuous scalar point function $\eta(\mathbf{r})$ is constructed for the entire surface by interpolation from the discrete values in (1.4–1). By definition, $\eta(\mathbf{r})$ has the values $\eta(\mathbf{r}_s)$ at the centers of the individual surface cells; it connects these continuously and smoothly at all intermediate points. This definition of $\eta(\mathbf{r})$ is meaningful if the body represented is sufficiently uniform so that η varies so slowly in tangential directions along the surface as to be sensibly constant over distances that are large compared with the lateral dimensions d_i of the cells. Since the thickness of the layer is of molecular magnitude, no condition of uniformity can or need be imposed in a direction normal to the surface. The dimensions of the surface density of charge are charge per unit area.

$$\eta(\mathbf{r}) \sim \frac{Q}{L^2} \quad \text{C/m}^2 \tag{1.4-2}$$

A static distribution of charge in a region constructed according to the free-charge or the bound-charge model may be described in terms of the scalar functions ρ and η to a degree of approximation that is adequate for all practical problems.

1.5 NUMBER DENSITIES

An alternative notation for describing the properties of a region is in terms of a *number density* which can be defined in any volume containing several different kinds of particles. For electrons, the number density at the center of the volume cell (given by the point $\mathbf{r}_i$) is defined as

$$n_e(\mathbf{r}_i) = \frac{N_{ei}}{\Delta\tau_i} \tag{1.5-1}$$

where N_{ei} is the number of electrons in the cell $\Delta\tau_i$. The continuous number density $n_e(\mathbf{r})$ may be obtained from the discrete values $n_e(\mathbf{r}_i)$ by interpolation.

If a surface representation is desired, the number density at the center of the surface cell (given by the point $\mathbf{r}_s$) may be defined as

$$n_e(\mathbf{r}_s) = \frac{N_{es}}{\Delta\tau_s} \tag{1.5-2}$$

where N_{es} is the number of electrons in the surface cell $\Delta\tau_s$. The continuous number density $n_{es}(\mathbf{r})$ is obtained by interpolation.

The number densities for positive charges are defined in a similar manner. If a volume representation is to be used, the number density at the center of the volume cell is

$$n_p(\mathbf{r}_i) = \frac{N_{pi}}{\Delta\tau_i} \tag{1.5-3}$$

where N_{pi} is the number of positive charges in $\Delta\tau_i$. Interpolation will yield the continuous density function $n_p(\mathbf{r})$. If the surface density is to be used, the number density at the center of the surface cell is given by

$$n_p(\mathbf{r}_s) = \frac{N_{ps}}{\Delta\tau_s} \tag{1.5-4}$$

where N_{ps} is the number of positive charges in $\Delta\tau_s$. The continuous function $n_{ps}(\mathbf{r})$ is obtained by interpolation.

The previously defined volume and surface densities of charge may be related to the number density for electrons as follows:

$$\rho_e(\mathbf{r}) = q_e n_e(\mathbf{r}) = -e n_e(\mathbf{r}) \tag{1.5–5}$$

$$\eta_e(\mathbf{r}) = q_e n_{es}(\mathbf{r}) = -e n_{es}(\mathbf{r}) \tag{1.5–6}$$

Similar relations are obtained for the positive charges:

$$\rho_p(\mathbf{r}) = q_p n_p(\mathbf{r}) = eZ n_p(\mathbf{r}) \tag{1.5–7}$$

$$\eta_p(\mathbf{r}) = q_p n_{ps}(\mathbf{r}) = eZ n_{ps}(\mathbf{r}) \tag{1.5–8}$$

The total volume and surface densities of charge for a region containing positive and negative particles may now be written:

$$\rho(\mathbf{r}) = \rho_e(\mathbf{r}) + \rho_p(\mathbf{r}) \tag{1.5–9}$$

$$\eta(\mathbf{r}) = \eta_e(\mathbf{r}) + \eta_p(\mathbf{r}) \tag{1.5–10}$$

The mass density may also be related to the number densities.

$$D_e(\mathbf{r}) = m_e n_e(\mathbf{r}), \qquad \text{for electrons} \tag{1.5–11}$$

$$D_p(\mathbf{r}) = m_p n_p(\mathbf{r}), \qquad \text{for protons} \tag{1.5–12}$$

$$D_n(\mathbf{r}) = m_n n_n(\mathbf{r}), \qquad \text{for neutrons} \tag{1.5–13}$$

The total mass density for a region containing positive and negative particles and neutrons is given by

$$D(\mathbf{r}) = D_e(\mathbf{r}) + D_p(\mathbf{r}) + D_n(\mathbf{r}) \tag{1.5–14}$$

The dimensions of the number density of charge are

$$n(\mathbf{r}) \sim \frac{1}{L^3} \qquad \mathrm{m}^{-3}$$

$$n_s(\mathbf{r}) \sim \frac{1}{L^2} \qquad \mathrm{m}^{-2}$$

1.6 ALTERNATIVE MODES OF REPRESENTATION

Up to this point the statistically stationary state has been characterized by the two continuous, slowly varying scalar point functions $\rho(\mathbf{r})$ and $\eta(\mathbf{r})$. The surface density η was introduced because a representation in terms of the volume density ρ alone is inadequate to represent conditions at a surface or boundary resulting from asymmetry in the direction normal to the surface. The definition of η required a simple change in the mode of subdivision of the region into elementary cells. The new mode of subdivision and the definition of ρ and η may be used for both free- and bound-charge models. But whereas in the free-charge model the surface conditions are *real surface effects* involving distinct layers of free charge that can be considered simply and naturally as separate from, and superimposed upon, volume phenomena, the same is not true of the bound-charge model. Here a condition ascribed specifically to the surface is actually a surface manifestation of a phenomenon existing throughout the interior. The appearance of a surface layer of charge is the result of a distortion and orientation *of all the atoms in the region.* And the charges on the surface are simply parts of the outermost layer of neutral but distorted atoms. From the physical point of view of the atomic model, the separate consideration of a thin surface sheet is not really appropriate or reasonable for the bound-charge model because the thin surface layer cuts off a part of the outer, closed shell of each atom. Hence, although a subdivision using surface cells and the separate definition of a surface density provides an adequate representation of the external properties of the bound-charge model from the mathematical point of view, an alternative, more appropriate representation is desirable from the point of view of the physical model. According to this, the entire surface effect is fundamentally a part of a volume phenomenon, and it is not merely plausible but logically necessary to provide an alternative representation entirely in terms of volume functions. Instead of changing the mode of subdivision and introducing separate surface cells and a separate surface density, an alternative procedure using the original subdivision into volume cells is required. The orientation-distortion effect in the interior, as well as its surface manifestation, must be represented by an additional function defined throughout the volume.

1.7 VOLUME DENSITY OF POLARIZATION

An alternative representation of the static state in terms of two volume functions instead of a volume and a surface function is designed specifically for the bound-charge model with a subdivision into volume cells only. The volume density of charge is defined to describe the average condition of total charge throughout the region just as before. It does not take account of the average separation and orientation of the statistical rest positions of positive and negative charge, so that a new function must be constructed for this purpose. Consider a region containing only closely bound charges which are exposed to the action of an external force that attracts negative and repels positive charge. Because there are no free charges in the region, there can be no general transfer of charge. Instead, the bound-charge groups associated with each atom or molecule are distorted and oriented in such a way that the statistical rest position of the entire negative charge in each group is moved away from coincidence with the rest position of the positive charge. Each bound group continues to be electrically neutral, but a statistical separation and orientation of its positive and negative charges takes place. Such a group may be described in terms of a statistically stationary positive charge q separated a distance d from a similar negative charge. A structure of this kind is called a *dipole*. Let a polar vector **d** be drawn from the statistical center of the negative charge to that of the positive charge. The polar vector

$$\mathbf{p} = q\mathbf{d} \qquad (1.7-1)$$

is a measure of the statistical separation of the rest positions of the positive and negative charge and their orientation in space as seen in Fig. 1.7–1. (The direction of a polar or ordinary vector differs from its opposite in a real physical sense. Thus the polar vector points from negative to positive charge where these two kinds of charge are physically different.) The polar vector **p** is called the *average polarization* of the bound group of charges in the atom or molecule; it is the polarization of a statistically equivalent dipole. The direction of the vector defines the axis of polarization; the magnitude of the vector is the electric moment.

   An entirely equivalent representation, which is more readily generalized to apply to a volume containing many charges, is illustrated in Fig. 1.7–2. An arbitrary origin is fixed at any convenient point near the rest positions of two equal and opposite charges. Let $\mathbf{d}_1$ and $\mathbf{d}_2$ be vectors drawn, respectively, from the origin to each of the two charges e_1 and e_2. The positive direction of each vector is from the origin to the charge, regardless of the sign of this latter. The polarization of the dipole composed of the two charges is defined by

$$\mathbf{p} = e_1\mathbf{d}_1 + e_2\mathbf{d}_2 \qquad (1.7-2)$$

Since e_1 and e_2 have been assumed to be equal in magnitude and opposite in sign, this relation may be written

$$\mathbf{p} = e_1(\mathbf{d}_1 - \mathbf{d}_2) = e_1\mathbf{d} \qquad (1.7-3)$$

It is therefore equivalent to (1.7–1) with e_1 replaced by q. In (1.7–3) it has been assumed that e_2 is the negative charge; that is, $e_2 = -e_1 = -e$, where e is positive.

   The notation used in (1.7–3) may be applied to define the polarization $\mathbf{p}(\mathbf{r}_i)$ of any number N of different charges in a volume element $\Delta\tau_i$ referred to the center of $\Delta\tau_i$, as shown in Fig. 1.7–3. The polarization $\mathbf{p}(\mathbf{r}_i)$ is given by the vector sum

$$\mathbf{p}(\mathbf{r}_i) = \sum_{j=1}^{N} e_j\mathbf{d}_j \qquad (1.7-4)$$

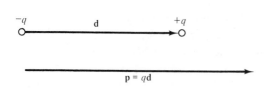

Figure 1.7–1   Representation of a dipole.

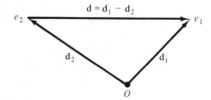

Figure 1.7–2   Equivalent representation of a dipole with unequal charges.

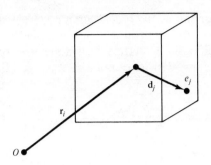

Figure 1.7-3 Location of charge e_j in volume element.

The vectors $\mathbf{d}_j$ are drawn from an origin at the center of the volume element $\Delta\tau_i$ to the individual charges e_j. In a region containing no free electrons, the $\Delta\tau_i$ elements may be so chosen that each one contains only complete atoms or molecules. With such a subdivision the volume density of charge is zero throughout. The electrical properties of the region are then characterized entirely by the polarization of the individual volume cells. Thus, if a vector $\mathbf{P}(\mathbf{r}_i)$ is defined at the center of each volume cell $\Delta\tau_i$ according to

$$\mathbf{P}(\mathbf{r}_i) = \frac{\mathbf{p}(\mathbf{r}_i)}{\Delta\tau_i} = \frac{\sum\limits_{j=1}^{N} e_j \mathbf{d}_j}{\Delta\tau_i} \tag{1.7-5}$$

a vector point function $\mathbf{P}(\mathbf{r})$ which is continuous and slowly varying throughout the region may be constructed by interpolation from the discrete values $\mathbf{P}(\mathbf{r}_i)$. If free charges are present in addition to the complete atoms, their distribution is described in terms of ρ. However, if ρ is not so slowly varying as to be sensibly constant over distances comparable with the dimensions of each $\Delta\tau_i$ cell, the nonuniformity of charge distribution contributes not only to $\rho(\mathbf{r}_i)$ but also to $\mathbf{p}(\mathbf{r}_i)$ defined at the center of the cell by (1.7-4). In order not to take this same variation into account twice, it must be subtracted from (1.7-5) if it is at all significant. Its value is easily determined as follows. Since ρ is a continuous function, the departure from a constant charge density at a point within the cell at a distance s from its center is given to a first approximation by $s(\partial\rho/\partial s)$. The derivative is evaluated at the center of the cell. The resulting contribution to the polarization of the entire cell referred to its center is obtained by multiplying this charge density by a vector $\mathbf{s}$ drawn from the center of the cell to the point and then integrating over $\Delta\tau_i$. Accordingly, if ρ is not sensibly constant, $\mathbf{P}(\mathbf{r}_i)$ must be defined as follows instead of by (1.7-5):

$$\mathbf{P}(\mathbf{r}_i) = \frac{1}{\Delta\tau_i}\left[\mathbf{p}(\mathbf{r}_i) - \int_{\Delta\tau_i} \mathbf{s}s\left(\frac{\partial\rho}{\partial s}\right)_i d\tau\right] \tag{1.7-6}$$

The integral in (1.7-6) with a vector in the integrand is a shorthand form for three integrals, one for each component. The function $\mathbf{P}(\mathbf{r}_i)$ in (1.7-6) evidently reduces to (1.7-5) if ρ is sensibly constant. A continuous function $\mathbf{P}(\mathbf{r})$ may be interpolated from the discrete values in (1.7-5) or (1.7-6) as required. $\mathbf{P}(\mathbf{r})$ is the vector volume density of polarization, or simply the *polarization vector*. A function, such as $\mathbf{P}(\mathbf{r})$, that assigns a vector to every point in a region is called a *vector point function*. Thus $\mathbf{P}(\mathbf{r})$ assigns to each point a vector that is a measure of the electric moment and its direction. The volume density of polarization has the dimensions of charge per unit area

$$P(\mathbf{r}) \sim \frac{Q}{L^2} \qquad C/m^2 \tag{1.7-7}$$

1.8 COMPARISON OF TWO REPRESENTATIONS

In describing the static-state properties of a region in terms of volume and surface densities of charge ρ and η, each of these two functions is defined in an entirely separate part of the region in which it alone bears the full responsibility. On the other hand, in the alternative representation in terms of volume densities of charge ρ and of polarization $\mathbf{P}$, both functions are defined in terms of the charges contained in the same volume elements. Both functions are defined throughout the whole region; to every point the function ρ assigns a scalar, the function $\mathbf{P}$ a vector. Since each point is characterized by two independently defined quantities, a question must arise as to what extent the two functions overlap and to what extent the entire representation depends on the mode of subdivision into volume cells. Furthermore, since the volume density of charge is unable to take account of asymmetrical conditions in the form of thin layers of charge at surfaces and boundaries, the volume density of polarization must in one way or another describe these.

Let a region that is under the influence of an external force be examined. Suppose, first, that it is divided into volume elements in such a way that each cell contains only complete and therefore neutral bound-charge groups (Fig. 1.8–1a). The volume density of charge is zero throughout and since the boundaries cut no dipoles, $\rho = 0$, $\mathbf{P} \neq 0$. If surface cells are used, dipoles at the ends will be cut by boundaries, $\eta \neq 0$, and $\mathbf{P}$ will be smaller than before because some dipoles are

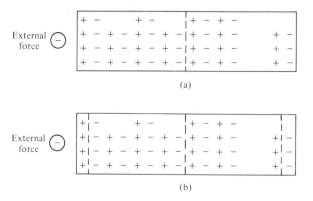

(a)

(b)

Figure 1.8–1 External force acting on a charged region.

cut. By careful location of the boundaries **P** can be made zero (Fig. 1.8–1b). Figure 1.8–2 shows an *electrically one-dimensional* section of a volume constructed so that the polarization is everywhere fixed in direction, but increasing in magnitude uniformly in a direction parallel to the axis of polarization. The volume density of polarization **P** is a uniformly increasing function in the direction from left to right. The three-dimensional picture of the volume under consideration is shown in Fig. 1.8–3. The volume V is a slab that extends from $x = 8$ to $x = 24$ and is infinite in the y and z directions. It is subdivided into cubical cells with $d_i = 4$ and $\Delta\tau_i = 4 \times 4 \times 4$. Because of the uniformity in the y and z directions, this may be simplified by choosing cells such that $\Delta\tau_i = 4 \times 1 \times 1$. ρ, **p**, and **P** may be calculated for each cell from the definitions

$$\rho_i = \frac{\sum e_j}{\Delta\tau_i}; \qquad \mathbf{p}_i = \sum e_j \mathbf{d}_j, \qquad \mathbf{P}_i = \frac{\sum e_j \mathbf{d}_j}{\Delta\tau_i}$$

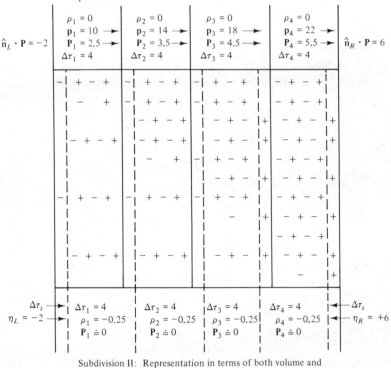

Subdivision I: Representation entirely in terms of the volume density of polarization **P**; $\rho = \eta = 0$. Note that doubly spaced dipoles are counted as two.

Subdivision II: Representation in terms of both volume and surface densities of charge, ρ and η; $\mathbf{P} \doteq 0$

Figure 1.8–2 Schematic diagram to show two possible modes of subdivision into volume cells in the static state.

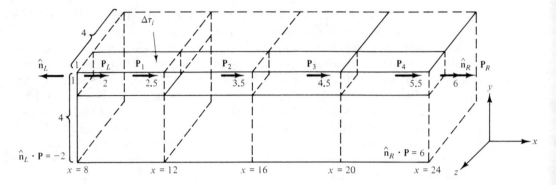

Figure 1.8-3   Representation of polarization.

and the representation in Fig. 1.8-2. Doubly spaced dipoles are counted as two in the volume cells. The functions for the four volume cells are

$$\rho_1 = 0; \qquad \rho_2 = 0; \qquad \rho_3 = 0; \qquad \rho_4 = 0$$

$$\mathbf{p}_1 = 10\hat{\mathbf{x}}; \qquad \mathbf{p}_2 = 14\hat{\mathbf{x}}; \qquad \mathbf{p}_3 = 18\hat{\mathbf{x}}; \qquad \mathbf{p}_4 = 22\hat{\mathbf{x}}$$

$$\mathbf{P}_1 = 2.5\hat{\mathbf{x}}; \qquad \mathbf{P}_2 = 3.5\hat{\mathbf{x}}; \qquad \mathbf{P}_3 = 4.5\hat{\mathbf{x}}; \qquad \mathbf{P}_4 = 5.5\hat{\mathbf{x}}$$

It is to be noted that the centers of the volume cells are at $x = 10$, 14, 18, and 22, respectively. The volume density of charge is zero since each volume element contains only bound charges and is therefore neutral. An interpolation for this mode of subdivision yields the continuous functions for subdivision I (Fig. 1.8-4):

$$\rho_I = 0; \qquad \mathbf{P}_I = 0.25x\hat{\mathbf{x}}, \qquad 8 \le x \le 24$$

$$\rho_I = 0; \qquad \mathbf{P}_I = 0, \qquad x < 8, \quad x > 24$$

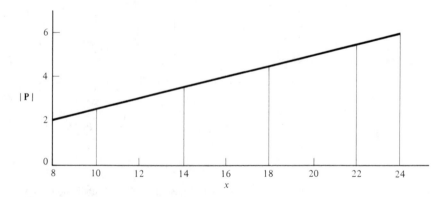

Figure 1.8-4 $|\mathbf{P}|$ as a function of x.

An extrapolation to the two surfaces results in

$$\hat{\mathbf{n}}_L \cdot \mathbf{P} = -\hat{\mathbf{x}} \cdot 2\hat{\mathbf{x}} = -2, \qquad \text{for the left-hand surface}$$

$$\hat{\mathbf{n}}_R \cdot \mathbf{P} = \hat{\mathbf{x}} \cdot 6\hat{\mathbf{x}} = 6, \qquad \text{for the right-hand surface}$$

The electrical properties of the volume V are completely described.

Now let the mode of subdivision be changed by separating a thin surface layer on each side with a boundary that cuts through the individual bound groups or atoms as shown in Fig. 1.8–2 (dashed lines). Each new volume element is found to be practically unpolarized but charged uniformly. Each surface element is charged. Hence ρ and η have nonvanishing values, while $\mathbf{P}$ is zero throughout. The functions, ρ, $\mathbf{p}$, and $\mathbf{P}$ may be determined for the new mode of subdivision II with both surface and volume cells, using as before,

$$\rho_i = \frac{\sum e_j}{\Delta \tau_i}; \qquad \mathbf{P}_i = \frac{\sum e_j \mathbf{d}_j}{\Delta \tau_i} = \frac{\mathbf{p}_i}{\Delta \tau_i}$$

With $\Delta \tau_i = 4$, it follows that

$$\rho_1 = -\tfrac{1}{4}; \qquad \rho_2 = -\tfrac{1}{4}; \qquad \rho_3 = -\tfrac{1}{4}; \qquad \rho_4 = -\tfrac{1}{4}$$

Similarly,

$$\mathbf{P}_1 = 0; \qquad \mathbf{P}_2 = 0; \qquad \mathbf{P}_3 = 0; \qquad \mathbf{P}_4 = 0$$

This follows from the fact that

$$\mathbf{p} = q_1 \mathbf{d}_1 + q_2 \mathbf{d}_2 + q_3 \mathbf{d}_3$$

$$= -ed_1(-\hat{\mathbf{x}}) + ed_1 \hat{\mathbf{x}} - e(2d_1)\hat{\mathbf{x}}$$

$$= 0$$

as shown in Fig. 1.8–5. The surface densities of charge on the left-hand and right-hand surfaces (η_L and η_R, respectively) are given by

$$\eta_L = \frac{-2}{1} = -2 = \hat{\mathbf{n}}_L \cdot \mathbf{P}_I$$

$$\eta_R = \frac{6}{1} = 6 = \hat{\mathbf{n}}_R \cdot \mathbf{P}_I \qquad (\Delta \tau_s = 1 \times 1)$$

An interpolation yields the following continuous functions for subdivision II:

$$\rho_{II} = -0.25; \qquad \mathbf{P}_{II} = 0; \qquad \eta_{IIL} = -2; \qquad \eta_{IIR} = 6$$

These functions describe completely the volume V for the second mode of subdivision.

If both of the modes of subdivision that have been described are discarded, and no attempt is made to have the boundaries cut or not cut through bound groups of charges, statistical conditions prevail along and across the boundaries of the

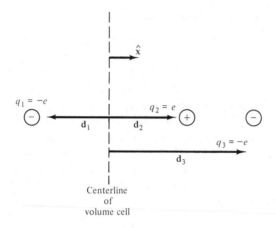

Figure 1.8–5   Dipole moment **p** for subdivision II.

Centerline
of
volume cell

volume cells just as in their interior. There will be, on the average, some bound groups near the boundaries of the cells completely within a cell, some partly in and partly out so as to be cut by the boundary line. On the average there will be more partly in and partly out groups on the right-hand boundary of each cell than on the left, because the number of polarized groups increases toward the right throughout the region. Thus the characteristics of the two special modes of subdivision prevail simultaneously. Each volume cell is both charged and polarized so that both ρ and **P** have nonvanishing values. At the surface η is likewise nonvanishing. The following questions must be answered: How can η, ρ, and **P** be combined to represent the magnitude, distribution, and orientation of charge throughout the volume and on all surfaces in a way that is unambiguous and independent of the mode of subdivision? Can this representation be extended to regions in which free charges as well as bound-charge groups are present? Let these questions be answered first for the interior in terms of ρ and **P** and later for the surfaces in terms of η and **P**.

1.9 ESSENTIAL VOLUME CHARACTERISTIC OF THE STATIC STATE: DIVERGENCE OF A VECTOR

The relationship between ρ and **P** in the interior of a region may be determined by examining what happens to the densities that describe the condition of charge in a typical volume cell (such as $\Delta\tau_2$ in Fig. 1.8–2) when the subdividing boundaries are shifted just enough to reduce the polarization in the cell to zero. Each dipole (or polarized atom or molecule) is assumed to be statistically fixed as shown one-dimensionally and schematically in Fig. 1.8–2. With subdivision I, there are only complete, neutral dipoles in $\Delta\tau_2$, so that $\rho_2 = 0$ (Fig. 1.9–1). When the boundary is moved to form subdivision II, four negative charges are included in $\Delta\tau_2$ on the right, while only three negative charges are excluded on the left. The net result is an addition of one negative charge. In subdivision I, no dipoles are cut by the boundaries of the cell. In subdivision II, three dipoles are cut by the left-hand boundary, four by the right-hand boundary. Therefore, three polarization vectors

$$\mathbf{p} = e\mathbf{d} \qquad (1.9\text{--}1)$$

pierce the left-hand boundary normally and pointing to the right, whereas four such vectors cross the right-hand boundary also directed to the right. In other words, there is an outwardly directed excess of one vector $\mathbf{p} = e\mathbf{d}$ pointing perpendicularly across the boundaries. Correspondingly, there is an excess of one negative charge in the cell, or a deficit of one positive charge. It follows from this highly simplified picture that there is a one-to-one correspondence between the number of negative charges appearing inside a volume cell in subdivision II and the excess of outwardly directed polarization vectors that pierce the cell walls normally in changing from I to II.

The volume function $\mathbf{P}$ measures the average density of polarization vectors due to individual dipoles or their equivalents in a small region about any point. Accordingly, the component of $\mathbf{P}$ directed along the outward or external normal to a closed surface at any point,

$$P_n = \hat{\mathbf{n}} \cdot \mathbf{P} \qquad (1.9\text{--}2)$$

is the magnitude of the average sum of the outwardly directed normal components of the elementary polarization vectors that pierce a unit area of the surface on which P_n is defined. Hence the surface integral

$$\int_{\Sigma} \hat{\mathbf{n}} \cdot \mathbf{P} \, d\sigma \qquad (1.9\text{--}3)$$

is a measure both of the number of elementary polarization vectors that pierce Σ normally and of the total positive charge that leaves (or the total negative charge that enters) the volume enclosed by the surface Σ ($\hat{\mathbf{n}}$ = unit external normal) when

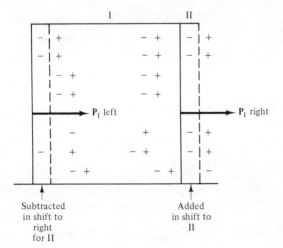

Subtracted
in shift to
right
for II

Added
in shift to
II

Figure 1.9–1   Change from subdivision I to subdivision II.

the mode of subdivision is changed to make **P** vanish. Accordingly, the net outflow of positive charge per unit volume in the change in subdivision is given by

$$
\begin{pmatrix} + \text{ charge removed or } - \text{ charge} \\ \text{added per unit volume in change} \\ \text{from subdivision I to subdivision II} \end{pmatrix} = \frac{\int_{\Sigma_i} \hat{\mathbf{n}} \cdot \mathbf{P}_\mathrm{I} \, d\sigma}{\Delta\tau_i} \qquad (1.9\text{--}4)
$$

Here Σ_i is the surface of $\Delta\tau_i$. The subscript I on **P** indicates that $\mathbf{P}_\mathrm{I}$ is determined using subdivision I. In subdivision II, $\mathbf{P}_\mathrm{II} = 0$. Alternatively,

$$
\begin{pmatrix} + \text{ charge added or } - \text{ charge removed} \\ \text{per unit volume in change from} \\ \text{subdivision I to subdivision II} \end{pmatrix} = \frac{-\int_{\Sigma_i} \hat{\mathbf{n}} \cdot \mathbf{P}_\mathrm{I} \, d\sigma}{\Delta\tau_i} \qquad (1.9\text{--}5)
$$

An added positive charge per unit volume in $\Delta\tau_i$ may be expressed as a volume density of bound charge $(\rho_{b\mathrm{II}})_i$ defined at the center of $\Delta\tau_i$. Thus

$$
(\rho_{b\mathrm{II}})_i = \frac{-\int_{\Sigma_i} \hat{\mathbf{n}} \cdot \mathbf{P}_\mathrm{I} \, d\sigma}{\Delta\tau_i} \qquad (1.9\text{--}6)
$$

A volume density of charge $(\rho_{b\mathrm{II}})_i$ due to polarization might be defined for each volume element and a continuous volume function interpolated from these discrete values. Since **P** is such an interpolated continuous function defined at every point, this is not necessary and the ordinary limit process of the calculus may be used. Thus

$$
\rho_{b\mathrm{II}} = \lim_{\Delta\tau \to 0} \frac{-\int_{\Sigma} \hat{\mathbf{n}} \cdot \mathbf{P}_\mathrm{I} \, d\sigma}{\Delta\tau} \qquad (1.9\text{--}7)
$$

The operation of taking the limit of the total outward normal flux of a vector such as **P** is called *evaluating the divergence of the vector.*

$$
\text{div } \mathbf{P} \equiv \nabla \cdot \mathbf{P} = \lim_{\Delta\tau \to 0} \frac{\int_{\Sigma} \hat{\mathbf{n}} \cdot \mathbf{P}_\mathrm{I} \, d\sigma}{\Delta\tau} \qquad (1.9\text{--}8)
$$

The divergence is a fundamental vector operator. With (1.9–8), (1.9–7) becomes

$$
\rho_{b\mathrm{II}} = -\nabla \cdot \mathbf{P}_\mathrm{I} \qquad (1.9\text{--}9)
$$

The electrical description of a bound-charge region may be written as follows:

$$
\begin{array}{lll}
\text{Subdivision I:} & \mathbf{P} = \mathbf{P}_\mathrm{I}; & \rho_{b\mathrm{I}} = 0 \\[4pt]
\text{Subdivision II:} & \mathbf{P}_\mathrm{II} = 0; & \rho_b = \rho_{b\mathrm{II}} = -\nabla \cdot \mathbf{P}_\mathrm{I}
\end{array} \qquad (1.9\text{--}10)
$$

The *interior* of a region containing only distorted, bound-charge atoms or statistically equivalent simple dipoles may be described either in terms of **P** alone with $\rho = 0$ using subdivision I or in terms of ρ alone with **P** $= 0$ using subdivision II.

If the region contains free charges in addition to distorted closed-shell atoms or other equivalent dipoles, the continuous functions **P** and ρ are constructed as defined above. If subdivision I is selected, none of the boundaries of the $\Delta\tau_i$ cells cuts through an atom or a dipole, so that polarization contributes nothing to ρ. Hence $\rho = \rho_f$, the volume density due to free charges only, and has a nonvanishing value. If subdivision II is chosen, the boundary surfaces of the $\Delta\tau_i$ cells are by definition so placed that they cut through enough dipoles to make the polarization in each interior cell vanish. The volume density of charge due to the free charges is independent of small shifts in the bounding surfaces, so that ρ_f will be the same as before. However, ρ, if determined according to definition, includes not only ρ_f but also ρ_b, due to the charges actually present in each cell from parts of dipoles cut by the boundaries. Hence the volume density of charge interpolated from the discrete values ρ_i is

$$\rho = \rho_f + \rho_b \tag{1.9-11}$$

But ρ_{bII} is equal to $-\mathbf{\nabla} \cdot \mathbf{P}_I$, so that $\rho_I - \mathbf{\nabla} \cdot \mathbf{P}_I$ is equal to ρ_{II}. If an intermediate mode of subdivision III is used in which, for example, only one-half of the dipoles necessary to reduce the polarization to zero are cut by the boundaries, $\rho_{bIII} = \frac{1}{2}\rho_{bII}$, $\mathbf{P}_{III} = \frac{1}{2}\mathbf{P}_I$; ρ_f is unchanged, and $\rho_{III} - \mathbf{\nabla} \cdot \mathbf{P}_{III}$ has the same value as ρ_{II} or as $\rho_I - \mathbf{\nabla} \cdot \mathbf{P}_I$. Thus

$$\bar{\rho} \equiv \rho - \mathbf{\nabla} \cdot \mathbf{P} \tag{1.9-12}$$

called the *essential volume density of charge* characteristic of the interior of a region, is *independent of the mode of subdivision* of that region into volume cells. If subdivision I is used, **P** has a nonvanishing value characteristic of the $\Delta\tau_i$ elements as constructed and ρ measures only ρ_f, the volume density of free charge. If subdivision II is used, **P** is zero and ρ measures a volume density equal to the sum of the volume densities of free charge ρ_f and of bound charge ρ_{bII}. If an intermediate subdivision is used, in particular, if statistical conditions obtain along the cell boundaries, **P** has a value smaller than $\mathbf{P}_I$ that depends on the distribution of dipoles, which are on the average partly in and partly out of the cells. ρ is then correspondingly larger than ρ_I. When integrated over a closed surface Σ, $-\int_\Sigma \hat{\mathbf{n}} \cdot \mathbf{P} \, d\sigma$ has a nonvanishing value only if **P** is not constant. If **P** is constant, the integral is zero everywhere and

$$\rho_{bII} = -\mathbf{\nabla} \cdot \mathbf{P}_I = 0 \tag{1.9-13}$$

In order to describe the condition of charge in the interior of a body, this may be divided into volume cells of correct size in any convenient way whatsoever. The discrete values ρ_i and $\mathbf{P}_i$ may be defined for the $\Delta\tau_i$ cells as constructed and the continuous functions ρ and **P** interpolated from them. The essential density $\rho - \mathbf{\nabla} \cdot \mathbf{P}$ is then a scalar point function defined throughout the region, continuous and slowly varying, and independent of the mode of subdivision. It characterizes the condition of charge at all points in the interior.

1.10 ESSENTIAL SURFACE CHARACTERISTIC OF THE STATIC STATE

In Sec. 1.9 it was shown that the function $(\rho - \nabla \cdot \mathbf{P})$ characterizes the interior of a region containing atoms constructed according to either the free-charge or the bound-charge model in a way that is independent of the mode of subdivision of the region into volume cells. In particular, it was demonstrated that this function is invariant to a change in subdivision from mode I to mode II. Mode I is a subdivision into volume cells only, with dividing lines drawn so that no atoms or molecules are cut by cell boundaries. Mode II is a subdivision into both volume and surface cells, with cell walls so placed that a thin layer of charges is treated separately around the surface and the average polarization of each cell in the interior is zero.

It has already been shown that a distribution of charge can be characterized completely using the second mode of subdivision in terms of a surface density of charge η and a volume density of charge ρ. The question now arises whether asymmetrical surface conditions (which are described in terms of ρ in subdivision II) as well as distributions of charge in the interior can be represented completely by the volume functions ρ and $\mathbf{P}$.

Both of these functions are interpolated from discrete values defined at the centers of thick interior cells. Moreover, since ρ is certainly unable to take account of thin layers of surface charge regardless of whether the surface condition is pictured as due to free charges or due to slices cut from polarized atoms or molecules, it follows that $\mathbf{P}$, although also a slowly varying volume function, must be able to describe surface conditions if a representation in terms of ρ and $\mathbf{P}$ alone is at all possible. This is evidently true only if the entire surface effect can be considered to be a part of an essentially volume phenomenon as in the bound-charge model, in which the distortion and orientation of atoms in the outermost layer of volume cells is assumed to be the same as in the interior. Accordingly, the slowly varying and continuous function $\mathbf{P}$ may be extrapolated to the edge of each volume cell along the surface. This extrapolated value, in particular its normal component $\hat{\mathbf{n}} \cdot \mathbf{P}$, is the characterization at the surface of a polarization effect throughout the interior. It must, therefore, replace η in describing surface conditions of charge using ρ and $\mathbf{P}$ above insofar as the bound-charge model is concerned.

In order to see how the normal component of $\mathbf{P}$ in subdivision I (volume cells only) plays the part of η in subdivision II (volume and surface cells), it is instructive to examine what happens to a thin layer of charges along both sides of the bounding surface between two electrically dissimilar regions when a change is made from subdivision I to subdivision II. Let each region be divided first into volume cells only according to scheme I with ρ_1 and $\mathbf{P}_1$ defined on one side and ρ_2 and $\mathbf{P}_2$ on the other side of the boundary. With only neutral polarizable units assumed to be present, $\rho_1 = \rho_2 = 0$. At the boundary, the normal components of the polarization vectors defined on each side are given by $\hat{\mathbf{n}}_1 \cdot \mathbf{P}_1$ and $\hat{\mathbf{n}}_2 \cdot \mathbf{P}_2$, with $\hat{\mathbf{n}}_1$ and $\hat{\mathbf{n}}_2$ outwardly directed unit normals referred to the volume indicated by the subscript. This is illustrated in Fig. 1.10–1.

The change to subdivision II is made by pulling apart the imaginary envelopes of the two volume cells $\Delta\tau_i$, one on each side of the surface, to leave room between them for the two thin surface cells of combined thickness $2d_c$, as shown in Fig. 1.10–1. Graphically, this simply means that all charges in a thin layer on each side of the boundary are removed from the volume cells $\Delta\tau_i$ as these are shifted away from the surface, and left in the newly constructed surface cells. As in the analogous case described for the interior, the total outward normal flux of the vector $\mathbf{P}$ across a closed surface measures the total positive charge (due to cut polarized units) that is moved out from within the surface as the mode of subdivision is changed from I to II.

The total positive charge that moves into the thin disk $2d_c\Delta\Sigma$ as this is formed is given by

$$-\int_{\Sigma} \hat{\mathbf{n}} \cdot \mathbf{P} \, d\sigma \qquad (1.10\text{–}1)$$

Here $\hat{\mathbf{n}}$ is the unit exterior normal to the surfaces enclosing the thin disk and Σ is its entire superficial area. In changing the subdivision from I to II by moving the volume cells apart, the surfaces $\Delta\Sigma_{1,2}$ which are parallel to the boundary between the regions are pierced by the charges that appear in the newly formed surface cells of combined volume $2d_c\Delta\Sigma$. The only significant contributions to this integral

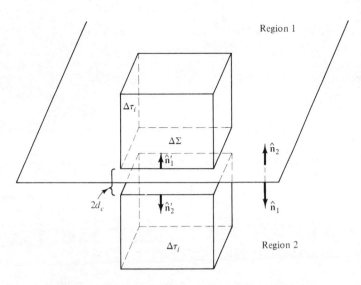

Figure 1.10–1   Volume and surface cells at a boundary.

come from the two equal areas $\Delta\Sigma_1$ and $\Delta\Sigma_2$. The thin edges are not similarly pierced, and they are in any case negligible in size compared with $\Delta\Sigma_1$ and $\Delta\Sigma_2$. Thus the integral

$$-\int_{\Sigma} \hat{\mathbf{n}} \cdot \mathbf{P} \, d\sigma = -\left(\int_{\Delta\Sigma_1} \hat{\mathbf{n}}_1' \cdot \mathbf{P}_1 \, d\sigma + \int_{\Delta\Sigma_2} \hat{\mathbf{n}}_2' \cdot \mathbf{P}_2 \, d\sigma \right) \qquad (1.10\text{--}2)$$

measures the total positive charge that was left in the two thin surface cells, each of volume $d_c\Delta\Sigma$, as these were formed by the moving apart of the volume cells $\Delta\tau_i$. Here $\hat{\mathbf{n}}_1'$ points *into* region 1, $\hat{\mathbf{n}}_2'$ *into* region 2, since both are external normals to the thin surface cells of combination volume $2d_c\Delta\Sigma$. The average positive charge per unit volume in these cells is obtained by dividing (1.10–2) by the volume $2d_c\Delta\Sigma$. Because $\mathbf{P}$ is a continuous function, it is permissible to pass to the limit and let $\Delta\Sigma$ approach zero to obtain the average density of bound charge in the surface layers due to contributions from cut, polarized units. With $\Delta\Sigma_1 = \Delta\Sigma_2 = \Delta\Sigma$ it is

$$(\rho_b)_{\text{av}} = \lim_{\Delta\Sigma\to 0} \left(\frac{\int_{\Delta\Sigma_1} \hat{\mathbf{n}}_1 \cdot \mathbf{P}_1 \, d\sigma}{2d_c\Delta\Sigma_1} + \frac{\int_{\Delta\Sigma_2} \hat{\mathbf{n}}_2 \cdot \mathbf{P}_2 \, d\sigma}{2d_c\Delta\Sigma_2} \right) \qquad (1.10\text{--}3)$$

Here $\hat{\mathbf{n}}_1 (= -\hat{\mathbf{n}}_1')$, $\hat{\mathbf{n}}_2 (= -\hat{\mathbf{n}}_2')$ are normals directed outwardly with respect to the region indicated by the subscript.

The average density of charge in the two layers on each side of the boundary is equal to one-half the sum of the individual densities defined in each region. Thus

$$(\rho_b)_{\text{av}} = \tfrac{1}{2}(\rho_{1b} + \rho_{2b}) \qquad (1.10\text{--}4)$$

A surface density of charge η may be defined as in (1.4–1):

$$\eta = \rho d_c \qquad (1.10\text{--}5a)$$

Accordingly,

$$d_c(\rho_{1b} + \rho_{2b}) = \eta_{1b} + \eta_{2b} \qquad (1.10\text{--}5b)$$

The multiplication of (1.10–3) by the constant $2d_c$ and the use of (1.10–4) and (1.10–5) lead to

$$\eta_{1b} + \eta_{2b} = \lim_{\Delta\Sigma\to 0} \left(\frac{\int_{\Delta\Sigma} \hat{\mathbf{n}}_1 \cdot \mathbf{P}_1 \, d\sigma}{\Delta\Sigma} + \frac{\int_{\Delta\Sigma} \hat{\mathbf{n}}_2 \cdot \mathbf{P}_2 \, d\sigma}{\Delta\Sigma} \right) \qquad (1.10\text{--}6)$$

The theorem of the mean for integrals permits writing

$$\int_{\Delta\Sigma} \hat{\mathbf{n}} \cdot \mathbf{P} \, d\sigma = \hat{\mathbf{n}} \cdot \overline{\mathbf{P}} \int_{\Delta\Sigma} d\sigma = \hat{\mathbf{n}} \cdot \overline{\mathbf{P}} \, \Delta\Sigma \qquad (1.10\text{--}7)$$

where $\hat{\mathbf{n}} \cdot \overline{\mathbf{P}}$ is a mean value at some point on $\Delta\Sigma$. Hence

$$\eta_{1b} + \eta_{2b} = \lim_{\Delta\Sigma \to 0} \left(\frac{\hat{\mathbf{n}}_1 \cdot \overline{\mathbf{P}}_1 \, \Delta\Sigma}{\Delta\Sigma} + \frac{\hat{\mathbf{n}}_2 \cdot \overline{\mathbf{P}}_2 \, \Delta\Sigma}{\Delta\Sigma} \right) \tag{1.10-8}$$

In passing to the limit, $\Delta\Sigma$ is shrunk down on the point at which $\hat{\mathbf{n}} \cdot \overline{\mathbf{P}}$ is defined. It is then no longer necessary to indicate a mean value because the surface has zero area. Let the following shorthand symbols be defined at every point along a surface of discontinuity:

$$\eta_b \equiv \eta_{1b} + \eta_{2b} \tag{1.10-9a}$$

$$\hat{\mathbf{n}} \cdot \mathbf{P} \equiv \hat{\mathbf{n}}_1 \cdot \mathbf{P}_1 + \hat{\mathbf{n}}_2 \cdot \mathbf{P}_2 \tag{1.10-9b}$$

With this notation (1.10-8) may be written in the following concise form with explicitly shown subscripts I and II to designate the mode of subdivision used in defining the densities:

$$\eta_{bII} = \hat{\mathbf{n}} \cdot \mathbf{P}_I \tag{1.10-10}$$

A value of the function η_b may be associated with every point along the surface separating regions 1 and 2. It measures the total surface density of charge in both regions associated with that point using subdivision II. It is important to bear in mind that η_b is a shorthand symbol for the sum of the two scalar functions η_{1b} and η_{2b} defined, respectively, in regions 1 and 2 at distances $d_c/2$ from each side of the boundary.

It is not difficult to see that the function $\eta + \hat{\mathbf{n}} \cdot \mathbf{P}$ is independent of the mode of subdivision. If only volume cells are used with a subdivision so devised that interior cells are all unpolarized, η describes the surface. If surface cells are used, but statistical conditions prevail across all cell boundaries so that the number of bound charges required to reduce the polarization to zero throughout the interior is not necessarily cut off by cell walls, η includes bound charges cut off by the walls of the surface cells while $\hat{\mathbf{n}} \cdot \mathbf{P}$ adds to η the contribution to the surface effect due to whatever volume polarization has not been canceled in the subdivision. It thus appears that the function $(\eta + \hat{\mathbf{n}} \cdot \mathbf{P})$ is indeed independent of the mode of subdivision into cells. It is a scalar point function that may be associated with every point on the boundary between two regions. It is called the *essential surface density of charge*. For the present it applies only to bodies containing no free charge. Let it be denoted by the symbol

$$\overline{\eta} \equiv \eta + \hat{\mathbf{n}} \cdot \mathbf{P} \tag{1.10-11}$$

Throughout the discussion above, it was assumed that the body or region was constructed according to the bound-charge model. This was necessary because all surface effects were looked upon as parts of a volume phenomenon in order that they could be described in terms of the normal component of the volume function $\mathbf{P}$ in the form $\hat{\mathbf{n}} \cdot \mathbf{P}$ wherever a subdivision into volume cells was used. When free charges are present, a distinct surface layer of charges may exist which is definitely not part of a volume phenomenon. Consequently, a description in terms of $\hat{\mathbf{n}} \cdot \mathbf{P}$ alone is apparently not possible, so that $\bar{\eta} \equiv \eta + \hat{\mathbf{n}} \cdot \mathbf{P}$ seems to be independent of the mode of subdivision only in the complete absence of free charges. If free charges are present, η includes a contribution η_f due to the free charges and a contribution η_b due to the bound charges contained in the surface cells ($\eta = \eta_f + \eta_b$). If the mode of subdivision is changed to volume cells only, η_b is represented by an equivalent value of $\hat{\mathbf{n}} \cdot \mathbf{P}$, whereas neither ρ nor $\mathbf{P}$ can take account of the free surface charge represented by η_f. Before concluding that ($\eta + \hat{\mathbf{n}} \cdot \mathbf{P}$) is useful (because of its independence of the mode of subdivision) only for the bound-charge model, the following must be considered. Given a pair of functions ρ_I and $\mathbf{P}_\mathrm{I}$ it is certainly mathematically possible to construct two different functions ρ_II and η_II which satisfy the pair of equations

$$\rho_\mathrm{II} = \rho_\mathrm{I} - \boldsymbol{\nabla} \cdot \mathbf{P}_\mathrm{I}; \qquad \eta_\mathrm{II} = \hat{\mathbf{n}} \cdot \mathbf{P}_\mathrm{I} \qquad (1.10\text{–}12)$$

The required values of ρ_II and η_II can be determined by solving these equations. In terms of physical models, the electrostatic properties of a body A are described completely by ρ_II and η_II with a subdivision into volume and surface cells, while the corresponding properties of a body B are described by ρ_I and $\mathbf{P}_\mathrm{I}$ with a subdivision into volume cells only. As a consequence of (1.10–12) the two bodies are electrostatically identical. Yet body A may contain an abundance of free charge in a surface layer while body B can have no free surface charge. It thus appears that every free-charge body described by ρ_II and η_II has a bound-charge counterpart described by ρ_I and $\mathbf{P}_\mathrm{I}$ that is indistinguishable from it in every way that can be described mathematically in terms of two continuous functions.

Two bodies that are electrostatically indistinguishable may be substituted for each other at will in all static-state conditions. Consequently, if every free-charge model is replaced mathematically by its bound-charge counterpart (whether this has a physical meaning or not), the function $\bar{\eta} \equiv \eta + \hat{\mathbf{n}} \cdot \mathbf{P}$ may be used to describe mathematically the surface conditions of both bound- and free-charge models.

This entire formulation may be criticized by arguing that the actual construction of the functions $\bar{\eta} \equiv \eta + \hat{\mathbf{n}} \cdot \mathbf{P}$ and $\bar{\rho} \equiv \rho - \boldsymbol{\nabla} \cdot \mathbf{P}$ is hopelessly intricate from the physical point of view. It involves sorting out the free charges and constructing ρ_f and η_f to describe their distribution; defining new functions ρ, η, and $\mathbf{P}$ to characterize the bound-charge counterpart of the free-charge distribution; and finally combining the functions so obtained with the corresponding functions constructed for the bound charges actually present. Such a criticism, however, overlooks the purpose of the formulation in terms of density functions. It is because it is impossible physically to construct any one of the functions ρ, η, and $\mathbf{P}$ as defined in terms of the distribution of countless millions of charges that the continuous functions ρ, η, and $\mathbf{P}$ have been introduced. It has been shown that the principal statistical properties of charge and its distribution can be described in a general way within a degree of approximation by the two functions $\bar{\rho}$ and $\bar{\eta}$. By building these functions into the mathematical model of electromagnetism and relating them to experimental analogues whenever possible, conclusions may be drawn regarding their structure in a variety of cases. It is then possible in any particular instance to interpret this structure in terms of a bound-charge model, a free-charge counterpart, or a combination of the two using any convenient and appropriate mode of subdivision. A choice can be made to depend entirely on physical reasons, because the mathematical specification of a given set of functions ρ, η, $\mathbf{P}$ does not demand a particular interpretation or mode of subdivision.

1.11 *SURFACE DENSITY OF POLARIZATION: DOUBLE LAYER*

The mathematical representation of surface distributions of charge by means of the essential surface density $\overline{\eta}$ does not attain the same degree of approximation that is achieved in the description of interior conditions in terms of the essential volume characteristic $\overline{\rho}$. This follows directly from the fact that two volume functions but only a single surface function have been provided. Thus, in the interior ρ describes the statistical distribution of discrete charges and $\mathbf{P}$ takes account of the distribution of polarized units corresponding to statistically simple dipoles. On the other hand, $\overline{\eta}$ describes only the statistical distribution of charges in a thin layer along the surface. The term $\mathbf{\hat{n}} \cdot \mathbf{P}$ which contributes to $\overline{\eta}$ must not be mistaken to be a separate surface function. Like η, it is able to describe only a surface distribution of simple charges, not a *surface distribution of dipoles*. Actually, none of the three functions ρ, η, $\mathbf{P}$ is suited to represent a distinctly surface layer of simply polarized atoms or molecules as illustrated in Fig. 1.11–1.

Such a thin surface double layer of charge is entirely plausible from the point of view of the atomic model. Certainly, the outermost layer of atoms at a surface is subject to asymmetrical forces that are not so much a result of the presence of separate charged bodies in the vicinity as of the essential one-sidedness of the intra-atomic forces themselves. The electron shells of the outermost atoms at a surface experience the repulsion of similar shells due to neighboring atoms only on three sides. A distortion of the outer shells of electrons on the exposed sides in the form of a shift relative to the positive nuclei might certainly be expected. Such a shift would naturally occur only in a direction normal to the boundary, and its effect would be precisely that of a double layer of charge of opposite sign or of a single layer of dipoles. If this effect exists, it is an intrinsic property of the atomic structure of the model in question. A surface polarization due to the action of external forces, while possible, is certainly negligible compared with an intrinsic surface polarization. A variation in forces due to relatively distant, externally situated charged bodies across the thickness d_c of a surface cell must be small compared

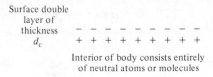

Surface double
layer of
thickness
d_c

Interior of body consists entirely
of neutral atoms or molecules

Figure 1.11–1 Surface polarization or double layer.

with the strong intra-atomic forces which are brought into play in establishing precisely that equilibrium which results in the intrinsic surface polarization.

The mathematical representation of a surface double layer, as pictured in the physical model, requires an additional density function $\mathbf{k}$ that can be defined in terms of $\mathbf{P}$ in a way quite analogous to the definition of η in terms of ρ. Thus the average density of polarization $\mathbf{P}_s$ at the center of a thin surface cell $\Delta\tau_s$ of thickness d_c and of volume $d_c\Delta\Sigma$ is

$$\mathbf{P}_s = \frac{\mathbf{p}_s}{d_c\Delta\Sigma_s} \tag{1.11-1}$$

A value $\mathbf{k}_{si}$ can be defined at the center of each surface cell according to

$$\mathbf{k}_{si} = d_c\mathbf{P}_s = \frac{d_c\mathbf{p}_s}{\Delta\tau_s} = \frac{\mathbf{p}_s}{\Delta\Sigma_s} \tag{1.11-2}$$

A continuous, slowly varying vector point function $\mathbf{k}_s$ can then be interpolated from the discrete values $\mathbf{k}_{si}$ and in this way defined throughout the surface shell. Since the surface asymmetry is entirely in a direction perpendicular to the boundary, the polarization vector $\mathbf{k}_s$ must be directed everywhere normal to the surface. Any polarization effect tangent to the surface differs in no essential way from the interior polarization, which is completely characterized by the volume function.

The surface function $\mathbf{k}_s$ plays no very significant part in the mathematical model of electromagnetism. In fact, its nature is such that it is difficult to devise experiments to involve it in an essential way. It is included here for completeness.

1.12 THE STATIC STATE AND ITS MATHEMATICAL MODEL: SUMMARY

The physical model of matter in the static state is constructed to conform to a picture of atoms and molecules built of positively and negatively charged units that are statistically at rest. Two special models are used. In the free-charge model an appreciable proportion of the charges experience no restraining constitutive forces over a time average to prevent them from moving freely under the action of a suitable external force. In the bound-charge model all the charges are constrained by strong forces of constitution to remain in definite and characteristic statistical groups, which can be distorted and oriented by external forces, but from which no charges can be removed over a time average. Models in which only a relatively small proportion of free electrons exist can be interpreted in terms of the combined properties of the extremes. It is not assumed that these models are necessarily counterparts of structures actually existing in nature. It is merely affirmed that close experimental analogues can be found for theoretical properties derived from these distinct types and from combinations of them. As an analytically convenient substitute for the intricate mechanism represented by a volume containing countless millions of atoms, two scalar point functions are defined throughout a charged region in terms of its average statistical properties. The atomic picture of discrete charges, each at a time-average location, is replaced by a mathematical model consisting of continuous functions defined at every point in a region or body. One of these functions characterizes the surface; the other the interior. The two functions $\bar{\eta} = \eta + \hat{n} \cdot \mathbf{P}$ and $\bar{\rho} = \rho - \nabla \cdot \mathbf{P}$ are the *essential electrical densities of charge*. They are expressed in terms of one vector and two scalar point functions. The scalar η is defined only in a thin surface layer; the scalar ρ and the vector $\mathbf{P}$ are defined throughout the volume. Mathematically, any two of the three functions are sufficient to represent uniquely the two general scalar characteristics $\bar{\rho}$ and $\bar{\eta}$ so that a given region may be characterized by a volume and a surface density of charge or by volume densities of charge and of polarization if in each case an appropriate mode of subdivision is used. In the first case, using ρ and η, surface and volume cells are required with the dividing boundaries of the cells so drawn that each cell is unpolarized. In the second case, only volume cells are needed and the dividing boundaries are so placed that no single atom is apportioned partly to one cell, partly to another. In any given atomic model one, or the other, or both of these special modes of subdivision may be quite unrealizable in a physically sensible way. For example, if a body is postulated which is composed entirely of atoms with closed electron shells, a subdivision into volume cells only and a representation in terms of ρ and $\mathbf{P}$ alone are physically reasonable, whereas a separate treatment of a surface layer and a subdivision of the interior that will cut through enough atoms to leave each cell unpolarized are not physically sensible. Exactly the reverse is true if a body is postulated which is composed of atoms with an abundance of free electrons. In this case there is no polarization due to distortion or orientation of atoms throughout the volume, so that a representation in terms of $\mathbf{P}$ is physically unreasonable. In both cases, however, it is possible to use an alternative model, not constructed as postulated, but which nevertheless exhibits exactly the same external electrostatic properties, and which is mathematically expressed in terms of a different pair of continuous functions. In other words, a body composed entirely of bound-charge atoms can be represented (insofar as its electrostatic properties are concerned) by a suitably constructed free-charge model using η and ρ. Or a region containing only free charges can be represented by an electrostatic counterpart composed entirely of bound-charge groups and characterized by ρ and $\mathbf{P}$. This does not identify the bound-charge model with the free-charge one, nor does it affirm that they have the same structural characteristics. It does state that a given distribution of bound charges has its electrostatic free-charge counterpart, and vice versa. For practical purposes, either the given distribution or its counterpart may be used since they are electrostatically indistinguishable. A body containing bound groups and a few free electrons may be described in terms of either model, looking upon it physically on the one hand as an almost bound-charge model, or on the other hand as an almost free-charge model.

Although η, ρ, and $\mathbf{P}$ (and also $\mathbf{k}_s$) are defined in terms of special physical models, the representation in terms of $\bar{\eta}$ and $\bar{\rho}$ (and $\mathbf{k}_s$) transcends in its generality any special case. In fact, the static state is characterized primarily by a mathematical model consisting of two (or three) functions and the restrictions of continuity and slow variation imposed on them. Any physical model whatsoever, the properties of which can be described in terms of such functions, is as satisfactory as any other from the analytical point of view. The theory of the static state and subsequently of its periodic variation in time is based primarily on the characterization in terms of these mathematical functions and only secondarily (and actually not necessarily) on the atomic models that were used in defining and describing them. A future discovery or observation which suggests a different kind of physical model from that described, but which leads by one way or another to a representation in terms of mathematical functions with the properties of $\bar{\eta}$ and $\bar{\rho}$ (and $\mathbf{k}_s$), will not change the mathematical picture of the static state as here defined. Any such change in the model must not be mistaken for a change in the experimentally observable physical world. For neither the physical nor the mathematical model is more than a tool for deriving theoretical quantities that have direct, and, hence, predictable experimental analogues. And neither the method nor the particular model that is used in deriving these fundamentally significant quantities is necessarily of particular or permanent importance in the experimentally observed description of the physical world.

STATIONARY STATES: THE STEADY STATE

1.13 THE STEADY STATE AND THE ATOMIC MODEL

The statistical representation of the static state led to the definition of average or statistical rest positions for the positive and negative charges contained in a body or region. It was specifically emphasized that the individual charges might move in irregular ways about these rest positions, but that their motion was of such a nature that the time-average properties of the dynamic model of a body were identical with those of the statistically static model.

The steady state is a generalization of the static state. It is characterized by a steady drift or circulation of electric charges relative to the statistically stationary rest positions that characterize the static condition. In terms of the dynamic picture, a steady average flow may be imagined superimposed on the random motions of the charges. This does not mean that at any instant all the charges are moving in the direction of the drift or circulation. It simply means that over a time average more charges of one sign move in the direction of the drift or circulation than in any other. In a region containing only one kind of charge, free-charge atoms or positive and negative ions, a steady drift might consist of an uninterrupted flow at constant velocity of one kind of charge in a definite direction, or of two kinds of charge in opposite directions relative to statistical rest positions fixed in the region. Such a drift is called a *convection current*. A special form of convection current is a steady drift of free electrons relative to statistically stationary nuclei. Such a drift is called a *conduction current.* Another possible kind of nonrandom motion of charges is a microscopic circulation of electrons about their respective nuclei. If the individual axes of rotation in the different atoms are oriented at random, the net effect of such intra-atomic rotation (insofar as a volume cell containing millions of them is concerned) is precisely that of random motion. On the other hand, if more of the elementary whirls have axes parallel to one definite direction than to any other, then a nonrandom condition prevails. A steady microscopic circulation of electrons about parallel axes is called a *magnetization current.*

In characterizing the steady state in terms of continuous functions, a method entirely analogous to that used in the static state is followed. Corresponding to the volume density of static charge ρ, a volume density of drifting charge or of convection current **J** is defined. In order to take into account possible asymmetrical conditions at the surface of a region, a surface density of drifting charge or of convection current **K** is constructed in close analogy with the surface density of charge η. A nonrandom surface flow of charges may be interpreted in two ways. Corresponding to the static accumulation of free charge at a surface, a motion of free charge along a surface may be imagined. Similarly, corresponding to the static orientation and distortion of bound-charge groups to give an effective static surface charge, it is possible to picture an orientation and alignment of elementary current whirls or circulations throughout a volume to produce an effective moving surface charge. In order to describe this volume effect in a physically more reasonable manner, a volume density of magnetization **M** is introduced in a way entirely analogous to the definition in the static state of the volume density of polarization **P**. Finally, by suitably combining the three steady-state functions in a manner resembling that used in combining the three static-state functions, a single surface function and a single volume function may be defined which are independent of the mode of subdivision, and which are, then, the essential characteristics of the steady state.

1.14 VOLUME DENSITY OF MOVING CHARGE (CONVECTION CURRENT)

Suppose that there is a continuous stream of electric charges through a typical volume element $\Delta\tau_i$ subject to the same size restrictions as the elements used in the static state. Such a stream may take the form of what is called a *convection current* in which all the charges in $\Delta\tau_i$ at any instant are a part of the drift; or it may be the special form of a convection current, called a *conduction current,* in which only free electrons are moving in a steady stream while the positive charges are statistically at rest. A function that is to characterize a current of either kind must take account of the magnitude of the average moving charge in each volume element and of its average velocity. If a typical charge e_j is moving with a mean drift velocity $\mathbf{v}_j$ relative to a fixed reference frame in a region, the *quantity of moving charge* is defined to be $e_j\mathbf{v}_j(\mathbf{r})$. The quantity of moving charge in the volume element $\Delta\tau_i$ (Fig. 1.14–1), in which there are N charges, is the vector sum or the resultant of the individual quantities of moving charge. It is

$$\sum_{j=1}^{N} e_j\mathbf{v}_j(\mathbf{r})$$

Let the average quantity of moving charge in a volume cell be called its volume density of drifting charge or *volume density of convection or conduction current*; let it be denoted by $\mathbf{J}(\mathbf{r}_i)$. For the cell $\Delta\tau_i$

$$\mathbf{J}(\mathbf{r}_i) \;=\; \frac{\displaystyle\sum_{j=1}^{N} e_j\mathbf{v}_j(\mathbf{r})}{\Delta\tau_i} \tag{1.14–1}$$

This is a vector defined at the center of the small volume cell; it is a measure of the average drift of electric charges through the cell both in magnitude and direction. A similar vector may be defined at the center of each volume cell throughout a body or region. By interpolating from these discrete vectors, a continuous *vector point function* $\mathbf{J}(\mathbf{r})$ may be constructed which, by definition, assumes the value $\mathbf{J}(\mathbf{r}_i)$ at the center of each volume cell $\Delta\tau_i$ and which smoothly connects these at all intermediate points.

As in the static case, a microscopic definition of the volume density of convection or conduction current is given by

$$\mathbf{J}(\mathbf{r}) \;=\; \sum_{j=1}^{N} e_j\mathbf{v}_j(\mathbf{r})\delta(|\mathbf{r} - \mathbf{r}_j|) \tag{1.14–2}$$

A somewhat different, although equivalent, definition of the volume density of current is in terms of the volume density ρ' of that part of the charge that is engaged in nonrandom motion with mean drift velocity $\mathbf{u}$. In the definition of $\mathbf{J}$ formulated above, no specific account was taken of charges engaged in nonrandom motion as distinct from charges that are statistically at rest. It is, however, clear that if a charge e is statistically at rest so that its rest position is fixed, its drift

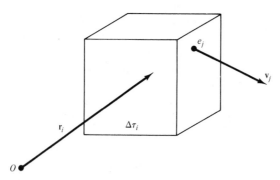

Figure 1.14–1   Volume element $\Delta\tau_i$ centered at $\mathbf{r}_i$ showing location and velocity of charge e_j.

velocity $\mathbf{v}$ is zero, and it contributes nothing to the volume density of current. If those charges in $\Delta\tau_i$ which are *not* statistically at rest are denoted by primes, it follows that

$$\mathbf{J}(\mathbf{r}_i) = \frac{\sum\limits_{j=1}^{N} e_j \mathbf{v}_j(\mathbf{r})}{\Delta\tau_i} = \frac{\sum\limits_{j=1}^{N'} e'_j \mathbf{v}_j(\mathbf{r})}{\Delta\tau_i} \qquad (1.14\text{--}3)$$

The average vector velocity of moving positive charge in $\Delta\tau_i$ is

$$\mathbf{u}(\mathbf{r}_i) = \frac{\sum\limits_{j=1}^{N'} e'_j \mathbf{v}_j(\mathbf{r})}{\sum\limits_{j=1}^{N'} |e'_j|} \qquad \left(\sum\limits_{j=1}^{N'} |e'_j| \neq 0\right) \qquad (1.14\text{--}4)$$

The volume density of charge $\rho'(\mathbf{r}_i)$ due to the *magnitude* of all charges engaged in drift motion in $\Delta\tau_i$ is defined in a way analogous to the definition of volume density of charge, that is,

$$\rho'(\mathbf{r}_i) = \frac{\sum\limits_{j=1}^{N'} |e'_j|}{\Delta\tau_i} \qquad (1.14\text{--}5)$$

With these two values substituted above,

$$\mathbf{J}(\mathbf{r}_i) = \mathbf{u}(\mathbf{r}_i)\rho'(\mathbf{r}_i) \qquad (1.14\text{--}6)$$

From discrete values of $\mathbf{J}(\mathbf{r}_i)$, $\mathbf{u}(\mathbf{r}_i)$, and $\rho'(\mathbf{r}_i)$ defined at the center of every volume element, a continuous scalar point function $\rho'(\mathbf{r})$, a continuous vector point function $\mathbf{u}(\mathbf{r})$, and a continuous vector point function $\mathbf{J}(\mathbf{r})$ are defined. These continuous functions are related by the expression

$$\mathbf{J}(\mathbf{r}) = \mathbf{u}(\mathbf{r})\rho'(\mathbf{r}) \qquad (1.14\text{--}7)$$

which defines the vector volume density of convection or conduction current $\mathbf{J}(\mathbf{r})$ in terms of the mean vector velocity of drifting charge $\mathbf{u}(\mathbf{r})$ and its volume density $\rho'(\mathbf{r})$. All three functions are continuous and slowly varying. The volume density of current differs from zero only when the density of charge in drift motion $\rho'(\mathbf{r})$ is not zero. If all the charges in the region are moving with the mean velocity $\mathbf{u}(\mathbf{r})$, then $\rho'(\mathbf{r}) = \rho(\mathbf{r})$ and $\mathbf{J}(\mathbf{r})$ gives the convection current density. If only electrons are moving with a mean drift velocity $\mathbf{u}(\mathbf{r})$, $\rho'(\mathbf{r})$ is the volume density of charge only for electrons in nonrandom motion and $\mathbf{J}(\mathbf{r})$ gives the conduction current density. The volume density of current has the dimensions of charge per unit time per unit area. The unit coulomb per second has the name *ampere* (A):

$$J \approx \frac{Q}{L^2 T} \qquad \text{A/m}^2 \qquad (1.14\text{--}8)$$

1.15 SURFACE DENSITY OF MOVING CHARGE (CONVECTION, CONDUCTION CURRENT)

As a result of asymmetrical conditions at the surface of a region or on the boundary between two dissimilar regions, the volume density of charge $\rho(\mathbf{r})$ is inadequate as a means of representing surface distributions of statistically stationary charge. In the same way and for similar reasons, the continuous slowly varying volume density of current $\mathbf{J}(\mathbf{r})$ is too insensitive to represent surface distributions of charge moving as a steady drift. Here, as before, the division of the region into volume cells alone is too coarse a mode of subdivision in the immediate vicinity of surfaces and boundaries to give proper weight to the rapid variations in current density that may exist there. This may be illustrated in terms of two relatively simple models which are analogous to the free-charge and bound-charge models in the static state. They are called the free-drift model and the spin model.

In the *free-drift model* there may be a thin layer of free electrons which is moving with a steady drift velocity along the surface while the charges in the interior are all statistically stationary so that the volume density of current $\mathbf{J}(\mathbf{r})$ as interpolated from the $\mathbf{J}(\mathbf{r}_i)$ of the volume cells $\Delta\tau_i$ is zero throughout the interior. For the volume cells along the surface, the $\mathbf{J}(\mathbf{r}_i)$ have nonvanishing values, but they are very small because the charges moving in a thin layer along the surface are averaged over the entire volume of the cell of which they occupy only a minute part. It is, therefore, necessary to treat a thin layer of cells along the surfaces separately.

In the *spin model*, the electrons form small intra-atomic whirls rotating in the same sense about mutually parallel axes. The net drift of charge through any volume element $\Delta\tau_i$ which contains a great many such microscopic whirls is zero for the element as a whole provided that the cell boundaries nowhere cut the paths of the individual whirls. Thus the volume density of current $\mathbf{J}(\mathbf{r}_i)$ as defined above is zero in every volume cell $\Delta\tau_i$, whether these are in the interior or along the surfaces. However, whenever the whirls are tangent to a surface or boundary, the net effect is that of a continuous sheet of current in a very thin layer in one direction. Thus a surface current can be constructed out of properly aligned elementary whirls existing throughout the volume. The surface sheet of current due to such interior whirls is no more taken into account by the function $\mathbf{J}(\mathbf{r})$ than is the surface sheet of charge due to distorted and oriented atoms represented by ρ in the analogous static case. Here, again, the difficulty may be overcome by treating a thin layer separately in terms of a specially constructed surface function.

A surface density of convection current may be defined much as was η by separating a very thin superficial layer of thickness d_c and dividing this up into surface cells $\Delta\tau_s$ (Fig. 1.15–1). A continuous two-dimensional vector point function $\mathbf{K}(\mathbf{r})$ is then interpolated from the discrete values,

$$\mathbf{K}(\mathbf{r}_s) = \frac{d_c \sum_{j=1}^{N} e_j \mathbf{v}_{js}}{\Delta\tau_s} \qquad (1.15–1)$$

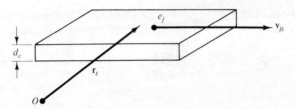

Figure 1.15–1 Surface cell.

defined at the center of $\Delta\tau_s$. The velocities $\mathbf{v}_{js}$ are all tangent to the surface or boundary. Alternatively, $\mathbf{K}(\mathbf{r})$ may be defined in terms of the surface density of the part of the charges (primed) that are engaged in nonrandom motion with a mean drift velocity $\mathbf{u}(\mathbf{r})$. That is,

$$\mathbf{K}(\mathbf{r}) = \mathbf{u}(\mathbf{r})\eta'(\mathbf{r}) = d_c\mathbf{u}(\mathbf{r})\rho'(\mathbf{r}) = d_c\mathbf{J}(\mathbf{r}) \qquad (1.15\text{–}2)$$

Here $\mathbf{u}(\mathbf{r})$ is constructed by interpolation from the $\mathbf{u}(\mathbf{r}_s)$ defined at the centers of the surface cells by an expression like (1.14–4); $\eta'(\mathbf{r})$ is interpolated from discrete values defined by

$$\eta'(\mathbf{r}_s) = \frac{d_c\sum_{j=1}^{N}|e_j'|}{\Delta\tau_s} \qquad (1.15\text{–}3)$$

$\mathbf{K}(\mathbf{r})$ has the dimensions of charge per unit time per unit length:

$$K \sim \frac{Q}{LT} \quad \text{A/m} \qquad (1.15\text{–}4)$$

It is necessary to require that the region represented be sufficiently uniform that the two-dimensional vector point function $\mathbf{K}(\mathbf{r})$ is so slowly varying as to be sensibly constant over lateral areas that are large compared with the largest dimension of the $\Delta\tau_s$ cells.

A steady-state distribution of charge moving in a region constructed according to either of the two steady-state models may be characterized in terms of the two, continuous slowly varying vector point functions $\mathbf{J}(\mathbf{r})$ and $\mathbf{K}(\mathbf{r})$.

1.16 NUMBER DENSITY REPRESENTATION OF MOVING CHARGES

The volume and surface densities of current described in Sec. 1.15 can be defined in terms of number densities. The volume density of moving negative charges, $\mathbf{J}_e(\mathbf{r})$, is given by

$$\mathbf{J}_e(\mathbf{r}) = q_e n_e(\mathbf{r})\mathbf{v}_e(\mathbf{r}) = -en_e(\mathbf{r})\mathbf{v}_e(\mathbf{r}) \qquad (1.16\text{–}1)$$

and the volume density of moving positive charges is given by

$$\mathbf{J}_p(\mathbf{r}) = q_p n_p(\mathbf{r})\mathbf{v}_p(\mathbf{r}) = eZn_p(\mathbf{r})\mathbf{v}_p(\mathbf{r}) \qquad (1.16\text{–}2)$$

$J_e(r)$ and $J_p(r)$ are interpolated functions like $J(r)$ in (1.14–7). The mean nonrandom velocity of the electrons is

$$v_e(r) = \frac{J_e(r)}{\rho_e(r)} = \frac{J_e(r)}{q_e n_e(r)} \qquad (1.16-3)$$

The mean nonrandom velocity of the positive charges is

$$v_p(r) = \frac{J_p(r)}{\rho_p(r)} = \frac{J_p(r)}{q_p n_p(r)} \qquad (1.16-4)$$

The volume density of current is defined as

$$J(r) = J_e(r) + J_p(r) = q_e n_e(r) v_e(r) + q_p n_p(r) v_p(r) \qquad (1.16-5)$$

The surface density of current $K(r)$ may be represented in a similar manner:

$$K(r) = K_e(r) + K_p(r) = q_e n_{es}(r) v_{es}(r) + q_p n_{ps}(r) v_{ps}(r) \qquad (1.16-6)$$

1.17 ALTERNATIVE MODES OF REPRESENTATION

Just as the static state was at first defined only in terms of surface and volume densities of charge using a special mode of subdivision appropriate to these functions, so the steady state has, up to this point, been represented entirely by surface and volume densities of drifting charge requiring the same special mode of subdivision into separate surface and volume cells. As in the static case, this mode of subdivision is physically reasonable for the free-drift model in which a separate and distinctly surface flow of free charges exists. On the other hand, the division into thin surface cells is not appropriate physically for the spin model because the boundaries of elementary cells are carefully and deliberately arranged to cut interior whirls (consisting of oriented groups of circulating electrons) so that in their steady circulation the charges continually pass from one cell into another and back. From the purely physical point of view, it is certainly artificial to call such parts of whirls at the surface a surface sheet of moving charge and to say that there is no circulation in the interior because of a clever allotment of parts of individual whirls to different volume cells in such a way that the net average circulation in each vanishes. This is illustrated roughly and schematically in Fig. 1.17–1. From the mathematical point of view, it is immaterial whether the subdivision used has a physical meaning or not, as long as it leads to a correct representation of the desired average effects. Nevertheless, a mathematical formulation that is closely adjusted to the physical model is often a valuable one. Hence, as in the static case, an alternative representation is introduced that is physically appropriate for the spin model and that describes a volume effect in terms of a volume function. Instead of separating the surface layer and defining a special density for it, only volume cells are used and a new volume density is defined throughout the region to take account of the elementary electronic whirls both in the interior and at the surface.

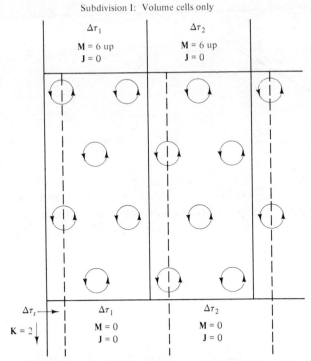

Subdivision I: Volume cells only

$\Delta\tau_1$

M = 6 up
J = 0

$\Delta\tau_2$

M = 6 up
J = 0

$\Delta\tau_s$

K = 2

$\Delta\tau_1$

M = 0
J = 0

$\Delta\tau_2$

M = 0
J = 0

Subdivision II: Surface and volume cells

Figure 1.17–1 Surface sheet of current caused by current whirls.

1.18 VOLUME DENSITY OF MAGNETIZATION

In the static state the volume density of polarization represents both in magnitude and in direction the average distortion and orientation of bound-charge groups. A simple model of a bound-charge atom consists of a closed shell of electrons associated with a positive nucleus. Under the action of suitable external forces, a separation of the average relative rest positions of the positive nucleus and the electrons may occur; the statistically equivalent representation of such a distorted atom is a dipole consisting of a positive charge separated a short distance from an equal negative charge. The distribution and orientation of statistical dipoles along a direction specified by the external force is called *polarization*.

In the steady state a model consisting of electrons rotating about an axis through an atom is an elementary magnet. Elementary magnets are produced by forces that cause the electrons in an atom to change their random orbits and circulate about a common axis. The orientation of such elementary magnets along parallel axes with a common direction of rotation will be called the *magnetization due to circulation*. Such an orientation may be superimposed on a random distribution. To characterize the magnetization of a volume element, it is necessary to specify the axis and the direction of rotation, the velocity of the whirling electron, and its distance from the axis.

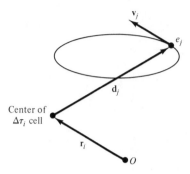

Figure 1.18–1   Orbital rotation of electron.

Consider first a single electron e_j moving in a circular path with a velocity $\mathbf{v}_j$ that is tangent to the circle. The distance from the center of the $\Delta\tau_i$ cell to the electron at any instant is $\mathbf{d}_j$ and the distance to the center of the $\Delta\tau_i$ cell from the origin is $\mathbf{r}_i$ (Fig. 1.18–1). The magnetization due to the circulation of orbital electrons of the $\Delta\tau_i$ cell is given by

$$\mathbf{m}_c(\mathbf{r}_i) = \tfrac{1}{2}\sum_{j=1}^{N} \mathbf{d}_j \times e_j\mathbf{v}_j \tag{1.18–1a}$$

$$= \tfrac{1}{2}\sum_{j=1}^{N} \overline{\mathbf{d}_j \times e_j\mathbf{v}_j} \tag{1.18–1b}$$

The instantaneous value (1.18–1a) and the time-average value (1.18–1b) of the magnetization due to circulation are equal to each other for statistically large numbers.

The vector volume density of magnetization of the circulating electrons in $\Delta\tau_i$ is defined by

$$\mathbf{M}_c(\mathbf{r}_i) = \frac{\mathbf{m}_c(\mathbf{r}_i)}{\Delta\tau_i}$$

$$= \frac{\tfrac{1}{2}\sum_{j=1}^{N} \mathbf{d}_j \times e_j\mathbf{v}_j}{\Delta\tau_i} \tag{1.18–2}$$

If a body or region is divided into volume cells $\Delta\tau_i$ and a vector $\mathbf{M}_c(\mathbf{r}_i)$ is defined at the center of each cell to describe the average orientation due to the circulation of the electrons, it is possible to interpolate from the discrete values $\mathbf{M}_c(\mathbf{r}_i)$ a continuous vector point function $\mathbf{M}_c(\mathbf{r})$ defined throughout the volume. This function is the *volume density of magnetization of the circulating electrons*. It assigns an axial vector to every point in the region or body in which it is defined, so that the direction of the vector is along the axis of magnetization in the positive sense according to the right-hand-screw convention, and the magnitude of the vector measures the average circulation of charge about the axis in a small region near the point.

There is, besides the magnetization of the electrons circulating in their orbits, the magnetization due to the *spin of the electrons*. The spin is the property of the electron whereby it rotates continuously about an axis of its own even while moving in an orbit around the positive nucleus. It is a quantum effect and cannot be explained by classical theories.

Consider a charge e of mass m_0 moving in a circle of radius $\mathbf{r}$ with a linear velocity $\mathbf{v}$ and an angular velocity $\omega = v/r$. Since we are dealing with speeds v which are much less than c, m_0 is the rest mass of the electron. When the electron moves in such a circular orbit, the angular momentum $\mathbf{L}$ is given by

$$|\mathbf{L}| = m_0 vr \qquad (1.18\text{--}3)$$

Although this is a nonrelativistic formula, it is a good approximation since $v \ll c$ (Fig. 1.18–2). The magnetic moment of the orbit is the product of the current I and the area of the orbit. The current I is the amount of charge passing a given point on the ring (formed by the orbit) per second:

$$I = \frac{ev}{2\pi r} \qquad (1.18\text{--}4)$$

since the electron makes $v/2\pi r$ rotations per second. With the area $A = \pi r^2$, the magnetic moment is given by

$$m = \frac{evr}{2} \qquad (1.18\text{--}5)$$

In vector notation

$$\mathbf{L} = \mathbf{r} \times m_0 \mathbf{v} \qquad (1.18\text{--}6a)$$

$$= \frac{m_0 A}{\pi} \boldsymbol{\omega} \qquad (1.18\text{--}6b)$$

$$\mathbf{m} = \tfrac{1}{2}\mathbf{r} \times e\mathbf{v} \qquad (1.18\text{--}7a)$$

$$= \frac{eA}{2\pi} \boldsymbol{\omega} \qquad (1.18\text{--}7b)$$

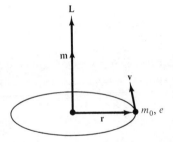

Figure 1.18–2   Electron moving in circular orbit.

Thus the magnetic moment in terms of the orbital angular momentum is

$$\mathbf{m} = \frac{e}{2m_0} \mathbf{L} \qquad (1.18\text{--}8)$$

Both $\mathbf{L}$ and $\mathbf{m}$ are directed perpendicular to the plane of the orbit and are in the same direction. It follows that the magnetic moment is constant in magnitude and direction whenever angular momentum is conserved. The factor $e/2m_0 = \gamma$ is called the *gyromagnetic ratio* or *orbital magnetomechanical ratio* of the electron. The relation (1.18–8) is true for orbital motion.

In certain crystalline structures such as ferromagnetic materials, the electron has a spin rotation which results in an angular momentum and a magnetic moment. These effects cannot be explained by classical theories as in orbital motion and are purely quantum effects. The ratio of $\mathbf{m}$ and $\mathbf{L}$ for the electron spin is twice as large as it is for the orbital motion of the spinning electron:

$$\mathbf{m}_{es} = \frac{e}{m_0} \mathbf{L}_{es} \qquad (1.18\text{--}9)$$

The two possible states of the spinning electron are given by $L_{es} = \pm \hbar/2$. Then

$$m_{es} = \pm \frac{e\hbar}{2m_0} \equiv \pm \mu_B \qquad (1.18\text{--}10)$$

where μ_B is the Bohr magneton and $\hbar = h/2\pi$. (Here h = Planck's constant = 6.61×10^{-34} J-s.) In general, the relation between the magnetic moment and the angular momentum can be written as follows:

$$\mathbf{m} = g\left(\frac{e}{2m_0}\right)\mathbf{L} \qquad (1.18\text{--}11)$$

where g, the Landé factor, is equal to 1 for orbital motion and 2 for spin.

The spin magnetomechanical ratio $\gamma_{es} = e/m_0$. The magnetic moment of the spin electron has been modified to allow for the fine structure by Schwinger:

$$m_{es} = \pm\left(1 + \frac{\alpha}{2\pi}\right)\frac{e\hbar}{2m_0} \qquad (1.18\text{--}12)$$

where $\alpha = e^2/4\pi\epsilon_0 c\hbar$ is the fine-structure constant. Then

$$\gamma_{es} = \left(1 + \frac{\alpha}{2\pi}\right)\frac{e}{m_0} \qquad (1.18\text{--}13)$$

The ratio of the gyromagnetic ratios $\gamma_{es}/\gamma \geq 2$. It is a characteristic of ferromagnetic materials that the electron spin moments are aligned parallel to one another.

The positive nucleus of an atom consisting of protons and neutrons also exhibits an intrinsic spin and is characterized by angular momentum and magnetic moment. The relationship is

$$\mathbf{m}_{ns} = g \left(\frac{e}{2M_0} \right) \mathbf{L}_{ns} \qquad (1.18\text{–}14)$$

where M_0 is the rest mass of the proton. In this case g is the nuclear g factor and is a number that is to be determined for each nucleus. The spin magnetic moment of the proton has a $g = 2(2.79)$. The neutron also has a spin magnetic moment whose value relative to its angular moment is $2(-1.93)$. The neutron has a magnetic moment like a rotating magnetic charge.

The total magnetization is the sum of the moments of the individual elements in a volume $\Delta\tau_i$, that is,

$$\mathbf{m}(\mathbf{r}_i) = \tfrac{1}{2} \sum_{j=1}^{N} \mathbf{d}_j \times e_j \mathbf{v}_j + \mathbf{m}_{es}(\mathbf{r}_i) + \mathbf{m}_{ns}(\mathbf{r}_i) \qquad (1.18\text{–}15)$$

The volume density of magnetization for $\Delta\tau_i$ is given by

$$\mathbf{M}(\mathbf{r}_i) = \frac{\mathbf{m}(\mathbf{r}_i)}{\Delta\tau_i} \qquad (1.18\text{–}16)$$

By defining similar quantities in each volume element and interpolating, a continuous vector point function, $\mathbf{M}(\mathbf{r})$, is defined throughout the volume. $\mathbf{M}(\mathbf{r})$ has the values $\mathbf{M}(\mathbf{r}_i)$ at the center of the volume cells and connects them smoothly. The dimensions of $\mathbf{M}$ are charge per unit length per unit time.

$$M \sim \frac{Q}{LT} \qquad \text{A/m} \qquad (1.18\text{–}17)$$

As in all preceding definitions of continuous densities, the body or region must be assumed to be sufficiently uniform that each cell differs only slightly from its neighbors, and the continuous interpolated function $\mathbf{M}(\mathbf{r})$ varies only very slowly over distances that are large compared with the dimensions of the volume cells. The definition of $\mathbf{M}$ implies in particular that the volume density of drifting charge $\mathbf{J}$ has an extremely small variation from cell to cell.

1.19 COMPARISON OF THE TWO REPRESENTATIONS

In describing the steady-state properties of a region in terms of a volume and a surface density of drifting charge and, alternatively, in terms of two volume densities, one of drifting charge and one of magnetization, a situation analogous to that encountered in static cases arises. In the mode of representation using surface and volume functions, the region is treated in two parts in each of which one function alone carries the burden of representation. In the alternative mode using volume functions only, both functions are defined in the entire region using the same volume cells. Since the volume density of drifting charge **J** (just as the volume density of charge ρ) is not sufficiently fine-grained to take account of surface conditions, it follows that the volume density of magnetization **M** must describe surface conditions of moving charge (much as **P** describes surface conditions of charge).

In examining the parts played by the three functions **J**, **K**, and **M** and their dependence on the mode of subdivision, it is possible to proceed by complete analogy with the static state and its three functions ρ, η, and **P**. If a given region containing only spin atoms is divided into volume cells in such a way that each cell contains only complete current whirls or elementary magnets, the volume density of current **J** is zero. There is no drift of charge across the boundaries of any volume element. Let the magnitude of the magnetization in the region increase uniformly in a direction normal to the fixed axis of magnetization. This condition is shown schematically and one-dimensionally in Fig. 1.19–1, where the number of elementary whirls increases steadily from left to right. The axis of magnetization is normal to the paper, pointing upward. A three-dimensional representation is given in Fig. 1.19–2.

The volume V is a slab that extends from $x = 8$ to $x = 24$ and is infinite in the y and z directions.

The functions for the four volume cells are as follows:

$$\mathbf{J}_1 = 0; \qquad \mathbf{J}_2 = 0; \qquad \mathbf{J}_3 = 0; \qquad \mathbf{J}_4 = 0$$

$$\mathbf{m}_1 = 10\hat{\mathbf{z}}; \qquad \mathbf{m}_2 = 14\hat{\mathbf{z}}; \qquad \mathbf{m}_3 = 18\hat{\mathbf{z}}; \qquad \mathbf{m}_4 = 22\hat{\mathbf{z}}$$

$$\mathbf{M}_1 = 2.5\hat{\mathbf{z}}; \qquad \mathbf{M}_2 = 3.5\hat{\mathbf{z}}; \qquad \mathbf{M}_3 = 4.5\hat{\mathbf{z}}; \qquad \mathbf{M}_4 = 5.5\hat{\mathbf{z}}$$

The continuous functions for subdivision I are obtained by interpolation:

$$\mathbf{J}_I = 0; \qquad \mathbf{M}_I = 0.25x\hat{\mathbf{z}}, \qquad 8 \le x \le 24$$

$$\mathbf{J}_I = 0; \qquad \mathbf{M}_I = 0, \qquad x < 8, \quad x > 24$$

If the continuous functions are extrapolated to the surfaces,

$$\mathbf{M}_{IL} = 0.25 \times 8\hat{\mathbf{z}} = 2\hat{\mathbf{z}}, \qquad \text{for the left-hand surface}$$

$$\mathbf{M}_{IR} = 0.25 \times 24\hat{\mathbf{z}} = 6\hat{\mathbf{z}}, \qquad \text{for the right-hand surface}$$

Furthermore,

$$\hat{\mathbf{n}}_L \times -\mathbf{M}_L = -\hat{\mathbf{x}} \times -\mathbf{M}_L = \hat{\mathbf{x}} \times \mathbf{M}_L = \hat{\mathbf{x}} \times 2\hat{\mathbf{z}} = -2\hat{\mathbf{y}}$$

$$\hat{\mathbf{n}}_R \times -\mathbf{M}_R = -\hat{\mathbf{x}} \times \mathbf{M}_R = -\hat{\mathbf{x}} \times 6\hat{\mathbf{z}} = 6\hat{\mathbf{y}}$$

as shown in Fig. 1.19–2.

Let the mode of subdivision be changed to volume and surface cells (subdivision II) as shown in Fig. 1.19–1 (dashed lines). Now some of the elementary

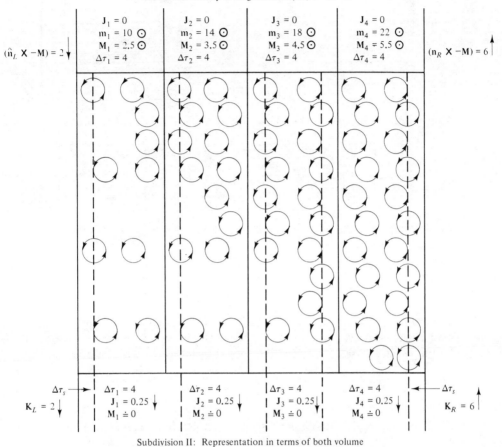

Subdivision I: Representation entirely in terms of
the volume density of magnetization, **M**; **J** = **K** = 0

Subdivision II: Representation in terms of both volume
and surface densities of moving charge, **J** and **K**; **M** $\doteq$ 0

Figure 1.19–1   Electrically one-dimensional schematic diagram to show two possible modes
of subdivision.

magnets near the surface are cut in two by the cell boundaries. The interior is then
subdivided by the dashed lines in such a way that the magnetization in each volume
cell is zero. With

$$\mathbf{J}_i = \frac{\sum e_j \mathbf{v}_j}{\Delta \tau_i}; \qquad \mathbf{M}_i = \frac{\sum \tfrac{1}{2}(\mathbf{d}_j \times e_j \mathbf{v}_j)}{\Delta \tau_i}$$

these functions have the following values for the four volume cells:

$$\mathbf{J}_1 = -\tfrac{1}{4}\hat{\mathbf{y}}; \qquad \mathbf{J}_2 = -\tfrac{1}{4}\hat{\mathbf{y}}; \qquad \mathbf{J}_3 = -\tfrac{1}{4}\hat{\mathbf{y}}; \qquad \mathbf{J}_4 = -\tfrac{1}{4}\hat{\mathbf{y}}$$

$$\mathbf{M}_1 = 0; \qquad\quad \mathbf{M}_2 = 0; \qquad\quad \mathbf{M}_3 = 0; \qquad\quad \mathbf{M}_4 = 0$$

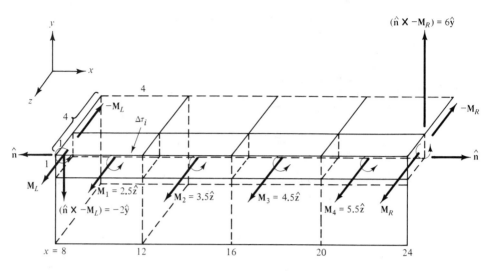

Figure 1.19-2 Representation of magnetization.

An interpolation results in the following continuous functions for subdivision II:

$$\mathbf{J}_{\mathrm{II}} = -\tfrac{1}{4}\hat{\mathbf{y}}; \qquad \mathbf{M}_{\mathrm{II}} = 0$$

Then in this mode of subdivision, the interior is characterized entirely by $\mathbf{J}$, whereas $\mathbf{M}$ characterized the interior in the mode of subdivision I. In the mode of subdivision II, the surface current is to be determined using $\mathbf{K}_s = d_c \sum e_j \mathbf{v}_{js}/\Delta\tau_s$, so that

$$\mathbf{K}_L = -\tfrac{2}{1}\hat{\mathbf{y}} = -2\hat{\mathbf{y}}, \qquad \text{on the left-hand surface}$$

$$\mathbf{K}_R = \tfrac{6}{1}\hat{\mathbf{y}} = 6\hat{\mathbf{y}}, \qquad \text{on the right-hand surface}$$

with $\Delta\tau_s = 1 \times 1$. The volume is now completely represented by $\mathbf{J}_{\mathrm{II}} = -0.25\hat{\mathbf{y}}$, $\mathbf{M}_{\mathrm{II}} = 0$, $\mathbf{K}_L = -2\hat{\mathbf{y}}$, and $\mathbf{K}_R = 6\hat{\mathbf{y}}$.

If neither of the modes of subdivision is used and no precautions are taken to have the cell walls cut or not cut through a certain number of atomic whirls, statistical conditions prevail along these walls as well as in the interior of the cells. Accordingly, there are always some atoms near the boundary completely in one volume cell, while others are cut by the boundary and are partly in one, partly in an adjacent cell. In the case considered here, the number of elementary whirls increases from left to right, so that on the average more of them are cut by the right than by the left boundary of each cell. The peculiarities of each of the two modes of subdivision exist simultaneously in the general picture, and all three functions $\mathbf{J}$, $\mathbf{K}$, and $\mathbf{M}$ have nonvanishing values. The situation is analogous to the static state in which all three functions ρ, η, and $\mathbf{P}$ had nonvanishing values for unrestricted subdivisions. It is to be expected by analogy that a suitable combination of $\mathbf{J}$, $\mathbf{M}$, and $\mathbf{K}$ will permit the definition of essential volume and surface characteristics that are independent of the mode of subdivision.

1.20 *ESSENTIAL VOLUME CHARACTERISTIC OF THE STEADY STATE: CURL OF A VECTOR*

A volume density of moving charge that is independent of the mode of subdivision may be obtained by a method analogous to that used in defining the function $\bar{\rho} = \rho - \nabla \cdot \mathbf{P}$ in the static state. That is, the representation of the distribution of moving charge in a typical volume cell is investigated when the cell walls are shifted from the one to the other of the two modes of subdivision illustrated in Fig. 1.19–1. With subdivision I each volume cell (e.g., $\Delta\tau_2$) contains only complete elementary magnets or whirls; none is cut by the boundaries. In moving the boundaries to subdivision II, in which $\mathbf{M} = 0$, three half-whirls are cut on the left and removed from the cell, while four half-whirls are cut on the right and added to the cell. The net result is an addition to the cell of one half-whirl representing a downwardly directed current of a certain magnitude. A similar result is observed in each volume cell. Hence the volume density of current $\mathbf{J}(\mathbf{r}_i)$, as defined in the usual way, has a nonvanishing value if this schematic picture is replaced by a distribution involving millions of current whirls. A continuous volume density of moving charge $\mathbf{J}$ can be interpolated for the entire region. It is clear from the schematic model that there is a direct correspondence between the number of elementary magnetization vectors $\mathbf{m} = \frac{1}{2}(\mathbf{d} \times e\mathbf{v})$ or current whirls cut by cell boundaries (in changing from subdivision I to subdivision II) and the volume density of current appearing in each volume cell in subdivision II. To include a *part* of a steady-state current whirl in a volume cell is exactly equivalent to adding an element of current parallel to that boundary which cuts the whirl.

It is necessary to define a quantity that is a measure of the number and orientation of the current whirls cut by a boundary and simultaneously of the equivalent current sheet due to these cut whirls on each side of the boundary. Such a quantity must resemble $\hat{\mathbf{n}} \cdot \mathbf{P}$ in the static state, which is a measure as well of the number of dipoles cut as of the amount of charge added to a volume cell in a change in subdivision. The vector $\mathbf{M}$ is a continuous function which measures the average density and direction of current whirls or of elementary magnetization vectors in a small region about any point. Since it is parallel to the axis of rotation of the circulating charges and a nonrandom axially directed drift is included in the function $\mathbf{J}$, $\mathbf{M}$ must be perpendicular to the plane of the motion of charges to be represented by $\mathbf{M}$. Charges moving parallel to the bounding surface of any volume cell must be moving perpendicular to the unit external normal $\hat{\mathbf{n}}$ to that surface. It follows that the circulating charges that are a part of a current whirl cut by a cell boundary in changing from subdivision I to II, and moving parallel to a boundary, must be perpendicular both to $\mathbf{M}$ and to an external normal to the surface. This does not require $\mathbf{M}$ to be perpendicular to $\hat{\mathbf{n}}$. A vector that has a direction perpendicular to $\mathbf{M}$ and $\hat{\mathbf{n}}$ and that is at the same time proportional to $\mathbf{M}$ is the vector product

$$\hat{\mathbf{n}} \times \mathbf{M} \qquad\qquad (1.20\text{–}1a)$$

Since $\hat{\mathbf{n}}$ is a unit vector, the magnitude of (1.20–1a) is

$$M \sin(n, M) = M \cos(s, M) \qquad (1.20\text{–}1b)$$

where $\mathbf{s}$ is a vector tangent to the surface. The magnitude of ($\hat{\mathbf{n}} \times \mathbf{M}$) is equal to the component of $\mathbf{M}$ parallel to the surface along which the circulation of charge takes place. An examination of Fig. 1.20–1 (which is drawn for the simplest case in which $\mathbf{M}$ is itself parallel to the surface) shows that in order to preserve the right-hand-screw convention as applied to current whirls of positive charge, it is necessary to use a vector equal in magnitude but opposite in direction to ($\hat{\mathbf{n}} \times \mathbf{M}$). Such a vector is

$$(\hat{\mathbf{n}} \times -\mathbf{M}) \qquad (1.20\text{–}2)$$

Figure 1.20–2 shows the orientation of the vector and the normal $\hat{\mathbf{n}}$ for the general case in which $\hat{\mathbf{n}}$ and $\mathbf{M}$ are not perpendicular to each other.

The magnitude of the vector ($\hat{\mathbf{n}} \times -\mathbf{M}$) is a measure of the average vector sum of the tangential components of the elementary magnetization vectors that pierce or lie in the plane of a unit area of the surface for which ($\hat{\mathbf{n}} \times -\mathbf{M}$) is defined. The direction of the vector ($\hat{\mathbf{n}} \times -\mathbf{M}$) is the same as the direction of motion along the surface of the positive charge in the elementary half-whirls (characterized by the magnetization vectors) on the side of the surface for which $\hat{\mathbf{n}}$ is an external normal. This is illustrated in Fig. 1.20–2, in which one current whirl is shown.

In order to determine the significance of the vector ($\hat{\mathbf{n}} \times -\mathbf{M}$) at a surface enclosing a volume cell, let the change in subdivision from I (volume cells only—no whirls cut) to II (surface and volume cells such that $\mathbf{M} = 0$) be examined in Fig. 1.20–1. If the boundaries of the volume cells are moved down half the diameter of a current whirl, the positive current sheet (composed of half-whirls) which is

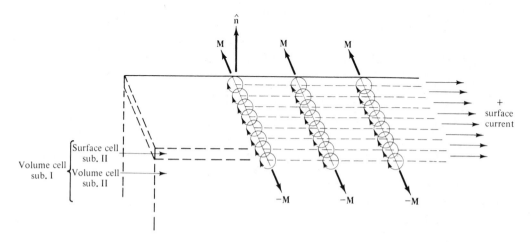

Figure 1.20–1 Surface current due to magnetization whirls.

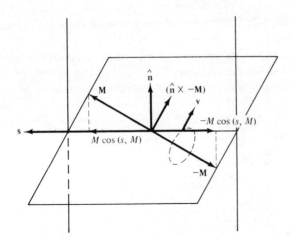

Figure 1.20–2 Orientation of magnetization vector and current whirl at a surface.

removed from this cell across each unit of its surface as the boundary is shifted down is measured by ($\hat{n}$ × $-\mathbf{M}$). This is equivalent to moving an identical, oppositely directed current sheet into the newly formed cell. That is, the sheet of positive current that appears in the volume cell per unit of surface in the change in subdivision is measured by $-(\hat{n} × -\mathbf{M})$ or ($\hat{n}$ × $\mathbf{M}$). Since $\mathbf{M}$ is a continuous function, the vector ($\hat{n}$ × $\mathbf{M}$) is defined at every point along the boundary of each volume cell in subdivision I. By integrating over the entire surface Σ of a volume cell

$$\int_{\Sigma} (\hat{n} × \mathbf{M_I})\, d\sigma \qquad (1.20\text{--}3)$$

a measure is obtained of the average number of positive half-current whirls that are cut and effectively added to the cell in the change in subdivision, and of the mean direction of their circulations as specified by their respective magnetization vectors. This integral is also a measure of the average positive current that is transferred across the boundaries of a volume cell as these are shifted to change from mode I to mode II. In (1.20–2) $\hat{n}$ is external to the cell, and $\mathbf{M}$ is necessarily the value in subdivision I, since $\mathbf{M} = 0$ in subdivision II. If this integral vanishes, the same number of current whirls is cut on each side so that there are no half-whirls downward added on one side that are not compensated by half-whirls downward lost on the other side, as shown in Fig. 1.17–1. If the integral does not vanish, the compensation is incomplete and a current is observed as in Fig. 1.19–1. The current per unit volume appearing in $\Delta\tau_k$ in subdivision II is

$$\frac{\int_{\Sigma} (\hat{n} × \mathbf{M_I})\, d\sigma}{\Delta\tau_k} \qquad (1.20\text{--}4)$$

In order to verify that (1.20–4) measures the positive current density, let a very simple case be studied in detail. Consider a region in which $\mathbf{M}$ is constant in

direction, vertically upward, but increases uniformly toward the right. A top view
of nine cubical volume elements of such a region is shown in Fig. 1.20–3.

The average value of $\mathbf{M}_k$ at the center of each cell as calculated using (1.18–16) is equivalent to a circulation around the cell boundaries as shown. In $\Delta\tau_1$, $3N$
charges are moving along each wall of area W^2; in $\Delta\tau_2$, $4N$ charges; in $\Delta\tau_3$, $5N$
charges.

Since the direction of $\mathbf{M}_k$ is known, it is merely necessary to calculate its
magnitude at the center of each cell using (1.18–15).

$$M_k = \frac{\displaystyle\sum_{\Delta\tau_k} \frac{1}{2} d_j e_j v_j \sin(d_j, v_j)}{W^3} = \frac{\displaystyle\sum_{\Delta\tau_k} e_j v_j}{4W^2} \qquad (1.20\text{–}5)$$

The second equality in (1.20–5) follows because

$$d_j \sin(d_j, v_j) = \frac{W}{2}$$

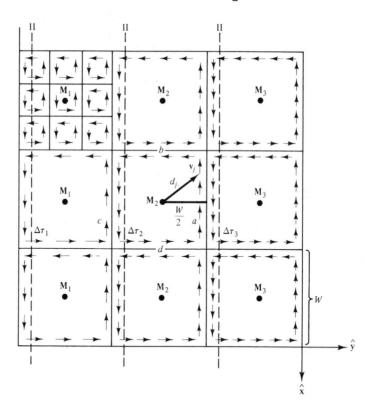

Figure 1.20–3 Magnetization and current along boundaries of volume cells. M is directed vertically
upward from the paper; its magnitude increases uniformly from left to right. Greater detail of the
typical circulation in each cell is shown in the upper left-hand corner. The boundaries for subdivision
II are shown in dashed lines.

However, $\frac{1}{4}\Sigma_{\Delta\tau_k} e_j v_j$ is the total current I_{ks} along each of the four sides in $\Delta\tau_k$ multiplied by the length of this side:

$$WI_{ks} = \frac{1}{4}\sum_{\Delta\tau_k} e_j v_j \qquad (1.20-6)$$

Therefore,

$$M_k = \frac{I_{ks}}{W} \qquad (1.20-7)$$

The magnitude of **M** on the boundary a between $\Delta\tau_2$ and $\Delta\tau_3$ is obtained by interpolation—in this case simple averaging—to be

$$M_a = \tfrac{1}{2}(M_2 + M_3) \qquad (1.20-8a)$$

Similarly, on the boundary c between $\Delta\tau_1$ and $\Delta\tau_2$,

$$M_c = \tfrac{1}{2}(M_1 + M_2) \qquad (1.20-8b)$$

On the boundaries b and d the value of M varies continuously. However,

$$M_b = M_d \qquad (1.20-9)$$

at opposite points.

One can now form (1.20–3). Since the magnitude of $(\hat{\mathbf{n}} \times \mathbf{M})$ on the two faces that are perpendicular to **M** is zero because $\hat{\mathbf{n}}$ and **M** are parallel,

$$\int_{\Sigma_2} \frac{\hat{\mathbf{n}} \times \mathbf{M}}{\Delta\tau_2}\, d\sigma = \frac{W^2}{W^3}\, (\hat{\mathbf{n}}_a \times \mathbf{M}_a + \hat{\mathbf{n}}_c \times \mathbf{M}_c)$$

$$+ \frac{1}{W^3} \int_{\Sigma = W^2} (\hat{\mathbf{n}}_b \times \mathbf{M}_b + \hat{\mathbf{n}}_d \times \mathbf{M}_d)\, d\sigma \qquad (1.20-10)$$

Here the $\hat{\mathbf{n}}$'s are unit external normals to $\Delta\tau_2$ on the sides indicated by the subscripts. Clearly,

$$\hat{\mathbf{n}}_c = -\hat{\mathbf{n}}_a; \qquad \hat{\mathbf{n}}_d = -\hat{\mathbf{n}}_b \qquad (1.20-11)$$

With (1.20–11) and (1.20–9),

$$\frac{\int_{\Sigma_2} (\hat{\mathbf{n}} \times \mathbf{M})\, d\sigma}{\Delta\tau_2} = \frac{1}{W}\, (M_a - M_c)\hat{\mathbf{x}} \qquad (1.20-12)$$

Here $\hat{\mathbf{x}}$ is a unit vector pointing toward the bottom of the page, as shown in Fig. 1.20–3. It specifies the direction of $\hat{\mathbf{n}}_a \times \mathbf{M}_a$ using the right-hand-screw convention. With (1.20–8a, b),

$$M_a - M_c = \tfrac{1}{2}(M_3 - M_1) \qquad (1.20-13a)$$

However,

$$M_2 = \tfrac{1}{2}(M_1 + M_3) \qquad (1.20-13b)$$

Hence

$$M_a - M_c = M_3 - M_2 \qquad (1.20\text{--}13c)$$

With (1.20–13c) and (1.20–7),

$$\frac{\int_{\Sigma_2} (\hat{\mathbf{n}} \times \mathbf{M})\, d\sigma}{\Delta\tau_2} = \frac{\hat{\mathbf{x}}(I_{3a} - I_{2a})}{W^2} \qquad (1.20\text{--}14a)$$

The expression on the right is the difference between the sheet of current I_{3a} directed down along side a in $\Delta\tau_3$ and I_{2a} flowing up along a in $\Delta\tau_2$. It measures the net flow in the direction $\hat{\mathbf{x}}$ along both sides of the boundary a between $\Delta\tau_3$ and $\Delta\tau_2$. If the mode of subdivision is changed to II as indicated by the dashed lines in Fig. 1.20–3, this net flow appears in $\Delta\tau_2$. Thus

$$\frac{\hat{\mathbf{x}}(I_{3a} - I_{2a})}{W^2} = \mathbf{J}_{m_2} \qquad (1.20\text{--}14b)$$

This simple example shows that (1.20–4) has the magnitude and the direction as well as the dimensions of a volume density of current.

The quantity

$$(\mathbf{J}_{m\mathrm{II}})_k = \frac{\int (\hat{\mathbf{n}} \times \mathbf{M}_\mathrm{I})\, d\sigma}{\Delta\tau_k} \qquad (1.20\text{--}15)$$

is the mean density of magnetization current in the cell $\Delta\tau_k$. If such a value is determined for each cell, a continuous volume density may be interpolated in the usual way. Since $\mathbf{M}$ is already such an interpolated function and is continuous throughout the region, it is permissible to pass to the limit directly and let the volume approach zero. The volume density of current that appears in the interior of a region as a result of changing the mode of subdivision from I to II is

$$\mathbf{J}_{m\mathrm{II}} = \lim_{\Delta\tau \to 0} \frac{\int (\hat{\mathbf{n}} \times \mathbf{M}_\mathrm{I})\, d\sigma}{\Delta\tau} \qquad (1.20\text{--}16)$$

An important shorthand symbol for defining the operation indicated on the right in (1.20–16) is the curl of a vector. It is defined by

$$\operatorname{curl} \mathbf{M} \equiv \nabla \times \mathbf{M} = \lim_{\Delta\tau \to 0} \frac{\int (\hat{\mathbf{n}} \times \mathbf{M})\, d\sigma}{\Delta\tau} \qquad (1.20\text{--}17)$$

and is itself a polar vector. The volume density of magnetization current $\mathbf{J}_m$ is a continuous vector point function defined throughout the body or region.

By a mere change in subdivision from I to II the description of the volume cells containing spin atoms only is altered as follows:

	Subdivision	
	I: Volume cells only; no cut spin atoms	II: Volume and surface cells; enough cut spin atoms to make $\mathbf{M}_{II} = 0$
$\mathbf{J} =$	0	$\mathbf{J}_m = \nabla \times \mathbf{M}_I$
$\mathbf{M} =$	$\mathbf{M}_I$	0

The interior of a region containing only spin-model atoms (describable by means of elementary magnets or whirls) may be characterized either in terms of $\mathbf{M}$ alone using subdivision I, or in terms of $\mathbf{J}$ alone using subdivision II. In the former case it is said to be magnetized.

If a region contains both free-drift and spin-model atoms, a representation of the interior using subdivision I (which does not cut through any of the elementary whirls) shows that each volume element is characterized by a volume density of magnetization $\mathbf{M}$ due to current whirls and by a volume density of current $\mathbf{J}_f$ due to the moving free charges. A volume density of magnetization current $\mathbf{J}_m$ due to the elementary magnets vanishes in subdivision I, since none of the whirls is cut by the cell boundaries. Hence $\mathbf{J} = \mathbf{J}_f$. When the same region is subdivided according to mode II, the interior is characterized by $\mathbf{M} = 0$ and by $\mathbf{J} = \mathbf{J}_f + \mathbf{J}_m$. Since $\mathbf{J}_f$ is slowly varying, it is also independent of small shifts in the location of cell boundaries, so that the same function is defined using either mode of subdivision. It has been shown that $\mathbf{J}_m$ in subdivision II is the same in magnitude and direction as $\nabla \times \mathbf{M}$ in subdivision I. Consequently, $(\mathbf{J} + \nabla \times \mathbf{M})$ has the same value in either mode of subdivision. If any intermediate subdivision is chosen in which the cell boundaries cut through *some* current whirls but not through enough to reduce the magnetization to zero, or if, in terms of the dynamic model, there is a statistical distribution along the boundaries so that there are always some whirls cut but not necessarily enough to make $\mathbf{M}$ vanish, both $\mathbf{M}$ and $\mathbf{J}_m$ have nonvanishing values. But $\nabla \times \mathbf{M}$ has a value precisely equal to the addition that would be made to $\mathbf{J}$ if a subdivision were chosen to reduce $\mathbf{M}$ to zero. Thus $(\mathbf{J} + \nabla \times \mathbf{M})$ is a vector point function that is independent of the mode of subdivision. It is the essential volume density of the steady state, analogous to the essential volume density of the static state $(\rho - \nabla \cdot \mathbf{P})$. Let it be denoted by

$$\bar{\mathbf{J}} \equiv \mathbf{J} + \nabla \times \mathbf{M} = \mathbf{J} - \nabla \times -\mathbf{M} \qquad (1.20\text{--}18)$$

Here the expression on the extreme right is written so that the sign agrees with that in $(\rho - \nabla \cdot \mathbf{P})$. It appears that $-\mathbf{M}$ and not $\mathbf{M}$ is analogous to $\mathbf{P}$.

1.21 ALTERNATIVE DEFINITION OF THE CURL OF A VECTOR

An alternative and often convenient definition of the curl of a vector is expressed in terms of the component of the curl in a definite direction or, in particular, in a direction normal to a given surface. Consider, for example, the component of $\nabla \times \mathbf{M}$ normal to the surface ΔS which cuts across a volume element $\Delta\tau$ dividing it into two parts. For simplicity, let the volume element $\Delta\tau$ be a small cube of side w (a volume element of any shape may be used), and let ΔS be a square drawn parallel to the front face as in Fig. 1.21–1. It is bounded by lines up and down on the two sides and across the top and bottom faces. Let $\hat{\mathbf{N}}$ be a unit normal to ΔS pointing out toward the front face from ΔS. The component of $\nabla \times \mathbf{M}$ [as defined by (1.20–17)] normal to ΔS is

$$\hat{\mathbf{N}} \cdot \nabla \times \mathbf{M} = \lim_{\Delta\tau \to 0} \frac{\int \hat{\mathbf{N}} \cdot (\hat{\mathbf{n}} \times \mathbf{M}) \, d\sigma}{\Delta\tau} \tag{1.21–1}$$

The double scalar and vector product may be rearranged as follows:

$$\hat{\mathbf{N}} \cdot \hat{\mathbf{n}} \times \mathbf{M} = \mathbf{M} \cdot \hat{\mathbf{N}} \times \hat{\mathbf{n}} \tag{1.21–2}$$

$\hat{\mathbf{N}}$ always points in the same direction; $\hat{\mathbf{n}}$ is an external normal to the surface of the cube, and hence points in different directions as $d\sigma$ moves from one face of the cube to another, in the course of integration. The vector product

$$\hat{\mathbf{s}}' \equiv \hat{\mathbf{N}} \times \hat{\mathbf{n}} \tag{1.21–3}$$

defines a unit vector which is normal to both $\hat{\mathbf{n}}$ and $\hat{\mathbf{N}}$. Since $\hat{\mathbf{N}}$ is fixed, $\hat{\mathbf{s}}'$ lies in the plane ΔS. In order to remain perpendicular to $\hat{\mathbf{n}}$ and satisfy the right-hand-screw relation of the vector product, $\hat{\mathbf{s}}'$ points along the bounding lines of ΔS in a counterclockwise sense as indicated in Fig. 1.21–1. The substitution of (1.21–3) in (1.21–1) gives

$$\hat{\mathbf{N}} \cdot \nabla \times \mathbf{M} = \lim_{\Delta\tau \to 0} \frac{\int_\Sigma (\mathbf{M} \cdot \hat{\mathbf{s}}') \, d\sigma}{\Delta\tau} \tag{1.21–4}$$

The integral in (1.21–4) includes no contributions from the front and back faces on which $\hat{\mathbf{s}}'$ vanishes because $\hat{\mathbf{N}}$ and $\hat{\mathbf{n}}$ are in the same or in opposite directions. Integration over the side, top, and bottom faces is equivalent to a sum of integrals

$$\oint_s \mathbf{M} \cdot d\mathbf{s}' \tag{1.21–5}$$

around the edges of each of an infinite family of parallel surfaces ΔS parallel to the xz plane. Let $d\sigma = ds' \, dy$; then

$$\hat{\mathbf{N}} \cdot \nabla \times \mathbf{M} = \lim_{\Delta\tau \to 0} \frac{\int_\Sigma (\mathbf{M} \cdot d\mathbf{s}') \, dy}{\Delta\tau} \tag{1.21–6}$$

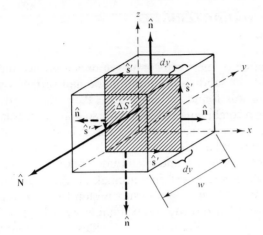

Figure 1.21–1 Cubical element to illustrate the alternative definition of the curl.

Since **M** is continuous, the theorem of the mean for integrals may be applied to the y integration. That is,

$$\hat{\mathbf{N}} \cdot \nabla \times \mathbf{M} = \lim_{\Delta\tau\to0} \frac{\overline{\oint_s (\mathbf{M} \cdot d\mathbf{s}')} \int dy}{\Delta\tau} \qquad (1.21\text{–}7)$$

Here the barred integral is a mean value for some contour s around one of the family of parallel surfaces ΔS. The integral $\int dy$ gives simply the thickness w of the cube; also $\Delta\tau = w\Delta S$. Hence (1.21–7) reduces to

$$\hat{\mathbf{N}} \cdot \nabla \times \mathbf{M} = \lim_{\Delta S\to0} \frac{\oint_s (\mathbf{M} \cdot d\mathbf{s}')}{\Delta S} \qquad (1.21\text{–}8)$$

The bar indicating a mean value is omitted, since in the limit $\Delta\tau$ approaches zero on the surface ΔS around which the integral has a mean value. The contour s traces the boundary of that surface. The positive direction of $\hat{\mathbf{s}}'$ around the contour obeys the right-hand-screw convention with respect to the normal $\hat{\mathbf{N}}$. The definition (1.21–8) of the component of the curl of a vector normal to a plane is often simpler than (1.20–17) to use and to visualize as an operation. For example, in rectangular coordinates

$$\nabla \times \mathbf{M} = \hat{\mathbf{x}} \lim_{\Delta y\Delta z\to0} \frac{\oint(\mathbf{M} \cdot d\mathbf{s})}{\Delta y\Delta z} + \hat{\mathbf{y}} \lim_{\Delta z\Delta x\to0} \frac{\oint(\mathbf{M} \cdot d\mathbf{s})}{\Delta z\Delta x}$$

$$+ \hat{\mathbf{z}} \lim_{\Delta y\Delta x\to0} \frac{\oint(\mathbf{M} \cdot d\mathbf{s})}{\Delta y\Delta x} \qquad (1.21\text{–}9)$$

1.22 ESSENTIAL SURFACE CHARACTERISTIC
OF THE STEADY STATE

The complete characterization of the steady state requires a surface function that is independent of the mode of subdivision into volume and surface cells. In the static state such a function was found to be $\eta + \hat{n} \cdot P$. In the steady state the analogous function must be a combination of K and M, so that the present situation with respect to M is analogous to that discussed for P in defining the surface characteristic of the static state. It has been shown (1.20–1b) that the vector product $(\hat{n} \times -M)$ is a measure of the direction and magnitude of positive charges moving parallel to the surface to which $\hat{n}$ is an external normal. At the boundary between two dissimilar regions, $(\hat{n} \times -M)$ describes in subdivision I what K represents in subdivision II, as can be verified directly by examining what happens to a thin layer containing parts of current whirls along both sides of a bounding surface between two regions when a change is made in the method of subdivision in each region from mode I to mode II. Consider first both regions divided according to mode I (Fig. 1.22–1). Each contains only volume cells, and both J and M may have nonvanishing values if both whirling and freely drifting charges are present. The value of J is due entirely to the drift of free charges, the value of M entirely to the circulation in spin atoms. Suppose, for a moment, that no free charges are present, so that $J = 0$; let M_1 be defined in region 1 on one side of the boundary, while M_2 is defined in region 2 on the other side. At the boundary the quantities $(\hat{n}_1 \times -M_1)$ and $(\hat{n}_2 \times -M_2)$ characterize the positive circulation on each side parallel to the boundary with normals outwardly directed with respect to the re-

Mode II: Obtained by moving cell wall to left   Mode I

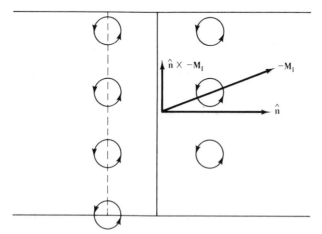

Figure 1.22–1 Change of mode of subdivision.

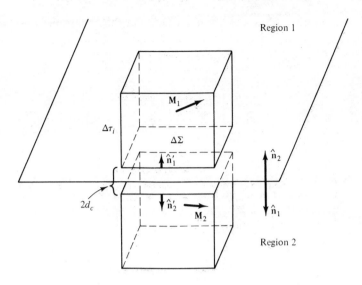

Figure 1.22–2 Volume and surface cells at a boundary.

gions. Consider two cubical volume cells one on each side of the boundary surface with adjacent faces $\Delta\Sigma_1$ and $\Delta\Sigma_2$. Let the mode of subdivision of the two regions be changed by pulling apart the two volume cells $\Delta\tau_1$ and $\Delta\tau_2$ (Fig. 1.22–2) to form two thin surface cells, of equal volume $d_c\Delta\Sigma_1$ and $d_c\Delta\Sigma_2$, one on each side of the boundary between the dissimilar regions. Each of the newly formed surface cells contains parts of current whirls still partly in $\Delta\tau_1$ and $\Delta\tau_2$. The number of current whirls that are cut and partly *added* to a given cell in the form of a positive current in changing the mode of subdivision is obtained by integrating the vector ($\hat{\mathbf{n}} \times \mathbf{M}$) over the enclosing surfaces as explained in conjunction with Fig. 1.20–2. If this integral is evaluated over the surface of the thin disk of thickness $2d_c$, significant contributions are obtained only from the large flat surfaces $\Delta\Sigma_1$ and $\Delta\Sigma_2$ because only these surfaces cut through the current whirls when the change in subdivision is made. Thus

$$\int_\Sigma (\hat{\mathbf{n}} \times \mathbf{M}) \, d\sigma = \int_{\Delta\Sigma_1} (\hat{\mathbf{n}}_1' \times \mathbf{M}_1) \, d\sigma + \int_{\Delta\Sigma_2} (\hat{\mathbf{n}}_2' \times \mathbf{M}_2) \, d\sigma \qquad (1.22\text{–}1)$$

where $\hat{\mathbf{n}}_1'$ points into region 1, $\hat{\mathbf{n}}_2'$ into region 2 since both must be *external* to the volume contained in the disk, $2d_c\Delta\Sigma$. Also, $\Delta\Sigma_1 = \Delta\Sigma_2$.

The average volume density of moving charge due to contributions from cut current whirls in these cells is

$$(\mathbf{J}_m)_{\text{av}} = \tfrac{1}{2}(\mathbf{J}_{1m} + \mathbf{J}_{2m}) \qquad (1.22\text{–}2)$$

If (1.22–1) is divided by the volume $2d_c\Delta\Sigma$ and $\Delta\Sigma$ is allowed to approach zero in the limit (since **M** is a continuous function this is permissible), the following continuous volume density of moving charge is defined:

$$\tfrac{1}{2}(\mathbf{J}_{1m} + \mathbf{J}_{2m}) =$$

$$-\lim_{2d_c\Delta\Sigma\to 0}\left[\frac{\displaystyle\int_{\Delta\Sigma_1}(\hat{\mathbf{n}}_1\times\mathbf{M}_1)\,d\sigma}{2d_c\Delta\Sigma_1} + \frac{\displaystyle\int_{\Delta\Sigma_2}(\hat{\mathbf{n}}_2\times\mathbf{M}_2)\,d\sigma}{2d_c\Delta\Sigma_2}\right] \qquad (1.22\text{–}3)$$

Here $\hat{\mathbf{n}}_1$ ($= -\hat{\mathbf{n}}'_1$), $\hat{\mathbf{n}}_2$ ($= -\hat{\mathbf{n}}'_2$) are normals directed outwardly with respect to the boundary surface between the regions 1 and 2. From (1.15–2), $\mathbf{J}_1 d_c$ is by definition the surface density of moving charge $\mathbf{K}_1$. Similarly, $\mathbf{K}_2$ is defined along the surface of region 2 by $\mathbf{J}_2 d_c$. Thus

$$\mathbf{K}_{1m} + \mathbf{K}_{2m} = \mathbf{J}_{1m}d_c + \mathbf{J}_{2m}d_c \qquad (1.22\text{–}4)$$

After (1.22–3) has been multiplied by $2d_c$ and combined with (1.22–4), the result is

$$\mathbf{K}_{1m} + \mathbf{K}_{2m} = -\lim_{\Delta\Sigma\to 0}\left[\frac{\displaystyle\int_{\Delta\Sigma_1}(\hat{\mathbf{n}}_1\times\mathbf{M}_1)\,d\sigma}{\Delta\Sigma_1} + \frac{\displaystyle\int_{\Delta\Sigma_2}(\hat{\mathbf{n}}_2\times\mathbf{M}_2)\,d\sigma}{\Delta\Sigma_2}\right] \qquad (1.22\text{–}5)$$

The application of the theorem of the mean for integrals to *each integral* in (1.22–5), leads to

$$\mathbf{K}_{1m} + \mathbf{K}_{2m} = -\lim_{\Delta\Sigma\to 0}\left[\frac{\overline{\hat{\mathbf{n}}_1\times\mathbf{M}_1}\displaystyle\int_{\Delta\Sigma_1}d\sigma}{\Delta\Sigma_1} + \frac{\overline{\hat{\mathbf{n}}_2\times\mathbf{M}_2}\displaystyle\int_{\Delta\Sigma_2}d\sigma}{\Delta\Sigma_2}\right] \qquad (1.22\text{–}6)$$

Here $\overline{\hat{\mathbf{n}}_1\times\mathbf{M}_1}$ and $\overline{\hat{\mathbf{n}}_2\times\mathbf{M}_2}$ are, respectively, mean values at points on the surfaces $\Delta\Sigma_1$ and $\Delta\Sigma_2$. In passing to the limit, $\Delta\Sigma_1$ and $\Delta\Sigma_2$ approach zero at the points where the mean values are defined, so that

$$\mathbf{K}_{1m} + \mathbf{K}_{2m} = -(\hat{\mathbf{n}}_1\times\mathbf{M}_1) - (\hat{\mathbf{n}}_2\times\mathbf{M}_2) \qquad (1.22\text{–}7)$$

Let the following shorthand notation be introduced:

$$\mathbf{K}_m \equiv \mathbf{K}_{1m} + \mathbf{K}_{2m} \qquad (1.22\text{–}8a)$$

$$\hat{\mathbf{n}}\times\mathbf{M} \equiv \hat{\mathbf{n}}_1\times\mathbf{M}_1 + \hat{\mathbf{n}}_2\times\mathbf{M}_2 \qquad (1.22\text{–}8b)$$

With this symbolism and with the mode of subdivision used in defining the densities explicitly indicated by the subscripts I and II, (1.22–7) becomes

$$\mathbf{K}_{m\mathrm{II}} = -(\hat{\mathbf{n}} \times \mathbf{M}_{\mathrm{I}}) \tag{1.22–9}$$

Just as for $\eta_{b\mathrm{II}}$ in the analogous case in the static state (1.10–10), a value for $\mathbf{K}_{m\mathrm{II}}$ may be associated with every point along the surface separating the two regions. It is a measure of the total surface density of moving charge in both regions associated with that point in terms of subdivision II.

It is easy to see that the function $(\mathbf{K} - \hat{\mathbf{n}} \times \mathbf{M})$ is independent of the mode of subdivision at least for a region containing no free charges by applying the same reasoning used in Sec. 1.10 in the analogous static case. This function is assigned the symbol $\overline{\mathbf{K}}$ and is named the essential surface density of the steady state. It is

$$\overline{\mathbf{K}} \equiv \mathbf{K} - \hat{\mathbf{n}} \times \mathbf{M} \equiv \mathbf{K} + (\hat{\mathbf{n}} \times -\mathbf{M}) \tag{1.22–10}$$

The expression on the extreme right is written in a form to bring out the analogy with the static-state function $\overline{\eta} = \eta + \hat{\mathbf{n}} \cdot \mathbf{P}$. It again appears that $-\mathbf{M}$ rather than $\mathbf{M}$ is the analogue of $\mathbf{P}$.

Because $(\hat{\mathbf{n}} \times \mathbf{M})$ can describe surface effects only if these are surface manifestations of a volume phenomenon, it cannot be used to characterize a sheet of drifting free charge. On the other hand, a description in terms of $\mathbf{J}_{\mathrm{I}}$ and $\mathbf{M}_{\mathrm{I}}$ can always be made mathematically equivalent to a description in terms of $\mathbf{J}_{\mathrm{II}}$ and $\mathbf{K}_{\mathrm{II}}$ if the following conditions are fulfilled:

$$\mathbf{J}_{\mathrm{II}} = \mathbf{J}_{\mathrm{I}} + \nabla \times \mathbf{M}_{\mathrm{I}} \tag{1.22–11}$$
$$\mathbf{K}_{\mathrm{II}} = -(\hat{\mathbf{n}} \times \mathbf{M}_{\mathrm{I}})$$

Since every free-drift model may be replaced by a mathematically equivalent spin model, the function

$$\overline{\mathbf{K}} = \mathbf{K} - (\hat{\mathbf{n}} \times \mathbf{M}) \tag{1.22–12}$$

may be used to describe mathematically the surface conditions of both spin and free-charge models. By building this function and the volume function $\mathbf{J}$ into the mathematical model of electromagnetism, it is possible to interpret them at will or for physical reasons in terms of a circulation model, a free-drift model, or a combination of these.

1.23 SURFACE DENSITY OF MAGNETIZATION

In order to complete the mathematical description of the steady state, an additional surface function to represent a surface distribution of current whirls might be defined. A surface magnetization is the steady-state analogue of a surface polarization or double layer in the static state. Mathematically, it is a simple matter to construct a steady-state function analogous to **k** in the static state. Whereas the function **k** plays only a very minor part in the coordination of the mathematical theory with experimental observations, no experimental effects have been observed that require a theoretical analogue that depends on a surface magnetization. Furthermore, it is difficult to conceive of conditions in terms of the atomic model that are sufficiently complex to produce surface magnetization. Accordingly, no surface density of magnetization is defined.

1.24 FORMAL ANALOGY BETWEEN THE STEADY STATE AND THE STATIC STATE

The physical and mathematical models of the steady state exhibit a complete parallelism in form, in development, and in interpretation with corresponding models in the static state. Atomic models describe two kinds of motion of the charges: a steady drift of free charges which are statistically at rest in the static state, and a circulation and orientation of charge. The latter is only formally a steady-state analogue of the distortion orientation of charge in the bound-charge static model. The two effects are parallel only in that both are concerned with an orientation. The distortion in the static state and the circulation in the steady state are quite unrelated, as are the external forces that produce them. In its mathematical model of continuous functions, the steady state is closely analogous to the static state. Both are characterized in terms of two continuous functions defined, one throughout the interior, the other on the surface of a region containing charges. In addition, the scalar product occurs in the static state wherever the vector product occurs in the steady state; the divergence of the vector appears in the static state wherever the curl of the vector appears in the steady state. This formal analogy between the two states is maintained throughout the mathematical development. Table 1.24–1 summarizes the definitions of important quantities in the two states and demonstrates the analogy between them.

1.25 MACROSCOPIC ELECTRICAL PROPERTIES OF MATTER: A SUMMARY OF THE STATIC AND STEADY STATES

According to atomic theory, all matter is composed of atoms and molecules, which, in turn, are formed by negatively and positively charged particles and neutral particles. Over distances that are large compared with the mean free path of these particles and times that are long compared with atomic events, it is possible to represent the approximate average properties of matter by continuous functions that vary slowly from point to point except near boundaries between different types of matter when rapid changes in the direction perpendicular to the boundary surface may be expected. Such functions are defined in terms of the statistical distributions

TABLE 1.24–1 SUMMARY OF ANALOGY BETWEEN STATIC AND STEADY STATES

Quantity	Static state	Steady state
Condition of charge	Dynamic: random motion	Dynamic: steady drift or circulation superimposed on random motion
	Static: statistically at rest	Static: steady drift or circulation relative to statistical rest positions
Atomic models	Free charge	Free drift
	Bound charge	Microscopic circulation
Elementary models	Dipole; $\mathbf{p} = e\mathbf{d}$ p defines electric moment; its direction is axis of polarization	Magnet;[a] $\mathbf{m} = \frac{1}{2}\mathbf{d} \times ev$ m defines magnetic moment; its direction is axis of magnetization
Density functions		
Volume density of:	Charge, ρ ($\rho_i = \Sigma e_j/\Delta\tau_i$)	Current, $\mathbf{J}$ ($\mathbf{J}_i = \Sigma e_j\mathbf{v}_j/\Delta\tau_i$)
Surface density of:	Charge, η ($\eta_s = \Sigma e_j d_c/\Delta\tau_s$)	Current, $\mathbf{K}$ ($\mathbf{K}_s = \Sigma e_j\mathbf{v}_j d_c/\Delta\tau_s$)
Volume density of:	Polarization, $\mathbf{P}$ ($\mathbf{P}_i = \Sigma e_j\mathbf{d}_j/\Delta\tau_i$ $= \mathbf{p}_i/\Delta\tau_i$)	Magnetization, $-\mathbf{M}$ ($\mathbf{M}_i = \Sigma\frac{1}{2}\mathbf{d}_j \times e_j\mathbf{v}_j/\Delta\tau_i$ $= \mathbf{m}_i/\Delta\tau_i$)
Essential volume density of:	Charge $\bar{\rho} \equiv \rho - \nabla \cdot \mathbf{P}$	Current $\bar{\mathbf{J}} \equiv \mathbf{J} + \nabla \times \mathbf{M}$
Essential surface density of:	Charge $\bar{\eta} \equiv \eta + \hat{\mathbf{n}} \cdot \mathbf{P}$	Current $\bar{\mathbf{K}} \equiv \mathbf{K} - \hat{\mathbf{n}} \times \mathbf{M}$
Operations	Scalar or dot product	Vector or cross product
	Divergence, $\nabla \cdot$	Curl, $\nabla \times$

[a] Electron spin also contributes to magnetization, but has no static-state analogue.

of the charges and their nonrandom velocities throughout the interior and along the boundaries of the region of interest. For most purposes it is sufficient to introduce six such functions to describe the average electrical properties. Of these, four represent the average conditions of charge and moving charge in the vicinity of each point in the interior and in a thin layer near a boundary, a fifth gives the average distributions and orientations of electric dipoles, and a sixth gives the microscopic circulations of electric charges, including spins (magnetic dipoles).

The definition of all density functions depends on the choice of volume and surface elements. They must be sufficiently large to contain enough charges so that statistical regularity obtains and sufficiently small so that the average property defined by them is sufficiently fine-grained for the desired macroscopic representation. In general, the selection of suitable interior and surface elements is straightforward in the case of solids and liquids, but not always possible for diffuse media such as ionized gases and plasmas. Once the regions under study have been subdivided into appropriate interior and surface cells, continuous density functions can be constructed by interpolation between the discrete values defined at the centers of the individual cells. Such a density is a function of the coordinates and of the time.

The elementary microscopic description of matter is in terms of the electron with an electrical charge $q_e = -e = -1.6 \times 10^{-19}$ C and a rest mass $m_e = 9.1 \times 10^{-31}$ kg; the positive particle or ion with an electrical charge $q_p = Ze$, where Z is an integer, and a mass m_p that is very great compared with m_e (for the proton $Z = 1$, $m_p = 1836.3m_e$); and the neutron with no electrical charge and a mass m_n that is also very large compared with m_e. If the interpolated average number of electrons per unit volume is $n_e(\mathbf{r})$, the continuous volume density of charge for electrons is $\rho_e(\mathbf{r}) = q_e n_e(\mathbf{r}) = -e n_e(\mathbf{r})$. The corresponding volume density of mass is $D_e(\mathbf{r}) = m_e n_e(\mathbf{r})$. Similarly, for positive ions with an interpolated number $n_p(\mathbf{r})$ per unit volume, the corresponding densities are $\rho_p(\mathbf{r}) = q_p n_p(\mathbf{r})$ and $D_p(\mathbf{r}) = m_p n_p(\mathbf{r})$. For neutrons, the densities are $\rho_n(\mathbf{r}) = 0$ and $D_n(\mathbf{r}) = m_n n_n(\mathbf{r})$. The total volume densities of charge and mass are $\rho(\mathbf{r}) = \rho_e(\mathbf{r}) + \rho_p(\mathbf{r})$ and $D(\mathbf{r}) = D_e(\mathbf{r}) + D_p(\mathbf{r}) + D_n(\mathbf{r})$. The volume density of charge $\rho(\mathbf{r})$ is a scalar point function that measures the average volume density of charge in coulombs per cubic meter in the neighborhood of every point in the interior of a region. By selecting cells that are of atomic thickness along boundaries, it is possible to define the average number of electrons or positive ions per unit area. If these are, respectively, $n_{es}(\mathbf{r})$ and $n_{ps}(\mathbf{r})$, the continuous surface densities of electrons and positive ions are $\eta_e(\mathbf{r}) = q_e n_{es}(\mathbf{r})$ and $\eta_p(\mathbf{r}) = q_p n_{ps}(\mathbf{r})$; the corresponding total surface density of charge in coulombs per square meter is $\eta(\mathbf{r}) = \eta_e(\mathbf{r}) + \eta_p(\mathbf{r})$, where $\mathbf{r}$ is a vector that ends at the surface. At a boundary between two media, it is usually convenient to define such a density in a thin layer on each side of the boundary.

If the mean nonrandom interpolated velocities of the electrons and positive charges are, respectively, $\mathbf{v}_e(\mathbf{r})$ and $\mathbf{v}_p(\mathbf{r})$, continuous volume and surface densities of moving charge or current may be defined as follows:

$$\mathbf{J}(\mathbf{r}) = q_e n_e(\mathbf{r})\mathbf{v}_e(\mathbf{r}) + q_p n_p(\mathbf{r})\mathbf{v}_p(\mathbf{r}) \tag{1.25-1}$$

$$\mathbf{K}(\mathbf{r}) = q_e n_{es}(\mathbf{r})\mathbf{v}_{es}(\mathbf{r}) + q_p n_{ps}(\mathbf{r})\mathbf{v}_{ps}(\mathbf{r}) \tag{1.25-2}$$

The continuous volume density of current $\mathbf{J}(\mathbf{r})$ in amperes per square meter is a vector point function that measures the average magnitude and direction of non-random flow of positive charges or their equivalent across each unit area in the interior of a region. Similarly, the surface density of current $\mathbf{K}(\mathbf{r})$ in amperes per meter is a vector point function that measures the average magnitude and direction tangent to a boundary of nonrandom flow of positive charges or their equivalent across each unit width of a surface or boundary layer of atomic thickness.

In a neutral unpolarized atom the time-average centers of the positive and negative charges coincide. Under the action of external forces, the center of the negative charges may be displaced slightly from the center of the positive charges. In this case the atom is said to be electrically polarized. When the atoms in a region are polarized, each volume element containing a great many such atoms is also polarized if the individual directions of polarization are determined by the same external force. This means that the time-average center of positive charge in each interior cell is displaced from the time-average center of the negative charge. If a vector drawn from the center of the negative charge to the center of the positive charge is denoted by $\mathbf{d}$, the polarization of the volume is $\mathbf{p} = Q_p\mathbf{d}$, where Q_p is the total positive charge. $Q_e = -Q_p$ is the total negative charge. The volume density of polarization $\mathbf{P}$ is the continuous vector point function that is obtained by interpolation from the discrete values of $\mathbf{p}/V$, where V is the volume of a typical interior cell. $\mathbf{P}(\mathbf{r})$ in coulombs per square meter is a measure of the number and orientation of electric dipoles per unit volume in the vicinity of any point.

If electric charges circulate about an axis through an atom or molecule or if the electron spins that characterize the atom do not cancel along some axis, the atom or molecule has a magnetic moment and is said to be *magnetically polarized* or *magnetized*. The intensity of magnetization due to the orbital electrons is determined by the number of circulating charges, their angular velocity about the axis, and their distance from the axis. Specifically, if a charge e travels around a circle of radius r with a velocity $\mathbf{v}$, the magnetization is defined by the axis vector $\mathbf{m}$ given by the vector product $\mathbf{m} = (\mathbf{r} \times e\mathbf{v})/2$. The magnetization of each uncanceled electron spin is equal to one Bohr magneton, $\mu_B = e\hbar/2m_e$, where $\hbar$ is Planck's constant divided by 2π. The volume density of magnetization $\mathbf{M}(\mathbf{r})$ in amperes per meter is the continuous vector point function that is obtained by interpolation from the vector sum of all contributions to the magnetic polarization per unit volume of interior cells. It specifies the magnitude and direction of the average magnetization per unit volume in the vicinity of every point in a region.

The volume densities $\rho(\mathbf{r})$, $\mathbf{J}(\mathbf{r})$, $\mathbf{P}(\mathbf{r})$, and $\mathbf{M}(\mathbf{r})$ and the surface densities $\eta(\mathbf{r})$ and $\mathbf{K}(\mathbf{r})$ are not all mutually independent but actually involve in their statistical definitions the manner in which the region to be described is subdivided into interior and surface cells. In the usual interpretation of these functions, it is implied that no charges that are parts of polarized or magnetized elements are included in two different cells. This means that only complete dipoles and current whirls are included in volume cells; thus all contributions to the densities $\rho(\mathbf{r})$, $\mathbf{J}(\mathbf{r})$, and $\eta(\mathbf{r})$ come from "free" charges and not from "bound" charges which are parts of neutral dipoles and self-contained circulations of current.

Since nonuniform distributions of dipoles and current whirls can contribute, respectively, to densities of charge and current, the actual effective volume and surface densities of charge are

$$\bar{\rho}(\mathbf{r}) \equiv \rho(\mathbf{r}) - \nabla \cdot \mathbf{P}(\mathbf{r}) \qquad \text{and} \qquad \bar{\eta}(\mathbf{r}) \equiv \eta(\mathbf{r}) + \hat{\mathbf{n}} \cdot \mathbf{P}(\mathbf{r}) \qquad (1.25\text{--}3)$$

where $\hat{\mathbf{n}}$ is the unit external normal. Similarly, the actual effective volume and surface densities of current are

$$\bar{\mathbf{J}}(\mathbf{r}) \equiv \mathbf{J}(\mathbf{r}) + \nabla \times \mathbf{M}(\mathbf{r}) \qquad \text{and} \qquad \bar{\mathbf{K}}(\mathbf{r}) \equiv \mathbf{K}(\mathbf{r}) - \hat{\mathbf{n}} \times \mathbf{M}(\mathbf{r}) \qquad (1.25\text{--}4)$$

NONSTATIONARY STATES

1.26 ESSENTIAL VOLUME CHARACTERISTIC OF THE UNSTEADY STATE: EQUATION OF CONTINUITY FOR ELECTRIC CHARGE

The static and steady states are both stationary states because the density functions that characterize conditions of statistical rest and of steady motion of electric charge at every point in a region are *independent of time*. In the nonstationary state the same density functions may be used to describe *instantaneous* distributions of charge and current, but they vary in time at every point. Moreover, as a consequence of the postulate of conservation of electric charge, definite relationships exist between the time rates of change of the densities of charge $\bar{\rho}$ and $\bar{\eta}$ and the densities of moving charge $\bar{\mathbf{J}}$ and $\bar{\mathbf{K}}$.

To determine the essential characteristics of the nonstationary state from the corresponding stationary quantities, consider a region containing electric charges for which all the four volume densities ρ, $\mathbf{P}$, $\mathbf{J}$, and $\mathbf{M}$ are defined in the usual way. The electrical properties of the region are described in terms of the essential densities

$$\bar{\rho} = \rho - \mathbf{\nabla} \cdot \mathbf{P} \tag{1.26-1a}$$

$$\bar{\mathbf{J}} = \mathbf{J} + \mathbf{\nabla} \times \mathbf{M} \tag{1.26-1b}$$

which are independent of the mode of subdivision into volume cells. In the static and steady states they are by definition constant in time.

Let an unsteady state exist in which the charges in the region move in a nonrandom way so that ρ and $\mathbf{P}$ are functions of the time. $\mathbf{M}$ and $\mathbf{J}$ may be constant or variable in time. Since $\partial\rho/\partial t$ and $\partial\mathbf{P}/\partial t$ are nonvanishing, the time rate of change of $\bar{\rho}$ in (1.26-1a) with an interchange of differentiation between the independent time and space variables is

$$\frac{\partial\bar{\rho}}{\partial t} = \frac{\partial\rho}{\partial t} - \mathbf{\nabla} \cdot \frac{\partial\mathbf{P}}{\partial t} \tag{1.26-2}$$

A variation in time of ρ means that the total charge contained in a volume cell $\Delta\tau$ increases or decreases or changes periodically as time passes. Since it has been postulated as a fundamental attribute of electricity that electric charge is conserved, so that no charges can be created or destroyed in $\Delta\tau$, it follows that electric charges must flow inward or outward across the closed bounding surface of $\Delta\tau$ if $\partial\rho/\partial t$ differs from zero. Such a flow may consist of free charges moving from volume cell to volume cell, or of parts of periodically distorted closed shells vibrating across the bounding surfaces. No contributions to such a flow come from oriented spin groups because the total charge at every instant in a given volume is unchanged by a variation in time of the number of current whirls or of their axes of rotation. With ρ defined to measure the average total charge per unit volume, it is clear that $\partial\rho/\partial t$ must measure the time rate of change of the average total charge in each unit of volume. This can be related to the volume density of current $\mathbf{J}$ in the following way.

Let Σ be the enclosing surface of a typical volume cell $\Delta\tau$. The postulate of conservation requires the rate of increase of the positive charge $\rho\Delta\tau$ in $\Delta\tau$ to equal the net positive charge entering across $\Delta\tau$ per unit time. (Negative charge leaving $\Delta\tau$ is equal in magnitude to positive charge entering it.) At any instant, the flow of positive charge across Σ into $\Delta\tau$ is given by the integral over Σ of the inward normal component of the volume density of moving charge $\mathbf{J}$. This integral is the negative of the total outward normal flux of the vector $\mathbf{J}$ across a closed surface:

$$\frac{\partial}{\partial t}(\rho\Delta\tau) = -\int_{\Sigma} \hat{\mathbf{n}}\cdot\mathbf{J}\,d\sigma \qquad (1.26\text{--}3)$$

where $\hat{\mathbf{n}}$ is the external normal to the surface Σ. Since $\Delta\tau$ is constant in time,

$$\frac{\partial}{\partial t}(\rho\Delta\tau) = \Delta\tau\,\frac{\partial\rho}{\partial t} \qquad (1.26\text{--}4)$$

Since ρ and $\mathbf{J}$ are continuous functions defined throughout the region, the simple limit process of the calculus may be used to obtain

$$\frac{\partial\rho}{\partial t} = -\lim_{\Delta\tau\to 0}\frac{\displaystyle\int_{\Sigma}\hat{\mathbf{n}}\cdot\mathbf{J}\,d\sigma}{\Delta\tau} = -\boldsymbol{\nabla}\cdot\mathbf{J} \qquad (1.26\text{--}5)$$

by definition of the divergence of a vector (1.9–8). The equation

$$\boldsymbol{\nabla}\cdot\mathbf{J} + \frac{\partial\rho}{\partial t} = 0 \qquad (1.26\text{--}6)$$

is called the *equation of continuity for electric charge*. It expresses the postulated principle of conservation of electricity in terms of the density functions of the mathematical model. It explicitly establishes a connection between the instantaneous steady-state volume density of drifting charge $\mathbf{J}$ and the instantaneous time rate of change of the volume density of charge ρ. If the value of $\partial\rho/\partial t$ given by the equation of continuity (1.26–6) is inserted in (1.26–2),

$$\frac{\partial\overline{\rho}}{\partial t} = -\boldsymbol{\nabla}\cdot\mathbf{J} - \boldsymbol{\nabla}\cdot\frac{\partial\mathbf{P}}{\partial t}$$

$$= -\boldsymbol{\nabla}\cdot\left(\mathbf{J} + \frac{\partial\mathbf{P}}{\partial t}\right)$$

or

$$\frac{\partial\overline{\rho}}{\partial t} + \boldsymbol{\nabla}\cdot\left(\mathbf{J} + \frac{\partial\mathbf{P}}{\partial t}\right) = 0 \qquad (1.26\text{--}7)$$

This is a formally generalized equation of continuity which has, in effect, been obtained from (1.26–6) simply by adding and subtracting the term $\nabla \cdot \partial \mathbf{P}/\partial t$. Its usefulness lies in the fact that the essential density of charge $\bar{\rho}$ is independent of the mode of subdivision into volume cells, whereas ρ is not. It follows that $(\mathbf{J} + \partial \mathbf{P}/\partial t)$ must characterize the moving charge in a way that is also independent of the mode of subdivision. This is true provided that there are no spin atoms, so that $\mathbf{M} = 0$. It has been shown in the steady state that if $\mathbf{M}$ differs from zero so that circulations of charge exist, $(\mathbf{J} + \nabla \times \mathbf{M})$ and not $\mathbf{J}$ alone is the characteristic of moving charge that is independent of the mode of subdivision into volume cells. In the general unsteady state in which $\mathbf{M}$ does not vanish, the essential characteristic of moving charge that is independent of the mode of subdivision may be obtained by noting that

$$\nabla \cdot \nabla \times \mathbf{M} = 0 \qquad (1.26\text{–}8)$$

so that $\nabla \times \mathbf{M}$ may be added to $(\mathbf{J} + \partial \mathbf{P}/\partial t)$ in (1.26–7) without disturbing the balance of the equation:

$$\frac{\partial \bar{\rho}}{\partial t} + \nabla \cdot \left(\mathbf{J} + \nabla \times \mathbf{M} + \frac{\partial \mathbf{P}}{\partial t} \right) = 0 \qquad (1.26\text{–}9)$$

The essential density of moving charge is

$$\overline{\rho_m \mathbf{v}} \equiv \mathbf{J} + \nabla \times \mathbf{M} + \frac{\partial \mathbf{P}}{\partial t} \qquad (1.26\text{–}10)$$

With (1.26–10), (1.26–9) becomes

$$\frac{\partial \bar{\rho}}{\partial t} + \nabla \cdot \overline{\rho_m \mathbf{v}} = 0 \qquad (1.26\text{–}11)$$

This is a symbolically convenient, generalized equation of continuity that differs only formally from (1.26–6) from which it may be obtained by adding and subtracting a term in $\mathbf{P}$ and adding a term in $\mathbf{M}$ which is identically zero. It is useful for obtaining the essential density of moving charge $\overline{\rho_m \mathbf{v}}$ in the nonstationary state. This is, at every instant, a continuous, slowly varying function of the space coordinates defined throughout the interior of a body or region. All its terms may be functions of time. The velocity $\bar{\mathbf{v}}$ is a continuous, slowly varying vector point function characterizing the average statistical velocity of moving charge including nonrandom motions of free charge and of closed-shell configurations of either the polarized or magnetized kind. $\bar{\rho}_m$ is the volume density associated with $\bar{\mathbf{v}}$; it includes all charges engaged in nonrandom motion.

The physical significance in terms of the atomic model of associating the term $\partial \mathbf{P}/\partial t$ with the convection or conduction current density $\mathbf{J}$ and the magnetization current density $\mathbf{J}_m = \nabla \times \mathbf{M}$ is readily visualized. Superficially, it is plausible that a more or less periodic reversal in the direction of polarization of a distorted and oriented group of bound charges, such as the closed shell of an atom or molecule, is equivalent to an alternating current. The actual explanation for the appearance of the term $\partial \mathbf{P}/\partial t$ in the essential characteristic of moving charge $\overline{\rho_m \mathbf{v}}$ is precisely the same as the explanation for the appearance of $-\nabla \cdot \mathbf{P}$ in the essential characteristic of charge $\overline{\rho}$. The contributions to $\partial \overline{\rho}/\partial t$ by the two terms $-\nabla \cdot \mathbf{J}$ and $-\nabla \cdot (\partial \mathbf{P}/\partial t)$, just as the contributions to $\overline{\rho}$ by ρ and $-\nabla \cdot \mathbf{P}$, depend on the mode of subdivision of a body into volume cells. If the boundaries of the cells are so placed that no dipoles are cut by them even when they experience periodically reversing distortion-orientation effects along a definite axis (subdivision I), then $(-\nabla \cdot \mathbf{P})$ is the contribution to $\overline{\rho}$ of a nonuniform polarization and $-\nabla \cdot (\partial \mathbf{P}/\partial t)$ is the contribution to $\partial \overline{\rho}/\partial t$ of a time rate of change of this polarization. This is equivalent to stating that $\partial \mathbf{P}/\partial t$ accounts for the contribution to the characteristic of moving charge of periodically reversing distortion orientations of bound groups. In the mode of subdivision I

$$\rho = \rho_f \qquad (1.26-12)$$

where ρ_f is the density of free charge and

$$\mathbf{J} = \mathbf{J}_f \qquad (1.26-13)$$

where $\mathbf{J}_f$ is the density of moving free charge since only the free charges contribute to the transfer across cell boundaries. The conservation of charge is described by

$$\frac{\partial \rho_f}{\partial t} + \nabla \cdot \mathbf{J}_f = 0 \qquad (1.26-14)$$

If the mode of subdivision is changed to mode II so that just enough dipoles are cut by the cell boundaries to make $\mathbf{P}$ zero everywhere, then $\nabla \cdot \mathbf{P}$ and $\partial \mathbf{P}/\partial t$ vanish. This corresponds to an alternating or periodically reversing current across the walls with a corresponding periodic change in the total charge in $\Delta\tau$. ρ and $\mathbf{J}$ calculated in the usual way differ from the values calculated in the first mode of subdivision precisely by the amounts,

$$\rho_{b\mathrm{II}} = -\nabla \cdot \mathbf{P}_{\mathrm{I}} \quad \text{and} \quad \mathbf{J}_{p\mathrm{II}} = \frac{\partial \mathbf{P}_{\mathrm{I}}}{\partial t} \qquad (1.26-15a)$$

In this case

$$\overline{\rho} = \rho_f + \rho_{b\mathrm{II}}$$

and

$$\qquad (1.26-15b)$$

$$\overline{\rho_m \mathbf{v}} = \mathbf{J}_f + \mathbf{J}_{p\mathrm{II}} = \mathbf{J}_f + \frac{\partial \mathbf{P}_{\mathrm{I}}}{\partial t}$$

where ρ_f is the average density of free charge and ρ_b is the average density of the parts of the dipoles that are cut and included within the boundaries of the cell. $\mathbf{J}_f$ represents the convection or conduction drifts of charge across the boundaries of cells and $\mathbf{J}_p$ the average motion of parts of vibrating dipoles across the same boundaries. The change in the mode of subdivision in the physical picture corresponds to a mere shift in the responsibility of representation in the mathematical model from one set of functions to another. The quantity $\mathbf{J}_p = \partial\mathbf{P}/\partial t$ is the volume density of polarization current due to parts of oscillating dipoles moving across cell walls. By appropriate choice this may be observed as $\mathbf{J}_p$ with $\mathbf{P} = 0$ or by $\partial\mathbf{P}/\partial t$ with $\mathbf{J}_p = 0$. Obviously, the effective alternating current due to oscillating dipoles must be included equally in the description independent of the location of cell walls.

For all modes of subdivision satisfying the restrictions on the size of individual cells, the two continuous functions

$$\bar{\rho} = \rho - \nabla \cdot \mathbf{P} \qquad (1.26\text{–}16a)$$

$$\overline{\rho_m\mathbf{v}} = \mathbf{J} + \nabla \times \mathbf{M} + \frac{\partial\mathbf{P}}{\partial t} \qquad (1.26\text{–}16b)$$

completely characterize the instantaneous conditions of charge and moving charge in the interior of a body or region to the degree of approximation here attempted.

1.27 ESSENTIAL SURFACE CHARACTERISTIC OF THE UNSTEADY STATE: SURFACE EQUATION OF CONTINUITY FOR ELECTRIC CHARGE

The equation of continuity, expressing the fundamental postulate of conservation of electric charge, is equally true along a surface, or more generally along the boundary between two electrically dissimilar regions, as in the interior. Its form is obtained by applying (1.26–6) to a thin surface layer on each side of the boundary between regions 1 and 2 (Fig. 1.27–1).

Consider a small rectangular surface element $\Delta\tau_s$ of thickness d_c on each side of a boundary and of area of base and top $\Delta\Sigma$. The combined volume of the double element is $\Delta\tau_s = 2d_c\Delta\Sigma$. Using the fundamental definition (1.9–8) of the divergence of a vector, (1.26–6) may be written in the form

$$\frac{\partial\rho}{\partial t} = \frac{-\lim\limits_{2d_c\Delta\Sigma\to 0}\int_\Sigma \hat{\mathbf{n}} \cdot \mathbf{J}\, d\sigma}{2d_c\Delta\Sigma} \qquad (1.27\text{–}1)$$

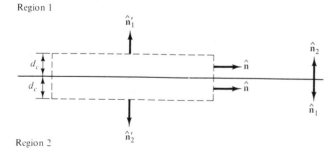

Region 1

Region 2

Figure 1.27–1 Surface layer at a boundary.

where $\hat{\mathbf{n}}$ is an external normal to the enclosing surface Σ of the element. Let Σ be considered in three parts. Σ_{dc} is the combined area in both regions 1 and 2 of the four narrow edges of the entire element of thickness $2d_c$; $\Delta\Sigma_1$ is the area of the top in region 1; $\Delta\Sigma_2$ is the area of the base in region 2. The surface integral in (1.27–1) can be written as the sum of three integrals over these three parts of the closed surface of the cell.

$$\frac{\partial \rho_{av}}{\partial t} = -\lim_{2d_c\Delta\Sigma\to 0} \left[\frac{\int_{\Sigma_{dc}} (\hat{\mathbf{n}}\cdot\mathbf{J}_{av})\,d\sigma}{2d_c\,\Delta\Sigma} + \frac{\int_{\Delta\Sigma_1} (\hat{\mathbf{n}}_1'\cdot\mathbf{J}_1)\,d\sigma}{2d_c\,\Delta\Sigma_1} + \frac{\int_{\Delta\Sigma_2} (\hat{\mathbf{n}}_2'\cdot\mathbf{J}_2)\,d\sigma}{2d_c\,\Delta\Sigma_2} \right] \tag{1.27–2}$$

$\hat{\mathbf{n}}_1'$ points out of $\Delta\tau_s$ from $\Delta\Sigma_1$ into region 1; $\hat{\mathbf{n}}_2'$ points across $\Delta\Sigma_2$ into region 2. So $\hat{\mathbf{n}}_1'$ and $\hat{\mathbf{n}}_2'$ represent the external normals to the cell boundaries.

In (1.27–2) ρ_{av} is the average volume density of charge in $2d_c\Delta\Sigma$. In terms of the volume densities in each region

$$\rho_{av} = \tfrac{1}{2}(\rho_1 + \rho_2) \tag{1.27–3}$$

Similarly, $\mathbf{J}_{av}$ in the integral over the narrow surface Σ_{dc} is the average value in the two regions:

$$\mathbf{J}_{av} = \tfrac{1}{2}(\mathbf{J}_1 + \mathbf{J}_2) \tag{1.27–4}$$

Using (1.27–3) and (1.27–4) in (1.27–2) and multiplying through by $2d_c$ gives

$$\frac{\partial}{\partial t}(d_c\rho_1 + d_c\rho_2) = -\lim_{2d_c\Delta\Sigma\to 0}\left[\frac{\int_{\Sigma_{dc}} \hat{\mathbf{n}}\cdot(d_c\mathbf{J}_1 + d_c\mathbf{J}_2)\,d\sigma}{2d_c\Delta\Sigma} \right.$$

$$\left. + \frac{\int_{\Delta\Sigma_1}(\hat{\mathbf{n}}_1'\cdot\mathbf{J}_1)\,d\sigma}{\Delta\Sigma_1} + \frac{\int_{\Delta\Sigma_2}(\hat{\mathbf{n}}_2'\cdot\mathbf{J}_2)\,d\sigma}{\Delta\Sigma_2} \right] \tag{1.27–5}$$

The surface density of charge (1.4–1) is

$$\eta = d_c\rho \tag{1.27–6}$$

The surface density of current (1.15–1) is

$$\mathbf{K} = d_c\mathbf{J} \tag{1.27–7}$$

From (1.27–6),

$$d_c\rho_1 + d_c\rho_2 = \eta_1 + \eta_2 \equiv \eta \tag{1.27–8}$$

Here η is the total surface density of charge associated with the boundary surface between the two regions. The use of (1.27–7) leads to

$$\hat{n} \cdot d_c J_1 + \hat{n} \cdot d_c J_2 = \hat{n} \cdot K_1 + \hat{n} \cdot K_2 \equiv \hat{n} \cdot K \qquad (1.27-9)$$

where K is the total surface density of current associated with the surface between the two regions. Since the normal in the first integral is directed across the narrow edges of the rectangular parallelepiped $\Delta \tau_s$, and hence is tangent to the bounding surface, it is clear that the current densities in (1.27–9) are parallel to the surface. With (1.27–8) and (1.27–9), (1.27–5) becomes

$$\frac{\partial \eta}{\partial t} = -\lim_{2d_c \Delta \Sigma \to 0} \left[\frac{\int_{\Sigma_{dc}} (\hat{n} \cdot K) \, d\sigma}{2d_c \Delta \Sigma} + \frac{\int_{\Delta \Sigma_1} (\hat{n}_1' \cdot J_1) \, d\sigma}{\Delta \Sigma_1} \right.$$

$$\left. + \frac{\int_{\Delta \Sigma_2} (\hat{n}_2' \cdot J_2) \, d\sigma}{\Delta \Sigma_2} \right] \qquad (1.27-10)$$

Here the first integral is the divergence of the two-dimensional vector K. The last two integrals may be transformed with the theorem of the mean.

$$\frac{\partial \eta}{\partial t} + \nabla \cdot K = -\lim_{\Delta \Sigma \to 0} \left(\hat{n}_1' \cdot J_1 \frac{\int_{\Delta \Sigma_1} d\sigma}{\Delta \Sigma_1} + \hat{n}_2' \cdot J_2 \frac{\int_{\Delta \Sigma_2} d\sigma}{\Delta \Sigma_2} \right) \qquad (1.27-11)$$

Upon performing the integration, passing to the limit, and introducing the exterior normals to the regions 1 and 2,

$$\hat{n}_1 = -\hat{n}_1'; \qquad \hat{n}_2 = -\hat{n}_2' \qquad (1.27-12)$$

the following equation, which contains both surface and volume functions, is obtained:

$$\frac{\partial \eta}{\partial t} + \nabla \cdot K - \hat{n} \cdot J = 0 \qquad (1.27-13)$$

where

$$\hat{n} \cdot J \equiv \hat{n}_1 \cdot J_1 + \hat{n}_2 \cdot J_2 \qquad (1.27-14)$$

For a surface effect involving a drift of charge along the surface only, (1.27–13) becomes

$$\frac{\partial \eta}{\partial t} + \nabla \cdot \mathbf{K} = 0 \qquad (1.27\text{--}15)$$

For a volume effect in which there is no motion of charge tangent to the boundary between two surfaces, (1.27–13) leads to the important boundary condition for volume density of current crossing a boundary:

$$\hat{\mathbf{n}} \cdot \mathbf{J} \equiv \hat{\mathbf{n}}_1 \cdot \mathbf{J}_1 + \hat{\mathbf{n}}_2 \cdot \mathbf{J}_2 = 0 \qquad (1.27\text{--}16)$$

If the surface and volume effects are independent, (1.27–15) and (1.27–16) are simultaneously and independently true.

The mixed surface equation (1.27–13) corresponds to the volume equation (1.26–6) in that it is not written in terms of essential densities that are independent of the mode of subdivision of the region into volume and surface cells. To generalize (1.27–13) to correspond to the volume equation (1.26–11), and in this way determine the essential surface characteristic of the nonstationary state, the procedure used in deriving (1.26–10) is followed. It involves adding and subtracting $\hat{\mathbf{n}} \cdot \partial \mathbf{P}/\partial t$, subtracting $\nabla \cdot (\hat{\mathbf{n}} \times \mathbf{M})$ and subtracting $\hat{\mathbf{n}} \cdot (\nabla \times \mathbf{M})$ in (1.27–13). The vector identity

$$\nabla \cdot (\hat{\mathbf{n}} \times \mathbf{M}) = \mathbf{M} \cdot \nabla \times \hat{\mathbf{n}} - \hat{\mathbf{n}} \cdot \nabla \times \mathbf{M}$$

is used. Since $\hat{\mathbf{n}}$ is a constant vector,

$$\nabla \times \hat{\mathbf{n}} = 0$$

and the identity reduces to

$$\nabla \cdot \hat{\mathbf{n}} \times \mathbf{M} + \hat{\mathbf{n}} \cdot \nabla \times \mathbf{M} = 0$$

Then

$$\frac{\partial}{\partial t} (\hat{\mathbf{n}} \cdot \mathbf{P}) - \hat{\mathbf{n}} \cdot \frac{\partial \mathbf{P}}{\partial t} - \nabla \cdot (\hat{\mathbf{n}} \times \mathbf{M}) - \hat{\mathbf{n}} \cdot (\nabla \times \mathbf{M}) = 0$$

and it may be added to (1.27–13) without altering the equation. A grouping and arrangement of terms yields for (1.27–13)

$$\frac{\partial}{\partial t} (\eta + \hat{\mathbf{n}} \cdot \mathbf{P}) + \nabla \cdot (\mathbf{K} - \hat{\mathbf{n}} \times \mathbf{M})$$

$$- \hat{\mathbf{n}} \cdot \left(\mathbf{J} + \nabla \times \mathbf{M} + \frac{\partial \mathbf{P}}{\partial t} \right) = 0 \qquad (1.27\text{--}17)$$

or

$$\frac{\partial \overline{\eta}}{\partial t} + \nabla \cdot \overline{\eta_m \mathbf{v}} - \hat{\mathbf{n}} \cdot \overline{\rho_m \mathbf{v}} = 0 \qquad (1.27\text{--}18)$$

where the essential volume and surface densities are defined as follows:

$$\bar{\rho} \equiv \rho - \nabla \cdot \mathbf{P} \tag{1.27--19}$$

$$\overline{\rho_m \mathbf{v}} \equiv \mathbf{J} + \nabla \times \mathbf{M} + \frac{\partial \mathbf{P}}{\partial t} \tag{1.27--20}$$

$$\bar{\eta} \equiv \eta + \hat{\mathbf{n}} \cdot \mathbf{P} \tag{1.27--21}$$

$$\overline{\eta_m \mathbf{v}} \equiv \overline{\mathbf{K}} \equiv \mathbf{K} - \hat{\mathbf{n}} \times \mathbf{M} \tag{1.27--22}$$

It follows from (1.27–22) that the essential surface characteristic of moving charge in the unsteady state is the same in form as the corresponding steady-state function. The notation $\overline{\eta_m \mathbf{v}}$ corresponds to the volume function $\overline{\rho_m \mathbf{v}}$. In the nonstationary state all terms in (1.27–22) are functions of time. By requiring the invariance of all densities in time, they reduce to the functions previously defined for the stationary states.

When it is desired to show explicitly that the densities are functions of the space coordinates and the time, the notation $\rho(\mathbf{r}, t)$, $\mathbf{J}(\mathbf{r}, t)$, and so on, is convenient.

1.28 EQUATION OF CONTINUITY FOR A CYLINDRICAL CONDUCTOR

A partial integration of the equation of continuity is readily carried out for a cylindrical region of length $2h$ and radius a. The following inequality is postulated:

$$a \ll h \tag{1.28--1}$$

The cylinder is oriented with its axis along the z axis of a cylindrical system of coordinates r, θ, z as shown in Fig. 1.28–1. Complete rotational symmetry is assumed to prevail. The cylinder will be identified later with a current-carrying conductor for which it will be shown that

$$\rho = 0; \quad \mathbf{P} = 0 \tag{1.28--2}$$

At this point it is sufficient to regard (1.28–2) as a special condition imposed on the cylinder. Subject to (1.28–2) the cylindrical region is characterized by

$$\bar{\rho} = 0; \quad \overline{\rho_m \mathbf{v}} = \mathbf{J}_f \tag{1.28--3a}$$

$$\bar{\eta} = \eta_f; \quad \overline{\eta_m \mathbf{v}} = 0 \tag{1.28--3b}$$

since $\mathbf{M} = 0$ in nonmagnetic materials and $\mathbf{K} = \mathbf{K}_f = 0$ when $\mathbf{J}_f \neq 0$.

Both volume and surface densities appear in (1.28–3a,b), so it is necessary to use a representation in terms of volume and surface cells. The equations of

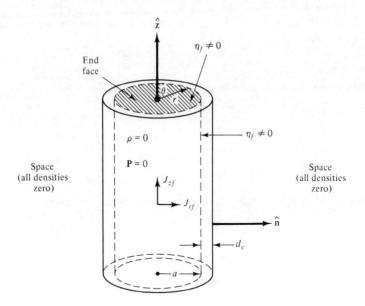

Figure 1.28–1   Cylindrical conductor.

continuity subject to (1.28–2) for the interior and the surface are

$$\boldsymbol{\nabla} \cdot \mathbf{J}_f = 0 \tag{1.28–4a}$$

$$\frac{\partial \eta_f}{\partial t} - \hat{\mathbf{n}} \cdot \mathbf{J}_f = 0 \tag{1.28–4b}$$

The current into the cylindrical surface layer of atomic thickness d_c from the interior is the radial component J_{rf} at

$$r = a - d_c$$

That is, in (1.28–4b),

$$\hat{\mathbf{n}} \cdot \mathbf{J}_f = \hat{\mathbf{n}}_1 \cdot \mathbf{J}_{1f} + \hat{\mathbf{n}}_2 \cdot \mathbf{J}_{2f} = \hat{\mathbf{n}}_1 \cdot \mathbf{J}_{1f}$$

since there is no moving charge outside the cylinder (region 2):

$$\hat{\mathbf{n}}_1 \cdot \mathbf{J}_{1f} = \hat{\mathbf{r}} \cdot \mathbf{J}_{1f} = J_{rf} \tag{1.28–5}$$

Then (1.28–4b) becomes

$$(J_{rf})_{r=a-d_c} = \frac{\partial \eta_f}{\partial t} \tag{1.28–6}$$

Also,

$$\boldsymbol{\nabla} \cdot \mathbf{J}_f = \frac{1}{r}\frac{\partial}{\partial r}(rJ_{rf}) + \frac{1}{r}\frac{\partial J_{\theta f}}{\partial \theta} + \frac{\partial J_{zf}}{\partial z} = 0 \tag{1.28–7}$$

which follows from (1.28–4a). Because rotational symmetry exists, it may be assumed that

$$J_{\theta f} = 0 \qquad\qquad (1.28\text{--}8)$$

Equation (1.28–7) now becomes

$$\frac{1}{r}\frac{\partial}{\partial r}(rJ_{rf}) = -\frac{\partial J_{zf}}{\partial z} \qquad\qquad (1.28\text{--}9a)$$

$$\frac{\partial}{\partial r}(rJ_{rf}) = -r\frac{\partial J_{zf}}{\partial z} \qquad\qquad (1.28\text{--}9b)$$

A partial integration of (1.28–9b) with respect to r gives

$$rJ_{rf}\Big|_0^{a-d_c} = -\int_0^{a-d_c} r\frac{\partial J_{zf}}{\partial z}\,dr$$

$$= -\frac{\partial}{\partial z}\int_0^{a-d_c} rJ_{zf}\,dr \qquad\qquad (1.28\text{--}10a)$$

The total axial current is defined to be

$$I_z \equiv \int_0^{a-d_c} 2\pi r J_{zf}\,dr \qquad\qquad (1.28\text{--}10b)$$

It follows that (1.28–10a) becomes

$$rJ_{rf}\Big|_0^{a-d_c} = -\frac{1}{2\pi}\frac{\partial}{\partial z}\int_0^{a-d_c} 2\pi r J_{zf}\,dr$$

or

$$(J_{rf})_{r=a-d_c} = \left(-\frac{1}{2\pi r}\frac{\partial I_z}{\partial z}\right)_{r=a-d_c} \qquad\qquad (1.28\text{--}11a)$$

$$= -\frac{1}{2\pi a}\frac{\partial I_z}{\partial z} = \frac{\partial \eta_f}{\partial t} \qquad\qquad (1.28\text{--}11b)$$

If the surface charge per unit length q is defined as

$$q = 2\pi a \eta_f \qquad\qquad (1.28\text{--}12)$$

(1.28–11b) reduces to the following equation of continuity for the cylindrical conductor:

$$\frac{\partial I_z}{\partial z} + \frac{\partial q}{\partial t} = 0 \qquad\qquad (1.28\text{--}13)$$

This could have been written down more quickly and without imposing (1.28–2) if the postulate of conservation of electric charge had been applied directly to a length dz of the cylinder. This was not done in order to bring clearly into the foreground the parts played by the several densities in a cylindrical region restricted by (1.28–2), (1.28–3b), and (1.28–8). It is clear from (1.28–9a) that even though a radial component of current does not appear in (1.28–13), such a component actually plays a fundamental part in maintaining a surface density of charge along the entire cylindrical surface of the conductor.

Near the ends of the conductor, specifically at $z = \pm(h - d_c)$, the total axial current entering the surface layer of thickness d_c of each end surface must satisfy the condition

$$(I_z)_{z = \pm(h - d_c)} = \frac{dQ_e}{dt} \tag{1.28–14}$$

$$Q_e = \int_0^{a - d_c} 2\pi r \eta_f \, dr, \quad \text{at } z = \pm(h - d_c) \tag{1.28–15}$$

Q_e represents the total surface charge on the end surface.

1.29 REPRESENTATION OF A CURRENT-CARRYING CONDUCTOR IN TERMS OF A VOLUME DENSITY OF POLARIZATION

Although a cylindrical conductor carrying an alternating current is characterized by axial and radial currents and a surface density of free charge only as shown in Sec. 1.28, it is sometimes convenient to represent the entire distribution by a physically fictitious but mathematically equivalent model characterized by $\mathbf{P}$ alone. Since $\mathbf{P}$ can under no circumstances describe surface currents, it follows that a description in terms of $\mathbf{P}$ alone is possible only if $\overline{\eta_m \mathbf{v}}$ vanishes. As in Sec. 1.28, the actual conductor is described completely by

$$\overline{\rho} = 0; \qquad \overline{\rho_m \mathbf{v}} = \mathbf{J}_f \tag{1.29–1a}$$

$$\overline{\eta} = \eta_f; \qquad \overline{\eta_m \mathbf{v}} = 0 \tag{1.29–1b}$$

It is now proposed to represent this in terms of a mathematically equivalent model using $\mathbf{P}$ alone. The essential densities as expressed in terms of $\mathbf{P}$ are

$$\overline{\rho} = -\nabla \cdot \mathbf{P}; \qquad \overline{\rho_m \mathbf{v}} = \frac{\partial \mathbf{P}}{\partial t} \tag{1.29–2a}$$

$$\overline{\eta} = \hat{\mathbf{n}} \cdot \mathbf{P}; \qquad \overline{\eta_m \mathbf{v}} = 0 \tag{1.29–2b}$$

It is necessary that

$$\mathbf{J}_f = \frac{\partial \mathbf{P}}{\partial t}; \qquad \eta_f = \hat{\mathbf{n}} \cdot \mathbf{P} \tag{1.29–3}$$

In order to make the representation (1.29–2a,b) and (1.29–3) equivalent to (1.29–1a,b) in the interior,

$$P_\theta = 0 \quad \text{(rotational symmetry)} \qquad (1.29\text{–}4a)$$

$$-\mathbf{\nabla} \cdot \mathbf{P}_I = \bar{\rho} = -\left[\frac{1}{r}\frac{\partial}{\partial r}(rP_r) + \frac{\partial P_z}{\partial z}\right] = 0 \qquad (1.29\text{–}4b)$$

$$\frac{\partial P_r}{\partial t} = J_{rf}; \qquad \frac{\partial P_z}{\partial t} = J_{zf} \qquad (1.29\text{–}4c)$$

For equivalence on the cylindrical surface,

$$(P_r)_{r=a} = \eta_f \text{ (cylinder)} \qquad (1.29\text{–}5)$$

For equivalence on the end surfaces,

$$(P_z)_{z=\pm h} = \eta_f \text{ (ends)} \qquad (1.29\text{–}6)$$

It is readily verified that these conditions and both equations of continuity are satisfied if

$$\frac{\partial \mathbf{P}}{\partial t} = \mathbf{J}_f \qquad (1.29\text{–}7)$$

and

$$I_z = \frac{\partial}{\partial t}\int_0^a 2\pi r P_z \, dr = \frac{\partial p_z}{\partial t} \qquad (1.29\text{–}8)$$

In (1.29–8),

$$p_z = \int_0^a 2\pi r P_z \, dr \qquad (1.29\text{–}9)$$

is the axial polarization or electric moment per unit length. It may be regarded as the axial component of an equivalent dipole for a unit length.

It is therefore possible to represent a distribution of free charge and current confined to a cylinder by a mathematically equivalent, but usually physically unavailable model using a distribution of the volume density of polarization.

1.30 REPRESENTATION OF A RING OF CURRENT BY A VOLUME DENSITY OF MAGNETIZATION

Consider a metal hoop of wire of rectangular cross section. The inner radius is a, the outer radius is b, and the thickness is d. It carries a total circulating current of magnitude I (Fig. 1.30–1).

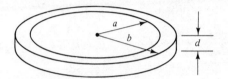

Figure 1.30–1 Metal loop.

If the current density is uniform, the electrical properties are represented completely by

$$\overline{\rho_m}\mathbf{v} = \mathbf{J'}; \qquad \overline{\eta_m}\mathbf{v} = 0, \quad \overline{\rho} = 0, \quad \overline{\eta} = 0 \text{ in the ring} \qquad (1.30\text{–}1)$$

$$\mathbf{J'} = \hat{\mathbf{\theta}}J'_\theta$$

$$= \hat{\mathbf{\theta}}\,\frac{I}{d(b - a)}, \qquad a \le r \le b, 0 \le z \le d$$

$$= 0, \qquad\qquad\qquad \text{elsewhere} \qquad\qquad (1.30\text{–}2)$$

All other densities are zero everywhere.

The problem is to replace this physical model with its mathematical representation in terms of $\mathbf{J'}$ by a physically fictitious but mathematically equivalent model using the volume density of magnetization $\mathbf{M''}$. Equations (1.30–1) and (1.30–2) now become

$$\overline{\rho_m}\mathbf{v} = \mathbf{\nabla} \times \mathbf{M''} = \mathbf{J'} = \hat{\mathbf{\theta}}\,\frac{I}{d(b - a)},$$

$$a \le r \le b, \quad 0 \le z \le d$$

$$= 0, \quad \text{elsewhere} \qquad\qquad (1.30\text{–}3)$$

$$(\mathbf{\nabla} \times \mathbf{M''})_\theta = \frac{\partial M''_r}{\partial z} - \frac{\partial M''_z}{\partial r}$$

$$= \frac{-\partial M''_z}{\partial r} \qquad\qquad (1.30\text{–}4)$$

if M''_r is taken to be zero.

Then, from (1.30–3),

$$\frac{\partial M''_z}{\partial r} = -\frac{I}{d(b - a)}, \qquad a \le r \le b, \quad 0, \le z \le d$$

$$= 0, \qquad\qquad\qquad \text{elsewhere} \qquad\qquad (1.30\text{–}5)$$

The integration of (1.30–5) results in

$$M_z'' = -\frac{I}{d(b-a)} r + C_1, \qquad a \le r \le b, \quad 0 \le z \le d$$

$$= C_2, \qquad\qquad\qquad \text{elsewhere} \tag{1.30–6}$$

The constants may be evaluated by requiring that M_z'' be continuous at $r = a$ and vanish for $r \ge b$.

$$\text{At } r = b: \qquad M_z'' = 0 = -\frac{Ib}{d(b-a)} + C_1$$

$$C_1 = \frac{Ib}{d(b-a)} \tag{1.30–7}$$

$$\text{For } a \le r \le b: \quad M_z'' = \frac{I(b-r)}{d(b-a)}$$

$$\text{For } 0 \le r \le a: \quad M_z'' = \frac{I}{d} \tag{1.30–8}$$

The ring of current is therefore equivalent to the magnetized disk with a constant

$$M_z = \frac{I}{d}, \qquad\qquad r \le a,$$

$$= \frac{I}{d}\frac{b-r}{b-a}, \qquad a \le r \le b$$

$$= 0, \qquad\qquad \text{elsewhere}$$

The magnetization of the entire disk is given by

$$m_z = d \int_0^b M_z 2\pi r \, dr$$

$$= I \int_0^a 2\pi r \, dr + \frac{I}{b-a} \int_a^b (b-r) 2\pi r \, dr$$

$$= I\pi a^2 + \frac{2\pi I}{b-a}\left[\frac{br^2}{2} - \frac{r^3}{3} \right]_a^b$$

$$= I\left\{ \pi a^2 + \frac{2\pi}{b-a}\left[\left(\frac{b^3}{2} - \frac{ba^2}{2} \right) - \left(\frac{b^3}{3} - \frac{a^3}{3} \right) \right] \right\}$$

$$= I\left[\pi a^2 + 2\pi\left(\frac{b^2}{6} + \frac{ab}{6} - \frac{a^2}{3} \right) \right] \tag{1.30–9}$$

If $b - a = c$, $b = a + c$ and

$$m_z = I\left[\pi a^2 + 2\pi\left(\frac{ac}{2} + \frac{c^2}{6}\right)\right]$$

$$= \pi I\left(a^2 + ac + \frac{c^2}{3}\right)$$

$$= \pi I r_e^2 \qquad (1.30\text{--}10)$$

where

$$r_e = a\left(1 + \frac{c}{a} + \frac{c^2}{3a^2}\right)^{1/2} \qquad (1.30\text{--}11)$$

Equation (1.30–10) may be written as

$$m_z = IS_e \qquad (1.30\text{--}12)$$

where S_e is the area enclosed by the loop.

If $b - a \to d$ (a current ring),

$$I = dK'_{f\theta} \qquad (1.30\text{--}13)$$

where $K'_{f\theta}$ is the free-charge current per unit width.

The magnetized disk may be represented as follows when it is subdivided into volume cells only (subdivision I):

$$\overline{\rho_m \mathbf{v}} = \mathbf{J}_{fI} + \nabla \times \mathbf{M}_I + \frac{\partial \mathbf{P}_I}{\partial t} = 0 \qquad (1.30\text{--}14a)$$

$$\mathbf{J}_{fI} = 0$$

$$\mathbf{P}_I = 0$$

$$\mathbf{M}_I = \hat{\mathbf{z}}M_{zI}$$

$$\overline{\eta_m \mathbf{v}} = \mathbf{K}_{fI} - (\hat{\mathbf{n}} \times \mathbf{M}_I) = -(\hat{\mathbf{n}} \times \mathbf{M}_I) \qquad (1.30\text{--}14b)$$

since $\mathbf{K}_{fI} = 0$ for a subdivision into volume cells.

If the mode of subdivision is now changed to include surface cells (subdivision II), $\mathbf{M} = 0$ and

$$\overline{\rho_m \mathbf{v}} = 0; \qquad \overline{\eta_m \mathbf{v}} = \mathbf{K}_{fII} \qquad (1.30\text{--}15)$$

For the two representations to be equivalent,

$$\mathbf{K}_{fII} = -\hat{\mathbf{n}} \times \mathbf{M}_I$$

$$= -\hat{\mathbf{r}} \times \hat{\mathbf{z}}M_{zI}$$

$$= \hat{\boldsymbol{\theta}}M_{zI} \qquad (1.30\text{--}16)$$

The magnetized disk can be described in terms of M_{zI} or $K_{f\theta II} = M_{zI}$ depending on the mode of subdivision. Let

$$K'_{f\theta} = \frac{I}{d} \text{ of the current ring } = K_{f\theta II} = M_{zI} \text{ of the magnetized disk}$$

By definition,

$$M_{zI} = \frac{m_{zI}}{\Delta\tau} = \frac{m_{zI}}{\pi b^2 d} \tag{1.30--17}$$

since M_{zI} is a constant. m_{zI} is the magnetization of the disk. Evidently, the current ring can be represented by a magnetized disk with total magnetic moment,

$$\mathbf{m} = \hat{z}M_z\pi b^2 d = \hat{z}IS \tag{1.30--18}$$

Now consider a magnetized disk of radius a and thickness d (Fig. 1.30--2). The electrical properties of the disk are completely characterized by

$$\mathbf{M} = \hat{z}M_z, \quad 0 \le r \le a, \quad 0 \le z \le d \tag{1.30--19}$$

where M_z is a constant. The electrical properties are described by the densities

$$\overline{\rho_m\mathbf{v}} = \mathbf{J}_I + \nabla \times \mathbf{M}_I + \frac{\partial\mathbf{P}_I}{\partial t} = 0 \tag{1.30--20a}$$

(since $\mathbf{J}_I = 0$, $\mathbf{M}_I$ = constant, $\partial\mathbf{P}_I/\partial t = 0$) and

$$\overline{\eta_m\mathbf{v}} = \mathbf{K}_I - \hat{n} \times \mathbf{M}_I$$

$$= -\hat{n} \times \mathbf{M}_I$$

$$= -\hat{r} \times \hat{z}M_{zI}$$

$$= \hat{\theta}M_{zI} \tag{1.30--20b}$$

(since $\mathbf{K}_I = 0$). In any new representation $\overline{\rho_m\mathbf{v}}$ and $\overline{\eta_m\mathbf{v}}$ must have the same values. A convenient choice is

$$\overline{\rho_m\mathbf{v}} = 0, \quad \text{with } \mathbf{J} = \mathbf{M} = \mathbf{P} = 0 \tag{1.30--21a}$$

$$\overline{\eta_m\mathbf{v}} = \mathbf{K}_{II} = \hat{\theta}M_{zI} \tag{1.30--21b}$$

This involves a change in the mode of subdivision from I (volume cells only) to II (surface and volume cells) with boundaries cutting enough current whirls to make

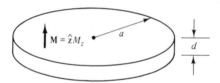

Figure 1.30--2   Magnetized disk.

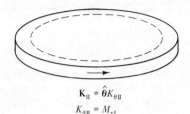

$$\mathbf{K}_{\text{II}} = \hat{\theta} K_{\theta\text{II}}$$
$$K_{\theta\text{II}} = M_{z\text{I}}$$

Figure 1.30–3   Surface current on magnetized disk.

$\mathbf{M}_{\text{II}} = 0$. $\mathbf{K}_{\text{II}}$ may be the surface effect of microscopic orbital motion or a sheet of free charges moving around the surface (Fig. 1.30–3).

The total current enclosing the ring is

$$I_{\text{II}} = dK_{\theta\text{II}} \tag{1.30–22}$$

It follows that

$$M_{z\text{I}} = K_{\theta\text{II}} = \frac{I_{\text{II}}}{d} \tag{1.30–23}$$

By definition

$$M_{z\text{I}} = \frac{m_{z\text{I}}}{\Delta\tau} = \frac{m_{z\text{I}}}{\pi b^2 d} \tag{1.30–24}$$

since $M_{z\text{I}}$ is constant. $m_{z\text{I}}$ is the magnetic moment of the entire disk. It follows that

$$m_{z\text{I}} = \pi b^2 d M_{z\text{I}} = \pi b^2 I = SI \tag{1.30–25}$$

where S is the area enclosed in the contour of current.

$$\mathbf{m} = \hat{z} M_z V = \hat{z} IS \tag{1.30–26}$$

where V is the volume in which $M_z \neq 0$. This is valid for any simply connected region. It may be concluded that a magnetized disk may be represented by an equivalent ring of current and such a ring may be represented by a magnetized disk.

PROBLEMS

1. With the aid of schematic diagrams, describe the orientation and distribution of simple dipoles in the following cases. It is assumed that only neutral bound-charge groups are present which may be represented by equivalent simple dipoles when they are distorted. In each case a suitable external force is presupposed. The values of $\mathbf{P}$ assume a subdivision into volume cells only; a is a constant.

   (a) A cube of side S placed with its center at the origin of coordinates and in which $P = P_x = ax$; $P_y = P_z = 0$.

   (b) A cylinder of height h and radius b in which $\mathbf{P} = a\mathbf{r}$.

$$r^2 = x^2 + y^2.$$

(c) A sphere of radius B in which $\mathbf{P} = a\mathbf{R}$.

$$R^2 = x^2 + y^2 + z^2.$$

(d) Calculate the values of ρ and η in each case.

2. Let the mode of subdivision in Problem 1 be changed into one using both volume and surface cells so chosen that $\mathbf{P} = 0$. Calculate ρ and η for the three cases (a), (b), and (c) in Problem 1 using the fundamental definition of divergence.

3. Repeat Problem 1 using the number density representation.

4. Derive $\nabla \cdot \mathbf{A}$ in cylindrical coordinates using the fundamental definition of the divergence.

5. Repeat Problem 4 for spherical coordinates.

6. Calculate $\nabla \cdot \mathbf{P}$ *using the appropriate coordinate form of the divergence* for each of three values of $\mathbf{P}$ given in Problem 1.

7. A right circular cylinder of radius a and height h is electrically polarized such that $\mathbf{P} = \hat{\mathbf{r}}br + \hat{\mathbf{z}}c$, where b and c are constants. Compute $\bar{\rho}$ for the interior and $\bar{\eta}$ for all surfaces.

8. A long cylindrical antenna of radius a has, at a given instant, a charge distribution given by $\rho = 0$, $\eta = (q/2\pi a) \sin kz$. $\mathbf{P} = 0$. Here q and k are independent of z. The ends of the antenna are at $kz = \pm \pi/2$. Express the charge distribution in terms of ρ and $\mathbf{P}$ alone. Show schematic diagrams for the two representations. (Neglect the small circular end surfaces.)

9. The electrostatic properties of a homogeneous sphere of radius a under the action of a symmetrical external force can be described by the following volume density functions using a subdivision into volume cells only. $\mathbf{P} = 5\mathbf{R}/R$; $\rho = 10/R$. Describe the same body in terms of ρ and η alone using an appropriately changed mode of subdivision. Which representation is to be preferred on the basis of simplicity? Of physical plausibility?

10. Derive $\nabla \times \mathbf{A}$ in cylindrical coordinates using the fundamental definition of the curl.

11. Repeat Problem 10 for spherical coordinates.

12. Derive the formula for $\nabla \times \mathbf{A}$ in Cartesian, cylindrical, and spherical coordinates using the alternative definition of the curl in terms of its component normal to a surface.

13. A section of a long cylindrical region of radius a is characterized in the steady state by the following densities using a subdivision into volume cells only. $\mathbf{M} = -4\hat{\theta}$; $\mathbf{J} = 4\hat{\mathbf{z}}/r$. Calculate the essential densities $\overline{\rho_m \mathbf{v}}$ and $\overline{\eta_m \mathbf{v}}$. Determine $\mathbf{J}$ and $\mathbf{K}$ for a subdivision into volume and surface cells such that $\mathbf{M} = 0$. (Consider only the cylindrical surface in calculating $\mathbf{K}$.) Show the directions of motion of the charges and the orientation of the density vectors for both representations by means of cross-sectional sketches. Which representation is to be preferred? Why?

14. A cylindrical bar of iron is completely characterized electrically by $\mathbf{M} = M_z\hat{\mathbf{z}}$. Here M_z is a constant and $\hat{\mathbf{z}}$ is a unit vector along the axis of the bar. Describe an equivalent representation in which $\mathbf{M} = 0$ using other density functions and a different mode of subdivision.

15. A hoop of flat copper lying in the r, θ plane has a small thickness d, a width w, and a mean radius R. It is characterized completely by a surface density of current $K_\theta \cos \omega t$

defined for both inner and outer surfaces of width w. Consider a representation entirely in terms of M_z. The equivalence in representation is to be valid only as observed from points outside the volume bounded by the hoop. That is, M_z may be defined at all points in this flat volume as well as in the copper. Characterize the loop antenna of n turns each carrying a current I in terms of an equivalent magnetic shell (i.e., in terms of M_z).

2

Mathematical
Description of Space
and of Simple Media

The fundamental purpose that a mathematical description of space must serve in the larger model of electromagnetism is to interconnect the density fields of matter. From the mathematical point of view, space consists of nothing more intricate than a coordinate system that assigns three numbers or coordinates to every point in order to relate it to an arbitrarily selected region. In certain regions, definitely located by these coordinates, the scalar and vector fields of the continuous densities characterizing matter have nonvanishing values. These regions define the geometrical positions of mathematical bodies. At all other points (i.e., in empty space) the density fields are zero. To interconnect scattered density fields, the mathematical model is extended to include space. This is accomplished by assigning two vectors to every point in space, including that which is empty and that which contains regions where the density fields are nonvanishing. The electrical structure of mathematical space is described in terms of two vector fields. To one of the two vector point functions, the *electric vector*, is assigned the symbol **E**; to the other, the *magnetic vector*, the symbol **B**. In a region in which **E** has a value at every point, an **E** field or an electric field is said to exist. In a region in which **B** has a value at every point, a **B** field or a magnetic field exists. The superposition of the two fields is called the *electromagnetic field*. Thus the electromathematical structure of all space is completely identified with the electromagnetic field. The definition of each of the two vectors **E** and **B** involves a numerical, experimentally determined proportionality constant with appropriate dimensions. These are the fundamental electric constant ϵ_0 and the fundamental magnetic constant μ_0. They are the factors with the aid of which numerical coordination is achieved between the mathematical model of electromagnetism and the world of experimental measurements. As such they play the role of universal constants. The electric constant is called the *permittivity of free space* or the dielectric constant of free space. The magnetic constant is the permeability of free space. Like the density fields, the electromagnetic field is a purely mathematical construct for which no *direct* experimental analogues are presumed to exist. It differs from the density fields in that it is not based on a physical model. No attempt is made to describe a mechanism to serve as a picture for the electromagnetic field in the way the atomic model serves the density fields. The electromagnetic field is thus a purely mathematical extension of the mathematical model of matter, not of its physical model.

The definition of the vectors **E** and **B** in terms of the continuous densities (which characterize the space occupied by matter) depends on a fundamental theorem in vector analysis. The theorem states that a vector field is uniquely determined if its divergence and curl are specified, and if the normal component of the field is known over a closed surface, or if the vector vanishes as $1/r^2$ at infinity.

The definition of the vectors **E** and **B** in terms of their respective divergences and curls is the second fundamental principle of electromagnetism. The first fundamental principle is the conservation of electric charge; it is mathematically expressed in the equation of continuity. The second principle (which contains the first) is expressed by a set of partial differential equations, the Maxwell–Lorentz field equations; these express the divergence and curl of the **E** and **B** vectors in terms of the density functions and the constants ϵ_0 and μ_0 as follows:

$$\epsilon_0 \, \boldsymbol{\nabla} \cdot \mathbf{E} = \bar{\rho} \tag{2.1-1}$$

$$\boldsymbol{\nabla} \times \mathbf{E} = -\dot{\mathbf{B}} \tag{2.1-2}$$

$$\mu_0^{-1} \boldsymbol{\nabla} \times \mathbf{B} = \overline{\rho_m \mathbf{v}} + \epsilon_0 \dot{\mathbf{E}} \tag{2.1-3}$$

$$\boldsymbol{\nabla} \cdot \mathbf{B} = 0 \tag{2.1-4}$$

The superscript dot is written for $\partial/\partial t$ and

$$\bar{\rho} = \rho - \boldsymbol{\nabla} \cdot \mathbf{P} \tag{2.1-5}$$

$$\overline{\rho_m \mathbf{v}} = \mathbf{J} + \boldsymbol{\nabla} \times \mathbf{M} + \dot{\mathbf{P}} \tag{2.1-6}$$

It is assumed that the region (or regions) that is characterized by $\bar{\rho}$ is as a whole at rest relative to the observer. The defining relations (2.1–1) to (2.1–4) describe the electromagnetic field completely in terms of the essential volume characteristics. It is important to note that the vectors **E** and **B** as defined in terms of the *average* or interpolated density functions $\bar{\rho}$ and $\overline{\rho_m \mathbf{v}}$ are themselves *average* values at points within matter. They do not define a "microscopic" or "local" field but rather a macroscopic field.

In the stationary states all time rates of change vanish, all functions are constant in time, and the field equations assume the completely symmetrical form

$$\epsilon_0 \nabla \cdot \mathbf{E} = \bar{\rho} \qquad (2.1-7)$$

$$\nabla \times \mathbf{E} = 0 \qquad (2.1-8)$$

$$\mu_0^{-1} \nabla \times \mathbf{B} = \bar{\mathbf{J}} \qquad (2.1-9)$$

$$\nabla \cdot \mathbf{B} = 0 \qquad (2.1-10)$$

This definition of the stationary electromagnetic field makes it possible to extend the analogy exhibited between the static and steady states in Chapter 1 as shown in Table 2.1–1.

TABLE 2.1–1

Quantity	Static state	Steady state
Electromagnetic field	Electric vector $\mathbf{E}$	Magnetic vector $\mathbf{B}$
	Electric constant ϵ_0	Magnetic constant μ_0^{-1}
Operation	Curl ($\nabla \times$)	Divergence ($\nabla \cdot$)

The dimensions of the electric and magnetic vectors and scalars are conveniently expressed in terms of the auxiliary dimensional symbol V and three of the four dimensional symbols already introduced. The dimensional equivalent of V in terms of Q, L, M, T is given by ML^2QT^{-4}. The unit associated with the symbol V is the volt.

$$E = \frac{V}{L} \qquad \frac{\text{volts}}{\text{meters}} \qquad (2.1-11)$$

With (2.1–11), (2.1–2) gives

$$B = \frac{VT}{L^2} \qquad \frac{\text{volt-seconds}}{\text{square meters}} \quad \text{or} \quad \frac{\text{webers}}{\text{square meters}} \quad \text{or} \quad \text{teslas} \qquad (2.1-12)$$

With (2.1–11) and (2.1–1) it follows that

$$\epsilon_0 = \frac{Q}{VL} \qquad \frac{\text{coulombs}}{\text{volt-meters}} \quad \text{or} \quad \frac{\text{farads}}{\text{meters}} \qquad (2.1-13)$$

The coulomb per volt is called the *farad*. Similarly, (2.1–3) and (2.1–12) lead to

$$\mu_0 = \frac{V^2}{QL} \qquad \frac{\text{volt-seconds}}{\text{ampere-meters}} \quad \text{or} \quad \frac{\text{henrys}}{\text{meters}} \qquad (2.1-14)$$

The volt-second per ampere or weber per ampere is called the *henry*.

Numerical values of the universal constants of ϵ_0 and μ_0 are obtained from suitably designed standard experiments.

$$\left.\begin{array}{l}\text{Electric constant}\\[2ex]\text{Permittivity of space}\end{array}\right\} \quad \begin{array}{l}\epsilon_0 = 8.854 \times 10^{-12}\\[2ex]= \dfrac{1}{36\pi} \times 10^{-9} \quad \dfrac{\text{farads}}{\text{meters}}\end{array} \qquad (2.1\text{–}15)$$

$$\left.\begin{array}{l}\text{Magnetic constant}\\[2ex]\text{Permeability of space}\end{array}\right\} \quad \begin{array}{l}\mu_0 = 1.257 \times 10^{-6}\\[2ex]= 4\pi \times 10^{-7} \quad \dfrac{\text{henrys}}{\text{meters}}\end{array} \qquad (2.1\text{–}16)$$

2.2 FIELD EQUATIONS AT A SURFACE: BOUNDARY CONDITIONS

A boundary surface is either the mathematical envelope between a charged region and space, where the density fields associated with the region vanish, or it is the mathematical envelope between two electrically different regions in contact, where the density fields associated with the two change abruptly. Conditions at the boundary between a charged region and space are obtained from those for two charged regions in contact by writing zero for one set of density fields.

Since the electromagnetic vectors $\mathbf{E}$ and $\mathbf{B}$ are defined in terms of all the volume densities, they cannot represent more rapid fluctuations in electrical conditions than can the densities themselves. Therefore, discontinuities in $\mathbf{E}$ and $\mathbf{B}$ can exist only at a boundary where an abrupt change from one set of densities to another occurs. In a thin layer of atomic thickness d_c on each side of a boundary are defined the essential surface densities of charge and moving charge $\bar{\eta} = \eta + \hat{\mathbf{n}} \cdot \mathbf{P}$ and $\overline{\eta_m \mathbf{v}} = \mathbf{K} - \hat{\mathbf{n}} \times \mathbf{M}$. Since the vectors $\mathbf{E}$ and $\mathbf{B}$ are defined in the interior in terms of the essential volume densities of charge and moving charge (current), it is to be expected that their behavior at a boundary is determined by the corresponding surface densities.

Let the equation

$$\epsilon_0 \, \boldsymbol{\nabla} \cdot \mathbf{E} = \rho - \boldsymbol{\nabla} \cdot \mathbf{P} \qquad (2.2\text{–}1)$$

be written for a boundary between regions 1 and 2 in which $\mathbf{E}_1$, $\mathbf{P}_1$, ρ_1 and $\mathbf{E}_2$, $\mathbf{P}_2$, ρ_2 are defined, respectively. The exterior unit normals to the regions are $\hat{\mathbf{n}}_1$ and $\hat{\mathbf{n}}_2$. From the definition (1.9–8) of the divergence of a vector, (2.2–1) may be written as follows for an element $\Delta\tau_s = 2d_c \, \Delta\Sigma$ between the two regions:

$$\lim_{2d_c\Delta\Sigma \to 0} \frac{\epsilon_0 \displaystyle\int_\Sigma (\hat{\mathbf{n}}' \cdot \mathbf{E}) \, d\sigma}{2d_c\Delta\Sigma} = \rho - \lim_{2d_c\Delta\Sigma \to 0} \frac{\displaystyle\int_\Sigma (\hat{\mathbf{n}}' \cdot \mathbf{P}) \, d\sigma}{2d_c\Delta\Sigma} \qquad (2.2\text{–}2)$$

Here $\hat{\mathbf{n}}'$ points *out* from the surface enclosing $\Delta\tau_s$.

The element of volume $2d_c\Delta\Sigma$ is the same thin disk, half in region 1 and half in region 2, used in defining the essential surface density of charge (1.10–11); as before, ρ is the average density of charge in the double cell

$$\rho = \tfrac{1}{2}(\rho_1 + \rho_2) \tag{2.2-3}$$

The two integrals in (2.2–2) are evaluated over the entire surface Σ of the element of volume $2d_c\Delta\Sigma$. This involves integration over the two parallel surfaces $\Delta\Sigma_1$ and $\Delta\Sigma_2$ and over the four narrow surfaces of width $2d_c$ which are perpendicular to the boundary. As has been stated before, surface effects that cannot be described by volume functions are due to rapid variations in distributions of charge in a direction *perpendicular* to the surface. Variations in distribution *parallel* to the surface do not differ from the ordinary slow variations in the interior and are, therefore, adequately represented by the volume functions. The integration over the narrow edges of width $2d_c$ thus contributes nothing to a distinctly surface effect of which complete account is taken in the integration over the surfaces $\Delta\Sigma$. The integrals are all four of the familiar form

$$\lim_{\Delta\Sigma\to0} \frac{\displaystyle\int_{\Delta\Sigma_1} \hat{\mathbf{n}}_1' \cdot \mathbf{P}_1 \, d\sigma}{\Delta\Sigma} = \lim_{\Delta\Sigma\to0} \frac{\overline{\hat{\mathbf{n}}_1' \cdot \mathbf{P}_1} \displaystyle\int d\sigma}{\Delta\Sigma} = \hat{\mathbf{n}}_1' \cdot \mathbf{P}_1 \tag{2.2-4}$$

The second integral in (2.2–4) is obtained using the theorem of the mean for integrals. The last step follows directly after allowing $\Delta\Sigma$ to approach zero. $(\hat{\mathbf{n}}_1' \cdot \mathbf{P}_1)$ is the value at the point where $\Delta\Sigma$ vanishes and at a distance d_c from the boundary. Let (2.2–3) and four integrals of the form (2.2–4) (with appropriately changed subscripts and with $\mathbf{E}$ written for $\mathbf{P}$ in two) be substituted in (2.2–2) and let this be multiplied through by $2d_c$. Next introduce the surface densities according to the definition

$$\eta_1 + \eta_2 = \rho_1 d_c + \rho_2 d_c \tag{2.2-5}$$

The exterior unit normals $\hat{\mathbf{n}}_1$ and $\hat{\mathbf{n}}_2$ to the large regions 1 and 2 are related to the outward normals $\hat{\mathbf{n}}_1'$ and $\hat{\mathbf{n}}_2'$ to $2d_c\Delta\Sigma$ by $\hat{\mathbf{n}}_1 = -\hat{\mathbf{n}}_1'$, $\hat{\mathbf{n}}_2 = -\hat{\mathbf{n}}_2'$. With this change (2.2–2) finally becomes

$$\epsilon_0\hat{\mathbf{n}}_1 \cdot \mathbf{E}_1 + \epsilon_0\hat{\mathbf{n}}_2 \cdot \mathbf{E}_2 = -(\eta_1 + \eta_2 + \hat{\mathbf{n}}_1 \cdot \mathbf{P}_1 + \hat{\mathbf{n}}_2 \cdot \mathbf{P}_2) \tag{2.2-6}$$

In shorthand notation, (2.2–6) may be written as

$$\epsilon_0\hat{\mathbf{n}} \cdot \mathbf{E} = -(\eta + \hat{\mathbf{n}} \cdot \mathbf{P}) = -\bar{\eta} \tag{2.2-7}$$

By the same reasoning, the equation

$$\nabla \cdot \mathbf{B} = 0 \tag{2.2-8}$$

has the following form at a boundary:

$$\hat{\mathbf{n}}_1 \cdot \mathbf{B}_1 + \hat{\mathbf{n}}_2 \cdot \mathbf{B}_2 \equiv \hat{\mathbf{n}} \cdot \mathbf{B} = 0 \tag{2.2-9}$$

If the third field equation

$$\mu_0^{-1} \nabla \times \mathbf{B} = \mathbf{J} + \nabla \times \mathbf{M} + \dot{\mathbf{P}} + \epsilon_0 \dot{\mathbf{E}} \qquad (2.2\text{--}10)$$

is expanded by introducing the definition for the curl (1.20–17) written for the small volume $\Delta\tau_s = 2d_c\Delta\Sigma$ at the boundary, it becomes

$$\lim_{2d_c\Delta\Sigma \to 0} \frac{\displaystyle\int_\Sigma (\hat{\mathbf{n}}' \times \mathbf{B})\, d\sigma}{2d_c\Delta\Sigma}$$

$$= \mu_0 \left[\mathbf{J} + \dot{\mathbf{P}} + \epsilon_0 \dot{\mathbf{E}} + \lim_{2d_c\Delta\Sigma \to 0} \frac{\displaystyle\int_\Sigma (\hat{\mathbf{n}}' \times \mathbf{M})\, d\sigma}{2d_c\Delta\Sigma} \right] \qquad (2.2\text{--}11)$$

If $\mathbf{J}_1$ and $\mathbf{J}_2$ are defined in the two regions, their average value in the combined surface cell is

$$\mathbf{J} = \tfrac{1}{2}(\mathbf{J}_1 + \mathbf{J}_2) \qquad (2.2\text{--}12)$$

Each of the two integrals includes integration over the parallel faces $\Delta\Sigma_1$ and $\Delta\Sigma_2$ and over the narrow surfaces of width $2d_c$ perpendicular to the boundary. Only integration over $\Delta\Sigma_1$ and $\Delta\Sigma_2$ contributes significantly to a distinctly surface effect caused by a rapid variation in the distribution of charge or current as the boundary is approached from the interior. Care must be exercised not to be misled by the fact that the vectors $\hat{\mathbf{n}} \times \mathbf{B}$ and $\hat{\mathbf{n}} \times \mathbf{M}$ are actually *parallel* to the boundary on $\Delta\Sigma_1$ and $\Delta\Sigma_2$. A rapid variation in these vectors due to asymmetries at the boundary must, nevertheless, occur in a direction *perpendicular* to the boundary. That is, a vector parallel to the boundary may change its direction or its magnitude very rapidly as the boundary is approached, while yet remaining parallel to it. Across the narrow surfaces of width $2d_c$, $\hat{\mathbf{n}} \times \mathbf{B}$ and $\hat{\mathbf{n}} \times \mathbf{M}$ can experience only the slow variation characteristic of the interior, and represented by $\nabla \times \mathbf{B}$ and $\nabla \times \mathbf{M}$. A typical one of four integrals evaluated over the surface $\Delta\Sigma$ is

$$\lim_{\Delta\Sigma_1 \to 0} \int_{\Delta\Sigma_1} \frac{(\hat{\mathbf{n}}_1' \times \mathbf{M}_1)\, d\sigma}{\Delta\Sigma_1} = \lim_{\Delta\Sigma_1 \to 0} \frac{\overline{\hat{\mathbf{n}}_1' \times \mathbf{M}_1} \displaystyle\int d\sigma}{\Delta\Sigma_1}$$

$$= \hat{\mathbf{n}}_1' \times \mathbf{M}_1 = -\hat{\mathbf{n}}_1 \times \mathbf{M}_1 \qquad (2.2\text{--}13)$$

The evaluation includes use of the theorem of the mean for integrals, allowing $\Delta\Sigma$ to approach zero where the mean value is defined, and introducing the external normal $\hat{\mathbf{n}}_1$ to region 1 by $\hat{\mathbf{n}}_1 = -\hat{\mathbf{n}}_1'$. Four integrals of the form (2.2–13) (with appropriately changed subscripts and with $\mathbf{B}$ written for $\mathbf{M}$ in two) and (2.2–12) are substituted in (2.2–11) and multiplied through by $2d_c$. The surface density of moving charge (current) $\mathbf{K}$ is introduced in the form

$$\mathbf{K}_1 + \mathbf{K}_2 = d_c\mathbf{J}_1 + d_c\mathbf{J}_2 \qquad (2.2-14)$$

It will be recalled that the surface function $\mathbf{K}$ is defined specifically because there may be rapid changes in the distribution of current as a boundary is approached from the interior. Consequently, a section of thickness d_c must be considered separately along all surfaces and boundaries where asymmetrical conditions prevail. A similar situation does not obtain for the functions $\mathbf{P}$ and $\mathbf{E}$. The former is a volume function defined by interpolation throughout the interior of each region; the latter is defined at all points in space and in bodies or charged regions in terms of the several *volume* densities. Both functions are continuous and slowly varying in the interior of each region. It follows, therefore, that the quantity $2d_c(\dot{\mathbf{P}} + \epsilon_0\dot{\mathbf{E}})$ defined for thin layers, each of thickness d_c and one in each region along their common boundary, will not differ significantly from the same quantity defined for two similar slices, one in the interior of each region. That is, the function $2d_c(\dot{\mathbf{P}} + \epsilon_0\dot{\mathbf{E}})$ describes nothing that is peculiar to the surface; its magnitude is necessarily extremely small because of the minuteness of d_c and the fact that $\dot{\mathbf{P}} + \epsilon_0\dot{\mathbf{E}}$ assumes no abnormally large values in the thin surface section on each side of the boundaries. Consequently, (2.2–11) may be written

$$\hat{\mathbf{n}}_1 \times \mathbf{B}_1 + \hat{\mathbf{n}}_2 \times \mathbf{B}_2 = -\mu_0(\mathbf{K}_1 + \mathbf{K}_2 - \hat{\mathbf{n}}_1 \times \mathbf{M}_1 - \hat{\mathbf{n}}_2 \times \mathbf{M}_2) \qquad (2.2-15)$$

All normals here are *external* to the regions indicated by the subscripts. In shorthand notation

$$\hat{\mathbf{n}} \times \mathbf{B} = -\mu_0(\mathbf{K} - \hat{\mathbf{n}} \times \mathbf{M}) = -\mu_0\overline{\eta_m\mathbf{v}} \qquad (2.2-16)$$

It follows by similar reasoning that the field equation

$$\nabla \times \mathbf{E} = -\dot{\mathbf{B}} \qquad (2.2-17)$$

leads to the following surface equation:

$$\hat{\mathbf{n}} \times \mathbf{E} = 0 \qquad (2.2-18)$$

The Maxwell–Lorentz equations written for surface effects on the boundary between regions 1 and 2 have the following form:

$$\epsilon_0\hat{\mathbf{n}}_1 \cdot \mathbf{E}_1 + \epsilon_0\hat{\mathbf{n}}_2 \cdot \mathbf{E}_2 = -(\eta_1 + \eta_2 + \hat{\mathbf{n}}_1 \cdot \mathbf{P}_1 + \hat{\mathbf{n}}_2 \cdot \mathbf{P}_2)$$

$$= -(\overline{\eta}_1 + \overline{\eta}_2) \qquad (2.2-19a)$$

$$\hat{\mathbf{n}}_1 \times \mathbf{E}_1 + \hat{\mathbf{n}}_2 \times \mathbf{E}_2 = 0 \qquad (2.2-19b)$$

$$\mu_0^{-1}(\hat{\mathbf{n}}_1 \times \mathbf{B}_1 + \hat{\mathbf{n}}_2 \times \mathbf{B}_2) = -(\mathbf{K}_1 + \mathbf{K}_2 + \hat{\mathbf{n}}_1 \times -\mathbf{M}_1 + \hat{\mathbf{n}}_2 \times -\mathbf{M}_2)$$

$$= -(\overline{\eta_m\mathbf{v}_1} + \overline{\eta_m\mathbf{v}_2}) \qquad (2.2-19c)$$

$$\hat{\mathbf{n}}_1 \cdot \mathbf{B}_1 + \hat{\mathbf{n}}_2 \cdot \mathbf{B}_2 = 0 \qquad (2.2-19d)$$

In shorthand

$$\epsilon_0 \hat{\mathbf{n}} \cdot \mathbf{E} = -(\eta + \hat{\mathbf{n}} \cdot \mathbf{P}) = -\overline{\eta} \tag{2.2-20a}$$

$$\hat{\mathbf{n}} \times \mathbf{E} = 0 \tag{2.2-20b}$$

$$\mu_0^{-1} \hat{\mathbf{n}} \times \mathbf{B} = -(\mathbf{K} + \hat{\mathbf{n}} \times -\mathbf{M}) = -\overline{\eta_m \mathbf{v}} \tag{2.2-20c}$$

$$\hat{\mathbf{n}} \cdot \mathbf{B} = 0 \tag{2.2-20d}$$

The interpretation of the boundary equations (2.2–19a–d) is not difficult. Equations (2.2–19a, d) apply to the normal components of the vectors **E** and **B**. Thus (2.2–19a) states that the normal component of the electric vector is discontinuous in crossing a boundary surface. The magnitude of the discontinuity is the essential surface characteristic of charge $\overline{\eta}$ divided by ϵ_0. In the same way (2.2–19d) states that the normal component of the magnetic vector is continuous across all boundaries. The interpretation of (2.2–19b, c) is more involved owing to the fact that the vector product of the external normal to the surface of a region and one of the field vectors actually does not specify any particular component of this field vector. It defines an axial vector that has the *magnitude of the tangential component* of the field vector at the surface and a direction normal to the plane formed by the field vector and the external normal. It follows at once that the magnitude of the discontinuity of the axial vector so defined is equal to the magnitude of the discontinuity in the tangential component of the field vector. Therefore, (2.2–19b) requires that *the tangential component of the electric vector be continuous in crossing all boundaries*, whereas (2.2–19c) requires *the tangential component of the magnetic vector to be discontinuous* by a magnitude equal to the essential surface density of current multiplied by μ_0. These boundary conditions are illustrated in Fig. 2.2–1. Table 2.2–1 illustrates the boundary conditions using shorthand notation as in (2.2–20).

The form of the boundary equations (2.2–19a–d) or (2.2–20a–d) is not changed in the stationary states. It is only necessary to require all functions to be

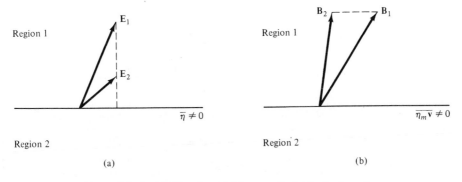

Figure 2.2–1   (a) Electric and (b) magnetic fields at a boundary.

TABLE 2.2–1

Vector	Component	Behavior at boundary	Magnitude of discontinuity
E	Normal	Discontinuous	$-\epsilon_0^{-1}\overline{\eta}$
E	Tangential	Continuous	0
B	Tangential	Discontinuous	$-\mu_0\overline{\eta_m \mathbf{v}}$
B	Normal	Continuous	0

invariant in time. The analogy between the static and steady states may be extended to include the following:

Quantity	Static state	Steady state
Operation	Vector product ($\times$)	Scalar product ($\cdot$)

2.3 THE FIELD EQUATIONS AS FUNDAMENTAL POSTULATES

The mathematical model of the electrical properties of matter is contained in the four essential characteristics that describe matter in terms of scalar and vector fields. Every point in the interior of a body is characterized by values of the two volume functions

$$\overline{\rho} = \rho - \nabla \cdot \mathbf{P} \tag{2.3–1}$$

$$\overline{\rho_m \mathbf{v}} = \mathbf{J} + \nabla \times \mathbf{M} + \dot{\mathbf{P}} \tag{2.3–2}$$

Every point on a boundary surface is characterized by values of the surface functions

$$\overline{\eta} = \eta + \hat{\mathbf{n}} \cdot \mathbf{P} \tag{2.3–3}$$

$$\overline{\eta_m \mathbf{v}} = \mathbf{K} - \hat{\mathbf{n}} \times \mathbf{M} \tag{2.3–4}$$

The mathematical model of the electrical properties of space is defined in terms of the two universal constants ϵ_0 and μ_0 and the two vector point functions **E** and **B** that are defined everywhere except in surface layers by

$$\epsilon_0 \nabla \cdot \mathbf{E} = \overline{\rho} \tag{2.3–5}$$

$$\nabla \times \mathbf{E} = -\dot{\mathbf{B}} \tag{2.3–6}$$

$$\mu_0^{-1} \nabla \times \mathbf{B} = \overline{\rho_m \mathbf{v}} + \epsilon_0 \dot{\mathbf{E}} \tag{2.3–7}$$

$$\nabla \cdot \mathbf{B} = 0 \tag{2.3–8}$$

$$\epsilon_0 \hat{\mathbf{n}} \cdot \mathbf{E} = -\overline{\eta} \tag{2.3–9}$$

$$\hat{\mathbf{n}} \times \mathbf{E} = 0 \tag{2.3–10}$$

$$\mu_0^{-1} \hat{\mathbf{n}} \times \mathbf{B} = -\overline{\eta_m \mathbf{v}} \tag{2.3–11}$$

$$\hat{\mathbf{n}} \cdot \mathbf{B} = 0 \tag{2.3–12}$$

The description of every point in the electromagnetic field is characterized by the nature and the distribution of the density fields that determine the essential volume and surface characteristics $\bar{\rho}$ and $\overline{\rho_m \mathbf{v}}$, $\bar{\eta}$ and $\overline{\eta_m \mathbf{v}}$. A knowledge of these densities is sufficient to define the field vectors $\mathbf{E}$ and $\mathbf{B}$. The final step in formulating a method for predicting and coordinating theoretical effects that are analogies of experimentally observable ones has been described. The mathematical densities identified with the electrical properties of matter have been fitted into a mathematical structure defined for all space called the electromagnetic field.

2.4 ALTERNATIVE FORMULATION OF FIELD EQUATIONS: AUXILIARY FIELD VECTORS AND CONSTANTS

An examination of the field and boundary equations in a purely formal way suggests a mathematically convenient rearrangement of the first and third equations. This consists in combining the terms according to their mathematical form rather than having them indicate explicitly that the divergence and curl of each of the vectors $\mathbf{E}$ and $\mathbf{B}$ are defined. Since ϵ_0 and μ_0 are constants, they may be written inside the operational symbols $(\nabla \cdot)$ and $(\nabla \times)$. Thus

$$\nabla \cdot (\epsilon_0 \mathbf{E} + \mathbf{P}) = \rho \tag{2.4-1}$$

$$\nabla \times (\mu_0^{-1}\mathbf{B} - \mathbf{M}) = \mathbf{J} + \frac{\partial}{\partial t}(\epsilon_0 \mathbf{E} + \mathbf{P}) \tag{2.4-2}$$

$$\hat{\mathbf{n}} \cdot (\epsilon_0 \mathbf{E} + \mathbf{P}) = -\eta \tag{2.4-3}$$

$$\hat{\mathbf{n}} \times (\mu_0^{-1}\mathbf{B} - \mathbf{M}) = -\mathbf{K} \tag{2.4-4}$$

For convenience in writing the following new symbols are defined:

$$\begin{aligned} \mathbf{D} &= \epsilon_0 \mathbf{E} + \mathbf{P} \\ \mathbf{H} &= \mu_0^{-1}\mathbf{B} - \mathbf{M} \end{aligned} \tag{2.4-5}$$

The vectors $\mathbf{D}$ and $\mathbf{H}$ are written simply as a shorthand notation. With no other qualifications, both vary with the mode of subdivision used in defining the density functions. At a later point a definite mode of subdivision must be selected, that is, one in which no bound groups are cut so that $\mathbf{P}$ and $-\mathbf{M}$ are maximum and $\rho = \rho_f$, $\mathbf{J} = \mathbf{J}_f$, $\eta = \eta_f$, $\mathbf{K} = \mathbf{K}_f$. At all points in space or in bodies where the $\mathbf{P}$ and $\mathbf{M}$ fields vanish, $\mathbf{D} = \epsilon_0 \mathbf{E}$, $\mathbf{H} = \mu_0^{-1}\mathbf{B}$. Since ϵ_0 and μ_0 are scalars, the vectors $\mathbf{D}$ and $\mathbf{H}$ point in the same direction and are proportional in magnitude, respectively, to the vectors $\mathbf{E}$ and $\mathbf{B}$ at all points where $\mathbf{P}$ and $-\mathbf{M}$ vanish or are not defined. On the other hand, in bodies where $\mathbf{P}$ and $-\mathbf{M}$ are nonvanishing, $\mathbf{D}$ and $\mathbf{H}$ are not in general proportional, respectively, to $\mathbf{E}$ and $\mathbf{B}$, nor do they point in the same direction.

In terms of the auxiliary vectors **D** and **H**, the Maxwell equations have a simpler appearance:

$$\nabla \cdot \mathbf{D} = \rho \qquad (2.4-6)$$

$$\nabla \times \mathbf{E} = -\dot{\mathbf{B}} \qquad (2.4-7)$$

$$\nabla \times \mathbf{H} = \mathbf{J} + \dot{\mathbf{D}} \qquad (2.4-8)$$

$$\nabla \cdot \mathbf{B} = 0 \qquad (2.4-9)$$

The auxiliary relations are

$$\mathbf{D} = \epsilon_0 \mathbf{E} + \mathbf{P} \qquad (2.4-10)$$

$$\mathbf{H} = \mu_0^{-1}\mathbf{B} - \mathbf{M} \qquad (2.4-11)$$

The surface equations in shorthand form are

$$\hat{\mathbf{n}} \cdot \mathbf{D} = -\eta \qquad (2.4-12)$$

$$\hat{\mathbf{n}} \times \mathbf{E} = 0 \qquad (2.4-13)$$

$$\hat{\mathbf{n}} \times \mathbf{H} = -\mathbf{K} \qquad (2.4-14)$$

$$\hat{\mathbf{n}} \cdot \mathbf{B} = 0 \qquad (2.4-15)$$

In dealing with models that have physically possible analogues it is never necessary to define a *surface* density of moving free charge. If **D** and **H** are defined using a subdivision that cuts no bound groups, η in (2.4–12) and **K** in (2.4–14) are, respectively, η_f and $\mathbf{K}_f$. Except in the physically unrealizable *but theoretically very important* case of a *perfect* free-charge model (perfect conductor), $\mathbf{K}_f$ is not required, so the right side of (2.4–14) is zero. The boundary conditions (2.4–12) to (2.4–15) are illustrated in Figs. 2.4–1 and 2.4–2 for the special case $\eta = \eta_f = 0$, $\mathbf{K} = \mathbf{K}_f = 0$. Media with **P** parallel to and in the same direction as **E** are said to be *dielectric*. Media with $-\mathbf{M}$ parallel to and in the same direction as **B** are *diamagnetic*. Media with $-\mathbf{M}$ parallel and directed opposite to **B** are *paramagnetic* when **M** is small and *ferromagnetic* when **M** is large.

Two useful auxiliary constants may be defined by combining the universal electric and magnetic constants ϵ_0 and μ_0.

Characteristic velocity of space $c = (\epsilon_0\mu_0)^{-1/2}$

$$= 2.998 \times 10^8$$

$$\simeq 3 \times 10^8 \text{ m/s} \qquad (2.4-16)$$

Characteristic resistance of space $\zeta_0 = (\mu_0\epsilon_0^{-1})^{1/2}$

$$= 376.7$$

$$\simeq 120\pi \text{ ohms } (\Omega) \qquad (2.4-17)$$

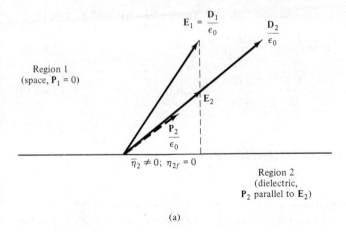

Region 1
(space, $P_1 = 0$)

$E_1 = \dfrac{D_1}{\epsilon_0}$

$\dfrac{D_2}{\epsilon_0}$

E_2

$\dfrac{P_2}{\epsilon_0}$

$\overline{\eta}_2 \neq 0; \; \eta_{2f} = 0$

Region 2
(dielectric,
P_2 parallel to E_2)

(a)

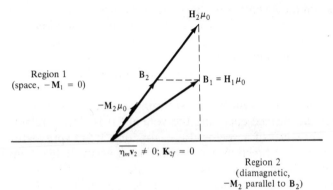

Region 1
(space, $-M_1 = 0$)

$H_2\mu_0$

B_2

$B_1 = H_1\mu_0$

$-M_2\mu_0$

$\overline{\eta_m v_2} \neq 0; \; K_{2f} = 0$

Region 2
(diamagnetic,
$-M_2$ parallel to B_2)

(b)

Figure 2.4–1   (a) Electric and (b) magnetic field vectors on both sides of a boundary.

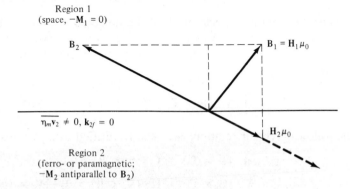

Region 1
(space, $-M_1 = 0$)

B_2

$B_1 = H_1\mu_0$

$\overline{\eta_m v_2} \neq 0, \; k_{2f} = 0$

$H_2\mu_0$

Region 2
(ferro- or paramagnetic;
$-M_2$ antiparallel to B_2)

Figure 2.4–2   Magnetic field vectors on both sides of a boundary.

The dimensions of c and ζ_0 are obtained directly from (2.1–13) and (2.1–14):

$$c = \frac{L}{T} \quad \frac{\text{meters}}{\text{seconds}} \tag{2.4–18}$$

$$\zeta_0 = \frac{VT}{Q} \quad \frac{\text{volts}}{\text{amperes}} \text{ or ohms} \tag{2.4–19}$$

2.5 INTEGRAL FORMS OF FIELD EQUATIONS: GENERAL THEOREMS

The Maxwell–Lorentz equations that define the electromagnetic field consist of four simultaneous, partial differential equations of the first order. With the aid of general integral theorems of the calculus, it is possible to transform them into integral relations that are often convenient in the solution of problems, especially those characterized by symmetry. In many cases, the boundary conditions are sufficiently simple so that explicit expressions for **E** and **B** in terms of the density functions can be obtained in this way. In order to establish the integral form of the field equations, it is first necessary to describe the fundamental integral theorems that are involved.

The *divergence theorem* is a theorem for transforming a volume integral into an integral evaluated over the surface enclosing the volume. It is closely related to the definition of the divergence of a vector. In vector form it is expressed in terms of a continuous vector point function **A** which defines a vector field. If any volume V is chosen in this field, it will be contained in a closed (mathematical) surface S. Let dV be an element of the volume, dS an element of the enclosing surface, and $\hat{\mathbf{n}}$ an external normal to the surface. The theorem is

$$\int_V \boldsymbol{\nabla} \cdot \mathbf{A} \, dV = \int_S \hat{\mathbf{n}} \cdot \mathbf{A} \, dS \tag{2.5–1}$$

The *curl theorem* is a special form of the divergence theorem. It is obtained from this by setting the vector **A** equal to the vector product $(\mathbf{B} \times \mathbf{C})$, where **C** is defined to be a constant vector. Upon applying the divergence theorem to this vector product,

$$\int_V \boldsymbol{\nabla} \cdot (\mathbf{B} \times \mathbf{C}) \, dV = \int_S \hat{\mathbf{n}} \cdot (\mathbf{B} \times \mathbf{C}) \, dS \tag{2.5–2}$$

The following vector relations are readily verified using Cartesian coordinates:

$$\hat{\mathbf{n}} \cdot (\mathbf{B} \times \mathbf{C}) = \mathbf{C} \cdot (\hat{\mathbf{n}} \times \mathbf{B}) \tag{2.5–3}$$

$$\boldsymbol{\nabla} \cdot (\mathbf{B} \times \mathbf{C}) = \mathbf{C} \cdot (\boldsymbol{\nabla} \times \mathbf{B}) - \mathbf{B} \cdot (\boldsymbol{\nabla} \times \mathbf{C}) \tag{2.5–4}$$

Since $\mathbf{C}$ is constant by definition, it follows that (2.5–4) reduces to

$$\mathbf{\nabla} \cdot (\mathbf{B} \times \mathbf{C}) = \mathbf{C} \cdot (\mathbf{\nabla} \times \mathbf{B}) \tag{2.5–5}$$

With (2.5–3) and (2.5–5) used in (2.5–2), the following relation is obtained:

$$\mathbf{C} \cdot \int_V (\mathbf{\nabla} \times \mathbf{B}) \, dV - \mathbf{C} \cdot \int_S (\hat{\mathbf{n}} \times \mathbf{B}) \, dS = 0 \tag{2.5–6}$$

Since the vector $\mathbf{C}$ is arbitrary and constant, the foregoing scalar product of the vector $\mathbf{C}$ and the vector defined by the difference between the two integrals vanishes for all orientations of the vector $\mathbf{C}$ only when the second member of the product vanishes identically. That is, when

$$\int_V (\mathbf{\nabla} \times \mathbf{B}) \, dV = \int_S (\hat{\mathbf{n}} \times \mathbf{B}) \, dS \tag{2.5–7}$$

This is the curl theorem. It is seen to be closely related to the definition of the curl of a vector.

Stokes's theorem is a theorem for transforming a surface integral over a cap- or cup-shaped surface into a line integral around the closed boundary of the surface. Consider any open cap- or cup-shaped surface S which may have any form whatsoever from a flat disk enclosed by the boundary line s to a deep balloon with only a narrow opening enclosed by the boundary line s. Let this surface be entirely in the field of a continuous vector point function $\mathbf{A}$. The theorem is

$$\int_{S(\text{cap})} \hat{\mathbf{n}} \cdot (\mathbf{\nabla} \times \mathbf{A}) \, dS = \oint_{s(\text{closed line})} (\mathbf{A} \cdot d\mathbf{s}) \tag{2.5–8}$$

The line integration around the closed boundary s is to be performed in a direction such that the right-hand-screw convention is satisfied with respect to the normal $\hat{\mathbf{n}}$ to the surface S. Stokes's theorem is seen to be closely related to the alternative definition of the curl of a vector in terms of its component normal to a surface.

Stokes's theorem and the divergence theorem may be applied to the four Maxwell–Lorentz equations to obtain integrals that are common in the technical literature. The shorthand notation using the auxiliary vectors $\mathbf{D}$ and $\mathbf{H}$ is used where convenient.

2.6 GAUSS'S THEOREM FOR THE E AND D VECTORS: INTEGRAL FORM OF $\mathbf{\nabla} \cdot \mathbf{D} = \rho$

Let both sides of the field equation

$$\mathbf{\nabla} \cdot \mathbf{D} = \rho \tag{2.6–1}$$

$$\mathbf{D} = \epsilon_0 \mathbf{E} + \mathbf{P}$$

be integrated over an arbitrary volume V enclosed by the surface S. Then

$$\int_V \mathbf{\nabla} \cdot \mathbf{D} \, dV = \int_V \rho \, dV \tag{2.6-2}$$

In any region in which the vector $\mathbf{D}$ is *continuous*, the divergence theorem may be used. Hence, when $\mathbf{D}$ is continuous throughout V and S, the divergence theorem applied to the left in (2.6-2) gives

$$\int_V \mathbf{\nabla} \cdot \mathbf{D} \, dV = \int_S (\hat{\mathbf{n}} \cdot \mathbf{D}) \, dS = \int_V \rho \, dV \tag{2.6-3}$$

The second integral in (2.6-3) is the total outward normal flux of the $\mathbf{D}$ vector across the closed surface S. This name does not imply that anything physically real is flowing out of the volume. It might be concluded in a perfectly general way from (2.6-3) that the normal $\mathbf{D}$ flux across any closed surface is equal to the total charge $Q = \int_V \rho \, dV$ contained within it. But the volume integral on the right in (2.6-3) only gives the contribution to the total charge due to the volume density ρ; it takes no account of possible surface distributions characterized by the density η which might be contained inside the enclosed envelope S. Since the $\mathbf{D}$ vector is discontinuous across surfaces or boundaries where η is nonvanishing, as seen in (2.4-12), the divergence theorem cannot be applied directly to such a region without first excluding the surfaces of discontinuity. Consequently, the integral (2.6-3) applies only to homogeneous volumes in which there are no discontinuities in $\mathbf{D}$.

To generalize (2.6-3) so that it may be applied to any region, suppose that the volume contained within S is composed of several dissimilar parts as indicated in Fig. 2.6-1. Let a surface of integration Σ be drawn around this composite region in such a way that it is entirely within a single homogeneous medium. In each of the regions 0, 1, 2, and 3, a different set of continuous densities ρ_i, $\mathbf{P}_i$ with appropriate subscripts is defined. On the boundaries between these regions, surface layers of charge may be described by the function η_{ij}. In the interior of each homogeneous region the following equation is true:

$$\mathbf{\nabla} \cdot \mathbf{D}_i = \rho_i \tag{2.6-4}$$

On the boundary surfaces between two dissimilar regions, such as 0 and 1, the following typical equation is true:

$$\hat{\mathbf{n}}_0 \cdot \mathbf{D}_0 + \hat{\mathbf{n}}_1 \cdot \mathbf{D}_1 = -(\eta_0 + \eta_1) \equiv -\eta_{01} \tag{2.6-5}$$

Here $\hat{\mathbf{n}}_0$ is an external normal to the region 0; $\hat{\mathbf{n}}_1$ is an external normal to region 1.

The boundary surfaces across which $\mathbf{D}$ is discontinuous are the envelopes S_{01}, S_{12}, S_{13}, and so on, as shown in Fig. 2.6-1. Let them be excluded from the region of integration by enclosing them in surfaces constructed both inside and outside the envelopes S. The volume V is thus divided into volumes V_j. The σ_i surfaces are to be drawn at small distances d_c (of atomic magnitude) from the surfaces S,

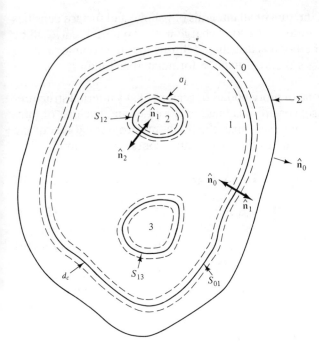

Figure 2.6–1   Regions with boundaries for obtaining Gauss's theorem for a discontinuous vector.

so that surface layers of charge defined by the surface density must be contained in the narrow regions each of thickness d_c on each side of the boundaries S. The expression for the total outward normal flux of the $\mathbf{D}$ vector across all the surfaces σ_i and the outside boundary Σ can now be written down. Since the surfaces σ_i are drawn very close to the surface S, no serious error is made by taking the integral over each side of S instead of over the two σ surfaces enclosing it. In the same way, the volume between pairs of σ surfaces is so small that it may be neglected in comparison with the rest, so that $\Sigma V_j = V$.

$$\int_{\Sigma} (\hat{\mathbf{n}}_0 \cdot \mathbf{D}_0)\, dS + \int_{S_{01}} [\hat{\mathbf{n}}_0 \cdot \mathbf{D}_0 + \hat{\mathbf{n}}_1 \cdot \mathbf{D}_1]\, dS_{01} + \sum_i \int_{S_{1i}} [\hat{\mathbf{n}}_1 \cdot \mathbf{D}_1$$

$$+ \hat{\mathbf{n}}_i \cdot \mathbf{D}_i]\, dS_{1i} = \sum_j \int_{V_j} \rho_j\, dV_j \qquad (2.6\text{–}6)$$

All normals in (2.6–6) are external to the regions indicated by subscripts. The substitution of the appropriate boundary conditions (2.6–5) in (2.6–6) for each pair of integrals leads to

$$\int_{\Sigma} (\hat{\mathbf{n}}_0 \cdot \mathbf{D}_0)\, dS = \left[\sum_j \int_{V_j} \rho_j\, dV_j + \int_{S_{01}} \eta_{01}\, dS_{01} + \sum_i \int_{S_{1i}} \eta_{1i}\, dS_{1i} \right]$$

$$= Q \qquad (2.6\text{–}7)$$

Q is defined by (2.6–7) to be the sum of all integrals of volume and surface densities of charge. Accordingly, it represents the total charge in the volume V enclosed by the surface Σ regardless of whether it is distributed throughout the volume or along surfaces or boundaries. Gauss's theorem may be formulated as follows:

Gauss's Theorem. If one or more bodies or regions of any internal structure, shape, or arrangement whatsoever are imagined enclosed in an envelope of integration Σ of entirely arbitrary shape and size, the total outward normal flux of the **D** vector across the surface is equal to the total charge contained within it

$$\int_{\Sigma} (\hat{\mathbf{n}} \cdot \mathbf{D}) \, dS = Q \tag{2.6–8}$$

$$\mathbf{D} = \epsilon_0 \mathbf{E} + \mathbf{P} \tag{2.6–9}$$

$$Q = \sum_j \int_{V_j} \rho_j \, dV_j + \sum_{ij} \int_{S_{ij}} \eta_{ij} \, dS_{ij} \tag{2.6–10}$$

If **P** is defined using a mode of subdivision that cuts through no bound groups as is customary when the **D** vector is used, Q in (2.6–10) is the total *free* charge within Σ and a subscript f is written on Q, ρ, and η. If Q and **D** vary in time, Gauss's theorem is true at every instant. If the field equation (2.6–1) is written in the form

$$\epsilon_0 \nabla \cdot \mathbf{E} = \bar{\rho}; \qquad \bar{\rho} = \rho - \nabla \cdot \mathbf{P} \tag{2.6–11}$$

the following alternative integral theorem is obtained:

$$\epsilon_0 \int_{\Sigma} (\hat{\mathbf{n}} \cdot \mathbf{E}) \, dS = \bar{Q} = \sum_j \int_{V_j} \bar{\rho}_j \, dV_j + \sum_{ij} \int_{S_{ij}} \bar{\eta}_{ij} \, dS_{ij} \tag{2.6–12}$$

2.7 THE AMPÈRE–MAXWELL THEOREM OF CIRCUITATION: INTEGRAL FORM OF $\nabla \times \mathbf{H} = \mathbf{J} + \dot{\mathbf{D}}$

The integration of both sides of the field equation

$$\nabla \times \mathbf{H} = \mathbf{J} + \frac{\partial \mathbf{D}}{\partial t} \tag{2.7–1}$$

with

$$\mathbf{H} = \mu_0^{-1} \mathbf{B} - \mathbf{M}$$
$$\mathbf{D} = \epsilon_0 \mathbf{E} + \mathbf{P} \tag{2.7–2}$$

over a cap surface S yields:

$$\int_{S(\text{cap})} \hat{\mathbf{N}} \cdot (\nabla \times \mathbf{H}) \, dS = \int_{S(\text{cap})} (\hat{\mathbf{N}} \cdot \mathbf{J}) \, dS + \frac{\partial}{\partial t} \int_{S(\text{cap})} (\hat{\mathbf{N}} \cdot \mathbf{D}) \, dS \tag{2.7–3}$$

When the vector **H** is continuous, Stokes's theorem applied to the integral on the left gives

$$\int_{S(\text{cap})} \hat{\mathbf{N}} \cdot (\boldsymbol{\nabla} \times \mathbf{H}) \, dS = \oint_{s(\text{line})} \mathbf{H} \cdot d\mathbf{s}$$

$$= \int_{S(\text{cap})} \hat{\mathbf{N}} \cdot \mathbf{J} \, dS + \frac{\partial}{\partial t} \int_{S(\text{cap})} \hat{\mathbf{N}} \cdot \mathbf{D} \, dS \quad (2.7\text{-}4)$$

The line integral in (2.7–4) is called the *circuitation* of the **H** vector, or simply the **H** circuitation around the closed contour s. The positive direction around the line is determined by the right-hand-screw convention referred to an arbitrarily directed normal $\hat{\mathbf{N}}$ to the surface S.

Since Stokes's theorem requires the vector to be continuous throughout the region of integration, special consideration must be given to the case where a cap surface is crossed by two or more dissimilar regions in contact because thin sheets of current of density **K** may then flow between them. If this is the case, the **H** vector is discontinuous across the boundary between the two regions, as seen in the boundary condition (2.4–14). To examine this situation, consider two electrically dissimilar regions or bodies 1 and 2 in contact along a boundary surface S_{12} (Fig. 2.7–1). Let region 1 be characterized by $\mathbf{H}_1$, $\mathbf{D}_1$, and $\mathbf{J}_1$; region 2 by $\mathbf{H}_2$, $\mathbf{D}_2$, and $\mathbf{J}_2$. For any closed contour in either region, (2.6–8) is valid. On the other hand, suppose a contour s to be drawn in such a way that it cuts the boundary surface S_{12}. Then any cap surface bounded by s is partly in region 1, partly in region 2. For simplicity, let the cap surface S be taken to be a plane perpendicular to the direction of the boundary surface S_{12} between the two regions. Let the line traced by the boundary surface where it intersects the plane S be s_{12}. This line cuts the plane cap surface S into two parts S_1 and S_2, so that $S_{\text{cap}} = S_1 + S_2$. The closed contour s bounding the plane is also cut into two parts s_1 and s_2, so that $s = s_1 + s_2$. S_1 is bounded by s_1 and one side of s_{12}, S_2 is bounded by s_2 and the other side of s_{12}. Let $\hat{\mathbf{N}}$ be a normal perpendicular to S and hence to s_{12}. $\hat{\mathbf{N}}$ is evidently parallel to the boundary surface S_{12} between the two regions where this is cut by the plane S.

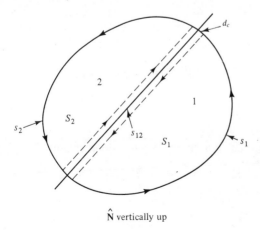

$\hat{\mathbf{N}}$ vertically up

Figure 2.7–1 Region with boundaries for obtaining the Ampère–Maxwell theorem for a discontinuous vector. Small arrows indicate the directions of integration around regions 1 and 2.

To form the circuitation of the **H** vector completely around the contour s, it is necessary to exclude the line of discontinuity s_{12} by drawing a line parallel to it on each side at a small distance d_c of atomic or molecular magnitude. The length of each of these two lines is essentially the same as that of s_{12}. A surface current of density **K**, if it exists, is necessarily confined to the strip of thickness $2d_c$ since **K** is defined only in this strip. Because each region is homogeneous, it is legitimate to form a line integral around S_1 and a second line integral around S_2. The path of integration for the first of these is along s_1 and s_{12} in a counterclockwise direction when viewed from above. This follows from the fact that the normal $\hat{\mathbf{N}}$ to the surface is directed vertically upward from the plane of the paper. Similarly, the path around S_2 will be counterclockwise around s_2 and s_{12}. The integrals may be taken along s_{12} instead of along lines parallel to s_{12} at distance d_c on each side because the surface contained between these lines is negligible. One of the two integrals is

$$\oint_{s_1 + s_{12}} \mathbf{H}_1 \cdot d\mathbf{s} = \int_{S_1} (\hat{\mathbf{N}} \cdot \mathbf{J}_1) \, dS + \frac{\partial}{\partial t} \int_{S_1} (\hat{\mathbf{N}} \cdot \mathbf{D}_1) \, dS \qquad (2.7\text{--}5)$$

The second integral is like (2.7–5) with subscript 1 written for 2 and 2 for 1. The sum of these two integrals is the same (neglecting the length d_c) as the counterclockwise integral completely around $S_1 + S_2 = S$, plus the clockwise, or minus the counterclockwise, integral around the boundary line s_{12} at a distance d_c from it. The latter integral is, of course, essentially equivalent to an integral along one side of s_{12} and back along the other side, since d_c is negligible. That is,

$$\oint_{s_1 + s_{12}} \mathbf{H}_1 \cdot d\mathbf{s} + \oint_{s_2 + s_{12}} \mathbf{H}_2 \cdot d\mathbf{s}$$

$$= \oint_{s_1 + s_2} \mathbf{H} \cdot d\mathbf{s} - \int_{s_{12}} \mathbf{H}_1 \cdot d\mathbf{s} - \int_{s_{12}} \mathbf{H}_2 \cdot d\mathbf{s} \qquad (2.7\text{--}6)$$

The negative sign before the last two integrals indicates that the direction of integration around the two sides of the line s_{12} has been changed from clockwise to counterclockwise. The value of **H** in the first integral on the right in (2.7–6) is taken to be $\mathbf{H}_1$ or $\mathbf{H}_2$ depending on in what region $d\mathbf{s}$ is.

With $s = s_1 + s_2$ and integration along both sides of s_{12} in the same direction instead of around s_{12}, (2.7–6) may be written as follows:

$$\oint_s \mathbf{H} \cdot d\mathbf{s} = \oint_{s_1 + s_{12}} \mathbf{H}_1 \cdot d\mathbf{s} + \oint_{s_1 + s_{12}} \mathbf{H}_2 \cdot d\mathbf{s} + \int_{s_{12}} (\mathbf{H}_1 - \mathbf{H}_2) \cdot d\mathbf{s} \qquad (2.7\text{--}7)$$

Let both sides of the boundary condition be multiplied scalarly by $\hat{\mathbf{N}}$; thus

$$\hat{\mathbf{N}} \cdot (\hat{\mathbf{n}}_1 \times \mathbf{H}_1) + \hat{\mathbf{N}} \cdot (\hat{\mathbf{n}}_2 \times \mathbf{H}_2) = -\hat{\mathbf{N}} \cdot \mathbf{K} \qquad (2.7\text{--}8)$$

Here $\hat{\mathbf{n}}_1$ and $\hat{\mathbf{n}}_2$ are external normals to the boundary between the regions 1 and 2. Since $\hat{\mathbf{n}}_2 = -\hat{\mathbf{n}}_1$, (2.7–8) can be written

$$\hat{\mathbf{N}} \cdot [\hat{\mathbf{n}}_1 \times (\mathbf{H}_1 - \mathbf{H}_2)] = -\hat{\mathbf{N}} \cdot \mathbf{K} \tag{2.7–9}$$

By cyclic permutation

$$(\mathbf{H}_1 - \mathbf{H}_2) \cdot (\hat{\mathbf{N}} \times \hat{\mathbf{n}}_1) = -\hat{\mathbf{N}} \cdot \mathbf{K} \tag{2.7–10}$$

But since $\hat{\mathbf{N}}$ and $\hat{\mathbf{n}}_1$ are not only normal to each other but also both perpendicular to s_{12}, it follows that the vector product $\hat{\mathbf{N}} \times \hat{\mathbf{n}}_1$ must be in the direction $d\mathbf{s}$ along s_{12}. With

$$\hat{\mathbf{N}} \times \hat{\mathbf{n}}_1 = -\frac{d\mathbf{s}}{|d\mathbf{s}|} \tag{2.7–11}$$

and the substitution in (2.7–10), the result is

$$(\mathbf{H}_1 - \mathbf{H}_2) \cdot d\mathbf{s} = (\hat{\mathbf{N}} \cdot \mathbf{K})\, ds \tag{2.7–12}$$

The use of (2.7–7) and (2.7–12) in (2.7–5) yields

$$\oint_s \mathbf{H} \cdot d\mathbf{s} = \sum_{j=1}^{2} \left[\int_{S_j} (\hat{\mathbf{N}} \cdot \mathbf{J})\, dS_j + \frac{\partial}{\partial t} \int_{S_j} (\hat{\mathbf{N}} \cdot \mathbf{D})\, dS_j \right]$$
$$+ \int_{s_{12}} (\hat{\mathbf{N}} \cdot \mathbf{K})\, ds \tag{2.7–13}$$

But

$$\sum_{j=1}^{2} \int_{S_j} (\hat{\mathbf{N}} \cdot \mathbf{J})\, dS_j + \int_{s_{12}} (\hat{\mathbf{N}} \cdot \mathbf{K})\, ds = I \tag{2.7–14}$$

is the total normal current crossing the cap surface S. If $\mathbf{M}$ is defined with a subdivision that cuts through no bound groups, as is customary when the $\mathbf{H}$ vector is used, I is the total moving *free* charge and the subscript f is written on $\mathbf{I}$, $\mathbf{J}$, and $\mathbf{K}$.

$$\oint_s \mathbf{H} \cdot d\mathbf{s} = I + \frac{\partial}{\partial t} \int_{S_{\text{cap}}} (\hat{\mathbf{N}} \cdot \mathbf{D})\, dS \tag{2.7–15}$$

with

$$S_{\text{cap}} = \sum_{j=1}^{2} S_j$$

These relations are easily generalized to include more than one surface of discontinuity. The Ampère–Maxwell circuitation theorem is expressed symbolically in (2.7–15). It states that the circuitation of the $\mathbf{H}$ vector around a closed contour s

is equal to the total normal current plus the time rate of change of the normal **D** flux across any cap surface bounded by the contour.

The right side of the Ampère–Maxwell relation consists of two terms of which the first is a current in the sense of moving charge. The second term was called the "displacement current" by Maxwell. Since, by definition, $\mathbf{D} = \epsilon_0 \mathbf{E} + \mathbf{P}$, the displacement current actually consists of two parts. These are

$$\frac{\partial}{\partial t} \int_S \hat{\mathbf{N}} \cdot \mathbf{D}\, dS = \epsilon_0 \int_S \hat{\mathbf{N}} \cdot \dot{\mathbf{E}}\, dS + \int_S \hat{\mathbf{N}} \cdot \dot{\mathbf{P}}\, dS \qquad (2.7\text{--}16)$$

It will be recalled that $\dot{\mathbf{P}}$ is the mathematical analogue of a motion of charge in the physical model due to the fluctuation of an orientation-distortion effect. It is called the *volume density of polarization current*. Thus the second integral on the right measures the total normal polarization current that traverses the cap surface S. It is properly called a current and represented by the symbol I_p. On the other hand, the first term on the right stands for the time rate of change of the **E** vector, which is not a current in the sense of moving charge in the physical model.

An alternative formulation of the circuitation theorem proceeds from the field equation

$$\mu_0^{-1} \boldsymbol{\nabla} \times \mathbf{B} = \overline{\rho_m \mathbf{v}} + \epsilon_0 \frac{\partial \mathbf{E}}{\partial t} \qquad (2.7\text{--}17)$$

instead of from (2.7–1). By the same formal reasoning,

$$\mu_0^{-1} \oint_s \mathbf{B} \cdot d\mathbf{s} = \overline{I} + \frac{\partial}{\partial t} \epsilon_0 \int_{S_{\text{cap}}} \hat{\mathbf{N}} \cdot \mathbf{E}\, dS \qquad (2.7\text{--}18)$$

with

$$\overline{I} = \int_{S_{\text{cap}}} (\hat{\mathbf{N}} \cdot \overline{\rho_m \mathbf{v}})\, dS + \sum_{ij} \int_{s_{ij}} (\hat{\mathbf{N}} \cdot \overline{\eta_m \mathbf{v}_{ij}})\, ds_{ij} \qquad (2.7\text{--}19)$$

$S_{\text{cap}} = \Sigma\, S_j$; s_{ij} is the boundary between S_i and S_j. In this formulation the entire moving charge or current is contained in the term $\overline{I}$.

2.8 FARADAY'S LAW OF CIRCUITATION: INTEGRAL FORM OF $\boldsymbol{\nabla} \times \mathbf{E} = -\partial\mathbf{B}/\partial t$

Let $\hat{\mathbf{N}}$ be a normal to a cap surface S bounded by a closed contour s. The components normal to S of the vectors defining each side of the second field equation (2.1–2) satisfy the relation

$$\hat{\mathbf{N}} \cdot \boldsymbol{\nabla} \times \mathbf{E} = -\frac{\partial}{\partial t}(\hat{\mathbf{N}} \cdot \mathbf{B}) \qquad (2.8\text{--}1)$$

If each side is integrated over the cap surface and Stokes's theorem is applied,

$$\int_{S(cap)} \hat{\mathbf{N}} \cdot (\boldsymbol{\nabla} \times \mathbf{E})\, dS = \oint_s \mathbf{E} \cdot d\mathbf{s} = -\frac{\partial}{\partial t} \int_{S(cap)} \hat{\mathbf{N}} \cdot \mathbf{B}\, dS \qquad (2.8\text{-}2)$$

If the cap surface is cut by one or more boundaries, the integral over the cap is replaced by a sum of integrals over the several parts of the cap. Faraday's law is given symbolically in (2.8–2). It may be stated as follows: The circuitation of the **E** vector in the positive direction around any closed contour in which **E** is continuous is equal to the negative time rate of change of the total normal **B** flux across any cap surface bounded by the contour. The positive direction is specified by the right-hand-screw convention referred to the arbitrarily directed normal $\hat{\mathbf{N}}$ to S.

2.9 GAUSS'S THEOREM FOR THE B VECTOR: INTEGRAL FORM OF $\boldsymbol{\nabla} \cdot \mathbf{B} = 0$

The use of (2.5–1) leads to

$$\boldsymbol{\nabla} \cdot \mathbf{B} = 0 \qquad (2.9\text{-}1)$$

$$\int_V \boldsymbol{\nabla} \cdot \mathbf{B}\, dV = \int_{S(closed)} \hat{\mathbf{n}} \cdot \mathbf{B}\, dS = 0$$

This equation states that the total outward normal flux of the **B** vector across any closed surface vanishes.

2.10 SUMMARY OF INTEGRAL FIELD EQUATIONS

In terms of the fundamental vectors the theorems are

(Gauss)
$$\epsilon_0 \int_{S(closed)} \hat{\mathbf{n}} \cdot \mathbf{E}\, dS = \overline{Q} \qquad (2.10\text{-}1)$$

(Faraday)
$$\oint_s \mathbf{E} \cdot d\mathbf{s} = -\frac{\partial}{\partial t} \int_{S(cap)} \hat{\mathbf{N}} \cdot \mathbf{B}\, dS \qquad (2.10\text{-}2)$$

(Ampère–Maxwell)
$$\mu_0^{-1} \oint_s \mathbf{B} \cdot d\mathbf{s} = \overline{I} + \frac{\partial}{\partial t} \epsilon_0 \int_{S(cap)} \hat{\mathbf{N}} \cdot \mathbf{E}\, dS \qquad (2.10\text{-}3)$$

(Gauss)
$$\int_{S(closed)} \hat{\mathbf{n}} \cdot \mathbf{B}\, dS = 0 \qquad (2.10\text{-}4)$$

where

$$\overline{Q} = \sum_j \int_{V_j} \overline{\rho}_j\, dV_j + \sum_{ij} \int_{S_{ij}(closed)} \overline{\eta}_{ij}\, dS_{ij} \qquad (2.10\text{-}5)$$

$$\overline{I} = \int_{S(cap)} (\hat{\mathbf{N}} \cdot \overline{\rho_m \mathbf{v}})\, dS + \sum_{ij} \int_{s_{ij}} (\hat{\mathbf{N}} \cdot \overline{\eta_m \mathbf{v}_{ij}})\, ds_{ij} \qquad (2.10\text{-}6)$$

S_{cap} is a cap surface with edge s; if surface currents cross it and divide it into parts S_j, S_{cap} is equivalent to $\Sigma_j\, S_j$. Also s_{ij} is a boundary between S_i and S_j; the sum includes the contour s if a surface current crosses this.

In terms of the auxiliary vectors

(Gauss)
$$\int_{S(closed)} \hat{\mathbf{n}} \cdot \mathbf{D}\, dS = Q \tag{2.10-7}$$

(Ampère–Maxwell)
$$\oint_s \mathbf{H} \cdot d\mathbf{s} = I + \frac{\partial}{\partial t}\int_{S(cap)} \hat{\mathbf{N}} \cdot \mathbf{D}\, dS \tag{2.10-8}$$

where

$$Q = \sum_j \int_{V_j} \rho_j\, dV_j + \sum_{ij} \int_{S_{ij}} \eta_{ij}\, dS_{ij} \tag{2.10-9}$$

$$I = \int_{S(cap)} (\hat{\mathbf{N}} \cdot \mathbf{J})\, dS + \sum_{ij} \int_{s_{ij}} (\hat{\mathbf{N}} \cdot \mathbf{K}_{ij})\, ds_{ij} \tag{2.10-10}$$

$$\mathbf{D} = \epsilon_0 \mathbf{E} + \mathbf{P}; \qquad \mathbf{H} = \mu_0^{-1}\mathbf{B} - \mathbf{M} \tag{2.10-11}$$

2.11 FIELD EQUATIONS IN SIMPLE MEDIA: PERMITTIVITY, PERMEABILITY

Maxwell's equations in their general forms (2.1–1) to (2.1–4) together with the definitions (2.1–5) and (2.1–6) of the essential densities, involve four volume densities: ρ, $\mathbf{P}$, $\mathbf{J}$, and $\mathbf{M}$. In many materials these densities are induced by an externally maintained electromagnetic field. The actual relations between any one of the densities and the exciting component of the field can be determined only from a detailed study of the atomic and molecular structure. This is beyond the scope of the present treatment. However, the general relationship between the densities and the field can be expressed formally as follows:

$$\mathbf{P} = \mathbf{P}(\mathbf{E}) = \mathbf{P}_0 + \epsilon_0\chi_e\mathbf{E} + \epsilon_0\chi_{e2}\,|E^2|\hat{\mathbf{E}} + \cdots \tag{2.11-1}$$

$$-\mathbf{M} = -\mathbf{M}(\mathbf{B}) = -\mathbf{M}_0 + \mu_0^{-1}\chi_m\mathbf{B} + \mu_0^{-1}\chi_{m2}\,|B^2|\,\hat{\mathbf{B}} + \cdots \tag{2.11-2}$$

$$\mathbf{J}_f = \mathbf{J}_f(\mathbf{E}) = \mathbf{J}_{f0} + \sigma\mathbf{E} + \sigma_2\,|E^2|\,\hat{\mathbf{E}} + \cdots \tag{2.11-3}$$

In these expressions each of the densities is represented by a series that includes a constant term independent of the exciting field, a linear term, and higher-order, nonlinear terms.

The constant terms represent, respectively, a permanent electric moment $\mathbf{P}_0$ such as exists in certain anisotropic crystals or that can be induced in waxes that are solidified in an electric field, a permanent magnetic moment $\mathbf{M}_0$ such as exists in ferromagnetic materials, and a permanent current $\mathbf{J}_{f0}$ as may exist in superconductors. Although such permanent densities are important in special circumstances, they are not of primary concern in major aspects of electromagnetic theory.

Of greatest importance in many applications of electromagnetism are the linear terms. Materials in which the linear term alone is a good approximation are known as simple media or linear media. These include most dielectrics in which the relation $\mathbf{P} = \epsilon_0 \chi_e \mathbf{E}$ is generally adequate, most diamagnetic and paramagnetic materials in which $-\mathbf{M} = \mu_0^{-1} \chi_m \mathbf{B}$,* and most conductors for which $\mathbf{J}_f = \sigma \mathbf{E}$ is an excellent approximation.†

The linear terms in (2.11–1) to (2.11–3) are not adequate in dielectrics when the electromagnetic field is of very high intensity as with laser beams in nonlinear optics, in all ferromagnetic materials that involve hysteresis phenomena, and in a number of conductors such as carbon.

When the linear terms are good approximations, the Maxwell equations are greatly simplified. In such media

$$\mathbf{D} = \epsilon_0 \mathbf{E} + \mathbf{P} = (1 + \chi_e)\epsilon_0 \mathbf{E} \tag{2.11–4}$$

$$\mathbf{H} = \mu_0^{-1}\mathbf{B} - \mathbf{M} = (1 + \chi_m)\mu_0^{-1}\mathbf{B} \tag{2.11–5}$$

$$\mathbf{J}_f = \sigma \mathbf{E} \tag{2.11–6}$$

In (2.11–4) and (2.11–5) the terms $(1 + \chi_e)$ and $(1 + \chi_m)$ are dimensionless quantities that depend only on the characteristic electric and magnetic structures of the models for which χ_e and χ_m are defined. Each of these quantities is conveniently represented by a symbol that stands for the properties of linear polarizability or linear magnetizability. They are

$$\epsilon_r = 1 + \chi_e \quad \text{or} \quad \chi_e = \epsilon_r - 1 \tag{2.11–7}$$

$$\mu_r^{-1} = 1 + \chi_m \quad \text{or} \quad \chi_m = \mu_r^{-1} - 1 \tag{2.11–8}$$

The quantity ϵ_r is the relative dielectric constant or relative permittivity of a linearly polarizable medium, the quantity μ_r is the relative permeability of a linearly magnetizable medium. Because the products $\epsilon_0 \epsilon_r$ and $\mu_0 \mu_r$ occur very frequently, it is convenient to introduce special symbols for them. Let

$$\epsilon \equiv \epsilon_0 \epsilon_r; \quad \mu \equiv \mu_0 \mu_r \tag{2.11–9}$$

---

* The magnetic susceptibility χ_m in this formula relates $-\mathbf{M}$ to $\mathbf{B}$. A more conventional representation relates $\mathbf{M}$ to $\mathbf{H}$ by the formula

$$\mathbf{M} = \chi_{mH}\mathbf{H} = -\mu_0^{-1}\chi_m\mathbf{B} = \mu_0^{-1}\frac{\chi_{mH}}{1 + \chi_{mH}}\mathbf{B}.$$

It follows that $\chi_m = -\chi_{mH}/(1 + \chi_{mH})$. Note that $\chi_m = -\chi_{mH}$ when $\chi_{mH} \ll 1$, an inequality that is often valid.

† A generalization of this relation to conductors that move with the velocity $\mathbf{v}$ is $\mathbf{J}_f = \sigma(\mathbf{E} + \mathbf{v} \times \mathbf{B})$.

The names absolute dielectric constant or permittivity for ϵ and absolute permeability for μ are used. In terms of these symbols,

$$\mathbf{D} = \epsilon_0 \epsilon_r \mathbf{E} = \epsilon \mathbf{E} \tag{2.11-10}$$

$$\mathbf{H} = \mu_0^{-1} \mu_r^{-1} \mathbf{B} = \mu^{-1} \mathbf{B} \tag{2.11-11}$$

$$\mathbf{P} = (\epsilon_r - 1)\epsilon_0 \mathbf{E} = (\epsilon - \epsilon_0)\mathbf{E} \tag{2.11-12}$$

$$-\mathbf{M} = (\mu_r^{-1} - 1)\mu_0^{-1} \mathbf{B} = (\mu^{-1} - \mu_0^{-1})\mathbf{B} \tag{2.11-13}$$

Since χ_e and χ_m and hence ϵ_r and μ_r are defined in terms of $\mathbf{P}$ and $-\mathbf{M}$ using a subdivision into volume cells that cuts no closely bound groups, it follows that there can be no contributions from cut parts of such groups either to the volume density of charge ρ or to the volume density of current $\mathbf{J}$. Therefore, these are limited to contributions due to free charge so that $\rho = \rho_f$; $\mathbf{J} = \mathbf{J}_f$. With (2.11–10) and (2.11–11) and the constitutive relation

$$\mathbf{J}_f = \sigma \mathbf{E} \tag{2.11-14}$$

the field equations assume the following forms, in which ρ_f is the only density function appearing explicitly:

$$\epsilon \nabla \cdot \mathbf{E} = \rho_f \tag{2.11-15}$$

$$\nabla \times \mathbf{E} = -\dot{\mathbf{B}} \tag{2.11-16}$$

$$\mu^{-1}(\nabla \times \mathbf{B}) = \sigma \mathbf{E} + \epsilon \dot{\mathbf{E}} \tag{2.11-17}$$

$$\nabla \cdot \mathbf{B} = 0 \tag{2.11-18}$$

The relations are valid in linearly polarizing, magnetizing, and conducting media. It is to be noted that if values of $\mathbf{P}$ and $-\mathbf{M}$ are determined using (2.11–12) and (2.11–13) and values of ϵ_r and μ_r are obtained from these equations in any particular case, the values so obtained *must* be interpreted in terms of a subdivision that cuts no closely bound polarized or magnetized groups.

Since $\mathbf{P}$ and $\mathbf{M}$ are defined entirely in terms of volume cells, there can be no contributions to surface densities of charge or current from cutoff surface slices of magnetized or polarized units. Any surface density must be due to free charge only. The surface equations or boundary conditions are

$$\epsilon_1 \hat{\mathbf{n}}_1 \cdot \mathbf{E}_1 + \epsilon_2 \hat{\mathbf{n}}_2 \cdot \mathbf{E}_2 = -\eta_{1f} - \eta_{2f} \quad \text{or} \quad \epsilon \hat{\mathbf{n}} \cdot \mathbf{E} = -\eta_f \tag{2.11-19}$$

$$\hat{\mathbf{n}}_1 \times \mathbf{E}_1 + \hat{\mathbf{n}}_2 \times \mathbf{E}_2 = 0 \quad \text{or} \quad \hat{\mathbf{n}} \times \mathbf{E} = 0 \tag{2.11-20}$$

$$\mu_1^{-1}(\hat{\mathbf{n}}_1 \times \mathbf{B}_1) + \mu_2^{-1}(\hat{\mathbf{n}}_2 \times \mathbf{B}_2) = -\mathbf{K}_{1f} - \mathbf{K}_{2f} \quad \text{or} \quad \mu^{-1}\hat{\mathbf{n}} \times \mathbf{B} = -\mathbf{K}_f \tag{2.11-21}$$

$$\hat{\mathbf{n}}_1 \cdot \mathbf{B}_1 + \hat{\mathbf{n}}_2 \cdot \mathbf{B}_2 = 0 \quad \text{or} \quad \hat{\mathbf{n}} \cdot \mathbf{B} = 0 \tag{2.11-22}$$

The simpler forms on the right are a shorthand form for the explicit expressions on the left. It may be surprising that surface densities appear in these equations when a subdivision into volume cells was used in defining **P** and **M**. This is due to the fact that a subdivision into surface and volume cells is retained for the free charges and a separate subdivision into volume cells only is used for the bound charges and current whirls.

The analogy between the static and steady states may be extended further to include the following constitutive parameters in the special case of linear media.

Quantity	Static state	Steady state
Constitutive parameters in simple models	Electric susceptibility, χ_e Relative permittivity, ϵ_r	Magnetic susceptibility, χ_m Reciprocal of relative permeability, μ_r^{-1}
	Absolute permittivity, ϵ	Reciprocal of absolute permeability, μ^{-1}

2.12 GENERAL DEFINITION OF A CONDUCTOR

A good conductor is characterized by a conductivity σ which is sufficiently large that a transient, nonrandom drifting of charge existing in a change from one stationary state to another is reduced to a negligibly small value in an experimentally insignificant length of time. The constitutive relation is then valid during all experimentally significant time intervals.

To define a homogeneous, linear conductor mathematically, it is necessary to establish a convenient criterion for the rapidity with which ρ_f vanishes or $\mathbf{J}_f$ becomes constant in the interior of a conducting medium after a change in the distribution of charge on the surface or in neighboring bodies. In the absence of detailed formulas for the constitutive parameters ϵ_r, μ_r, and σ in terms of atomic and molecular structure, such a definition is possible only if it is assumed that time lags in polarization, in magnetization, and in current responses to changes in **E** and **B** are smaller than an experimentally observable interval t_{ex}. That is, the definition of a conductor includes the requirement that the constitutive relations

$$\mathbf{P} = (\epsilon_r - 1)\epsilon_0\mathbf{E}; \qquad -\mathbf{M} = (\mu_r^{-1} - 1)\mu_0^{-1}\mathbf{B}; \qquad \mathbf{J}_f = \sigma\mathbf{E} \qquad (2.12\text{-}1)$$

be true at all times within experimental observation. Subject to (2.12–1) the field equations assume the form

$$\epsilon\nabla \cdot \mathbf{E} = \rho_f \qquad (2.12\text{-}2a)$$

$$\mu^{-1}(\nabla \times \mathbf{B}) = \sigma\mathbf{E} + \epsilon\dot{\mathbf{E}} \qquad (2.12\text{-}2b)$$

Upon taking the divergence of (2.12–2b) and recalling that the operation $(\nabla \cdot \nabla \times)$ acting on any vector yields zero, one obtains

$$\nabla \cdot (\sigma\mathbf{E} + \epsilon\dot{\mathbf{E}}) = 0 \qquad (2.12\text{-}3)$$

The use of (2.12–2a) to eliminate **E** yields the following linear differential equation in ρ_f:

$$\dot{\rho}_f + \frac{\sigma}{\epsilon} \rho_f = 0 \tag{2.12–4}$$

The solution of this equation is

$$\rho_f = \rho_0 e^{-(\sigma/\epsilon)t} \tag{2.12–5}$$

Here ρ_0 is the value of ρ_f at $t = 0$. From (2.12–5) it is clear that ρ_f reduces to a negligibly small value in a time that depends on the ratio ϵ/σ. For convenience, let this quantity be called—after Maxwell—the time of relaxation and denoted by

$$T_R = \frac{\epsilon}{\sigma} \tag{2.12–6}$$

The time of relaxation is the time required for ρ_f to reduce to $1/e$ of its initial value ρ_0. It is noteworthy that it is independent of μ. It follows from (2.12–5) that when

$$\sigma = 0; \qquad \rho_f = \rho_0 = \text{constant} \tag{2.12–7}$$

$$\sigma \doteq \infty; \qquad \rho_f \doteq 0 \tag{2.12–8}$$

In a perfect nonconductor as defined by (2.12–7), the volume density of charge is everywhere constant. Since closely bound groups such as atoms and molecules are electrically neutral as a whole, it is in most cases correct to set the constant ρ_0 in (2.12–7) equal to zero. In a *perfect* conductor as defined by (2.12–8), the volume density of charge vanishes at all times.

If it is assumed that for all practical purposes it is adequate for ρ_f to reduce to 1 percent of its initial value in a time that is short compared with an experimentally observable interval, then a conductor can be defined by the following conditions:

$$e^{-t/T_R} \ll 1 \quad \text{or} \quad e^{t/T_R} \geq 100, \qquad t \ll t_{\text{ex}} \tag{2.12–9}$$

Upon taking the logarithm of (2.12–9) and noting that $e^{4.6} = 100$,

$$T_R \equiv \frac{\epsilon}{\sigma} \leq \frac{t}{4.6} \quad \text{or} \quad \frac{\sigma}{\epsilon} \geq \frac{4.6}{t}, \qquad t \ll t_{\text{ex}} \tag{2.12–10}$$

By assigning a definite time t that is short compared with an experimentally significant interval t_{ex} in any particular case, a conductor is defined by (2.12–10) to be a model in which the ratio of material parameters σ/ϵ is greater than a specified magnitude. Let this definition be expressed by the following shorthand symbolism, which implies that a time limit has been established and that (2.12–1) and (2.12–10) are both true. A conductor is defined symbolically by

$$\sigma \gg 0 \tag{2.12–11a}$$

If (2.12–11a) is true, $\rho_f \sim 0$ within the required degree of accuracy during all significant intervals of time. In periodic phenomena, the time of relaxation T_R may be required to be negligible compared with $T_p/2\pi$, where T_p is the period. With

$$T_p = \frac{1}{f} = \frac{2\pi}{\omega}$$

(2.12–11a) becomes more specifically

$$T_R \ll \frac{T_p}{2\pi}; \qquad \frac{\omega\epsilon}{\sigma} \ll 1 \qquad\qquad (2.12\text{–}11b)$$

In terms of the field equation (2.12–2b), the definition of a conductor (2.12–10) is equivalent to requiring that

$$\sigma \mathbf{E} \gg \epsilon \dot{\mathbf{E}} \qquad\qquad (2.12\text{–}12)$$

Hence, in a conductor (2.12–2b) reduces to

$$\mu^{-1}\nabla \times \mathbf{B} \doteq \sigma \mathbf{E} \qquad\qquad (2.12\text{–}13)$$

With (2.12–1), (2.12–12) is equivalent to

$$\mathbf{J}_f \gg \dot{\mathbf{P}} + \epsilon_0 \dot{\mathbf{E}} \qquad\qquad (2.12\text{–}14)$$

With (2.12–1) and (2.12–8), (2.12–2a) gives

$$\nabla \cdot \mathbf{E} = 0 \qquad\qquad (2.12\text{–}15a)$$

so that with $\epsilon \mathbf{E} = \mathbf{P} + \epsilon_0 \mathbf{E}$

$$\nabla \cdot \mathbf{P} = 0 \qquad\qquad (2.12\text{–}15b)$$

Since $\mathbf{P}$ is thus seen to play no significant part in the definitions of $\mathbf{E}$ and $\mathbf{B}$ in a conductor, it may be concluded that although no proof has been given that $\mathbf{P}$ necessarily vanishes in a conductor, its effect is certainly negligibly small, so that it is as well to write

$$\mathbf{P} \doteq 0 \qquad\qquad (2.12\text{–}16)$$

If a nonstationary state prevails in which there is a periodically varying distribution of charge so that

$$\mathbf{E}_{\text{inst}} = \mathbf{E}\cos(\omega t + \theta_E); \qquad \mathbf{B}_{\text{inst}} = \mathbf{B}\cos(\omega t + \theta_B) \qquad (2.12\text{–}17)$$

a time lag in the instantaneous values of polarization, magnetization, and current density may be expressed by writing (2.12–1) in the form

$$\mathbf{P}(t) = \tilde{\chi}_e \epsilon_0 \mathbf{E}\cos(\omega t + \theta_E - \theta_P) \qquad\qquad (2.12\text{–}18a)$$

$$-\mathbf{M}(t) = \tilde{\chi}_m \mu_0^{-1} \mathbf{B}\cos(\omega t + \theta_B - \theta_M) \qquad\qquad (2.12\text{–}18b)$$

$$\mathbf{J}_f(t) = \tilde{\sigma} \mathbf{E}\cos(\omega t + \theta_E - \theta_J) \qquad\qquad (2.12\text{–}18c)$$

Here the parameters $\tilde{\chi}_e$, $\tilde{\chi}_m$, $\tilde{\sigma}$ as well as the phase angles θ_P, θ_M, θ_J are functions of ω. The definition of a good conductor in the periodic state includes the following conditions:

$$\theta_P \ll 1; \qquad \theta_M \ll 1; \qquad \theta_J \ll 1$$
$$\tilde{\chi}_e \doteq \chi_e; \qquad \tilde{\chi}_m \doteq \chi_m; \qquad \tilde{\sigma} \doteq \sigma \qquad (2.12\text{–}19)$$

With the substitution of (2.12–17) in (2.12–2b) and in (2.12–13) [to which (2.12–2b) reduces by definition of a conductor], the following two equations are obtained:

$$\nabla \times \mathbf{B} \cos(\omega t + \theta_B) = \sigma\mu\mathbf{E}\left[\cos(\omega t + \theta_E) - \frac{\omega\epsilon}{\sigma}\sin(\omega t + \theta_E)\right] \quad (2.12\text{–}20)$$

$$\nabla \times \mathbf{B} \cos(\omega t + \theta_B) \doteq \sigma\mu\mathbf{E}\cos(\omega t + \theta_E) \qquad (2.12\text{–}21)$$

With the trigonometric formula

$$a \sin x + b \cos x = (a^2 + b^2)^{1/2}\cos\left(x - \tan^{-1}\frac{a}{b}\right) \qquad (2.12\text{–}22)$$

(2.12–20) becomes

$$\nabla \times \mathbf{B} \cos(\omega t + \theta_B)$$
$$= \sigma\mu\mathbf{E}\left[1 + \left(\frac{\omega\epsilon}{\sigma}\right)^2\right]^{1/2}\cos\left(\omega t + \theta_E + \tan^{-1}\frac{\omega\epsilon}{\sigma}\right) \qquad (2.12\text{–}23)$$

Subject to (2.12–10), (2.12–23) must reduce to (2.12–21). A more convenient, though equivalent definition of a conductor for the harmonic state than (2.12–10) may be formulated by requiring (2.12–11b) to be satisfied, that is,

$$\frac{\omega\epsilon}{\sigma} = \omega T_R \ll 1 \qquad (2.12\text{–}24)$$

Subject to (2.12–24),

$$\left[1 + \left(\frac{\omega\epsilon}{\sigma}\right)^2\right]^{1/2} \doteq 1; \qquad \tan^{-1}\frac{\omega\epsilon}{\sigma} \doteq \frac{\omega\epsilon}{\sigma} \ll 1 \qquad (2.12\text{–}25)$$

Subject to (2.12–25), (2.12–23) is equivalent to (2.12–21) to the desired degree of approximation.

The definitions of conductors, semiconductors, and nonconductors may be summarized as follows:

Conductor: $\sigma/\epsilon \geq 4.6/t$ (shorthand form, $\sigma \gg 0$), where t is a time that is short compared with experimentally measurable intervals. For harmonic time dependence, $\sigma/\epsilon \gg \omega$, where $\omega = 2\pi f$ and f is the highest frequency. In a conductor, the constitutive relations

$$\mathbf{P} = (\epsilon_r - 1)\epsilon_0\mathbf{E}; \qquad -\mathbf{M} = (\mu_r^{-1} - 1)\mu_0^{-1}\mathbf{B}; \qquad \mathbf{J}_f = \sigma\mathbf{E}$$

are true at all times within experimental observation; $\mathbf{P}$ is always negligibly small; ρ_f vanishes.

Semiconductor: $\sigma > 0$, $\mathbf{J}_f \to \sigma\mathbf{E}$, $\rho_f \to 0$; the constitutive relations are obeyed after an observable time interval or with an observable time lag.

Nonconductor: $\sigma = 0$, $\mathbf{J}_f = 0$ for all values of $\mathbf{E}$; $\rho_f = \text{const} (= 0)$.

In simple conductors and nonconductors (in which it may be assumed that $\rho_f = 0$) the field equations and the corresponding boundary equations for media 1 and 2 are

$$\nabla \cdot \mathbf{E} = 0 \tag{2.12--26a}$$

$$\nabla \times \mathbf{E} = -\dot{\mathbf{B}} \tag{2.12--26b}$$

$$\nabla \times \mathbf{B} = \mu(\sigma\mathbf{E} + \epsilon\dot{\mathbf{E}}) \tag{2.12--26c}$$

$$\nabla \cdot \mathbf{B} = 0 \tag{2.12--26d}$$

$$\epsilon_1\hat{\mathbf{n}}_1 \cdot \mathbf{E}_1 + \epsilon_2\hat{\mathbf{n}}_2 \cdot \mathbf{E}_2 = -\eta_{1f} - \eta_{2f} \tag{2.12--27a}$$

$$\hat{\mathbf{n}}_1 \times \mathbf{E}_1 + \hat{\mathbf{n}}_2 \times \mathbf{E}_2 = 0 \tag{2.12--27b}$$

$$\mu_1^{-1}\hat{\mathbf{n}}_1 \times \mathbf{B}_1 + \mu_2^{-1}\hat{\mathbf{n}}_2 \times \mathbf{B}_2 = -\mathbf{K}_{1f} - \mathbf{K}_{2f} \tag{2.12--27c}$$

$$\hat{\mathbf{n}}_1 \cdot \mathbf{B}_1 + \hat{\mathbf{n}}_2 \cdot \mathbf{B}_2 = 0 \tag{2.12--27d}$$

Because ρ and $\mathbf{P}$ vanish in their interior, the electrical properties of conductors may be represented mathematically in terms of the surface density η_f alone instead of in terms of $\bar{\rho}$ and $\bar{\eta}$. In the solution of most boundary-value problems involving practically available conductors and frequencies, no surface layers of current (moving free charge) which are so thin that they cannot be described adequately in terms of $\mathbf{J}_f$ are encountered. Therefore, $\mathbf{K}_f$ is not required and need not be defined. It follows that a representation entirely in terms of η_f, $\mathbf{J}_f$, and $\mathbf{M}$ is possible. This is not true of problems involving the idealized case of perfect conductors with $\sigma \to \infty$. For these a surface function $\mathbf{K}_f$ rather than a representation in terms of an equivalent $\hat{\mathbf{n}} \times \mathbf{M}$ is often desirable because values of $\mathbf{M}$ actually associated physically with spin atoms may be encountered also. The function $\mathbf{K}_f$ need be used only in problems involving infinite conductivity but is often a useful approximation for very good conductors. For ordinary conductors and boundaries between two such conductors numbered 1 and 2

$$\nabla \cdot \mathbf{E} = 0 \tag{2.12--28a}$$

$$\nabla \times \mathbf{E} = -\dot{\mathbf{B}} \tag{2.12--28b}$$

$$\nabla \times \mathbf{B} = \mu\sigma\mathbf{E} \tag{2.12--28c}$$

$$\nabla \cdot \mathbf{B} = 0 \tag{2.12--28d}$$

where $\mathbf{E}$, $\mathbf{B}$, σ, and μ have the subscript 1 or 2 and, on the boundary,

$$\epsilon_1 \hat{\mathbf{n}}_1 \cdot \mathbf{E}_1 + \epsilon_2 \hat{\mathbf{n}}_2 \cdot \mathbf{E}_2 = -\eta_{1f} - \eta_{2f} \qquad (2.12\text{--}29\text{a})$$

$$\hat{\mathbf{n}}_1 \times \mathbf{E}_1 + \hat{\mathbf{n}}_2 \times \mathbf{E}_2 = 0 \qquad (2.12\text{--}29\text{b})$$

$$\mu_1^{-1}(\hat{\mathbf{n}}_1 \times \mathbf{B}_1) + \mu_2^{-1}(\hat{\mathbf{n}}_2 \times \mathbf{B}_2) = 0 \qquad (2.12\text{--}29\text{c})$$

$$\hat{\mathbf{n}}_1 \cdot \mathbf{B}_1 + \hat{\mathbf{n}}_2 \cdot \mathbf{B}_2 = 0 \qquad (2.12\text{--}29\text{d})$$

In nonconductors free charges are present in negligible amounts or not at all so that the entire representation may be expressed in terms of $\mathbf{P}$ and $\mathbf{M}$ alone. The following sets of equations are valid, respectively, in nonconductors and on the boundaries between nonconductors 1 and 2:

$$\nabla \cdot \mathbf{E} = 0 \qquad (2.12\text{--}30\text{a})$$

$$\nabla \times \mathbf{E} = -\dot{\mathbf{B}} \qquad (2.12\text{--}30\text{b})$$

$$\nabla \times \mathbf{B} = \mu\epsilon\dot{\mathbf{E}} \qquad (2.12\text{--}30\text{c})$$

$$\nabla \cdot \mathbf{B} = 0 \qquad (2.12\text{--}30\text{d})$$

$$\epsilon_1 \hat{\mathbf{n}}_1 \cdot \mathbf{E}_1 + \epsilon_2 \hat{\mathbf{n}}_2 \cdot \mathbf{E}_2 = 0 \qquad (2.12\text{--}31\text{a})$$

$$\hat{\mathbf{n}}_1 \times \mathbf{E}_1 + \hat{\mathbf{n}}_2 \times \mathbf{E}_2 = 0 \qquad (2.12\text{--}31\text{b})$$

$$\mu_1^{-1}(\hat{\mathbf{n}}_1 \times \mathbf{B}_1) + \mu_2^{-1}(\hat{\mathbf{n}}_2 \times \mathbf{B}_2) = 0 \qquad (2.12\text{--}31\text{c})$$

$$\hat{\mathbf{n}}_1 \cdot \mathbf{B}_1 + \hat{\mathbf{n}}_2 \cdot \mathbf{B}_2 = 0 \qquad (2.12\text{--}31\text{d})$$

Since the right-hand members of the boundary conditions (2.12–31a–d) are all zero, the following relations for the magnitudes of $\mathbf{E}$ and $\mathbf{B}$ on the two sides of the boundary may be written. All directions are referred to the external normal to region 1. The notation $(\hat{\mathbf{n}}, \mathbf{E})$ in the argument of a trigonometric function means the angle between $\hat{\mathbf{n}}$ and $\mathbf{E}$.

$$\epsilon_1 E_1 \cos(\hat{\mathbf{n}}_1, \mathbf{E}_1) = \epsilon_2 E_2 \cos(\hat{\mathbf{n}}_1, \mathbf{E}_2) \quad \text{or} \quad \epsilon_1 E_{1n} = \epsilon_2 E_{2n} \qquad (2.12\text{--}32\text{a})$$

$$E_1 \sin(\hat{\mathbf{n}}_1, \mathbf{E}_1) = E_2 \sin(\hat{\mathbf{n}}_1, \mathbf{E}_2) \quad \text{or} \quad E_{1t} = E_{2t} \qquad (2.12\text{--}32\text{b})$$

$$\mu_1^{-1} B_1 \sin(\hat{\mathbf{n}}_1, \mathbf{B}_1) = \mu_2^{-1} B_2 \sin(\hat{\mathbf{n}}_1, \mathbf{B}_2) \quad \text{or} \quad \mu_1^{-1} B_{1t} = \mu_2^{-1} B_{2t} \qquad (2.12\text{--}33\text{a})$$

$$B_1 \cos(\hat{\mathbf{n}}_1, \mathbf{B}_1) = B_2 \cos(\hat{\mathbf{n}}_1, \mathbf{B}_2) \quad \text{or} \quad B_{1n} = B_{2n} \qquad (2.12\text{--}33\text{b})$$

From these equations, the following general ratios are obtained:

$$\frac{\tan(\hat{\mathbf{n}}_1, \mathbf{E}_1)}{\tan(\hat{\mathbf{n}}_1, \mathbf{E}_2)} = \frac{\epsilon_1}{\epsilon_2} \tag{2.12–34a}$$

$$\frac{\cot(\hat{\mathbf{n}}_1, \mathbf{B}_1)}{\cot(\hat{\mathbf{n}}_1, \mathbf{B}_2)} = \frac{\mu_2}{\mu_1} \tag{2.12–34b}$$

$$\frac{E_2}{E_1} = \left[\frac{\epsilon_1^2}{\epsilon_2^2} + \left(1 - \frac{\epsilon_1^2}{\epsilon_2^2}\right)\sin^2(\hat{\mathbf{n}}_1, \mathbf{E}_1)\right]^{1/2} \tag{2.12–35a}$$

$$\frac{B_2}{B_1} = \left[\frac{\mu_2^2}{\mu_1^2} + \left(1 - \frac{\mu_2^2}{\mu_1^2}\right)\cos^2(\hat{\mathbf{n}}_1, \mathbf{B}_1)\right]^{1/2} \tag{2.12–35b}$$

It is sometimes convenient to represent a distribution of moving free charges entirely in terms of $\mathbf{P}$ and $-\mathbf{M}$. This is possible if the free-charge model is first replaced by a *mathematically equivalent* bound-charge model which is characterized by the same essential densities. These are given by

$$\overline{\rho_m \mathbf{v}} = \nabla \times \mathbf{M} + \dot{\mathbf{P}}; \qquad \overline{\rho} = -\nabla \cdot \mathbf{P} \tag{2.12–36}$$

$$\overline{\eta_m \mathbf{v}} = -(\hat{\mathbf{n}} \times \mathbf{M}); \qquad \overline{\eta} = \hat{\mathbf{n}} \cdot \mathbf{P}$$

Once this change in model and representation has been made, a relative dielectric constant (permittivity) and a relative permeability can be defined in the usual way so that (2.12–30a–d) and (2.12–31a–d) apply. It is well to note that the definitions of permittivity and permeability imply a *linear relation between* $\mathbf{P}$ *and* $\mathbf{E}$ and *between* $-\mathbf{M}$ *and* $\mathbf{B}$ and therefore a bound-charge model of special structure. The fact that a bound-charge model can always be constructed to be equivalent to a *particular* distribution of moving free charge does not mean that the equivalence is necessarily maintained under the action of diverse external forces. This certainly would not be the case in general, especially if linear relations between $\mathbf{P}$ and $\mathbf{E}$ and $-\mathbf{M}$ and $\mathbf{B}$ are assumed to be maintained. A conducting medium does not behave like a dielectric, even though it is possible to represent a particular distribution of free charge in terms of an equivalent polarization.

In the general form (2.12–26a–d) for simple media the first-order equations are readily converted to second-order equations in which the variables are separated. Thus with (2.12–26b, c)

$$\nabla \times \nabla \times \mathbf{E} + \mu\sigma\dot{\mathbf{E}} + \mu\epsilon\ddot{\mathbf{E}} = 0 \tag{2.12–37}$$

$$\nabla \times \nabla \times \mathbf{B} + \mu\sigma\dot{\mathbf{B}} + \mu\epsilon\ddot{\mathbf{B}} = 0 \tag{2.12–38}$$

2.13 THE TIME-INDEPENDENT MAXWELL EQUATIONS: PERIODIC TIME DEPENDENCE

The time-dependent Maxwell–Lorentz equations are

$$\nabla \cdot \mathbf{E}(\mathbf{r}, t) = \epsilon_0^{-1}\bar{\rho}(\mathbf{r}, t) \qquad (2.13-1)$$

$$\nabla \times \mathbf{E}(\mathbf{r}, t) = -\dot{\mathbf{B}}(\mathbf{r}, t) \qquad (2.13-2)$$

$$\nabla \times \mathbf{B}(\mathbf{r}, t) = \mu_0[\overline{\rho_m \mathbf{v}}(\mathbf{r}, t) + \epsilon_0\dot{\mathbf{E}}(\mathbf{r}, t)] \qquad (2.13-3)$$

$$\nabla \cdot \mathbf{B}(\mathbf{r}, t) = 0 \qquad (2.13-4)$$

where the dependence of the electromagnetic vectors and the densities on the time and the space coordinates is explicitly shown. Time-independent equations can be obtained in a very general manner in terms of Fourier transforms. Let

$$\mathbf{E}(\mathbf{r}, t) = \frac{1}{2\pi} \int_{-\infty}^{\infty} \mathbf{E}(\mathbf{r}, \omega)e^{j\omega t} \, d\omega \qquad (2.13-5)$$

$$\mathbf{B}(\mathbf{r}, t) = \frac{1}{2\pi} \int_{-\infty}^{\infty} \mathbf{B}(\mathbf{r}, \omega)e^{j\omega t} \, d\omega \qquad (2.13-6)$$

$$\bar{\rho}(\mathbf{r}, t) = \frac{1}{2\pi} \int_{-\infty}^{\infty} \bar{\rho}(\mathbf{r}, \omega)e^{j\omega t} \, d\omega \qquad (2.13-7)$$

$$\overline{\rho_m \mathbf{v}}(\mathbf{r}, t) = \frac{1}{2\pi} \int_{-\infty}^{\infty} \overline{\rho_m \mathbf{v}}(\mathbf{r}, \omega)e^{j\omega t} \, d\omega \qquad (2.13-8)$$

The associated inverse transforms are

$$\mathbf{E}(\mathbf{r}, \omega) = \int_{-\infty}^{\infty} \mathbf{E}(\mathbf{r}, t)e^{-j\omega t} \, dt \qquad (2.13-9)$$

$$\mathbf{B}(\mathbf{r}, \omega) = \int_{-\infty}^{\infty} \mathbf{B}(\mathbf{r}, t)e^{-j\omega t} \, dt \qquad (2.13-10)$$

$$\bar{\rho}(\mathbf{r}, \omega) = \int_{-\infty}^{\infty} \bar{\rho}(\mathbf{r}, t)e^{-j\omega t} \, dt \qquad (2.13-11)$$

$$\overline{\rho_m \mathbf{v}}(\mathbf{r}, \omega) = \int_{-\infty}^{\infty} \overline{\rho_m \mathbf{v}}(\mathbf{r}, t)e^{-j\omega t} \, dt \qquad (2.13-12)$$

The substitution of (2.13–5) to (2.13–8) in (2.13–1) to (2.13–4) leads to

$$\int_{-\infty}^{\infty} [\nabla \cdot \mathbf{E}(\mathbf{r}, \omega) - \epsilon_0^{-1}\overline{\rho}(\mathbf{r}, \omega)]e^{j\omega t}\, d\omega = 0 \quad (2.13\text{–}13)$$

$$\int_{-\infty}^{\infty} [\nabla \times \mathbf{E}(\mathbf{r}, \omega) + j\omega\mathbf{B}(\mathbf{r}, \omega)]e^{j\omega t}\, d\omega = 0 \quad (2.13\text{–}14)$$

$$\int_{-\infty}^{\infty} \{\nabla \times \mathbf{B}(\mathbf{r}, \omega) - \mu_0[\overline{\rho_m\mathbf{v}}(\mathbf{r}, \omega) - j\omega\epsilon_0\mathbf{E}(\mathbf{r}, \omega)]\}e^{j\omega t}\, d\omega = 0 \quad (2.13\text{–}15)$$

$$\int_{-\infty}^{\infty} \nabla \cdot \mathbf{B}(\mathbf{r}, \omega)e^{j\omega t}\, d\omega = 0 \quad (2.13\text{–}16)$$

If the Fourier transform of a function is zero, the function itself must be zero. Hence

$$\nabla \cdot \mathbf{E}(\mathbf{r}, \omega) = \epsilon_0^{-1}\overline{\rho}(\mathbf{r}, \omega) \qquad\qquad (2.13\text{–}17)$$

$$\nabla \times \mathbf{E}(\mathbf{r}, \omega) = -j\omega\mathbf{B}(\mathbf{r}, \omega) \qquad\qquad (2.13\text{–}18)$$

$$\nabla \times \mathbf{B}(\mathbf{r}, \omega) = \mu_0[\overline{\rho_m\mathbf{v}}(\mathbf{r}, \omega) + j\omega\epsilon_0\mathbf{E}(\mathbf{r}, \omega)] \qquad (2.13\text{–}19)$$

$$\nabla \cdot \mathbf{B}(\mathbf{r}, \omega) = 0 \qquad\qquad (2.13\text{–}20)$$

These are the time-independent Maxwell equations. The associated boundary conditions in shorthand form are

$$\hat{\mathbf{n}} \cdot \mathbf{E}(\mathbf{r}, \omega) = -\epsilon_0^{-1}\overline{\eta}(\mathbf{r}, \omega) \qquad\qquad (2.13\text{–}21)$$

$$\hat{\mathbf{n}} \times \mathbf{E}(\mathbf{r}, \omega) = 0 \qquad\qquad (2.13\text{–}22)$$

$$\hat{\mathbf{n}} \times \mathbf{B}(\mathbf{r}, \omega) = -\mu_0\overline{\eta_m\mathbf{v}}(\mathbf{r}, \omega) \qquad\qquad (2.13\text{–}23)$$

$$\hat{\mathbf{n}} \cdot \mathbf{B}(\mathbf{r}, \omega) = 0 \qquad\qquad (2.13\text{–}24)$$

If solutions of (2.13–17) to (2.13–20) are obtained as functions of the coordinates and the frequency, the time-dependent solution is obtained with the help of the Fourier transforms (2.13–5) to (2.13–8).

A very simple time dependence is obtained with the help of the delta function when

$$\mathbf{E}(\mathbf{r}, \omega) = \tfrac{1}{2}\mathbf{E}(\mathbf{r}, \omega_1) \int_{-\infty}^{\infty} e^{-j(\omega - \omega_1)t}\, dt + \tfrac{1}{2}\mathbf{E}^*(\mathbf{r}, \omega_1) \int_{-\infty}^{\infty} e^{-j(\omega + \omega_1)t}\, dt$$

$$= \pi[\mathbf{E}(\mathbf{r}, \omega_1)\delta(\omega - \omega_1) + \mathbf{E}^*(\mathbf{r}, \omega_1)\delta(\omega + \omega_1)] \qquad (2.13\text{–}25)$$

is substituted in the Fourier transform (2.13–5). Thus, with (2.13–25),

$$\mathbf{E}(\mathbf{r}, t) = \frac{1}{2\pi} \int_{-\infty}^{\infty} \mathbf{E}(\mathbf{r}, \omega)e^{j\omega t}\, d\omega = \tfrac{1}{2}\mathbf{E}(\mathbf{r}, \omega_1)e^{j\omega_1 t} + \tfrac{1}{2}\mathbf{E}^*(\mathbf{r}, \omega_1)e^{-j\omega_1 t}$$

or

$$\mathbf{E}(\mathbf{r}, t) = \text{Re } [\mathbf{E}(\mathbf{r}, \omega_1)e^{j\omega_1 t}] \tag{2.13-26}$$

It is readily verified that the substitution into (2.13–1) to (2.13–4) of the simple time dependence (2.13–26) with

$$\mathbf{B}(\mathbf{r}, t) = \text{Re } [\mathbf{B}(\mathbf{r}, \omega_1)e^{j\omega_1 t}] \tag{2.13-27}$$

$$\bar{\rho}(\mathbf{r}, t) = \text{Re } [\bar{\rho}(\mathbf{r}, \omega_1)e^{j\omega_1 t}] \tag{2.13-28}$$

$$\overline{\rho_m \mathbf{v}}(\mathbf{r}, t) = \text{Re } [\overline{\rho_m \mathbf{v}}(\mathbf{r}, \omega_1)e^{j\omega_1 t}] \tag{2.13-29}$$

yields (2.13–17) to (2.13–20) directly. In this case solutions of (2.13–17) to (2.13–20) subject to (2.13–21) to (2.13–24) need only be multiplied by $e^{j\omega_1 t}$ and the real part ,taken. This gives the solution of (2.13–1) to (2.13–4). Note that complex scalars like $\bar{\rho}(\mathbf{r}, \omega)$ are not vectors. In (2.13–28), $\bar{\rho}(\mathbf{r}, t)$ is a real instantaneous value. It is obtained from the complex scalar $\bar{\rho}(\mathbf{r}, \omega)$. This can be written as follows:

$$\bar{\rho}(\mathbf{r}, \omega) = \bar{\rho}_r(\mathbf{r}, \omega) + j\bar{\rho}_i(\mathbf{r}, \omega) = |\bar{\rho}(\mathbf{r}, \omega)| \, e^{j\theta_\rho(\mathbf{r},\omega)} \tag{2.13-30a}$$

where

$$|\bar{\rho}(\mathbf{r}, \omega)| = [\bar{\rho}_r^2(\mathbf{r}, \omega) + \bar{\rho}_i^2(\mathbf{r}, \omega)]^{1/2}; \quad \theta_\rho(\mathbf{r}, \omega) = \tan^{-1} \frac{\bar{\rho}_i(\mathbf{r}, \omega)}{\bar{\rho}_r(\mathbf{r}, \omega)} \tag{2.13-30b}$$

It follows from (2.13–28) that

$$\bar{\rho}(\mathbf{r}, t) = \text{Re } [|\bar{\rho}(\mathbf{r}, \omega)| \, e^{j[\omega_1 t + \theta_\rho(\mathbf{r},\omega)]}]$$

$$= |\bar{\rho}(\mathbf{r}, \omega)| \cos [\omega_1 t + \theta_\rho(\mathbf{r}, \omega)] \tag{2.13-31}$$

For a vector

$$\mathbf{E}(\mathbf{r}, \omega) = \mathbf{E}_r(\mathbf{r}, \omega) + j\mathbf{E}_i(\mathbf{r}, \omega) = |\mathbf{E}(\mathbf{r}, \omega)| \, e^{j\theta_E(\mathbf{r},\omega)} \tag{2.13-32}$$

$$|\mathbf{E}(\mathbf{r}, \omega)| = [\mathbf{E}_r^2(\mathbf{r}, \omega) + \mathbf{E}_i^2(\mathbf{r}, \omega)]^{1/2}; \quad \theta_E(\mathbf{r}, \omega) = \tan^{-1} \frac{\mathbf{E}_i(\mathbf{r}, \omega)}{\mathbf{E}_r(\mathbf{r}, \omega)} \tag{2.13-33}$$

In (2.13–32), $\mathbf{E}(\mathbf{r}, \omega)$ is a complex vector; $\mathbf{E}_r(\mathbf{r}, \omega)$ and $\mathbf{E}_i(\mathbf{r}, \omega)$ are real vectors. From (2.13–26),

$$\mathbf{E}(\mathbf{r}, t) = \text{Re}[|\mathbf{E}(\mathbf{r}, \omega)| \, e^{j[\omega_1 t + \theta_E(\mathbf{r}, \omega)]}]$$

$$= |\mathbf{E}(\mathbf{r}, \omega)| \cos [\omega_1 t + \theta_E(\mathbf{r}, \omega)] \tag{2.13-34}$$

In (2.13–33), $\mathbf{E}(\mathbf{r}, t)$ is the real instantaneous value. The quantity $|\mathbf{E}(\mathbf{r}, \omega)| \, e^{j[\omega_1 t + \theta_E(\mathbf{r}, \omega)]}$ is a complex instantaneous value.

Tabulation of the general time-independent equations:

Equations for interior:

$$\epsilon_0 \nabla \cdot \mathbf{E} = \bar{\rho} \tag{2.13-35a}$$

$$\nabla \times \mathbf{E} = -j\omega \mathbf{B} \tag{2.13-35b}$$

$$\mu_0^{-1} \nabla \times \mathbf{B} = \overline{\rho_m \mathbf{v}} + j\omega\epsilon_0 \mathbf{E} \tag{2.13-35c}$$

$$\nabla \cdot \mathbf{B} = 0 \tag{2.13-35d}$$

Boundary conditions:

$$\epsilon_0\, \hat{\mathbf{n}}_1 \cdot \mathbf{E}_1 + \epsilon_0 \hat{\mathbf{n}}_2 \cdot \mathbf{E}_2 = -\bar{\eta}_1 - \bar{\eta}_2 \tag{2.13-36a}$$

$$\hat{\mathbf{n}}_1 \times \mathbf{E}_1 + \hat{\mathbf{n}}_2 \times \mathbf{E}_2 = 0 \tag{2.13-36b}$$

$$\mu_0^{-1}(\hat{\mathbf{n}}_1 \times \mathbf{B}_1) + \mu_0^{-1}(\hat{\mathbf{n}}_2 \times \mathbf{B}_2) = -\overline{\eta_m \mathbf{v}_1} - \overline{\eta_m \mathbf{v}_2} \tag{2.13-36c}$$

$$\hat{\mathbf{n}}_1 \cdot \mathbf{B}_1 + \hat{\mathbf{n}}_2 \cdot \mathbf{B}_2 = 0 \tag{2.13-36d}$$

Density functions:

$$\bar{\rho} = \rho - \nabla \cdot \mathbf{P} \tag{2.13-37a}$$

$$\overline{\rho_m \mathbf{v}} = \mathbf{J} + \nabla \times \mathbf{M} + j\omega \mathbf{P} \tag{2.13-37b}$$

$$\bar{\eta} = \eta + \hat{\mathbf{n}} \cdot \mathbf{P} \tag{2.13-38a}$$

$$\overline{\eta_m \mathbf{v}} = \mathbf{K} - \hat{\mathbf{n}} \times \mathbf{M} \tag{2.13-38b}$$

Equations of continuity:

$$\nabla \cdot \overline{\rho_m \mathbf{v}} + j\omega\bar{\rho} = 0 \tag{2.13-39a}$$

$$\nabla \cdot (\overline{\eta_m \mathbf{v}_1} + \overline{\eta_m \mathbf{v}_2}) + j\omega(\bar{\eta}_1 + \bar{\eta}_2) - (\hat{\mathbf{n}}_1 \cdot \overline{\rho_m \mathbf{v}_1}) - (\hat{\mathbf{n}}_2 \cdot \overline{\rho_m \mathbf{v}_2}) = 0 \tag{2.13-39b}$$

2.14 COMPLEX FIELD EQUATIONS IN SIMPLE MEDIA

In Sec. 2.11 it is shown that Maxwell's time-dependent equations assume a simple form for homogeneous isotropic media in which the following linear relations obtain:

$$\mathbf{P}(\mathbf{r}, t) = \epsilon_0 \chi_e \mathbf{E}(\mathbf{r}, t) = \epsilon_0(\epsilon_r - 1)\mathbf{E}(\mathbf{r}, t) \tag{2.14-1}$$

$$-\mathbf{M}(\mathbf{r}, t) = \mu_0^{-1} \chi_m \mathbf{B}(\mathbf{r}, t) = \mu_0^{-1}(\mu_r^{-1} - 1)\mathbf{B}(\mathbf{r}, t) \tag{2.14-2}$$

$$\mathbf{J}_f(\mathbf{r}, t) = \sigma \mathbf{E}(\mathbf{r}, t) \tag{2.14-3}$$

In these relations the proportionality constants are assumed to be independent of the space coordinates and of the time. This means that any change in the electric or magnetic field is accompanied *instantaneously* by a proportional change in the volume densities of polarization, magnetization, and conduction current. If the field varies periodically, the related density varies proportionally and without time lag. Since this type of response is an idealization of what is observed in the physical world, it is necessary to provide a less restrictive theory. This can be accomplished quite easily for periodically varying fields.

Suppose that

$$\mathbf{P}(\mathbf{r}, t) = \epsilon_0 \chi_e \left[\mathbf{E}(\mathbf{r}, t) + C_1 \frac{\partial \mathbf{E}(\mathbf{r}, t)}{\partial t} + C_2 \frac{\partial^2 \mathbf{E}(\mathbf{r}, t)}{\partial t^2} + \cdots \right] \qquad (2.14\text{-}4)$$

$$-\mathbf{M}(\mathbf{r}, t) = \mu_0^{-1} \chi_m \left[\mathbf{B}(\mathbf{r}, t) + D_1 \frac{\partial \mathbf{B}(\mathbf{r}, t)}{\partial t} + D_2 \frac{\partial^2 \mathbf{B}(\mathbf{r}, t)}{\partial t^2} + \cdots \right] \qquad (2.14\text{-}5)$$

$$\mathbf{J}_f(\mathbf{r}, t) = \sigma \left[\mathbf{E}(\mathbf{r}, t) + S_1 \frac{\partial \mathbf{E}(\mathbf{r}, t)}{\partial t} + S_2 \frac{\partial^2 \mathbf{E}(\mathbf{r}, t)}{\partial t^2} + \cdots \right] \qquad (2.14\text{-}6)$$

It is assumed that χ_e, χ_m, and σ are constant in time and independent of the space coordinates. The electromagnetic vectors in (2.14-4) to (2.14-6) are arbitrary functions of the time. For periodic time $\mathbf{E}(\mathbf{r}, t) = \text{Re}\,[\mathbf{E}(\mathbf{r})e^{j\omega t}]$, $\mathbf{B}(\mathbf{r}, t) = \text{Re}\,[\mathbf{B}(\mathbf{r})e^{j\omega t}]$, where $\mathbf{E}(\mathbf{r})$ and $\mathbf{B}(\mathbf{r})$ are complex, it follows that

$$\mathbf{P}(\mathbf{r}, t) = \text{Re}\,[\mathbf{P}(\mathbf{r})e^{j\omega t}]; \qquad \mathbf{P}(\mathbf{r}) = \epsilon_0(\epsilon_r' - j\epsilon_r'' - 1)\mathbf{E}(\mathbf{r}) \qquad (2.14\text{-}7)$$

$$-\mathbf{M}(\mathbf{r}, t) = -\text{Re}\,[\mathbf{M}(\mathbf{r})e^{j\omega t}]; \qquad -\mathbf{M}(\mathbf{r}) = \mu_0^{-1}[(\mu_r' - j\mu_r'')^{-1} - 1]\mathbf{B}(\mathbf{r}) \qquad (2.14\text{-}8)$$

$$\mathbf{J}_f(\mathbf{r}, t) = \text{Re}\,[\mathbf{J}_f(\mathbf{r})e^{j\omega t}]; \qquad \mathbf{J}_f(\mathbf{r}) = (\sigma' - j\sigma'')\mathbf{E}(\mathbf{r}) \qquad (2.14\text{-}9)$$

where $\mathbf{P}(\mathbf{r})$, $\mathbf{M}(\mathbf{r})$, and $\mathbf{J}_f(\mathbf{r})$ are, in general, complex and where the now complex constitutive parameters are

$$\epsilon_0(\epsilon_r' - j\epsilon_r'') - \epsilon_0 = \epsilon_0(\epsilon_r - 1)(1 + j\omega C_1 - \omega^2 C_2 \cdots) \qquad (2.14\text{-}10)$$

$$\mu_0^{-1}(\mu_r' - j\mu_r'')^{-1} - \mu_0^{-1} = \mu_0^{-1}(\mu_r^{-1} - 1)(1 + j\omega D_1 - \omega^2 D_2 \cdots) \qquad (2.14\text{-}11)$$

$$\sigma' - j\sigma'' = \sigma(1 + j\omega S_1 - \omega^2 S_2 \cdots) \qquad (2.14\text{-}12)$$

Note that the complex auxiliary vectors are given by

$$\mathbf{D}(\mathbf{r}) = \epsilon_0 \mathbf{E}(\mathbf{r}) + \mathbf{P}(\mathbf{r}) = \epsilon_0(\epsilon_r' - j\epsilon_r'')\mathbf{E}(\mathbf{r}) \qquad (2.14\text{-}13)$$

$$\mathbf{H}(\mathbf{r}) = \mu_0^{-1}\mathbf{B}(\mathbf{r}) - \mathbf{M}(\mathbf{r}) = \mu_0^{-1}(\mu_r' - j\mu_r'')^{-1}\mathbf{B}(\mathbf{r}) \qquad (2.14\text{-}14)$$

Evidently, $\mathbf{D}(\mathbf{r})$ and $\mathbf{E}(\mathbf{r})$, $\mathbf{H}(\mathbf{r})$ and $\mathbf{B}(\mathbf{r})$ differ not only in amplitude but also in phase.

In general, the symbols ϵ, μ, and σ, which are necessarily real when used in the time-dependent equations, will be assumed to denote the following complex quantities whenever they occur in the time-independent equations with the assumed time dependence $e^{j\omega t}$:

$$\epsilon = \epsilon_0 \epsilon_r = \epsilon_0(\epsilon_r' - j\epsilon_r'')$$

$$\mu^{-1} = \mu_0^{-1}\mu_r^{-1} = \mu_0^{-1}(\mu_r' - j\mu_r'')^{-1} \qquad \sigma = \sigma' - j\sigma'' \qquad (2.14\text{-}15)$$

This dual meaning of the same symbols is in common use and should cause no confusion. The same convention also applies to the density functions and the field vectors for which the same symbols are used for the real quantities in the time-dependent equations and for the generally complex quantities in the time-independent equations. For simple media with the time dependence $e^{j\omega t}$, the latter are

$$\epsilon \nabla \cdot \mathbf{E} = \rho_f \qquad (2.14\text{-}16a)$$

$$\nabla \times \mathbf{E} = -j\omega \mathbf{B} \qquad (2.14\text{-}16b)$$

$$\mu^{-1}\nabla \times \mathbf{B} = (\sigma + j\omega\epsilon)\mathbf{E} \qquad (2.14\text{-}17a)$$

$$\nabla \cdot \mathbf{B} = 0 \qquad (2.14\text{-}17b)$$

The boundary conditions are

$$\epsilon_1(\hat{\mathbf{n}}_1 \cdot \mathbf{E}_1) + \epsilon_2(\hat{\mathbf{n}}_2 \cdot \mathbf{E}_2) = -\eta_{1f} - \eta_{2f} \qquad (2.14\text{-}18a)$$

$$(\hat{\mathbf{n}}_1 \times \mathbf{E}_1) + (\hat{\mathbf{n}}_2 \times \mathbf{E}_2) = 0 \qquad (2.14\text{-}18b)$$

$$\mu_1^{-1}(\hat{\mathbf{n}}_1 \times \mathbf{B}_1) + \mu_2^{-1}(\hat{\mathbf{n}}_2 \times \mathbf{B}_2) = -\mathbf{K}_{1f} - \mathbf{K}_{2f} \qquad (2.14\text{-}19a)$$

$$(\hat{\mathbf{n}}_1 \cdot \mathbf{B}_1) + (\hat{\mathbf{n}}_2 \cdot \mathbf{B}_2) = 0 \qquad (2.14\text{-}19b)$$

The equations of continuity for electric charge are

$$\sigma \nabla \cdot \mathbf{E} + j\omega\rho_f = 0 \qquad (2.14\text{-}20a)$$

$$\sigma_1(\hat{\mathbf{n}}_1 \cdot \mathbf{E}_1) + \sigma_2(\hat{\mathbf{n}}_2 \cdot \mathbf{E}_2) - \nabla \cdot (\mathbf{K}_{1f} + \mathbf{K}_{2f}) - j\omega(\eta_{1f} + \eta_{2f}) = 0 \qquad (2.14\text{-}20b)$$

The elimination of ρ_f from (2.14–16a) and (2.14–20a) leads to

$$\nabla \cdot \mathbf{E} = 0 \qquad (2.14\text{-}21a)$$

Similarly, with (2.14–18a) and (2.14–20b), and the absence of perfect conductors ($\mathbf{K}_f = 0$),

$$(\sigma_1 + j\omega\epsilon_1)(\hat{\mathbf{n}}_1 \cdot \mathbf{E}_1) + (\sigma_2 + j\omega\epsilon_2)(\hat{\mathbf{n}}_2 \cdot \mathbf{E}_2) = 0 \qquad (2.14\text{-}21b)$$

These equations apply to all simple media with a linear time lag.

The complex factor $\sigma + j\omega\epsilon$ which appears in (2.14–17a) and in (2.14–21b) can be expanded as follows:

$$\sigma + j\omega\epsilon = \sigma' - j\sigma'' + j\omega(\epsilon' - j\epsilon'')$$

$$= (\sigma' + \omega\epsilon'') + j\omega\left(\epsilon' - \frac{\sigma''}{\omega}\right) \qquad (2.14\text{-}22)$$

where the real quantities $\sigma' + \omega\epsilon''$ and $\epsilon' - \sigma''/\omega$ are, respectively, the real effective conductivity σ_e and the real effective permittivity, ϵ_e. They are

$$\sigma_e \equiv \sigma' + \omega\epsilon''; \qquad \epsilon_e = \epsilon_0\epsilon_{er} = \epsilon_0\left(\epsilon_r' - \frac{\sigma''}{\omega\epsilon_0}\right) \qquad (2.14\text{-}23)$$

With this notation

$$\sigma + j\omega\epsilon = \sigma_e + j\omega\epsilon_e = j\omega\epsilon_e(1 - jp_e) = j\omega\bar{\epsilon} \qquad (2.14\text{--}24a)$$

where

$$p_e \equiv \frac{\sigma_e}{\omega\epsilon_e} \qquad (2.14\text{--}24b)$$

is the *loss tangent* and

$$\bar{\epsilon} \equiv \epsilon - \frac{j\sigma}{\omega} = \epsilon_e - \frac{j\sigma_e}{\omega} \qquad (2.14\text{--}24c)$$

is a generalized permittivity that includes all properties relating to both polarization and conduction. The effective conductivity in (2.14–23) includes a term σ' due to free charge moving in an imperfect conductor and a term $\omega\epsilon''$ due to time lags in polarization in a dielectric. In good conductors, the latter is negligible; in good dielectrics operated at frequencies near those leading to molecular or atomic resonances or relaxation phenomena, the term $\omega\epsilon''$ is large and σ' usually small. The effective relative permittivity ϵ_{er} involves a term ϵ_r' due to ordinary polarization and a term $-\sigma''/\omega\epsilon_0$ due to a time lag in convection current response. In good dielectrics, the term $-\sigma''/\omega\epsilon_0$ is negligible compared with ϵ_r', so that the loss tangent becomes $p_e = \omega\epsilon''/\omega\epsilon' = \epsilon_r''/\epsilon_r'$, which is independent of the frequency unless ϵ_r' and ϵ_r'' are themselves functions of the frequency.

The real and imaginary parts of the generalized complex permittivity $\bar{\epsilon}$ as defined in (2.14–24c) are not independent. With $\bar{\epsilon} = \epsilon_r - j\epsilon_i$, where $\epsilon_r = \epsilon_e$, $\epsilon_i = \sigma_e/\omega$, the interrelation is contained in the following Kronig–Kramers relations:

$$\epsilon_r(\omega) - \epsilon_r(\infty) = \frac{2}{\pi} \int_0^\infty \frac{\omega'\epsilon_i(\omega')}{\omega'^2 - \omega^2} \, d\omega' \qquad (2.14\text{--}25a)$$

$$\epsilon_i(\omega) = -\frac{2\omega}{\pi} \int_0^\infty \frac{\epsilon_r(\omega') - \epsilon_r(\infty)}{\omega'^2 - \omega^2} \, d\omega' \qquad (2.14\text{--}25b)$$

In ionized gases or plasmas such as found in the ionosphere, the constitutive relation between the electric field and the volume density of current J_f due to moving electrons is given approximately by (2.14–9) with

$$\sigma' \sim \frac{n_e e^2}{m_e} \frac{v}{\omega^2 + v^2}; \qquad \sigma'' \sim \frac{n_e e^2}{m_e} \frac{\omega}{\omega^2 + v^2} \qquad (2.14\text{--}26)$$

where e is the charge and m_e the mass of the electron, n_e is the number density of electrons, and v is the collision frequency. When collisions are neglected, the following simple forms are obtained with (2.14–24c) in (2.14–23):

$$\sigma_e \sim 0; \qquad \epsilon_{er} \sim 1 - \frac{\sigma''}{\omega\epsilon_0} = 1 - \frac{e^2 n_e}{\omega^2 \epsilon_0 m_e} \qquad (2.14\text{--}27)$$

Evidently, depending on the frequency in $\omega = 2\pi f$, the real effective relative permittivity ϵ_{er} can be positive as in ordinary dielectrics and underdense plasmas, negative as in overdense plasmas, or zero. The zero value occurs at the *plasma frequency* $f_p = \omega_p/2\pi$, where

$$\omega_p = \left[\frac{e^2 n_e}{\epsilon_0 m_e}\right]^{1/2} \tag{2.14-28}$$

With (2.14–24c) the field and boundary equations become

$$\boldsymbol{\nabla} \cdot \mathbf{E} = 0 \tag{2.14-29a}$$

$$\boldsymbol{\nabla} \times \mathbf{E} = -j\omega\mathbf{B} \tag{2.14-29b}$$

$$\boldsymbol{\nabla} \times \mathbf{B} = j\omega\mu\tilde{\epsilon}\mathbf{E} \tag{2.14-30a}$$

$$\boldsymbol{\nabla} \cdot \mathbf{B} = 0 \tag{2.14-30b}$$

$$\tilde{\epsilon}_1(\hat{\mathbf{n}}_1 \cdot \mathbf{E}_1) + \tilde{\epsilon}_2(\hat{\mathbf{n}}_2 \cdot \mathbf{E}_2) = 0 \tag{2.14-31a}$$

$$\hat{\mathbf{n}}_1 \times \mathbf{E}_1 + \hat{\mathbf{n}}_2 \times \mathbf{E}_2 = 0 \tag{2.14-31b}$$

$$\mu_1^{-1}(\hat{\mathbf{n}}_1 \times \mathbf{B}_1) + \mu_2^{-1}(\hat{\mathbf{n}}_2 \times \mathbf{B}_2) = 0 \tag{2.14-32a}$$

$$\hat{\mathbf{n}}_1 \cdot \mathbf{B}_1 + \hat{\mathbf{n}}_2 \cdot \mathbf{B}_2 = 0 \tag{2.14-32b}$$

The associated second-order equations are

$$\boldsymbol{\nabla} \times \boldsymbol{\nabla} \times \mathbf{E} - k^2\mathbf{E} = 0 \tag{2.14-33}$$

$$\boldsymbol{\nabla} \times \boldsymbol{\nabla} \times \mathbf{B} - k^2\mathbf{B} = 0 \tag{2.14-34}$$

where k in

$$k^2 \equiv \omega^2\mu\tilde{\epsilon} \tag{2.14-35}$$

is the complex wave number.

2.15 IMPRESSED AND INTRINSIC ELECTRIC FIELDS

A steady, periodically varying or transient flow of electric charge in a simple medium requires at some location an active charge-separating region or generator. Such regions are usually localized and provided with terminals which are kept oppositely charged by internal forces that may, for example, be electrochemical, electromechanical, thermoelectric, or photoelectric. These forces can be represented by an equivalent impressed or intrinsic electric field $\mathbf{E}^e$ that is continuously maintained between the terminals in a direction opposite to that of the electric field due to the currents and charges. The volume density of current in such a region can be represented by the relation

$$\mathbf{J}_f(\mathbf{r}, t) = \sigma[\mathbf{E}(\mathbf{r}, t) + \mathbf{E}^e(\mathbf{r}, t)] \tag{2.15-1}$$

Outside the region, $E^e(\mathbf{r}, t) = 0$ and (2.15–1) reduces to (2.14–3). If the impressed field varies periodically, (2.15–1) becomes

$$\mathbf{J}_f(\mathbf{r}, t) = \text{Re } [\mathbf{J}_f(\mathbf{r})e^{j\omega t}] \qquad (2.15\text{–}2)$$

$$\mathbf{J}_f(\mathbf{r}) = \sigma[\mathbf{E}(\mathbf{r}) + \mathbf{E}^e(\mathbf{r})] \qquad (2.15\text{–}3)$$

where, in general, $\sigma = \sigma' - j\sigma''$. This generalized constitutive relation may be used instead of the simpler one (2.14–9) whenever it is necessary to indicate expressly the existence of an intrinsic or impressed electric field in a localized region.

It is often convenient to assume an electric current to be excited by one (or more) charge-separating regions maintained between adjacent surfaces across a conductor at specified cross sections. Such a region is called a *slice generator* or a *point generator* if the problem is one-dimensional. Such a generating region can be further idealized to be maintained across the infinitely close circular edges of adjoining tubular conductors. If these lie along the z axis, the impressed electric field is defined as follows:

$$E_z^e(z, t) = \delta(z)V^e(z, t) \qquad (2.15\text{–}4)$$

where $V^e(z, t)$ is an electromotive force (emf) and $\delta(z)$ is the delta function. Such a fictitious but theoretically useful generator is known as a δ-*function generator*. With the properties of the δ function, (2.15–4) gives

$$\int_{-h}^{h} E_z^e(z, t) \, dz = \int_{-h}^{h} \delta(z)V^e(z, t) \, dz = V^e(0, t) \qquad (2.15\text{–}5)$$

when integrated over the length of a conductor. Clearly, a time-varying emf with magnitude $V^e(0, t)$ is maintained at $z = 0$ along the conductor.

2.16 GENERALIZED COEFFICIENTS OF SIMPLE MEDIA

The complex quantity k which occurs in the second-order equations (2.14–33) and (2.14–34) is of great importance in electromagnetic theory. When the medium is air with $\bar{\epsilon} = \epsilon_0$, $\mu = \mu_0$, it reduces to

$$k_0 = \omega(\mu_0\epsilon_0)^{1/2} = \frac{\omega}{c} = \frac{2\pi f}{c} = \frac{2\pi}{\lambda_0} \qquad (2.16\text{–}1)$$

This follows with (2.4–16). Evidently, since c is a velocity equal to 3×10^8 m/s, f is the frequency in seconds and λ_0 is the wavelength in meters, it follows that k_0 is dimensionally a reciprocal length measured in meters^{-1}. It is known as the *wave number*.

When the medium is characterized by $\bar{\epsilon}$ and μ, the wave number is complex and given by

$$k = \omega(\mu\bar{\epsilon})^{1/2} = \beta - j\alpha = k_0(N - j\chi) \qquad (2.16\text{--}2)$$

where β is the real phase constant, α is the real attenuation constant, N is the real index of refraction, and χ is the real extinction coefficient. Corresponding to the velocity c in air, v is a phase velocity defined by

$$v = \frac{\omega}{\beta} = \frac{c}{N} \qquad (2.16\text{--}3)$$

To evaluate the real and imaginary parts of k, it is convenient to make use of the loss tangent $p_e = \sigma_e/\omega\epsilon_e$ and to assume that μ is real, which is generally true of nonferromagnetic media. Thus with

$$k_e \equiv k_0\epsilon_{er}^{1/2} = \omega(\mu\epsilon_e)^{1/2} \qquad (2.16\text{--}4)$$

it follows that

$$k = \beta - j\alpha = \begin{cases} k_e(1 - jp_e)^{1/2} = k_e[f(p_e) - jg(p_e)], & \epsilon_{er} > 0 \quad (2.16\text{--}5a) \\[2mm] (-j\omega\mu\sigma_e)^{1/2}, & \epsilon_{er} = 0 \\[2mm] -j|k_e|(1 + j|p_e|)^{1/2} = |k_e|[g(|p_e|) - jf(|p_e|)], & \epsilon_{er} < 0 \quad (2.16\text{--}5b) \end{cases}$$

The condition $\epsilon_{er} > 0$ applies to all solids and liquids and to *underdense plasmas* for which $0 < \epsilon_{er} < 1$. For *overdense plasmas*, $\epsilon_{er} < 0$. The intermediate condition, $\epsilon_{er} = 0$, applies to good conductors such as metals and to plasmas very near the plasma frequency. The functions $f(p)$ and $g(p)$ are defined as follows:

$$f(p) = [\tfrac{1}{2}(\sqrt{1 + p^2} + 1)]^{1/2} = \cosh\left(\tfrac{1}{2}\sinh^{-1} p\right) \qquad (2.16\text{--}6a)$$

$$g(p) = [\tfrac{1}{2}(\sqrt{1 + p^2} - 1)]^{1/2} = \sinh\left(\tfrac{1}{2}\sinh^{-1} p\right) \qquad (2.16\text{--}6b)$$

They are tabulated in Appendix II. Note that

$$\frac{\alpha}{\beta} = \frac{g(p_e)}{f(p_e)} < 1, \quad \epsilon_{er} > 0; \qquad \frac{\alpha}{\beta} = 1, \quad \epsilon_{er} = 0$$

$$\frac{\alpha}{\beta} = \frac{f(|p_e|)}{g(|p_e|)} > 1, \quad \epsilon_{er} < 0 \qquad (2.16\text{--}7)$$

For some purposes it is useful to define a characteristic velocity v_e and a characteristic or wave impedance ζ_e for a simple medium. Thus by analogy with the constants

$$c = (\mu_0\epsilon_0)^{-1/2} = 3 \times 10^8 \text{ m/s}$$

$$\zeta_0 = \left(\frac{\mu_0}{\epsilon_0}\right)^{1/2} = 120\pi \doteq 376.7 \text{ ohms}$$

it is possible to define

$$v_e = (\mu\epsilon_e)^{-1/2} = c(\mu_r\epsilon_{er})^{-1/2} \tag{2.16-8}$$

$$\zeta_e = \left(\frac{\mu}{\epsilon_e}\right)^{1/2} = \zeta_0\left(\frac{\mu_r}{\epsilon_{er}}\right)^{1/2} \tag{2.16-9}$$

These constants play a similar role in electromagnetic phenomena in linear dielectrics as do c and ζ_0 in space.

Poor conductors are defined by

$$p_e^2 \ll 1 \tag{2.16-10}$$

so that

$$f(p_e) \doteq 1; \qquad g(p_e) \doteq \frac{p_e}{2} \tag{2.16-11}$$

$$\beta \doteq k_e = \frac{\omega}{v_e}; \qquad v \doteq v_e \tag{2.16-12}$$

$$\alpha \doteq k_e\frac{\sigma_e}{2\omega\epsilon_e} = \tfrac{1}{2}\sigma_e\zeta_e \tag{2.16-13}$$

Note that

$$p_e = \frac{2\alpha}{\beta} \tag{2.16-14}$$

so that

$$\left(\frac{\alpha}{\beta}\right)^2 \ll 1 \tag{2.16-15}$$

Good conductors are defined by

$$p_e \gg 1 \tag{2.16-16}$$

and

$$\epsilon_{er} = 0 \tag{2.16-17}$$

so that

$$f(p_e) \doteq g(p_e) \doteq \left(\frac{p_e}{2}\right)^{1/2} \tag{2.16-18}$$

and

$$\beta \doteq \alpha \doteq k_e\left(\frac{p_e}{2}\right)^{1/2} = \left(\frac{\omega\sigma_e\mu}{2}\right)^{1/2} \tag{2.16-19}$$

The reciprocal of α is dimensionally a length that is called the *skin depth* or *skin thickness* in good conductors. It is

$$d_s = \left(\frac{2}{\omega\mu\sigma_e}\right)^{1/2} \tag{2.16–20}$$

In terms of the skin depth,

$$\beta \sim \alpha \sim \frac{1}{d_s} \tag{2.16–21}$$

2.17 AN APPLICATION OF THE MAXWELL–LORENTZ EQUATIONS: THE REFLECTION AND TRANSMISSION OF PLANE ELECTROMAGNETIC WAVES AT THE SURFACE OF THE EARTH

An instructive and useful application of the field equations is to the determination of the electromagnetic field in the earth when a plane wave is normally incident on its surface. It is to be anticipated that the incident wave will be partly reflected back into the air, partly transmitted into the earth. As shown in Fig. 2.17–1, the upper half-space (region 0, air) is characterized by the wave number $k_0 = \omega/c$, where $c = (\mu_0\epsilon_0)^{-1/2} \sim 3 \times 10^8$ m/sec is the velocity of electromagnetic waves (including light). The wave impedance in air is $\zeta_0 = (\mu_0/\epsilon_0)^{1/2} \sim 120\pi$ ohms. The lower half-space (region 1, earth) has the complex wave number

$$k_1 = \beta_1 - j\alpha_1 = k_0[\epsilon_{1er}(1 - jp_{1e})]^{1/2} = \omega(\mu_1\bar{\epsilon}_1)^{1/2} \tag{2.17–1}$$

The generalized complex permittivity is $\bar{\epsilon}_1 = \epsilon_{e1} - j\sigma_{e1}/\omega$, ϵ_{e1} is the real effective permittivity, σ_{e1} the real effective conductivity, and $p_{e1} = \sigma_{e1}/\omega\epsilon_{e1}$ the real effective loss tangent. In the following the subscript e is omitted for simplicity. It is assumed that $\mu_1 = \mu_0$. The real phase constant β_1 and attenuation constant α_1 can be

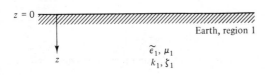

Air, region 0
ϵ_0, μ_0
k_0, ζ_0

$z = 0$

Earth, region 1

z

$\tilde{\epsilon}_1$, μ_1
k_1, ζ_1

Figure 2.17–1   Air–earth half-spaces.

expressed in terms of the functions $f(p_1)$ and $g(p_1)$ defined in (2.16–6a) and (2.16–6b). With (2.16–4) and (2.16–5a)

$$\beta_1 = k_0 \sqrt{\epsilon_{1r}} f(p_1) = \frac{\sigma_1 \zeta_0}{\sqrt{\epsilon_{1r}}} \left[\frac{f(p_1)}{p_1} \right] \tag{2.17–2}$$

$$\alpha_1 = k_0 \sqrt{\epsilon_{1r}} g(p_1) = \frac{\sigma_1 \zeta_0}{\sqrt{\epsilon_{1r}}} \left[\frac{g(p_1)}{p_1} \right] \tag{2.17–3}$$

Graphs of $f(p)$, $g(p)$, $f(p)/p$, and $g(p)/p$ as functions of p are given in Fig. 2.17–2. The wave impedance of the medium is

$$\zeta_1 = \frac{\zeta_0}{\sqrt{\epsilon_{1r}} (1 - jp_1)^{1/2}} = \frac{\zeta_0 [f(p_1) + jg(p_1)]}{\sqrt{\epsilon_{1r}} (1 + p_1^2)^{1/2}} \tag{2.17–4}$$

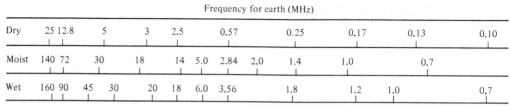

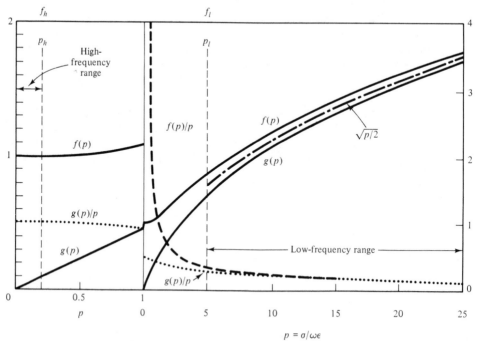

Figure 2.17–2 Functions $f(p)$ and $g(p)$ and related quantities.

The plane bounding the two regions is defined by $z = 0$ with the positive z axis directed downward into the earth. A plane electromagnetic wave normally incident in air on the boundary surface is represented by

$$E_x^i(z, t) = \text{Re } [E_x^i(z, \omega)e^{j\omega t}]$$

$$= \text{Re } [E_x^i(0, \omega)e^{j\omega(t - z/c)}] \tag{2.17–5a}$$

The complex amplitude is

$$E_x^i(z, \omega) = E_x^i(0, \omega)e^{-jk_0 z} \tag{2.17–5b}$$

$E_x^i(0, \omega)$ is the magnitude of the incident electric field at the surface ($z = 0$). The associated incident magnetic field has the complex amplitude

$$B_y^i(z, \omega) = \frac{E_x^i(z, \omega)}{c} = \frac{k_0}{\omega} E_x^i(z, \omega) \tag{2.17–5c}$$

The reflected wave has the real instantaneous value

$$E_x^r(z, t) = \text{Re } [E_x^r(z, \omega)e^{j\omega t}]$$

$$= \text{Re } [E_x^r(0, \omega)e^{j\omega(t + z/c)}] \tag{2.17–6a}$$

The complex amplitude is

$$E_x^r(z, \omega) = E_x^r(0, \omega)e^{+jk_0 z} \tag{2.17–6b}$$

since the reflected wave travels in the negative z direction. The associated reflected magnetic field has the complex amplitude

$$B_y^r(z, \omega) = -\frac{E_x^r(z, \omega)}{c} = -\frac{k_0}{\omega} E_x^r(z, \omega) \tag{2.17–6c}$$

Equation (2.17–6c) is easily obtained from the field equations.

The transmitted wave has the real instantaneous value

$$E_x^t(z, t) = \text{Re } [E_x^t(z, \omega)e^{j\omega t}]$$

$$= \text{Re } [E_x^t(0, \omega)e^{j(\omega t - k_1 z)}] \tag{2.17–7a}$$

The complex amplitude is therefore

$$E_x^t(z, \omega) = E_x^t(0, \omega)e^{-jk_1 z} \tag{2.17–7b}$$

The associated transmitted magnetic field has the complex amplitude

$$B_y^t(z, \omega) = \frac{k_1}{\omega} E_x^t(z, \omega) \tag{2.17–7c}$$

The boundary between the two regions is at the plane defined by $z = 0$. In the absence of surface distributions of charge or current on the boundary, the boundary conditions (2.2–20b,c) may be used in the form

$$\hat{n}_1 \times \mathbf{E}_1 + \hat{n}_2 \times \mathbf{E}_2 = 0 \tag{2.17–8a}$$

$$\hat{n}_1 \times \mathbf{B}_1 + \hat{n}_2 \times \mathbf{B}_2 = 0 \tag{2.17–8b}$$

at $z = 0$. These lead to the following equations:

$$E_x^i(0, \omega) + E_x^r(0, \omega) = E_x^t(0, \omega) \tag{2.17–9a}$$

$$B_y^i(0, \omega) + B_y^r(0, \omega) = B_y^t(0, \omega) \tag{2.17–9b}$$

With (2.17–5c), (2.17–6c) and (2.17–7c), (2.17–9b) is reduced to

$$k_0[E_x^i(0, \omega) - E_x^r(0, \omega)] = k_1 E_x^t(0, \omega) \tag{2.17–9c}$$

The solutions of these equations are

$$E_x^r(0, \omega) = E_x^i(0, \omega) \frac{k_0 - k_1}{k_0 + k_1} \tag{2.17–10a}$$

$$E_x^t(0, \omega) = E_x^i(0, \omega) \frac{2k_0}{k_0 + k_1} \tag{2.17–10b}$$

The real instantaneous values of the reflected and transmitted waves follow:

$$E_x^r(z, t) = E_x^i(0, \omega) \, |\Gamma_r| \cos(\omega t + k_0 z + \psi_r) \tag{2.17–11a}$$

where the complex reflection coefficient for the electric field is

$$\Gamma_r = \frac{k_0 - k_1}{k_0 + k_1} = |\Gamma_r| \, e^{j\psi_r} \tag{2.17–11b}$$

and

$$E_x^t(z, t) = E_x^i(0, \omega) \, |\Gamma_t| \, e^{-\alpha_1 z} \cos(\omega t - \beta_1 z + \psi_t) \tag{2.17–12a}$$

where the complex transmission coefficient for the electric field is

$$\Gamma_t = \frac{2k_0}{k_0 + k_1} = |\Gamma_t| \, e^{j\psi_t} \tag{2.17–12b}$$

Equation (2.17–11a) represents a plane wave traveling upward after reflection. Equation (2.17–12a) represents a plane wave traveling downward into the earth. It has a real phase velocity $v_p = \omega/\beta_1$ and an amplitude that is exponentially attenuated.

With (2.17–1) to (2.17–3) it is readily shown that in general

$$\Gamma_r = \frac{1 - \sqrt{\epsilon_{1r}} \, [f(p_1) - jg(p_1)]}{1 + \sqrt{\epsilon_{1r}} \, [f(p_1) - jg(p_1)]} \tag{2.17–13a}$$

$$\Gamma_t = \frac{2}{1 + \sqrt{\epsilon_{1r}} \, [f(p_1) - jg(p_1)]} \tag{2.17–13b}$$

It is seen that the relation $E_x^i(0, \omega) = E_x^t(0, \omega) - E_x^r(0, \omega)$ is now satisfied by (2.17–13a,b) since it can be reduced to the form

$$1 = \Gamma_t - \Gamma_r \qquad (2.17\text{–}14)$$

The amplitudes and phase angles of the reflection and transmission coefficients are

$$|\Gamma_r| = \frac{\{[1 - \epsilon_{1r}f^2(p_1) - \epsilon_{1r}g^2(p_1)]^2 + 4\epsilon_{1r}g^2(p_1)\}^{1/2}}{\{[1 + \sqrt{\epsilon_{1r}}\,f(p_1)]^2 + \epsilon_{1r}g^2(p_1)\}^{1/2}} \qquad (2.17\text{–}15a)$$

$$\psi_r = \tan^{-1}\frac{2\sqrt{\epsilon_{1r}}\,g(p_1)}{[1 - \epsilon_{1r}f^2(p_1) - \epsilon_{1r}g^2(p_1)]} \qquad (2.17\text{–}15b)$$

and

$$|\Gamma_t| = \frac{2}{\{[1 + \sqrt{\epsilon_{1r}}\,f(p_1)]^2 + \epsilon_{1r}g^2(p_1)\}^{1/2}} \qquad (2.17\text{–}16a)$$

$$\psi_t = \tan^{-1}\frac{\sqrt{\epsilon_{1r}}\,g(p_1)}{1 + \sqrt{\epsilon_{1r}}\,f(p_1)} \qquad (2.17\text{–}16b)$$

The attenuation of the field as the wave travels in the conducting earth is of particular practical interest. The ratio of the field at depth $z = d$ to the transmitted field at the surface is given by (2.17–7b)

$$\frac{E_x^t(d, \omega)}{E_x^t(0, \omega)} = e^{-jk_1 d} = e^{-\alpha_1 d}e^{-j\beta_1 d} \qquad (2.17\text{–}17)$$

The magnitude of the ratio, which is simply $e^{-\alpha_1 d}$, is shown in Fig. 2.17–3 as a function of frequency in megahertz with a depth d in meters as the parameter for three different types of earth.

The ratios of transmitted or reflected fields at arbitrary distances d from the surface to the incident field at the surface are of interest. The ratios $E_x^t(d, \omega)/E_x^i(0, \omega)$ and $E_x^r(d, \omega)/E_x^i(0, \omega)$ are known as the steady-state transfer functions. They satisfy the equations

$$G_t(d, \omega) = G_R(d, \omega) + jG_I(d, \omega) \qquad (2.17\text{–}18a)$$

$$= |G_t(d, \omega)|\, e^{j\Phi_t(d,\omega)} \qquad (2.17\text{–}18b)$$

$$= |\Gamma_t|\, e^{j(\psi_t - k_1 d)} \qquad (2.17\text{–}18c)$$

$$G_r(d, \omega) = |G_r(d, \omega)|\, e^{j\Phi_r(d,\omega)} = |\Gamma_r|\, e^{j(\psi_r + k_0 d)} \qquad (2.17\text{–}19a)$$

$$G_i(d, \omega) = e^{-jk_0 d} \qquad (2.17\text{–}19b)$$

The steady-state properties of both the transmitted and reflected waves are determined by the complex transfer functions $G_t(d, \omega)$ and $G_r(d, \omega)$. These functions are now examined in the low- and high-frequency ranges.

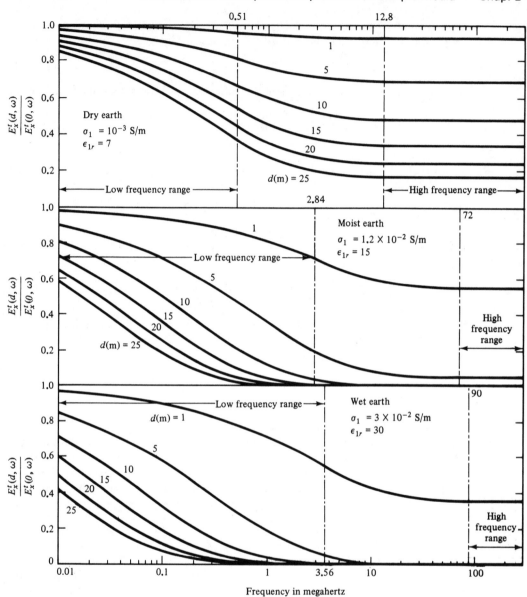

Figure 2.17–3 Magnitude of the ratio of the electric field at depth d (meters) to the electric field at the surface as a function of the frequency.

The low-frequency range is defined by the inequality

$$p_1^2 \gg 1 \qquad \text{or} \qquad p_1^2 \geq 25 \qquad\qquad (2.17\text{–}20a)$$

which is equivalent to

$$\omega \le \frac{0.2\sigma_1}{\epsilon_1} \quad \text{or} \quad f \le f_l \tag{2.17-20b}$$

where

$$f_l = 3560\left(\frac{\sigma_1}{\epsilon_{1r}}\right) \quad \text{MHz} \tag{2.17-20c}$$

is the upper limit of the low-frequency range. In this range it is readily shown that $f(p_1) \doteq g(p_1) = \sqrt{p_1/2}$ (Fig. 2.17–2). Then

$$\sqrt{\epsilon_{1r}}\, f(p_1) \doteq \sqrt{\epsilon_{1r}}\, g(p_1) = \left(\frac{\sigma_1}{2\omega\epsilon_0}\right)^{1/2} \tag{2.17-21}$$

and

$$|\Gamma_t| \doteq 2\sqrt{\frac{\omega\epsilon_0}{\sigma_1}}; \quad \psi_t \doteq \tan^{-1}\frac{(\sigma_1/2\omega\epsilon_0)^{1/2}}{1 + (\sigma_1/2\omega\epsilon_0)^{1/2}} \doteq \frac{\pi}{4} \tag{2.17-22}$$

$$|\Gamma_r| \doteq 1; \quad \psi_r = \tan^{-1}\left[-2\left(\frac{2\omega\epsilon_0}{\sigma_1}\right)^{1/2}\right] \tag{2.17-23}$$

It is easily seen that the transmission coefficient is zero at zero frequency and the incident wave is totally reflected. Furthermore,

$$\beta_1 \doteq \alpha_1 \doteq k_0\left(\frac{\epsilon_{1r}p_1}{2}\right)^{1/2} = \left(\frac{\omega\mu_0\sigma_1}{2}\right)^{1/2} = \frac{1}{d_s} \tag{2.17-24}$$

where d_s is the skin depth. It follows that, in the low-frequency range,

$$|G_t(d, \omega)| \doteq 2\left(\frac{\omega\epsilon_0}{\sigma_1}\right)^{1/2} \exp[-d(\omega\mu_0\sigma_1/2)^{1/2}] \tag{2.17-25a}$$

$$\psi_t - \beta_1 d = \Phi_t(d, \omega) \doteq \frac{\pi}{4} - d\left(\frac{\omega\mu_0\sigma_1}{2}\right)^{1/2} \tag{2.17-25b}$$

It is seen from (2.17–25a) that $|G_t(d, \omega)|$ increases with frequency owing to the term $\omega^{1/2}$ and simultaneously decreases due to the exponential attenuation. The maximum of $|G_t(d, \omega)|$ (provided that it exists in the specified low-frequency range) occurs when

$$d\left(\frac{\omega\mu_0\sigma_1}{2}\right)^{1/2} = 1 \quad \text{or} \quad f = \frac{1}{\pi d^2 \mu_0 \sigma_1} \tag{2.17-26}$$

The associated maximum value is

$$|G_t(d,\,\omega)|_{max} = \frac{2}{d\sigma_1}\left(\frac{2\epsilon_0}{\mu_0}\right)^{1/2} e^{-1} = \frac{2.76}{d\sigma_1} \times 10^{-3} \qquad (2.17\text{--}27)$$

These relations are valid only in the range $0 \le f \le f_l$.

The high frequency range is defined by the inequality

$$p_1^2 \ll 1; \qquad p_1^2 \le 0.04 \qquad (2.17\text{--}28a)$$

which is equivalent to

$$\omega \ge \frac{5\sigma_1}{\epsilon_1} \qquad \text{or} \qquad f \ge f_h \qquad (2.17\text{--}28b)$$

where

$$f_h = 9 \times 10^4 \left(\frac{\sigma_1}{\epsilon_{1r}}\right) \qquad \text{MHz} \qquad (2.17\text{--}28c)$$

is the lower limit of the high-frequency range. In this range, $f(p_1) \doteq 1$, $g(p_1) \doteq p_1/2 = \sigma_1/2\omega\epsilon_1$ (Fig. 2.17–2), so that

$$|\Gamma_t| \doteq \frac{2}{1 + \sqrt{\epsilon_{1r}}}; \qquad \psi_t \doteq \frac{\sigma_1}{2\omega\epsilon_0\,(\sqrt{\epsilon_{1r}} + \epsilon_{1r})} \ll 1 \qquad (2.17\text{--}29)$$

$$|\Gamma_r| \doteq \frac{1 - \epsilon_{1r}}{1 + \sqrt{\epsilon_{1r}}}; \qquad \psi_r \doteq \tan^{-1}\frac{\sigma_1}{2\omega\epsilon_0\,\sqrt{\epsilon_{1r}}\,(1 - \epsilon_{1r})} \qquad (2.17\text{--}30)$$

$$\beta_1 \doteq \omega(\mu_0\epsilon_1)^{1/2}; \qquad \alpha_1 \doteq \frac{\sigma_1}{2}\left(\frac{\mu_0}{\epsilon_1}\right)^{1/2} \qquad (2.17\text{--}31)$$

In this range, the attenuation constant α_1 is independent of the frequency and the wave number β_1 is linear in the frequency. The function $|G_t(d,\,\omega)|$ is given by

$$|G_t(d,\,\omega)| \doteq \frac{2}{1 + \sqrt{\epsilon_{1r}}} \exp[(-\sigma_1 d/2)(\mu_0/\epsilon_1)^{1/2}] \qquad (2.17\text{--}32)$$

$$\psi_t - \beta_1 d = \Phi_t(d,\,\omega)$$

$$\doteq \frac{\sigma_1}{2\omega\epsilon_0\,(\sqrt{\epsilon_{1r}} - \epsilon_{1r})} - \omega d(\mu_0\epsilon_1)^{1/2}$$

$$\doteq -\omega d(\mu_0\epsilon_1)^{1/2} \qquad (2.17\text{--}33)$$

At sufficiently high frequencies, the magnitude of the transfer function for transmission is a constant independent of the frequency.

The behavior of the steady-state transfer function (transmission) has been evaluated numerically for three practically important media:

1. Dry earth

$$\epsilon_{1r} = 7; \quad \sigma_1 = 10^{-3} \text{ S/m}; \quad p_1 = 2.58 \times 10^6/f$$

$$f_l = 0.51 \text{ MHz}; \quad f_h = 12.8 \text{ MHz}$$

2. Moist earth

$$\epsilon_{1r} = 15; \quad \sigma_1 = 1.2 \times 10^{-2} \text{ S/m}; \quad p_1 = 14.4 \times 10^6/f$$

$$f_l = 2.84 \text{ MHz}; \quad f_h = 72 \text{ MHz}$$

3. Wet earth

$$\epsilon_{1r} = 30; \quad \sigma_1 = 3 \times 10^{-2} \text{ S/m}; \quad p_1 = 18.0 \times 10^6/f$$

$$f_l = 3.56 \text{ MHz}; \quad f_h = 90 \text{ MHz}$$

The general formulas must be used for the intermediate range. Figures 2.17–4, 2.17–5, and 2.17–6 show the behavior of the transmitted field for these cases.

2.18 AN APPLICATION OF THE MAXWELL–LORENTZ FIELD EQUATIONS: THE REFLECTION AND TRANSMISSION OF A PLANE ELECTROMAGNETIC PULSE AT THE SURFACE OF THE EARTH

If the electromagnetic disturbance normally incident on the surface of the earth is a pulse with a planar front, the spectrum of frequencies involved in the reflection and transmission extends from zero to infinity. Depending on the distribution of frequencies in this incident pulse and the electrical properties of the earth, it is partly reflected and partly transmitted. A description of the transmitted pulse at various depths is of interest in conjunction with the shielding characteristics of the earth. These are of two types: (1) reflection at the surface so that only a fraction of the incident field enters the earth, and (2) attenuation with depth of the part that is transmitted. Since the frequency characteristics of reflection at the surface and attenuation in the course of transmission are quite different, it may be anticipated that the shape of the pulse transmitted into the earth is quite different from that of the incident pulse.

Let the electromagnetic field incident on the surface of the earth at $z = 0$ consist of a simple Gaussian pulse with the form

$$E_x^i(z, t) = E_x^i(0, 0) \exp\left[-\frac{(t - z/c)^2}{2t_1^2}\right] \qquad (2.18–1)$$

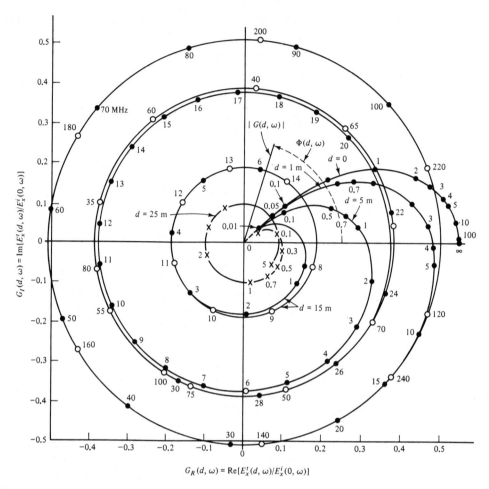

Figure 2.17–4 Transfer function $G_r(d, \omega) = E_x^t(d, \omega)/E_z^i(0,\omega)$ for dry earth ($\varepsilon_r = 7$, $\sigma = 10^{-3}$ S/m) in the complex plane. The parameter is the frequency in megahertz. The depth d is in meters.

Here t_1 is a constant parameter that is a measure of the pulse width in time and $z_1 = ct_1$ is the corresponding pulse width in air. The distribution (2.18–1) is illustrated in Fig. 2.18–1. At $z = 0$, this gives $E_x^i(0, t)/E_x^i(0, 0)$ as a function of time; at $t = 0$, it gives $E_x^i(z, 0)/E_x^i(0, 0)$ as a function of the normalized distance z/c. At $t = t_1$, the pulse is reduced to $e^{-1/2} = 0.606$ of its maximum at $t = 0$. The pulse is reduced to half-amplitude in the time $1.176t_1$.

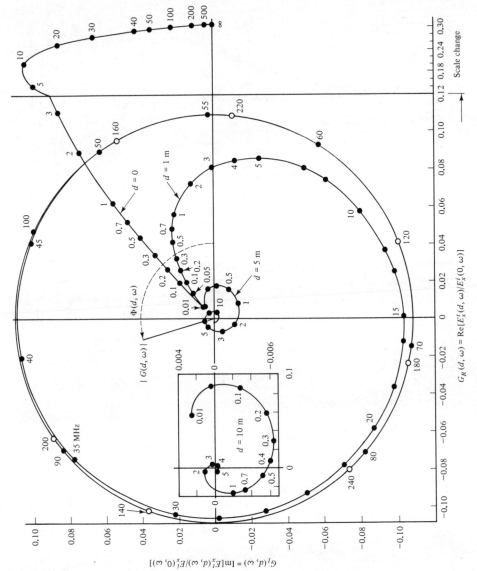

Figure 2.17–5 Like Fig. 2.17-4 for wet earth ($\varepsilon_r = 30$, $\sigma = 3 \times 10^{-2}$ S/m).

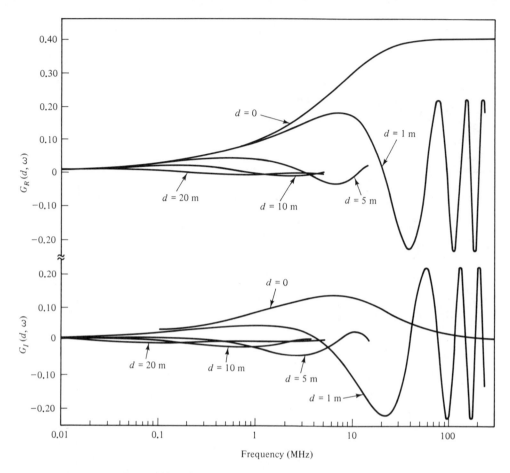

Figure 2.17–6 Real and imaginary parts of transfer function $G(d, \omega) = E_x^t(d, \omega)/E_x^i(0, \omega)$ for moist earth ($\varepsilon_r = 15$, $\sigma = 1.2 \times 10^{-2}$ S/m).

The spectrum contained in (2.18–1) is given by the Fourier transform

$$E_x^i(z, \omega) = \int_{-\infty}^{\infty} E_x^i(z, t)e^{-j\omega t}\, dt \qquad (2.18\text{–}2a)$$

$$= E_x^i(0, 0) \int_{-\infty}^{\infty} \exp\left[-\frac{(t - z/c)^2}{2t_1^2} \right] e^{-j\omega t}\, dt \qquad (2.18\text{–}2b)$$

The integral in (2.18–2b) may be evaluated using a table of integral transforms. It is

$$E_x^i(z, \omega) = E_x^i(0, 0) \exp\left(\frac{-j\omega z}{c} \right) t_1(2\pi)^{1/2} \exp\left(\frac{-\omega^2 t_1^2}{2} \right) \qquad (2.18\text{–}2c)$$

At $z = 0$, that is, at the boundary between air and the dissipative medium,

$$E_x^i(0, \omega) = E_x^i(0, 0)t_1(2\pi)^{1/2} \exp\left(\frac{-\omega^2 t_1^2}{2}\right) \qquad (2.18\text{--}2d)$$

The reflected and transmitted fields are given by

$$E_x^r(z, \omega) = E_x^i(0, \omega)G_r(z, \omega) \qquad (2.18\text{--}3a)$$

$$E_x^t(z, \omega) = E_x^i(0, \omega)G_t(z, \omega) \qquad (2.18\text{--}3b)$$

where $G_r(z, \omega)$ and $G_t(z, \omega)$ are the steady-state transfer functions defined in (2.17–19a) and (2.17–18c). The instantaneous fields at a distance z due to the entire Gaussian pulse are obtained by an inverse Fourier transform:

$$E_x^r(z, t) = \frac{1}{2\pi} \int_{-\infty}^{\infty} E_x^r(z, \omega)e^{j\omega t}\, d\omega$$

$$= \frac{1}{2\pi} \int_{-\infty}^{\infty} E_x^i(0, \omega)G_r(z, \omega)e^{j\omega t}\, d\omega \qquad (2.18\text{--}4a)$$

$$E_x^t(z, t) = \frac{1}{2\pi} \int_{-\infty}^{\infty} E_x^t(z, \omega)e^{j\omega t}\, d\omega$$

$$= \frac{1}{2\pi} \int_{-\infty}^{\infty} E_x^i(0, \omega)G_t(z, \omega)e^{j\omega t}\, d\omega \qquad (2.18\text{--}4b)$$

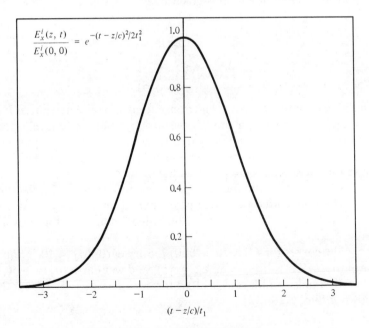

Figure 2.18–1 Incident Gaussian pulse.

$E_x^r(z, t)$ and $E_x^t(z, t)$ as well as $E_x^i(0, \omega)$ are real quantities. Hence the integrals may be written as follows:

$$E_x^r(z, t) = \frac{1}{2\pi} \int_{-\infty}^{\infty} E_x^i(0, \omega) |G_r(z, \omega)| \cos[\omega t + \Phi_r(z, \omega)] d\omega \quad (2.18\text{–}5a)$$

$$E_x^t(z, t) = \frac{1}{2\pi} \int_{-\infty}^{\infty} E_x^i(0, \omega) |G_t(z, \omega)| \cos[\omega t + \Phi_t(z, \omega)] d\omega \quad (2.18\text{–}5b)$$

Since the integral is even in ω and, with (2.18–2d), the reflected and transmitted fields become

$$E_x^r(z, t) = 2E_x^i(0, 0) \frac{t_1}{(2\pi)^{1/2}} \left[\int_0^{\infty} |G_r(z, \omega)| \cos \Phi_r(z, \omega) \exp\left(\frac{-\omega^2 t_1^2}{2}\right) \right.$$

$$\left. \times \cos \omega t \, d\omega - \int_0^{\infty} |G_r(z, \omega)| \sin \Phi_r(z, \omega) \exp\left(\frac{-\omega^2 t_1^2}{2}\right) \sin \omega t \, d\omega \right] \quad (2.18\text{–}6a)$$

$$E_x^t(z, t) = 2E_x^i(0, 0) \frac{t_1}{(2\pi)^{1/2}} \left[\int_0^{\infty} |G_t(z, \omega)| \cos \Phi_t(z, \omega) \exp\left(\frac{-\omega^2 t_1^2}{2}\right) \right.$$

$$\left. \times \cos \omega t \, d\omega - \int_0^{\infty} |G_t(z, \omega)| \sin \Phi_t(z, \omega) \exp\left(\frac{-\omega^2 t_1^2}{2}\right) \sin \omega t \, d\omega \right] \quad (2.18\text{–}6b)$$

The incident field at $z = 0$ is given by

$$E_x^i(0, t) = E_x^i(0, 0) \frac{t_1}{(2\pi)^{1/2}} \int_{-\infty}^{\infty} \exp\left(\frac{-\omega^2 t_1^2}{2}\right) \cos \omega t \, d\omega$$

$$= E_x^i(0, 0) \exp\left(\frac{-t^2}{2t_1^2}\right) \quad (2.18\text{–}7)$$

This is an even function of t. The transmitted and reflected pulses have both positive and negative parts even though the incident pulse is entirely positive. Since the field is totally reflected at zero frequency, the transmitted pulse must have equal positive and negative parts so that the average is zero. The shape of the transmitted pulse for $z \geq 0$ differs greatly from the incident pulse at $z \leq 0$. The reflected pulse at $z = 0$ resembles the incident pulse since there is total reflection at $\omega = 0$.

The transmitted field may be evaluated numerically for different values of z by taking the upper limit at $\omega_c = 2.6t_1$ instead of infinity; this introduces a negligible error.

The curves for $E_x^t(z, t)/E_x^i(0, 0)$ for dry earth ($\sigma_1 = 10^{-3}$ S/m, $\epsilon_{1r} = 7$), moist earth ($\sigma_1 = 1.2 \times 10^{-2}$ S/m, $\epsilon_{1r} = 15$), and wet earth ($\sigma_1 = 3 \times 10^{-2}$ S/m, $\epsilon_{1r} = 30$) with a pulse width of $t_1 = 1$ μs for $z = d = 0, 1, 5, 10, 15, 20,$ and 25 m are shown in Fig. 2.18–2.

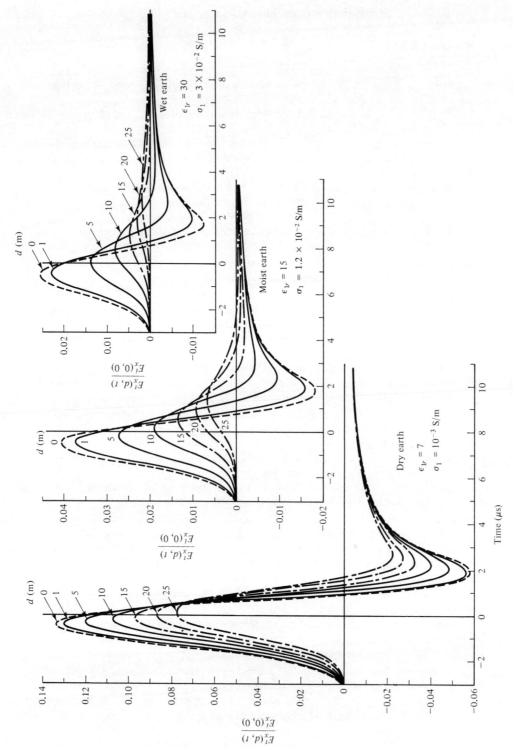

Figure 2.18–2 Transmitted part of a Gaussian pulse at time t (microseconds) and depth d (meters) in dry earth, moist earth, and wet earth.

PROBLEMS

1. Write the boundary conditions for the **E** vector and the **B** vector at the boundaries between two regions.
   (a) Region 1 is a good conductor described in terms of η_f, $\mathbf{J}_f$, and **M**; region 2 is free space.
   (b) Regions 1 and 2 are both dielectrics described in terms of different values of **P**.
   (c) Regions 1 and 2 are both simply polarizing dielectrics characterized by different dielectric constants.
   (d) Region 1 is a perfect conductor described in terms of η_f and $\mathbf{K}_f$; region 2 is a simply polarizing but imperfect dielectric described in terms of a dielectric constant and $\mathbf{J}_f$.
   (e) Region 1 is a good conductor described in terms of $\mathbf{J}_f$ and **M** and region 2 is a simply polarizing dielectric described in terms of a dielectric constant.

2. Prove that the field equations are consistent with the principle of conservation of electric charge as formulated in the equation of continuity.

3. A very long coaxial cable consists of a solid inner conductor of radius a and outer conductor of inner radius b_1 and outer radius b_2. The cable is parallel to the z axis. The space between the two conductors is air. There is a uniform current density J_z on the inner conductor and a uniform current density $-J'_z$ on the outer conductor such that the total currents in the inner and outer conductors are the same in magnitude. Determine **B** and **H** for all radii from zero to points completely outside the cable in terms of the total current in the inner conductor. Write the boundary conditions in coordinate form at each boundary.

4. Determine the effect on **B** and **H** in Problem 3 if the medium between the conductors is now characterized by a volume density of magnetization $-\mathbf{M}$ that is proportional to **B** in magnitude and is
   (a) A diamagnetic material [**B** and $-\mathbf{M}$ are in the same direction and the proportionality constant is $\mu_0^{-1}(\mu_r^{-1} - 1)$ with $\mu_r^{-1} > 1$].
   (b) A paramagnetic material [**B** and $-\mathbf{M}$ are opposite in direction and the proportionality constant is $\mu_0^{-1}(\mu_r^{-1} - 1)$ with $\mu_r^{-1} < 1$].

5. Consider the cable of Problem 3. Now assume that there is a uniform surface charge density η on the inner conductor and a uniform surface charge density $-\eta'$ on the inner surface of the outer conductor. The rotationally symmetric distributions on the outer and inner conductors are such that the total charge per unit length on outer and inner conductors is the same in magnitude. Determine **E** and **D** for all radii from zero to points completely outside the cable in terms of the total charge per unit length on the inner conductor. Write the boundary conditions in coordinate form at each boundary.

6. Determine the effect on **E** and **D** in Problem 5 if there is a dielectric medium between the two conductors that is characterized by a volume density of polarization **P**. **P** is proportional to **E** and in the same direction. $\epsilon_0(\epsilon_r - 1)$ is the proportionality factor.

7. An infinitely long transmission line consists of two parallel wires each of radius 0.5 cm separated by a distance of 25 cm between centers. The two wires carry direct currents in opposite directions; the current in wire 1 is 2 A, that in wire 2 is 1 A. Beginning with the integral form of the field equations, determine **B** at a distance of 15 cm from wire

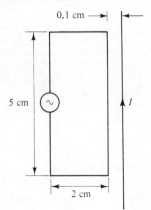

0.1 cm ⟶

5 cm

I

2 cm

Figure P2–8

2 and 20 cm from wire 1 in a cross-sectional plane. Show the directions of current and field vector on a diagram.

8. The magnitude of the quantity $\oint (\mathbf{E} \cdot d\mathbf{s})$ around the small rectangle of wire in Fig. P2–8 is determined by measurement to be 8 mV. What is the magnitude of the alternating current I opposite the center of the loop? The frequency is 20 MHz. The resistance of the ammeter is 0.1 Ω.

9. The xy plane is the boundary between a good (but not perfect) conductor (region 1) and a perfect dielectric (region 2). Each is a simple medium with ϵ_r, μ_r, and σ assumed known.
    (a) Assume that $\mathbf{E}$ has only an x component and $\mathbf{B}$ only a y component. Further assume that both $\mathbf{E}$ and $\mathbf{B}$ have a periodic time dependence. What are the partial differential equations for $\mathbf{E}$ and $\mathbf{B}$ both in the dielectric and conductor? Obtain the boundary conditions.
    (b) Assuming $\mathbf{E}$ and $\mathbf{B}$ to be determined from the solutions of the equations subject to the boundary conditions in part (a), write down expressions for ρ, η, $\mathbf{P}$, $\mathbf{J}$, $\mathbf{K}$, and $-\mathbf{M}$ in terms of $\mathbf{E}$ and $\mathbf{B}$. If any are zero, state the reasons.

10. Consider a very long coaxial transmission line that has a free charge of q coulombs per unit length on the surface of the inner conductor and an opposite charge of $-q$ coulombs per unit length on the inner surface of the outer conductor. Suppose that the region between the two conductors is then half-filled with a dielectric material ($\epsilon = \epsilon_0 \epsilon_r$) as shown in Fig. P2–10. If the total essential density of surface charge $\bar{\eta}$ at $r = r_1$ is assumed to be uniform and $\bar{\rho} = 0$ in all space, find the $\mathbf{E}$ and $\mathbf{D}$ fields in all space in terms of q. Determine if the assumption of a uniform distribution of $\bar{\eta}$ about the inner conductor violates any boundary conditions. What is $\mathbf{P}$ in the region $r_1 < r < r_2$? Does $\mathbf{P}$ violate any of the essential density equations? Sketch the free surface charge distribution about the center conductor.

11. A coaxial transmission line carries a free surface current of I amperes (dc) on the inner conductor and a current $-I$ in the opposite direction on the outer conductor. A non-conducting wedge of magnetic material ($\mu = \mu_0 \mu_r$) is then placed between the inner and outer conductor as shown in Fig. P2–11. Can the boundary conditions on the $\mathbf{B}$

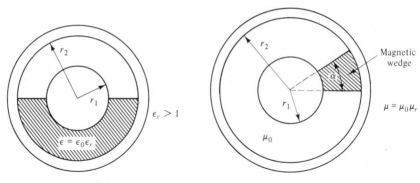

Figure P2-10   **Figure P2-11**

and **H** fields be satisfied by assuming that $\overline{\eta_m \mathbf{v}}$ is constant at $r = r_1$ and $\overline{\rho_m \mathbf{v}} = 0$ for $r_2 > r > r_1$? Find **B** and **H** under this assumption. What is **M** in the magnetic material? Sketch the total free surface current as a function of position along the circumference of the inner conductor.

12. (a) An iron toroid (region 2) with a circular cross section is completely characterized by the volume density of magnetization $\mathbf{M} = \hat{\boldsymbol{\theta}} M_\theta$. The surrounding region 1 is air. If $\mathbf{B}_1 = 0$, determine $\mathbf{B}_2$, $\mathbf{H}_2$, and $\mathbf{H}_1$.
    (b) A gap of length d ($d \ll a$) is cut across the toroid so that the mean length of the iron is $L = 2\pi b - d$, where b is the mean radius of the toroid. (It is assumed that $b \gg a$, where a is the cross-sectional radius.) If the iron is still characterized by $\mathbf{M} = \hat{\boldsymbol{\theta}} M_\theta$, determine B_θ and H_θ in the iron and in the gap, both at radius b. (Use integral forms of Maxwell's equations.)

13. Determine whether the following materials may be treated as good conductors at $f = 60$ Hz, 6 kHz, 6 MHz, 900 MHz, and 10 GHz.
    (a) Salt water: $\sigma = 4$ S/m; $\epsilon_r = 80$
    (b) Distilled water: $\sigma = 2 \times 10^{-4}$ S/m; $\epsilon_r = 81$
    (c) Copper: $\sigma = 5.8 \times 10^7$ S/m; $\epsilon_r = 1$
    (d) Dry earth: $\sigma = 1.2 \times 10^{-2}$ S/m; $\epsilon_r = 15$
    (e) Wet earth: $\sigma = 3 \times 10^{-2}$ S/m; $\epsilon_r = 30$

14. A plane wave with complex amplitude $\mathbf{E}^{\text{inc}}(\mathbf{r}) = \hat{\mathbf{x}} E_x^{\text{inc}}(0) e^{-jk_0 z}$ is incident on the plane boundary surface at $z = 0$ between air and an imperfect dielectric (ϵ, μ_0, σ) (Fig. P2-14). Obtain the resulting total instantaneous fields $\mathbf{E}(\mathbf{r}, t)$ and $\mathbf{B}(\mathbf{r}, t)$ in the air and the

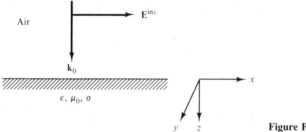

Figure P2-14

imperfect dielectric. With the aid of diagrams interpret these results in terms of traveling and standing waves.

15. Discuss how the traveling and standing waves of Problem 14 will differ when the imperfect dielectric is:
    (a) A perfect dielectric.
    (b) Seawater: $\sigma \epsilon_0 / \epsilon \doteq 5/80 = 1/16$.
    (c) A plasma where $\epsilon < 0$.
    (d) A perfect conductor.

16. An antenna is to be protected from the elements by placing it inside a radome. The dome is constructed of a dielectric material of dielectric constant ϵ_d and thickness t. The interior of the dome is filled with a dielectric foam $\epsilon = \epsilon_f$. The problem is to determine ϵ_d and t so as to minimize the reflection by the dome of the signal emitted by the antenna.

    Approximate the problem by considering a plane wave normally incident on an infinite dielectric sheet separating the two regions of dielectric constants ϵ_0 and ϵ_f. In this approximation the reflected wave can be made zero by properly choosing ϵ_d and t (Fig. P2–16).

17. Show that the reflection coefficients for a plane wave incident on a plane boundary between air (region 1) and an imperfect dielectric (region 2) at an angle θ_1 with the normal may be expressed as follows:

$$f_{er} = f_{er}'' + jf_{er}' = \frac{N^2 \cos \theta_1 - \sqrt{N^2 - \sin^2 \theta_1}}{N^2 \cos \theta_1 + \sqrt{N^2 - \sin^2 \theta_1}}$$

$$f_{mr} = f_{mr}'' + jf_{mr}' = \frac{\cos \theta_1 - \sqrt{N^2 - \sin^2 \theta_1}}{\cos \theta_1 + \sqrt{N^2 - \sin^2 \theta_1}}$$

where $N = k/k_0 = \sqrt{\bar{\epsilon}/\epsilon_0} = \sqrt{\epsilon_{er}(1 - jp_e)}$ is the (complex) index of refraction if it is assumed that $\mu_2 = \mu_0$. The real effective relative permittivity and conductivity of region 2 are, respectively, $\epsilon_e = \epsilon_{er}\epsilon_0$ and σ_e; $p_e = \sigma_e/\omega\epsilon_e$.

18. Justify the statement: An incident electric field that is linearly polarized in an arbitrary direction in air is reflected from a plane imperfect dielectric as an elliptically polarized field with the shape of the ellipse depending on ϵ_{er} and σ_e and on the angle of incidence θ_1. Is the reflected field ever linearly polarized when the dielectric is imperfect? When the dielectric is perfect?

19. Consider the special case where a plane wave with electric-field vector $\mathbf{E}$ normal to the plane of incidence is obliquely incident on a perfectly conducting plane (Fig. P2–19).

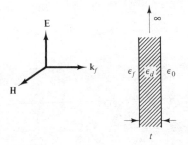

Figure P2–16

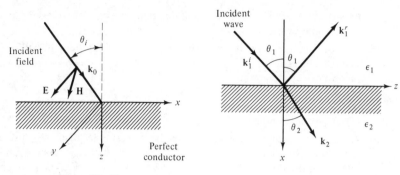

Figure P2–19   **Figure P2–20**

Describe the combination of incident and reflected waves in terms of traveling and standing waves.

20. Using the Fresnel equations derived in Problem 17 and the conditions

$$\epsilon_1 > \epsilon_2; \qquad \sqrt{\frac{\epsilon_1}{\epsilon_2}}\, \sin \theta_1 > 1$$

show that the total internal reflection of a plane wave in medium 1 will occur for both polarizations of the incident electric field (i.e., $\mathbf{E}^i$ in the plane of incidence and normal to it). Both media are perfect dielectrics (Fig. P2–20).

Vector and Scalar
Potential Functions

3.1 *GRADIENT OF A SCALAR FIELD*

A continuous mathematical function that assigns a scalar to every point in a region is called a *scalar point function*. Such a function may characterize a certain property of a physical model such as volume density of charge, volume density of mass, temperature, humidity, pressure, and elevation above sea level.

An important property of a scalar point function is its space rate of change. Consider, for example, a scalar point function ϕ that is continuous with its derivatives. The simplest form for this function is that of a constant. If ϕ is everywhere constant, the rate of change of ϕ at every point and in every direction vanishes.

$$\phi = C; \qquad \frac{d\phi}{ds} = 0 \qquad (3.1–1)$$

If the function varies so that a different number is assigned to each point, $d\phi/ds$ measures the rate of increase of the function ϕ in the direction in which ds is taken. It may vary not only from one location to another in the region, but also in different directions relative to a single point. Thus the space rate of change of a scalar point function itself defines a continuous function that assigns not one magnitude to every point, but a *magnitude to every direction* at that point. That is, the property associated with the space rate of change of a scalar point function is characterized not only by magnitude but also by direction. Therefore, the magnitude $d\phi/ds$ at a particular point and in a particular direction (as specified by the orientation of the element ds) is the component of a vector in the direction of ds. The component G_s of a vector $\mathbf{G}$ in a given direction specified by the unit vector $\hat{\mathbf{s}}$ is the scalar product

$$G_s = \hat{\mathbf{s}} \cdot \mathbf{G} \tag{3.1-2}$$

Hence $d\phi/ds$ is the component of a vector $\mathbf{G}$ in the direction of ds specified by the unit vector $\hat{\mathbf{s}}$. That is,

$$\frac{d\phi}{ds} = G_s = \hat{\mathbf{s}} \cdot \mathbf{G} \tag{3.1-3}$$

This component vanishes in a direction in which ϕ remains constant. In such a direction

$$\frac{d\phi}{ds} = 0 = \hat{\mathbf{s}} \cdot \mathbf{G} \tag{3.1-4}$$

Since $\hat{\mathbf{s}}$ cannot vanish and $\mathbf{G}$ is zero only if $d\phi/ds$ vanishes in all directions, which is assumed not to be the case, there is no way to satisfy the right side of (3.1-4) except by requiring $\mathbf{G}$ to be perpendicular to $\hat{\mathbf{s}}$ when this points in a direction in which ϕ is constant. Therefore, the vector $\mathbf{G}$, characterizing the space rate of change of ϕ, is always in a direction normal to surfaces along which ϕ is constant. The maximum value of $d\phi/ds$ is in a direction that makes the cosine of the angle between $\hat{\mathbf{s}}$ and $\mathbf{G}$ equal to unity. That is,

$$\left(\frac{d\phi}{ds}\right)_{max} = |\mathbf{G}| = G \tag{3.1-5}$$

Hence the vector $\mathbf{G}$ defines both the magnitude and the direction of the maximum space rate of increase of ϕ. It is directed perpendicular to the surfaces $\phi = $ constant. The vector point function $\mathbf{G}$ which defines the direction and magnitude of the maximum space rate of increase of a scalar point function is called the *gradient of the scalar point function*. The operation of taking the derivative of a scalar function in the direction in which the derivative is a maximum is assigned the symbol ∇:

$$\left(\frac{d}{ds}\right) \text{ in the direction of the maximum} \equiv \nabla \tag{3.1-6}$$

Accordingly,

$$\left(\frac{d\phi}{ds}\right)_{max} = |\nabla\phi| = G \tag{3.1-7}$$

$$\mathbf{G} = \nabla\phi \tag{3.1-8}$$

The symbol ∇ is a differential operator that specifies differentiation with respect to the space coordinates in the direction in which the derivative has its maximum value. More precisely, the operator ∇ operating on a scalar point function ϕ produces a vector point function $\nabla\phi$, the direction of which at every point is that of the greatest space rate of increase of ϕ and the magnitude of which measures the maximum rate of increase $(d\phi/ds)_{max}$ of that function. Alternatively, $\nabla\phi$ is a vector which has for a component in any direction the rate of change $d\phi/ds$ in that direction.

The gradient of a scalar field may be defined in a way that is analogous to the definition of the divergence and the curl of a vector field. Thus

$$\nabla\phi = \lim_{\Delta\tau\to 0} \frac{\int_\Sigma \hat{\mathbf{n}}\phi \, d\sigma}{\Delta\tau} \qquad (3.1\text{--}9)$$

Here $\hat{\mathbf{n}}$ is an external normal to the surface Σ enclosing the volume $\Delta\tau$. It is easily shown that this definition is equivalent to that given above. Let the volume $\Delta\tau$ be chosen so that two sides always lie along surfaces on which ϕ is constant, while the other four sides are perpendicular to these surfaces as in Fig. 3.1–1. Everywhere on the top surface ΔS_1, $\phi = \phi_1$; everywhere on the bottom surface ΔS_2, $\phi = \phi_2$ with $\phi_1 > \phi_2$; on the four sides ΔS_i, ϕ varies continuously from ϕ_1 to ϕ_2. Let the mean value of ϕ on each side be $\overline{\phi}$. With this notation and assuming $\Delta\tau$ to be sufficiently small, (3.1–9) becomes

$$\nabla\phi = \lim_{\Delta\tau\to 0} \frac{\hat{\mathbf{n}}_1\phi_1\Delta S_1 + \hat{\mathbf{n}}_2\phi_2\Delta S_2 + \sum_{i=3}^{6} \overline{\phi}_i\hat{\mathbf{n}}_i\Delta S_i}{\Delta\tau} \qquad (3.1\text{--}10)$$

As the volume is made smaller and smaller, $\Delta S_1 \to \Delta S_2$ and the opposite pairs of surfaces ΔS_i become equal. The average values $\overline{\phi}$ on opposite side surfaces ΔS_i also become equal, so that $\hat{\mathbf{n}}_i\overline{\phi}_i\Delta S_i$ for opposite sides are alike in magnitude and opposite in sign. Hence the sum in (3.1–10) vanishes in the limit, leaving

$$\nabla\phi = \lim_{\Delta\tau\to 0} \frac{\hat{\mathbf{n}}_1\Delta S_1(\phi_1 - \phi_2)}{\Delta\tau}$$

$$= \lim_{\Delta\tau\to 0} \frac{\hat{\mathbf{n}}_1\Delta S_1\Delta\phi}{\Delta S_1\Delta s} \qquad (3.1\text{--}11)$$

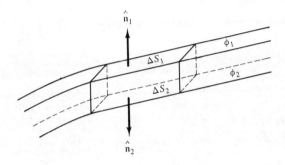

Figure 3.1–1 Parallelepiped in a scalar field.

In (3.1–11), Δs is the distance between the parallel upper and lower surfaces. Finally,

$$\nabla \phi = \hat{\mathbf{n}}_1 \lim_{\Delta s \to 0} \frac{\Delta \phi}{\Delta s} = \hat{\mathbf{n}}_1 \frac{d\phi}{ds} \tag{3.1-12}$$

Here $\hat{\mathbf{n}}_1$ points in a direction perpendicular to a surface of constant ϕ at the point where $d\phi/ds$ is defined and in the direction of increasing ϕ. The form (3.1–9) is convenient for expressing the gradient in various systems of orthogonal coordinates.

As an example of the gradient, let $\phi = h$ measure the height above sea level at every point on a mountain range. Let $s = (x^2 + y^2)^{1/2}$ measure the distance in the horizontal plane. The lines $\phi = h = $ const. or $d\phi/ds = 0$ are level contours that neither rise nor descend. The direction specified by $\nabla \phi$ at any point gives the direction of steepest ascent from that point, and the magnitude of $\nabla \phi$ is the numerical value of the steepest upward slope, or the tangent of the largest angle of upward inclination. On the other hand, $-\nabla \phi$ is directed opposite to $\nabla \phi$ at every point so that it gives the direction and magnitude of steepest downward slope. If $\nabla \phi = 0$ at a point, there is no direction of steepest rise, the slope in all directions is zero. This is true on a level plateau or on the uppermost part of a rounded mountain peak. If water is poured out at any point, it starts to flow in the direction of $-\nabla \phi$.

3.2 GREEN'S THEOREM

Green's theorem is an integral theorem involving the gradient. It may be obtained directly from the divergence theorem by setting

$$\mathbf{A} = u\nabla v \tag{3.2-1}$$

with u and v arbitrary scalar point functions. Thus

$$\int_V \nabla \cdot (u\nabla v)\, dV = \int_S (\hat{\mathbf{n}} \cdot u\nabla v)\, dS \tag{3.2-2}$$

With the vector relation

$$\nabla \cdot (\phi \mathbf{A}) = \phi \nabla \cdot \mathbf{A} + \mathbf{A} \cdot \nabla \phi \tag{3.2-3}$$

the integral on the left may be transformed into two integrals. The integral on the right may be changed in form by noting that

$$\hat{\mathbf{n}} \cdot \nabla v = \nabla_n v = \frac{\partial v}{\partial n} \tag{3.2-4}$$

With (3.2–3) and (3.2–4), (3.2–2) becomes

$$\int_V u\nabla \cdot \nabla v\, dV + \int_V (\nabla u \cdot \nabla v)\, dV = \int_S u\frac{\partial v}{\partial n}\, dS \tag{3.2-5}$$

By definition of the operator $\mathbf{\nabla}^2$,

$$\mathbf{\nabla}^2 v \equiv \mathbf{\nabla} \cdot \mathbf{\nabla} v \tag{3.2-6}$$

Hence

$$\int_V u \mathbf{\nabla}^2 v \; dV = \int_S u \frac{\partial v}{\partial n} \; dS - \int_V (\mathbf{\nabla} u \cdot \mathbf{\nabla} v) \; dV \tag{3.2-7}$$

This is one form of Green's theorem. A second form, called *Green's symmetrical theorem*, may be obtained from (3.2–7) by interchanging u and v and subtracting the expression so obtained from (3.2–7). The result is

$$\int_V (u \mathbf{\nabla}^2 v - v \mathbf{\nabla}^2 u) \; dV = \int_S \left(u \frac{\partial v}{\partial n} - v \frac{\partial u}{\partial n} \right) dS \tag{3.2-8}$$

3.3 POTENTIAL FUNCTIONS

Definition.   A *scalar potential* is a scalar point function the negative gradient of which is a vector point function. In other words, any scalar point function ϕ is a scalar potential if $-\mathbf{\nabla}\phi$ defines a vector point function $\mathbf{K}$. A surface on which ϕ is constant is called an *equipotential surface*.

A vector point function is not necessarily the gradient of a scalar potential. The class of vector functions that can be derived as the negative gradients of scalar potentials is limited by the following theorem:

Theorem.   A necessary and sufficient condition that a vector field $\mathbf{K}$ possess a *scalar* potential ϕ is that

$$\mathbf{\nabla} \times \mathbf{K} = 0 \tag{3.3-1}$$

If this is true, the vector $\mathbf{K}$ is called a *potential vector* and its field is *irrotational*. It may be derived from its associated scalar potential by setting

$$\mathbf{K} = -\mathbf{\nabla}\phi \tag{3.3-2}$$

Definition.   A *vector potential* is a vector point function the curl of which is also a vector point function. In other words, a vector point function $\mathbf{A}$ is a vector potential if $\mathbf{\nabla} \times \mathbf{A}$ defines a second vector point function $\mathbf{C}$.

The class of vector functions that can be derived from vector potentials is limited by the following theorem.

Theorem.   A necessary and sufficient condition that a vector field $\mathbf{C}$ possess a vector potential $\mathbf{A}$ is that

$$\mathbf{\nabla} \cdot \mathbf{C} = 0 \tag{3.3-3}$$

If (3.3–3) is true, **C** is a *solenoidal* vector; its field is *rotational*. It may be derived from a vector potential by setting

$$\mathbf{C} = \nabla \times \mathbf{A} \tag{3.3-4}$$

Any single-valued vector point function **S** that together with its derivatives is finite and continuous and vanishes at infinity can be derived from a scalar potential ϕ and a vector potential **A** in a form known as the *Helmholtz theorem*:

$$\mathbf{S} = -\nabla\phi + \nabla \times \mathbf{A} \tag{3.3-5}$$

3.4 EQUATIONS OF D'ALEMBERT, POISSON, AND LAPLACE

The direct determination of **E** and **B** by solving the Maxwell–Lorentz equations is usually difficult. The solution is often facilitated by defining potential functions in terms of the field vectors, because it is possible to transform the four first-order field equations into two second-order equations in the potential functions that are formally integrable in a general way. From the integrals so obtained, and the definitions of potential functions, it is possible to calculate **E** and **B** and, from these and the boundary conditions, the distributions of current and charge. Frequently, current and charge are more conveniently determined directly from the potential functions than from the field vectors.

An examination of the four field equations discloses that the magnetic vector **B** is solenoidal because its divergence vanishes. Hence **B** can be derived from a vector potential **A** defined by

$$\nabla \times \mathbf{A} = \mathbf{B} \tag{3.4-1}$$

The vector point function **A** defined (incompletely) in (3.4–1) is of fundamental importance especially in the theory of antennas. It is called the *magnetic vector potential*.

To define a scalar potential it is necessary to find a vector with vanishing curl. From the symmetry of electric and magnetic quantities, the second field equation should be used for this purpose, since this is the electric analogue of the magnetic fourth equation. The second equation is

$$\nabla \times \mathbf{E} = -\dot{\mathbf{B}} \tag{3.4-2}$$

This is easily transformed with (3.4–1) to read

$$\nabla \times (\mathbf{E} + \dot{\mathbf{A}}) = 0 \tag{3.4-3}$$

The vector $(\mathbf{E} + \dot{\mathbf{A}})$ is a potential vector because its curl vanishes. It can be derived from a scalar potential ϕ defined by

$$-\nabla\phi = \mathbf{E} + \dot{\mathbf{A}} \tag{3.4-4}$$

This is also a fundamentally important relation. The scalar potential defined by (3.4–4) is called the *electric scalar potential*. It is evident that if the scalar and vector potentials φ and **A** are known, the electromagnetic vectors **E** and **B** may be calculated directly from

$$\mathbf{E} = -\nabla\phi - \dot{\mathbf{A}} \tag{3.4–5}$$

$$\mathbf{B} = \nabla \times \mathbf{A} \tag{3.4–6}$$

With the scalar and vector potentials defined, the next step is to eliminate **E** and **B** from the field equations. It is to be noted that whereas the scalar φ is defined completely (except for an additive constant) by (3.4–4), the vector **A** is not defined completely by (3.4–1). To define a vector, both its curl and divergence must be specified. Nothing has been said so far about the divergence of **A**, and to this extent at least the vector **A** is still arbitrary.

Since the potential functions φ and **A** are defined in terms of two of the field equations, they must still be made to satisfy the other two, namely,

$$\epsilon_0\nabla \cdot \mathbf{E} = \overline{\rho} \tag{3.4–7}$$

$$\mu_0^{-1}\nabla \times \mathbf{B} = \overline{\rho_m\mathbf{v}} + \epsilon_0\dot{\mathbf{E}} \tag{3.4–8}$$

Direct substitution of (3.4–5) and (3.4–6) in (3.4–7) and (3.4–8) gives

$$\nabla \cdot \nabla\phi + \nabla \cdot \dot{\mathbf{A}} = \frac{-\overline{\rho}}{\epsilon_0} \tag{3.4–9}$$

$$\nabla \times \nabla \times \mathbf{A} = \mu_0(\overline{\rho_m\mathbf{v}} - \epsilon_0\nabla\dot{\phi} - \epsilon_0\ddot{\mathbf{A}}) \tag{3.4–10}$$

The symbol ∇^2 stands for two different operations, depending on whether it operates on a scalar φ or a vector **A**. It is called the Laplacian operator and with (3.2–6) it is defined in the two cases as follows:

$$\nabla^2\phi \equiv \nabla \cdot \nabla\phi \tag{3.4–11}$$

$$\nabla^2\mathbf{A} \equiv \nabla\nabla \cdot \mathbf{A} - \nabla \times \nabla \times \mathbf{A} \tag{3.4–12}$$

With this symbolism (3.4–9) and (3.4–10) become

$$\nabla^2\phi + \nabla \cdot \dot{\mathbf{A}} = \frac{-\overline{\rho}}{\epsilon_0} \tag{3.4–13}$$

$$\nabla^2\mathbf{A} - \nabla\nabla \cdot \mathbf{A} = \mu_0(-\overline{\rho_m\mathbf{v}} + \epsilon_0\nabla\dot{\phi} + \epsilon_0\ddot{\mathbf{A}}) \tag{3.4–14}$$

Since the vector potential **A** is still not completely defined, it is possible to assign any convenient value to $\nabla \cdot \mathbf{A}$. The value that is chosen is that which separates the variables **A** and φ in (3.4–13) and (3.4–14). Let the definition of **A** be completed (except for an additive constant vector) by writing

$$\nabla \times \mathbf{A} = \mathbf{B} \tag{3.4-15}$$

$$\nabla \cdot \mathbf{A} = -\epsilon_0\mu_0 \frac{\partial \phi}{\partial t} = -\frac{1}{c^2}\frac{\partial \phi}{\partial t} \tag{3.4-16}$$

where $c = (\epsilon_0\mu_0)^{-1/2}$. Equation (3.4–16) is known as the *Lorentz condition* when written in the form

$$\nabla \cdot \mathbf{A} + \epsilon_0\mu_0 \frac{\partial \phi}{\partial t} = 0 \tag{3.4-17}$$

It closely resembles the generalized equation of continuity of electric charge

$$\nabla \cdot \overline{\rho_m\mathbf{v}} + \frac{\partial \overline{\rho}}{\partial t} = 0 \tag{3.4-18}$$

As will be seen and subject to (3.4–17), $\mathbf{A}$ is defined in terms of $\overline{\rho_m\mathbf{v}}$ and ϕ in terms of $\overline{\rho}$. The resemblance between (3.4–17) and (3.4–18) is evidently more than formal. Equation (3.4–17) corresponds to an equation of continuity for potentials. If it is satisfied, (3.4–13) and (3.4–14) reduce to

$$\nabla^2\phi - \mu_0\epsilon_0\ddot{\phi} = \frac{-\overline{\rho}}{\epsilon_0} \tag{3.4-19}$$

$$\nabla^2\mathbf{A} - \mu_0\epsilon_0\ddot{\mathbf{A}} = -\mu_0\overline{\rho_m\mathbf{v}} \tag{3.4-20}$$

These are the fundamental *d'Alembert equations* governing the potential fields ϕ and $\mathbf{A}$.

In the stationary states (3.4–19) and (3.4–20) reduce to a form known as *Poisson's equation*.

$$\nabla^2\phi = \frac{-\overline{\rho}}{\epsilon_0} \tag{3.4-21}$$

$$\nabla^2\mathbf{A} = -\mu_0\overline{\mathbf{J}} \tag{3.4-22}$$

The definitions for the potential functions in the stationary states are

$$-\nabla\phi = \mathbf{E} \tag{3.4-23}$$

$$\nabla \times \mathbf{A} = \mathbf{B}$$

$$\nabla \cdot \mathbf{A} = 0 \tag{3.4-24}$$

Since (3.4–21) involves only the essential volume density of charge $\overline{\rho} = \rho - \nabla\cdot\mathbf{P}$ and the electric constant ϵ_0, it is a static-state equation. It is analogous to the steady-state equation (3.4–22) that involves the essential volume density of moving charge $\overline{\mathbf{J}} = \mathbf{J} + \nabla \times \mathbf{M}$ and the magnetic constant μ_0. The formal analogy between the static and steady states may be extended as follows:

Quantity	Static state	Steady state
Potential function	Electric scalar potential, ϕ	Magnetic vector potential, **A**
Operation	∇^2 (operating on a scalar)	∇^2 (operating on a vector)

A special form of Poisson's equation called *Laplace's equation* is

$$\nabla^2\phi = 0 \qquad (3.4\text{--}25)$$

$$\nabla^2\mathbf{A} = 0 \qquad (3.4\text{--}26)$$

3.5 BOUNDARY CONDITIONS FOR POTENTIAL FUNCTIONS

The general boundary conditions for the electromagnetic field vectors are given in Chapter 2. They may be written as follows at the boundary between two media 1 and 2:

$$\hat{\mathbf{n}}_1 \cdot \mathbf{E}_1 + \hat{\mathbf{n}}_2 \cdot \mathbf{E}_2 = \frac{-\overline{\eta}}{\epsilon_0} \qquad (3.5\text{--}1\text{a})$$

$$\hat{\mathbf{n}}_1 \times \mathbf{E}_1 + \hat{\mathbf{n}}_2 \times \mathbf{E}_2 = 0 \qquad (3.5\text{--}1\text{b})$$

$$\hat{\mathbf{n}}_1 \times \mathbf{B}_1 + \hat{\mathbf{n}}_2 \times \mathbf{B}_2 = -\mu_0\overline{\eta_m\mathbf{v}} \qquad (3.5\text{--}1\text{c})$$

$$\hat{\mathbf{n}}_1 \cdot \mathbf{B}_1 + \hat{\mathbf{n}}_2 \cdot \mathbf{B}_2 = 0 \qquad (3.5\text{--}1\text{d})$$

The shorthand notation $\overline{\eta} = \overline{\eta}_1 + \overline{\eta}_2$; $\overline{\eta_m\mathbf{v}} = \overline{\eta_m\mathbf{v}}_1 + \overline{\eta_m\mathbf{v}}_2$ is used. Normals are external to the region indicated by the subscript. The electromagnetic vectors in (3.5–1a–d) can be replaced by the potential functions using (3.4–5) and (3.4–6), that is,

$$\mathbf{E} = -\nabla\phi - \dot{\mathbf{A}} \qquad (3.5\text{--}2)$$

$$\mathbf{B} = \nabla \times \mathbf{A} \qquad (3.5\text{--}3)$$

With

$$\hat{\mathbf{n}} \cdot \nabla\phi = \frac{\partial\phi}{\partial n} \qquad (3.5\text{--}4)$$

and the substitution of (3.5–2) in (3.5–1a), the following equation results:

$$\frac{\partial\phi_1}{\partial n_1} + \frac{\partial\phi_2}{\partial n_2} + \hat{\mathbf{n}}_1 \cdot \dot{\mathbf{A}}_1 + \hat{\mathbf{n}}_2 \cdot \dot{\mathbf{A}}_2 = \frac{\overline{\eta}}{\epsilon_0} \qquad (3.5\text{--}5)$$

Before the potential functions are substituted for the electric field in (3.5–1b), let (3.5–1b) be multiplied vectorially by a unit vector $\hat{s}$ tangent to the surface of discontinuity as illustrated in Fig. 3.5–1. Then

$$\hat{s} \times \hat{n}_1 \times E_1 + \hat{s} \times \hat{n}_2 \times E_2 = 0 \tag{3.5–6}$$

With the use of the general vector formula

$$A \times B \times C = B(A \cdot C) - C(A \cdot B) \tag{3.5–7a}$$

and since $\hat{s}$ and $\hat{n}$ are mutually perpendicular, it follows that $\hat{s} \cdot \hat{n} = 0$ and

$$\hat{s} \times \hat{n} \times E = \hat{n}(\hat{s} \cdot E) \tag{3.5–7b}$$

Hence (3.5–1b) is equivalent to

$$\hat{n}_1(\hat{s} \cdot E_1) + \hat{n}_2(\hat{s} \cdot E_2) = 0 \tag{3.5–8}$$

Since $\hat{n}_1 = -\hat{n}_2$, (3.5–8) becomes

$$\hat{s} \cdot E_1 - \hat{s} \cdot E_2 = 0 \tag{3.5–9}$$

The substitution of (3.5–2) in (3.5–9) and the use of (3.5–4) with $\hat{s}$ written for $\hat{n}$ yield

$$\frac{\partial \phi_1}{\partial s} - \frac{\partial \phi_2}{\partial s} + \hat{s} \cdot \dot{A}_1 - \hat{s} \cdot \dot{A}_2 = 0 \tag{3.5–10}$$

Equations (3.5–5) and (3.5–10) are satisfied if, at the boundary,

$$A_1 - A_2 = 0 \tag{3.5–11}$$

$$\left(\frac{\partial \phi}{\partial n_1}\right)_1 + \left(\frac{\partial \phi}{\partial n_2}\right)_2 = \frac{\overline{\eta}}{\epsilon_0} \tag{3.5–12}$$

$$\left(\frac{\partial \phi}{\partial s}\right)_1 - \left(\frac{\partial \phi}{\partial s}\right)_2 = 0 \tag{3.5–13}$$

These boundary conditions, although not as general as possible, are consistent with (3.5–1a, b). For all boundaries characterized only by $\overline{\eta}$ and $\overline{\eta_m v}$ it is possible to write

$$\phi_1 - \phi_2 = 0 \tag{3.5–14}$$

Equations (3.5–11) and (3.5–14) are not correct if surface densities of magnetization and polarization are defined. However, most practical problems do not involve these densities.

In order to determine the boundary conditions for the normal and tangential derivatives of the vector potential by substituting (3.5–3) in (3.5–1c,d), it is necessary to express $\nabla \times A$ in terms of its components parallel and perpendicular to the boundary surface. This may be done in terms of a set of rectangular axes defined by the unit vectors $\hat{n}$, $\hat{s}$, $\hat{p}$ arranged to form a right-handed system. As

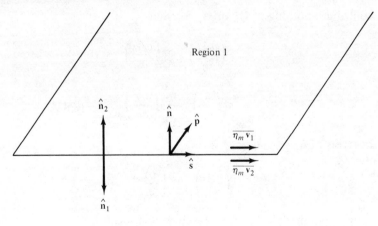

Figure 3.5–1   Rectangular coordinates at a boundary.

shown in Fig. 3.5–1, $\hat{\mathbf{n}} = \hat{\mathbf{n}}_2 = -\hat{\mathbf{n}}_1$ is perpendicular to the boundary, $\hat{\mathbf{s}}$ is parallel to the surface and in the *direction of* $\overline{\eta_m \mathbf{v}}$, and $\hat{\mathbf{p}}$ is parallel to the surface and perpendicular to $\hat{\mathbf{s}}$. The curl may be expressed in the form

$$
\nabla \times \mathbf{A} = \begin{vmatrix} \hat{\mathbf{n}} & \hat{\mathbf{s}} & \hat{\mathbf{p}} \\ \dfrac{\partial}{\partial n} & \dfrac{\partial}{\partial s} & \dfrac{\partial}{\partial p} \\ A_n & A_s & A_p \end{vmatrix} \tag{3.5–15}
$$

Equation (3.5–15) is now substituted for **B** in (3.5–1c):

$$
\hat{\mathbf{n}}_1 \times \left\{ \hat{\mathbf{n}}_1 \left[\left(\frac{\partial A_p}{\partial s} - \frac{\partial A_s}{\partial p} \right)_1 \right] - \hat{\mathbf{s}} \left[\left(\frac{\partial A_p}{\partial n} - \frac{\partial A_n}{\partial p} \right)_1 \right] + \hat{\mathbf{p}} \left[\left(\frac{\partial A_s}{\partial n} - \frac{\partial A_n}{\partial s} \right)_1 \right] \right\}
$$

$$
+ \hat{\mathbf{n}}_2 \times \left\{ \hat{\mathbf{n}}_2 \left[\left(\frac{\partial A_p}{\partial s} - \frac{\partial A_s}{\partial p} \right)_2 \right] - \hat{\mathbf{s}} \left[\left(\frac{\partial A_p}{\partial n} - \frac{\partial A_n}{\partial p} \right)_2 \right] + \hat{\mathbf{p}} \left[\left(\frac{\partial A_s}{\partial n} - \frac{\partial A_n}{\partial s} \right)_2 \right] \right\}
$$

$$
= -\mu_0 \overline{\eta_m \mathbf{v}} \tag{3.5–16a}
$$

The terms $\hat{\mathbf{n}} \times \hat{\mathbf{n}} = 0$ for both regions; the terms $\hat{\mathbf{n}} \times \hat{\mathbf{s}}$ are perpendicular to $\hat{\mathbf{s}}$, which is by definition in the direction of $\overline{\eta_m \mathbf{v}}$, so they must vanish; the terms $\hat{\mathbf{n}} \times \hat{\mathbf{p}} = -\hat{\mathbf{s}}$ are parallel to $\overline{\eta_m \mathbf{v}}$. Consequently, (3.5–16a) reduces to

$$
\hat{\mathbf{s}} \left(\frac{\partial A_s}{\partial n} - \frac{\partial A_n}{\partial s} \right)_1 - \hat{\mathbf{s}} \left(\frac{\partial A_s}{\partial n} - \frac{\partial A_n}{\partial s} \right)_2 = -\mu_0 \overline{\eta_m \mathbf{v}} \tag{3.5–16b}
$$

It will be seen that **A** is always parallel to the moving charge from which it is calculated. Since $\overline{\eta_m \mathbf{v}}$ in (3.5–16a,b) is by definition in the direction of $\hat{\mathbf{s}}$, it can contribute nothing to A_n or A_p, so that

$$\left(\frac{\partial A_n}{\partial s}\right)_1 - \left(\frac{\partial A_n}{\partial s}\right)_2 = 0 \qquad (3.5\text{–}17)$$

$$\hat{s}\left(\frac{\partial A_s}{\partial n}\right)_1 - \hat{s}\left(\frac{\partial A_s}{\partial n}\right)_2 = -\mu_0\overline{\eta_m}\mathbf{v} \qquad (3.5\text{–}18)$$

In terms of $\hat{\mathbf{n}}_1 = -\hat{\mathbf{n}} = -\hat{\mathbf{n}}_2$,

$$\left(\frac{\partial A_s}{\partial n_1}\right)_1 + \left(\frac{\partial A_s}{\partial n_2}\right)_2 = \mu_0\overline{\eta_m}v \qquad (3.5\text{–}19)$$

with A_s parallel to $\overline{\eta_m}\mathbf{v}$. The terms $\hat{\mathbf{n}} \times \hat{\mathbf{s}}$ result in

$$\left(\frac{\partial A_n}{\partial p}\right)_1 - \left(\frac{\partial A_n}{\partial p}\right)_2 = \left(\frac{\partial A_p}{\partial n_1}\right)_1 + \left(\frac{\partial A_p}{\partial n_2}\right)_2 = 0 \qquad (3.5\text{–}20)$$

with A_p perpendicular to $\overline{\eta_m}\mathbf{v}$ but tangent to the boundary.

The substitution of (3.5–15) in (3.5–1d) results in terms involving $\hat{\mathbf{n}} \cdot \hat{\mathbf{n}}$, $\hat{\mathbf{n}} \cdot \hat{\mathbf{s}}$, and $\hat{\mathbf{n}} \cdot \hat{\mathbf{p}}$. Since $\hat{\mathbf{n}} \cdot \hat{\mathbf{s}}$ and $\hat{\mathbf{n}} \cdot \hat{\mathbf{p}}$ are zero, (3.5–1d) yields

$$\left(\frac{\partial A_p}{\partial s} - \frac{\partial A_s}{\partial p}\right)_1 - \left(\frac{\partial A_p}{\partial s} - \frac{\partial A_s}{\partial p}\right)_2 = 0 \qquad (3.5\text{–}21)$$

Since A_s and A_p are independent, (3.5–21) may be separated to give

$$\left(\frac{\partial A_p}{\partial s}\right)_1 - \left(\frac{\partial A_p}{\partial s}\right)_2 = 0 \qquad (3.5\text{–}22)$$

$$\left(\frac{\partial A_s}{\partial p}\right)_1 - \left(\frac{\partial A_s}{\partial p}\right)_2 = 0 \qquad (3.5\text{–}23)$$

In summary, the following boundary conditions obtain for the scalar potential:

$$\phi_1 - \phi_2 = 0 \qquad (3.5\text{–}24)$$

$$\epsilon_0\left(\frac{\partial\phi}{\partial n_1}\right)_1 + \epsilon_0\left(\frac{\partial\phi}{\partial n_2}\right)_2 = \overline{\eta} \equiv \overline{\eta}_1 + \overline{\eta}_2 \qquad (3.5\text{–}25)$$

$$\left(\frac{\partial\phi}{\partial s}\right)_1 - \left(\frac{\partial\phi}{\partial s}\right)_2 = 0 \qquad (3.5\text{–}26)$$

That is, ϕ and its tangential derivative are continuous across a boundary; its normal derivative is discontinuous by $\overline{\eta}/\epsilon_0$.

The boundary conditions for the vector potential are

$$\mathbf{A}_1 - \mathbf{A}_2 = 0 \tag{3.5-27}$$

$$\left(\frac{\partial A_s}{\partial n_1}\right)_1 + \left(\frac{\partial A_s}{\partial n_2}\right)_2 = \mu_0 \overline{\eta_m \upsilon} \equiv \mu_0 (\overline{\eta_m \upsilon_1} + \overline{\eta_m \upsilon_2}) \tag{3.5-28}$$

$$\left(\frac{\partial A_p}{\partial n_1}\right)_1 + \left(\frac{\partial A_p}{\partial n_2}\right)_2 = 0 \tag{3.5-29}$$

$$\left(\frac{\partial A_{n,s}}{\partial p}\right)_1 - \left(\frac{\partial A_{n,s}}{\partial p}\right)_2 = 0 \tag{3.5-30}$$

$$\left(\frac{\partial A_{n,p}}{\partial s}\right)_1 - \left(\frac{\partial A_{n,p}}{\partial s}\right)_2 = 0 \tag{3.5-31}$$

The vector potential and the tangential derivative of its tangential component are continuous across a boundary; the normal derivative of its tangential component in the direction of $\overline{\eta_m \mathbf{v}}$ is discontinuous by $\mu_0 \overline{\eta_m \upsilon}$; the normal derivative of its tangential component at right angles to $\overline{\eta_m \mathbf{v}}$ is continuous. More briefly, only the normal derivative of the component of the vector potential *parallel* to the surface current is discontinuous.

3.6 EQUATIONS OF D'ALEMBERT, POISSON, AND LAPLACE IN SIMPLE MEDIA

The potential equations that were obtained in Sec 3.4 involve the essential densities $\overline{\rho} = \rho - \nabla \cdot \mathbf{P}$ and $\overline{\rho_m \mathbf{v}} = \mathbf{J} + \nabla \times \mathbf{M} + \dot{\mathbf{P}}$. The solution of these equations for ϕ and $\mathbf{A}$ thus assumes a knowledge of all the volume densities ρ, $\mathbf{P}$, $\mathbf{J}$, and $\mathbf{M}$. In media in which the constitutive parameters ϵ_r, μ_r, and σ may be used, considerable simplification is achieved by reducing the number of density functions that must be known. In writing equations for simple media it is advantageous to consider two cases separately. The first of these involves linearly polarizing and magnetizing media so that the parameters ϵ_r and μ_r appear and the densities $\mathbf{P}$ and $-\mathbf{M}$ do not. However, no specific assumptions are made regarding the volume density of convection current $\mathbf{J}_f$ so that it appears explicitly. The second formulation is like the first in assuming simply polarizing and magnetizing media, but it assumes simply conducting media as well, so that all three parameters ϵ_r, μ_r, and σ appear and $\mathbf{P}$, $-\mathbf{M}$, and $\mathbf{J}_f$ do not.

In simply polarizing and magnetizing media, the first and third field equations have the following form:

$$\epsilon \nabla \cdot \mathbf{E} = \rho_f \tag{3.6-1}$$

$$\mu^{-1} \nabla \times \mathbf{B} = \mathbf{J}_f + \epsilon \dot{\mathbf{E}} \tag{3.6-2}$$

If these equations are compared with (3.4–7) and (3.4–8), they are found to be formally exactly like the more general ones. It is only necessary to write

$$
\begin{array}{ccc}
\epsilon & \text{for} & \epsilon_0 \\
\mu & \text{for} & \mu_0 \\
\rho_f & \text{for} & \bar{\rho} \\
\mathbf{J}_f & \text{for} & \overline{\rho_m \mathbf{v}}
\end{array}
\qquad (3.6\text{–}3)
$$

By defining the scalar and vector potentials according to

$$
-\nabla\phi = \mathbf{E} + \dot{\mathbf{A}} \qquad (3.6\text{–}4)
$$

$$
\nabla \times \mathbf{A} = \mathbf{B} \qquad (3.6\text{–}5a)
$$

$$
\nabla \cdot \mathbf{A} = -\epsilon\mu \frac{\partial\phi}{\partial t} \qquad (3.6\text{–}5b)
$$

equations corresponding to (3.4–19) and (3.4–20) are obtained:

$$
\nabla^2\phi - \epsilon\mu\ddot{\phi} = -\frac{\rho_f}{\epsilon} \qquad (3.6\text{–}6)
$$

$$
\nabla^2\mathbf{A} - \epsilon\mu\ddot{\mathbf{A}} = -\mu\mathbf{J}_f \qquad (3.6\text{–}7)
$$

The corresponding steady-state equations are

$$
\nabla^2\phi = -\frac{\rho_f}{\epsilon} \qquad (3.6\text{–}8)
$$

$$
\nabla^2\mathbf{A} = -\mu\mathbf{J}_f \qquad (3.6\text{–}9)
$$

The second form of the field equations in simple media is derived from

$$
\epsilon\nabla \cdot \mathbf{E} = \rho_f \qquad (3.6\text{–}10)
$$

$$
\mu^{-1}\nabla \times \mathbf{B} = \sigma\mathbf{E} + \epsilon\dot{\mathbf{E}} \qquad (3.6\text{–}11)
$$

The following definitions of the scalar and vector potentials:

$$
-\nabla\phi = \mathbf{E} + \dot{\mathbf{A}} \qquad (3.6\text{–}12)
$$

$$
\nabla \times \mathbf{A} = \mathbf{B} \qquad (3.6\text{–}13a)
$$

$$
\nabla \cdot \mathbf{A} = -\mu(\sigma\phi + \epsilon\dot{\phi}) \qquad (3.6\text{–}13b)
$$

lead to the relations

$$\nabla^2\phi - \sigma\mu\dot{\phi} - \epsilon\mu\ddot{\phi} = -\frac{\rho_f}{\epsilon} \tag{3.6-14}$$

$$\nabla^2\mathbf{A} - \sigma\mu\dot{\mathbf{A}} - \epsilon\mu\ddot{\mathbf{A}} = 0 \tag{3.6-15}$$

These equations differ from the previous ones in that terms in $\dot{\phi}$ and $\dot{\mathbf{A}}$ appear together with the conductivity σ. Since it has been shown that ρ_f vanishes in conductors and is a constant which may be set equal to zero in nonconductors, the following symmetrical equations are usually valid:

$$\nabla^2\phi - \sigma\mu\dot{\phi} - \epsilon\mu\ddot{\phi} = 0 \tag{3.6-16}$$

$$\nabla^2\mathbf{A} - \sigma\mu\dot{\mathbf{A}} - \epsilon\mu\ddot{\mathbf{A}} = 0 \tag{3.6-17}$$

The Lorentz condition that must be satisfied is

$$\nabla \cdot \mathbf{A} + \sigma\mu\phi + \epsilon\mu\dot{\phi} = 0 \tag{3.6-18}$$

Depending on the particular problem to be investigated, one of the three forms of the potential equations may be selected. It is to be noted that the last form involves no volume densities and only the three constitutive parameters.

The boundary conditions for the potential functions at a boundary between two simple media are obtained as in Sec. 3.5:

$$\phi_1 - \phi_2 = 0 \tag{3.6-19}$$

$$\epsilon_1\left(\frac{\partial\phi}{\partial n_1}\right)_1 + \epsilon_2\left(\frac{\partial\phi}{\partial n_2}\right)_2 = \eta_{1f} + \eta_{2f} \tag{3.6-20}$$

$$\left(\frac{\partial\phi}{\partial s}\right)_1 - \left(\frac{\partial\phi}{\partial s}\right)_2 = 0 \tag{3.6-21}$$

$$A_{1s} - A_{2s} = 0; \qquad \mu_1^{-1}A_{1n} - \mu_2^{-1}A_{2n} = 0 \tag{3.6-22}$$

$$\mu_1^{-1}\left(\frac{\partial A_s}{\partial n_1}\right)_1 + \mu_2^{-1}\left(\frac{\partial A_s}{\partial n_2}\right)_2 = K_{1f} + K_{2f} \tag{3.6-23}$$

$$\mu_1^{-1}\left(\frac{\partial A_p}{\partial n_1}\right)_1 + \mu_2^{-1}\left(\frac{\partial A_p}{\partial n_2}\right)_2 = 0 \tag{3.6-24}$$

$$\left(\frac{\partial A_{n,s,p}}{\partial p}\right)_1 - \left(\frac{\partial A_{n,s,p}}{\partial p}\right)_2 = 0 \tag{3.6-25}$$

$$\mu_1^{-1}\left(\frac{\partial A_{n,s}}{\partial s}\right)_1 - \mu_2^{-1}\left(\frac{\partial A_{n,s}}{\partial s}\right)_2 = 0; \qquad \left(\frac{\partial A_p}{\partial s}\right)_1 - \left(\frac{\partial A_p}{\partial s}\right)_2 = 0 \tag{3.6-26}$$

The field equations defining the vectors **E** and **B** have been replaced by two mutually independent potential equations written both in the general form and in the special forms suited to simple media. The advantage of the potential formulation is a mathematical one.

3.7 THE POTENTIAL EQUATIONS WITH PERIODIC TIME DEPENDENCE IN FREE SPACE AND IN SIMPLE MEDIA

When the density functions vary periodically in time according to $\bar{\rho}(t) = \bar{\rho}e^{j\omega t}$, $\overline{\rho_m \mathbf{v}}(t) = \overline{\rho_m \mathbf{v}}e^{j\omega t}$, where now $\bar{\rho}$ and $\overline{\rho_m \mathbf{v}}$ are complex, the potential functions also have this form, namely, $\phi(t) = \phi e^{j\omega t}$, $\mathbf{A}(t) = \mathbf{A}e^{j\omega t}$. In this case the complex potential functions satisfy the following equations in free space:

$$\nabla^2 \phi + k_0^2 \phi = -\frac{\bar{\rho}}{\epsilon_0} \tag{3.7-1a}$$

$$\nabla^2 \mathbf{A} + k_0^2 \mathbf{A} = -\mu_0 \overline{\rho_m \mathbf{v}} \tag{3.7-1b}$$

where $k_0 = \omega(\mu_0 \epsilon_0)^{1/2} = \omega/c$. The Lorentz condition becomes

$$\nabla \cdot \mathbf{A} + \frac{jk_0^2}{\omega} \phi = 0 \tag{3.7-1c}$$

The electromagnetic field vectors are related to the potential functions as follows:

$$\mathbf{E} = -\nabla\phi - j\omega\mathbf{A} = -\frac{j\omega}{k_0^2} (\nabla\nabla \cdot \mathbf{A} + k_0^2 \mathbf{A}) \tag{3.7-2a}$$

$$\mathbf{B} = \nabla \times \mathbf{A} \tag{3.7-2b}$$

In simple media, the equations for the potential functions are

$$\nabla^2 \phi + k^2 \phi = 0 \tag{3.7-3a}$$

$$\nabla^2 \mathbf{A} + k^2 \mathbf{A} = 0 \tag{3.7-3b}$$

where $k = \omega(\mu\bar{\epsilon})^{1/2}$. The Lorentz condition is

$$\nabla \cdot \mathbf{A} + j\frac{k^2}{\omega} \phi = 0 \tag{3.7-3c}$$

The relations for the electromagnetic vectors are given by

$$\mathbf{E} = -\nabla\phi - j\omega\mathbf{A} = -\frac{j\omega}{k^2} (\nabla\nabla \cdot \mathbf{A} + k^2 \mathbf{A}) \tag{3.7-4a}$$

$$\mathbf{B} = \nabla \times \mathbf{A} \tag{3.7-4b}$$

3.8 THE SOLUTIONS OF THE POTENTIAL EQUATIONS: HELMHOLTZ'S INTEGRALS

The solutions of equations (3.7–1a,b) may be written in different forms, but it is generally advantageous to select those which satisfy (3.7–1c). The form of solution that is particularly useful in solving transmission-line and antenna problems is a particular integral called the *Helmholtz integral*. The Helmholtz integrals for (3.7–1a,b) are derived as follows. First consider the equation

$$\nabla^2 v + k_0^2 v = 0 \tag{3.8–1}$$

where v is the scalar point function. Let this equation be written in spherical coordinates (R, Θ, Φ) (see Fig. 3.8–1) under the assumption that v is independent of Θ and Φ. Then (3.8–1) reduces to

$$\frac{1}{R^2} \frac{d}{dR} \left(R^2 \frac{dv}{dR} \right) + k_0^2 v = 0 \tag{3.8–2}$$

which is equivalent to

$$\frac{d^2}{dR^2} (Rv) + k_0^2 (Rv) = 0 \tag{3.8–3}$$

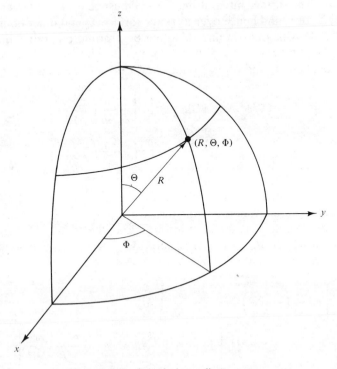

Figure 3.8–1   Spherical coordinate system.

A particular integral of (3.8–3) and hence of (3.8–1) is

$$v = \frac{e^{-jk_0R}}{R} \tag{3.8-4}$$

as may be verified by direct substitution.

Now let ϕ and v be two scalar point functions that are finite and continuous with their first and second derivatives in a volume τ enclosed in a surface Σ_τ. According to Green's symmetrical theorem (3.2–8),

$$\int_\tau (\phi \nabla^2 v - v \nabla^2 \phi) d\tau = \int_{\Sigma_\tau} \left(\phi \frac{\partial v}{\partial n} - v \frac{\partial \phi}{\partial n} \right) d\sigma \tag{3.8-5}$$

Let v in (3.8–5) be identified with v in (3.8–4) noting that R is the distance from a fixed point $P(x, y, z)$ where ϕ is to be determined, to a variable point of integration $P'(x', y', z')$. Both points are within τ. If (3.8–4) is to be substituted in (3.8–5), it is necessary to exclude temporarily the point P, because as P' is moved throughout τ, $1/R$ becomes infinite at $R = 0$. Therefore, let the point P be enclosed in a small sphere of radius a, and let (3.8–5) be formed for the region τ_a which is the same as τ excluding the volume of the small sphere. The surface of this sphere is $\Sigma_a = 4\pi a^2$. The surface integrals in (3.8–5) must be evaluated over the complete boundary Σ_τ of τ_a and hence over its outer surface Σ_0 and inner boundary Σ_a. Thus, using (3.8–4) with (3.8–1) and denoting by a prime quantities that are functions only of the point $P'(x', y', z')$ which locates the element of integration as this is moved throughout τ_a and on its boundaries, (3.8–5) becomes

$$-\int_{\tau_a} \frac{e^{-jk_0R}}{R} (\nabla^2 \phi' + k_0^2 \phi') d\tau' = \int_{\Sigma_0} \left[\phi' \frac{\partial}{\partial n} \left(\frac{e^{-jk_0R}}{R} \right) - \frac{e^{-jk_0R}}{R} \frac{\partial \phi'}{\partial n} \right] d\sigma'$$

$$+ \int_{\Sigma_a} \left[\phi' \frac{\partial}{\partial n} \left(\frac{e^{-jk_0R}}{R} \right) - \frac{e^{-jk_0R}}{R} \frac{\partial \phi'}{\partial n} \right] d\sigma' \tag{3.8-6}$$

The distance $R = [(x - x')^2 + (y - y')^2 + (z - z')^2]^{1/2}$ is a function of both primed and unprimed coordinates. On the surface $\Sigma_a = 4\pi a^2$, the external normal to the volume τ_a is directed inward along the radius. Hence, on this surface $\partial/\partial n = -\partial/\partial R$ and $d\sigma = a^2 d\Omega$, where $d\Omega$ is an element of solid angle. The integral over Σ_a in (3.8–6) becomes

$$\int_{\Sigma_a} \left[\phi' \frac{\partial}{\partial n} \left(\frac{e^{-jk_0R}}{R} \right) - \frac{e^{-jk_0R}}{R} \frac{\partial \phi'}{\partial n} \right] d\sigma'$$

$$= \int_{\Sigma_a} \left[\phi' \left(\frac{1}{R^2} + \frac{jk_0}{R} \right) e^{-jk_0R} + \frac{e^{-jk_0R}}{R} \frac{\partial \phi'}{\partial R} \right]_{R=a} a^2 d\Omega' \tag{3.8-7}$$

Since $\partial\phi'/\partial R$ is finite in τ, (3.8–7) reduces to the following simple form in the limit as a is allowed to approach zero:

$$\lim_{a\to 0}\int_{\Sigma_a}\left[\phi'\frac{\partial}{\partial n}\left(\frac{e^{-jk_0R}}{R}\right) - \frac{e^{-jk_0R}}{R}\frac{\partial\phi'}{\partial n}\right]d\sigma'$$

$$= \lim_{a\to 0}\int_{\Sigma_a}\phi'\,d\Omega'$$

$$= 4\pi\phi \qquad (3.8\text{–}8)$$

In (3.8–8) $d\Omega'$ is an element of solid angle. The function ϕ depends only on the coordinates (x, y, z) of the point P. With (3.8–8) substituted in (3.8–6), the resulting equation is

$$4\pi\phi = -\int_{\tau}\frac{e^{-jk_0R}}{R}(\nabla^2\phi' + k_0^2\phi')\,d\tau'$$

$$-\int_{\Sigma_0}\left[\phi'\frac{\partial}{\partial n}\left(\frac{e^{-jk_0R}}{R}\right) - \frac{e^{-jk_0R}}{R}\frac{\partial\phi'}{\partial n}\right]d\sigma' \qquad (3.8\text{–}9)$$

With the use of (3.7–1a) in the volume integral in (3.8–9) and after the expansion of the integrand of the surface integral, (3.8–9) becomes

$$\phi = \frac{1}{4\pi\epsilon_0}\int_{\tau}\frac{\bar{\rho}'}{R}e^{-jk_0R}\,d\tau'$$

$$+ \frac{1}{4\pi}\int_{\Sigma_0}\left[R\left(\frac{\partial\phi'}{\partial n} + jk_0\phi'\frac{\partial R}{\partial n}\right) + \phi'\frac{\partial R}{\partial n}\right]\frac{e^{-jk_0R}}{R^2}\,d\sigma' \qquad (3.8\text{–}10)$$

This is the complete expression for the scalar potential if, as has been assumed, ϕ and its first derivative are continuous throughout the region τ. The boundary conditions on ϕ and its derivatives, (3.5–19) to (3.5–21), require ϕ to be continuous but its normal derivative at a boundary with a nonvanishing surface charge to be discontinuous. If such boundaries exist within τ, they must be excluded in the integration. This is done by enclosing such a boundary between two parallel surfaces each at a constant distance d_c (atomic dimension) from the boundary. The volume integral in (3.8–10) is then carried out over all volumes excluding only the narrow slices of thickness $2d_c$ between parallel surfaces Σ_i enclosing boundaries where $\partial\phi/\partial n$ is discontinuous. Additional surface integrals appear on the right in (3.8–10). They have the form

$$\int_{\Sigma_i}\left[\left(\frac{\partial\phi'}{\partial n_1}\right)_1 + \left(\frac{\partial\phi'}{\partial n_2}\right)_2\right]\frac{e^{-jk_0R}}{R}\,d\sigma' \qquad (3.8\text{–}11)$$

where the subscripts 1 and 2 are used to designate the regions on each side of the boundary; $\hat{n}_1$ and $\hat{n}_2$ are external normals to the regions indicated by the subscript at the boundary. The normal derivatives are taken on each side of the boundary. Equation (3.5–25) may be written for complex amplitudes

$$\left(\frac{\partial \phi}{\partial n_1}\right)_1 + \left(\frac{\partial \phi}{\partial n_2}\right)_2 = \frac{\bar{\eta}}{\epsilon_0} \tag{3.8–12}$$

With (3.8–12) substituted in (3.8–11), and (3.8–11) included in (3.8–10), the resulting expression for ϕ is

$$\phi = \frac{1}{4\pi\epsilon_0}\left(\int_\tau \frac{\bar{\rho}'}{R} e^{-jk_0R}\, d\tau' + \int_\Sigma \frac{\bar{\eta}'}{R} e^{-jk_0R}\, d\sigma'\right) + \frac{1}{4\pi}\int_{\Sigma_0} u'\, d\sigma' \tag{3.8–13}$$

For convenience in writing, only a single volume integral and a single surface integral are shown. In general, the volume integral is to be evaluated over all regions in which $\bar{\rho}$ differs from zero, the surface integral over all boundaries where $\bar{\eta}$ is defined. The following shorthand is used:

$$u' \equiv \left[R\left(\frac{\partial \phi'}{\partial n} + jk_0\phi' \frac{\partial R}{\partial n}\right) + \phi' \frac{\partial R}{\partial n}\right] \frac{e^{-jk_0R}}{R^2} \tag{3.8–14}$$

Since each of the Cartesian components of **A** satisfies an equation like (3.7–1a), each must have a solution of the form (3.8–13). If these are combined vectorially and use is made of the boundary conditions (3.5–27) to (3.5–31) for complex amplitudes, the integral solution of (3.7–1b) is

$$\mathbf{A} = \frac{\mu_0}{4\pi}\left(\int_\tau \frac{\overline{\rho_m \mathbf{v}'}}{R} e^{-jk_0R}\, d\tau' + \int_\Sigma \frac{\overline{\eta_m \mathbf{v}'}}{R} e^{-jk_0R}\, d\sigma'\right)$$

$$+ \frac{1}{4\pi}\int_{\Sigma_0} \mathbf{u}'_A\, d\sigma' \tag{3.8–15}$$

with

$$\mathbf{u}'_A = \left[R\left(\frac{\partial \mathbf{A}'}{\partial n} + jk_0 \mathbf{A}' \frac{\partial R}{\partial n}\right) + \mathbf{A}' \frac{\partial R}{\partial n}\right] \frac{e^{-jk_0R}}{R^2} \tag{3.8–16}$$

In applying (3.8–13) and (3.8–15) to the solution of electromagnetic problems, it is necessary to know u' and $\mathbf{u}'_A$ on the envelope Σ_0 that encloses the regions in which periodically varying distributions of charge and current are maintained. The location of this envelope is quite arbitrary and may be chosen wherever convenient. If it is possible to locate it in such a way that both u' and $\mathbf{u}'_A$ vanish everywhere on its surface, the last integral in (3.8–13) and (3.8–15) vanishes and the calculation of scalar and vector potentials reduces to integrations over such volumes and surfaces within Σ_0 in and on which nonvanishing densities of periodically varying charge and current exist. This is true whenever the surface Σ_0 can be chosen in such a way that on it both ϕ and **A** are zero at all points. Both ϕ and **A** vanish with **E** and **B**, and the amplitude of a periodically varying electromagnetic field reduces to negligibly small values in the interior of a sufficiently good conductor. Therefore, if a periodically varying distribution of charge is maintained in a cavity in a good conductor, it may be assumed that u' and $\mathbf{u}'_A$ are both zero if Σ_0 is entirely in the conductor.

For the general solution of electromagnetic problems that involve radiating sources in space, the last integrals in (3.8–13) and (3.8–15) are required to vanish by the imposition of a *radiation condition* in the general form

$$\lim_{R \to \infty} R \left(\frac{\partial \phi}{\partial R} + jk_0 \phi \right) = 0 \tag{3.8–17}$$

and a similar relation for the vector potential. This condition is equivalent to a requirement that the potentials approach infinity as outward-traveling waves of the form e^{-jk_0R}/R. The radiation condition is usually combined with the "condition for regularity at infinity"

$$\lim_{R \to \infty} (R\phi) \text{ is finite} \tag{3.8–18}$$

and a similar condition for the vector potential.

The radiation condition can be shown to be a consequence of the principle of causality, which states that a signal cannot be observed before it has been generated and emitted.

The solutions of (3.7–1a, b) can be shown to be unique provided that (3.8–17) and (3.8–18) are satisfied. They are given by

$$\phi = \frac{1}{4\pi\epsilon_0} \left[\int_\tau \frac{\overline{\rho}'}{R} e^{-jk_0R} \, d\tau' + \int_\Sigma \frac{\overline{\eta}'}{R} e^{-jk_0R} \, d\sigma' \right] \tag{3.8–19}$$

$$\mathbf{A} = \frac{\mu_0}{4\pi} \left[\int_\tau \frac{\overline{\rho_m \mathbf{v}'}}{R} e^{-jk_0R} \, d\tau' + \int_\Sigma \frac{\overline{\eta_m \mathbf{v}'}}{R} e^{-jk_0R} \, d\sigma' \right] \tag{3.8–20}$$

The integration is to be carried out over all regions and surfaces where nonvanishing density functions are defined. Since the electromagnetic vectors and the potential functions are by definition computed *only* from charges and moving charges, all such solutions that are physically significant must be due to distributions of charge and current somewhere in space. Hence, if the integrations in (3.8–19) and (3.8–20) are assumed to be taken over all regions with nonvanishing values of the density functions, (3.8–19) and (3.8–20) are the general solutions. Equations (3.8–19) and (3.8–20) are used to determine ϕ and $\mathbf{A}$ due to charges and currents in a particular region such as an antenna. This problem is discussed in later sections.

An especially simple form of the vector potential is that of an idealization known as an *infinitesimal dipole*. This is essentially extensionless and has an electric dipole moment

$$\mathbf{p} = \int_\tau \mathbf{P}' \, d\tau' \tag{3.8–21}$$

It is obtained from (3.8–20) with $\overline{\rho_m \mathbf{v}} = j\omega\mathbf{P}$ and with $R \sim R_0$, the distance to the center of the dipole. Thus

$$\mathbf{A} = \frac{j\omega\mu_0\mathbf{p}}{4\pi} \frac{e^{-jk_0R_0}}{R_0} \tag{3.8–22}$$

Since $p = 2Qh$ where $2h$ is the length of the dipole and Q the periodically varying charge at one end, and since $j\omega Q = I$, the current generated by the moving charge, it follows that

$$j\omega\mathbf{p} = 2h\mathbf{I} \tag{3.8–23}$$

so that

$$\mathbf{A} = \frac{\mu_0 h\mathbf{I}}{2\pi} \frac{e^{-jk_0R_0}}{R_0} \tag{3.8–24}$$

3.9 SKIN EFFECT AND INTERNAL IMPEDANCE

The transverse distribution of current in rotationally symmetric cylindrical conductors depends on the field outside the conductor primarily in terms of the field on its surface. This makes it possible to determine in approximate but very general terms the circuit properties per unit length of such a conductor for a simple geometry and then use these in the determination of the complete characteristics of closed and quasi-closed electric circuits, transmission lines, and antennas.

Consider a circular conducting cylinder of radius a that extends along the z axis of a system of cylindrical coordinates ρ, θ, z. It is contained in a cylindrical cavity of radius b surrounded by an infinite conducting medium. A rotationally symmetric electric field with the complex amplitude $E_z(a)$ is maintained on its surface $\rho = a$. The transverse distribution of the axially directed field $E_z(\rho)$ and of the volume density of current $J_z(\rho)$ for $\rho \leq a$ can be determined from the general equation for the vector potential in a simple medium, that is, from

$$\nabla^2\mathbf{A} + k^2\mathbf{A} = 0 \tag{3.9–1}$$

with

$$k = \omega(\mu\bar{\epsilon})^{1/2} \tag{3.9–2a}$$

$$\bar{\epsilon} = \epsilon_0 \left(\epsilon_{er} - j\frac{\sigma_e}{\omega\epsilon_0} \right) \tag{3.9–2b}$$

Equation (3.9–1) can be expressed in the cylindrical coordinates (ρ, θ, z) and, since there is rotational symmetry,

$$\frac{\partial\mathbf{A}}{\partial\theta} = 0; \qquad A_\theta = 0 \tag{3.9–3}$$

Furthermore, by definition,

$$\nabla^2 \mathbf{A} = \nabla\nabla \cdot \mathbf{A} - \nabla \times \nabla \times \mathbf{A} \tag{3.9–4}$$

The expansion of (3.9–4) in cylindrical coordinates and its substitution into (3.9–1) result in the following equations:

$$\frac{\partial^2 A_\rho}{\partial z^2} + \frac{\partial}{\partial\rho} \frac{1}{\rho} \frac{\partial}{\partial\rho} (\rho A_\rho) + k^2 A_\rho = 0 \tag{3.9–5}$$

$$\frac{\partial^2 A_z}{\partial z^2} + \frac{1}{\rho} \frac{\partial}{\partial\rho} \rho \frac{\partial A_z}{\partial\rho} + k^2 A_z = 0 \tag{3.9–6}$$

The solution of (3.9–5) is obtained by the method of separation of variables. $A_z(\rho, z)$ is written as follows:

$$A_z(\rho, z) = f_z(z)F_z(\rho) \tag{3.9–7}$$

where $f_z(z)$ is a dimensionless function of z alone and $F_z(\rho)$ is a function of ρ alone. The substitution of (3.9–7) in (3.9–6) yields the equation

$$F_z \frac{\partial^2 f_z}{\partial z^2} + f_z \frac{1}{\rho} \frac{\partial}{\partial\rho} \frac{\partial F_z}{\partial\rho} + k^2 f_z F_z = 0 \tag{3.9–8}$$

When (3.9–8) is written in the form

$$\frac{1}{f_z} \frac{\partial^2 f_z}{\partial z^2} + k^2 = -\frac{1}{F_z} \frac{1}{\rho} \frac{\partial}{\partial\rho} \rho \frac{\partial F_z}{\partial\rho} \tag{3.9–9}$$

it is seen that the left side is a function of z alone and the right side is a function of ρ alone. Hence the two sides of (3.9–9) can be equal to each other only if both are equal to a constant that may be multivalued. If this constant is denoted by the symbol κ^2 and

$$\gamma^2 = \kappa^2 - k^2 \tag{3.9–10}$$

then

$$\frac{\partial^2 f_z}{\partial z^2} - \gamma^2 f_z = 0 \tag{3.9–11}$$

$$\frac{1}{\rho} \frac{\partial}{\partial\rho} \rho \frac{\partial F_z}{\partial\rho} + \kappa^2 F_z = 0 \tag{3.9–12}$$

Equation (3.9–12) is differentiated to get

$$\frac{\partial^2 F_z}{\partial\rho^2} + \frac{1}{\rho} \frac{\partial F_z}{\partial\rho} + \kappa^2 F_z = 0 \tag{3.9–13}$$

A further transformation may be obtained with the new independent variable

$$x = \kappa\rho \tag{3.9–14a}$$

With it

$$\frac{\partial}{\partial x} = \frac{\partial}{\partial \rho}\frac{d\rho}{dx} = \frac{1}{\kappa}\frac{\partial}{\partial \rho} \tag{3.9-14b}$$

$$\frac{\partial^2}{\partial x^2} = \frac{\partial}{\partial x}\left(\frac{1}{\kappa}\frac{\partial}{\partial \rho}\right) = \frac{1}{\kappa^2}\frac{\partial^2}{\partial \rho^2} \tag{3.9-14c}$$

Equation (3.9–13) is now written in terms of the new independent variable

$$\frac{\partial^2 F_z}{\partial x^2} + \frac{1}{x}\frac{\partial F_z}{\partial x} + F_z = 0 \tag{3.9-15}$$

Equation (3.9–15) is known as the *Fourier equation*.

Solutions for (3.9–5) may also be written in the form

$$A_\rho(\rho, z) = f_\rho(z)F_\rho(\rho) \tag{3.9-16}$$

With the same procedure as that used for $A_z(\rho, z)$,

$$\frac{\partial^2 f_\rho}{\partial z^2} - \gamma^2 f_\rho = 0 \tag{3.9-17a}$$

$$\frac{d}{d\rho}\frac{1}{\rho}\frac{\partial}{\partial \rho}(\rho F_\rho) + \kappa^2 F_\rho = 0 \tag{3.9-17b}$$

where κ is not necessarily the same constant as before. Differentiation yields

$$\frac{\partial^2}{\partial \rho^2}F_\rho + \frac{1}{\rho}\frac{\partial F_\rho}{\partial \rho} + \left(\kappa^2 - \frac{1}{\rho^2}\right)F_\rho = 0 \tag{3.9-17c}$$

or

$$\frac{\partial^2}{\partial x^2}F_\rho + \frac{1}{x}\frac{\partial F_\rho}{\partial x} + \left(1 - \frac{1}{x^2}\right)F_\rho = 0 \tag{3.9-17d}$$

Equations (3.9–15) and (3.9–17d) are special cases of the general *Bessel equation*

$$\frac{\partial^2 F}{\partial x^2} + \frac{1}{x}\frac{\partial F}{\partial x} + \left(1 - \frac{n^2}{x^2}\right)F = 0 \tag{3.9-18}$$

Solutions for $A(\rho, z)$ may be obtained in terms of solutions of the Bessel equation.

Electromagnetic Field and Boundary Conditions for a Cylindrical Conductor

The electromagnetic vectors in a simple medium can be calculated from the complex vector potential with the formulas

$$\mathbf{E} = \frac{-j\omega}{k^2}(\mathbf{\nabla\nabla}\cdot\mathbf{A} + k^2\mathbf{A}) \tag{3.9-19a}$$

$$\mathbf{B} = \mathbf{\nabla}\times\mathbf{A} \tag{3.9-19b}$$

With (3.9–1) and (3.9–4),

$$\mathbf{E} = \frac{-j\omega}{k^2} \nabla \times \nabla \times \mathbf{A} \tag{3.9–20}$$

at points in a simple medium. Hence

$$E_\rho = \frac{j\omega}{k^2} \left[\frac{\partial}{\partial z} \left(\frac{\partial A_\rho}{\partial z} - \frac{\partial A_z}{\partial \rho} \right) \right]; \qquad B_\rho = 0 \tag{3.9–21a}$$

$$E_\theta = 0; \qquad B_\theta = \frac{\partial A_\rho}{\partial z} - \frac{\partial A_z}{\partial \rho} \tag{3.9–21b}$$

$$E_z = \frac{-j\omega}{k^2} \left[\frac{1}{\rho} \frac{\partial}{\partial \rho} \rho \left(\frac{\partial A_\rho}{\partial z} - \frac{\partial A_z}{\partial \rho} \right) \right]; \qquad B_z = 0 \tag{3.9–21c}$$

Since all components of the field can be expressed in terms of either A_ρ or A_z alone, it is usually possible and convenient to set one or the other of the two components equal to zero without loss of generality. The preferred choice is likely to be the component most closely related to the dominant component of current: A_z for J_z, A_ρ for J_ρ. In the long cylindrical conductor only the axial current is of interest, so that it is appropriate to choose A_z and set $A_\rho = 0$. It then follows that

$$E_\rho = \frac{-j\omega}{k^2} \frac{\partial^2 A_z}{\partial z \partial \rho}; \qquad B_\rho = 0 \tag{3.9–22a}$$

$$E_\theta = 0; \qquad B_\theta = -\frac{\partial A_z}{\partial \rho} \tag{3.9–22b}$$

$$E_z = \frac{j\omega}{k^2} \frac{1}{\rho} \frac{\partial}{\partial \rho} \rho \left(\frac{\partial A_z}{\partial \rho} \right) = -j\omega \frac{\kappa^2}{k^2} A_z; \qquad B_z = 0 \tag{3.9–22c}$$

The general boundary conditions for the tangential components of $\mathbf{E}$ and $\mathbf{B}$ at the boundary between region 1, a cylindrical conductor with radius a, and region 2, air, have the following form:

$$E_{1z}(a) = E_{2z}(a); \qquad \mu_1^{-1} B_{1\theta}(a) = \mu_2^{-1} B_{2\theta}(a) \tag{3.9–23}$$

They may be expressed in terms of A_z as follows:

$$\frac{\kappa_1^2}{k_1^2} A_{1z}(a) = \frac{\kappa_2^2}{k_2^2} A_{2z}(a); \qquad \frac{1}{\mu_1} \left[\frac{\partial A_{1z}(\rho)}{\partial \rho} \right]_{\rho=a} = \frac{1}{\mu_2} \left[\frac{\partial A_{2z}(\rho)}{\partial \rho} \right]_{\rho=a} \tag{3.9–24}$$

Region 2 (air) has a radius b; beyond it is a conducting region 3 that extends to infinity. At the boundary $\rho = b$, between regions 2 and 3, relations corresponding to (3.9–23) obtain. These are the equations from which the separation constants κ_1 and κ_2 must be determined. Before this can be done expressions for A_z in the conductor ($\rho \leq a$) and outside the conductor, $\rho \geq a$, must be obtained.

Since the cylinder, region 1, $\rho \leq a$, and the outside infinite region 3, $\rho \geq b$, are good conductors and the medium between them, region 2, is air, the following forms of the wave numbers are applicable:

$$k_3 = k_1 = \beta_1 - j\alpha_1 = (-j\omega\mu_1\sigma_1)^{1/2} = (1 - j)\left(\frac{\omega\mu_1\sigma_1}{2}\right)^{1/2} \qquad (3.9\text{--}25)$$

It is assumed that $\mu_3 = \mu_1$ is real.

$$k_2 = k_0 = \omega(\mu_0\epsilon_0)^{1/2} = \frac{\omega}{c} \qquad (3.9\text{--}26)$$

Solutions for $A_z(\rho, z)$ in terms of $f_z(z)$ and $F_z(\rho)$ as obtained from (3.9–11) and (3.9–13) that are appropriate to the three regions are

$$A_{1z}(\rho, z) = D_1 J_0(\kappa_1\rho) \exp(-\gamma_1 z), \qquad\qquad \rho \leq a \qquad (3.9\text{--}27)$$

$$A_{2z}(\rho, z) = [D_2 J_0(\kappa_2\rho) + D_3 N_0(\kappa_2\rho)] \exp(-\gamma_2 z), \qquad a \leq \rho \leq b \quad (3.9\text{--}28)$$

$$A_{3z}(\rho, z) = D_4 H_0(\kappa_3\rho) \exp(-\gamma_3 z), \qquad\qquad \rho \geq b \qquad (3.9\text{--}29)$$

The indicated choice of Bessel functions which are solutions of (3.9–18) is made so that A_z does not become infinite either at $\rho = 0$ in the inner conductor or at $\rho = \infty$ in the outer conducting medium. The superscript (1) or (2) must be added to $H_0(\kappa_3\rho)$ in (3.9–29) according to which one vanishes at $\rho = \infty$. Note that for a wave traveling in the positive z direction in all three regions it is necessary that $\gamma_1 = \gamma_2 = \gamma_3$ or, since $k_3 = k_1$,

$$\sqrt{\kappa_1^2 - k_1^2} = \sqrt{\kappa_2^2 - k_2^2} \qquad (3.9\text{--}30)$$

When (3.9–27) to (3.9–29) are substituted in (3.9–24) and in the corresponding expressions at $\rho = b$, the following equations are readily obtained:

$$\frac{\mu_1\kappa_1}{k_1^2} \frac{J_0(\kappa_1 a)}{J_0'(\kappa_1 a)} = \frac{\mu_2\kappa_2}{k_2^2} \frac{J_0(\kappa_2 a) + CN_0(\kappa_2 a)}{J_0'(\kappa_2 a) + CN_0'(\kappa_2 a)} \qquad (3.9\text{--}31)$$

$$\frac{\mu_2\kappa_2}{k_2^2} \frac{J_0(\kappa_2 b) + CN_0(\kappa_2 b)}{J_0'(\kappa_2 b) + CN_0'(\kappa_2 b)} = \frac{\mu_1\kappa_1}{k_1^2} \frac{H_0(\kappa_1 b)}{H_0'(\kappa_1 b)} \qquad (3.9\text{--}32)$$

where $C = D_4/D_3$ is an arbitrary constant and where the primes denote differentiation with respect to the argument. The only solution of these equations that is significant for an inner conductor of sufficiently small cross section so that $k_2 a = k_0 a \leq 1$ is the one valid for very good conductors for which k_1^2 and κ_1^2 are both very large. This solution is readily obtained from (3.9–31) and (3.9–32) in the limit of very large values of $\sigma_3 = \sigma_1$, so that $k_1 = \sqrt{-j\omega\mu_1\sigma_1}$ is also very large. Since $\gamma_1 = \sqrt{\kappa_1^2 - k_1^2}$ must remain finite even when $k_1 \to \infty$, it is clear that κ_1 must also be very large. That is,

$$\kappa_1 \sim k_1 \sim \sqrt{-j\omega\mu_1\sigma_1} = |k_1|e^{-j\pi/4}; \qquad |k_1| = \sqrt{\omega\mu_1\sigma_1} \qquad (3.9\text{--}33)$$

It follows that the left side of (3.9–31) and the right side of (3.9–32) are very small with k_1^2 in the denominator. This requires κ_2 to be very small, approaching zero in the limit $\sigma_1 \to \infty$. With $\gamma_1 = \sqrt{\kappa_1^2 - k_1^2} = \sqrt{\kappa_2^2 - k_2^2}$, it is evident that

$$\gamma_1 \sim jk_2 = jk_0 \tag{3.9-34}$$

It is readily verified with the use of the small argument approximations of the Bessel functions $[J_0(x) \sim 1, J_0'(x) \sim -x/2, N_0(x) \sim (2/\pi)[\ln{(x/2)} + 0.5772], N_0'(x) = 2/\pi x]$ that the limit $\kappa_2 \to 0$ is consistent with (3.9–31) and (3.9–32) in a manner that is *independent* of both C and the outer radius b. This means that the solution (3.9–27) for the interior of the inner conductor is independent of the presence of the outer conductor. That is,

$$A_{1z}(\rho, z) \sim D_1 J_0(\kappa_1 \rho) e^{-jk_{0}z} \tag{3.9-35}$$

for any cylindrical conductor that is excited to preserve complete rotational symmetry. This formula is readily expressed in the form

$$A_{1z}(\rho) = A_{1z}(a) \frac{J_0(\kappa_1 \rho)}{J_0(\kappa_1 a)} \tag{3.9-36}$$

If $E_z(a)$ is the z component of the electric field at the surface $\rho = a$ of the conductor, it follows from (3.9–22c) that

$$E_{1z}(\rho) = E_{1z}(a) \frac{J_0(\kappa_1 \rho)}{J_0(\kappa_1 a)} \tag{3.9-37}$$

Also, since the volume density of current is $J_z = \sigma E_z$, it follows that

$$J_{1z}(\rho) = J_{1z}(a) \frac{J_0(\kappa_1 \rho)}{J_0(\kappa_1 a)} \tag{3.9-38}$$

The solutions for the vector potential, electric field, and current in the interior of the conductor involve Bessel functions of the form $J_0(j^{-1/2}y)$ with $y = |k_1|\rho$ real and $|k_1| = \sqrt{\omega\mu_1\sigma_1}$. The Bessel functions can be expressed in magnitude-angle form:

$$J_0(j^{-1/2}y) = M_0(y) e^{j\theta_0(y)} \tag{3.9-39}$$

The functions $M_0(y)$ and $\theta_0(y)$ are tabulated in Appendix III. It follows that

$$\frac{A_z(\rho)}{A_z(a)} = \frac{E_z(\rho)}{E_z(a)} = \frac{J_z(\rho)}{J_z(a)} = \frac{M_0(|k_1|\rho)}{M_0(|k_1|a)} e^{-j[\theta_0(|k_1|a) - \theta_0(|k_1|\rho)]} \tag{3.9-40}$$

These relations characterize completely the distribution of the *axial* components of the vector potential, electric field, and current density in any cross section of the conductor in terms of the values at the surface, $\rho = a$. When $|k_1|\rho$ is sufficiently large,

$$M_0(y) \doteq \frac{1}{(2\pi y)^{1/2}} e^{y/\sqrt{2}}, \qquad y \geq 10 \tag{3.9-41a}$$

$$\theta_0(y) \doteq \frac{y}{\sqrt{2}} - \frac{\pi}{8}, \qquad y \geq 10 \tag{3.9-41b}$$

Then

$$\frac{A_z(\rho)}{A_z(a)} = \frac{E_z(\rho)}{E_z(a)} = \frac{J_z(\rho)}{J_z(a)} = \sqrt{\frac{a}{\rho}}\, e^{-\alpha_1(a-\rho)} e^{-j\beta_1(a-\rho)} \tag{3.9-42}$$

Since $(a - \rho)$ measures the distance s radially in from the surface and

$$\alpha_1 = \beta_1 = \sqrt{\omega\mu_1\sigma_1/2} = \frac{1}{d_s} \tag{3.9-43}$$

where d_s is the *skin depth*,

$$\frac{A_z(s)}{A_z(s = 0)} = \frac{E_z(s)}{E_z(s = 0)} = \frac{J_z(s)}{J_z(s = 0)} = \sqrt{\frac{a}{a-s}}\, e^{-s/d_s} e^{-js/d_s} \tag{3.9-44}$$

The skin depth d_s is the radial distance from the surface at which J_z is reduced to $1/e$ of its value at the surface provided that this distance is negligibly small compared with the radius a.

Curves showing the magnitude of the ratio of volume density of current $J_z(\rho)$ at radius ρ to that at radius a and the relative phase are given in Fig. 3.9–1. For the larger values of $|k_1|a$, the phase reverses many times from the surface of the conductor to the axis so that the real current density is characterized by concentric rings of current alternately in opposite directions. The real instantaneous current density $J_z(\rho, t)$ is given by

$$\frac{J_z(\rho, t)}{J_z(a)} = \frac{M_0(|k_1|\rho)}{M_0(|k_1|a)} \cos\left[\omega t + \theta_0(|k_1|\rho) - \theta_0(|k_1|a)\right] \tag{3.9-45}$$

The total current $I_z = \int_0^a J_z(\rho)2\pi\rho\, d\rho$ is readily evaluated and expressed in terms of $E_z(a) = J_z(a)/\sigma_1$. The ratio of the axial electric field at the surface of a given cross section of the conductor to the total current across that cross section, that is, $E_z(a)/I_z(a)$, is a complex quantity called the *internal impedance per unit length of the conductor*. It is denoted by $z^i = r^i + jx^i$; where r^i is the internal or ohmic resistance per unit length, x^i is the internal reactance per unit length. Then

$$z^i = r^i + jx^i = \frac{E_z(a)}{I_z(a)}$$

$$= \frac{1}{\pi a^2\sigma_1}\left(\frac{k_1 a}{2}\right)\frac{J_0(k_1 a)}{J_1(k_1 a)}$$

$$= \frac{1}{\pi a^2\sigma_1}\left(\frac{|k_1|a}{2}\right)\frac{M_0(|k_1|a)}{M_1(|k_1|a)}\, e^{-j[\theta_1(|k_1|a) - \theta_0(|k_1|a) - 3\pi/4]} \tag{3.9-46a}$$

with $k_1 = \beta_1 - j\alpha_1 = j^{-1/2}|k_1| = e^{-j\pi/4}\sqrt{\omega\mu_1\sigma_1}$.

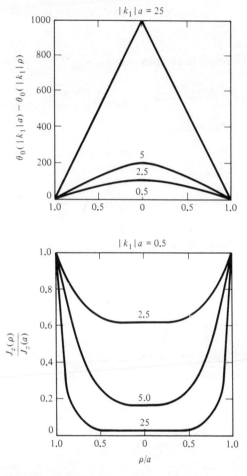

Figure 3.9–1 Relative amplitude and phase angle of the volume density of axial current in a cylindrical conductor for four values of $|k_1|a$.

The requirement that $|k_1|a$ be small is the equivalent in the conductor of the condition for the near zone. In this case,

$$r^i \doteq r_0 = \frac{1}{\pi a^2 \sigma_1}; \qquad x^i = 0; \qquad (|k_1|a)^2 \ll 4 \qquad (3.9\text{–}46b)$$

r_0 is the dc resistance per unit length. The complex frequency factor for the conductor is

$$F = \frac{|k_1|a}{2} \frac{M_0(|k_1|a)}{M_1(|k_1|a)} \, e^{-j[\theta_1(|k_1|a) - \theta_0(|k_1|a) - 3\pi/4]} \qquad (3.9\text{–}46c)$$

The internal impedance may therefore be written as follows:

$$z^i = r_0 F \tag{3.9-47}$$

For small values of $|k_1|a$ but not as small as assumed in (3.9–46b),

$$r^i \doteq r_0 = \frac{1}{\pi a^2 \sigma_1}; \qquad x^i \doteq \frac{\omega \mu_1}{8\pi}; \qquad (|k_1|a)^4 \ll 192 \tag{3.9-48}$$

If $|k_1|a$ is sufficiently large,

$$z^i = r^i + jx^i \tag{3.9-49}$$

$$= \frac{1+j}{2\pi a} \sqrt{\frac{\omega \mu_1}{2\sigma_1}}, \qquad |k_1|a \geq 10 \tag{3.9-50}$$

This is known as the *Rayleigh formula*. For small arguments, the density of current is sensibly constant and the resistance per unit length is inversely proportional to the area of cross section. For large arguments, on the other hand, most of the current is confined to a relatively thin layer near the surface and the impedance per unit length is inversely proportional to the circumference.

If the argument $|k_1|a$ is sufficiently large to permit the use of the asymptotic formulas for the Bessel functions, the total axial current $I_z(a)$ is distributed in a layer sufficiently near the circumference to allow the definition of a *quasi-surface current* K_z' which is equal to the magnetic H-field at the surface of the cylindrical conductor. Thus

$$K_z' = H_\theta(a) = \frac{I_z(a)}{2\pi a} \tag{3.9-51}$$

The ratio

$$Z^s = \frac{E_z(a)}{H_\theta(a)} = 2\pi a z^i \tag{3.9-52}$$

is called the *surface impedance*. When $|k_1|a$ is large,

$$Z^s = R^s + jX^s = (1+j)\sqrt{\frac{\omega \mu_1}{2\sigma_1}} = \frac{1+j}{d_s \sigma_1} \tag{3.9-53}$$

The surface impedance depends only on the material parameters μ_1 and σ_1 and the frequency. Since Z^s is independent of the radius of the conductor, this can be allowed to approach infinity so that the cylinder becomes an infinite plane sheet. The surface impedance (3.9–53) is a good approximation for metal sheets that are thick compared with the skin depth d_s. It may be concluded that the surface impedance expresses a relation between the tangential and mutually perpendicular components of the electric and magnetic fields at the surface of any highly conducting sheet with a thickness that is great compared with the skin depth. The relation (3.9–52) is sometimes called the *impedance boundary condition*.

The current distribution and internal impedance per unit length of a tubular conductor of inner radius b and outer radius c can be calculated in a manner similar to that for a solid conductor. If the thickness of the wall is large compared with the skin depth [i.e., $(c - b) \geq 4d_s$], the density at radial distances $s = c - r$ from the outer surface such that $(r - b)/d_s \geq 4$ is given by

$$\frac{J_z(s)}{J_z(s = 0)} = \sqrt{\frac{c}{c - s}}\, e^{-s/d_s} e^{-js/d_s} \tag{3.9-54}$$

Equation (3.9–54) is of the same form as (3.9–44) for a solid conductor of the same radius. The internal impedance is given by

$$z^i = \frac{E_z(c)}{I_z(c)} = \frac{k_1}{2\pi c \sigma_1} \frac{1 + j}{\sqrt{2}} \tag{3.9-55}$$

The electric field $E_z(b) = E_i$ in the space inside a metal tube when $|k_1|b \geq 10$ can be shown to be:

$$E_z(b) = E_i = \frac{J_z(s = 0)}{\sigma_1}\, 2 \sqrt{\frac{c}{b}}\, e^{-(c-b)/d_s} e^{-j(c-b)/d_s} \tag{3.9-56}$$

The field E_o at the outer surface is obtained with $r = c$ or $s = 0$,

$$E_z(c) = E_o = \frac{J_z(s = 0)}{\sigma_1} \tag{3.9-57}$$

The ratio of the field inside the tube to the field in space just outside the tube is

$$\frac{E_i}{E_o} = 2 \sqrt{\frac{c}{b}}\, e^{-(c-b)/d_s} e^{-j(c-b)/d_s} \tag{3.9-58}$$

The magnitude of the ratio in (3.9–58) is

$$\left| \frac{E_i}{E_o} \right| = 2 \sqrt{\frac{c}{b}}\, e^{-(c-b)/d_s} \tag{3.9-59}$$

Equation (3.9–59) illustrates the principle underlying one aspect of *electromagnetic shielding*. By making the ratio $(c - b)/d_s$ sufficiently large (or d_s sufficiently small), E_i, the field in the space enclosed by the metal tube, can be made as small as desired. At high frequencies, materials such as copper, brass, or aluminum make very effective shields. At low frequencies, very thick copper or a ferromagnetic alloy is required. Note that (3.9–59) applies equally well (approximately) to plane sheets of metal. A second aspect of electromagnetic shielding involves the reduction in magnitude of the electric field on the surface of the conductor by reflection or scattering. The reflection coefficient for a plane wave normally incident on a half-space is given by (2.17–11b). It shows that as the frequency is reduced to zero, the reflection coefficient approaches -1. At zero frequency a metal surface is a perfect shield.

3.10 APPLICATION OF THE POTENTIAL FUNCTIONS TO THE DERIVATION OF THE INTEGRAL EQUATION FOR THE CURRENT IN A CYLINDRICAL ANTENNA

An instructive and subsequently useful application of the potential functions is to the formulation of the equation that governs the distribution of current along a center-driven, electrically thin tubular conductor that lies along the z axis of a system of cylindrical coordinates (ρ, θ, z) as shown in Fig. 3.10–1. The surface of the cylindrical antenna is at $\rho = a$ and $-h \leq z \leq h$. An antenna is usually driven by a voltage maintained across its terminals by a coaxial or two-wire transmission line. When the cross section of the line is electrically sufficiently small, the driving voltage can be approximated by an emf concentrated across a narrow slice of the

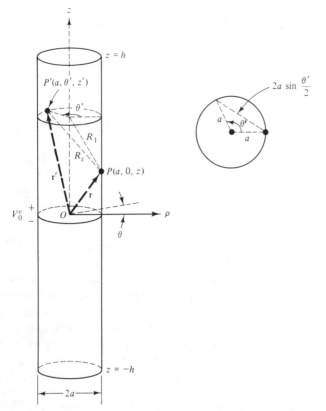

Figure 3.10–1 Cylindrical antenna.

antenna. Such a generator can be represented by a delta function with active electric field $V_0^e\delta(z)$.

The antenna is approximated by a perfectly conducting cylinder; so there are only surface densities of current $K(z, t)$ and charge $\eta(z, t)$. The boundary condition on the tangential component of an electric field requires that it vanish at the surface of a perfect conductor. Then

$$E_z(\mathbf{r}, t) = -V_0^e(t)\delta(z), \qquad \text{when } \rho = a, \quad -h \leq z \leq h \qquad (3.10\text{-}1)$$

with $\mathbf{r} = \boldsymbol{\rho} + \mathbf{z}$ as shown in Fig. 3.10–1. If the generator is assumed to oscillate at a single frequency,

$$V_0^e(t) = \text{Re}\,(V_0^e e^{j\omega t}); \qquad E_z(\mathbf{r}, t) = \text{Re}\,[E_z(\mathbf{r})e^{j\omega t}] \qquad (3.10\text{-}2)$$

Equation (3.7–4a) gives the relationship between the vector potential function $\mathbf{A}$ and the electric field vector $\mathbf{E}$:

$$\mathbf{E}(\mathbf{r}) = \frac{-j\omega}{k_0^2}\,[\nabla\nabla \cdot \mathbf{A}(\mathbf{r}) + k_0^2\mathbf{A}(\mathbf{r})] \qquad (3.10\text{-}3a)$$

On the surface of the antenna where $\rho = a$, $-h \leq z \leq h$, this reduces to

$$E_z(\mathbf{r}) = \frac{-j\omega}{k_0^2}\left[\frac{\partial^2 A_z(\mathbf{r})}{\partial z^2} + k_0^2 A_z(\mathbf{r})\right] = -V_0^e\delta(z) \qquad (3.10\text{-}3b)$$

The z component (axial component) of the vector potential on the surface of the cylinder satisfies the equation

$$\frac{\partial^2 A_z(\mathbf{r})}{\partial z^2} + k_0^2 A_z(\mathbf{r}) = \frac{-jk_0^2}{\omega}\,V_0^e\delta(z) \qquad (3.10\text{-}4)$$

This is a second-order partial differential equation the solution of which consists of a complementary function and a particular integral. The complementary function is the solution of the homogeneous equation

$$\frac{\partial^2 A_z(\mathbf{r})}{\partial z^2} + k_0^2 A_z(\mathbf{r}) = 0 \qquad (3.10\text{-}5)$$

and is given by

$$A_{zc} = \frac{-j}{c}\,(C_1 \cos k_0 z + C_2 \sin k_0 z) \qquad (3.10\text{-}6a)$$

where C_1 and C_2 are constants of integration that remain to be evaluated. A particular integral is

$$A_{zp} = \frac{V_0^e}{2\omega}\,k_0 e^{-jk_0|z|} \qquad (3.10\text{-}6b)$$

as can be verified by substitution into (3.10–4).

It follows that the general solution of (3.10–4) is

$$A_z(z) = \frac{V_0^e k_0}{2\omega} e^{-jk_0|z|} - \frac{j}{c}(C_1 \cos k_0 z + C_2 \sin k_0 z) \qquad (3.10\text{–}7)$$

Since the delta-function generator is at the center of the antenna, the vector potential is even in z, so that

$$A_z(-z) = A_z(z) \qquad (3.10\text{–}8)$$

This requires that

$$C_2 = 0 \qquad (3.10\text{–}9)$$

and

$$A_z(z) = \frac{-j}{c}\left(C \cos k_0 z + \frac{V_0^e}{2} \sin k_0|z|\right) \qquad (3.10\text{–}10)$$

where $C = C_1 + j(V_0^e/2)$ is a new arbitrary constant that must be evaluated from the boundary conditions. The corresponding expressions for the scalar potential may be obtained from the Lorentz condition in the one-dimensional form

$$\phi(z) = \frac{j\omega}{k_0^2}\frac{\partial A_z(z)}{\partial z} = C \sin k_0 z + \frac{V_0^e}{2} \cos k_0 z, \qquad 0 \le z \le h \qquad (3.10\text{–}11)$$

$$\phi(z) = C \sin k_0 z - \frac{V_0^e}{2} \cos k_0 z, \qquad\qquad -h \le z < 0 \quad (3.10\text{–}12)$$

Note that

$$\phi(-z) = -\phi(z) \qquad (3.10\text{–}13)$$

and

$$\phi(z) - \phi(-z) = 2\phi(z) = 2C \sin k_0 z + V_0^e \cos k_0 z, \qquad 0 \le z \le h \qquad (3.10\text{–}14)$$

The emf V_0^e of the delta-function generator is therefore defined by

$$V_0^e = \lim_{z \to 0}[\phi(z) - \phi(-z)] = \lim_{z \to 0} 2\phi(z) \qquad (3.10\text{–}15)$$

The axial (z) component of the vector potential on the surface of the perfectly conducting tube has been expressed in terms of the source V_0^e in (3.10–10). The general solution for the vector potential is given in (3.8–20). The antenna is a nonmagnetic perfect conductor, so the current is entirely on the surface and the solution reduces to

$$\mathbf{A} = \hat{\mathbf{z}}\frac{\mu_0}{4\pi}\int_\Sigma \frac{K_z(z')}{R} e^{-jk_0 R}\, d\sigma' \qquad (3.10\text{–}16)$$

If the total rotationally symmetric current is $I_z(z')$ and the radius of the antenna is a,

$$I_z(z') = 2\pi a K_z(z') \qquad (3.10\text{–}17a)$$

and the charge per unit length is

$$q(z') = 2\pi a\eta(z') = \frac{j}{\omega} \frac{\partial I_z(z')}{\partial z'} \qquad (3.10\text{–}17b)$$

With (3.10–17a, b) in (3.10–16), this becomes

$$A_z(z) = \frac{\mu_0}{4\pi} \int_{-h}^{h} I_z(z')\, dz' \int_{-\pi}^{\pi} \frac{e^{-jk_0 R_s}}{R_s} \frac{d\theta'}{2\pi}, \qquad \rho = a \qquad (3.10\text{–}18)$$

$\mathbf{r}$ is the vector drawn from the origin at the center of the antenna to a point $P(a, \theta, z)$ on the surface where A_z is calculated. The element of integration $d\sigma' = a\, d\theta'\, dz'$ is located at the point $P'(a, \theta', z')$ and the vector $\mathbf{r}'$ is drawn from the origin to this point. Then the distance between $\mathbf{r}$ and $\mathbf{r}'$ is given by

$$|\mathbf{r} - \mathbf{r}'| = R_s = [(x - x')^2 + (y - y')^2 + (z - z')^2]^{1/2} \qquad (3.10\text{–}19)$$

With the transformation from rectangular to cylindrical coordinates: $x = \rho \cos \theta$, $y = \rho \sin \theta$, $z = z$, (3.10–19) reduces to

$$|\mathbf{r} - \mathbf{r}'| = R_s = \left[(z - z')^2 + \left(2a \sin \frac{\theta'}{2} \right)^2 \right]^{1/2} \qquad (3.10\text{–}20)$$

If the kernel

$$\mathcal{K}_s(z, z') = \int_{-\pi}^{\pi} \frac{e^{-jk_0 R_s}}{R_s} \frac{d\theta'}{2\pi} \qquad (3.10\text{–}21)$$

is introduced into (3.10–18), it may be written as follows:

$$A_z(z) = \frac{\mu_0}{4\pi} \int_{-h}^{h} I_z(z')\mathcal{K}_s(z, z')\, dz' \qquad (3.10\text{–}22)$$

Since (3.10–10) and (3.10–18) both represent the vector potential, the following integral equation is obtained:

$$\int_{-h}^{h} I_z(z')\mathcal{K}_s(z, z')\, dz' = \frac{-4\pi j}{\zeta_0} \left(C \cos k_0 z + \frac{V_0^e}{2} \sin k_0|z| \right) \qquad (3.10\text{–}23)$$

with $\zeta_0 = (\mu_0\epsilon_0)^{1/2} = 120\pi$ ohms. Here $I_z(z')$ is the total current on the cylindrical conductor which satisfies the condition

$$k_0 a \ll 1 \qquad (3.10\text{–}24)$$

An excellent approximation of the kernel $\mathcal{K}_s(z, z')$ is obtained when R_s is replaced by its average value, $R = [(z - z')^2 + a^2]^{1/2}$. The new approximate kernel is

$$\mathcal{K}_s(z, z') = \frac{e^{-jk_0 R}}{R} = \frac{\cos k_0 R}{R} - j \frac{\sin k_0 R}{R} \qquad (3.10\text{–}25)$$

An important property of this kernel when $k_0 a \ll 1$ is that its real part, $\mathcal{K}_{sR}/k_0 = (\cos k_0 R)/k_0 R$, has a very sharp and high peak of magnitude $1/k_0$ at $z' = z$. The imaginary part, $\mathcal{K}_{sI}/k_0 = -(\sin k_0 R)/k_0 R$ has no such peak and approaches the value -1 at $z' = z$.

3.11 THE INTEGRAL EQUATION FOR THE CURRENT IN A THIN CYLINDRICAL ANTENNA IMMERSED IN A DISSIPATIVE MEDIUM

The derivation of the integral equation (3.10–23) for the current in a tubular conductor in air is readily extended to obtain the corresponding equation for the same antenna when the surrounding medium is an imperfect dielectric such as earth, lake, or seawater, as shown in Fig. 3.11–1. Such a medium (region 1) is well approximated by the linear constitutive relations which involve the generally complex permittivity $\epsilon_1 = \epsilon_1' - j\epsilon_1''$, complex conductivity $\sigma_1 = \sigma_1' - j\sigma_1''$, and a real value of the permeability $\mu_1 \sim \mu_0$. These are conveniently combined into the real effective values, $\epsilon_{1e} = \epsilon_1' - \sigma_1''/\omega$, $\sigma_{1e} = \sigma_1' + \omega\epsilon_1''$, which are contained in the complex wave number

$$k_1 = \beta_1 - j\alpha_1 = \omega(\mu_1\bar{\epsilon}_1)^{1/2} = \omega\left[\mu_1\left(\epsilon_{1e} - \frac{j\sigma_{1e}}{\omega}\right)\right]^{1/2} \tag{3.11–1}$$

and the wave impedance

$$\zeta_1 = \left(\frac{\mu_1}{\bar{\epsilon}_1}\right)^{1/2} = \frac{\omega\mu_1}{k_1} \tag{3.11–2}$$

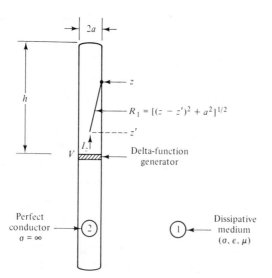

Figure 3.11–1 Dipole in a dissipative medium.

For the antenna in a dissipative medium, these quantities appear in place of the real parameters

$$k_0 = \omega(\mu_0\epsilon_0)^{1/2} = \frac{\omega}{c}; \quad \zeta_0 = \left(\frac{\mu_0}{\epsilon_0}\right)^{1/2} \tag{3.11-3}$$

The differential equation for $A_z(z)$ is like that for the antenna in air but with k_0 replaced by k_1 and ζ_0 by ζ_1. The boundary condition for $E_z(\mathbf{r}, t)$ on the surface of its antenna is like (3.10–1) and the integral equation is

$$\int_{-h}^{h} I_z(z')\mathcal{K}_s(z, z')\, dz'$$

$$= \frac{-4\pi j}{\zeta_1} \left(C \cos k_1 z + \frac{V_0^e}{2} \sin k_1|z| \right), \quad -h \leq z \leq h \tag{3.11-4}$$

Here the approximate kernel is

$$\mathcal{K}_s(z, z') = \frac{e^{-jk_1 R}}{R} = \frac{e^{-\alpha_1 R} \cos \beta_1 R}{R} - j\frac{e^{-\alpha_1 R} \sin \beta_1 R}{R} \tag{3.11-5}$$

The presence of the real exponential factor $e^{-\alpha_1 R}$ in the kernel makes a significant difference in the solution of (3.11–4), as explained in a later chapter.

As for the antenna in air, the total axial current is $I_z(z') = 2\pi a K_z(z')$. But the equation for the charge per unit length $q(z') = 2\pi a \eta(z')$ is related to $I_z(z')$ in a more complicated manner, owing to the presence of a radial current $I_\rho(z')$ into the surrounding medium. The equation of continuity for surface currents at the surface of a perfect conductor (region 2) bounded by a simple medium is

$$\nabla \cdot \mathbf{K}_{2f} + j\omega\eta_{2f} - \hat{\mathbf{n}}_1 \cdot \mathbf{J}_{1f} = 0 \tag{3.11-6}$$

For the cylindrical antenna, and with the subscript f omitted:

$$\frac{\partial K_{2z}(z)}{\partial z} + j\omega[\eta_2(z) + \eta_1(z)] - \hat{\mathbf{n}}_1 \cdot \mathbf{J}_1(z) = 0, \quad \rho = a \tag{3.11-7}$$

When multiplied by $2\pi a$ and with $I_{2z}(z) = 2\pi a K_{2z}(z)$, $q(z) = 2\pi a \eta(z)$, and $I_{1\rho}(z) = 2\pi a J_{1\rho}(z)$, (3.11–7) becomes

$$\frac{\partial I_{2z}(z)}{\partial z} + j\omega[q_2(z) + q_1(z)] + I_{1\rho}(z) = 0 \tag{3.11-8}$$

This equation shows that the rate of decrease of the axial current $I_{2z}(z')$ on the conductor involves not only the charging of the surface $\rho = a$ but also a radial current into the surrounding medium. The axial current along the antenna oscillates by charging the surface; then a part of this charge leaks off radially in the current $I_{1\rho}(z')$. Specifically, the two equations contained in (3.11–8) are

$$\frac{\partial I_{2z}(z)}{\partial z} + j\omega q_2(z) = 0 \tag{3.11-9}$$

$$j\omega q_1(z) + I_{1\rho}(z) = 0 \tag{3.11-10}$$

The relation between $q_1(z)$ and $q_2(z)$ is readily established with the general boundary condition on the normal component of **E**, which takes the following form on the surface of a perfect conductor:

$$\epsilon_1 E_{1\rho} = \eta_{1f} + \eta_{2f} \tag{3.11-11}$$

and

$$j\omega\eta_{1f} = -J_{1\rho} = -\sigma_1 E_{1\rho} \tag{3.11-12}$$

It follows that

$$\frac{q_1(z')}{q_2(z')} = \frac{\eta_1}{\eta_2} = -\frac{\sigma_1}{\sigma_1 + j\omega\epsilon_1} = -\frac{\sigma_1' - j\sigma_1''}{\sigma_{1e} + j\omega\epsilon_{1e}} \tag{3.11-13}$$

Evidently, if the integral equation (3.11–4) is solved for the axial current $I_{2z}(z)$, the charge per unit length associated with this current can be obtained from (3.11–9). The charge per unit length $q_1(z)$ can then be calculated from (3.11–13) and this can be substituted in (3.11–10) to determine the radial current $I_{1\rho}(z')$ that enters the surrounding medium from the surface at $\rho = a$.

3.12 THE INTEGRAL EQUATION FOR THE CURRENT IN AN INSULATED ANTENNA

When a dipole or monopole antenna made of highly conducting material is coated with a uniform layer of dielectric material and immersed in a homogeneous isotropic medium that may be a dielectric or conductor, the antenna is characterized by two quite different sets of properties that depend on the relative magnitudes of the wave numbers that characterize the dielectric layer and the ambient medium. If the wave number of the dielectric sheath has a larger magnitude than the wave number of the infinite medium (as for polystyrene-coated wire in air), total internal reflection occurs in the dielectric sheath and a surface wave travels along it. If, on the other hand, the magnitude of the wave number of the ambient medium significantly exceeds that of the insulating sleeve (as for a Styrofoam-coated wire in seawater), the antenna behaves like a generalized transmission line. Depending on the thickness of the insulating sheath, the transfer of power to the ambient medium per unit length—either to be radiated through it if it is a good dielectric such as fresh water at a high frequency or dissipated in it near the antenna if it is a good conductor such as seawater—can be made large or small. The wave number of the dielectric-coated surface-wave antenna is quite different from that of the insulated, transmission-line-type antenna.

    The integral equation of interest here is for an insulated antenna in an ambient medium with a wave number that is much greater in magnitude than that of the insulating sheath. For this a solution is possible that has a simple transmission-line

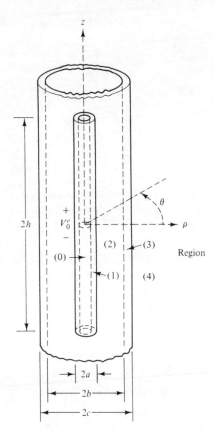

Figure 3.12–1 Insulated antenna driven by a delta-function generator.

form which will be discussed in a later chapter. The structure of the insulated antenna is shown in Fig. 3.12–1. It consists of a highly conducting thin-walled tubular cylinder of radius a (region 1) and wall thickness $(a_+ - a_-)$ enclosed in a sheath of insulating material with outer radius b (region 2) characterized by $\mu_2 = \mu$, ϵ_2, σ_2; the related wave number is $k_2 = \omega[\mu(\epsilon_2 + i\sigma_2/\omega)]^{1/2}$ and the wave impedance is $\zeta_2 = \omega\mu/k_2$. The same material fills the interior (region 0) of the thin-walled open-ended tube. If the insulating material is a fluid, it may be contained in a thin-walled glass or plastic tube with inner radius b and outer radius c (region 3) with the constitutive parameters $\mu_3 = \mu$, ϵ_3, σ_3. The wave number is $k_3 = \omega[\mu(\epsilon_3 + i\sigma_3/\omega)]^{1/2}$ and the wave impedance is $\zeta_3 = \omega\mu/k_3$. Outside the insulating regions 2 and 3 is an infinite homogeneous and isotropic ambient medium (region 4) characterized by $\mu_4 = \mu$, ϵ_4, σ_4 and by $k_4 = \omega[\mu(\epsilon_4 + i\sigma_4/\omega)]^{1/2}$ and $\zeta_4 = \omega\mu/k_4$. It is assumed that all materials are nonmagnetic, so that with $n = 1, 2, 3,$ or 4, $\mu_n = \mu = \mu_0$. Both ϵ_n and σ_n can be complex but the real effective permittivities ϵ_{en} and conductivities σ_{en} are defined so that $\epsilon_n + i\sigma_n/\omega = \epsilon_{en} + i\sigma_{en}/\omega$.

The analysis that leads to the simple transmission-line-like solution is valid only when the following inequalities are satisfied:

$$|k_4^2| \gg |k_2^2|; \qquad |k_4^2| \gg |k_3^2| \tag{3.12-1a}$$

which may be written quantitatively as

$$|k_4| \geq 3|k_2|; \qquad |k_4| \geq 3|k_3| \tag{3.12-1b}$$

It is also assumed that the following inequalities are satisfied:

$$|k_2 a| < |k_2 b| \ll 1; \qquad |k_3 b| < |k_3 c| \ll 1 \tag{3.12-1c}$$

The axis of the tube coincides with the z axis. At $z = 0$, a delta-function generator maintains the electric field $E_z^e = -V_0^e \delta(z)$ with emf V_0^e. The axial current I_z is related to the rotationally symmetric surface current density K_z by the relation $I_z = 2\pi a K_z$. Since there is rotational symmetry, the nonvanishing cylindrical components of the electromagnetic field in each of the several regions are $E_{n\rho}$, E_{nz}, and $E_{n\theta}$, where $n = 0, 2, 3, 4$. They satisfy Maxwell's equations, which may be solved by introducing Fourier transforms. The algebra is formidable and is not carried out here. The integral equation for the current in the inner conductor of the insulated antenna is formally like that for a bare antenna, that is,

$$\int_{-h}^{h} I(z') \mathcal{K}(z - z') \, dz' = C \cos k_L z + i \frac{V_0^e}{2\zeta_2} \sin k_L |z| \tag{3.12-2}$$

Note that the time dependence $e^{-i\omega t}$ instead of $e^{j\omega t}$ has been assumed in order to agree with notation conventional in the theory of the insulated antenna. Conversion is simple since $j = -i$. The kernel $\mathcal{K}(z - z')$ is well approximated by

$$\mathcal{K}(z - z') = \frac{k_L}{4\pi^2 k_2} \int_{-\infty}^{\infty} I_0(a\zeta) K_0(a\zeta)$$

$$\times \left[1 - \frac{I_0(a\zeta) K_0(b\zeta)}{I_0(b\zeta) K_0(a\zeta)} \right] e^{-i\zeta(z - z')} \, d\zeta \tag{3.12-3}$$

where I_0 and K_0 are the modified Bessel functions. Although formally complicated, this kernel has the important property that it peaks sharply in *both* its real and imaginary parts at $z = z'$. A sample graph of $\mathcal{K}(z)$ is shown in Fig. 3.12–2. An approximate formula for the wave number k_L is

$$k_L = k_2 \left[1 + \frac{H_0^{(1)}(k_4 b)}{k_4 b H_1^{(1)}(k_4 b) \ln (b/a)} \right]^{1/2} \tag{3.12-4}$$

where $H_0^{(1)}(k_4 b)$ and $H_1^{(1)}(k_4 b)$ are the Hankel functions of the first kind and orders 0 and 1. The associated characteristic impedance is

$$Z_c = \frac{\omega \mu_0 k_L}{2\pi k_2^2} \ln \left(\frac{b}{a} \right) \tag{3.12-5}$$

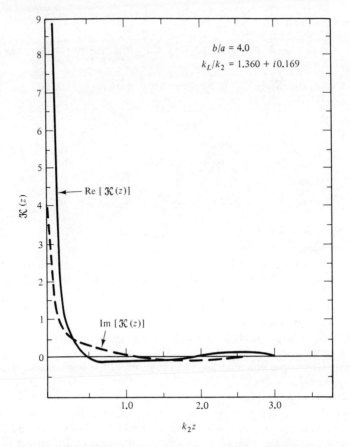

Figure 3.12–2   Kernel $K(z)$ as a function of k_2z. Relative permittivity $\varepsilon_{r4} = 3.8$, loss tangent $p_4 = 0.08$; $\varepsilon_{r2} = 1.0$, $p_2 = 0$; frequency $f = 380$ MHz; $k_2a = 2.53 \times 10^{-2}$.

PROBLEMS

1. Show that the scalar potential in the stationary state,

$$\phi = \frac{1}{4\pi\epsilon_0} \left(\int_\tau \frac{\bar{\rho}'}{R} d\tau' + \int_\Sigma \frac{\bar{\eta}'}{R} d\sigma' \right)$$

is equivalent to

$$\phi = \frac{1}{4\pi\epsilon_0} \left\{ \int_\tau \frac{\rho'}{R} d\tau' + \int_\Sigma \frac{\eta'}{R} d\sigma' + \int_\tau \left[\mathbf{\nabla}' \left(\frac{1}{R} \right) \cdot \mathbf{P}' \right] d\tau' \right\}$$

2. Show that the vector potential in the stationary state,

$$\mathbf{A} = \frac{\mu_0}{4\pi} \left(\int_\tau \frac{\bar{\mathbf{J}}'}{R} d\tau' + \int_\Sigma \frac{\bar{\mathbf{K}}'}{R} d\sigma' \right)$$

is equivalent to

$$\mathbf{A} = \frac{\mu_0}{4\pi} \left\{ \int_\tau \frac{\mathbf{J'}}{R} d\tau' + \int_\Sigma \frac{\mathbf{K'}}{R} d\sigma' + \int_\tau \left[\mathbf{\nabla'} \left(\frac{1}{R} \right) \times -\mathbf{M'} \right] d\tau' \right\}$$

3. A spherical charge distribution has the density $\rho(r)$ given by

$$\rho(r) = \begin{cases} \rho_0 r^2, & 0 \le r \le \dfrac{R}{4} \\[2mm] \rho_0(R - r)^2, & \dfrac{R}{4} \le r \le R \\[2mm] 0, & r > R \end{cases}$$

The charge distribution is in free space. Determine the electrostatic scalar potential everywhere.

4. The Hertzian dipole is equivalent to a pair of small metal spheres or circular metal disks each of radius b that are connected by a short wire of length $2h$ and radius a such that $a \ll b \ll h \ll \lambda$ (wavelength). The two spheres or disks are charged alternately with a total charge $+Q$ or $-Q$ by a periodically varying current of uniform amplitude I maintained in the wire.

(a) Derive expressions for the scalar and vector potential functions using the Helmholtz integrals.

(b) Express the potential functions in terms of the polarization vector (the dipole moment) $p = 2hQ$.

(c) Calculate the electromagnetic field from the scalar and vector potential functions.

5. (a) Calculate and plot the cross-sectional distribution of current in a copper wire of radius $a = 0.5$ mm at frequencies of 60 Hz, 1.5 MHz, and 3 GHz.

(b) Calculate the internal resistance r^i and reactance x^i per unit length for the above wire at frequencies of 60 Hz, 150 MHz, and 3 GHz.

6. Calculate the internal resistance r^i and reactance x^i per unit length of a coaxial cable consisting of a silvered copper inner conductor of radius 0.4 mm and a tinned copper outer conductor of inner radius 0.265 mm. σ for silver $= 6.14 \times 10^7$ S/m, σ for tin $= 1.87 \times 10^7$ S/m. The frequency is 3 GHz.

7. Calculate the thickness of copper, aluminum, and sheet steel at which the amplitude of the electric field tangent to a plane boundary is reduced to 1 percent of the value at the boundary for 1 MHz and 1 GHz.

$$\sigma \text{ for copper} = 5.8 \times 10^7 \text{ S/m}$$

$$\sigma \text{ for aluminum} = 3.54 \times 10^7 \text{ S/m}$$

$$\sigma \text{ for sheet steel} = 1 \times 10^7 \text{ S/m}$$

$$\mu_r \text{ for sheet steel} = 10^3$$

8. Determine the skin depth and surface impedance of:

(a) Salt water at 60 MHz

$$\sigma = 4.3 \text{ S/m}; \quad \epsilon_r = 80; \quad \mu_r = 1$$

(b) Brass at 3 GHz

$$\sigma = 1.5 \times 10^7 \text{ S/m}; \qquad \epsilon_r = 1; \qquad \mu_r = 1$$

(c) Wet earth at 10 MHz

$$\sigma = 3 \times 10^{-2} \text{ S/m}; \qquad \epsilon_r = 10; \qquad \mu_r = 1$$

9. Calculate the ratio of the electric field just outside and just inside an infinitely long aluminum tube of wall thickness 2 mm and radius 8 cm at 1.5 MHz and 2.25 GHz.

10. A copper wire (region 1) of radius a is surrounded by a polyfoam (air) cylinder (region 2) of outer radius b. The infinite region 3 ($r > b$) is an imperfect conductor. A periodically varying axial electric field with amplitude $E_z(b)$ is maintained at $r = b$ in the dielectric (Fig. P3–10).

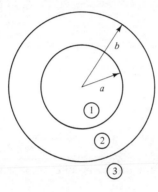

Figure P3–10.

(a) Write down the general formula for the volume density of current $J_z(r, t)$ for $r \geq b$. Specialize this for $b\sqrt{\omega\mu_3\sigma_3} \geq 10$. With the aid of a diagram illustrate the propagation of waves of axial current radially outward from $r = b$.

(b) Determine the total axial current I_z in region 3 in terms of $E_z(b)$.

(c) Obtain expressions for the impedance per unit length of the outer conductor defined by

$$z^i = \frac{E_z(b)}{I_z} = r^i + jx^i$$

in general, for $b\sqrt{\omega\mu_3\sigma_3} \ll 1$, and for $b\sqrt{\omega\mu_3\sigma_3} \geq 10$. Compare these values with the corresponding one of z^i for the inner conductor.

(d) Determine z^i for the inner and outer conductors if $a = 0.2$ cm, $b = 0.5$ cm, with region 1 copper ($\sigma = 5.65 \times 10^7$ S/m), region 3 salt water ($\epsilon_r = 80$, $\sigma = 5$ S/m), at an angular frequency $\omega = 10^5$.

(e) Determine z^i for the outer conductor if this is copper.

4

Electromagnetic Force and Energy

Macroscopic electromagnetic theory is concerned primarily with the solution of the Maxwell–Lorentz equations subject to the boundary conditions associated with a particular configuration of conductors and dielectrics. Of interest are the electromagnetic field and the associated distributions of current and charge. Another important aspect of electromagnetism is its relation with macroscopically non-electrical phenomena which occur in the generation and observation of electric fields and currents. Electromechanical and electrochemical devices are of particular importance. The relations between electromagnetic quantities and corresponding quantities in mechanics and chemistry occur primarily through the concepts of force and energy.

4.1 DEFINITION OF ELECTROMAGNETIC FORCE AND TORQUE

The repulsion–attraction effect between stationary and moving charges is one of the fundamental postulates of the atomic model. In terms of the physical model of bodies and regions composed of billions of widely separated charges in motion, the problem of attraction and repulsion between the charges in separate bodies is hardly different from the problem of the attractions and repulsions between the charges of a single homogeneous body. However, with the physical model of each body replaced by a small number of continuous density functions, the interaction of charges in different bodies becomes specifically a problem of formulating the interaction between the density fields characterizing these bodies. The density of mass is one of the density fields associated with every body. Relative motions and conditions of equilibrium of such density fields are described by the mathematical model of mechanics which is entirely adequate in predicting experimental analogues for *uncharged* bodies. Since many electrical measurements involve experiments using mechanical elements, a connection between the mathematical models of mechanics and electromagnetics must be provided. The force vector is defined in mechanics to determine the conditions of motion or of equilibrium between the density of mass fields or simply between masses. This vector, together with the standard of mass, is defined in terms of the equation of motion of the center of mass of a body,

$$\mathbf{F}_M = \frac{d}{dt}(m\mathbf{v}) \tag{4.1-1}$$

where $m = \int_\tau D \, d\tau$ and $\mathbf{v}$ is the velocity of the center of mass.

Consider a body characterized not only by the mechanical property of mass but also by the electrical properties of charge and distribution of charge. Any condition of equilibrium or of motion of the body may, in general, be formulated in terms of mechanical forces and electrical repulsion-attraction effects. It is convenient to define an electromagnetic force to represent the latter and so complete the mathematical model of the electrical properties of matter in space by relating it to the mathematical model of mechanics. The electromagnetic force vector must take account of the attraction-repulsion effects of bodies containing stationary and moving charges in terms of the density functions of all bodies in the electromagnetic field.

The electromagnetic force $\mathbf{F}$ and torque $\mathbf{T}$ acting on a body are defined so that the following conditions of equilibrium are satisfied for a body at rest:

$$\mathbf{F}_M + A_e\mathbf{F} = 0 \tag{4.1-2a}$$

$$\mathbf{T}_M + A_e\mathbf{T} = 0 \tag{4.1-2b}$$

Here $\mathbf{F}_M$ and $\mathbf{T}_M$ are the resultant mechanical force and torque acting on the body, and the factor A_e is a constant that depends on the relation between the electrical and mechanical units of force. It may be defined either as a fundamental constant to be determined experimentally as the mechanical equivalent of electricity, or it may be assigned an arbitrary dimensionless numerical value. The latter alternative is chosen in the practical system of units in which A_e is dimensionless and equal to 1. If there is no mechanical force to balance the electrical force, the body is accelerated. A mechanical force of inertia may then be written in the familiar form

$$\mathbf{F}_M = -\frac{d}{dt}(m\mathbf{v}) \tag{4.1-3}$$

Here $\mathbf{v}$ is the velocity of the center of mass of the body of mass m. With $A_e = 1$, the force equation (4.1-2a) becomes the following equation of motion:

$$\mathbf{F} - \frac{d}{dt}(m\mathbf{v}) = 0 \tag{4.1-4}$$

The definition of the electromagnetic force $\mathbf{F}$ which is to be used in (4.1–2a) may be looked upon as a third fundamental postulate of electrodynamics.

The electromagnetic force $\mathbf{F}$ and vector torque $\mathbf{T}$ acting on a volume τ characterized by the densities $\bar{\rho}$, $\bar{\eta}$, $\overline{\rho_m \mathbf{v}}$, $\overline{\eta_m \mathbf{v}}$ are defined to be

$$\mathbf{F} = \int_\tau (\bar{\rho}\mathbf{E} + \overline{\rho_m \mathbf{v}} \times \mathbf{B})\, d\tau + \int_\Sigma (\bar{\eta}\mathbf{E} + \overline{\eta_m \mathbf{v}} \times \mathbf{B})\, d\sigma \qquad (4.1-5)$$

$$\mathbf{T} = \int_\tau \mathbf{r} \times d\mathbf{F}_\tau + \int_\Sigma \mathbf{r} \times d\mathbf{F}_\sigma \qquad (4.1-6)$$

In (4.1–6) $\mathbf{r}$ is the vector from an arbitrary origin to the element $d\tau$ or $d\sigma$; $d\mathbf{F}_\tau$ is the integrand of the volume integral, and $d\mathbf{F}_\sigma$ the integrand of the surface integral in (4.1–5). In (4.1–5) and (4.1–6) $\mathbf{E}$ and $\mathbf{B}$ are calculated from the field equations in terms of the continuous volume and surface densities of all bodies and regions *exterior* to τ. If τ is at rest relative to the observer, the densities $\bar{\rho}$, $\bar{\eta}$, $\overline{\rho_m \mathbf{v}}$, and $\overline{\eta_m \mathbf{v}}$ are the essential characteristics of charge and moving charge already defined as follows:

$$\begin{aligned} \bar{\rho} &\equiv \rho - \nabla \cdot \mathbf{P}; & \overline{\rho_m \mathbf{v}} &\equiv \mathbf{J} + \nabla \times \mathbf{M} + \dot{\mathbf{P}} \\ \bar{\eta} &\equiv \eta + \hat{\mathbf{n}} \cdot \mathbf{P}; & \overline{\eta_m \mathbf{v}} &\equiv \mathbf{K} - \hat{\mathbf{n}} \times \mathbf{M} \end{aligned} \qquad (4.1-7)$$

Since these densities are obtained by interpolation from averages defined at the center of each volume or surface cell, the values of $\mathbf{E}$ and $\mathbf{B}$, and hence of $\mathbf{F}$ calculated from them, have a meaning only at points that are far from the density fields contributing to $\mathbf{E}$ and $\mathbf{B}$ as compared with the dimensions of the volume cells used in defining them. That is, the electromagnetic force acting on a volume τ_1 due to charges and currents in a second region τ_2 can be calculated from (4.1–5) only if each point of τ_1 is very much farther from every point in τ_2 than the dimension of the typical volume cell in τ_2. In defining the continuous densities in τ_2 by interpolation, the average electrical properties of each volume cell are, in effect, assumed to be concentrated at its center. The implication is that the densities correctly represent the average properties only if it may be assumed without serious error that all charges in each volume cell are at the same distance from the point where $\mathbf{E}$ and $\mathbf{B}$ are calculated as is the center of the cell in question. The restriction is not important insofar as the mutual interaction of bodies as a whole is concerned.

4.2 DYNAMICAL EQUATION

The electromagnetic force $\mathbf{F}$ is defined by (4.1–5) for use in (4.1–2a). Since this applies to conditions of equilibrium of bodies as a whole, the integrals in (4.1–5) are necessarily evaluated over the entire volume and surface of the body which is acted on by the mechanical force $\mathbf{F}_M$. The resultant electromagnetic force $\mathbf{F}$ acting on a body as a whole can be considered to be the vector sum of a number of components. For example, the components acting on different sections of the body may be considered separately. That is, the component of electromagnetic force acting on each of a large number of small volumes into which the body is subdivided may be calculated. By vector addition, the same resultant force is obtained.

Since the densities are by definition continuous functions in any body or region, it is quite legitimate to calculate the average electromagnetic force acting on any volume, however small, provided only that it is a part of a larger region in which the density functions are properly defined. Equation (4.1–5) cannot be used to calculate the electromagnetic force on a region that is not much larger than a typical volume cell, and which is not a part of a more extensive, homogeneous body, because statistical conditions cannot be assumed to prevail and no density functions can be defined. However, if the electromagnetic field due to all bodies surrounding a region in which volume and surface densities are properly defined is sufficiently slowly varying so that it may be assumed to be sensibly constant over the entire region, a simplification is possible. In this case $\mathbf{E}$ and $\mathbf{B}$ may be removed from under the sign of integration in (4.1–5) and

$$\mathbf{F} = q\mathbf{E} + q_m\mathbf{v} \times \mathbf{B} \tag{4.2–1}$$

with

$$q = \int_\tau \bar{\rho}\, d\tau + \int_\Sigma \bar{\eta}\, d\sigma$$

$$q_m\mathbf{v} = \int_\tau \overline{\rho_m\mathbf{v}}\, d\tau + \int_\Sigma \overline{\eta_m\mathbf{v}}\, d\sigma \tag{4.2–2}$$

The integrals of vectors are shorthand forms that can be evaluated in any convenient system of coordinates with the integrands expressed in component form. The integrations are carried out for each component separately and the results are combined into a new vector.

It is immaterial whether the total charge q in τ is determined by first constructing continuous density functions and integrating, or by direct summation of the charges. This is also true for $q_m\mathbf{v}$. It follows that (4.2–1) may be used for as small a region as desired if q and $q_m\mathbf{v}$ are calculated by direct summation. If q and q_m coincide so that all charges are in motion as in an electron stream, (4.2–1) reduces to the form

$$\mathbf{F} = q(\mathbf{E} + \mathbf{v} \times \mathbf{B}) \tag{4.2–3}$$

This expression may be used to define the electromagnetic force acting on a small group of charges or even on a single charge moving with nonrandom velocity $\mathbf{v}$, if $\mathbf{E}$ and $\mathbf{B}$ are sensibly constant over its extension, and if it is at a distance from all charged regions which is large compared with the dimensions of the volume cells used in constructing the densities entering into the definitions of $\mathbf{E}$ and $\mathbf{B}$.

Due to the foregoing restrictions, (4.2–3) cannot be used to calculate the electromagnetic force acting on a charge in the interior of a body or region. It may be used to calculate the component of electromagnetic force on an interior charge due to the entire charge distribution *outside* a sphere drawn around the charge and with radius large compared with a typical volume cell. The force due to the charge distribution within this sphere, the *local force*, cannot be calculated using (4.2–3). For certain symmetrical distributions, it can be shown to be zero. In other cases, well-founded assumptions have to be made.

     If the entire electromagnetic force **F** acting on a charge moving with mean velocity **v** is given by (4.2–3) and no other forces are involved, the charge is accelerated. If its mass is m, the mechanical force resisting acceleration is

$$\mathbf{F}_M = -\frac{d}{dt}(m\mathbf{v}) \tag{4.2–4}$$

so that the equation of motion is

$$q(\mathbf{E} + \mathbf{v} \times \mathbf{B}) = \frac{d}{dt}(m\mathbf{v}) \tag{4.2–5}$$

When the charge is an electron, $q = e = -|e|$ and (4.2–5) is called the *dynamical equation of the electron*.

4.3 CONCEPT OF ENERGY

The force equation of Sec. 4.2 may be expressed in a different and frequently useful form by defining two *energy functions*. The basis for the definition is the law of conservation of energy. The concept of energy follows from the definition of mechanical force:

$$\mathbf{F}_M = m\frac{d\mathbf{v}}{dt} \tag{4.3–1}$$

A study of the line integral of force, called *work*, together with the relation

$$\frac{d\mathbf{v}}{dt} = \frac{d\mathbf{v}}{ds}\frac{ds}{dt}$$

$$= \frac{d\mathbf{v}}{ds}v \tag{4.3–2}$$

leads to the concept of energy:

$$\int_{s_0}^{s} \mathbf{F}_M \cdot d\mathbf{s} = \tfrac{1}{2}mv^2 - \tfrac{1}{2}mv_0^2 \tag{4.3–3}$$

The left side in (4.3–3) defines the work done by the force $\mathbf{F}_M$ in the motion of the mass m from s_0 to s; the right side is a change in the *kinetic energy*. The left side of (4.3–3) can be integrated directly when $\mathbf{F}_M \cdot d\mathbf{s}$ is a perfect differential.

This is true when $\mathbf{F}_M$ is a potential vector so that it can be derived from a scalar potential function V. Thus, provided that $\nabla \times \mathbf{F}_M = 0$,

$$\mathbf{F}_M \cdot d\mathbf{s} = -dV \tag{4.3-4}$$

where V is a scalar function. The integration yields

$$V + \tfrac{1}{2}mv^2 = V_0 + \tfrac{1}{2}mv_0^2 \tag{4.3-5}$$

The left- and right-hand sides are constant for any motion of m and (4.3–5) becomes a statement of the law of conservation of energy. The mathematical quantity $V + \tfrac{1}{2}mv^2$ is fundamentally involved in any motion of the mass m in a "conservative mechanical system" and is invariant in time. This constant is the *total energy of the system* and is denoted by W. The scalar potential function V is the *potential energy*; $\tfrac{1}{2}mv^2$ is the *kinetic energy*.

4.4 DEFINITION OF THE ELECTROMAGNETIC ENERGY FUNCTION AND THE ENERGY-TRANSFER FUNCTION

A function U, the *electromagnetic energy*, must satisfy the equation

$$\frac{dW}{dt} + A_q \frac{dQ}{dt} + A_e \frac{dU}{dt} + A_q T_q + A_e T = 0 \tag{4.4-1}$$

in a closed region involving a representation in terms of the mathematical models of macroscopic mechanics, thermodynamics, and electromagnetism. W is the mechanical energy (or the mechanical equivalent of any other energy function); Q is the thermal energy; T_q is the thermal transfer function of the enclosing surface; A_q is the mechanical equivalent of heat. It is presumed that a function U, the *electromagnetic energy*, and a function T, the *electromagnetic energy-transfer function* of the enclosing boundary, can be found to satisfy this equation in such a way that A_e is a numerical constant. It may be called the *mechanical equivalent of electromagnetic energy*. In the practical system of units, the energy functions are so defined that A_e is dimensionless and equal to 1. It is assumed in writing (4.4–1) that each energy-transfer function and the time rate of change of each energy function carries its own sign according as the forces contributing to it do active (positive) work or resist its performance (do negative work). The interconnection between electromagnetism and mechanics might be established in terms of the power equation (4.4–1) by suitably defining the electromagnetic energy function U. Actually, a connection has already been established by constructing the electromagnetic force vector $\mathbf{F}$ to satisfy the vector force equation

$$\mathbf{F}_M + \mathbf{F} = 0 \tag{4.4-2}$$

where $\mathbf{F}_M$ is the resultant mechanical force and $\mathbf{F}$ was defined in Sec. 4.1. The definition of electromagnetic energy and energy-transfer functions must be consistent with this equation and with the definition of electromagnetic force on which it depends.

The equilibrium between electrical and mechanical forces acting on the charges in an element of volume $d\tau$ is expressed in

$$d\mathbf{F}_M + d\mathbf{F} = 0 \qquad (4.4-3)$$

Let (4.4–3) be multiplied scalarly by the mean nonrandom velocity $\mathbf{u}$ of the free charges in $d\tau$ that are actually engaged in nonrandom motion and integrated over the volume τ,

$$\int_\tau d\mathbf{F}_M \cdot \mathbf{u} + \int_\tau d\mathbf{F} \cdot \mathbf{u} = 0 \qquad (4.4-4)$$

The first integral on the left may be interpreted to be the sum of two terms by resolving $d\mathbf{F}_M$ into an active, charge-separating force $d\mathbf{F}_M'$ and an oppositely directed "frictional" or resisting force $d\mathbf{F}_M''$. The active force $d\mathbf{F}_M'$ is nonvanishing only in charge-separating regions, of which at least one is assumed to be present within τ. The frictional force $d\mathbf{F}_M''$ resists the acceleration of the nonrandom motion and leads to an increase in random or heat motion. Accordingly, all positive work is done on the charges by $d\mathbf{F}_M'$ in the charge-separating region against the "thermal" forces $d\mathbf{F}_M''$ and the electromagnetic force $d\mathbf{F}$ which acts wherever there are charges.

$$\int_\tau d\mathbf{F}_M' \cdot \mathbf{u} = \frac{dW}{dt} \qquad (4.4-5)$$

$$\int_\tau d\mathbf{F}_M'' \cdot \mathbf{u} = -A_q \frac{dQ}{dt} \qquad (4.4-6)$$

W is the total mechanical, chemical, or other energy function associated with the charge-separating region. It has a positive time rate of change because it is associated with forces that do work on the moving free charges in τ. Accordingly, W decreases in time. Q is the thermal energy function associated with the moving free charges in the region τ, which is assumed to be thermally isolated for simplicity, so that $T_q = 0$ and Q increases in time. The time rate of change of Q is negative because it is associated with forces against which work is done. With $T_q = 0$, the sum of (4.4–5) and (4.4–6) is equal to the first integral in (4.4–4). Now (4.4–4) becomes

$$\frac{dW}{dt} - A_q \frac{dQ}{dt} + \int_\tau d\mathbf{F} \cdot \mathbf{u} = 0 \qquad (4.4-7)$$

It is desired to define an energy function U within τ and, if necessary, an energy-transfer function T on its boundary surface Σ_τ such that

$$\int_\tau d\mathbf{F} \cdot \mathbf{u} = -\frac{dU}{dt} - T \qquad (4.4-8)$$

with dU/dt and T positive. If this can be done, (4.4–7) and (4.4–1) will have the same form.

If surfaces of discontinuity that are characterized by surface densities of current are excluded by enclosing them in surfaces S_j so that the volume τ is divided into several volumes τ_i such that $\tau = \Sigma_i \tau_i$, the electromagnetic force acting on free charges engaged in nonrandom motion in an element $d\tau_i$ is

$$d\mathbf{F}_i = (\rho_f' \mathbf{E} + \mathbf{J}_f \times \mathbf{B})\, d\tau_i \qquad (4.4\text{--}9)$$

The force acting on free charges engaged in nonrandom motion in an element $d\sigma_j$ of a boundary surface S_j within Σ_τ is

$$d\mathbf{F}_j = (\eta_f' \mathbf{E} + \mathbf{K}_f \times \mathbf{B})\, d\sigma_j \qquad (4.4\text{--}10)$$

Furthermore,

$$\mathbf{J}_f = \rho_f' \mathbf{u}; \qquad \mathbf{K}_f = \eta_f' \mathbf{u} \qquad (4.4\text{--}11)$$

with ρ_f' and η_f', respectively, volume and surface densities of free charge moving with mean nonrandom velocity $\mathbf{u}$.

The substitution of (4.4–9) and (4.4–10) in the integral in (4.4–8) gives

$$\int_\tau d\mathbf{F} \cdot \mathbf{u} = \sum_i \int_{\tau_i} [\rho_f'(\mathbf{E} \cdot \mathbf{u}) + (\mathbf{J}_f \times \mathbf{B}) \cdot \mathbf{u}]\, d\tau_i$$

$$+ \sum_j \int_{S_j} [\eta_f'(\mathbf{E} \cdot \mathbf{u}) + (\mathbf{K}_f \times \mathbf{B}) \cdot \mathbf{u}]\, d\sigma_j \qquad (4.4\text{--}12)$$

with

$$(\mathbf{J}_f \times \mathbf{B}) \cdot \mathbf{u} = \rho_f'(\mathbf{u} \times \mathbf{B}) \cdot \mathbf{u} = \rho_f' \mathbf{B} \cdot (\mathbf{u} \times \mathbf{u}) = 0 \qquad (4.4\text{--}13)$$

since $\mathbf{u} \times \mathbf{u} = 0$. A similar product involving $\mathbf{K}_f$ also vanishes. With (4.4–11), (4.4–12) becomes

$$\int_\tau d\mathbf{F} \cdot \mathbf{u} = \sum_i \int_{\tau_i} (\mathbf{J}_f \cdot \mathbf{E})\, d\tau_i + \sum_j \int_{S_j} (\mathbf{K}_f \cdot \mathbf{E})\, d\sigma_j \qquad (4.4\text{--}14)$$

The integrals on the right in (4.4–14) may be transformed by expressing $\mathbf{J}_f$ and $\mathbf{K}_f$ in terms of the electromagnetic vectors using their fundamental definitions. They must be expressed either in terms of $\mathbf{E}$, $\mathbf{B}$, $\mathbf{P}$, and $\mathbf{M}$ or in terms of $\mathbf{E}$, $\mathbf{B}$, $\mathbf{D}$, and $\mathbf{H}$. The latter representation is more commonly used. Furthermore, since $\mathbf{J}_f$ is involved explicitly, a mode of subdivision in interpreting $\mathbf{P}$ and $\mathbf{M}$ or $\mathbf{D}$ and $\mathbf{H}$ is implied that cuts through no bound groups. The following equations are used:

$$\nabla \times \mathbf{H} = \mathbf{J}_f + \dot{\mathbf{D}} \qquad (4.4\text{--}15)$$

$$\hat{\mathbf{n}} \times \mathbf{H} = -\mathbf{K}_f \qquad (4.4\text{--}16)$$

Then

$$\mathbf{J}_f \cdot \mathbf{E} = (\nabla \times \mathbf{H}) \cdot \mathbf{E} - \dot{\mathbf{D}} \cdot \mathbf{E} \tag{4.4-17}$$

$$\mathbf{K}_f \cdot \mathbf{E} = -(\hat{\mathbf{n}} \times \mathbf{H}) \cdot \mathbf{E}$$

$$= -(\mathbf{H} \times \mathbf{E}) \cdot \hat{\mathbf{n}}$$

$$= (\mathbf{E} \times \mathbf{H}) \cdot \hat{\mathbf{n}} \tag{4.4-18}$$

Substitution in (4.4–14) gives

$$\int_\tau d\mathbf{F} \cdot \mathbf{u} = \sum_i \int_{\tau_i} [(\nabla \times \mathbf{H}) \cdot \mathbf{E} - \dot{\mathbf{D}} \cdot \mathbf{E}] \, d\tau_i$$

$$+ \sum_j \int_{S_j} [\hat{\mathbf{n}} \cdot (\mathbf{E} \times \mathbf{H})] \, d\sigma_j \tag{4.4-19}$$

The right sides of (4.4–14) and (4.4–19) are mathematically equivalent, but they differ in a fundamental way. Whereas (4.4–14) involves integration only over those *parts of the volume* τ which contain moving free charges (i.e., conductors), (4.4–19) involves integration over the *entire volume* τ. This is a consequence of the fact that the electromagnetic vectors are defined throughout space and, as long as only mathematical significance is attached to energy and to the electromagnetic field, no other comment is required. On the other hand, if energy is to be regarded as a material substance, the change from (4.4–14) to (4.4–19) is of fundamental significance because the "energy substance" apparently confined to conductors within Σ_τ in (4.4–14) may be interpreted as distributed throughout space in (4.4–19). *It is in passing from (4.4–14) to (4.4–19) that the idea of localizing and "storing" energy in space is made abruptly possible.*

The first term on the right in (4.4–19) may be transformed with the aid of the vector identity

$$\nabla \cdot (\mathbf{E} \times \mathbf{H}) = \mathbf{H} \cdot (\nabla \times \mathbf{E}) - \mathbf{E} \cdot (\nabla \times \mathbf{H}) \tag{4.4-20a}$$

which gives

$$\mathbf{E} \cdot (\nabla \times \mathbf{H}) = -\nabla \cdot (\mathbf{E} \times \mathbf{H}) + \mathbf{H} \cdot (\nabla \times \mathbf{E}) \tag{4.4-20b}$$

With the Maxwell equation

$$\nabla \times \mathbf{E} = -\dot{\mathbf{B}} \tag{4.4-21}$$

the volume integral in (4.4–19) becomes

$$-\int_{\tau_i} [\nabla \cdot (\mathbf{E} \times \mathbf{H}) + (\mathbf{H} \cdot \dot{\mathbf{B}}) + (\dot{\mathbf{D}} \cdot \mathbf{E})] \, d\tau_i \tag{4.4-22}$$

Let it be assumed that the boundary Σ_τ enclosing the region τ is not a surface of discontinuity. The divergence theorem can then be applied to the volume τ or the volumes τ_i within Σ_τ after properly enclosing boundaries where discontinuities exist within surfaces S_j:

$$\sum_i \int_{\tau_i} \mathbf{\nabla} \cdot (\mathbf{E} \times \mathbf{H}) \, d\tau_i = \sum_j \int_{S_j + \Sigma_\tau} \mathbf{\hat{n}} \cdot (\mathbf{E} \times \mathbf{H}) \, d\sigma \qquad (4.4\text{--}23)$$

After the formation of (4.4–19) it is noted that all integrals over surfaces of discontinuity S_j cancel to leave

$$\int_\tau d\mathbf{F} \cdot \mathbf{u} = -\sum_i \int_{\tau_i} [(\mathbf{H} \cdot \mathbf{\dot{B}}) + (\mathbf{\dot{D}} \cdot \mathbf{E})] \, d\tau_i - \int_{\Sigma_\tau} \mathbf{\hat{n}} \cdot (\mathbf{E} \times \mathbf{H}) \, d\sigma \qquad (4.4\text{--}24)$$

If the expression on the right side of (4.4–24) is compared with the right side of (4.4–8), it is clear that energy functions U and T may be defined to satisfy the following relations. The sum of integrals over the volumes τ_i is written as a single integral over $\tau = \Sigma_i \, \tau_i$ for simplicity and because surface currents are not required for practically available conductors, so that no discontinuities can occur. The analysis was carried out including possible surface currents because the idealized case of perfect conductors where all currents are surface currents is often assumed in engineering problems. In such cases, the sum of integrals must be written for the single integral, but no change in interpretation is required.

$$\frac{dU}{dt} \equiv \int_\tau [(\mathbf{H} \cdot \mathbf{\dot{B}}) + (\mathbf{\dot{D}} \cdot \mathbf{E})] \, d\tau \qquad (4.4\text{--}25)$$

$$T \equiv \int_{\Sigma_\tau} [\mathbf{\hat{n}} \cdot (\mathbf{E} \times \mathbf{H})] \, d\sigma \qquad (4.4\text{--}26)$$

These are the conventional definitions for the time rate of change of an electromagnetic energy function U and for an electromagnetic energy-transfer function T. Because (4.4–25) includes a term involving only the vectors $\mathbf{E}$ and $\mathbf{D}$ associated with electric phenomena and another involving only the vectors $\mathbf{B}$ and $\mathbf{H}$ associated with magnetic phenomena, separate time rates of change of electric and magnetic energies U_E and U_M are defined:

$$\frac{dU_E}{dt} \equiv \int_\tau (\mathbf{\dot{D}} \cdot \mathbf{E}) \, d\tau; \qquad \frac{dU_M}{dt} \equiv \int_\tau (\mathbf{H} \cdot \mathbf{\dot{B}}) \, d\tau \qquad (4.4\text{--}27)$$

A vector $\mathbf{S}$ called the *Poynting vector* is defined as follows:

$$\mathbf{S} \equiv \mathbf{E} \times \mathbf{H} \qquad (4.4\text{--}28)$$

so that T is the total outward normal flux of the Poynting vector:

$$T = \int_{\Sigma_\tau} (\mathbf{\hat{n}} \cdot \mathbf{S}) \, d\sigma \qquad (4.4\text{--}29)$$

From the point of view of general electromagnetic theory, the electromagnetic energy functions defined in (4.4–25) and (4.4–26) are not attractive because it is not possible to express them entirely in terms of the fundamental vectors **E** and **B**. The appearance of **D** and **H** in the energy functions, as well as the original assumption that only free charges are involved in the force equation, not only implies a definite mode of subdivision but presupposes that time variations in polarization and magnetization lead to no increase in thermal energy. If the forces that oppose periodic variations in polarization and magnetization lead to an increase in the random motions of charges and groups of charges so that a rise in the thermal energy associated with τ occurs, this is not included in dQ/dt in (4.4–7). It is actually contained in dU/dt, so the electromagnetic function U as defined includes what is recognized as thermal energy unless there is no increase in heat associated with changes in time of polarization and magnetization. It is only in the special case in which thermal energy is increased exclusively as a result of free charges moving in imperfect conductors that U is strictly an *electromagnetic* energy function.

If the auxiliary vectors **D** and **H** are expanded in accordance with their definitions,

$$\mathbf{D} \equiv \epsilon_0 \mathbf{E} + \mathbf{P}; \qquad \mathbf{H} \equiv \mu_0^{-1}\mathbf{B} - \mathbf{M} \tag{4.4–30}$$

equations (4.4–25) and (4.4–29) have a more complicated but also a more fundamental form:

$$\frac{dU}{dt} = \int_\tau \left[\frac{d}{dt}\left(\tfrac{1}{2}\mu_0^{-1}B^2 + \tfrac{1}{2}\epsilon_0 E^2\right) - (\mathbf{M} \cdot \dot{\mathbf{B}}) + (\dot{\mathbf{P}} \cdot \mathbf{E}) \right] d\tau \tag{4.4–31}$$

$$\mathbf{S} = \mu_0^{-1}(\mathbf{E} \times \mathbf{B}) - (\mathbf{E} \times \mathbf{M}) \tag{4.4–32}$$

The dimensions of the energy functions determined from (4.4–25) and (4.4–26) are as follows:

$$U \approx \frac{Q}{LT}\frac{VT}{L^2}L^3 \approx QV \qquad \text{coulomb-volts or joules} \tag{4.4–33}$$

$$T \approx \frac{V}{L}\frac{Q}{LT}L^2 \approx \frac{QV}{T} \qquad \text{volt-amperes or watts} \tag{4.4–34}$$

The coulomb-volt has the name joule; the joule per second is the watt. U is measured in joules, T in watts. Since the factor A_e in (4.4–1) is by definition dimensionless and equal to unity, the mechanical joule and watt and the electrical joule and watt are the same. The auxiliary dimension V in volts can be expressed directly in terms of Q, L, M, and T. Since mechanical energy W has the dimensions

$$W \approx \frac{ML^2}{T^2} \qquad \text{joules} \tag{4.4–35}$$

it follows that

$$V \approx \frac{ML^2}{T^2Q} \qquad \frac{\text{joules}}{\text{coulombs}} \quad \text{or} \quad \text{volts} \tag{4.4–36}$$

In electrical problems it is usually more convenient to retain the auxiliary dimension V than to introduce its equivalent in terms of Q, L, M, and T.

4.5 ENERGY FUNCTIONS IN SPACE AND IN SIMPLE MEDIA

If the region enclosed by the boundary Σ_τ consists exclusively of space and simple media, the energy functions have a more attractive form. In space where all densities vanish, $\mathbf{D} = \epsilon_0 \mathbf{E}$ and $\mathbf{H} = \mu_0^{-1}\mathbf{B}$:

$$\frac{dU}{dt} = \frac{d}{dt}\int_\tau \tfrac{1}{2}(\mu_0^{-1}B^2 + \epsilon_0 E^2)\,d\tau \tag{4.5-1}$$

$$T = \int_{\Sigma_\tau} (\hat{\mathbf{n}} \cdot \mathbf{S})\,d\sigma; \qquad \mathbf{S} = \mu_0^{-1}\mathbf{E} \times \mathbf{B} \tag{4.5-2}$$

Explicit definition of the electromagnetic energy functions is suggested by (4.5-1):

$$U = U_M + U_E = \int_\tau \tfrac{1}{2}\mu_0^{-1}B^2\,d\tau + \int_\tau \tfrac{1}{2}\epsilon_0 E^2\,d\tau \tag{4.5-3}$$

In simply polarizing and magnetizing media the relations

$$\mathbf{P} = (\epsilon_r - 1)\epsilon_0\mathbf{E}; \qquad -\mathbf{M} = (\mu_r^{-1} - 1)\mu_0^{-1}\mathbf{B} \tag{4.5-4}$$

are valid. This implies an instantaneous response in polarization and magnetization to changes in $\mathbf{E}$ and $\mathbf{B}$ and *no* associated increase in random or heat motion. The relations (4.5-4) give

$$\mathbf{D} = \epsilon_r\epsilon_0\mathbf{E} = \epsilon\mathbf{E}; \qquad \mathbf{H} = (\mu_r\mu_0)^{-1}\mathbf{B} = \mu^{-1}\mathbf{B} \tag{4.5-5}$$

The energy functions in simple media are

$$U = U_M + U_E = \int_\tau \tfrac{1}{2}\mu^{-1}B^2\,d\tau + \int_\tau \tfrac{1}{2}\epsilon E^2\,d\tau \tag{4.5-6}$$

$$T = \int_{\Sigma_\tau} \hat{\mathbf{n}} \cdot \mathbf{S}\,d\sigma; \qquad \mathbf{S} = \mu^{-1}(\mathbf{E} \times \mathbf{B}) \tag{4.5-7}$$

In a simply conducting region, by definition

$$\mathbf{J}_f = \sigma\mathbf{E} \tag{4.5-8}$$

or, if a charge-separating region characterized by an impressed or intrinsic electric field $\mathbf{E}^e$ (which by definition is equal in magnitude and opposite in direction to the electric field required to prevent the separation) is involved,

$$\mathbf{J}_f = \sigma(\mathbf{E} + \mathbf{E}^e) \tag{4.5-9}$$

Since surface densities of moving free charge are not needed to describe simple conductors with *finite* conductivity,

$$\mathbf{K}_f = 0 \tag{4.5-10}$$

If (4.5–9) and (4.5–10) are used in (4.4–14), it follows that

$$\int_\tau d\mathbf{F} \cdot \mathbf{u} = \int_\tau \sigma[(\mathbf{E} + \mathbf{E}^e) \cdot \mathbf{E}]\, d\tau \tag{4.5-11}$$

This may be expanded into

$$\int_\tau d\mathbf{F} \cdot \mathbf{u} = \int_\tau \sigma[(\mathbf{E} + \mathbf{E}^e) \cdot (\mathbf{E} + \mathbf{E}^e)]\, d\tau - \int_\tau \sigma[(\mathbf{E} + \mathbf{E}^e) \cdot \mathbf{E}^e]\, d\tau \tag{4.5-12}$$

With (4.5–9) this gives

$$\int_\tau d\mathbf{F} \cdot \mathbf{u} = \int_\tau \frac{J_f^2}{\sigma}\, d\tau - \int_\tau (\mathbf{J}_f \cdot \mathbf{E}^e)\, d\tau \tag{4.5-13}$$

Substitution in (4.4–7) leads to

$$\int_\tau d\mathbf{F}_M \cdot \mathbf{u} = \frac{dW}{dt} - A_q \frac{dQ}{dt}$$

$$= \int_\tau (\mathbf{J}_f \cdot \mathbf{E}^e)\, d\tau - \int_\tau \frac{J_f^2}{\sigma}\, d\tau \tag{4.5-14}$$

The first term on the right can now be identified with the power supplied to the region τ by nonelectrical forces in a charge-separating region; the second term is the power dissipated as heat. The first integral on the right in (4.5–14) vanishes except in a charge-separating region where $\mathbf{E}^e$ differs from zero. With (4.5–14), (4.5–6), and (4.5–7), the general power equation for simple media may be written in the alternative form

$$\int_\tau (\mathbf{J}_f \cdot \mathbf{E}^e)\, d\tau - \int_\tau \frac{J_f^2}{\sigma}\, d\tau - \frac{d}{dt} \int_\tau \tfrac{1}{2}(\mu^{-1}B^2 + \epsilon E^2)\, d\tau$$

$$- \int_{\Sigma_\tau} \mu^{-1}\hat{\mathbf{n}} \cdot (\mathbf{E} \times \mathbf{B})\, d\sigma = 0 \tag{4.5-15}$$

The power equation (4.5–15) and the associated energy functions (4.5–6) and (4.5–7) as written for simple media are useful in the solution of problems of many types, such as radiation from antennas and dissipation in wave guides. The first integral in (4.5–15) measures the time rate of decrease of the mechanical (or other nonelectrical) energy function W associated with τ. The second integral is the time rate of increase of the thermal energy function Q associated with moving *free* charges in τ. The third integral defines the time rate of increase of the electromagnetic energy function U associated with τ. The surface integral measures the time rate of increase of energy functions associated with all regions outside τ due to the action of electromagnetic forces on charges outside Σ_τ. If the first integral is zero in a particular case, it follows that a nonvanishing and nontransient current $\mathbf{J}_f$ can exist in τ only if the surface integral is negative or the external normal $\hat{\mathbf{n}}$ is reversed. This means that the electromagnetic forces due to moving charges *outside* τ are doing work on the charges *in* τ to maintain the current.

If energy is assumed to be a physical substance endowed with properties that admit of its localization and distribution in charge-filled or empty regions and of its flow from one region to another, the foregoing interpretation of the power equation is not only modified but also greatly specialized in its meaning. Thus, if the integral (4.5–6) defines the electromagnetic energy "stored" in τ, the integrand when written for simple media or for space as

$$\tfrac{1}{2}\mu^{-1}B^2 + \tfrac{1}{2}\epsilon E^2; \qquad \tfrac{1}{2}\mu_0^{-1}B^2 + \tfrac{1}{2}\epsilon_0 E^2 \tag{4.5-16}$$

must specify its spatially distributed density. (The quantities $\tfrac{1}{2}\mu^{-1}B^2$ and $\tfrac{1}{2}\mu_0^{-1}B^2$ are *magnetic energy densities*; $\tfrac{1}{2}\epsilon E^2$ and $\tfrac{1}{2}\epsilon_0 E^2$ *electric energy densities*.)

With the energy substance assumed distributed throughout space with the density specified by (4.5–16), the energy-transfer function T is interpreted to measure the flow of energy across the enclosing boundary Σ_τ. If the flow is outward, as indicated by an outwardly directed normal, the energy density outside τ is increasing, that inside is decreasing. If the flow is inward, as indicated by an inwardly directed normal, the reverse is true. Furthermore, since T measures the *total* flow of energy across Σ_τ, it is concluded that the integrand in (4.5–7) *must* measure the flow of energy across *each element* $d\sigma$ of the surface. Accordingly, the Poynting vector $\mathbf{S}$ is interpreted to define the direction and magnitude of the actual flow of energy across a unit area at every point. This conclusion is not mathematically justified. The surface integral in (4.5–7) was originally obtained by the application of the divergence theorem to the volume integral; thus

$$\int_\tau \nabla \cdot \mathbf{S}\, d\tau = \int_{\Sigma_\tau} (\hat{\mathbf{n}} \cdot \mathbf{S})\, d\sigma \tag{4.5-17}$$

This theorem has a meaning *only if the integration is extended over a completely closed surface*. It does not admit of an integration over only *part* of a closed surface. It follows that from the mathematical point of view no meaning can be attached to an integral like that on the right in (4.5–17) if the integration is carried out over a surface that is not closed, or to the integrand itself. This is made especially clear by the fact that any integration over a specified part of a closed surface can be made to have *any desired value* by merely adding to $\mathbf{S}$ a suitably defined solenoidal vector $\mathbf{C}$. By definition, a solenoidal vector satisfies the condition $\nabla \cdot \mathbf{C} = 0$, so that

$$\int_\tau \nabla \cdot (\mathbf{S} + \mathbf{C})\, d\tau = \int_{\Sigma_\tau} (\hat{\mathbf{n}} \cdot \mathbf{S})\, d\sigma \tag{4.5-18}$$

There is no change in the surface integral over the closed surface Σ_τ and there is no reason, either mathematical or physical, that makes it necessary or even reasonable to prefer the vector $\mathbf{S}$ to the vector $\mathbf{S} + \mathbf{C}$ because both lead to the same power equation. The "amount of energy" that is assumed to "flow" across a surface which is only a part of the closed surface Σ_τ is arbitrary and not at all uniquely specified by $\int_\Sigma (\hat{\mathbf{n}} \cdot \mathbf{S}) \, d\sigma$. All values of $\int_\Sigma \hat{\mathbf{n}} \cdot (\mathbf{S} + \mathbf{C}) \, d\sigma$ are equally reasonable with $\mathbf{C}$ *any solenoidal vector whatsoever.*

4.6 COMPLEX ENERGY FUNCTIONS

The specialization of the energy formulation to complex harmonic functions of time must begin with the power equation

$$\frac{dW}{dt} - A_q \frac{dQ}{dt} - \frac{dU}{dt} - T = 0 \tag{4.6-1}$$

It is assumed in writing (4.6–1) that the forces contributing to W in a closed but not isolated region τ do positive work and so supply energy to the region, while all other forces within and outside the region do negative work (i.e., receive energy). In (4.6–1), all derivative and transfer functions are positive. Equation (4.6–1) expresses the relation between the *instantaneous* time rates of change of mechanical, thermal, and electromagnetic energy functions and the instantaneous electromagnetic energy-transfer function T defined for a thermally isolated region τ and its closed boundary Σ_τ. If the field vectors $\mathbf{E}$ and $\mathbf{B}$ and the auxiliary vectors $\mathbf{D}$ and $\mathbf{H}$ vary harmonically in time, the energy functions and the transfer function, which are defined in terms of $\mathbf{E}$, $\mathbf{B}$, $\mathbf{D}$, and $\mathbf{H}$, must also have periodic time variations. The instantaneous time rates of change of the energy functions are not of primary interest. Their average values over a longer interval are of practical importance. Such an interval includes many periods T_p and, since each period is like every other one, a time average over a single period is the same as the time average over a great many whole periods. A fraction of a period in a sufficiently long time contributes only a negligible amount to the average.

The time average of any function $X(t)$ over a period T_p is given by

$$\langle X(t) \rangle = \frac{1}{T_p} \int_0^{T_p} X(t) \, dt \tag{4.6-2}$$

The time-average value is indicated by the symbol $\langle \cdot \rangle$. Thus for

$$X(t) = X \cos(\omega t + \theta_x) \tag{4.6-3}$$

$$\langle X(t) \rangle = \frac{1}{T_p} \int_0^{T_p} X \cos(\omega t + \theta_x) \, dt = 0$$

$$\langle X^2(t) \rangle = \frac{1}{T_p} \int_0^{T_p} X^2 \cos^2(\omega t + \theta_x) \, dt$$

$$= \frac{1}{\omega T_p} \int_0^{2\pi} X^2 \cos^2(x + \theta_x) \, dx = \frac{X^2}{2} \tag{4.6-4}$$

$$\frac{d}{dt} \langle X^2(t) \rangle = 0 \tag{4.6-5}$$

If $Y(t) = Y \cos(\omega t + \theta_y)$, the time-average product $\langle X(t)Y(t) \rangle$ becomes

$$\langle X(t)Y(t) \rangle = \frac{1}{T_p} \int_0^{T_p} XY \cos(\omega t + \theta_y) \cos(\omega t + \theta_x) \, dt$$

$$= \tfrac{1}{2} XY \cos(\theta_x - \theta_y) \qquad (4.6\text{--}6)$$

In the complex notation the instantaneous values of the desired real functions are given by the real parts of the following complex instantaneous functions:

$$X(t) = X e^{j\omega t} = |X| e^{j\theta_x} e^{j\omega t} \qquad (4.6\text{--}7)$$

$$Y(t) = Y e^{j\omega t} = |Y| e^{j\theta_y} e^{j\omega t} \qquad (4.6\text{--}8)$$

The complex conjugates of the functions are

$$X^*(t) = X^* e^{-j\omega t} = |X| e^{-j\theta_x} e^{-j\omega t} \qquad (4.6\text{--}9a)$$

$$Y^*(t) = Y^* e^{-j\omega t} = |Y| e^{-j\theta_y} e^{-j\omega t} \qquad (4.6\text{--}9b)$$

It follows that

$$\langle X^2(t) \rangle = \tfrac{1}{2}|X|^2 = \tfrac{1}{2} X(t)X^*(t) = \tfrac{1}{2} XX^* \qquad (4.6\text{--}10)$$

$$\langle X(t)Y(t) \rangle = \tfrac{1}{2}|XY| \cos(\theta_x - \theta_y) = \mathrm{Re}(\tfrac{1}{2} XY^*) \qquad (4.6\text{--}11)$$

A time-average power equation may be derived from

$$\int_\tau \langle \mathbf{J}_f(t) \cdot \mathbf{E}(t) \rangle \, d\tau + \int_\Sigma \langle \mathbf{K}_f(t) \cdot \mathbf{E}(t) \rangle \, d\sigma$$

$$= \mathrm{Re}\left\{ \frac{1}{2}\left[\int_\tau (\mathbf{J}_f^* \cdot \mathbf{E}) \, d\tau + \int_\Sigma (\mathbf{K}_f^* \cdot \mathbf{E}) \, d\sigma \right] \right\} \qquad (4.6\text{--}12)$$

To expand the right side of (4.6–12), it is necessary to substitute for $\mathbf{J}_f^*$ and $\mathbf{K}_f^*$ from the equations

$$\nabla \times \mathbf{H}^* = \mathbf{J}_f^* - j\omega \mathbf{D}^* \qquad (4.6\text{--}13)$$

$$\hat{\mathbf{n}} \times \mathbf{H}^* = -\mathbf{K}_f^* \qquad (4.6\text{--}14)$$

to obtain

$$\frac{1}{2} \int_\tau [\mathbf{E} \cdot (\nabla \times \mathbf{H}^*) + j\omega(\mathbf{E} \cdot \mathbf{D}^*)] \, d\tau - \sum_i \frac{1}{2} \int_{S_i} (\hat{\mathbf{n}} \times \mathbf{H}^*) \cdot \mathbf{E} \, d\sigma_i \qquad (4.6\text{--}15)$$

Since

$$\nabla \cdot (\mathbf{E} \times \mathbf{H}^*) = \mathbf{H}^* \cdot (\nabla \times \mathbf{E}) - \mathbf{E} \cdot (\nabla \times \mathbf{H}^*) \qquad (4.6\text{--}16)$$

and

$$\nabla \times \mathbf{E} = -j\omega \mathbf{B} \qquad (4.6\text{--}17)$$

(4.6–15) can be written as follows:

$$j \frac{\omega}{2} \int_\tau [(\mathbf{E} \cdot \mathbf{D}^*) - (\mathbf{B} \cdot \mathbf{H}^*)] \, d\tau - \frac{1}{2} \int_\tau \nabla \cdot (\mathbf{E} \times \mathbf{H}^*) \, d\tau$$

$$+ \sum_i \frac{1}{2} \int_{S_i} \hat{\mathbf{n}} \cdot (\mathbf{E} \times \mathbf{H}^*) \, d\sigma_i \qquad (4.6\text{–}18)$$

After the application of the divergence theorem to the second integral (expanded, if necessary, into a sum over separate regions where the functions $\mathbf{E}$ and $\mathbf{H}$ are continuous), the integrals over surfaces of discontinuity cancel the corresponding integrals from the last term in (4.6–18) and only the integral over the enclosing surface Σ_τ remains. Thus

$$\frac{1}{2} \left[\int_\tau (\mathbf{J}_f^* \cdot \mathbf{E}) \, d\tau + \int_{\Sigma_\tau} (\mathbf{K}_f^* \cdot \mathbf{E}) \, d\sigma \right]$$

$$= -\frac{1}{2} j\omega \int_\tau [(\mathbf{B} \cdot \mathbf{H}^*) - (\mathbf{E} \cdot \mathbf{D}^*)] \, d\tau - \frac{1}{2} \int_{\Sigma_\tau} \hat{\mathbf{n}} \cdot (\mathbf{E} \times \mathbf{H}^*) \, d\sigma \qquad (4.6\text{–}19)$$

This is a complex power equation that can be separated into two real equations. The equation given by the real parts in (4.6–19) is the desired time-average power equation; that given by the imaginary parts is an additional true equation.

As in the general case of unrestricted time dependence, the complex power equation assumes an attractive form particularly in space and in simple media with no time lags in polarization, magnetization, or current responses. In such media, by definition

$$\mathbf{J}_f = \sigma(\mathbf{E} + \mathbf{E}^e); \qquad \mathbf{D} = \epsilon\mathbf{E}; \qquad \mathbf{H} = \mu^{-1}\mathbf{B} \qquad (4.6\text{–}20a)$$

with σ, ϵ, and μ real. Also,

$$\mathbf{K}_f = 0 \qquad (4.6\text{–}20b)$$

$$\mathbf{J}_f^* \cdot \mathbf{E} = \sigma(\mathbf{E}^* + \mathbf{E}^{e*}) \cdot \mathbf{E} = \frac{J_f^2}{\sigma} - \mathbf{J}_f^* \cdot \mathbf{E}^e \qquad (4.6\text{–}21a)$$

$$\mathbf{B} \cdot \mathbf{H}^* = \mu^{-1}B^2 \qquad (4.6\text{–}21b)$$

$$\mathbf{E} \cdot \mathbf{D}^* = \epsilon E^2 \qquad (4.6\text{–}21c)$$

The complex Poynting vector is

$$\mathbf{S} = \frac{1}{2}(\mathbf{E} \times \mathbf{H}^*) = \hat{\mathbf{S}}|S|e^{j\theta_s} = \hat{\mathbf{S}}|S|(\cos\theta_s + j\sin\theta_s) \qquad (4.6\text{–}21d)$$

The substitution from (4.6–21a–c) into (4.6–19), with $\mathbf{J}_f$ assumed to be in phase with $\mathbf{E}^e$, leads to:

$$\frac{1}{2}\int_\tau \frac{J_f^2}{\sigma}\, d\tau \; - \; \frac{1}{2}\int_\tau \mathbf{J}_f \cdot \mathbf{E}^e\, d\tau$$

$$= \; -j\frac{\omega}{2}\int_\tau (\mu^{-1}B^2 - \epsilon E^2)\, d\tau \; - \int_{\Sigma_\tau} \hat{\mathbf{n}}\cdot\hat{\mathbf{S}}|S|e^{j\theta_s}\, d\sigma \qquad (4.6\text{–}22)$$

The separation into real and imaginary parts yields two equations. The real power equation is

$$\frac{1}{2}\int_\tau (\mathbf{J}_f \cdot \mathbf{E}^e)\, d\tau \; - \; \frac{1}{2}\int_\tau \frac{J_f^2}{\sigma}\, d\tau \; - \int_{\Sigma_\tau} \hat{\mathbf{n}}\cdot\hat{\mathbf{S}}|S|\cos\theta_s\, d\sigma = 0 \qquad (4.6\text{–}23)$$

An auxiliary equation, which may be called the *reactive power equation,* is

$$\frac{\omega}{2}\int_\tau (\mu^{-1}B^2 - \epsilon E^2)\, d\tau \; + \int_{\Sigma_\tau} \hat{\mathbf{n}}\cdot\hat{\mathbf{S}}|S|\sin\theta_s\, d\sigma = 0 \qquad (4.6\text{–}24)$$

It is now possible to define time-average electric and magnetic energy functions $\langle U_E(t)\rangle$ and $\langle U_M(t)\rangle$ and a complex energy-transfer function $T = T_r + jT_i$, the real part of which is the time-average energy-transfer function. Thus

$$\langle U_M(t)\rangle = \frac{\mu^{-1}}{4}\int_\tau B^2\, d\tau; \qquad \langle U_E(t)\rangle = \frac{\epsilon}{4}\int_\tau E^2\, d\tau \qquad (4.6\text{–}25)$$

$$T_r = \langle T(t)\rangle = \operatorname{Re}\!\left(\int_{\Sigma_\tau} \hat{\mathbf{n}}\cdot\mathbf{S}\, d\sigma\right) = \int_{\Sigma_\tau} \hat{\mathbf{n}}\cdot\hat{\mathbf{S}}|S|\cos\theta_s\, d\sigma \qquad (4.6\text{–}26)$$

$$T_i = \operatorname{Im}\!\left(\int_{\Sigma_\tau} \hat{\mathbf{n}}\cdot\mathbf{S}\, d\sigma\right) = \int_{\Sigma_\tau} \hat{\mathbf{n}}\cdot\hat{\mathbf{S}}|S|\sin\theta_s\, d\sigma \qquad (4.6\text{–}27)$$

Since the power equation (4.6–23) does not involve the time-average electric and magnetic energy functions, it must be concluded that

$$\langle U(t)\rangle = \langle U_M(t)\rangle + \langle U_E(t)\rangle = \text{constant} \qquad (4.6\text{–}28)$$

Whenever

$$T_i = 0 \qquad (4.6\text{–}29)$$

which is certainly true when *one* or both of the following conditions are satisfied *at all points* on the surface Σ_τ,

$$\theta_s = \theta_H - \theta_E = 0 \qquad (4.6\text{–}30)$$

$$\hat{\mathbf{n}}\cdot\mathbf{S} = 0 \qquad (4.6\text{–}31)$$

it follows from (4.6–24) that

$$\langle U_M(t)\rangle = \langle U_E(t)\rangle \qquad (4.6\text{–}32)$$

and

$$\langle U(t) \rangle = 2 \langle U_M(t) \rangle = 2 \langle U_E(t) \rangle \qquad (4.6\text{--}33)$$

It is found that (4.6–29) is true in many practically important problems.

It is possible to take account of time lags in polarization and current response in sufficiently specialized physical models by noting that if the conductivity and the dielectric constant are complex, (4.6–13) may be written as follows using (2.14–17a) and (2.14–24a):

$$\begin{aligned} \nabla \times \mathbf{H}^* &= (\sigma^* - j\omega\epsilon^*)\mathbf{E}^* \\ &= (\sigma_e - j\omega\epsilon_e)\mathbf{E}^* \end{aligned} \qquad (4.6\text{--}34)$$

The expression on the right-hand side of (4.6–34) is identically the same as that for no time lags except that the real effective conductivity and dielectric constant replace σ and ϵ. If it is assumed that the forces required to maintain a periodically varying complex polarization current $j\omega \mathbf{P}e^{j\omega t}$ are the same as those required to maintain an equivalent complex current of free charge $\mathbf{J}e^{j\omega t}$, so that $\sigma_e\mathbf{E}^*$ is assumed to be equivalent in all respects involving the equilibrium of forces to $\sigma\mathbf{E}^*$ with $\sigma_e = \sigma$, then (4.6–23), (4.6–24), and (4.6–25) may be modified by writing a subscript e on both σ and ϵ. Since $\sigma_e = \sigma' + \omega\epsilon''$, it follows, if $\mathbf{E} + \mathbf{E}^e$ is maintained constant with $\mathbf{E}^e$ directed opposite to $\mathbf{E}$ in charge-separating regions, that the rate of change of the average heat-energy function Q given by

$$A_q \frac{d}{dt} \langle Q(t) \rangle = \frac{1}{2} \int_\tau \frac{J_f^2}{\sigma} \, d\tau = \frac{1}{2} \int_\tau \sigma(E + E^e)^2 \, d\tau \qquad (4.6\text{--}35)$$

is increased by $\frac{1}{2}\int_\tau \omega\epsilon''(E + E^e)^2 \, d\tau$ owing to polarization lag when $\sigma_e = \sigma' + \omega\epsilon''$ is written for σ. The time-average electric energy function

$$\langle U_e(t) \rangle = \frac{\epsilon}{4} \int_\tau E^2 \, d\tau \qquad (4.6\text{--}36)$$

is decreased when $\epsilon_e = \epsilon' - \sigma''/\omega$ is written for ϵ.

4.7 A POWER EQUATION IN TERMS OF SCALAR AND VECTOR POTENTIALS

A simple and useful alternative power equation may be formulated using the scalar and vector potentials. The combination of the relations

$$\mathbf{E} = -\nabla\phi - \dot{\mathbf{A}} \qquad (4.7\text{--}1)$$

and

$$\mathbf{J}_f = \sigma(\mathbf{E} + \mathbf{E}^e) \qquad (4.7\text{--}2)$$

leads to the following equation:

$$\mathbf{E}^e = \frac{\mathbf{J}_f}{\sigma} + \nabla\phi + \dot{\mathbf{A}} \tag{4.7-3}$$

When (4.7–3) is multiplied scalarly by $\mathbf{J}_f$ and integrated over an arbitrary volume τ, the result is

$$\int_\tau \mathbf{J}_f \cdot \mathbf{E}^e \, d\tau = \int_\tau \frac{J_f^2}{\sigma} \, d\tau + \int_\tau \mathbf{J}_f \cdot \nabla\phi \, d\tau + \int_\tau \mathbf{J}_f \cdot \dot{\mathbf{A}} \, d\tau \tag{4.7-4a}$$

which may also be written as follows:

$$\int_\tau \mathbf{J}_f \cdot \mathbf{E}^e \, d\tau = \int_\tau \frac{J_f^2}{\sigma} \, d\tau - \int_\tau \mathbf{J}_f \cdot \mathbf{E} \, d\tau \tag{4.7-4b}$$

With the vector identity

$$\nabla \cdot (\phi\mathbf{J}_f) = \phi(\nabla \cdot \mathbf{J}_f) + \mathbf{J}_f \cdot (\nabla\phi) \tag{4.7-5}$$

and the divergence theorem

$$\int_\tau \nabla \cdot (\phi\mathbf{J}_f) \, d\tau = \int_\Sigma \phi(\hat{\mathbf{n}} \cdot \mathbf{J}_f) \, d\Sigma \tag{4.7-6}$$

(4.7–4a) can be transformed into

$$\int_\tau (\mathbf{J}_f \cdot \mathbf{E}^e) \, d\tau = \int_\tau \frac{J_f^2}{\sigma} \, d\tau - \int_\tau \phi(\nabla \cdot \mathbf{J}_f) \, d\tau$$
$$+ \int_\tau (\mathbf{J}_f \cdot \dot{\mathbf{A}}) \, d\tau + \int_\Sigma \phi(\hat{\mathbf{n}} \cdot \mathbf{J}_f) \, d\Sigma \tag{4.7-7}$$

It is to be noted that Σ includes Σ_τ, a surface completely enclosing τ, and all additional surfaces within τ across which $\mathbf{J}_f$ (or ϕ) is discontinuous. If the equation of continuity for free charge

$$\nabla \cdot \mathbf{J}_f + \dot{\rho}_f = 0 \tag{4.7-8}$$

is substituted into (4.7–7), it becomes

$$\int_\tau (\mathbf{J}_f \cdot \mathbf{E}^e) \, d\tau - \int_\tau \frac{J_f^2}{\sigma} \, d\tau - \int_\tau \dot{\rho}_f\phi \, d\tau - \int_\tau (\mathbf{J}_f \cdot \dot{\mathbf{A}}) \, d\tau$$
$$- \int_\Sigma \phi(\hat{\mathbf{n}} \cdot \mathbf{J}_f) \, d\Sigma = 0 \tag{4.7-9}$$

This equation resembles (4.5–15), that is,

$$\int_\tau (\mathbf{J}_f \cdot \mathbf{E}^e) \, d\tau - \int_\tau \frac{J_f^2}{\sigma} \, d\tau - \frac{d}{dt} \int_\tau \tfrac{1}{2}(\mu^{-1}B^2 + \epsilon E^2) \, d\tau$$
$$- \int_{\Sigma_\tau} \mu^{-1}\hat{\mathbf{n}} \cdot (\mathbf{E} \times \mathbf{B}) \, d\Sigma \tag{4.7-10}$$

The first two terms in (4.7–9) and (4.7–10) are exactly the same. The next two in each equation are one electric and one magnetic volume integral. The last term is a surface integral including the enclosing boundary. There is, however, a fundamental difference between (4.7–9) and (4.7–10). Whereas (4.7–10) involves integration over all space where $\mathbf{E}$ and $\mathbf{B}$ are nonvanishing, (4.7–9) involves integration *only over that part of space where electric currents and charges differ from zero.* *All* the terms in (4.7–9) vanish in free space. Thus an interpretation of (4.7–9) in terms of "energy" distributed in completely evacuated space is impossible. In particular, the surface integral certainly does not suggest or permit an interpretation in terms of a "flow of energy *in space*" across Σ_τ. Actually, the surface integral vanishes everywhere except in conductors.

A further transformation of (4.7–9) is possible with the surface equation of continuity,

$$\nabla \cdot \mathbf{K}_f + \dot{\eta}_f - \hat{\mathbf{n}} \cdot \mathbf{J}_f = 0 \tag{4.7–11}$$

When perfect conductors are excluded, $\mathbf{K}_f$ need not be defined since $\mathbf{J}_f$ is adequate. Hence (4.7–11) reduces to

$$\hat{\mathbf{n}} \cdot \mathbf{J}_f = \dot{\eta}_f \tag{4.7–12}$$

and (4.7–9) becomes

$$\int_\tau (\mathbf{J}_f \cdot \mathbf{E}^e) \, d\tau - \int_\tau \frac{J_f^2}{\sigma} \, d\tau - \int_\tau \dot{\rho}_f \phi \, d\tau$$

$$- \int_\tau (\mathbf{J}_f \cdot \dot{\mathbf{A}}) \, d\tau - \int_\Sigma \dot{\eta}_f \phi \, d\Sigma = 0 \tag{4.7–13}$$

This is the general equation.

As shown in Sec. 2.13, ρ_f vanishes in nonconductors and conductors if the time dependence is periodic. In this important case, (4.7–13) reduces to

$$\int_\tau (\mathbf{J}_f \cdot \mathbf{E}^e) \, d\tau - \int_\tau \frac{J_f^2}{\sigma} \, d\tau - \int_\tau (\mathbf{J}_f \cdot \dot{\mathbf{A}}) \, d\tau - \int_\Sigma \dot{\eta}_f \phi \, d\sigma = 0 \tag{4.7–14}$$

This is a power equation expressed in terms of scalar and vector potentials.

4.8 ENERGY-TRANSFER FUNCTION FOR THE SURFACE OF A CYLINDRICAL ANTENNA

The complex transfer function $T = T_r + jT_i$ is defined by

$$T = \int_{\Sigma_\tau} \hat{\mathbf{n}} \cdot \mathbf{S} \, d\sigma \tag{4.8–1}$$

where $d\sigma$ is an element of any completely closed surface Σ_τ on which $\hat{\mathbf{n}}$ is an external unit normal. **S** is the complex Poynting vector defined in space by

$$\mathbf{S} = \tfrac{1}{2}\mu_0^{-1}\mathbf{E} \times \mathbf{B}^* \tag{4.8-2}$$

The real part T_r of the integral (4.8–1) is a measure of the total transfer of power from within Σ_τ. It is necessarily independent of the shape and size of the envelope Σ_τ provided that the energy functions defined within Σ_τ are unchanged. In particular, T_r has the same value if Σ_τ is a great sphere enclosing an antenna completely in its far zone, or if Σ_τ is a cylindrical envelope only an infinitesimal amount larger than the actual surface of the antenna. T_i, on the other hand, is not necessarily independent of the shape and location of Σ_τ, so that it may not be concluded that because T_i vanishes when Σ_τ is a sphere in the far zone, it will also vanish when Σ_τ is the cylindrical surface of the antenna itself.

On the cylindrical envelope $\rho = a$, and in terms of the cylindrical coordinates ρ, θ, z with the z axis along the antenna,

$$\mathbf{\hat{n}} \cdot \mathbf{S} = S_\rho = \tfrac{1}{2}\mu_0^{-1}(\mathbf{E} \times \mathbf{B}^*)_\rho = \tfrac{1}{2}\mu_0^{-1}(E_\theta B_z^* - E_z B_\theta^*) \tag{4.8-3}$$

On the end surfaces at $z = \pm h$,

$$\mathbf{\hat{n}} \cdot \mathbf{S} = \pm S_z = \pm\tfrac{1}{2}\mu_0^{-1}(\mathbf{E} \times \mathbf{B}^*)_z = \pm\tfrac{1}{2}\mu_0^{-1}(E_\rho B_\theta^* - E_\theta B_\rho^*) \tag{4.8-4}$$

The upper sign is for the top of the antenna, the lower sign is for the bottom.

For a cylindrical antenna, with a rotationally symmetric emf,

$$E_\theta = 0; \qquad B_\rho = 0 \tag{4.8-5}$$

so that

$$S_r = -\tfrac{1}{2}\mu_0^{-1}E_z B_\theta^* \tag{4.8-6}$$

$$S_z = \pm\tfrac{1}{2}\mu_0^{-1}E_\rho B_\theta^* \tag{4.8-7}$$

It follows that

$$T = -\tfrac{1}{2}\mu_0^{-1}\int_{-h}^{h}(E_z B_\theta^*)_{\rho=a}2\pi a\ dz + \tfrac{1}{2}\mu_0^{-1}\int_0^a (E_\rho B_\theta^*)_{z=h}2\pi\rho\ d\rho$$

$$-\tfrac{1}{2}\mu_0^{-1}\int_0^a (E_\rho B_0^*)_{z=-h}2\pi\rho\ d\rho \tag{4.8-8}$$

If the area of the end surfaces is sufficiently small compared with the cylindrical surface so that it is correct to write

$$2\pi a^2 \ll 2\pi ah \tag{4.8-9}$$

or

$$a \ll h \tag{4.8-10}$$

the contribution to T by the last two integrals in (4.8–8) is negligible. This follows because the very small radial currents on the ends cannot lead to an average tangential electromagnetic field at the ends that equals the average tangential field on the cylindrical sides due to the axial current. Accordingly, a good approximation subject to (4.8–10) is

$$T \doteq -\tfrac{1}{2}\mu_0^{-1} \int_{-h}^{h} (E_z B_\theta^*)_{\rho=a} 2\pi a \, dz \tag{4.8–11}$$

It is possible to express B_θ in terms of the total current using the Ampère–Maxwell theorem of circuitation (2.10–3). For a single surface this is

$$\mu_0^{-1} \oint \mathbf{B} \cdot d\mathbf{s} = \bar{I} + j\omega\epsilon_0 \int_{S(\text{cap})} (\hat{\mathbf{N}} \cdot \mathbf{E}) \, dS \tag{4.8–12}$$

with

$$\bar{I} = \int_{S(\text{cap})} (\hat{\mathbf{N}} \cdot \overline{\rho_m \mathbf{v}}) \, dS + \int_s (\hat{\mathbf{N}} \cdot \overline{\eta_m \mathbf{v}}) \, ds \tag{4.8–13}$$

Let $S(\text{cap})$ be the cross section of the cylindrical conductor at any point along its length, and s the circumference. In the conductor (not perfect)

$$(\hat{\mathbf{N}} \cdot \overline{\rho_m \mathbf{v}}) = J_{zf}; \qquad \hat{\mathbf{N}} \cdot \overline{\eta_m \mathbf{v}} = K_{zf} = 0; \qquad \bar{I} = I_z \tag{4.8–14}$$

By definition of a good conductor,

$$J_{zf} = \sigma E_z \gg \omega\epsilon_0 E_z \tag{4.8–15}$$

Since the antenna is rotationally symmetrical, (4.8–12) becomes

$$\mu_0^{-1} \oint B_\theta a \, d\theta = \mu_0^{-1} 2\pi a B_\theta = I_z \tag{4.8–16}$$

With (4.8–16), (4.8–11) becomes

$$T \doteq -\tfrac{1}{2} \int_{-h}^{h} (E_z)_{\rho=a} I_z^* \, dz \tag{4.8–17}$$

If the antenna is symmetrical and center-driven so that

$$I(z) = I(-z) \tag{4.8–18}$$

$$T = -\tfrac{1}{2} \int_0^h (E_z)_{\rho=a} I_z^*(z) \, dz + \tfrac{1}{2} \int_0^{-h} (E_z)_{\rho=a} I_z^*(-z) \, dz \tag{4.8–19}$$

With z substituted for $-z$ in the second integral, this becomes like the first integral. Then

$$T = -\int_0^h (E_z)_{\rho=a} I_z^*(z) \, dz \tag{4.8–20}$$

The real part of (4.8–20) gives the total power transferred from a cylindrical region of length $2h$ and radius a carrying a total current I_z. $(E_z)_{\rho=a}$ is the tangential component of the electric field at the surface of the cylinder. At all points along

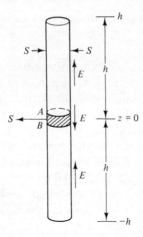

Figure 4.8–1   Electric field and Poynting vector on the cylindrical surface of a good but not perfect conductor with a generator between A and B.

the part of the conducting surface that contains no charge-separating agency (A to h and B to $-h$ in Fig. 4.8–1), $(E_z)_{\rho=a}$ and I_z are in the same direction. [The phase difference between $(E_{z\,\text{inst}})_{\rho=a}$ and $I_{z\,\text{inst}}$ is in no case greater than 45°.] At points along the conducting surface of the charge-separating region, $(E_z)_{\rho=a}$ and I_z are oppositely directed. It follows from (4.8–20) that integration over the length of the conductor not including the length AB contributes a negative value to T, while the integral over the cylindrical surface of AB gives a positive value. It is interesting to note that that part of the integration which is carried out over the surface of the antenna proper, not including the surface of the generating region, yields only the rate of increase of heat energy in the conductor, not at all the rate at which energy functions outside the conductor increase due to radiation. This is given entirely by that part of the integration which is carried out over the surface enclosing the charge-separating region. Since no particular significance can be attached to integration of the normal component of the Poynting vector over a *part* of a closed surface, it is not legitimate to attempt to localize the "flow of energy" on the basis of the meaningless parts of a meaningful integral.

An erroneous and peculiarily inconsistent analysis based on (4.8–20) proceeds as follows.

1. The conductor is assumed to be perfect.
2. A sinusoidal distribution of current is *assumed* along the antenna. (This is an incorrect assumption except for an infinitely thin, perfect conductor.)
3. The electric field $(E_z)_{\rho=a}$ is calculated from the assumed current. A *nonvanishing* value is obtained in direct contradiction to postulate 1, which requires a vanishing field.
4. The field so obtained is substituted in (4.8–20) to determine T and ultimately the input impedance of the antenna.

It is obvious that the impedance of an antenna cannot be obtained by this sequence of errors and contradictions.

The Poynting vector on the cylindrical surface of the conductor is given by (4.8–6). If the conductor is perfect, E_z is zero and hence S_r vanishes at all points along the antenna. If the conductor is not perfect, E_z has a small value and S_r is negative (i.e., **S** points radially *into* the conductor). Between the circumferences at A and B of the antenna near $z = 0$ (Fig. 4.8–1), an impressed field $\mathbf{E}^e$ is maintained; it is in the direction of the current, while the electric field **E** (which does not include $\mathbf{E}^e$) is in the opposite direction. It follows that S_r is positive between A and B so that **S** is outwardly directed. An interpretation based on the assumption that the Poynting vector specifies the magnitude and direction of flow of "spatially distributed energy" leads to the following conclusions. No "energy" leaves the antenna or enters it if the conductor is perfect. "Energy" flows radially into the antenna from the "stored energy" in space if the conductor is imperfect to account for the increase in thermal energy in the conductor. "Energy" is "radiated" outward into space from the charge-separating region AB (Fig. 4.8–1).

If the antenna is driven from a two-conductor line which is in turn driven from a generator or charge-separating region AB (Fig. 4.8–2) and the Poynting vector is determined at the surface of the conductors including those of the line, the following is learned. If the conductors are perfect, the tangential electric field vanishes everywhere along all conductors so that there is no outwardly directed component. If the conductors are imperfect, the Poynting vector has a small inward

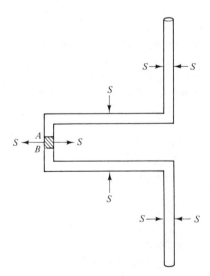

Figure 4.8–2 Poynting vector at the surfaces of antenna and transmission line. A generator is between A and B.

component, as shown in Fig. 4.8–2. The Poynting vector has a large outward component only between the terminals A and B.

It might be concluded from this discussion that since no "energy" leaves the antenna in either case the moving charges in the antenna can play no part whatsoever in maintaining currents in neighboring or distant conductors. This conclusion is in direct contradiction to the postulate that all moving charges exert forces and hence do work on one another. Since no "energy" leaves the antenna *if the Poynting vector correctly localizes its flow*, the antenna is apparently quite unnecessary as long as the same potential difference is maintained between the surfaces A and B. In fact, if the antenna in Fig 4.8–2 is replaced by a lumped impedance consisting of a suitable resistance and reactance that constitute an identical load at the end of the line, the Poynting vector is unchanged, and directed outward between A and B, indicating that the same "energy flow into space" is maintained by the generating region in both cases. On the other hand, the long antenna is capable of "receiving" only a very small inward "flow of energy from space," whereas the lumped impedance "receives from space" practically all the energy that is "transferred to space" from the generator. It is to be noted that the long transmission line seems to have nothing to do with the outward "flow of energy" from the charge-separating region AB nor with the inward "flow of energy from space" to either the antenna or the lumped impedance. According to the Poynting vector, energy does not travel in the conductors of the line, but in the surrounding space.

The discussion above further emphasizes that nothing of value is gained by attaching *arbitrary* physical significance to the Poynting vector or to electromagnetic or other energy functions.

PROBLEMS

1. Show that the electromagnetic force defined by (4.1–5) may be expanded into

$$\mathbf{F} = \int_\tau (\rho\mathbf{E} + \mathbf{J} \times \mathbf{B})\, d\tau + \int_\Sigma (\eta\mathbf{E} + \mathbf{K} \times \mathbf{B})\, d\sigma + \int_\tau (\mathbf{P} \cdot \nabla)\mathbf{E}\, d\tau$$
$$+ \int_\tau (\mathbf{M} \cdot \nabla)\mathbf{B}\, d\tau + \int_\tau (\mathbf{P} \times \mathbf{B})\, d\tau$$

2. The general definition of the electromagnetic vector torque acting on a body is

$$\mathbf{T} = \int_\tau \mathbf{r} \times d\mathbf{F}_\tau + \int_\Sigma \mathbf{r} \times d\mathbf{F}_\sigma$$

where

$$d\mathbf{F}_\tau = (\overline{\rho}\mathbf{E} + \overline{\rho_m\mathbf{v}} \times \mathbf{B})\, d\tau; \qquad d\mathbf{F}_\sigma = (\overline{\eta}\mathbf{E} + \overline{\eta_m\mathbf{v}} \times \mathbf{B})\, d\sigma$$

Show that this is equivalent to

$$\mathbf{T} = \int_\tau (\mathbf{r} \times \rho\mathbf{E}) \, d\tau + \int_\Sigma (\mathbf{r} \times \eta\mathbf{E}) \, d\sigma + \int_\tau [\mathbf{r} \times (\mathbf{J} \times \mathbf{B})] \, d\tau$$

$$+ \int_\Sigma [\mathbf{r} \times (\mathbf{K} \times \mathbf{B})] \, d\sigma + \int_\tau [\mathbf{r} \times (\dot{\mathbf{P}} \times \mathbf{B})] \, d\tau$$

$$+ \int_\tau [\mathbf{r} \times (\mathbf{P} \cdot \nabla)\mathbf{E}] \, d\tau + \int_\tau [\mathbf{r} \times (\mathbf{M} \cdot \nabla)\mathbf{B}] \, d\tau$$

$$+ \int_\tau \{\mathbf{r} \times [\mathbf{M} \times (\nabla \times \mathbf{B})]\} \, d\tau + \int_\tau (\mathbf{P} \times \mathbf{E}) \, d\tau + \int_\tau (\mathbf{M} \times \mathbf{B}) \, d\tau$$

The first five terms are known as body torques, the sixth term is the torque on an electric dipole in a nonuniform field, the seventh and eighth are torques on a magnetic dipole in a nonuniform field, the ninth term is the torque on an electric dipole in a uniform field, and the tenth is the torque on a magnetic dipole in a uniform field.

3. A group of n charged conducting bodies and a surrounding space-charge density ρ is confined to a finite domain in free space. Show that the electrostatic energy of the system

$$U_e = \frac{\epsilon_0}{2} \int_V |E|^2 \, d\tau$$

may be written as

$$\tfrac{1}{2} \sum_{j=1}^{n} \phi_j Q_j + \tfrac{1}{2} \int_V \phi\rho \, d\tau$$

where ϕ_j is the potential of the jth conductor and Q_j is its total charge. Use Green's theorem,

$$\int_V \psi\nabla^2\phi = \int_s \psi \frac{\partial\phi}{\partial n} - \int_V \nabla\psi \cdot \nabla\phi$$

4. Show that the pressure on an infinite conducting plane with a uniform surface charge density η is $\eta^2/2\epsilon_0$ by equating the work done per unit area in moving the plane to the change in energy of the electrostatic field.

5. An uncharged conducting sphere of small radius is connected to a fine wire which extends to infinity. At time $t = 0$ a uniform and constant current I begins to flow in the wire and to charge the sphere. Calculate the scalar potential ϕ at a distance R from the sphere as a function of time for *all times*. Plot ϕ as

(a) A function of R at a given time.
(b) A function of time for a given R.

Could the electric field for this problem be obtained by merely taking the gradient of ϕ? Explain.

6. Two thin, straight parallel wires are separated by a distance a and each carries a current I in the same direction. Compute the force per unit length on one of the wires, using

the integral form of $\nabla \times \mathbf{H} = \mathbf{J} + \dot{\mathbf{D}}$ and the force equation

$$\mathbf{F} = \int_V (\bar{\rho}\mathbf{E} + \overline{\rho_m \mathbf{v}} \times \mathbf{B})\, d\tau$$

Is the force attractive or repulsive?

7. (a) Two electrodes lie in the yz plane at $x = 0$ and $x = b$. q_x is the charge per unit area in a slice of unit thickness at x, M_x the mass of this charge, u_x its nonrandom velocity, and a_x its nonrandom acceleration. Use the field and force equations to obtain the following relations:

$$\frac{dJ_x}{dx} = 0; \qquad J_x = q_x u_x + \epsilon_0 \frac{dE_x}{dt}$$

$$q_x E_x = M_x \frac{du_x}{dt} = M_x a_x$$

(b) Use the results of part (a) to show that

$$J_x = \epsilon_0 \frac{dE_x}{dt}$$

$$\frac{da_x}{dt} = \frac{eJ_x}{m\epsilon_0}$$

where e is the charge on the electron and m is the mass of the electron.

5

Electromagnetic Waves in Unbounded Regions

5.1 THE GENERAL ELECTROMAGNETIC FIELD

The Helmholtz integrals that define the potential field in terms of the essential densities of charge and moving charge are

$$\phi = \frac{1}{4\pi\epsilon_0} \left(\int_\tau \frac{\overline{\rho'}}{R} e^{-jk_0R} \, d\tau' + \int_\Sigma \frac{\overline{\eta'}}{R} e^{-jk_0R} \, d\Sigma' \right) \tag{5.1-1a}$$

$$\mathbf{A} = \frac{\mu_0}{4\pi} \left(\int_\tau \frac{\overline{\rho_m \mathbf{v'}}}{R} e^{-jk_0R} \, d\tau' + \int_\Sigma \frac{\overline{\eta_m \mathbf{v'}}}{R} e^{-jk_0R} \, d\Sigma' \right) \tag{5.1-1b}$$

The relationships between the potential functions and the electric and magnetic vectors for periodic time dependence are

$$\mathbf{E} = -\nabla\phi - j\omega\mathbf{A} \tag{5.1-2a}$$

$$\mathbf{B} = \nabla \times \mathbf{A} \tag{5.1-2b}$$

The substitution of (5.1–1a,b) in (5.1–2a,b) leads to explicit formulas for $\mathbf{E}$ and $\mathbf{B}$ that are integral solutions of the Maxwell equations that define these vectors.

It is important to note that the potentials and the field vectors are functions of the unprimed coordinates of the point P (the field point). The gradient (∇) and curl ($\nabla \times$) operations in (5.1–2a,b) involve the unprimed coordinates and not the primed variables of integration that locate the volume element $d\tau'$ and the surface element $d\Sigma'$ at P' (the source point). (When the operators ∇, $\nabla \cdot$, and $\nabla \times$ apply to the primed variables, they are written with a prime: ∇', $\nabla' \cdot$, and $\nabla' \times$. R is the distance between P and P' and is therefore a function of both the primed and unprimed coordinates.) The density functions characterizing the region of integration are functions of the primed variables only, as seen in (5.1–1a,b). It follows that

$$\nabla \overline{\rho'} = 0; \qquad\qquad \nabla \overline{\eta'} = 0 \qquad\qquad (5.1\text{–}3a)$$

$$\nabla \times \overline{\rho_m \mathbf{v}'} = 0; \qquad \nabla \times \overline{\eta_m \mathbf{v}'} = 0 \qquad\qquad (5.1\text{–}3b)$$

The ∇ and $(\nabla \times)$ operators may be applied under the sign of integration in (5.1–1a,b) whenever the integrals are regular and this is the case when the point P is not in the region of integration τ. If the point P is in τ, the potentials usually cannot be evaluated because the density functions are not known everywhere in an extended region. Therefore, P is assumed to be outside τ.

Before the evaluation of $\nabla \phi$ and $\nabla \times \mathbf{A}$ is carried out, some vector relations must be discussed. If ψ is a scalar function of R,

$$\nabla \psi = \frac{\partial \psi}{\partial R} \nabla R = \frac{\partial \psi}{\partial R} \hat{\mathbf{R}} \qquad\qquad (5.1\text{–}4)$$

where $\hat{\mathbf{R}}$ is a unit vector. It follows from the definition of the gradient that

$$|\nabla R| = \left(\frac{\partial R}{\partial R} \right)_{\text{max}} = 1 \qquad\qquad (5.1\text{–}5)$$

The direction of the maximum rate of increase of $\mathbf{R}$ is outward and is specified by the unit vector $\hat{\mathbf{R}}$. If ψ is a scalar and $\mathbf{C}$ a vector, the following vector relation is readily proved:

$$\nabla \times (\psi \mathbf{C}) = \psi \nabla \times \mathbf{C} + \nabla \psi \times \mathbf{C} \qquad\qquad (5.1\text{–}6)$$

It is now convenient to apply (5.1–4) or (5.1–6) to (5.1–1a,b) with

$$\psi = \frac{e^{-jk_0 R}}{R} \qquad\qquad (5.1\text{–}7a)$$

and

$$\mathbf{C} = \overline{\rho_m \mathbf{v}'} \qquad\qquad (5.1\text{–}7b)$$

Then

$$\nabla \psi = -\hat{\mathbf{R}} e^{-jk_0 R} \left(\frac{1}{R^2} + \frac{jk_0}{R} \right) \qquad\qquad (5.1\text{–}8)$$

With (5.1–3b) and (5.1–8)

$$\nabla \times (\psi \mathbf{C}) = (\nabla \psi) \times \mathbf{C}$$

$$= -\hat{\mathbf{R}} \times \overline{\rho_m \mathbf{v}'} e^{-jk_0 R} \left(\frac{1}{R^2} + \frac{jk_0}{R} \right) \qquad\qquad (5.1\text{–}9)$$

With (5.1–3) to (5.1–9) the volume integrals in (5.1–2a,b) are

$$-\nabla\phi = -\frac{1}{4\pi\epsilon_0}\int_\tau \nabla(\overline{\rho'}\psi)\, d\tau'$$

$$= -\frac{1}{4\pi\epsilon_0}\int_\tau \overline{\rho'}\nabla\psi\, d\tau'$$

$$= \frac{1}{4\pi\epsilon_0}\int_\tau \hat{R}\overline{\rho'}e^{-jk_0R}\left(\frac{1}{R^2} + \frac{jk_0}{R}\right) d\tau' \tag{5.1–10}$$

$$\nabla \times \mathbf{A} = \frac{\mu_0}{4\pi}\int_\tau \nabla \times (\psi\mathbf{C})\, d\tau'$$

$$= -\frac{\mu_0}{4\pi}\int_\tau (\hat{R} \times \overline{\rho_m\mathbf{v'}})e^{-jk_0R}\left(\frac{1}{R^2} + \frac{jk_0}{R}\right) d\tau' \tag{5.1–11}$$

With

$$\frac{1}{\mu_0\epsilon_0} = c^2; \qquad k_0 = \frac{\omega}{c} \tag{5.1–12}$$

it follows that

$$j\omega\mathbf{A} = \frac{1}{4\pi\epsilon_0}\int_\tau \overline{\rho_m\mathbf{v'}}e^{-jk_0R}\frac{jk_0}{Rc}\, d\tau' \tag{5.1–13}$$

When (5.1–10), (5.1–11), and (5.1–13) are substituted into the expressions (5.1–2a,b) for **E** and **B**, the general integrals for the electromagnetic field are obtained. They are

$$\mathbf{E} = \frac{1}{4\pi\epsilon_0}\int_\tau e^{-jk_0R}\left[\frac{\hat{R}\overline{\rho'}}{R^2} + \left(\overline{\rho'}\hat{R} - \frac{\overline{\rho_m\mathbf{v'}}}{c}\right)\frac{jk_0}{R}\right] d\tau' + S_E \tag{5.1–14a}$$

$$\mathbf{B} = -\frac{\mu_0}{4\pi}\int_\tau e^{-jk_0R} (\hat{R} \times \overline{\rho_m\mathbf{v'}})\left(\frac{1}{R^2} + \frac{jk_0}{R}\right) d\tau' + S_B \tag{5.1–14b}$$

S_E and S_B are the surface integrals which are obtained from the volume integrals in each case by writing η for ρ and Σ for τ.

The electromagnetic field at points in space due to the currents and charges on a cylindrical conductor with small cross sections and with its axis along the z axis of a coordinate system may be obtained from (5.1–14a,b). This is accomplished by carrying out the integrations over the small cross section and circumference of the conductor with R taken to be the constant distance to the axis. The conductor is assumed to extend from $z = -h$ to $z = +h$ along the z axis and to have a radius a. Then

$$\mathbf{E} = \frac{1}{4\pi\epsilon_0} \int_{-h}^{h} e^{-jk_0 R} \left[\frac{\hat{\mathbf{R}}}{R^2} q' + \frac{jk_0}{R} \left(\hat{\mathbf{R}} q' - \hat{\mathbf{z}} \frac{I_z'}{c} \right) \right] dz' \qquad (5.1\text{--}15a)$$

$$\mathbf{B} = -\frac{\mu_0}{4\pi} \int_{-h}^{h} e^{-jk_0 R} (\hat{\mathbf{R}} \times \hat{\mathbf{z}}) I_z' \left(\frac{1}{R^2} + \frac{jk_0}{R} \right) dz' \qquad (5.1\text{--}15b)$$

where

$$q' = \int_S \overline{\rho'} \, dS' + \int_s \overline{\eta'} \, ds' \qquad (5.1\text{--}16a)$$

$$I_z' = \int_S \overline{\rho_m v_z'} \, dS' + \int_s \overline{\eta_m v_z'} \, ds' \qquad (5.1\text{--}16b)$$

The first integral in (5.1–16a,b) is taken over the cross section $S = \pi a^2$ of the conductor, the second integral around its circumference $s = 2\pi a$; q' and I_z' are the amplitudes, respectively, of the total charge per unit length and the total axial conduction current. Contributions due to currents directed radially in the conductor are negligible.

The instantaneous values of $\mathbf{E}$ and $\mathbf{B}$ are obtained by introducing the time factor $e^{j\omega t}$ and selecting the real part. It is therefore necessary to refer all phases to a single quantity which may be conveniently chosen to be the current. The complex amplitude q' may be expressed in terms of I_z' with the aid of the equation of continuity written for the cylindrical conductor in the one-dimensional form

$$\frac{dI'}{dz'} + j\omega q' = 0 \qquad (5.1\text{--}17)$$

or

$$q' = \frac{j}{\omega} \frac{dI'}{dz'} = \frac{j}{k_0 c} \frac{dI_z'}{dz'} \qquad (5.1\text{--}18)$$

In order to obtain the instantaneous or time-dependent electric and magnetic fields it is necessary to multiply (5.1–15a,b) by $e^{j\omega t}$ on both sides and to select the real parts as the desired solutions. These are generated by the following time-dependent currents and charges per unit length

$$I_z'(t) = \text{Re}\,(I_z' e^{j\omega t}) = I_z' \cos \omega t \qquad (5.1\text{--}19)$$

$$q'(t) = \text{Re}\,(q' e^{j\omega t}) = -\frac{1}{\omega} \frac{dI_z'}{dz'} \sin \omega t \qquad (5.1\text{--}20)$$

The electric and magnetic fields at all points outside the cylindrical conductor are

$$
\mathbf{E}(t) = -\frac{1}{4\pi\epsilon_0} \int_{-h}^{h} \left[\frac{\hat{\mathbf{R}}}{Rc} \frac{dI'_z}{dz'} \cos(\omega t - k_0 R) \right.
$$

$$
\left. + \left(\frac{\hat{\mathbf{R}}}{k_0 c R^2} \frac{dI'_z}{dz'} - \hat{\mathbf{z}} \frac{k_0 I'_z}{cR} \right) \sin(\omega t - k_0 R) \right] dz' \tag{5.1-21a}
$$

$$
\mathbf{B}(t) = -\frac{\mu_0}{4\pi} \int_{-h}^{h} (\hat{\mathbf{R}} \times \hat{\mathbf{z}}) I'_z \left[\frac{1}{R^2} \cos(\omega t - k_0 R) - \frac{k_0}{R} \sin(\omega t - k_0 R) \right] dz' \tag{5.1-21b}
$$

The vector $\mathbf{E}$ has a component along $\mathbf{R}$ and a second component parallel to $\hat{\mathbf{z}}$ and, hence, parallel to the current in the conductor. $\mathbf{B}$ is everywhere perpendicular to both $\mathbf{R}$ and $\hat{\mathbf{z}}$ and so is directed along tangents to circles drawn with the conductor as the common axis. Therefore, $\mathbf{E}$ and $\mathbf{B}$ are *everywhere mutually perpendicular* and the equiphase surfaces of $\mathbf{E}$ and of $\mathbf{B}$ due to the periodically varying charge and current *in an element dz* of the conductor are spherical shells. This follows from the fact that at a given instant the integrals in (5.1–21a,b) are constant if R is constant.

The general electromagnetic field is too complicated to allow a simple geometrical interpretation. However, component parts of the field can be considered and described in simple form.

5.2 *INDUCTION AND RADIATION FIELDS*

It is convenient to study the general electromagnetic field defined by (5.1–14a,b) in two parts. Let each of the vectors $\mathbf{E}$ and $\mathbf{B}$ be expressed as the sum of two components as follows:

$$
\mathbf{E} = \mathbf{E}^i + \mathbf{E}^r; \qquad \mathbf{B} = \mathbf{B}^i + \mathbf{B}^r \tag{5.2-1}
$$

The components with the superscript i define what is called the *induction field*. This is given by

$$
\mathbf{E}^i = \frac{1}{4\pi\epsilon_0} \int_\tau \hat{\mathbf{R}} \frac{\overline{\rho'}}{R^2} e^{-jk_0 R} \, d\tau' + S^i_E \tag{5.2-2a}
$$

$$
\mathbf{B}^i = -\frac{\mu_0}{4\pi} \int_\tau \frac{\hat{\mathbf{R}} \times \overline{\rho_m \mathbf{v}'}}{R^2} e^{-jk_0 R} \, d\tau' + S^i_B \tag{5.2-2b}
$$

The surface integrals S^i_E and S^i_B are obtained by writing η for ρ and Σ for τ in the volume integrals. The components in (5.2–1) with the superscript r define the *radiation field*. It is given by

$$
\mathbf{E}^r = \frac{1}{4\pi\epsilon_0} \int_\tau e^{-jk_0 R} \frac{jk_0}{R} \left(\overline{\rho'} \, \hat{\mathbf{R}} - \frac{\overline{\rho_m \mathbf{v}'}}{c} \right) d\tau' + S^r_E \tag{5.2-3a}
$$

$$
\mathbf{B}^r = -\frac{\mu_0}{4\pi} \int_\tau e^{-jk_0 R} \frac{jk_0}{R} (\hat{\mathbf{R}} \times \overline{\rho_m \mathbf{v}'}) \, d\tau' + S^r_B \tag{5.2-3b}
$$

The induction field is proportional to $1/R^2$, the radiation field to k_0/R. The ratio of these two factors is k_0R. It follows that with any given k_0 and with sufficiently small values of R (i.e., $\mathbf{E}$ and $\mathbf{B}$ calculated at points sufficiently near the periodically varying charge distribution), the induction field is so large compared with the radiation field that the latter may be neglected. On the other hand, for R very large, the reverse is true; the radiation field predominates and the induction field may be disregarded because it is insignificant.

5.3 INDUCTION OR NEAR ZONE: COULOMB'S LAW AND BIOT–SAVART FORMULA

Let the *induction* or *near zone* be defined to be that region in the immediate vicinity of the volume on and in which charge and current densities are nonvanishing which satisfies the inequality defining the quasi-stationary state:

$$k_0R \ll 1 \tag{5.3-1}$$

Consequently,

$$E^i \gg E^r; \qquad B^i \gg B^r \tag{5.3-2}$$

and

$$\mathbf{E} \doteq \mathbf{E}^i; \qquad \mathbf{B} \doteq \mathbf{B}^i \tag{5.3-3}$$

As a consequence of (5.3–1),

$$e^{-jk_0R} = 1 - jk_0R - \frac{(k_0R)^2}{2} \cdots \doteq 1 \tag{5.3-4}$$

This is equivalent to assuming quasi-instantaneous action at a sufficiently short distance. With (5.3–4), the induction field is identified with the field of the quasi-stationary state. It is

$$k_0R \ll 1 \begin{cases} \mathbf{E} \doteq \mathbf{E}^i = \dfrac{1}{4\pi\epsilon_0}\left(\displaystyle\int_\tau \hat{\mathbf{R}}\,\frac{\overline{\rho'}}{R^2}\,d\tau' + \int_\Sigma \hat{\mathbf{R}}\,\frac{\overline{\eta'}}{R^2}\,d\Sigma'\right) & (5.3\text{-}5a) \\[4mm] \mathbf{B} \doteq \mathbf{B}^i = -\dfrac{\mu_0}{4\pi}\left(\displaystyle\int_\tau \frac{\hat{\mathbf{R}} \times \overline{\rho_m\mathbf{v}'}}{R^2}\,d\tau' + \int_\Sigma \frac{\hat{\mathbf{R}} \times \overline{\eta_m\mathbf{v}'}}{R^2}\,d\Sigma'\right) & (5.3\text{-}5b) \end{cases}$$

In the stationary state, with all densities constant in time,

$$\mathbf{E} = \frac{1}{4\pi\epsilon_0}\left(\int_\tau \hat{\mathbf{R}}\,\frac{\overline{\rho'}}{R^2}\,d\tau' + \int_\Sigma \hat{\mathbf{R}}\,\frac{\overline{\eta'}}{R^2}\,d\Sigma'\right) \tag{5.3-6a}$$

$$\mathbf{B} = -\frac{\mu_0}{4\pi}\left(\int_\tau \frac{\hat{\mathbf{R}} \times \overline{\mathbf{J}'}}{R^2}\,d\tau' + \int_\Sigma \frac{\hat{\mathbf{R}} \times \overline{\mathbf{K}'}}{R^2}\,d\Sigma'\right) \tag{5.3-6b}$$

The electric field of the stationary state is the *electrostatic field*; the magnetic field of the stationary state is the *magnetostatic field*. Since $\omega = 0$ in the stationary states, the radiation fields vanish and the entire stationary field is the induction field. At low frequencies and for all significant distances, $\omega R/c$ is so small that the quasi-stationary or near-zone field is the entire field.

A fundamentally important static-state relation is obtained by specializing (5.3–6a). Let the distance from any point P where $\mathbf{E}$ is calculated to the center of a volume τ be R_0, a distance that is large compared with the maximum dimension of τ. Then

$$
\mathbf{E} = \frac{1}{4\pi\epsilon_0} \left(\int_\tau \hat{\mathbf{R}} \frac{\overline{\rho}'}{R^2} \, d\tau' + \int_\Sigma \hat{\mathbf{R}} \frac{\overline{\eta}'}{R^2} \, d\Sigma' \right)
$$

$$
= \frac{\hat{\mathbf{R}}_0}{4\pi\epsilon_0 R_0^2} \left(\int_\tau \overline{\rho}' \, d\tau' + \int_\Sigma \overline{\eta}' \, d\Sigma' \right) = \frac{\hat{\mathbf{R}}_0 q}{4\pi\epsilon_0 R_0^2} \qquad (5.3\text{–}7)
$$

where q is the total charge within and on the surface of τ. The electrostatic force acting on a second volume V characterized by essential densities $\overline{\rho}$ and $\overline{\eta}$ is obtained from the general definition of the electromagnetic force $\mathbf{F}$ (4.1–5).

$$
\mathbf{F}_e = \int_V \overline{\rho}\mathbf{E} \, dV + \int_S \overline{\eta}\mathbf{E} \, dS \qquad (5.3\text{–}8)
$$

If the dimensions of V are small compared with R, $\mathbf{E}$ is constant over V, so that

$$
\mathbf{F}_e = \mathbf{E} \left(\int_V \overline{\rho} \, dV + \int_S \overline{\eta} \, dS \right) = \mathbf{E}Q \qquad (5.3\text{–}9)
$$

where Q is the total charge within and on V. Upon combining (5.3–7) and (5.3–9), we obtain

$$
\mathbf{F}_e = \hat{\mathbf{R}}_0 \frac{qQ}{4\pi\epsilon_0 R_0^2} \qquad (5.3\text{–}10)
$$

In this expression R_0 is the distance between the centers of the two volumes τ and V. $\mathbf{F}_e$ is the electrostatic force acting on V due to τ provided that all parts of the two volumes are separated by distances that are large compared with the dimensions of the volumes V and τ. If the two volumes are symmetrically charged spheres, (5.3–10) is valid without such a restriction, so that V need not be small in terms of R_0. The relation (5.3–10) is *Coulomb's law*. It is an important special form of the electromathematical model that is actually contained in the field and force equations as a particular case. Coulomb's law is a fundamental postulate in a more conventional formulation of electrodynamics.

If the near-zone relations (5.3–5a,b) are written for a cylindrical conductor in space, they assume the following form:

$$k_0 R \ll 1 \begin{cases} \mathbf{E} \doteq \mathbf{E}^i = \dfrac{1}{4\pi\epsilon_0} \displaystyle\int_{-h}^{h} \dfrac{q'}{R^2} \hat{\mathbf{R}} \, dz' & \text{(5.3--11a)} \\[3em] \mathbf{B} \doteq \mathbf{B}^i = -\dfrac{\mu_0}{4\pi} \displaystyle\int_{-h}^{h} (\hat{\mathbf{R}} \times \hat{\mathbf{z}}) \dfrac{I'_z}{R^2} \, dz' & \text{(5.3--11b)} \end{cases}$$

Equation (5.3–11b) may be written in the equivalent form

$$d\mathbf{B}^i = \frac{\mu_0 I_s}{4\pi} \frac{d\mathbf{s} \times \hat{\mathbf{R}}}{R^2} \tag{5.3--12}$$

The relation (5.3–12) is the *Biot–Savart formula*. The direction of the magnetic field due to the current I_s in the element $d\mathbf{s}$ of a wire is perpendicular to the plane containing $\mathbf{R}$ and $d\mathbf{s}$. $\mathbf{R}$ is a vector drawn from $d\mathbf{s}$ to the point where the field is computed. The magnitude of the field is

$$dB^i = \frac{\mu_0 I_s}{4\pi} \frac{\sin (R, s)}{R^2} \, ds \tag{5.3--13}$$

where (R, s) is the angle between $\mathbf{s}$ and $\mathbf{R}$.

5.4 RADIATION OR FAR ZONE

The *radiation* or *far zone* is defined by

$$k_0 R \gg 1 \tag{5.4--1}$$

As a consequence of (5.4–1),

$$E^r \gg E^i; \qquad B^r \gg B^i \tag{5.4--2}$$

and

$$\mathbf{E} \doteq \mathbf{E}^r; \qquad \mathbf{B} \doteq \mathbf{B}^r \tag{5.4--3}$$

The radiation-zone field is given by

$$k_0 R \gg 1 \begin{cases} \mathbf{E} \doteq \mathbf{E}^r = \dfrac{-jk_0}{4\pi\epsilon_0} \displaystyle\int_{\tau} \left(\hat{\mathbf{R}}\overline{\rho'} - \dfrac{\overline{\rho_m \mathbf{v}'}}{c} \right) \dfrac{e^{-jk_0 R}}{R} \, d\tau' + S_E^r & \text{(5.4--4)} \\[3em] \mathbf{B} \doteq \mathbf{B}^r = \dfrac{-jk_0\mu_0}{4\pi} \displaystyle\int_{\tau} (\hat{\mathbf{R}} \times \overline{\rho_m \mathbf{v}'}) \dfrac{e^{-jk_0 R}}{R} \, d\tau' + S_B^r & \text{(5.4--5)} \end{cases}$$

The corresponding equations for a cylindrical conductor in space are

$$
k_0 R \gg 1
\begin{cases}
\mathbf{E} \doteq \mathbf{E}^r = \dfrac{jk_0}{4\pi\epsilon_0} \displaystyle\int_{-h}^{h} \left(\hat{\mathbf{R}} q' - \hat{\mathbf{z}} \dfrac{I_z'}{c} \right) \dfrac{e^{-jk_0 R}}{R}\, dz' & (5.4\text{--}6) \\[4mm]
\mathbf{B} \doteq \mathbf{B}^r = \dfrac{-jk_0\mu_0}{4\pi} \displaystyle\int_{-h}^{h} (\hat{\mathbf{R}} \times \hat{\mathbf{z}}) I_z' \dfrac{e^{-jk_0 R}}{R}\, dz' & (5.4\text{--}7)
\end{cases}
$$

The electromagnetic field in the radiation zone has the same simple form as the potential field. A wave picture of spherical equiphase surfaces expanding with the constant velocity c is appropriate to describe the electromagnetic radiation field due to the periodically varying charges and currents in an element dz'. The magnetic waves are transverse; the electric waves may be both transverse and longitudinal.

It is seen that the general electromagnetic field may be divided into a near or induction-zone field, and a far or radiation-zone field. All points in the former are by definition sufficiently near to the moving charges so that an *instantaneous* action at a distance is a good approximation to the fundamentally correct *retarded* action at a distance. In the near zone it is not convenient or illuminating to use a wave picture because in the vicinity of the periodically varying charges and currents the equiphase surfaces of $\mathbf{B}$ and the components of $\mathbf{E}$ do not move with a constant velocity. The phase velocity is very high near the moving charges and the approximation (5.3–4) is equivalent to assuming an infinite phase velocity. The integrals that define the approximate electromagnetic field in the induction or near zone are usually interpreted using geometrical pictures that correctly describe exactly only the stationary states in which all densities are constant in time. The equivalence between quantities in the stationary states and the corresponding complex amplitudes in the quasi-stationary state makes it possible to use the same geometrical pictures for amplitudes of periodic phenomena as for stationary magnitudes, and this is equivalent to assuming an instantaneous action at a distance.

In the radiation zone, the simplest form of retarded action at a distance is valid. The equiphase surfaces expand with a *constant* velocity and the electric and magnetic fields can be described in terms of a simple wave picture. The radiation zone is therefore frequently called the *wave zone*. The induction zone and the radiation zone do not overlap and an intermediate zone exists between them. This is the region between $k_0 R \ll 1$ and $k_0 R \gg 1$. The electromagnetic field in the intermediate zone has neither the stationary-state properties of the near-zone field nor the wave properties of the far-zone field. The field in this region can only be represented by the integrals (5.1–14a,b) of the general electromagnetic field.

5.5 SPHERICAL AND PLANE ELECTROMAGNETIC WAVES

The field in the radiation zone of a cylindrical conductor of length $2h$ in space as given in (5.4–6) and (5.4–7) is

$$\mathbf{E}^r = \frac{jk_0}{4\pi\epsilon_0} \int_{-h}^{h} \left(\hat{\mathbf{R}}q' - \hat{\mathbf{z}}\frac{I_z'}{c} \right) \frac{e^{-jk_0R}}{R} \, dz' \tag{5.5–1}$$

and

$$\mathbf{B}^r = \frac{-j\mu_0 k_0}{4\pi} \int_{-h}^{h} (\hat{\mathbf{R}} \times \hat{\mathbf{z}})I_z' \frac{e^{-jk_0R}}{R} \, dz' \tag{5.5–2}$$

If the conductor is oriented in space along an arbitrarily directed s axis, (5.5–1) and (5.5–2) describe the radiation field with z replaced by s as shown in Fig. 5.5–1. In the radiation zone, the distance R_0 from the origin at the center of the antenna to the point where the field is to be calculated satisfies the inequality

$$R_0 \gg h \tag{5.5–3a}$$

and

$$k_0 R_0 \gg 1 \tag{5.5–3b}$$

Therefore, the vector $\mathbf{R}$ drawn from the point P' locating the element of integration ds' in the conductor to the point P where the field is to be calculated (Fig. 5.5–1) may be written as follows:

$$\mathbf{R} = \mathbf{R}_0 - \mathbf{s}' \tag{5.5–4a}$$

$$R = [R_0^2 + s'^2 - 2(\mathbf{R}_0 \cdot \mathbf{s}')]^{1/2} \tag{5.5–4b}$$

As a result of (5.5–3a), it follows that

$$R \doteq R_0 \tag{5.5–5}$$

in all factors that cannot vanish as R increases. These include all quantities except the arguments of the exponential functions. In these the first-order small terms

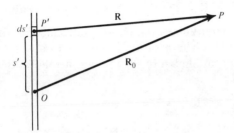

Figure 5.5–1 Section of antenna with the origin of coordinates at O.

must be retained. That is,

$$R \doteq R_0 - (\hat{\mathbf{R}}_0 \cdot \mathbf{s}') = R_0 - s' \cos(R_0, s') \tag{5.5-6}$$

where $\hat{\mathbf{R}}_0$ is a unit vector pointing from O to P and (R_0, s') is the angle between $\mathbf{s}'$ and $\hat{\mathbf{R}}_0$.

With (5.5–5) in amplitude factors and (5.5–6) in arguments, the relations (5.5–1) and (5.5–2) become

$$\mathbf{E}^r = \frac{jk_0}{4\pi\epsilon_0} \frac{e^{-jk_0R_0}}{R_0} \int_{-h}^{h} \left(\hat{\mathbf{R}}_0 q' - \frac{\hat{\mathbf{s}}I_s'}{c} \right) e^{jk_0(\hat{\mathbf{R}}_0 \cdot \mathbf{s}')} \, ds' \tag{5.5-7a}$$

$$\mathbf{B}^r = \frac{-j\mu_0 k_0}{4\pi} \frac{e^{-jk_0R_0}}{R_0} \int_{-h}^{h} (\hat{\mathbf{R}}_0 \times \hat{\mathbf{s}}) I_s' e^{jk_0(\hat{\mathbf{R}}_0 \cdot \mathbf{s}')} \, ds' \tag{5.5-7b}$$

where $\hat{\mathbf{s}}$ is a unit vector along the axis of the antenna. Equation (5.5–7b) can be written in the same form as the potential functions if $e^{-jk_0R_0}$ is moved under the sign of integration and recombined with $e^{jk_0(\hat{\mathbf{R}}_0 \cdot \mathbf{s}')}$ to give e^{-jk_0R}:

$$\mathbf{B}^r = \frac{-j\mu_0 k_0 (\hat{\mathbf{R}}_0 \times \hat{\mathbf{s}})}{4\pi} \int_{-h}^{h} \frac{I_s'}{R} e^{-jk_0R} \, ds' \tag{5.5-8}$$

Each element of current $I_s' \, ds'$ may be associated with a radially expanding spherical train of equiphase surfaces or waves moving with the constant velocity c. Since the vector potential $\mathbf{A}$ may be written in the form

$$\mathbf{A} = \frac{\mu_0 \hat{\mathbf{s}}}{4\pi} \int_{-h}^{h} \frac{I_s'}{R} e^{-jk_0R} \, ds' \tag{5.5-9}$$

and it has been shown that

$$\mathbf{E} = -\frac{j\omega}{k_0^2} (\boldsymbol{\nabla}\boldsymbol{\nabla} \cdot \mathbf{A} + k_0^2\mathbf{A})$$

it follows with (3.7–3b) and (3.4–12) that

$$\mathbf{E}^r = \frac{jk_0}{4\pi\epsilon_0 c} \frac{e^{-jk_0R_0}}{R_0} \int_{-h}^{h} (\hat{\mathbf{R}}_0 \times \hat{\mathbf{R}}_0 \times \hat{\mathbf{s}}) I_s' e^{jk_0(\hat{\mathbf{R}}_0 \cdot \mathbf{s}')} \, ds'$$

$$= c(\mathbf{B}^r \times \hat{\mathbf{R}}_0) \tag{5.5-10}$$

If a system of polar coordinates R_0, Θ, Φ is introduced with its origin at the center of the antenna and with Θ measured from its axis which lies along the z axis of a Cartesian system, as shown in Fig. 5.5–2, it follows that in (5.5–7b), z may be written for s and

$$(\hat{\mathbf{R}}_0 \cdot \mathbf{s}') = z' \cos\Theta \tag{5.5-11}$$

Also,

$$-(\hat{\mathbf{R}}_0 \times \hat{\mathbf{s}}) = (\hat{\mathbf{s}} \times \hat{\mathbf{R}}_0) = \hat{\boldsymbol{\Phi}} \sin\Theta \tag{5.5-12}$$

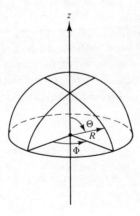

Figure 5.5–2   Polar coordinates.

and

$$\hat{\mathbf{R}}_0 \times \hat{\mathbf{R}}_0 \times \hat{\mathbf{s}} = -(\hat{\mathbf{R}}_0 \times \hat{\boldsymbol{\Phi}}) \sin\Theta = \hat{\boldsymbol{\Theta}} \sin\Theta \qquad (5.5\text{–}13)$$

Then

$$\mathbf{B}^r = \hat{\boldsymbol{\Phi}} B^r_{\Phi}; \qquad \mathbf{E}^r = \hat{\boldsymbol{\Theta}} E^r_{\Theta} \qquad (5.5\text{–}14)$$

The components of the far-zone field of an antenna oriented along the z axis may be written as follows:

$$E^r_R = 0; \qquad\qquad\qquad\qquad B^r_R = 0$$

$$E^r_{\Theta} = f(h, \Theta)\, \frac{e^{-jk_0R_0}}{R_0} = F(\Theta, R_0); \qquad B^r_{\Theta} = 0$$

$$E^r_{\Phi} = 0; \qquad\qquad\qquad B^r_{\Phi} = \frac{f(h, \Theta)}{c}\, \frac{e^{-jk_0R_0}}{R_0} = \frac{F(\Theta, R_0)}{c} \qquad (5.5\text{–}15)$$

with

$$f(h, \Theta) = \frac{j\omega\mu_0}{4\pi} \int_{-h}^{h} I'_z \, e^{jk_0z'\cos\Theta} \sin\Theta \, dz' \qquad (5.5\text{–}16\text{a})$$

$$F(\Theta, R_0) = f(h, \Theta)\, \frac{e^{-jk_0R_0}}{R_0} \qquad (5.5\text{–}16\text{b})$$

Equations (5.5–15) and (5.5–16a,b) define the radiation-zone field of an antenna of half-length h in space. The function $F(\Theta, R_0)$ satisfies the spherical wave equation

$$\frac{\partial^2}{\partial R_0^2}\,(R_0F) + k_0^2(R_0F) = 0 \qquad (5.5\text{–}17)$$

The function F describes the transverse spherical waves expanding radially with a constant velocity c. The equiphase surfaces are separated by a constant distance $\lambda_0 = 2\pi/k_0$.

In most of the problems involving the radiation-zone field of an antenna, interest lies primarily in the field in a relatively small part of space. That is, it is frequently necessary to deal only with **E** and **B** in a region that is small compared with the distance R_0 from a point P_0 at its origin in the distant antenna. If the region is identified with a volume with a radial thickness $R' \ll R_0$, the distance $R = OP$ to any point P in V is approximately given by

$$R \doteq R_0 \qquad\qquad (5.5\text{--}18)$$

in amplitude factors. In arguments of periodic functions, (5.5–18) may not be used. Let the direction cosines of the radial line $R_0 = OP_0$ be expressed in terms of the rectangular coordinates x, y, z with origin also at O. The direction cosines of R_0 are

$$l_0 = \cos (R_0, x); \qquad m_0 = \cos (R_0, y); \qquad n_0 = \cos (R_0, z) \qquad (5.5\text{--}19a)$$

The components of R_0 along the coordinate axes are (x_0, y_0, z_0) given by

$$x_0 = R_0 l_0; \qquad y_0 = R_0 m_0; \qquad z_0 = R_0 n_0 \qquad (5.5\text{--}19b)$$

It follows that

$$R_0 = x_0 l_0 + y_0 m_0 + z_0 n_0 \qquad\qquad (5.5\text{--}20)$$

Any line s lying parallel to OP_0 has the same direction cosines as R_0. If the coordinates of its end point are x, y, z,

$$s = x l_0 + y m_0 + z n_0 \qquad\qquad (5.5\text{--}21)$$

Therefore, the direction of any radial vector $R = OP$ terminating at P in V is the same as that of $R_0 = OP_0$. If the coordinates of P are x, y, z,

$$R \doteq s = x l_0 + y m_0 + z n_0 \qquad\qquad (5.5\text{--}22)$$

With (5.5–18) in amplitude factors and (5.5–21) in arguments, it follows that

$$F \doteq K e^{-jk_0 s} \qquad\qquad (5.5\text{--}23)$$

where

$$K = \frac{f(\Theta)}{R} \doteq \frac{f(\Theta)}{R_0} = \text{constant} \qquad (5.5\text{--}24)$$

If the time factor is included,

$$F e^{j\omega t} \doteq K e^{j(\omega t - k_0 s)} \qquad\qquad (5.5\text{--}25)$$

This expression characterizes the approximate distribution of **E** and **B** in any volume V at a large distance from the antenna. It is readily interpreted in terms of a simple wave picture. The relation (5.5–21) defines a plane at right angles to s and at a distance s from the origin. Equation (5.5–25) may be described using a picture of plane equiphase surfaces at right angles to s and traveling along s with a constant velocity c. That is, the arcs of radially expanding spherical equiphase surfaces defined by (5.5–16b) which pass through the volume V may be assumed to be approximately plane and of constant amplitude provided that the distance to V from the source is sufficiently great and the solid angle subtended by it is small. Equation (5.5–25) satisfies the plane-wave equation

$$\frac{d^2F}{ds^2} + k_0^2 F = 0 \qquad (5.5\text{–}26)$$

which is derived from (5.5–17) with the conditions (5.5–18) and (5.5–22).

In problems involving specifically the field in a volume V in the wave zone and not directly the antenna or its location, it is convenient to use a set of rectangular coordinates as shown in Fig. 5.5–3. The x axis lies along the line OP_0 so that it coincides with the spherical coordinate R_0; the y axis is tangent to and in the direction of the coordinate Φ_0 at P_0; the z axis is tangent to and in the direction $-\Theta_0$ at P_0. With this choice, x, y, z are a right-handed system and it follows that the electromagnetic field at P_0 and approximately throughout V is given by

$$E_\Theta^r \doteq -E_z^r; \qquad B_\Phi^r \doteq B_y^r \qquad (5.5\text{–}27)$$

Also, $s = x$. With this notation, the radiation-zone field in a volume V is

$$-E_z^r(t) = cB_y^r(t) = Ke^{j(\omega t - k_0 x)} \qquad (5.5\text{–}28)$$

This expression may be interpreted in terms of plane transverse waves normal to the x axis and traveling along this with a constant velocity c.

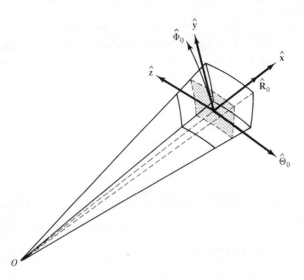

Figure 5.5–3 Rectangular coordinates related to spherical coordinates.

5.6 POLARIZED ELECTROMAGNETIC WAVES

The electromagnetic field in a sufficiently small region V in the radiation zone of the antenna is defined approximately by the following relations:

$$\mathbf{E}(t) = -\hat{\mathbf{z}}E_z^r e^{j\omega t} = -\hat{\mathbf{z}}K e^{j(\omega t - k_0 x)} \qquad (5.6-1a)$$

$$\mathbf{B}(t) = \hat{\mathbf{y}}B_y^r e^{j\omega t} = \hat{\mathbf{y}}\frac{K}{c}e^{j(\omega t - k_0 x)} \qquad (5.6-1b)$$

The positive x axis lies along the radial vector $\mathbf{R}_0 = OP_0$ locating the center P_0 of the region V. From the real part of (5.6–1a,b), it can be concluded that $\mathbf{E}(t)$ and $\mathbf{B}(t)$ at every point change continuously and periodically in length as time passes, but that they are directed parallel, respectively, to the z and y axes at the point P_0 and approximately parallel to these axes throughout the volume V. The $\mathbf{E}$ field is therefore linearly polarized along the z axis, the $\mathbf{B}$ field is linearly polarized along the y axis. Both fields are polarized in the plane at right angles to the direction of propagation of the equiphase surfaces or wave fronts. All the statements made so far are valid regardless of the location of V relative to the antenna, provided only that it is in the radiation zone. The amplitude factor K in (5.6–1a,b) depends both on the spherical angle Θ_0 locating the vector $\mathbf{R}_0 = OP_0$ and on the length of R_0. But for all values of Θ_0 and of R_0 (such that $k_0 R_0 \gg 1$), the electromagnetic field at P_0 and approximately throughout V always lies in a plane at right angles to OP_0. Consequently, a change in the orientation of the antenna at O can at most alter the amplitude K or rotate the axes of polarization of $\mathbf{E}$ and $\mathbf{B}$ in this plane. Therefore, the electric and magnetic fields in the neighborhood of any point in the radiation zone of a dipole antenna are always linearly polarized along mutually perpendicular axes that lie in a plane at right angles to the line joining the point with the center of the antenna.

If the electromagnetic field in V is not that of a single antenna but of a combination of antennas grouped at O and oriented relative to one another in any way whatsoever (e.g., two antennas at right angles to each other), the resultant electromagnetic field in V due to the entire array is still in a plane at right angles to OP_0. If the x axis is along OP_0 as before and the y and z axes are fixed in any positions that make a right-handed system, then, in the most general case, $\mathbf{E}(t)$ and $\mathbf{B}(t)$ may have components with different complex amplitudes along the y and z axes. Assuming that only a single frequency is involved, $\mathbf{E}(t)$ and $\mathbf{B}(t)$ may be expressed as follows:

$$\mathbf{E}(t) \doteq -(\hat{\mathbf{z}}E_z^r + \hat{\mathbf{y}}E_y^r)e^{j\omega t} \qquad (5.6-2a)$$

$$\mathbf{B}(t) \doteq (\hat{\mathbf{y}}B_y^r + \hat{\mathbf{z}}B_z^r)e^{j\omega t} \qquad (5.6-2b)$$

where

$$-E_z^r = cB_y^r = Ke^{-jk_0 x} \qquad (5.6-3a)$$

$$-E_y^r = cB_z^r = Ne^{-jk_0 x} \qquad (5.6-3b)$$

The complex amplitudes K and N may be written in terms of the real amplitudes (a and b) and phase angles (g and p). Thus

$$K = ae^{-jg}; \qquad N = be^{-jp} \tag{5.6-4}$$

With (5.6–3a,b) and (5.6–4), the real parts of (5.6–2a,b) become

$$-E_z^r(t) = cB_y^r(t) = a \cos(\omega t - k_0 x - g) = a \cos \psi \tag{5.6-5a}$$

$$-E_y^r(t) = cB_z^r(t) = b \cos(\omega t - k_0 x - p) = b \cos(\psi - \delta) \tag{5.6-5b}$$

where ψ and δ are given by

$$\psi \equiv \omega t - k_0 x - g \tag{5.6-6}$$

$$\delta \equiv p - g \tag{5.6-7}$$

It can be verified that the following expressions are true:

$$\frac{E_z^2(t)}{a^2} + \frac{E_y^2(t)}{b^2} - \frac{2E_y(t)E_z(t)\cos\delta}{ab} = \sin^2\delta \tag{5.6-8a}$$

$$\frac{B_y^2(t)}{a^2} + \frac{B_z^2(t)}{b^2} - \frac{2B_y(t)B_z(t)\cos\delta}{ab} = \frac{\sin^2\delta}{c^2} \tag{5.6-8b}$$

These equations define the loci of the ends of the vectors $\mathbf{E}(t)$ and $\mathbf{B}(t)$ as ψ varies. At a given distance x from the source, the argument ψ can change only in time. Hence (5.6–8a,b) give the time variation in length and orientation of $\mathbf{E}(t)$ and $\mathbf{B}(t)$ at all points in V that lie in the yz plane. The relations (5.6–8a,b) are the equations of ellipses in the general case in which the major and minor axes do not lie along the coordinate axes as shown in Fig. 5.6–1. The major axis of the ellipse characterizing $\mathbf{E}(t)$ is along the minor axis of the ellipse belonging to $\mathbf{B}(t)$. Under all conditions, the ellipses lie in the yz plane if the x axis is chosen to pass through the origin of the periodically varying charge in the antennas and through the center

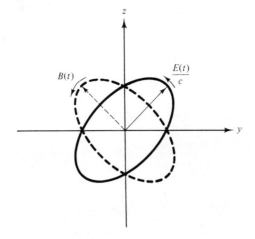

Figure 5.6–1   Instantaneous radiation fields.

of the volume V in which the field is calculated. The orientation of the ellipses in the yz plane and their eccentricity depend on the magnitudes of the amplitudes a and b of the components of $\mathbf{E}(t)$ and $\mathbf{B}(t)$ along the coordinate axes and on the phase difference δ between these components. The most general electromagnetic field in a small region in the radiation zone of a group of antennas operating at a single frequency is elliptically polarized in a plane at right angles to the line joining the source with the region.

Since (5.6–8a,b) gives the locus of the ends of $\mathbf{E}(t)$ and $\mathbf{B}(t)$ in general, it must include the case of linear polarization, which has been shown to characterize the radiation-zone field of a *single* antenna. When the components of $\mathbf{E}(t)$ [or $\mathbf{B}(t)$] are mutually in phase with $\delta = 0$, or exactly opposite in phase with $\delta = \pi$, (5.6–8a) becomes

$$\frac{E_z(t)}{a} = \pm \frac{E_y(t)}{b} \tag{5.6-9}$$

This is a linear relation between $E_y(t)$ and $E_z(t)$ so that the ellipse is degenerate and a straight line with an orientation that depends on the amplitudes a and b of the components. If $b = 0$, $E_y(t) = 0$ and the axis of polarization is the z axis. If $a = b$, the axis of polarization is the 45° line.

As a result of the fact that $\mathbf{E}(t)$ and $\mathbf{B}(t)$ are elliptically polarized in general, there is always a preferred direction for $\mathbf{E}(t)$ and one at right angles to this for $\mathbf{B}(t)$ unless the ellipse degenerates into a circle. This is true when the components of $\mathbf{E}(t)$ [or those of $\mathbf{B}(t)$] have the same amplitudes and differ in phase by $\pi/2$. Thus, if $a = b$ and $\delta = \pm\pi/2$, (5.6–8a,b) reduce to

$$E_y^2(t) + E_z^2(t) = b^2 \tag{5.6-10a}$$

$$B_y^2(t) + B_z^2(t) = \frac{b^2}{c^2} \tag{5.6-10b}$$

These are the equations of circles in the yz plane.

In summary, the electric and magnetic fields in a sufficiently small volume V in the radiation zone of a monochromatic source are in general elliptically polarized. The ellipse, the circle, or the straight line always lies in the plane at right angles to the direction of propagation of the approximately plane waves that pass through V.

5.7 RADIATION FUNCTIONS IN THE FAR ZONE OF AN ANTENNA

The vector potential in the radiation zone of an antenna was shown in (5.5–9) to be

$$\mathbf{A}^r = \frac{\hat{\mathbf{s}}\mu_0}{4\pi} \frac{e^{-jk_0R_0}}{R_0} \int_{-h}^{h} I_s' e^{jk_0(\hat{\mathbf{R}}_0 \cdot \mathbf{s}')} \, ds' \tag{5.7-1}$$

where $\mathbf{R}_0$ is the vector distance from the center of the antenna (O) to the point P where the field is to be calculated and $\hat{\mathbf{s}}$ is the unit vector along the axis of the antenna. A comparison of (5.7–1) and (5.5–7a,b) shows that

$$\mathbf{B}^r = -jk_0(\hat{\mathbf{R}}_0 \times \mathbf{A}^r) \tag{5.7-2}$$

$$\mathbf{E}^r = c(\mathbf{B}^r \times \hat{\mathbf{R}}_0) = -j\omega[(\hat{\mathbf{R}}_0 \times \mathbf{A}^r) \times \hat{\mathbf{R}}_0]$$

$$= j\omega[\hat{\mathbf{R}}_0 \times (\hat{\mathbf{R}}_0 \times \mathbf{A}^r)] \tag{5.7-3}$$

The entire electromagnetic field in the far zone is computed from the vector potential.

If the origin of the spherical coordinate system is located at the center of the antenna,

$$\mathbf{A}^r = \hat{\mathbf{R}}_0 A_R^r + \hat{\mathbf{\Theta}} A_\Theta^r + \hat{\mathbf{\Phi}} A_\Phi^r \tag{5.7-4}$$

Since

$$\hat{\mathbf{R}}_0 \times \hat{\mathbf{R}}_0 = 0; \quad \hat{\mathbf{R}}_0 \times \hat{\mathbf{\Theta}} = \hat{\mathbf{\Phi}}; \quad \hat{\mathbf{R}}_0 \times \hat{\mathbf{\Phi}} = -\hat{\mathbf{\Theta}} \tag{5.7-5}$$

it follows that

$$\mathbf{B}^r = -jk_0(\hat{\mathbf{\Phi}} A_\Theta^r - \hat{\mathbf{\Theta}} A_\Phi^r) \tag{5.7-6}$$

$$\mathbf{E}^r = -j\omega(\hat{\mathbf{\Theta}} A_\Theta^r + \hat{\mathbf{\Phi}} A_\Phi^r) \tag{5.7-7}$$

If the axis of the antenna lies along the z axis so that

$$\mathbf{A}^r = \hat{\mathbf{z}} A_z^r \tag{5.7-8}$$

$$A_\Phi^r = 0; \quad A_\Theta^r = -A_z^r \sin \Theta \tag{5.7-9}$$

In this case,

$$\mathbf{B}^r = -jk_0 A_\Theta^r \hat{\mathbf{\Phi}} \tag{5.7-10}$$

$$\mathbf{E}^r = -j\omega A_\Theta^r \hat{\mathbf{\Theta}} \tag{5.7-11}$$

The general expression (5.7–1) for the vector potential in the radiation zone of a cylindrical antenna that has its axis along the direction $\hat{\mathbf{s}}$ may be written as follows:

$$\mathbf{A}^r = \hat{\mathbf{s}} A_s^r; \quad A_s^r = \left(\frac{\zeta_0 I_p}{4\pi\omega} \frac{e^{-jk_0R_0}}{R_0} \right) \left[k_0 \int_{-h}^{h} f(s') e^{jk_0(\hat{\mathbf{R}}_0 \cdot \mathbf{s}')} \, ds' \right] \tag{5.7-12}$$

I_p is the complex amplitude of the current at a convenient but quite arbitrary reference point p in the antenna. Reference currents usually selected are either the input current at the origin $s' = 0$ or the maximum current whatever its location. $f(s')$ is a dimensionless function which is complex in general.

$$I_s' = I_p f(s') \tag{5.7-13}$$

The components A_Θ^r and A_Φ^r, which are alone important in computing the electromagnetic field in the far zone, may be referred to any conveniently oriented set of polar coordinates R_0, Θ, Φ.

$$A_\Theta^r = (\hat{\Theta} \cdot \hat{s})A_s^r; \qquad A_\Phi^r = (\hat{\Phi} \cdot \hat{s})A_s^r \qquad (5.7\text{--}14)$$

The complex Poynting vector in space is defined as follows:

$$\mathbf{S} = \frac{1}{2\mu_0} \mathbf{E} \times \mathbf{B}^* \qquad (5.7\text{--}15)$$

where $\mathbf{B}^*$ is the complex conjugate of $\mathbf{B}$. With (5.7–6), (5.7–7), and $\hat{\Theta} \times \hat{\Phi} = \hat{\mathbf{R}}_0$, $\zeta_0 = (\mu_0/\epsilon_0)^{1/2}$,

$$\mathbf{S} = \frac{\omega^2}{2\zeta_0} (A_\Theta^r A_\Theta^{r*} + A_\Phi^r A_\Phi^{r*})\hat{\mathbf{R}}_0$$

$$= \frac{\omega^2}{2\zeta_0} [(A_\Theta^r)^2 + (A_\Phi^r)^2]\hat{\mathbf{R}}_0 \qquad (5.7\text{--}16)$$

It follows that the complex Poynting vector in the far zone is real and directed radially outward on each spherical surface of constant phase.

The complex electromagnetic energy-transfer function

$$T = T_r + jT_i$$

is real if $\mathbf{S}$ is real:

$$T = T_r = \int_{\Sigma_\tau} (\hat{\mathbf{n}} \cdot \mathbf{S}) \, d\sigma; \quad T_i = 0 \qquad (5.7\text{--}17)$$

When Σ_τ is the surface of a great sphere

$$T = R_0^2 \int_0^{2\pi} \int_0^{\pi} S \sin \Theta \, d\Theta \, d\Phi \qquad (5.7\text{--}18)$$

With (5.7–16),

$$T = \frac{\omega^2 R_0^2}{2\zeta_0} \int_0^{2\pi} \int_0^{\pi} [(A_\Theta^r)^2 + (A_\Phi^r)^2] \sin \Theta \, d\Theta \, d\Phi \qquad (5.7\text{--}19)$$

T is a measure of the total power transferred over a time average from the moving charges within a sphere of radius R_0 that contribute to A_Θ^r and A_Φ^r to moving charges outside this sphere.

The function $[(A_\Theta^r)^2 + (A_\Phi^r)^2]$ contains factors that are not functions of Θ and Φ. They can be removed from under the sign of integration to leave a dimensionless function in the integrand. This is the *space radiation function*, $K_p^2(\Theta, \Phi)$, defined by

$$K_p^2(\Theta, \Phi) = \left(\frac{4\pi\omega R_0}{\zeta_0 I_p}\right)^2 [(A_\Theta^r)^2 + (A_\Phi^r)^2]$$

$$= \left(\frac{4\pi\omega R_0}{\zeta_0}\right)^2 \left(\frac{1}{I_p I_p^*}\right)(A_\Theta^r A_\Theta^{r*} + A_\Phi^r A_\Phi^{r*}) \qquad (5.7\text{–}20)$$

The subscript on K is the same as on I since its value depends on the choice of reference current. The substitution of (5.7–20) in (5.7–19) yields

$$T = \frac{I_p^2 \zeta_0}{32\pi^2} \int_0^{2\pi} \int_0^\pi K_p^2(\Theta, \Phi) \sin\Theta \, d\Theta \, d\Phi \qquad (5.7\text{–}21)$$

In the case of a single antenna with its axis along the z axis about which Φ is measured, there is rotational symmetry with respect to the coordinate Φ so that $K_p^2(\Theta, \Phi)$ reduces to $K_p^2(\Theta)$. Then

$$T = \frac{I_p^2 \zeta_0}{16\pi} \int_0^\pi K_p^2(\Theta) \sin\Theta \, d\Theta \qquad (5.7\text{–}22)$$

The function T in (5.7–21) or (5.7–22) is a measure of the time-average rate of decrease of energy functions within the great sphere of radius R_0 and of the corresponding time-average rate of increase of energy functions outside the sphere. It is the total time-average transferred or radiated power. In a dissipationless medium it is independent of the radius of the sphere provided that R_0 is large enough so that all points of the sphere are in the far zone of the currents contributing to the field.

The Poynting vector may be expressed in terms of the space radiation function

$$\mathbf{S} = \frac{I_p^2 \zeta_0}{32\pi^2 R_0^2} K_p^2(\Theta, \Phi)\hat{\mathbf{R}}_0 \qquad (5.7\text{–}23)$$

The radiated power divided by the square of the rms reference current is dimensionally a resistance, the external or radiation resistance R_p^e referred to the current I_p.

$$R_p^e = \frac{2T}{I_p^2} = \frac{\zeta_0}{16\pi^2} \int_0^{2\pi} \int_0^\pi K_p^2(\Theta, \Phi) \sin\Theta \, d\Theta \, d\Phi \qquad (5.7\text{–}24)$$

If I_p is the current at the driving point of the antenna, R_p^e is the input resistance which can be measured. If I_p is not the current at the driving point, R_p^e is not a measurable quantity.

5.8 DIRECTIVITY AND GAIN

The directional characteristics of an antenna in the far zone are conveniently expressed by their absolute directivity D. This is defined by

$$D = \frac{P_m}{P} \tag{5.8-1}$$

where P is the power actually radiated by the antenna, and P_m is the power that would be radiated from a physically fictitious *omnidirectional* antenna that radiates the same power. In terms of the space radiation function $K_p^2(\Theta, \Phi)$,

$$D = \frac{K_p^2(\Theta_m, \Phi_m) \int_0^{2\pi} \int_0^\pi \sin\Theta \, d\Theta \, d\Phi}{\int_0^{2\pi} \int_0^\pi K_p^2(\Theta, \Phi) \sin\Theta \, d\Theta \, d\Phi}$$

$$= \frac{30 K_p^2(\Theta_m, \Phi_m)}{R_p^e} \tag{5.8-2}$$

The space radiation function is a maximum in the direction (Θ_m, Φ_m).

In comparing the properties of antennas it is often convenient to use the ideal half-wave dipole in air as a standard of comparison. This has an assumed current distribution of the form $I_s = I_0 \cos k_0 s$ and a radiation function $K_0(\Theta) = \cos[(\pi/2) \cos\Theta] \csc\Theta$, so that its directivity is $D = 1.64$ with $\Theta_m = \pi/2$. Also, its input resistance is $R_0^e = 73$ ohms.

The relative directivity D_r of an antenna is defined as

$$D_r = \frac{D}{1.64} \tag{5.8-3}$$

The gain in decibels of an antenna of absolute directivity D referred to a half-wave dipole is

$$G \text{ (dB)} = 10 \log_{10} D_r \tag{5.8-4}$$

5.9 THE INFINITESIMAL OR HERTZIAN DIPOLE

The vector potential of a z-directed infinitesimal dipole with the electric moment $j\omega p_{1z} = 2hI_z$ is obtained from (3.8–24) to be

$$A_z = \frac{\mu_0 h I_z}{2\pi} \frac{e^{-jkR}}{R} \tag{5.9-1}$$

where in air $k = k_0 = \omega(\mu_0\epsilon_0)^{1/2}$, in a general dissipative medium $k = k_1 = \beta_1 - j\alpha_1 = \omega[\mu_1(\epsilon_1 - j\sigma_1/\omega)]^{1/2}$. In the cylindrical coordinates (ρ, θ, z), $R = (\rho^2 + $

$z^2)^{1/2}$ since rotational symmetry obtains. The electromagnetic field of the dipole is readily evaluated from the general relations $\mathbf{E} = -\nabla\phi - j\omega\mathbf{A}$, $\mathbf{B} = \nabla \times \mathbf{A}$ with the Lorentz condition $\nabla \cdot \mathbf{A} + j(k^2/\omega)\phi = 0$. Since $\mathbf{A} = \hat{\mathbf{z}}A_z$, the cylindrical components are given by

$$\mathbf{E} = \frac{-j\omega}{k^2}\left(\hat{\boldsymbol{\rho}}\frac{\partial}{\partial\rho} + \hat{\mathbf{z}}\frac{\partial}{\partial z}\right)\left(\frac{\partial A_z}{\partial z}\right) - j\omega A_z; \qquad \mathbf{B} = -\hat{\boldsymbol{\theta}}\frac{\partial A_z}{\partial\rho} \qquad (5.9\text{--}2)$$

When (5.9–1) is substituted in (5.9–2), the complete field is

$$E_\rho = \frac{-j\omega\mu h I_z}{2\pi k^2}e^{-jkR}\left(\frac{-k^2}{R} + \frac{3jk}{R^2} + \frac{3}{R^3}\right)\frac{\rho z}{R^2} \qquad (5.9\text{--}3)$$

$$E_z = \frac{-j\omega\mu h I_z}{2\pi k^2}e^{-jkR}\left[\frac{k^2(R^2 - z^2)}{R^3} + \frac{jk(3z^2 - R^2)}{R^4} + \frac{3z^2 - R^2}{R^5}\right] \qquad (5.9\text{--}4)$$

$$B_\theta = \frac{\mu h I_z}{2\pi}e^{-jkR}\left(\frac{jk}{R} + \frac{1}{R^2}\right)\frac{\rho}{R} \qquad (5.9\text{--}5)$$

These formulas are readily converted to the spherical coordinates (R, Θ, Φ), with the relations $E_\Theta = E_\rho\cos\Theta - E_z\sin\Theta$, $E_R = E_\rho\sin\Theta + E_z\cos\Theta$, $B_\Phi = B_\theta$.

$$E_\Theta = \frac{j\omega\mu h I_z}{2\pi k^2}e^{-jkR}\left(\frac{k^2}{R} - \frac{jk}{R^2} - \frac{1}{R^3}\right)\sin\Theta \qquad (5.9\text{--}6)$$

$$E_R = \frac{-j\omega\mu h I_z}{2\pi k^2}e^{-jkR}\left(\frac{2jk}{R^2} + \frac{2}{R^3}\right)\cos\Theta \qquad (5.9\text{--}7)$$

$$B_\Phi = \frac{\mu h I_z}{2\pi}e^{-jkR}\left(\frac{jk}{R} + \frac{1}{R^2}\right)\sin\Theta \qquad (5.9\text{--}8)$$

The radiation or far field at $|kR| \gg 1$ obtained from these formulas is

$$E_\Theta^r = \frac{j\omega\mu h I_z}{2\pi}\frac{e^{-jkR}}{R}\sin\Theta; \qquad B_\Phi^r = \frac{j\mu k h I_z}{2\pi}\frac{e^{-jkR}}{R}\sin\Theta \qquad (5.9\text{--}9)$$

In air, $k = k_0$ and the amplitudes decrease as $1/R$; in a dissipative medium, $k = k_1 = \beta_1 - j\alpha_1$ and the amplitudes decrease as $e^{-\alpha_1 R}/R$. Note that $E_\Theta^r = (k/\omega)B_\Phi^r$. For the dipole in air $k/\omega = k_0/\omega = c = 3 \times 10^8$ m/s and E_Θ^r and B_Φ^r are in phase; in a dissipative medium $k = k_1$ is complex and they are not in phase.

The space radiation function and the radiation resistance of the Hertzian dipole in air can be evaluated from (5.7–20) and (5.7–24). With (5.9–1), $A_\Theta = A_z\sin\Theta$, so that

$$K_p(\Theta) = 2k_0 h\sin\Theta; \qquad R_p^e = 80k_0^2 h^2 \quad \text{ohms} \qquad (5.9\text{--}10)$$

Also, with (5.8–1) to (5.8–4) and with $\Theta_m = \pi/2$,

$$D = 1.5; \qquad D_r = 0.91; \qquad G\text{ (dB)} = -0.39 \qquad (5.9\text{--}11)$$

PROBLEMS

1. A Hertzian dipole (which is an infinitesimally small current element) is equivalent to a pair of small metal spheres which are alternately charged with a total charge $+Q$ and $-Q$ by a periodically varying current of uniform amplitude I maintained in the wire. Using the Helmholtz integrals of Chapter 3, show that the scalar and vector potentials at any point P at a distance R_0 from the center of the dipole (see Figure P5–1) are given by

$$\phi = \frac{1}{4\pi\epsilon_0} e^{-jk_0R_0} \left(\frac{jk_0}{R} + \frac{1}{R_0^2} \right) (\hat{\mathbf{R}}_0 \cdot \mathbf{p})$$

$$\mathbf{A} = \frac{\mu_0}{4\pi} j\omega\mathbf{p} \frac{e^{-jk_0R_0}}{R_0}$$

$$\mathbf{p} = \hat{s}2hQ = -\hat{s}j\frac{2hI}{\omega}$$

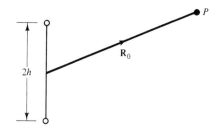

Figure P5–1

2. Starting from Fig. P5–1 and

$$\mathbf{E} = -\nabla\phi - j\omega\mathbf{A}$$

$$\mathbf{B} = \nabla \times \mathbf{A}$$

   show that the electromagnetic field of a Hertzian dipole is given by

$$\mathbf{E} = \frac{e^{-jk_0R_0}}{4\pi\epsilon_0} \left\{ -\frac{k_0^2}{R_0} [\hat{\mathbf{R}}_0 \times (\hat{\mathbf{R}}_0 \times \mathbf{p})] + j\frac{k_0}{R_0^2} [3\hat{\mathbf{R}}_0(\hat{\mathbf{R}}_0 \cdot \mathbf{p}) - \mathbf{p}] \right.$$

$$\left. + \frac{1}{R_0^3} [3\hat{\mathbf{R}}_0(\hat{\mathbf{R}}_0 \cdot \mathbf{p}) - \mathbf{p}] \right\}$$

$$\mathbf{B} = -\frac{j\omega\mu_0}{4\pi} e^{-jk_0R_0} (\hat{\mathbf{R}}_0 \times \mathbf{p}) \left(\frac{1}{R_0^2} + j\frac{k_0}{R_0} \right)$$

3. Determine the components of $\mathbf{E}$ and $\mathbf{B}$ in Problem 2 in
   (a) Cartesian coordinates.
   (b) Spherical coordinates.

4. (a) Calculate the electromagnetic field of the Hertzian dipole in the far zone and plot the field pattern in a polar and a rectangular plot.
   (b) Determine the Poynting vector for the Hertzian dipole.

5. The axial distribution of current in a center-driven antenna of half-length h that satisfies the condition $k_0 h \ll 1$ is given by

$$I_z = I_0 \left(1 - \frac{|z|}{h}\right)$$

The axis of the antenna coincides with the z axis of a coordinate system; its center, the driving point, is at $z = 0$.

(a) Determine $\mathbf{A}$, $\mathbf{E}$, and $\mathbf{B}$.
(b) Determine $\mathbf{A}$, $\mathbf{E}$, and $\mathbf{B}$ in the far zone in free space.
(c) Determine the Poynting vector.
(d) Plot the field pattern in polar form.

6

General Theorems
of Electromagnetic Theory
and Their Applications

Most problems in electromagnetic theory are advantageously formulated directly in terms of the Maxwell–Lorentz equations and the related boundary conditions. However, there are some general situations which are conveniently treated with the help of certain general theorems that provide the means for a more rapid solution or for special insights. Among those that are most generally useful are the theorem of images, the reciprocal theorem, and the theory of electrodynamic similitude.

6.1 IMAGE FIELDS

A straight conductor extends in space from C_a to D_a at an arbitrary angle above the mathematical plane $z = 0$ as shown in Fig. 6.1–1. It carries a periodically varying axial current distributed in an unspecified way. Periodically varying concentrations of charge exist in appropriate distribution as required to satisfy the equation of continuity of electric charge. Below the plane $z = 0$ is an identical second (or image) conductor extending in space from C_i to D_i and arranged to be the exact geometrical image of the first conductor *except in one respect*. All currents and charges, while the same in magnitude at image points, are opposite in direction and sign, respectively. For example, at the point shown in Fig. 6.1–1, the current $\mathbf{I}_a$ is directed from C_a to D_a in the first conductor (subscript a), whereas at the image point the current $\mathbf{I}_i$ in the image conductor (subscript i) is the same in magnitude as $\mathbf{I}_a$ but in the opposite direction with respect to image points (i.e., from D_i to C_i). It is to be noted that if the two conductors are parallel to the plane $z = 0$ as in Fig. 6.1–2a, the reversed current in the image makes *currents at image points* in the two conductors *opposite in direction*; if the two conductors are collinear and perpendicular to the plane $z = 0$ as in Fig. 6.1–2b, the reversed current in the image makes *currents at image points* in the two conductors *the same in direction*. The lower conductor in Fig. 6.1–2a is sometimes called an *image of the upper conductor with current in the opposite direction*; the lower conductor in Fig. 6.1–2b is the *image of the upper conductor with current in the same direction*. It is to be noted that, actually both are images of the actual conductors with currents *reversed* with respect to corresponding image points. It follows directly from the equation of continuity that a reversal of current in the image involves a change in sign of the charge in the image. Accordingly, if there is a concentration of *positive* charge near a given point in the upper conductor, there is an equal concentration of *negative* charge near the image point in the lower conductor.

The complex amplitude of the resultant vector potential $\mathbf{A}$ at any point P (Fig. 6.1–1) is the vector sum of the complex amplitudes of the potentials due to the currents in the individual conductors. Thus for the two conductors,

$$\mathbf{A} = \mathbf{A}_a + \mathbf{A}_i = \frac{\mu_0}{4\pi} \left(\hat{\mathbf{s}}_a \int_{D_a}^{C_a} \frac{I_a'}{R_a} e^{-jk_0 R_a} \, ds_a' + \hat{\mathbf{s}}_i \int_{C_i}^{D_i} \frac{I_i'}{R_i} e^{-jk_0 R_i} \, ds_i' \right) \qquad (6.1\text{–}1)$$

If the point P where $\mathbf{A}$ is calculated is on the plane $z = 0$, $R_a = R_i$ and the two integrals in (6.1–1) are the same. The directions of $\mathbf{A}_a$ and $\mathbf{A}_i$ are those of $\hat{\mathbf{s}}_a$ and $-\hat{\mathbf{s}}_i$, where $\hat{\mathbf{s}}_a$ and $\hat{\mathbf{s}}_i$ are, respectively, unit vectors in the directions $C_a D_a$ and $C_i D_i$,

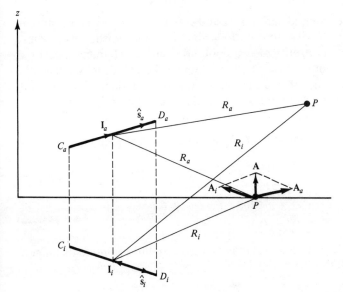

Figure 6.1–1 Antenna $C_a D_a$ with image $C_i D_i$ with reversed current.

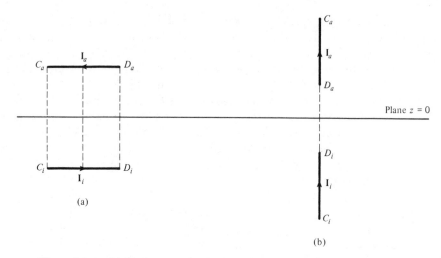

Figure 6.1–2   (a) Horizontal and (b) vertical antennas $C_a D_a$ with images $C_i D_i$ with reversed currents.

as shown in Fig. 6.1–1. The resultant vector potential **A** is necessarily perpendicular to the plane $z = 0$ at all points.

The scalar potential at any point P is

$$\phi = \phi_a + \phi_i$$

$$= \frac{1}{4\pi\epsilon_0} \left(\int_{D_a}^{C_a} \frac{q_a'}{R_a} e^{-jk_0R_a} \, ds_a' + \int_{C_i}^{D_i} \frac{q_i'}{R_i} e^{-jk_0R_i} \, ds_i' \right) \qquad (6.1\text{–}2)$$

Since $q_a' = -q_i'$, the two integrals are equal and opposite in sign at all points in the plane $z = 0$ where $R_a = R_i$. Therefore, the scalar potential vanishes everywhere on this plane, and the plane itself is an equipotential surface characterized by $\phi = 0$. If ϕ is positive at a point (x, y, z) slightly above the plane, it has an equal negative value at a corresponding point $(x, y, -z)$ below the plane. The gradient of ϕ, $\nabla\phi$, is perpendicular to the plane since this is an equipotential surface.

The electromagnetic field at any point P may be computed from

$$\mathbf{E} = -\nabla\phi - j\omega\mathbf{A}$$

$$= -j \frac{\omega}{k_0^2} \left(\nabla\nabla \cdot \mathbf{A} + k_0^2\mathbf{A} \right) \qquad (6.1\text{–}3)$$

$$\mathbf{B} = \nabla \times \mathbf{A} \qquad (6.1\text{–}4)$$

At any point on the plane $z = 0$, **E** and **B** may be resolved into components tangent to the plane (subscript s) and components perpendicular to the plane (subscript n for downward normal).

$$E_n = -\frac{\partial \phi}{\partial n} - j\omega A_n = -2\left(\frac{\partial \phi_a}{\partial n} + j\omega A_{an}\right) = 2E_{an} \qquad (6.1-5)$$

$$E_s = -\frac{\partial \phi}{\partial s} - j\omega A_s = 0 \qquad (6.1-6)$$

$$B_s = (\nabla \times \mathbf{A})_s = 2(\nabla \times \mathbf{A}_a)_s = 2B_{as} \qquad (6.1-7)$$

$$B_n = (\nabla \times \mathbf{A})_n = \lim_{\Delta\Sigma \to 0} \frac{\oint A_s \, ds}{\Delta\Sigma} = 0 \qquad (6.1-8)$$

The vanishing of B_n is due to the fact that there is no component A_s of $\mathbf{A}$ tangent to the plane $z = 0$ in which the contour of integration and $\Delta\Sigma$ in (6.1–8) must lie. The electromagnetic field in the upper half-space (i.e., with $z \geq 0$) is defined by (6.1–3) and (6.1–4) with (6.1–1) and (6.1–2). The electric field is perpendicular and the magnetic field is tangent to the plane $z = 0$.

Let the lower conductor in Fig. 6.1–1 be imagined removed and the entire half-space below $z = 0$ filled with a *perfectly* conducting medium, as in Fig. 6.1–3. In this medium, and hence for $z \leq 0$, the electromagnetic field must be zero. Above the perfectly conducting plane at $z = 0$, the potential functions and the electromagnetic field must be determined from the currents and charges in the upper conductor (which is exactly as before) and from the surface densities of

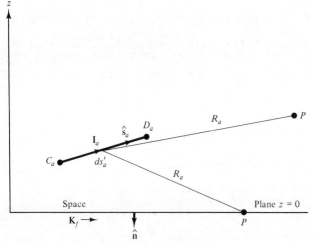

Figure 6.1–3   Antenna $C_a D_a$ over a perfectly conducting half-space.

current and charge on the conducting plane. Thus

$$\mathbf{A} = \mathbf{A}_a + \mathbf{A}_{\text{plane}}$$

$$= \frac{\mu_0 \hat{\mathbf{s}}_a}{4\pi} \int_{C_a}^{D_a} \frac{I'_a}{R_a} e^{-jk_0 R_a} ds'_a + \frac{\mu_0}{4\pi} \int_{\text{plane}} \frac{\mathbf{K}'_f}{R} e^{-jk_0 R} d\sigma' \qquad (6.1\text{-}9)$$

$$\phi = \phi_a + \phi_{\text{plane}}$$

$$= \frac{1}{4\pi\epsilon_0} \int_{C_a}^{D_a} \frac{q'_a}{R_a} e^{-jk_0 R_a} ds'_a + \frac{1}{4\pi\epsilon_0} \int_{\text{plane}} \frac{\eta'_f}{R} e^{-jk_0 R} d\sigma' \qquad (6.1\text{-}10)$$

Here $\mathbf{A}_a$ and ϕ_a are the same as in (6.1–1) and (6.1–2) because I'_a and q'_a are the same. The boundary conditions at $z = 0$ for the electromagnetic field calculated from (6.1–9) and (6.1–10) using (6.1–3) and (6.1–4) are

$$\hat{\mathbf{n}} \cdot \mathbf{E} = \frac{-\eta_f}{\epsilon_0}; \qquad \hat{\mathbf{n}} \times \mathbf{E} = 0 \qquad (6.1\text{-}11)$$

$$\hat{\mathbf{n}} \times \mathbf{B} = -\mu_0 \mathbf{K}_f; \qquad \hat{\mathbf{n}} \cdot \mathbf{B} = 0 \qquad (6.1\text{-}12)$$

Here $\mathbf{E}$ and $\mathbf{B}$ are the electric and magnetic fields in space just above the conducting plane; $\hat{\mathbf{n}}$ is an external normal to the upper half-space; η_f and $\mathbf{K}_f$ are surface densities of free charge and current in the conducting plane. In terms of the magnitudes of the normal and tangential components, (6.1–11) and (6.1–12) may be written as follows:

$$E_n = \frac{-\eta_f}{\epsilon_0}; \qquad E_s = 0 \qquad (6.1\text{-}13)$$

$$B_s = -\mu_0 K_f; \qquad B_n = 0 \qquad (6.1\text{-}14)$$

If the boundary conditions (6.1–13) and (6.1–14) at the perfectly conducting plane $z = 0$ are compared with the field (6.1–5) through (6.1–8) at the mathematical plane $z = 0$, it is seen that they can be made to coincide provided that it is required that the following conditions be true:

$$\frac{-\eta_f}{\epsilon_0} = 2E_{an} = -2 \left(\frac{\partial \phi_a}{\partial n} + j\omega A_{an} \right) \qquad (6.1\text{-}15)$$

$$-\mu_0 K_f = 2B_{an} = 2(\nabla \times \mathbf{A}_a)_s \qquad (6.1\text{-}16)$$

Subject to (6.1–15) and (6.1–16), the electromagnetic field calculated from $\mathbf{A}$ and ϕ as given in (6.1–1) and (6.1–2) is the same for $z \geq 0$ as the electromagnetic field calculated from $\mathbf{A}$ and ϕ as given in (6.1–9) and (6.1–10) subject to (6.1–11) and (6.1–12). This follows from the fact that the two solutions satisfy the same differential equations (the Maxwell–Lorentz equations) in the entire upper half-space ($z \geq 0$) and the same conditions at the boundary ($z = 0$). Since any solution that satisfies the differential equations and the boundary conditions is the unique solution no matter how obtained, it follows that the assumed conditions (6.1–15) and (6.1–16) that lead to the unique solution must be correct themselves.

The following conclusions constitute the *theorem of images*. The electromagnetic vectors at any point P for which $z \geq 0$ in Fig. 6.1–3 are the same as at the corresponding point in Fig. 6.1–1. It is assumed that the conductor $C_a D_a$ is identical physically and electrically in the two cases and that $C_i D_i$ in Fig. 6.1–1 is a geometrical image of $C_a D_a$ but with currents reversed in direction and with charges altered in sign. After $\mathbf{E}$ and $\mathbf{B}$ have been determined for $z \geq 0$ in the single-conductor, infinite-conducting-plane problem of Fig. 6.1–3 by solving the conductor-with-image problem of Fig. 6.1–1, the surface densities η_f and $\mathbf{K}_f$ in the problem of Fig. 6.1–3 may be determined using (6.1–11) and (6.1–12). The field vectors $\mathbf{E}$ and $\mathbf{B}$ due to the surface distributions of current and charge on the perfectly conducting plane in Fig. 6.1–3 are the same at all points in the upper half-space ($z \geq 0$) as the fields due to currents and charges in the image conductor in Fig. 6.1–1. This is not true of the vector potential, as is clear from the directions of $\mathbf{A}_i$ and $\mathbf{A}_{\text{plane}}$ in (6.1–1) and (6.1–9).

The theorem of images permits the substitution of a relatively simple problem involving only a conductor and its image in space for a rather difficult one involving a single conductor over a perfectly conducting infinite plane. This is possible only for points on and above this plane. The theorem is easily generalized to any combination of conductors over a perfectly conducting infinite plane. If the plane is a good conductor ($\sigma_e / \omega \epsilon_e \gg 1$) rather than a perfect one ($\sigma_e = \infty$), the solutions obtained by assuming the conductor perfect are usually good approximations. If the conducting surface is not infinite or not plane, the boundary conditions are much more complicated and the method of images is strictly not applicable. In some instances, moderately good approximations are obtained by applying the method of images even when the conducting surface is finite and not plane. An example is the approximate solution of problems involving antennas over the surface of the earth at frequencies that are sufficiently low so that over moist earth or salt water $\sigma_e / \omega \epsilon_e$ is very large. The method of images must be applied with great care in practical problems involving conducting surfaces of *finite* extent. A finite plane, even if quite large is usually not a satisfactory substitute for an infinite plane and there may be errors in assuming it to be so.

The theorem of images is readily applied to a monopole antenna erected vertically over a highly conducting half-space, as shown in Fig. 6.1–4. It is assumed that the monopole is base-driven by a delta-function generator with emf $\frac{1}{2} V_0^e$ and is of length h and radius a. By the theorem of images, it is equivalent to a dipole of length $2h$ center-driven by a delta-function generator with emf V_0^e. The current distribution on the monopole is the same as that on half of a dipole. The input impedance of the monopole is half that of the dipole and the input admittance is twice the input admittance of the dipole.

The antenna of height h erected over a conducting half-space has the same far-field pattern as the corresponding isolated antenna of height $2h$. Practically, it is often convenient to construct the antenna over a ground plane or to erect it over

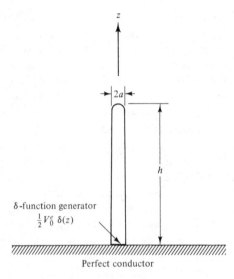

Figure 6.1–4 Base-driven monopole over perfectly conducting ground screen.

the earth. Figure 6.1–5 shows the field pattern of a very thin monopole of length $h = \lambda/4$ over various conductors. The theorem of images may also be applied when there are two or more antennas, such as a collinear array when the antennas are erected one above the other or an array of parallel elements, which is discussed in Chapter 8. Figure 6.1–6 shows the field pattern of a half-wave dipole at a height d over a conducting plane.

6.2 RAYLEIGH–CARSON RECIPROCAL THEOREM

The Rayleigh–Carson reciprocal theorem has wide applications in the general theory of electric circuits especially those involving antennas. It is derived subject only to the condition that the essential density of moving charge $\overline{\rho_m \mathbf{v}}$ be linearly

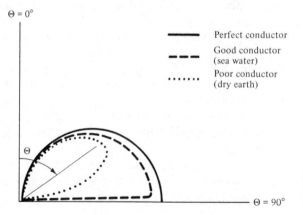

Figure 6.1–5 Vertical field pattern of thin antenna ($h = \lambda/4$) over a conducting plane.

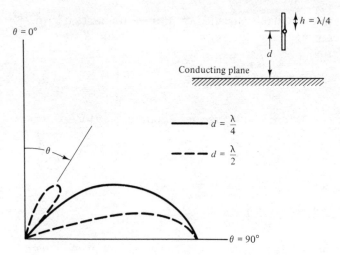

Figure 6.1–6 Field pattern of vertical center-driven antenna of half-length $h = \lambda/4$ over a conducting plane.

related to the electric field **E**. It can be shown that this is true in all simply polarizing, magnetizing, and conducting media for which

$$\mathbf{P} = (\epsilon_r - 1)\epsilon_0\mathbf{E}; \qquad -\mathbf{M} = (\mu_r^{-1} - 1)\mu_0^{-1}\mathbf{B}; \qquad \mathbf{J}_f = \sigma(\mathbf{E} + \mathbf{E}^e) \qquad (6.2\text{–}1)$$

Subject to (6.2–1),

$$\overline{\rho_m\mathbf{v}} = Y\mathbf{E} + \mu_r\sigma\mathbf{E}^e$$

where

$$Y \equiv \mu_r[\sigma_e + j\omega\epsilon_0(\epsilon_{er} - \mu_r^{-1})] \qquad (6.2\text{–}2)$$

$\mathbf{E}^e$ is an impressed field independent of $\overline{\rho_m\mathbf{v}}$ and $\mathbf{E}$ in the regions under study. It is clear that wherever (6.2–1) is true, the density of moving charge $\overline{\rho_m\mathbf{v}}$ is linearly related to the electric field.

Since the electric field is expressed in terms of vector and scalar potentials by the fundamental relation

$$\mathbf{E} = -\nabla\phi - j\omega\mathbf{A} \qquad (6.2\text{–}3)$$

the following relation is obtained from (6.2–2) and (6.2–3):

$$\frac{\overline{\rho_m\mathbf{v}}}{Y} - \frac{\mu_r\sigma\mathbf{E}^e}{Y} = -\nabla\phi - j\omega\mathbf{A} \qquad (6.2\text{–}4)$$

In practical problems only imperfect conductors are involved and surface densities of moving free charge are not required, so $\mathbf{K}_f = 0$. The vector potential can be represented entirely in terms of the volume density $\overline{\rho_m\mathbf{v}}$:

$$\mathbf{A} = \frac{\mu}{4\pi}\int_\tau \frac{\overline{\rho_m\mathbf{v}}}{R} e^{-jk_0R} \, d\tau \qquad (6.2\text{–}5)$$

Integration is with respect to the variables locating $d\tau$. Primes are omitted in (6.2–5) to avoid confusion when primes and double primes are introduced later.

Let a vector $\mathbf{G}$ be defined by

$$\mathbf{G} = \frac{\mu_r \sigma}{Y} \mathbf{E}^e - \nabla\phi \tag{6.2-6}$$

If the nonvanishing values of $\mathbf{E}^e$ are maintained in a good conductor for which $Y \doteq \mu_r \sigma$ where σ is real,

$$\mathbf{G} = \mathbf{E}^e - \nabla\phi \tag{6.2-7}$$

With (6.2–4) and (6.2–5) substituted in (6.2–6) or (6.2–7),

$$\mathbf{G} = \frac{\overline{\rho_m \mathbf{v}}}{Y} + \frac{j\omega\mu}{4\pi} \int_\tau \frac{\overline{\rho_m \mathbf{v}}}{R} e^{-jk_0 R} \, d\tau \tag{6.2-8}$$

Equation (6.2–8) is a vector integral equation in $\overline{\rho_m \mathbf{v}}$. The kernel of the integral is $e^{-jk_0 R}/R$. It is symmetric with respect to any two points in space such as (x_1, y_1, z_1) and (x_2, y_2, z_2) since the distance between the two points

$$R = [(x_1 - x_2)^2 + (y_1 - y_2)^2 + (z_1 - z_2)^2]^{1/2}$$

is invariant to an interchange of subscripts. As a consequence of this symmetry, the theory of integral equations allows the following general theorem to be written.

Theorem.   Any two pairs of functions $\overline{\rho_m \mathbf{v}}'$ and $\mathbf{G}'$, $\overline{\rho_m \mathbf{v}}''$ and $\mathbf{G}''$, which are functions of the space coordinates and which satisfy the integral equation (6.2–8), obey the following reciprocal relation:

$$\int_\tau (\overline{\rho_m \mathbf{v}}' \cdot \mathbf{G}'') \, d\tau = \int_\tau (\overline{\rho_m \mathbf{v}}'' \cdot \mathbf{G}') \, d\tau \tag{6.2-9}$$

The expansion of $\mathbf{G}$ with (6.2–6) and the observation that significant values of $\overline{\rho_m \mathbf{v}}$ occur only in good conductors where $Y = \sigma\mu_r$ so that (6.2–6) reduces to (6.2–7) lead to

$$\int_\tau (\overline{\rho_m \mathbf{v}}' \cdot \mathbf{E}^{e''}) \, d\tau - \int_\tau (\overline{\rho_m \mathbf{v}}'' \cdot \mathbf{E}^{e'}) \, d\tau$$

$$= \int_\tau (\overline{\rho_m \mathbf{v}}' \cdot \nabla\phi'') \, d\tau - \int_\tau (\overline{\rho_m \mathbf{v}}'' \cdot \nabla\phi') \, d\tau \tag{6.2-10}$$

The integrals on the right may be transformed using the vector identity

$$\nabla \cdot (\mathbf{C}\psi) = \psi\nabla \cdot \mathbf{C} + \mathbf{C} \cdot \nabla\psi \tag{6.2-11}$$

and the divergence theorem

$$\int_\tau \nabla \cdot \mathbf{C} \, d\tau = \int_\Sigma \hat{\mathbf{n}} \cdot \mathbf{C} \, d\sigma \tag{6.2-12}$$

For example, the first integral on the right in (6.2–10) becomes

$$\int_\tau (\overline{\rho_m \mathbf{v}'} \cdot \boldsymbol{\nabla}\phi'') \, d\tau = \int_\tau \boldsymbol{\nabla} \cdot (\phi'' \overline{\rho_m \mathbf{v}'}) \, d\tau - \int_\tau \phi'' \boldsymbol{\nabla} \cdot (\overline{\rho_m \mathbf{v}'}) \, d\tau$$

$$= \int_\Sigma \phi'' (\hat{\mathbf{n}} \cdot \overline{\rho_m \mathbf{v}'}) \, d\sigma - \int_\tau \phi'' \boldsymbol{\nabla} \cdot (\overline{\rho_m \mathbf{v}'}) \, d\tau \quad (6.2\text{–}13)$$

The surface integral in (6.2–13) vanishes if the surface Σ is chosen to cross no regions in which $\overline{\rho_m \mathbf{v}'}$ differs from zero. With the equation of continuity

$$\boldsymbol{\nabla} \cdot \overline{\rho_m \mathbf{v}} + j\omega\bar{\rho} = 0 \quad (6.2\text{–}14)$$

(6.2–13) becomes

$$\int_\tau (\overline{\rho_m \mathbf{v}'} \cdot \boldsymbol{\nabla}\phi'') \, d\tau = j\omega \int_\tau \phi'' \bar{\rho}' \, d\tau \quad (6.2\text{–}15)$$

The scalar potential satisfies the general wave equation

$$\boldsymbol{\nabla}^2\phi + k_0^2\phi = \frac{-\bar{\rho}}{\epsilon_0} \quad (6.2\text{–}16)$$

With (6.2–16) solved for $\bar{\rho}$ substituted in (6.2–15), the following equation is obtained:

$$\int_\tau (\overline{\rho_m \mathbf{v}'} \cdot \boldsymbol{\nabla}\phi'') \, d\tau = -j\omega\epsilon_0 \int_\tau \phi'' (\boldsymbol{\nabla}^2\phi' + k_0^2 \, \phi') \, d\tau \quad (6.2\text{–}17)$$

Similarly,

$$\int_\tau (\overline{\rho_m \mathbf{v}''} \cdot \boldsymbol{\nabla}\phi') \, d\tau = -j\omega\epsilon_0 \int_\tau \phi' (\boldsymbol{\nabla}^2\phi'' + k_0^2 \, \phi'') \, d\tau \quad (6.2\text{–}18)$$

With (6.2–17) and (6.2–18) in (6.2–10), it follows that

$$\int_\tau (\overline{\rho_m \mathbf{v}'} \cdot \mathbf{E}^{e''}) \, d\tau - \int_\tau (\overline{\rho_m \mathbf{v}''} \cdot \mathbf{E}^{e'}) \, d\tau$$

$$= j\omega\epsilon_0 \int_\tau (\phi' \boldsymbol{\nabla}^2\phi'' - \phi'' \boldsymbol{\nabla}^2\phi') \, d\tau \quad (6.2\text{–}19)$$

The right side can be transformed using Green's symmetrical theorem. It becomes

$$j\omega\epsilon_0 \int_\Sigma \left(\phi' \frac{\partial\phi''}{\partial n} - \phi'' \frac{\partial\phi'}{\partial n} \right) d\sigma \quad (6.2\text{–}20)$$

with n an external normal to the enclosing surface Σ. If this is chosen to be a spherical shell of radius R that is very large compared with the greatest dimension between regions in which $\mathbf{E}^e$ differs from zero, then

$$\frac{\partial\phi}{\partial n} = \frac{\partial\phi}{\partial R} \quad (6.2\text{–}21)$$

on this shell. Furthermore, the scalar potential ϕ on the surface of a sufficiently large shell in the far zone has the form

$$\phi = C \frac{e^{-jk_0R}}{R} \tag{6.2-22}$$

where C is independent of R. With (6.2–21) and in the limit as $R \to \infty$,

$$\frac{\partial \phi}{\partial n} = \frac{\partial \phi}{\partial R} = -jk_0\phi \tag{6.2-23}$$

If (6.2–23) is written for ϕ' and ϕ'' and substituted in (6.2–20), this vanishes. Therefore, the right side of (6.2–19) is zero and

$$\int_\tau (\overline{\rho_m \mathbf{v}'} \cdot \mathbf{E}^{e''}) \, d\tau = \int_\tau (\overline{\rho_m \mathbf{v}''} \cdot \mathbf{E}^{e'}) \, d\tau \tag{6.2-24}$$

These integrals are to be evaluated over all regions where both $\mathbf{E}^e$ and $\overline{\rho_m \mathbf{v}}$ differ from zero.

In applying (6.2–24) to electric circuits and antennas, it is usually possible to assume that regions where $\mathbf{E}^e$ differs from zero are equivalent to short distances between a finite number of pairs of surfaces each of small area. Each of the pairs constitutes the terminal A and B of a generator or its equivalent across which there is a total current

$$I_s = \int_S \overline{\rho_m v_s} \, dS \tag{6.2-25}$$

due to the field $\mathbf{E}^e$ maintained between A and B. These terminals are sufficiently close together so that the current I_s has the same amplitude at A, at B, and in the generator between A and B. Let $\mathbf{E}^e$ be a mean value on each surface between n pairs of terminals A and B:

$$\int_\tau (\overline{\rho_m \mathbf{v}'} \cdot \mathbf{E}^{e''}) \, d\tau = \sum_{j=1}^{n} \left(\hat{s} I_s' \cdot \int_A^B \mathbf{E}^{e''} \, ds \right)_j \tag{6.2-26}$$

Let the positive direction of current for each set of terminals be from B to A. Also let

$$V_{AB}^e = \int_A^B E^e \, ds \tag{6.2-27}$$

be the driving potential maintained across each pair of terminals by the n generators. Then

$$\int_\tau (\overline{\rho_m \mathbf{v}'} \cdot \mathbf{E}^{e''}) \, d\tau = \sum_{j=1}^{n} I_j' V_{ABj}^{e''} \tag{6.2-28}$$

With (6.2–28) and an expression like it but with primes and double primes interchanged, (6.2–24) becomes

$$\sum_{j=1}^{n} I'_j V^{e''}_{ABj} = \sum_{j=1}^{n} I''_j V^{e'}_{ABj} \tag{6.2–29}$$

This is the final expression. The theorem it expresses may be summarized as follows:

Statement I: A set of driving potential differences $V^{e'}_1, \ldots, V^{e'}_j, \ldots, V^{e'}_n$ maintained across n pairs of terminals in a network of conductors or an array of antennas produces a set of currents $I'_1, \ldots, I'_j, \ldots, I'_n$ at these terminals.

Statement II: A different set of driving potential differences $V^{e''}_1, \ldots, V^{e''}_j, \ldots, V^{e''}_n$ maintained across the *same* n terminals produces the set of currents $I''_1, \ldots, I''_j, \ldots, I''_n$.

Theorem.   The driving potential differences and the currents in the two cases are related by the following reciprocal relation:

$$\sum_{j=1}^{n} I'_j V^{e''}_j = \sum_{j=1}^{n} I''_j V^{e'}_j$$

Condition. The total moving charge in the entire region is everywhere linearly related to the electric field, so that

$$\overline{\rho_m \mathbf{v}} \sim \mathbf{E}$$

The proportionality constant may be complex.

The reciprocal theorem may be expressed nonmathematically as follows: If a generator with an emf or driving potential difference of complex amplitude V^e between its terminals maintains a current of complex amplitude I through a load connected between any other pair of terminals in the same or in a coupled network, the current in the load is unaltered if load and generator are interchanged provided that the impedances connected between each pair of terminals are the same in both cases and the generator maintains the same emf.

In applying the reciprocal theorem it is usually convenient to consider the currents due to only one driving potential difference at a time, since it is possible to add algebraically the individually determined currents due to several voltages. This follows from the fact that the differential equations involved are linear with constant coefficients. Suppose that when $V^{e'}_j$ is applied at the terminals j, a current I'_i exists at terminals i, and when $V^{e''}_i$ is applied at terminals i, a current I''_j exists at terminals j. The reciprocal theorem reduces to the important form

$$I''_j V^{e'}_j = I'_i V^{e''}_i \tag{6.2–30}$$

Further simplification in (6.2–30) results if the same potential difference is applied in the one case across the terminals j as in the other case across the terminals i.

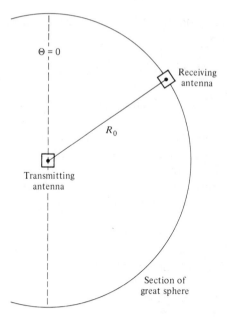

Figure 6.3–1 Arrangement of antennas to illustrate reciprocity.

When

$$V_j^{e'} = V_i^{e''} \tag{6.2–31}$$

it follows that the reciprocal theorem becomes simply

$$I_j'' = I_i' \tag{6.2–32}$$

6.3 APPLICATION OF THE RAYLEIGH–CARSON RECIPROCAL THEOREM

An important and useful application of the Rayleigh–Carson reciprocal theorem is to the determination of the directional properties of a dipole antenna (No. 1) when used for reception from the known field pattern of the same antenna when used for transmission (Fig. 6.3–1). The far field of a center-driven dipole antenna of length $2h_1$ at the radial distance R_0 is

$$E_{\Theta 1}^r = j60 I_{10} \frac{e^{-jk_0 R_0}}{R_0} F_0(\Theta_1, k_0 h_1) \tag{6.3–1}$$

where $F_0(\Theta_1, k_0 h_1)$ is the far-field pattern and

$$I_{10} = \frac{V_{10}^e}{Z_{10} + Z_{1g}} \tag{6.3–2}$$

is the current at the driving point; Z_{10} is the self-impedance of the antenna, V_{10}^e is the emf and Z_{1g} the impedance of the generator. In order to observe the field pattern $F_0(\Theta_1, k_0h_1)$, a center-loaded receiving dipole (Fig. 6.3–1) is moved around the transmitting antenna tangent to the meridian of a great sphere with radius R_0. The current I_{20} in the load is proportional to the electric field $E_{\Theta1}$ along the axis of the antenna. It can be expressed in the form

$$I_{20} = \frac{-2h_{e2}(\pi/2)E_{\Theta1}}{Z_{20} + Z_{2L}} \tag{6.3-3}$$

where Z_{20} is the self-impedance of the antenna and Z_{2L} is the load impedance. The proportionality parameter $2h_{e2}(\pi/2)$ is known as the *effective length* of the receiving antenna. Note that the antenna is moved so that the angle Θ_2 between its axis and the direction of the radius R_0 to the center of the distant transmitter is always $\pi/2$. When (6.3–1) with (6.3–2) is substituted in (6.3–3), the result is

$$I_{20} = \frac{-2h_{e2}(\pi/2)}{Z_{20} + Z_{2L}} \left[\frac{j60V_{10}^e}{Z_{10} + Z_{1g}} \frac{e^{-jk_0R_0}}{R_0} F_0(\Theta_1, k_0h_1) \right] \tag{6.3-4}$$

As the receiving antenna is moved to vary Θ_1 from 0 to $\pi/2$, I_{20} varies in a manner proportional to $F_0(\Theta_1, k_0h_1)$.

Now let the generator at the center of antenna 1 be interchanged with the load at the center of antenna 2. As antenna 2, now the transmitter, is moved along the meridian of the great sphere just as before, the current in the load of antenna 1 is given by

$$I_{10} = \frac{-2h_{e1}(\Theta_1)}{Z_{10} + Z_{1L}} \left[\frac{j60V_{20}^e}{Z_{20} + Z_{2g}} \frac{e^{-jk_0R_0}}{R_0} F_0(\pi/2, k_0h_2) \right] \tag{6.3-5}$$

This expression is like (6.3–4), with the $F_0(\Theta_1, k_0h_1)$ replaced by $F_0(\pi/2, k_0h_2)$, where h_2 is the half-length of antenna 2 and $F_0(\pi/2, k_0h_2)$ is the far field of antenna 2 in the direction $\Theta_2 = \pi/2$. Similarly, $h_{e2}(\pi/2)$ in (6.3–4) is replaced by $h_{e1}(\Theta_1)$ since angle Θ_1 between the axis of antenna 1 and the direction along R_0 to antenna 2 varies as antenna 2 is moved.

The reciprocal theorem as applied to two antennas reduces to

$$I_{10}'V_{10}^{e''} = I_{20}''V_{20}^{e'} \tag{6.3-6}$$

where I_{10}' is the current maintained in the load of antenna 1 (used as a receiver) by the driving voltage $V_{20}^{e'}$ at the center of antenna 2 (used as a transmitter) and I_{20}'' is the current maintained in the load of antenna 2 (used as receiver) by the driving voltage $V_{10}^{e''}$ at the center of antenna 1 (used as transmitter). If the two driving voltages are made equal $V_{10}^{e''} = V_{20}^{e'}$, it follows that

$$I_{10}' = I_{20}'' \tag{6.3-7}$$

Let (6.3–4) be condition prime with $I_{20} = I_{20}'$, $V_{10}^e = V_{10}^{e'}$; let (6.3–5) be condition double prime with $I_{10} = I''_{10}$, $V^e_{20} = V^{e''}_{20}$. Then with $V^{e''}_{20} = V^{e'}_{10}$, (6.3–7) gives

$$\frac{h_{e1}(\Theta_1)}{Z_{10} + Z_{1L}} \frac{F_0(\pi/2, k_0 h_2)}{Z_{20} + Z_{2g}} = \frac{h_{e2}(\pi/2)}{Z_{20} + Z_{2L}} \frac{F_0(\Theta_1, k_0 h_1)}{Z_{10} + Z_{1g}} \tag{6.3–8}$$

If $Z_{1L} = Z_{1g}$, $Z_{2L} = Z_{2g}$, and both sides are multiplied by k_0 to make them dimensionless:

$$k_0 h_{e1}(\Theta_1) F_0(\pi/2, k_0 h_2) = k_0 h_{e2}(\pi/2) F_0(\Theta_1, k_0 h_1) \tag{6.3–9}$$

This is satisfied if, in general,

$$k_0 h_e(\Theta) = F_0(\Theta, k_0 h) \tag{6.3–10}$$

for each antenna. Thus the dimensionless electrical effective half-length of the antenna when used for reception is equal to the far-field pattern of the same antenna when used for transmission. This means that the current in the central load of a receiving dipole is given by

$$I_{10} = \frac{2 h_{e1}(\Theta_1) E_{\Theta 2}}{Z_{10} + Z_{1L}} \tag{6.3–11}$$

where $h_{e1}(\Theta_1) = F_0(\Theta_1, k_0 h_1)/k_0$ and $F_0(\Theta_1, k_0 h_1)$ is a factor in (6.3–1). Thus the directional properties of an antenna are the same for transmission and reception. Note that for an electrically short antenna with $k_0 h < 1$,

$$k_0 h_e(\Theta) = F_0(\Theta, k_0 h) = \tfrac{1}{2} k_0 h \sin \Theta \tag{6.3–12}$$

Also, $h_e(\pi/2) = h/2$.

6.4 ELECTRODYNAMICAL SIMILITUDE AND THE THEORY OF MODELS

The experimental study of electromagnetic phenomena is often simplified if scale models of convenient size are used to simulate a full-sized system. The system under study may be either too small or too large to be convenient for use in the laboratory and a suitable scale model would facilitate the study.

Let the scale model have physical dimensions that differ from those of the actual system by a factor n, which may be less than or greater than 1 according to the size convenient for use in the laboratory. The electromagnetic field at all points in an isotropic homogeneous medium can be determined from the equations for the scalar and vector potentials, that is, from

$$\nabla^2 \phi + k^2 \phi = 0 \tag{6.4–1a}$$

$$\nabla^2 \mathbf{A} + k^2 \mathbf{A} = 0 \tag{6.4–1b}$$

where $k^2 = \omega^2\mu(\epsilon - j\sigma/\omega)$. The corresponding equations are true at all points in a modeled medium. They are

$$\nabla_m^2\phi_m + k_m^2\phi_m = 0 \qquad (6.4\text{--}2a)$$

$$\nabla_m^2\mathbf{A}_m + k_m^2\mathbf{A}_m = 0 \qquad (6.4\text{--}2b)$$

where $k_m^2 = \omega_m^2\mu_m(\epsilon_m - j\sigma_m/\omega_m)$. The electrical properties of the model are required to be the same as the original except for a change in scale. It follows that the scalar potential ϕ_m and vector potential $\mathbf{A}_m$ in the medium about the model should differ from the original potentials ϕ and $\mathbf{A}$ by, at most, a constant factor m. Furthermore, since all lengths in the model are to be changed by a factor $n = r_m/r$, where r_m and r are characteristic lengths, and since the operator ∇_m^2 is equivalent to a second derivative with respect to the space coordinates

$$\nabla_m^2 = \frac{\nabla^2}{n^2} \qquad (6.4\text{--}3)$$

(6.4–2a) for the scalar potential becomes

$$\nabla^2\phi + n^2k_m^2\phi = 0 \qquad (6.4\text{--}4)$$

after ϕ_m has been multiplied by m and $m\phi_m$ replaced by ϕ. If the model is to be electrically equivalent to the original system, (6.4–4) and (6.4–2a,b) must be identical. It follows that

$$k_m^2 = \frac{k^2}{n^2} \qquad \text{or} \qquad k_m r_m = kr \qquad (6.4\text{--}5)$$

That is, the electrical length (which is complex in general) must be kept invariant. In free space, $k = k_0 = \omega/c$, where c is the velocity of light. Equation (6.4–5) is then equivalent to

$$\omega_m = \frac{\omega}{n} \qquad \text{or} \qquad f_m = \frac{f}{n} \qquad (6.4\text{--}6a)$$

where f is the frequency. Alternatively,

$$f_m r_m = fr \qquad (6.4\text{--}6b)$$

The same results are obtained with the equations for the vector potential $\mathbf{A}$.

The theory of scaling may be extended to a more general linear isotropic homogeneous medium. In this case, with k complex, (6.4–5) becomes $k_m^2 r_m^2 = \omega_m^2 r_m^2\mu_m(\epsilon_m - j\sigma_m/\omega_m) = k^2r^2 = \omega^2r^2\mu(\epsilon - j\sigma/\omega)$, where ϵ and σ stand for the real effective values ϵ_e and σ_e. This is equivalent to two conditions:

$$\omega_m^2 r_m^2\mu_m\epsilon_m = \omega^2r^2\mu\epsilon; \qquad \omega_m r_m^2\mu_m\sigma_m = \omega r^2\mu\sigma \qquad (6.4\text{--}7)$$

This theory of models has useful applications. For example, in living organisms, it is at times desirable to insert antennas for heating purposes as in hy-

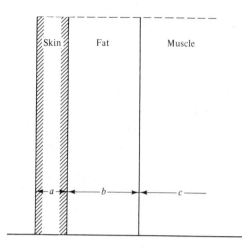

Figure 6.4–1 Section of three-layered half-space.

perthermia treatment of cancer or for use as transponders. The layers of skin, fat, and muscle with thicknesses a, b, and c are shown in Fig. 6.4–1. Such a system may be modeled in the laboratory by using three containers with solutions separated by thin plastic walls that will not modify the dielectric properties. One of the frequencies used in biomedical applications is 2.24 GHz. At this frequency the constitutive parameters for skin, fat, and muscle are as follows:

	ϵ_r	σ (S/m)
Skin	42.5	1.88
Fat	6	0.144
Muscle	48.5	1.81

where ϵ_r is the relative permittivity and σ is the conductivity. A convenient frequency for use in the laboratory is 600 MHz. Since the materials are nonmagnetic ($\mu = \mu_0$), (6.4–7) reduces to

$$\omega_m^2 \epsilon_m r_m^2 = \omega^2 \epsilon r^2 \tag{6.4–8a}$$

$$\omega_m \sigma_m r_m^2 = \omega \sigma r^2 \tag{6.4–8b}$$

At 600 MHz the frequency ratio is $f_m/f = \omega_m/\omega = 0.27$. The layers of skin and muscle can be approximated by a saltwater solution which has a permittivity of about $\epsilon_{mr} = 80$ at $f = 600$ MHz. For skin (6.4–8a) gives

$$\frac{a_m}{a} = \frac{\omega}{\omega_m} \sqrt{\frac{\epsilon}{\epsilon_m}} = \frac{1}{0.27} \sqrt{\frac{42.5}{80}} = 3.0; \qquad a_m = 3a$$

For muscle,

$$\frac{c_m}{c} = \frac{1}{0.27}\sqrt{\frac{48.5}{80}} = 2.9; \qquad c_m = 2.9a$$

The substitution of these results in (6.4–8b) yields for skin,

$$\frac{\sigma_m}{\sigma} = \frac{\omega a^2}{\omega_m a_m^2} = \frac{1}{0.27} \times \frac{1}{9} = 0.41; \qquad \sigma_m = 0.77 \text{ S/m}$$

For muscle,

$$\frac{\sigma_m}{\sigma} = \frac{1}{0.27} \times \frac{1}{8.4} = 0.44; \qquad \sigma_m = 0.795 \text{ S/m}$$

The layer of fat can be modeled by a layer of clean sand moistened with fresh water or very slightly salty water in proportions to obtain $\epsilon_r \sim 10$, so that from (6.4–8a),

$$\frac{b_m}{b} = \frac{1}{0.27}\sqrt{\frac{6}{10}} = 2.9; \qquad b_m = 2.9b$$

and, from (6.4–8b),

$$\frac{\sigma_m}{\sigma} = \frac{1}{0.27} \times \frac{1}{8.4} = 0.44; \qquad \sigma_m = 0.063 \text{ S/m}$$

From these calculations, the laboratory model should have layers that are roughly three times as thick as those in the living organism with the skin modeled by a layer of salt water with $\epsilon_r = 80$ and $\sigma = 0.77$ S/m, the layer of fat by a sand–freshwater mixture with $\epsilon_r = 10$ and $\sigma = 0.063$ S/m, and the muscle by salt water with $\epsilon_r = 80$ and $\sigma = 0.795$ S/m. Measurements made on properly scaled antennas in the model can be converted directly to the smaller antennas in the living organism.

PROBLEMS

1. A monopole is the upper half of a Hertzian dipole erected vertically over a perfectly conducting half-space. Determine its vector potential, electromagnetic field, Poynting vector, space radiation function, and radiation resistance and compare with those of the Hertzian dipole.

2. A broadcast antenna is erected vertically over moist earth ($\sigma = 10^{-2}$ S/m). Its length is $5\lambda/8$ and the frequency is 1.5 MHz. Laboratory measurements are to be made on a model at 60 MHz. What must be the length of the model and the dielectric constant and conductivity of the material to be used to simulate the moist earth? Is it physically realizable?

3. A center-driven antenna of half-length $h = \lambda/4$ has an input resistance at resonance of 70 Ω. The voltage applied at the terminals is 100 V rms at 300 MHz.
   (a) Calculate the rms electric field at a distance of 2 km from the antenna, $\Theta = 60°$.
   (b) Determine the power transferred to the matched load of a receiving antenna placed parallel to the field of part (a) (i.e., at 2 km from the transmitting antenna with $\Theta = 60°$). The receiving antenna is like the transmitting antenna. Assume its effective half-length to be $h_e = \lambda/2\pi$. Note that the effective electrical half-length of a receiving antenna is given by

   $$k_0 h_e(\Theta) = \frac{\cos (k_0 h \cos \Theta) - \cos k_0 h}{\sin \Theta \sin k_0 h}$$

   where Θ is the orientation of the receiving antenna with respect to the wave front or surface of constant phase of the electric field of the transmitting antenna.

4. A half-wave dipole receiving antenna has a conjugate matched load. It is due north of a distant transmitting antenna that maintains a vertically polarized electric field of 1 mV/m at the receiving antenna. The receiving antenna is inclined 36.9° toward the south. The plane containing the receiving antenna and the line joining it to the distant transmitter is inclined 45° from the vertical. Determine the power in the load of the receiving antenna if $\Omega = 2 \ln (2h/a) = 10$, $h = \lambda_0/4$. The frequency is 10 MHz.

5. A receiving antenna for which $\Omega = 2 \ln (2h/a) = 15$ is placed in a circularly polarized electric field of magnitude 100 μV/m. The plane of polarization is the yz plane; $f = 10$ MHz.
   (a) Determine the orientation of the antenna and its half-length h for which the maximum open-circuit voltage is maintained across the load terminals at the center. What is this maximum voltage?
   (b) What will be the voltage across the terminals of a receiver connected as a load to the terminals of the antenna in part (a) if its impedance is $Z_L = 100,000 + jX_L$ and X_L is adjusted for maximum voltage across R_L?
   (c) What will be the voltage across Z_L if the conditions of part (b) obtain but the antenna is inclined at 53° from the xy plane?

6. A receiving antenna of half-length $h = \lambda_0/4$ and radius such that $\Omega = 2 \ln (2h/a) = 20$ is center-loaded by an impedance $Z_L = 4.5 + j60 \, \Omega$. The current in Z_L is zero when the antenna lies along a horizontal east-west axis. As the antenna is rotated in the vertical plane containing north and south, a maximum current of 60 μA is observed in Z_L when the antenna is vertical, a minimum value of 20 μA when the antenna is horizontal. $f = 10$ MHz.
   (a) What is the polarization of the electric field at the receiving antenna?
   (b) What is its magnitude in volts per meter? Give maximum and minimum values.
   (c) In what direction is the distant transmitter?

7. In a system provided for underground communication, the frequency is 5 kHz. The lithosphere is assumed to consist of a region of rock of low conductivity between more highly conducting layers at the surface and the mantle. The ϵ_{er} is assumed to be 5 for all three layers. The loss tangents are as follows: surface layer, 0.12; rock, 1; mantle, 5. The thickness of the layers is 25 km, 5 km, and an infinite half-space for the mantle. If an operating frequency of 115 MHz is to be used in the scale model, calculate the constitutive parameters of the materials used in the model and the dimensions.

7

Scattering and Diffraction of Plane Waves

The oscillating electric currents maintained by generators in antennas and other radiating systems over a wide band of frequencies induce similarly oscillating currents in surrounding more or less distant bodies and matter-filled regions. The interaction is quantitatively accurately described in terms of trains of electromagnetic waves that travel outward from the source and, in passing over the obstacles, induce currents in them. These, in turn, generate a secondary electromagnetic field known as the *scattered* or *reradiated field*. Where it penetrates the geometrical shadow it is known as the *diffracted field*. Its nature depends on the electrical properties, shape, and orientation of the obstacle relative to the incident field. The actual calculation of the scattered field and the induced currents that generate it constitutes an important application of Maxwell's equations and the associated boundary conditions. Among the analytically solvable problems, the scattering and diffraction by a conducting or totally absorbing (black) half-plane is among the most important and instructive. It has interesting applications in many aspects of electromagnetism, ranging from radio to optical frequencies.

7.1 REFLECTION BY A CONDUCTING PLANE: ACTUAL AND VIRTUAL SOURCES

As an introduction to the problem of the half-plane, it is useful to review the reflection of plane electromagnetic waves incident on a perfectly conducting infinite plane at an arbitrary angle. Let the xy plane ($z = 0$) be the perfectly conducting surface on which a plane electromagnetic wave is incident in the half-space $z \le 0$ as shown in Fig. 7.1–1. The linearly polarized electric field is parallel to the x axis. The magnetic field is parallel to the y axis. The direction of propagation is specified by the vector wave number **k**, which makes an angle θ^i with the xz plane. The incident electric field at (ρ, θ) in cylindrical and at (y, z) in Cartesian coordinates is given by

$$E_x^i = Ee^{ik\rho\cos(\theta - \theta^i)} = Ee^{ik(y\sin\theta^i + z\cos\theta^i)} \tag{7.1–1}$$

where the time dependence $e^{-i\omega t}$ is understood.

The boundary condition on the total field—incident plus scattered—is

$$E_x(z = 0) = 0 = E_x^i(z = 0) + E_x^s(z = 0) \tag{7.1–2}$$

for all values of x and y. The appropriate scattered field that satisfies this condition is

$$E_x^s = -Ee^{ik\rho\cos(\pi - \theta - \theta^i)} = -Ee^{ik(y\sin\theta^i - z\cos\theta^i)} \tag{7.1–3}$$

The total field is

$$E_x = E_x^i + E_x^s = 2iEe^{iky\sin\theta^i}\sin(kz\cos\theta^i); \quad z \le 0 \tag{7.1–4}$$

The time-dependent value is the real part of (7.1–4) after multiplication by $e^{-i\omega t}$. It is

$$E_x(t) = 2E\sin(kz\cos\theta^i)\sin(\omega t - ky\sin\theta^i) \tag{7.1–5}$$

This represents a standing wave in the direction of the negative z axis. It is given by $\sin(kz\cos\theta^i)$ and is independent of the time. The apparent wavelength of the

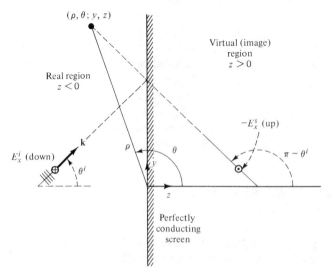

Figure 7.1–1 Plane electromagnetic wave incident on a perfectly conducting screen.

standing-wave pattern is

$$\lambda_z = \frac{2\pi}{k \cos \theta^i} \tag{7.1-6}$$

which reduces to the free-space wavelength at normal incidence when $\theta^i = 0$. In the direction of the coordinate y (7.1–5) represents a traveling or progressive wave given by $\sin(\omega t - ky \sin \theta^i) = \sin \omega(t - y/v_y)$, where the phase velocity in the y direction is

$$v_y = \frac{\omega}{k \sin \theta^i} = \frac{c}{\sin \theta^i} \tag{7.1-7}$$

This reduces to $v_y = \infty$ when $\theta^i = 0$ at normal incidence, and to $v_y = c$ at grazing incidence with $\theta^i = \pi/2$.

The complete field in the region of interest $(z < 0)$ is made up of the incident field and a reflected field. This latter is identically the same as the field of a fictitious image or virtual source of plane waves in the region $z > 0$ in the absence of the conducting screen. The superposition of the fields from the actual and virtual sources provides a total field that satisfies the boundary condition $E_x(z = 0) = 0$. Note that the superposition of the two fields is physically meaningful only in the region $z \leq 0$. The actual field behind the perfectly conducting screen at $z = 0$ is zero. The virtual field due to the image source in the region $z \geq 0$ is a part of the mathematical formalism of the theorem of images.

7.2 FORMULATION OF THE PROBLEM OF THE CONDUCTING HALF-PLANE

The structure of interest is shown in Fig. 7.2–1. It consists of an infinitely thin, perfectly conducting screen that occupies the half-plane defined by $z = 0$, $y \geq 0$, so that its finite edge coincides with the x axis. A plane electromagnetic wave is incident on the screen in a direction such that the x axis (the edge of the screen) lies in a wave front. This means that the entire electromagnetic field is independent of x since all wave fronts extend from $x = -\infty$ to $x = \infty$. Also, $\partial/\partial x = 0$. With these conditions, Maxwell's equations in Cartesian coordinates and with the time dependence $e^{-i\omega t}$ reduce to the following two mutually independent groups:

$$\text{(I)} \quad \frac{\partial B_z}{\partial y} - \frac{\partial B_y}{\partial z} + \frac{ik^2}{\omega} E_x = 0 \tag{7.2-1}$$

$$\frac{\partial E_x}{\partial z} - i\omega B_y = 0 \tag{7.2-2}$$

$$\frac{\partial E_x}{\partial y} + i\omega B_z = 0 \tag{7.2-3}$$

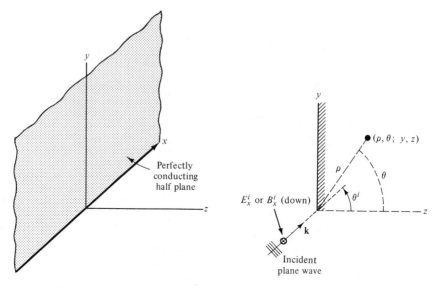

Figure 7.2–1 Perfectly conducting half-plane and incident plane wave.

$$\text{(II)}\quad \frac{\partial E_z}{\partial y} - \frac{\partial E_y}{\partial z} - i\omega B_x = 0 \tag{7.2–4}$$

$$\frac{\partial B_x}{\partial z} + \frac{ik^2}{\omega} E_y = 0 \tag{7.2–5}$$

$$\frac{\partial B_x}{\partial y} - \frac{ik^2}{\omega} E_z = 0 \tag{7.2–6}$$

Differentiation and substitution within each group lead to

$$\text{(I)}\quad \frac{\partial^2 E_x}{\partial y^2} + \frac{\partial^2 E_x}{\partial z^2} + k^2 E_x = 0 \tag{7.2–7}$$

$$B_y = -\frac{i}{\omega}\frac{\partial E_x}{\partial z}; \qquad B_z = \frac{i}{\omega}\frac{\partial E_x}{\partial y} \tag{7.2–8}$$

$$\text{(II)}\quad \frac{\partial^2 B_x}{\partial y^2} + \frac{\partial^2 B_x}{\partial z^2} + k^2 B_x = 0 \tag{7.2–9}$$

$$E_y = \frac{i\omega}{k^2}\frac{\partial B_x}{\partial z}; \qquad E_z = -\frac{i\omega}{k^2}\frac{\partial B_x}{\partial y} \tag{7.2–10}$$

If (7.2–7) or (7.2–9) is solved for E_x or B_x, the other two associated components are determined by differentiation.

The boundary conditions for the two problems are simply:

(I) $E_x(z = 0) = 0,$ on the screen   (7.2–11)

(II) $E_y(z = 0) = 0$ or $\left(\dfrac{\partial B_x}{\partial z}\right)_{z=0} = 0,$ on the screen   (7.2–12)

Problems of type (I) in which the dependent variable E_x vanishes on the boundary are known as *Dirichlet problems*. Problems of type (II) in which the normal derivative of the dependent variable B_x vanishes on the boundary are known as *Neumann problems*.

The solution of (7.2–7) and (7.2–9) subject to the boundary conditions (7.2–11) and (7.2–12) can be carried out with the help of a generalization of the method of images. Let this be introduced with a function of the cylindrical coordinates ρ, θ which has the form $F = F[\rho, \sin(\theta/2)]$, where F is to be *single-valued* in each argument, ρ and $\sin(\theta/2)$. Actually, at each point in the yz plane F has *two* distinct values that correspond to the angles θ and $\theta + 2\pi$. However, F can be made a single-valued function if the yz plane is regarded as two planes superimposed on each other in the following manner. Both planes, the upper and the lower, are cut along a semi-infinite straight line at $\theta = \pi/2$. The left side of the cut of the upper plane is joined to the right side of the cut of the lower plane. It now requires two complete turns to move from $\theta = -7\pi/2$ to $\theta = \pi/2$ back to the starting point. This two-sheeted plane is known as a *Riemann surface*. Over it, $F[\rho, \sin(\theta/2)]$ is a single-valued function of ρ and $\sin(\theta/2)$. The half-plane screen divides the yz-Riemann surface into two parts that are completely separated. Specifically, when the cut and the screen coincide, one side of the screen is at $\theta = \pi/2$, the other side at $\theta = -4\pi + \pi/2 = -7\pi/2$. Let real and virtual ranges be defined as follows:

$$\text{Real range:} \quad -\frac{3\pi}{2} \le \theta \le \frac{\pi}{2}$$

$$\text{Virtual range:} \quad -\frac{7\pi}{2} \le \theta \le -\frac{3\pi}{2}$$

(7.2–13)

7.3 SINGLE-VALUED SOLUTIONS ON THE TWO-SHEETED RIEMANN SURFACE

The equation of interest has the form

$$\frac{\partial^2 f}{\partial y^2} + \frac{\partial^2 f}{\partial z^2} + k^2 f = 0$$

(7.3–1)

where $f = E_x$ subject to $f = 0$ on the screen or $f = B_x$ subject to $\partial f/\partial z = 0$ on the screen. A solution of this equation is sought in the semiparabolic coordinates

u and v defined as follows:

$$u = z = \rho \cos \theta; \qquad v = \rho - z = \rho(1 - \cos \theta) \qquad (7.3\text{--}2)$$

It follows that

$$\rho = v(1 - \cos \theta)^{-1} \qquad (7.3\text{--}3)$$

For any given v, this is the equation of a parabola with the directrix at $z = -v/2$ and the focus at $z = v/2$.

The coordinates u and v are related to the half-angle $\theta/2$ by

$$v = 2\rho \sin^2 \frac{\theta}{2}; \qquad u = \rho\left(1 - 2 \sin^2 \frac{\theta}{2}\right) \qquad (7.3\text{--}4)$$

The change in variables expressed by (7.3–2) leads to quite long formulas for $\partial^2/\partial z^2$ and $\partial^2/\partial y^2$. With these the transformed equation is

$$\frac{\partial^2 f}{\partial u^2} + \frac{2v}{u + v}\left(\frac{\partial^2 f}{\partial v^2} - \frac{\partial^2 f}{\partial u \, \partial v}\right) + \frac{1}{u + v}\frac{\partial f}{\partial v} + k^2 f = 0 \qquad (7.3\text{--}5)$$

The variables in (7.3–5) are not separable, but it is possible to find a particular integral that has the form

$$f = U(u)V(v) \qquad (7.3\text{--}6)$$

With (7.3–6), (7.3–5) can be expressed as follows:

$$V(U'' + k^2 U) + \frac{2v}{u + V}\left(-U'V' + UV'' + \frac{1}{2v} UV'\right) = 0 \qquad (7.3\text{--}7)$$

where the primes denote differentiation with respect to the argument u for U and v for V. This equation is satisfied by

$$U'' + k^2 U = 0 \qquad (7.3\text{--}8)$$

$$\frac{-U'}{U} + \frac{V''}{V'} + \frac{1}{2v} = 0 \qquad (7.3\text{--}9)$$

A particular integral of (7.3–8) is

$$U = e^{iku} = e^{ik\rho \cos \theta} \qquad (7.3\text{--}10)$$

which corresponds to the incident plane wave at normal incidence and with unit amplitude.

The substitution of (7.3–10) into (7.3–9) gives

$$\frac{V''}{V'} + \frac{1}{2v} = \frac{U'}{U} = ik \qquad (7.3\text{--}11)$$

The equation to be solved is

$$\frac{V''}{V'} + \frac{1}{2v} - ik = 0 \qquad (7.3\text{--}12)$$

which is satisfied by the particular integral

$$V = C_1 \int_{C_2}^{v} \frac{e^{ikw}}{\sqrt{w}}\, dw \tag{7.3-13}$$

With (7.3–10) and (7.3–13) in (7.3–6), it follows that

$$f = UV = C_1 e^{iku} \int_{C_2}^{v} \frac{e^{ikw}}{\sqrt{w}}\, dw \tag{7.3-14}$$

Now let $w = \pi s^2/2k$. This gives

$$f = C_1' e^{iku} \int_{C_2'}^{\pm\sqrt{2kv/\pi}} e^{i\pi s^2/2}\, ds$$

$$= C_1'\, e^{ik\rho\cos\theta} \int_{C_2'}^{\pm 2\sqrt{k\rho/\pi}\,\sin(\theta/2)} e^{i\pi s^2/2}\, ds$$

$$= C_1' e^{ik\rho\cos\theta}\left[F\!\left(\pm 2\sqrt{\frac{k\rho}{\pi}}\,\sin\frac{\theta}{2} \right) - F(C_2') \right] \tag{7.3-15}$$

The function f is two-valued on the real plane, single-valued on the two-sheeted Riemann surface.

The constants C_1' and C_2' and the sign of the upper limit can be determined from the requirement that f be a plane wave at infinite radial distance, $\rho = \infty$. In the real plane, $-3\pi/2 \le \theta \le 0$, $f \to e^{ik\rho\cos\theta}$, which is the incident field, as $\rho \to \infty$. This is obtained with the $-$ sign in the upper limit, which is then positive in the range $-2\pi \le \theta \le 0$ in which $\sin(\theta/2)$ is negative. Since

$$\int_{-\infty}^{\infty} e^{i\pi s^2/2}\, ds = \sqrt{2}\, e^{i\pi/4} \tag{7.3-16}$$

it is possible to set $C_1' = e^{-i\pi/4}/\sqrt{2}$, $C_2' = -\infty$, and obtain $f = e^{ik\rho\cos\theta}$ as $\rho \to \infty$. With these values

$$f = e^{ik\rho\cos\theta}\,\frac{e^{-i\pi/4}}{\sqrt{2}}\left[F\!\left(-2\sqrt{\frac{k\rho}{\pi}}\,\sin\frac{\theta}{2} \right) - F(-\infty) \right] \tag{7.3-17}$$

where

$$F(w) \equiv \int_0^w e^{i\pi s^2/2}\, ds; \qquad F(\pm\infty) \equiv \pm\frac{e^{i\pi/4}}{\sqrt{2}} = \pm\frac{1+i}{2} \tag{7.3-18}$$

is the Fresnel integral.*

---

*An alternative definition is obtained with the substitution $t = \pi s^2/2$. It is

$$F(W) = \int_0^W \frac{e^{it}}{\sqrt{2\pi t}}\, dt$$

with $W = \pi w^2/2$.

The particular integral (7.3–17) is a solution of the equation (7.3–1), but so far no boundary conditions have been imposed. Before these are applied, it is useful to examine the nature of the solution in its present form. A first step is a review of the properties of the Fresnel integral with real argument.

7.4 THE FRESNEL INTEGRAL AND THE CORNU SPIRAL

The Fresnel integral (7.3–18) is complex with the real and imaginary parts given by

$$F(w) = C(w) + iS(w) \tag{7.4–1a}$$

where

$$C(w) = \int_0^w \cos \frac{\pi s^2}{2} \, ds; \qquad S(w) = \int_0^w \sin \frac{\pi s^2}{2} \, ds \tag{7.4–1b}$$

Note that

$$F(-w) = -F(w) \tag{7.4–2}$$

Graphs of the integrands, $\cos (\pi s^2/2)$ and $\sin (\pi s^2/2)$, are shown in Fig. 7.4–1a and of the integrals $C(w)$ and $S(w)$ in Fig. 7.4–1b. Numerical values are given in Table 7.4–1. A graph of $F(w)$ in the complex plane is shown in Fig. 7.4–1c. This is known as the *Cornu spiral*. The argument w is measured along the spiraling contour from the origin. It is positive in the first quadrant, negative in the third quadrant.

At $w = 0$, $F(0) = 0$. Here the tangent to the curve is horizontal. Near $w = 0$, $S(w) \sim \pi w^3/6$, $C(w) \sim w$. At $w = \infty$, $F(\infty) = (1 + i)/2$, $|F(\infty)| = \sqrt{2}/2$. At

TABLE 7.4–1 The Fresnel Integral $C(w) + iS(w)$

w	$C(w)$	$S(w)$	w	$C(w)$	$S(w)$
0.0	0	0	3.0	0.606	0.496
0.2	0.200	0.004	3.2	0.466	0.593
0.4	0.398	0.033	3.4	0.438	0.430
0.6	0.581	0.111	3.6	0.588	0.492
0.8	0.723	0.249	3.8	0.448	0.566
1.0	0.780	0.438	4.0	0.498	0.420
1.2	0.715	0.623	4.2	0.542	0.563
1.4	0.543	0.714	4.4	0.438	0.462
1.6	0.366	0.639	4.6	0.567	0.516
1.8	0.334	0.451	4.8	0.434	0.497
2.0	0.488	0.343	5.0	0.564	0.499
2.2	0.636	0.456	5.2	0.439	0.497
2.4	0.555	0.620	5.4	0.557	0.514
2.6	0.389	0.550	5.6	0.452	0.470
2.8	0.468	0.392	5.8	0.530	0.546

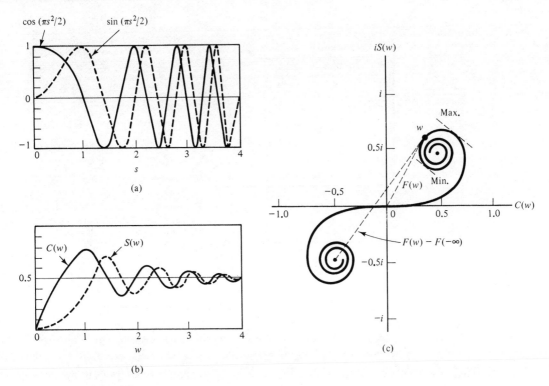

Figure 7.4–1 Fresnel integrals: (a) integrands; (b) real and imaginary integrals; (c) Cornu spiral.

$w = -\infty$, $F(-\infty) = -(1 + i)/2$. Note that on the graph, $F(w)$ is the line from the origin to the point w on the spiral contour. As w increases, the length of this line oscillates. The quantity $F(w) - F(-\infty)$ is the line drawn from the point w to the point $w = -\infty$.

7.5 APPLICATION OF THE PARTICULAR INTEGRAL TO THE ABSORBING (BLACK) HALF-PLANE: NORMAL INCIDENCE

It is useful to consider first the important special case of a plane wave normally incident on the half-plane, as shown schematically in Fig. 7.5–1. In the sense of geometrical optics, the illuminated sector is defined by $-3\pi/2 < \theta < 0$, the shadow region is $0 < \theta \leq \pi/2$, and the shadow boundary is the plane $\theta = 0$. The particular integral for normal incidence is (7.3–17), that is,

$$f_1 = e^{ik\rho\cos\theta} \frac{e^{-i\pi/4}}{\sqrt{2}} \left[F\left(-2\sqrt{\frac{k\rho}{\pi}} \sin\frac{\theta}{2} \right) - F(-\infty) \right] \qquad (7.5–1)$$

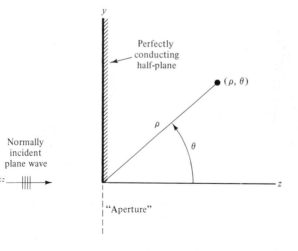

Figure 7.5–1 Plane wave normally incident on conducting half-plane.

As $\rho \rightarrow \infty$,

$$f_1 = \begin{cases} e^{ikz} & \text{for } -3\pi/2 < \theta < 0 \text{ with sin } (\theta/2) \text{ negative} \qquad (7.5\text{–}2a) \\ 0 & \text{for } 0 < \theta < \pi/2 \text{ with sin } (\theta/2) \text{ positive} \qquad (7.5\text{–}2b) \end{cases}$$

These forms of the solution actually correspond to geometrical optics in the absence of reflection (i.e., for a perfectly absorbing "black screen"). The incident field on arriving at the screen passes from the real to the virtual space and disappears (is absorbed in the screen).

The half-plane $\theta = -\pi/2$ is the "aperture" through which the field passes. For it, $\cos \theta = 0$, $\sin (\theta/2) = -\sqrt{2}/2$, so that

$$f_1 = \frac{e^{-i\pi/4}}{\sqrt{2}} \left[F\left(\sqrt{\frac{2k\rho}{\pi}}\right) - F(-\infty) \right] \qquad (7.5\text{–}3)$$

Behind the screen, $\theta = \pi/2$ and

$$f_1 = \frac{e^{-i\pi/4}}{\sqrt{2}} \left[F\left(-\sqrt{\frac{2k\rho}{\pi}}\right) - F(-\infty) \right] \qquad (7.5\text{–}4)$$

These are the rigorous solutions for an incident unpolarized (scalar) electromagnetic field in the plane $z = 0$ for a black, perfectly absorbing screen. It gives a complete and accurate picture of optical diffraction. A graph of the function f as given by (7.5–3) and (7.5–4) in the "aperture" and behind the screen is shown in Fig. 7.5–2. It represents the amplitude of the complete field for a perfectly absorbing screen. It decays rapidly and smoothly in the shadow region, and oscillates with decreasing amplitude about the constant field of geometrical optics in the aperture.

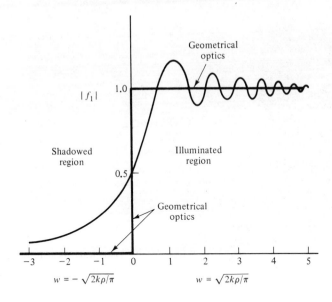

Figure 7.5–2 Field $|f_1| = (1/\sqrt{2})$ $[F(w) - F(-\infty)]$ for "black" screen; normally incident field.

In front of the screen $\theta = -3\pi/2$, $\cos \theta = 0$, and $\sin (\theta/2) = -\sqrt{2}/2$, so that

$$f_1 = \frac{e^{-i\pi/4}}{\sqrt{2}} \left[F\left(\sqrt{\frac{2k\rho}{\pi}} \right) - F(-\infty) \right] \tag{7.5–5}$$

which is identical with (7.5–3) for the field in the aperture. Since the screen is perfectly absorbing (nonreflecting), it is to be expected that it behaves just like the empty space of the aperture.

7.6 GENERAL INCIDENCE ON A CONDUCTING HALF-PLANE: E_x AND B_x POLARIZATIONS

The geometry for a plane wave incident on the half-plane at an arbitrary angle of incidence θ^i is shown in Fig. 7.6–1. The same solution as for normal incidence with $\theta - \theta^i$ substituted for θ satisfies the wave equation (7.3–1). The new, more general solution is

$$f_1(\rho, \theta) = e^{ik\rho \cos(\theta - \theta^i)} \frac{e^{-i\pi/4}}{\sqrt{2}} [F(s) - F(-\infty)] \tag{7.6–1}$$

where

$$s = -2\sqrt{\frac{k\rho}{\pi}} \sin \frac{1}{2}(\theta - \theta^i) \tag{7.6–2}$$

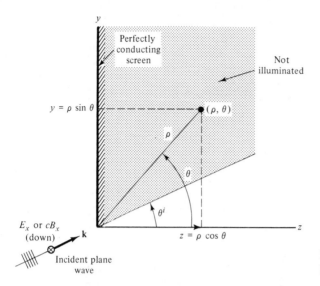

Figure 7.6–1   Plane wave incident on half-plane at an arbitrary angle.

The illuminated sector is

$$-\frac{3\pi}{2} \leq \theta \leq \theta^i \qquad \text{or} \qquad -\frac{1}{2}\left(\frac{3\pi}{2} + \theta^i\right) \leq \frac{1}{2}(\theta - \theta^i) \leq 0 \qquad (7.6\text{-}3)$$

In this sector $\sin[(\theta - \theta^i)/2]$ is negative so that s in (7.6-2) is positive. In the limit, $\rho \to \infty$, $F(s \to \infty) - F(-\infty) = \sqrt{2}\, e^{i\pi/4}$, so that

$$f_1(\rho \to \infty, \theta) \to e^{ik\rho\cos(\theta - \theta^i)} = e^{ik(y\sin\theta^i + z\cos\theta^i)} \qquad (7.6\text{-}4)$$

which is the incident field.

The shadowed sector is

$$\theta^i \leq \theta \leq \frac{\pi}{2} \qquad \text{or} \qquad 0 \leq \frac{1}{2}(\theta - \theta^i) \leq \frac{1}{2}\left(\frac{\pi}{2} - \theta^i\right) \qquad (7.6\text{-}5)$$

In this sector $\sin[(\theta - \theta^i)/2]$ is positive, so that s is negative. In the limit as $\rho \to \infty$, $F(s \to -\infty) - F(-\infty) = 0$, so that $f_1(\rho \to \infty, \theta) \to 0$.

It is now convenient to impose the boundary conditions in order to determine the scattered field when the incident linearly polarized electromagnetic field is in the presence of a perfectly conducting half-plane. The solution $f_1(\rho, \theta)$ satisfies the differential equation and has the desired behavior at $\rho \to \infty$. It does not satisfy either of the following boundary conditions for the two polarizations:

$$\text{(I)} \quad f = E_x = 0, \qquad \text{on the screen}, \qquad \theta = \frac{\pi}{2}, \ -\frac{3\pi}{2} \qquad (7.6\text{-}6)$$

$$\text{(II)} \quad \frac{\partial f}{\partial n} = \frac{\partial B_x}{\partial z} = 0, \qquad \text{on the screen} \qquad (7.6\text{-}7)$$

In order that $f_1(\rho, \theta)$ satisfy one (or the other) of these boundary conditions, a reflected field can be introduced. This must be generated in the virtual plane by a virtual or image source at infinity. It must enter the real plane *through* the screen and so act as the reflected field. Such a field is obtained with the substitution of $-(\pi + \theta^i)$ for θ^i in $f_1(\rho, \theta)$, so that

$$\cos(\theta - \theta^i) \rightarrow -\cos(\theta + \theta^i); \sin \tfrac{1}{2}(\theta - \theta^i) \rightarrow \cos \tfrac{1}{2}(\theta + \theta^i) \qquad (7.6\text{--}8)$$

The reflected field is then given by

$$f_2(\rho, \theta) = e^{-ik\rho\cos(\theta + \theta^i)} \frac{e^{-i\pi/4}}{\sqrt{2}} [F(t) - F(-\infty)] \qquad (7.6\text{--}9)$$

where

$$t = -2\sqrt{\frac{k\rho}{\pi}} \cos \tfrac{1}{2}(\theta + \theta^i) \qquad (7.6\text{--}10)$$

It is now readily verified that the boundary conditions (7.6–6) and (7.6–7) are satisfied by the following combinations of the incident and reflected fields and that they have the correct behavior at infinity:

$$\left.\begin{array}{c} E_x \\ cB_x \end{array}\right\} = f_1(\rho, \theta) \mp f_2(\rho, \theta) \qquad (7.6\text{--}11)$$

where $f_1(\rho, \theta)$ is given by (7.6–1) with (7.6–2) and $f_2(\rho, \theta)$ is given by (7.6–9) with (7.6–10). These are the final solutions, respectively, for E_x and cB_x. The other components are obtained by differentiation. They are

$$\left.\begin{array}{c} cB_y \\ E_y \end{array}\right\} = (f_2 \pm f_1)\cos \theta^i \mp \frac{e^{i(k\rho - \pi/4)}}{\sqrt{2\pi k\rho}} [\sin \tfrac{1}{2}(\theta - \theta^i) \pm \cos \tfrac{1}{2}(\theta + \theta^i)] \qquad (7.6\text{--}12)$$

$$\left.\begin{array}{c} cB_z \\ E_z \end{array}\right\} = (f_2 \mp f_1)\sin \theta^i \mp \frac{e^{i(k\rho - \pi/4)}}{\sqrt{2\pi k\rho}} [\cos \tfrac{1}{2}(\theta + \theta^i) \mp \sin \tfrac{1}{2}(\theta - \theta^i)] \qquad (7.6\text{--}13)$$

Note that on the screen defined by $\theta = \pi/2$ and $\theta = -3\pi/2$, $f_1 = f_2$ and $\sin[(\theta - \theta^i)/2] - \cos[(\theta + \theta^i)/2] = 0$, so that $E_x = E_y = 0$, as required by the boundary conditions.

With $F(-\infty) = -F(\infty) = -e^{i\pi/4}/\sqrt{2}$, it can be shown that the two terms in $F(-\infty)$ in (7.6–11) can be combined to give

$$\left.\begin{array}{c} E_x \\ cB_x \end{array}\right\} = e^{iky\sin\theta^i} \left\{ \left[\begin{array}{c} i\sin(kz\cos\theta^i) \\ \cos(kz\cos\theta^i) \end{array} \right] \right.$$

$$\left. + \frac{e^{-i\pi/4}}{\sqrt{2}} [F(s)e^{ikz\cos\theta^i} \mp F(t)e^{-ikz\cos\theta^i}] \right\} \qquad (7.6\text{--}14)$$

Here the first term on the right is one-half the field with a full conducting plane, as can be verified by comparison with (7.1–4).

7.7 SPECIAL RANGES OF THE TOTAL FIELD

The principal components E_x and cB_x for the two polarizations are conveniently studied in the three regions illustrated in Fig. 7.7–1. As indicated, region I is illuminated by both the incident and reflected fields, region II by the incident field alone, and region III by neither field directly. Let the fields be examined successively in the three regions beginning with the shadow, region III.

Region III.   The shadowed region, $\theta^i \leq \theta \leq \pi/2$; the diffracted field. In this range, $0 \leq \theta - \theta^i \leq (\pi/2) - \theta^i$, so that $0 \leq \sin \frac{1}{2}(\theta - \theta^i) \leq \sin [(\pi/4) - (\theta^i/2)]$ and $-2\sqrt{k\rho/\pi} \sin [(\pi/4) - (\theta^i/2)] \leq s \leq 0$, where $s = -2\sqrt{k\rho/\pi} \sin \frac{1}{2}(\theta - \theta^i)$. Similarly, $\theta^i \leq \frac{1}{2}(\theta + \theta^i) \leq (\pi/4) + (\theta^i/2)$, so that $-2\sqrt{k\rho/\pi} \cos \theta^i \leq t \leq -2\sqrt{k\rho/\pi} \cos [(\pi/4) + (\theta^i/2)]$, where $t = -2\sqrt{k\rho/\pi} \cos \frac{1}{2}(\theta + \theta^i)$. At normal incidence, $\theta^i = 0$, and the ranges of s and t are

$$-\sqrt{\frac{2k\rho}{\pi}} \leq s \leq 0; \qquad -2\sqrt{\frac{k\rho}{\pi}} \leq t \leq -\sqrt{\frac{2k\rho}{\pi}} \tag{7.7-1}$$

When $\sqrt{2k\rho/\pi} \geq 1$, $|f_1(\rho, \theta)| = (1/\sqrt{2}) \, | \, F(s) - F(-\infty)|$ is significant since it is measured from $F(0) = 0$, whereas $|f_2(\rho, \theta)| = (1/\sqrt{2})|F(t) - F(-\infty)|$ is very small since it is measured between adjacent points in the spiral. This means that except where $\sqrt{2k\rho/\pi} < 1$ or $k\rho < \pi/2$, the field in the shadowed region is due almost entirely to $f_1(\rho, \theta)$ and is essentially the same as for the "black" screen as given by Fig. 7.5–2. Furthermore, the fields E_x and cB_x differ negligibly. Thus, in region III with $k\rho \leq \pi/2$,

$$E_x \sim cB_x \sim f_1(\rho, \theta) = \frac{e^{-i\pi/4}}{\sqrt{2}} \, e^{ik\rho \cos(\theta - \theta^i)}[F(s) - F(-\infty)] \tag{7.7-2}$$

where $s = -2\sqrt{k\rho/\pi} \sin \frac{1}{2}(\theta - \theta^i)$.

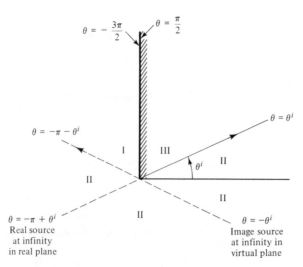

Figure 7.7–1   Regions of illumination and shadow in real plane. I, Illuminated by real and virtual sources; II, illuminated by real source only; III, not illuminated by either source, shadow.

It can be shown that for $|s|$ and $|t|$ both large and $\theta \neq \theta^i$,

$$\left.\begin{array}{c} E_x \\ cB_x \end{array}\right\} \doteq \frac{e^{i[(\pi/4) + k\rho]}}{2\sqrt{2\pi k\rho}} \left[\frac{1}{\sin \frac{1}{2}(\theta - \theta^i)} \mp \frac{1}{\cos \frac{1}{2}(\theta + \theta^i)} \right]$$

$$\sim \frac{e^{i\pi/4}}{2\sqrt{2\pi k}} \frac{e^{ik\rho}}{\sqrt{\rho}} \frac{1}{\sin \frac{1}{2}(\theta - \theta^i)}; \qquad \theta \neq \theta^i, \quad \theta < \frac{\pi}{2} \qquad (7.7-3)$$

Since the amplitude decreases slowly as θ increases from near θ^i, it is clear that the field penetrates quite far into the geometrical shadow. Note that the radial dependence $e^{ik\rho}/\sqrt{\rho}$ is that of a cylindrical wave traveling outward from the edge of the screen. This edge is an apparent line source for the field in the shadow.

Directly behind the screen, $\theta = \pi/2, 0 \leq \rho \leq \infty$, it is easily shown that

$$E_x = f_1\left(\rho, \frac{\pi}{2}\right) - f_2\left(\rho, \frac{\pi}{2}\right) = 0 \qquad (7.7-4)$$

$$cB_x = f_1\left(\rho, \frac{\pi}{2}\right) + f_2\left(\rho, \frac{\pi}{2}\right)$$

$$= \frac{2e^{-i\pi/4}}{\sqrt{2}} e^{ik\rho \sin \theta^i} \left\{ F\left[-\sqrt{\frac{2k\rho}{\pi}} \left(\cos \frac{\theta^i}{2} - \sin \frac{\theta^i}{2} \right) \right] - F(-\infty) \right\} \qquad (7.7-5)$$

At normal incidence $\theta^i = 0$ and

$$cB_x = \sqrt{2}\, e^{-i\pi/4} \left[F\left(-\sqrt{\frac{2k\rho}{\pi}} \right) - F(-\infty) \right] \qquad (7.7-6)$$

This is the same as for a "black" screen but multiplied by 2.

Region II. Illuminated by the incident field, not illuminated by the reflected field; $-\pi - \theta^i < \theta < \theta^i$. This range includes the "aperture" defined by $\theta = -\pi/2, 0 \leq \rho \leq \infty$, where the field is given by

$$\left.\begin{array}{c} E_x \\ cB_x \end{array}\right\} = f_1\left(\rho, -\frac{\pi}{2}\right) \mp f_2\left(\rho, -\frac{\pi}{2}\right)$$

$$= \frac{e^{-i\pi/4}}{\sqrt{2}} e^{-ik\rho \sin \theta^i} \left\{ F\left[\sqrt{\frac{2k\rho}{\pi}} \left(\cos \frac{\theta^i}{2} + \sin \frac{\theta^i}{2} \right) \right] \right.$$

$$\left. \mp F\left[-\sqrt{\frac{2k\rho}{\pi}} \left(\cos \frac{\theta^i}{2} + \sin \frac{\theta^i}{2} \right) \right] - \begin{array}{c} 0 \\ 2F(-\infty) \end{array} \right\} \qquad (7.7-7)$$

This is simplified greatly with normal incidence, that is, with $\theta^i = 0$. Thus, since

$$F(-\infty) = -F(\infty),$$

$$E_x = \sqrt{2}\, e^{-i\pi/4} F\left(-\sqrt{\frac{2k\rho}{\pi}}\right); \qquad cB_x = 1 \qquad (7.7\text{--}8)$$

Elsewhere in region II no simplification of the general formula occurs except when $k\rho \gg 1$. Where this is true and except near the boundaries $\theta = \theta^i$ and $\theta = -\pi - \theta^i$,

$$f_1 \sim e^{ik\rho\cos(\theta-\theta^i)} \text{ (the incident field)}; \qquad f_2 \sim 0 \qquad (7.7\text{--}9)$$

It follows that

$$E_x \sim cB_x \sim e^{ik\rho\cos(\theta-\theta^i)} \qquad (7.7\text{--}10)$$

Near the shadow boundary, $\theta^i = 0$,

$$E_x \sim cB_x \sim e^{ik\rho\cos\theta} \qquad (7.7\text{--}11)$$

which is like the field with a "black" screen.

Region I.   Illuminated by the incident and reflected waves; $-3\pi/2 \le \theta \le -\pi - \theta^i$. There is no simplification of the general formula in this range except when $k\rho \gg 1$. In this case

$$f_1(\rho, \theta) \sim e^{ik\rho\cos(\theta-\theta^i)}; \qquad f_2(\rho, \theta) = e^{-ik\rho\cos(\theta+\theta^i)} \qquad (7.7\text{--}12)$$

It follows that

$$E_x = 2ie^{iky\sin\theta^i}\sin(kz\cos\theta^i) \qquad (7.7\text{--}13)$$

$$cB_x = 2e^{iky\sin\theta^i}\cos(kz\cos\theta^i) \qquad (7.7\text{--}14)$$

These represent waves traveling in the y direction, that is, parallel to the screen, with the phase velocity

$$v_y = \frac{c}{\sin\theta^i} \qquad (7.7\text{--}15)$$

and standing waves in the $-z$ direction, that is, perpendicular to the screen, with the wavelength

$$\lambda_z = \frac{\lambda}{\cos\theta^i} \qquad (7.7\text{--}16)$$

where $\lambda = 2\pi/k = c/f$ is the free-space wavelength. This behavior is just like that for the infinite screen.

7.8 THE SURFACE CURRENT ON THE CONDUCTING HALF-PLANE: NORMAL INCIDENCE

The surface density of current **K** on the perfectly conducting half-plane is obtained from the boundary condition

$$\mu_0^{-1}(\hat{\mathbf{n}} \times \mathbf{B}) = -\mathbf{K} \tag{7.8-1}$$

On the illuminated front surface $\theta = -3\pi/2$, $\hat{\mathbf{n}} = \hat{\mathbf{z}}$; on the back surface in the shadow $\theta = \pi/2$; $\hat{\mathbf{n}} = -\hat{\mathbf{z}}$. It follows that

$$\mathbf{K}\left(-\frac{3\pi}{2}\right) = -\mu_0^{-1}\hat{\mathbf{z}} \times \mathbf{B}\left(-\frac{3\pi}{2}\right); \quad \mathbf{K}\left(\frac{\pi}{2}\right) = \mu_0^{-1}\hat{\mathbf{z}} \times \mathbf{B}\left(\frac{\pi}{2}\right) \tag{7.8-2}$$

On the infinitely thin perfectly conducting screen only the total current is significant. It is given by

$$\mathbf{K}_t = \mathbf{K}\left(-\frac{3\pi}{2}\right) + \mathbf{K}\left(\frac{\pi}{2}\right) \tag{7.8-3}$$

When E_x (B_x) is parallel to the edge, B_y (B_x) is the tangential component of the magnetic field. It follows that for

$$E_x \text{ polarization: } \quad \mathbf{K}\left(-\frac{3\pi}{2}\right) = \hat{\mathbf{x}}\mu_0^{-1}B_y\left(-\frac{3\pi}{2}\right); \quad \mathbf{K}\left(\frac{\pi}{2}\right) = -\hat{\mathbf{x}}\mu_0^{-1}B_y\left(\frac{\pi}{2}\right) \tag{7.8-4}$$

$$B_x \text{ polarization: } \quad \mathbf{K}\left(-\frac{3\pi}{2}\right) = -\hat{\mathbf{y}}\mu_0^{-1}B_x\left(-\frac{3\pi}{2}\right); \quad \mathbf{K}\left(\frac{\pi}{2}\right) = \hat{\mathbf{y}}\mu_0^{-1}B_x\left(\frac{\pi}{2}\right) \tag{7.8-5}$$

The corresponding total currents are

$$E_x \text{ polarization: } \quad \mathbf{K}_t = \hat{\mathbf{x}}\mu_0^{-1}\left[B_y\left(-\frac{3\pi}{2}\right) - B_y\left(\frac{\pi}{2}\right)\right] \tag{7.8-6}$$

$$B_x \text{ polarization: } \quad \mathbf{K}_t = -\hat{\mathbf{y}}\mu_0^{-1}\left[B_x\left(-\frac{3\pi}{2}\right) - B_x\left(\frac{\pi}{2}\right)\right] \tag{7.8-7}$$

where, for normal incidence ($\theta^i = 0$),

$$cB_y(\theta) = f_1(\rho, \theta) + f_2(\rho, \theta) - \frac{e^{i(k\rho - \pi/4)}}{\sqrt{2\pi k\rho}}\left[\sin\left(\frac{\theta}{2}\right) + \cos\left(\frac{\theta}{2}\right)\right] \tag{7.8-8}$$

$$cB_x(\theta) = f_1(\rho, \theta) + f_2(\rho, \theta) \tag{7.8-9}$$

In practice, infinitely thin, perfectly conducting screens are not available, but a thin sheet of metal is a good approximation. Usually it is thick compared to the skin depth so that the surface currents on the two sides are the distinct and physically separated currents given by (7.8–4) and (7.8–5). With (7.6–11) and (7.6–12) the currents $K = \hat{x}K_x$ with E_x parallel to the edge of the screen are:

$$\zeta_0 K_x\left(-\frac{3\pi}{2}\right) = 2f_1\left(-\frac{3\pi}{2}\right) + \frac{e^{i(k\rho - \pi/4)}}{\sqrt{\pi k\rho}}$$

$$= \sqrt{2}\, e^{-i\pi/4}\left[F\left(\sqrt{\frac{2k\rho}{\pi}}\right) - F(-\infty)\right] + \frac{e^{i(k\rho - \pi/4)}}{\sqrt{\pi k\rho}} \qquad (7.8\text{--}10)$$

$$= \begin{cases} 2 & \text{as } \rho \to \infty \\ \dfrac{e^{i(k\rho - \pi/4)}}{\sqrt{\pi k\rho}} & \text{as } \rho \to 0 \end{cases} \qquad (7.8\text{--}11)$$

$$\zeta_0 K_x\left(\frac{\pi}{2}\right) = -2f_1\left(\frac{\pi}{2}\right) + \frac{e^{i(k\rho - \pi/4)}}{\sqrt{\pi k\rho}}$$

$$= -\sqrt{2}\, e^{-i\pi/4}\left[F\left(-\sqrt{\frac{2k\rho}{\pi}}\right) - F(-\infty)\right] + \frac{e^{i(k\rho - \pi/4)}}{\sqrt{\pi k\rho}} \qquad (7.8\text{--}12)$$

$$= \begin{cases} 0 & \text{as } \rho \to \infty \\ \dfrac{e^{i(k\rho - \pi/4)}}{\sqrt{\pi k\rho}} & \text{as } \rho \to 0 \end{cases} \qquad (7.8\text{--}13)$$

Similarly, the currents $\mathbf{K} = \hat{y}K_y$ when B_x is parallel to the edge of the screen are

$$\zeta_0 K_y\left(-\frac{3\pi}{2}\right) = -2f_1\left(-\frac{3\pi}{2}\right) = -\sqrt{2}\, e^{-i\pi/4}\left[F\left(\sqrt{\frac{2k\rho}{\pi}}\right) - F(-\infty)\right] \qquad (7.8\text{--}14)$$

$$= \begin{cases} -2 & \text{as } \rho \to \infty \\ -1 & \text{as } \rho \to 0 \end{cases} \qquad (7.8\text{--}15)$$

$$\zeta_0 K_y\left(\frac{\pi}{2}\right) = 2f_1\left(\frac{\pi}{2}\right) = \sqrt{2}\, e^{-i\pi/4}\left[F\left(-\sqrt{\frac{2k\rho}{\pi}}\right) - F(-\infty)\right] \qquad (7.8\text{--}16)$$

$$= \begin{cases} 0 & \text{as } \rho \to \infty \\ 1 & \text{as } \rho \to 0 \end{cases} \qquad (7.8\text{--}17)$$

These currents for the two polarizations are quite different. When E_x is parallel to the edge, the surface currents are also parallel to the edge. When B_x is parallel to the edge, the surface currents are perpendicular to the edge, oppositely directed at equal distances from the edge, and equal in magnitude at the edge ($\rho = 0$). This means that the surface current approaches the edge downward on the illuminated side, proceeds around the edge, and then continues upward on the shadowed side. The magnitude of the surface current $\zeta_0 K_y$ is shown in Fig. 7.8–1.

The amplitude $|\zeta_0 K_x|$ when E_x is parallel to the edge has distributions like $|\zeta_0 K_y|$ when B_x is parallel to the edge for that part of the current given by the Fresnel integral terms. This is oppositely directed on the two sides of the screen.

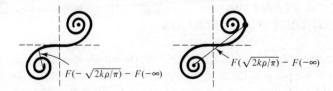

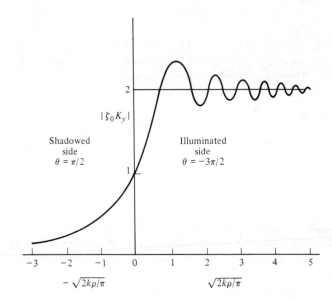

Figure 7.8–1 Surface current $\zeta_0 K_y$ on the two sides of the screen when B_x is parallel to the edge; normal incidence.

$\zeta_0 K_x$ has an additional term, $e^{i(k\rho - \pi/4)}/\sqrt{\pi k\rho}$, which is a radially outward traveling wave on both sides of the screen and originating at the edge. This acts like a line source.

The total currents for the two polarizations are

$$\zeta_0 K_{xt} = \sqrt{2}\, e^{-i\pi/4}\left[F\left(\sqrt{\frac{2k\rho}{\pi}}\right) - F\left(-\sqrt{\frac{2k\rho}{\pi}}\right)\right] + \frac{2e^{i(k\rho - \pi/4)}}{\sqrt{\pi k\rho}} \qquad (7.8\text{–}18)$$

$$= \begin{cases} 2 & \text{as } \rho \to \infty \\[2mm] \dfrac{2e^{i(k\rho - \pi/4)}}{\sqrt{\pi k\rho}} & \text{as } \rho \to 0 \end{cases} \qquad (7.8\text{–}19)$$

$$\zeta_0 K_{yt} = -\sqrt{2}\, e^{-i\pi/4}\left[F\left(\sqrt{\frac{2k\rho}{\pi}}\right) - F\left(-\sqrt{\frac{2k\rho}{\pi}}\right)\right] \qquad (7.8\text{–}20)$$

$$= \begin{cases} -2 & \text{as } \rho \to \infty \\ 0 & \text{as } \rho \to 0 \end{cases} \qquad (7.8\text{–}21)$$

7.9 GENERAL INCIDENCE OF A PLANE WAVE
ON A HALF-PLANE: ARBITRARY POLARIZATION

The solutions obtained in Secs. 7.6 to 7.8 for the field scattered by a conducting half-plane apply specifically to polarizations of the incident field with either the electric or the magnetic field parallel to the edge of the screen. In the actual analysis this coincides with the x axis of a Cartesian system of coordinates. Also, it lies in a wave front or surface of constant phase. This means that the plane of incidence, which is perpendicular to the wave fronts and contains the vector wave number $\mathbf{k}$, is the yz plane; $\mathbf{k} = \hat{\mathbf{y}}k_y + \hat{\mathbf{z}}k_z$.

The solution when the electric field is $\mathbf{E} = \hat{\mathbf{x}}E_x = \mathbf{E}_\perp$—where the subscript $\perp$ refers to orientation with respect to the plane of incidence—includes the magnetic field $\mathbf{B}_\parallel = \hat{\mathbf{y}}B_y - \hat{\mathbf{z}}B_z$. The associated complex Poynting vector is

$$\mathbf{S} = \frac{1}{2\mu_0}\,\mathbf{E} \times \mathbf{B}^* = \frac{1}{2\mu_0}\,\mathbf{E}_\perp \times \mathbf{B}_\parallel^* = \frac{1}{2\mu_0}\,(\hat{\mathbf{z}}E_xB_y^* + \hat{\mathbf{y}}E_xB_z^*) \qquad (7.9\text{-}1)$$

The solution when the magnetic field is $\mathbf{B} = \hat{\mathbf{x}}B_x = \mathbf{B}_\perp$ includes the electric field $\mathbf{E}_\parallel = -\hat{\mathbf{y}}E_y + \hat{\mathbf{z}}E_z$. The complex Poynting vector is

$$\mathbf{S} = \frac{1}{2\mu_0}\,\mathbf{E} \times \mathbf{B}^* = \frac{1}{2\mu_0}\,\mathbf{E}_\parallel \times \mathbf{B}_\perp = \frac{1}{2\mu_0}\,(\hat{\mathbf{z}}E_yB_x^* - \hat{\mathbf{y}}E_zB_x^*) \qquad (7.9\text{-}2)$$

Within the requirement that the edge of the half-plane lie in a wave front and, hence, perpendicular to the plane of incidence, a general solution for arbitrary angles of incidence and polarization is readily constructed simply as the superposition of the two solutions for $\mathbf{E}_\perp\mathbf{B}_\parallel$ and $\mathbf{E}_\parallel\mathbf{B}_\perp$. This is evident from Fig. 7.9–1, which shows a plane wave characterized by $\mathbf{E} = \mathbf{E}_\perp + \mathbf{E}_\parallel$ and magnetic field $\mathbf{B} = \mathbf{B}_\parallel + \mathbf{B}_\perp$.

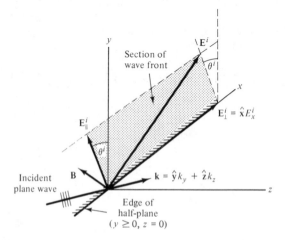

Figure 7.9–1 Arbitrarily incident plane wave.

PROBLEMS

1. An infinite perfectly conducting cylinder of radius a has its axis along the z coordinate of a cylindrical coordinate system, (ρ, θ, z) . A plane electromagnetic wave with the magnetic field parallel to the z axis is incident on the cylinder.

    Given the free-space Green's function for a line source,

$$G = \frac{i}{4} H_0^{(1)}(k|\boldsymbol{\rho} - \boldsymbol{\rho}'|)$$

   and Green's symmetrical theorem, show with all steps that the total magnetic field outside the cylinder is given by

$$B_z(\rho, \theta) = B_0 \left[e^{ik\rho\cos\theta} - \sum_{m=0}^{\infty} \epsilon_m i^m \frac{J_m'(ka)}{H_m^{(1)'}(ka)} H_m^{(1)}(k\rho) \cos m\theta \right]$$

   where ϵ_m is the Neumann number defined by $\epsilon_m = 1, m = 0; \epsilon_m = 2, m \neq 0$. Also show that the surface current on the cylinder is

$$K_\theta(\theta) = \frac{-i2B_0}{\mu_0 \pi ka} \sum_{m=0}^{\infty} \epsilon_m i^m \frac{\cos m\theta}{H_m^{(1)'}(ka)}$$

   and that the *geometrical-optics current* is

$$|K_\theta(\theta)| = 2|H_z^{\text{inc}}(\theta)|$$

   where $H_z^{\text{inc}}(\theta)$ is the value on the surface of the cylinder, $\rho = a$. *Note:* The potential at all points in space due to a unit source is known as the *time-dependent Green's function* and represented by $G(\mathbf{R}, t)$, which is a solution of the equation

$$\nabla^2 G(\mathbf{R}, t) - \frac{1}{c^2} \frac{\partial^2}{\partial t^2} G(\mathbf{R}, t) = -\delta(\mathbf{R})\delta(t)$$

   with $\mathbf{R} = \mathbf{r} - \mathbf{r}'$. The free-space Green's function is given by $G(\mathbf{R}) = e^{ik_0 R}/4\pi R$ for periodic time dependence.

2. When the cross-sectional dimensions of a scattering obstacle are small compared to the wavelength and the point of observation is very far from the obstacle (i.e., for the cylinder, $ka \ll 1, k\rho \gg 1$), the scattering is known as *Rayleigh scattering*.

    Show that for the conducting cylinder, the leading terms in the relevant fields are

$$\frac{E_z(\rho, \theta)}{E_0^i} \doteq \sqrt{\frac{\pi}{2k\rho}} e^{i(k\rho + \pi/4)} \left(\ln \frac{\gamma ka}{2} - \frac{i\pi}{2} \right)^{-1}$$

$$\frac{B_z(\rho, \theta)}{B_0^i} \doteq -\sqrt{\frac{\pi}{2k\rho}} e^{i(k\rho + \pi/4)} (ka)^2 \left(\frac{1}{2} - \cos \theta \right)$$

   respectively, when E_z^{inc} and B_z^{inc} are parallel to the axis. ($\ln \gamma = 0.5772$.)

3. Show that the total scattering cross sections for Rayleigh scattering for the two cases are

$$\frac{\sigma_D}{4a} = \frac{\pi^2}{ka[\pi^2 + 4\ln^2(\gamma ka/2)]}, \qquad \text{when } E_z \text{ is parallel to the axis}$$

$$\frac{\sigma_N}{4a} = \frac{3\pi^2(ka)^3}{16}, \qquad \text{when } B_z \text{ is parallel to the axis}$$

Note that the last formula includes contributions from the $(ka)^4$ term which is not included in the formula for the field in Problem 2. Correspondingly, in terms of the general series representation, contributions come from the $m = 1$ as well as the $m = 0$ terms.

4. The magnetic field near a perfectly conducting half-plane defined by $z = 0$, $y \geq 0$ is given by

$$H_x = f_1(\rho, \theta) + f_2(\rho, \theta)$$

where

$$f_1(\rho, \theta) = e^{ik\rho\cos\theta}\frac{e^{-i\pi/4}}{\sqrt{2}}[F(s) - F(-\infty)]; \qquad s = -2\sqrt{\frac{k\rho}{\pi}}\sin\frac{\theta}{2}$$

$$f_2(\rho, \theta) = e^{-ik\rho\cos\theta}\frac{e^{-i\pi/4}}{\sqrt{2}}|F(t) - F(-\infty)|; \qquad t = -2\sqrt{\frac{k\rho}{\pi}}\cos\frac{\theta}{2}$$

when the incident field is $H_x^{\text{inc}} = e^{ik\rho\cos\theta}$ (Fig. P7–4).

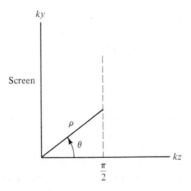

Figure P7–4.

(a) Calculate and plot the magnitude and phase of $H_x = |H_x|e^{i\psi}$ along the plane $kz = \pi/2$. Include at least the first maximum and minimum in the illuminated region and a reasonable distance in the shadow. Note that

$$F(w) = C(w) + iS(w)$$

where $C(w)$ and $S(w)$ are the Fresnel integrals which are tabulated.* Also show $f_1(\rho, \theta)$.

---

* E. Jahnke and F. Emde, *Tables of Functions* (New York: Dover Publications, Inc., 1945).

(b) Determine the surface density of current on both sides of the screen (i.e., at $\theta = -3\pi/2$ and $\theta = \pi/2$) and the total current.

Note: If preferred, this problem may be evaluated by writing a program for the computer. In this case, H_x may be evaluated along several planes corresponding to different distances kz over a wider range of y to include several maxima and minima.

Theory of the Linear Antenna and Antenna Arrays

Current-carrying radiating structures are generally made of metal. They can have many forms, including the spheroid, the bicone, and the thin tubular cylinder. Each of these has characteristic advantages in the application of electromagnetic theory when used as a single isolated element. When two or more are to be combined into directive arrays, the spheroidal or biconical symmetry is violated and the thin tubular cylinder has significant advantages. The integral equation for the current in the thin tubular cylinder is derived in Chapter 3. Its solution is the subject of this chapter.

8.1 EQUATION FOR THE CURRENT

The integral equation for the current in a cylindrical dipole antenna is [(3.10−22) and (3.10−23)]

$$\frac{4\pi}{\mu_0} A_z(z) = \int_{-h}^{h} I_z(z') \frac{e^{-jk_0R}}{R} \, dz'$$

$$= -j\frac{4\pi}{\zeta_0} (C_1 \cos k_0z + \tfrac{1}{2}V_0^e \sin k_0|z|) \qquad (8.1-1)$$

where V_0^e is the emf of the delta-function generator, $\zeta_0 = \sqrt{\mu_0/\epsilon_0} \sim 120\pi$ ohms and $R = [(z - z')^2 + a^2]^{1/2}$. The length of the dipole is $2h$ and its radius is a (Fig. 8.1–1). There is no simple rigorous method of solution for the current $I_z(z)$ and the admittance $I_z(0)/V_0^e$ in (8.1–1). Several approximate methods have been formulated; of these only a few are sufficiently simple and accurate to permit the subsequent evaluation of the electromagnetic field maintained by the current in the antenna. A convenient method involves the replacement of the integral equation (8.1–1) by an algebraic equation that is approximately equivalent to it.

To obtain the algebraic equation for the current, it is necessary to rearrange (8.1–1). The integrand consists of the unknown current $I_z(z)$, which vanishes at $z = \pm h$, is continuous at $z = 0$, and is even in z so that $I_z(-z) = I_z(z)$, and the kernel $K(z, z')$ given by

$$K(z, z') = K_R(z, z') + jK_I(z, z') = \frac{e^{-jk_0R}}{R} \tag{8.1–2}$$

with

$$K_R(z, z') = \frac{\cos k_0R}{R}; \qquad K_I(z, z') = -\frac{\sin k_0R}{R} \tag{8.1–3}$$

The dimensionless quantities $K_R(z, z')/k_0$ and $K_I(z, z')/k_0$ are shown in Fig. 8.1–2 as functions of $k_0|z - z'|$. A comparison of the two functions in the lower figure shows that they behave quite differently. $K_R(z, z')/k_0$ has a sharp high peak at $z = z'$. Its magnitude at this point is equal to $1/k_0a$ and since it is postulated that $k_0a \ll 1$ (a very thin antenna), this magnitude is very large compared with 1. $K_I(z, z')/k_0$, however, varies only slowly with $k_0|z - z'|$ and is always less than 1.

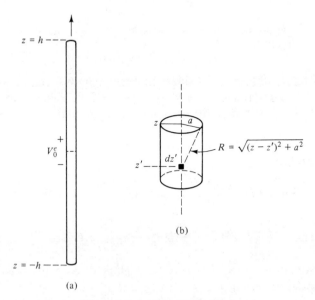

Figure 8.1–1   Cylindrical antenna.

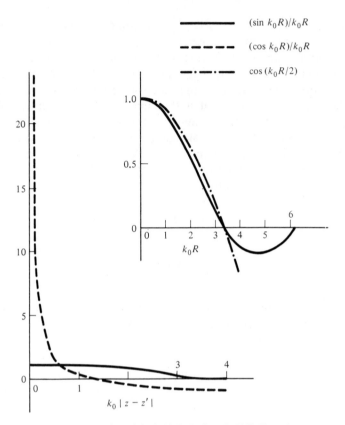

Figure 8.1–2   Functions $(\sin k_0 R)/k_0 R$, $(\cos k_0 R)/k_0 R$, and $\cos (k_0 R/2)$.

The upper figure in Fig. 8.1–2 shows that $(\sin k_0 R)/k_0 R$ can be well approximated by $\cos (k_0 R/2)$ in the range $0 \le k_0|z - z'| \le 5\pi/4$; the value of $\cos (k_0 R/2)$ is hardly affected if the small quantity $k_0 a$ is neglected and $k_0 R$ is approximated by $k_0|z - z'|$.

The following approximations may, therefore, be written for the two parts of the integral in (8.1–1):

$$J_R(h, z) = \int_{-h}^{h} I_z(z') \frac{\cos k_0 R}{R} \, dz' = \Psi_1(z) I_z(z) \doteq \Psi_1 I_z(z) \tag{8.1–4}$$

$$J_I(h, z) = -\int_{-h}^{h} I_z(z') \frac{\sin k_0 R}{R} \, dz' = -k_0 \int_{-h}^{h} I_z(z') \cos \frac{k_0(z - z')}{2} \, dz' \tag{8.1–5}$$

The approximation in (8.1–4) may be explained as follows. The kernel is quite small except at or very near $z = z'$, where it rises to a very large value. It is clear, therefore, that the current near $z' = z$ is significant primarily in determining the value of the integral at z. Hence the integral is approximately proportional to $I(z)$.

The proportionality constant is Ψ_1 and is to be determined where $I_z(z)$ is a maximum.

The integral in (8.1–5) may be transformed as follows:

$$J_I(h, z) = -k_0 \int_{-h}^{h} I_z(z') \cos \frac{k_0(z - z')}{2} \, dz'$$

$$= -k_0 \int_0^h I_z(z') \left[\cos \frac{k_0(z - z')}{2} + \cos \frac{k_0(z + z')}{2} \right] dz'$$

$$= -2k_0 \cos \frac{k_0 z}{2} \int_0^h I_z(z') \cos \frac{k_0 z'}{2} \, dz'$$

For antennas for which $k_0 h \leq 5\pi/4$,

$$J_I(h, z) \doteq J_I(h, 0) \cos \frac{k_0 z}{2}$$

$$J_I(h, 0) = -2k_0 \int_0^h I_z(z') \cos \frac{k_0 z'}{2} \, dz' \tag{8.1–6}$$

The approximation in (8.1–4) may be further refined using the fact that while the integral on the left becomes quite small at the ends of the antenna where $z = \pm h$, the right-hand side vanishes identically at these points since $I_z(\pm h) = 0$. Therefore, a better approximation than (8.1–4) is the following:

$$J_R(h, z) - J_R(h, h) = \int_{-h}^{h} I_z(z')[K_R(z, z') - K_R(h, z')] \, dz'$$

$$\doteq \Psi_2 I_z(z) \tag{8.1–7}$$

where Ψ_2 is a new constant.

8.2 MODIFIED EQUATION FOR THE CURRENT

The integral equation (8.1–1) must now be modified so that (8.1–7) may be used. When $(4\pi/\mu_0)A_z(h)$ is subtracted from both sides of (8.1–1), the result is

$$\frac{4\pi}{\mu_0} [A_z(z) - A_z(h)] = \int_{-h}^{h} I_z(z') K_d(z, z') \, dz'$$

$$= -j \frac{4\pi}{\zeta_0} [C_1 \cos k_0 z + \tfrac{1}{2} V_0^e \sin k_0 |z| + U] \tag{8.2–1}$$

where

$$U = \frac{-j\zeta_0}{4\pi} \int_{-h}^{h} I_z(z') K(h, z') \, dz' \tag{8.2–2}$$

and the difference kernel is

$$K_d(z, z') = K(z, z') - K(h, z') \qquad (8.2\text{-}3)$$

The constant C_1 in (8.2–1) can now be expressed in terms of U and V_0^e by setting $z = h$. The left side of (8.2–1) vanishes and the right side yields

$$C_1 = \frac{-(\frac{1}{2}V_0^e \sin k_0 h + U)}{\cos k_0 h} \qquad (8.2\text{-}4)$$

The substitution of this value of C_1 in (8.2–1) yields the following modified integral equation for the current:

$$\int_{-h}^{h} I_z(z')K_d(z, z')\,dz'$$

$$= \frac{j4\pi}{\zeta_0 \cos k_0 h} \left[\frac{1}{2}V_0^e \sin k_0(h - |z|) + U(\cos k_0 z - \cos k_0 h)\right] \qquad (8.2\text{-}5)$$

This equation together with (8.1–6) and (8.1–7) can be reduced to an algebraic form to facilitate the solution.

8.3 REDUCTION OF THE INTEGRAL EQUATION TO AN EQUIVALENT ALGEBRAIC EQUATION AND ITS SOLUTION

With the use of (8.1–6) and (8.1–7) the integral in (8.2–5) may be approximated as follows:

$$\int_{-h}^{h} I_z(z')K_d(z, z')\,dz' \doteq I_z(z)\Psi_2 + jJ_l(h, 0)\left(\cos \frac{k_0 z}{2} - \cos \frac{k_0 h}{2}\right) \qquad (8.3\text{-}1)$$

The substitution of (8.3–1) into (8.2–5) gives

$$I_z(z)\Psi_2 = -jJ_l(h, 0)\left(\cos \frac{k_0 z}{2} - \cos \frac{k_0 h}{2}\right)$$

$$+ j\frac{4\pi}{\zeta_0 \cos k_0 h}\left[\frac{1}{2}V_0^e \sin k_0(h - |z|) + U(\cos k_0 z - \cos k_0 h)\right]$$

Evidently a zero-order form of the current is given by

$$I_z(z) \doteq [I_z(z)]_0$$

$$= I_V\left[\sin k_0(h - |z|) + T_U(\cos k_0 z - \cos k_0 h) + T_D\left(\cos \frac{k_0 z}{2} - \cos \frac{k_0 h}{2}\right)\right] \qquad (8.3\text{-}2)$$

The complex coefficients are

$$I_V = \frac{j2\pi V_0^e}{\Psi_2 \zeta_0 \cos k_0 h} \tag{8.3-3a}$$

$$T_U = \frac{2U}{V_0^e} \tag{8.3-3b}$$

$$T_D = \frac{jJ_1(h, 0)}{\Psi_2 I_V} \tag{8.3-3c}$$

Equation (8.3–2) suggests that an approximation of the current consists of the superposition of three terms each of which represents a different distribution. The first term is a simple sinusoid which is the component of the current maintained directly by the generator; it does not take account of the effects of coupling between the different parts of the antenna or of radiation. The currents induced by the interaction between charges moving in widely separated sections of the antenna are given by the second and third terms in (8.3–2). The second term, which is the shifted cosine, is the current maintained by that part of the interaction that is equivalent to a constant field acting in phase at all points along the antenna. The third term, the shifted cosine with half-angle arguments, is the correction that takes account of the phase lag introduced by a retarded interaction instead of an instantaneous one.

Conventionally, only the first term, the sinusoidal distribution, is assumed to be an adequate approximation of the current in a thin cylindrical antenna. The three-term expression greatly improves the approximation since it takes account of the very important contributions from coupling among the parts of the antenna and from radiation.

The coefficients Ψ_2, $J_1(h, 0)$, and U can all be evaluated. Actually, it is sufficient to determine only the new coefficients I_V, T_U, and T_D defined in (8.3–3a–c). When (8.3–2) is substituted into (8.2–5), the following integrals involve the real part of the difference kernel $K_{dR}(z, z')$. Their approximate evaluation is accomplished with (8.1–7):

$$\int_{-h}^{h} \sin k_0(h - |z'|)K_{dR}(z, z')\, dz' \doteq \Psi_{dR} \sin k_0(h - |z|) \tag{8.3-4a}$$

$$\int_{-h}^{h} (\cos k_0 z' - \cos k_0 h)K_{dR}(z, z')\, dz' \doteq \Psi_{dUR}(\cos k_0 z - \cos k_0 h) \tag{8.3-4b}$$

$$\int_{-h}^{h} \left(\cos \frac{k_0 z'}{2} - \cos \frac{k_0 h}{2}\right) K_{dR}(z, z')\, dz' \doteq \Psi_{dDR}\left(\cos \frac{k_0 z}{2} - \cos \frac{k_0 h}{2}\right) \tag{8.3-4c}$$

The constants Ψ_{dR}, Ψ_{dUR}, and Ψ_{dDR} are evaluated later.

The integrals obtained with the imaginary part $K_{dI}(z, z')$ of the difference kernel are approximated by applying (8.1–6). The resulting expressions are

$$\int_{-h}^{h} \sin k_0(h - |z'|) K_{dI}(z, z')\, dz' \doteq \Psi_{dI}\left(\cos\frac{k_0 z}{2} - \cos\frac{k_0 h}{2}\right) \qquad (8.3-5a)$$

$$\int_{-h}^{h} (\cos k_0 z' - \cos k_0 h) K_{dI}(z, z')\, dz' \doteq \Psi_{dUI}\left(\cos\frac{k_0 z}{2} - \cos\frac{k_0 h}{2}\right) \qquad (8.3-5b)$$

$$\int_{-h}^{h} \left(\cos\frac{k_0 z'}{2} - \cos\frac{k_0 h}{2}\right) K_{dI}(z, z')\, dz' \doteq \Psi_{dDI}\left(\cos\frac{k_0 z}{2} - \cos\frac{k_0 h}{2}\right) \qquad (8.3-5c)$$

The three constants Ψ_{dI}, Ψ_{dUI}, and Ψ_{dDI} are evaluated later. When (8.3–2) is substituted in (8.2–2), the result is

$$U = \frac{-j\zeta_0 I_V}{4\pi}[\Psi_V(h) + T_U\Psi_U(h) + T_D\Psi_D(h)] \qquad (8.3-6)$$

where

$$\Psi_V(h) = \int_{-h}^{h} \sin k_0(h - |z'|) K(h, z')\, dz' \qquad (8.3-7a)$$

$$\Psi_U(h) = \int_{-h}^{h} (\cos k_0 z' - \cos k_0 h) K(h, z')\, dz' \qquad (8.3-7b)$$

$$\Psi_D(h) = \int_{-h}^{h} \left(\cos\frac{k_0 z'}{2} - \cos\frac{k_0 h}{2}\right) K(h, z')\, dz' \qquad (8.3-7c)$$

After (8.3–4a–c) and (8.3–5a–c) are used to reduce the integral on the left of (8.2–5) to a sum of terms, the entire integral equation is replaced by an algebraic equation consisting of three distributions $\sin k_0(h - |z|)$, $\cos (k_0 z) - \cos (k_0 h)$, and $\cos (k_0 z/2) - \cos (k_0 h/2)$. This algebraic equation is

$$\left(I_V\Psi_{dR} - j\frac{2\pi V_0^e}{\zeta_0 \cos k_0 h}\right)\sin k_0(h - |z|)$$

$$+ \left(I_V T_U \Psi_{dUR} - \frac{j4\pi U}{\zeta_0 \cos k_0 h}\right)(\cos k_0 z - \cos k_0 h)$$

$$+ I_V(j\Psi_{dI} + j\Psi_{dUI}T_U + \Psi_{dD}T_D)\left(\cos\frac{k_0 z}{2} - \cos\frac{k_0 h}{2}\right) = 0 \qquad (8.3-8)$$

where $\Psi_{dD} = \Psi_{dDR} + j\Psi_{dDI}$.

Equation (8.3–8) is satisfied for all values of z when the coefficient of each of the distributions vanishes. This requires that

$$I_V = \frac{j2\pi V_0^e}{\zeta_0 \Psi_{dR} \cos k_0 h} \qquad (8.3-9a)$$

$$T_U[\Psi_{dUR}\cos k_0 h - \Psi_U(h)] - T_D\Psi_D(h) = \Psi_V(h) \tag{8.3-9b}$$

$$T_U\Psi_{dUI} - jT_D\Psi_{dD} = -\Psi_{dI} \tag{8.3-9c}$$

Equations (8.3–9b,c) are readily solved for T_U and T_D, with the result

$$T_U = Q^{-1}[\Psi_V(h)\Psi_{dD} - j\Psi_D(h)\Psi_{dI}] \tag{8.3-10a}$$

$$T_D = -jQ^{-1}\{\Psi_{dI}[\Psi_{dUR}\cos k_0 h - \Psi_U(h)] + \Psi_V(h)\Psi_{dUI}\} \tag{8.3-10b}$$

$$Q = \Psi_{dD}[\Psi_{dUR}\cos k_0 h - \Psi_U(h)] + j\Psi_D(h)\Psi_{dUI} \tag{8.3-11}$$

The Ψ functions in (8.3–10a,b) and (8.3–11) are defined at the value of z associated with the maximum of the current distribution function using (8.3–4a–c) and (8.3–5a–c). In the range of interest for this analysis, $k_0 h \le 5\pi/4$. The maximum of $\sin k_0(h - |z|)$ is at $z = 0$ when $k_0 h \le \pi/2$ but at $z = h - \lambda/4$ when $k_0 h \ge \pi/2$. The maxima of $(\cos k_0 z - \cos k_0 h)$ and $\cos(k_0 z/2) - \cos(k_0 h/2)$ are at $z = 0$. Hence the following definitions are appropriate:

$$\Psi_{dR} = \Psi_{dR}(z_m) \begin{cases} z_m = 0, & k_0 h \le \dfrac{\pi}{2} \\[2mm] z_m = h - \dfrac{\lambda}{4}, & k_0 h > \dfrac{\pi}{2} \end{cases} \tag{8.3-12}$$

$$\Psi_{dR}(z) = \csc k_0(h - |z|) \int_{-h}^{h} \sin k_0(h - |z'|)$$
$$\times [K_R(z, z') - K_R(h, z')]\, dz' \tag{8.3-13}$$

$$\Psi_{dUR} = (1 - \cos k_0 h)^{-1} \int_{-h}^{h} (\cos k_0 z' - \cos k_0 h)$$
$$\times [K_R(0, z') - K_R(h, z')\, dz' \tag{8.3-14}$$

$$\Psi_{dD} = \left(1 - \cos\frac{k_0 h}{2}\right)^{-1} \int_{-h}^{h} \left(\cos\frac{k_0 z'}{2} - \cos\frac{k_0 h}{2}\right)$$
$$\times [K(0, z') - K(h, z')]\, dz' \tag{8.3-15}$$

$$\Psi_{dI} = \left(1 - \cos\frac{k_0 h}{2}\right)^{-1} \int_{-h}^{h} \sin k_0(h - |z'|)$$
$$\times [K_I(0, z') - K_I(h, z')]\, dz' \tag{8.3-16}$$

$$\Psi_{dUI} = \left(1 - \cos\frac{k_0 h}{2}\right)^{-1} \int_{-h}^{h} (\cos k_0 z' - \cos k_0 h)$$
$$\times [K_I(0, z') - K_I(h, z')]\, dz' \tag{8.3-17}$$

These integrals are readily evaluated for any particular antenna with the help of the computer. They may also be expressed in terms of the following integrals:

$$C_a(h, z) = \int_{-h}^{h} \cos k_0 z' \frac{e^{-jk_0R}}{R} dz' \tag{8.3-18a}$$

$$S_a(h, z) = \int_{-h}^{h} \sin k_0|z'| \frac{e^{-jk_0R}}{R} dz' \tag{8.3-18b}$$

$$E_a(h, z) = \int_{-h}^{h} \frac{e^{-jk_0R}}{R} dz' \tag{8.3-18c}$$

Each of these can be expanded in terms of the generalized sine, cosine, and exponential integral functions which are tabulated.

The approximation for the current in an isolated cylindrical antenna with $k_0h \le 5\pi/4$ and $k_0a \ll 1$ is given by

$$I_z(z) = \frac{j2\pi V_0^e}{\zeta_0 \Psi_{dR} \cos k_0 h} \left[\sin k_0(h - |z|) + T_U(\cos k_0 z - \cos k_0 h) \right.$$
$$\left. + T_D \left(\cos \frac{k_0 z}{2} - \cos \frac{k_0 h}{2} \right) \right] \tag{8.3-19}$$

The associated driving-point admittance evaluated at $z = 0$ is

$$Y_0 = \frac{j2\pi}{\zeta_0 \Psi_{dR} \cos k_0 h} \left[\sin k_0 h + T_U(1 - \cos k_0 h) \right.$$
$$\left. + T_D \left(1 - \cos \frac{k_0 h}{2} \right) \right] \tag{8.3-20}$$

Figure 8.3-1 shows the current distribution for a full-wave antenna ($k_0 h = \pi$).

When $k_0 h = \pi/2$, (8.3-19) and (8.3-20) are indeterminate; however, they can be rearranged to yield

$$I_z(z) = \frac{-j2\pi V_0^e}{\zeta_0 \Psi_{dR}} \left[\sin k_0 |z| - \sin k_0 h + T'_U(\cos k_0 z - \cos k_0 h) \right.$$
$$\left. - T'_D \left(\cos \frac{k_0 z}{2} - \cos \frac{k_0 h}{2} \right) \right] \tag{8.3-21}$$

$$Y_0 = \frac{j2\pi}{\zeta_0 \Psi_{dR}} \left[\sin k_0 h - T'_U(1 - \cos k_0 h) + T'_D \left(1 - \cos \frac{k_0 h}{2} \right) \right] \tag{8.3-22}$$

with

$$T'_U = -\frac{T_U + \sin k_0 h}{\cos k_0 h}; \qquad T'_D = \frac{T_D}{\cos k_0 h} \tag{8.3-23}$$

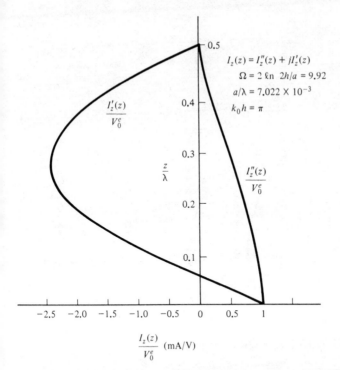

$$I_z(z) = I_z''(z) + jI_z'(z)$$
$$\Omega = 2 \ \text{\r{l}n} \ 2h/a = 9.92$$
$$a/\lambda = 7.022 \times 10^{-3}$$
$$k_0 h = \pi$$

$$\frac{I_z(z)}{V_0^e} \ (\text{mA/V})$$

Figure 8.3–1 Current in upper half of full-wave dipole.

Equations (8.3–21) and (8.3–22) are useful near $k_0 h = \pi/2$. Figure 8.3–2 shows the current distribution for the half-wave antenna.

When the antenna is electrically short ($k_0 h < 1$), the trigonometric functions can be expanded in series and only the leading terms retained. Then

$$I_z(z) = \frac{j2\pi V_0^e}{\zeta_0 \Psi_{dR}} \left[k_0 h \left(1 - \frac{|z|}{h} \right) + \frac{1}{2} k_0^2 h^2 T \left(1 - \frac{z^2}{h^2} \right) \right] \qquad (8.3\text{--}24)$$

This distribution has triangular and parabolic components. The admittance for an electrically short antenna is

$$Y_0 = \frac{j2\pi}{\zeta_0 \Psi_{dR}} \left(k_0 h + \frac{1}{2} k_0^2 h^2 T \right) \qquad (8.3\text{--}25)$$

where $T = T_U + T_D/4$.

The three-term approximation does not yield quite accurate values for the susceptance since the simple trigonometric functions cannot adequately account for the rapid change in current near the driving point when the antenna is not near resonance. However, the introduction of higher-order terms makes the formulation more complicated. The error in the susceptance can be corrected by including a suitable lumped susceptance in a corrective network at the junction. Such a network is necessary in practice to take account of the terminal-zone effects near and at the junction between the feeding line and the antenna.

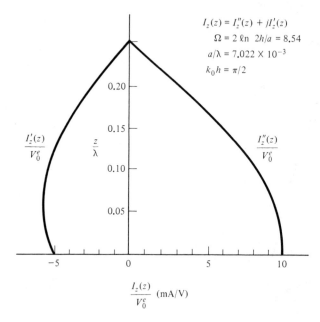

$$I_z(z) = I''_z(z) + jI'_z(z)$$
$$\Omega = 2 \ln \; 2h/a = 8.54$$
$$a/\lambda = 7.022 \times 10^{-3}$$
$$k_0 h = \pi/2$$

$$\frac{I_z(z)}{V^e_0} \;\; (\text{mA/V})$$

Figure 8.3–2 Current in upper half of half-wave dipole.

8.4 AN APPROXIMATE TWO-TERM THEORY

For most purposes, a somewhat simpler two-term approximation of the current is useful. The difference between the functions $F_{0z} = \cos k_0 z - \cos k_0 h$ and $H_{0z} = \cos(k_0 z/2) - \cos(k_0 h/2)$ is small and the formulation is simplified by consolidating the two terms. If F_{0z} is substituted everywhere for H_{0z}, the expression for the current in (8.3–19) is approximated as follows when $k_0 h \leq 5\pi/4$:

$$I_z(z) = \frac{j2\pi V^e_0}{\zeta_0 \Psi_{dR} \cos k_0 h} [\sin k_0(h - |z|) + T(\cos k_0 z - \cos k_0 h)] \qquad (8.4–1)$$

where $T = T_U + T_D$ and T_U, T_D are defined in (8.3–10a,b). The function T then has the form

$$T = \frac{\Psi_V(h) - j\Psi_{dI} \cos k_0 h}{\Psi_{dU} \cos k_0 h - \Psi_U(h)} \qquad (8.4–2)$$

The substitutions $\Psi_{dD} = \Psi_{dU}$ and $\Psi_D(h) = \Psi_U(h)$ are used since $F_{0z} \doteq H_{0z}$.

Near $k_0 h = \pi/2$ the approximate expression for the current is

$$I_z(z) = \frac{-j2\pi V^e_0}{\zeta_0 \Psi_{dR}} [\sin k_0 |z| - \sin k_0 h + T'(\cos k_0 z - \cos k_0 h)] \qquad (8.4–3)$$

where

$$T' = -\frac{T + \sin k_0 h}{\cos k_0 h} \qquad (8.4–4)$$

Since $\Psi_U(h) = \Psi_V(h) = C_a(h, h)$, T' reduces simply to

$$T' = \frac{\Psi_{dU} - j\Psi_{dI} - S_a(\lambda/4, \lambda/4) + E_a(\lambda/4, \lambda/4)}{C_a(\lambda/4, \lambda/4)} \qquad (8.4\text{--}5)$$

for $k_0 h = \pi/2$. The functions $C_a(h, h)$, $S_a(h, h)$ and $E_a(h, h)$ are defined in (8.3–18a,b,c).

As with the three-term theory, the quadrature component of the current near the driving point is not adequately represented by simple trigonometric functions. The same correction used in the three-term approximation must be used here to obtain accurate quantitative agreement with measured values of the susceptance.

8.5 THE COMPLETE FIELD OF MULTIPLE HALF-WAVE ANTENNAS IN CONFOCAL COORDINATES

The electromagnetic field of a thin antenna at all distances may be expressed in an illuminating form at half-lengths h given by

$$h = n\frac{\lambda_0}{4}; \qquad k_0 h = \frac{n\pi}{2}, \qquad n \text{ odd} \qquad (8.5\text{--}1)$$

The distribution of current is well approximated by

$$I_z = I_0 \cos k_0 z \qquad (8.5\text{--}2)$$

as shown in Fig. 8.5–1. The associated vector potential is

$$\mathbf{A} = \hat{z}A_z = \hat{z}\,\frac{\mu_0}{4\pi} \int_{-h}^{h} I_z(z') \frac{e^{-jk_0 R}}{R}\, dz' \qquad (8.5\text{--}3)$$

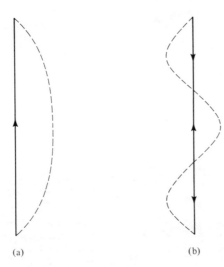

(a)

(b)

Figure 8.5–1 Current distribution on a thin center-driven antenna: (a) $n = 1$; (b) $n = 3$.

In the cylindrical coordinates (ρ, θ, z) the distances from a point z' on the upper and lower halves of a center-driven antenna to an arbitrary point $P(x, y, z)$ are given by

$$R \equiv R_1 = [(z - z')^2 + \rho^2]^{1/2} \tag{8.5-4a}$$

$$R_2 = [(z + z')^2 + \rho^2]^{1/2} \tag{8.5-4b}$$

and the distances from P to the upper and lower ends of the antenna and to its center are given by

$$R_{1h} = [(h - z)^2 + \rho^2]^{1/2} \tag{8.5-5a}$$

$$R_{2h} = [(h + z)^2 + \rho^2]^{1/2} \tag{8.5-5b}$$

$$R_0 = (z^2 + \rho^2)^{1/2} \tag{8.5-5c}$$

These distances are shown in Fig. 8.5-2. The expression for A_z may be written in the following equivalent form:

$$A_z = \frac{\mu_0}{4\pi} \int_0^h I_0 \cos k_0 z' \left(\frac{e^{-jk_0R_1}}{R_1} + \frac{e^{-jk_0R_2}}{R_2} \right) dz' \tag{8.5-6}$$

The magnetic field is readily evaluated from the vector potential with the relation

$$\mathbf{B} = \nabla \times \mathbf{A} \tag{8.5-7}$$

Since $A_\rho = A_\theta = 0$, it follows that the components B_ρ and B_z are zero and B_θ is given by

$$B_\theta = \frac{-\partial A_z}{\partial \rho} \tag{8.5-8}$$

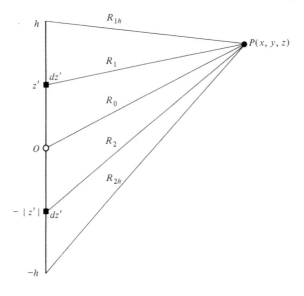

Figure 8.5-2 Distances to the field point $P(x, y, z)$ from the center-driven antenna.

The differentiation of (8.5–6) according to (8.5–8) gives

$$B_\theta = \frac{-\mu_0 I_0}{4\pi} \frac{\partial}{\partial \rho} \left[\int_0^h \cos k_0 z' \left(\frac{e^{-jk_0R_1}}{R_1} + \frac{e^{-jk_0R_2}}{R_2} \right) dz' \right] \tag{8.5-9}$$

The integral in (8.5–9) is readily evaluated (Appendix VI) and the magnetic field is

$$B_\rho = 0 \tag{8.5-10a}$$

$$B_\theta = \frac{j\mu_0 I_0}{4\pi\rho} \left(e^{-jk_0R_{1h}} + e^{-jk_0R_{2h}} \right) \tag{8.5-10b}$$

$$B_z = 0 \tag{8.5-10c}$$

The electric field at points in space where all currents vanish can be evaluated from the Maxwell equation:

$$\mathbf{E} = \frac{-j\omega}{k_0^2} \nabla \times \mathbf{B} \tag{8.5-11}$$

Since

$$\mathbf{B} = \hat{\boldsymbol{\theta}} B_\theta; \qquad B_\rho = 0; \qquad B_z = 0 \tag{8.5-12}$$

$$(\nabla \times \mathbf{B})_\rho = -\frac{1}{\rho} \frac{\partial}{\partial z} (\rho B_\theta) \tag{8.5-13a}$$

$$(\nabla \times \mathbf{B})_\theta = 0 \tag{8.5-13b}$$

$$(\nabla \times \mathbf{B})_z = \frac{1}{\rho} \frac{\partial}{\partial \rho} (\rho B_\theta) \tag{8.5-13c}$$

From (8.5–10b)

$$\rho B_\theta = \frac{j\mu_0 I_0}{4\pi} \left(e^{-jk_0R_{1h}} + e^{-jk_0R_{2h}} \right) \tag{8.5-14}$$

In carrying out the differentiations required in (8.5–13a,c) it is to be noted that

$$\frac{\partial R_{1h}}{\partial z} = \frac{z - h}{R_{1h}}; \qquad \frac{\partial R_{2h}}{\partial z} = \frac{z + h}{R_{2h}} \tag{8.5-15}$$

and

$$\frac{\partial R_{1h}}{\partial \rho} = \frac{\rho}{R_{1h}}; \qquad \frac{\partial R_{2h}}{\partial \rho} = \frac{\rho}{R_{2h}} \tag{8.5-16}$$

With (8.5–15) and (8.5–16)

$$\frac{\partial}{\partial z} (\rho B_\theta) = \frac{\mu_0 k_0 I_0}{4\pi} \left(\frac{z - h}{R_{1h}} e^{-jk_0R_{1h}} + \frac{z + h}{R_{2h}} e^{-jk_0R_{2h}} \right) \tag{8.5-17}$$

$$\frac{\partial}{\partial \rho}(\rho B_\theta) = \frac{\mu_0 k_0 I_0 \rho}{4\pi}\left(\frac{e^{-jk_0 R_{1h}}}{R_{1h}} + \frac{e^{-jk_0 R_{2h}}}{R_{2h}}\right) \tag{8.5-18}$$

(8.5–13a–c), (8.5–17), and (8.5–18) may now be used to form the cylindrical components of the electric field at all points in space:

$$E_\rho = \frac{jI_0\zeta_0}{4\pi\rho}\left(\frac{z-h}{R_{1h}}e^{-jk_0 R_{1h}} + \frac{z+h}{R_{2h}}e^{-jk_0 R_{2h}}\right) \tag{8.5-19a}$$

$$E_\theta = 0 \tag{8.5-19b}$$

$$E_z = \frac{-jI_0\zeta_0}{4\pi}\left(\frac{e^{-jk_0 R_{1h}}}{R_{1h}} + \frac{e^{-jk_0 R_{2h}}}{R_{2h}}\right) \tag{8.5-19c}$$

where $\zeta_0 = (\mu_0/\epsilon_0)^{1/2}$.

An infinitely thin antenna of half-length

$$h = \frac{n\lambda_0}{4}; \qquad k_0 h = \frac{n\pi}{2}, \qquad n \text{ even} \tag{8.5-20}$$

with a distribution of current

$$I_z = I_0 \sin k_0 z \tag{8.5-21}$$

as shown in Fig. 8.5–3 also oscillates in a natural mode. Such an antenna cannot, however, be center-driven, so it has to be driven asymmetrically if driven by a single generator. If two properly phased generators are used, a symmetrical distribution is possible. This distribution is equivalent to that in two antennas each of half-length $h = n\lambda_0/4$, n odd, placed end to end. The field of each half of such a combination is given by (8.5–10a–c) and (8.5–19a–c) with an origin at the center of each half and with $2h$ the length of each half. If the origin is shifted to the center

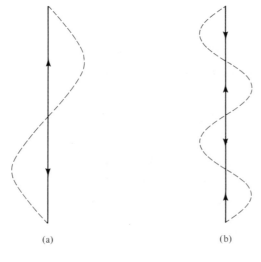

(a)   (b)

Figure 8.5–3 Current distribution on an antenna driven with equal and opposite voltages located $\lambda/4$ from the ends: (a) $n = 2$; (b) $n = 4$.

of the entire structure and $2h$ is its full length, the field of the upper half is given by (8.5–10a–c) and (8.5–19a–c) with R_0 written for R_{2h}. The current is reversed in the lower half and its field is given by (8.5–10a–c) and (8.5–19a–c) with a negative sign and R_0 written for R_{1h}. If the two fields are combined, the resultant field for $h = n\lambda_0/4$, n even, is

$$B_\rho = 0 \tag{8.5–22a}$$

$$B_\theta = \frac{j\mu_0 I_0}{4\pi\rho}\left(e^{-jk_0R_{1h}} - e^{-jk_0R_{2h}}\right) \tag{8.5–22b}$$

$$B_z = 0 \tag{8.5–22c}$$

$$E_\rho = \frac{jI_0\zeta_0}{4\pi\rho}\left(\frac{z-h}{R_{1h}}e^{-jk_0R_{1h}} - \frac{z+h}{R_{2h}}e^{-jk_0R_{2h}}\right) \tag{8.5–23a}$$

$$E_\theta = 0 \tag{8.5–23b}$$

$$E_z = \frac{-jI_0\zeta_0}{4\pi}\left(\frac{e^{-jk_0R_{1h}}}{R_{1h}} - \frac{e^{-jk_0R_{2h}}}{R_{2h}}\right) \tag{8.5–23c}$$

Since (8.5–22a–c) and (8.5–23a–c) differ from (8.5–10a–c) and (8.5–19a–c) only in a negative sign, the two sets of fields may be treated together.

The time-dependent values of the components of the electromagnetic field given in (8.5–10a–c), (8.5–19a–c), (8.5–22a–c), and (8.5–23a–c) are obtained by multiplying both sides by $e^{j\omega t}$ and taking the real part, assuming I_0 to be real. The solution corresponds to a time dependence

$$I_0(t) = I_0 \cos \omega t \tag{8.5–24}$$

The time-dependent values are

$$B_\rho(t) = 0 \tag{8.5–25a}$$

$$B_\theta(t) = \frac{-\mu_0 I_0}{4\pi\rho}\left[\sin(\omega t - k_0R_{1h}) \pm \sin(\omega t - k_0R_{2h})\right] \tag{8.5–25b}$$

$$B_z(t) = 0 \tag{8.5–25c}$$

$$E_\rho(t) = \frac{-I_0\zeta_0}{4\pi\rho}\left[\frac{z-h}{R_{1h}}\sin(\omega t - k_0R_{1h}) \pm \frac{z+h}{R_{2h}}\sin(\omega t - k_0R_{2h})\right] \tag{8.5–26a}$$

$$E_\theta(t) = 0 \tag{8.5–26b}$$

$$E_z(t) = \frac{I_0\zeta_0}{4\pi}\left[\frac{1}{R_{1h}}\sin(\omega t - k_0R_{1h}) \pm \frac{1}{R_{2h}}\sin(\omega t - k_0R_{2h})\right] \tag{8.5–26c}$$

The upper sign is for n odd, the lower sign for n even in $h = n\lambda_0/4$.

The complete electromagnetic field of the dipole is advantageously expressed in prolate spheroidal instead of cylindrical coordinates so that it can be described in terms of mathematical surfaces of simple geometric shape on all points of which one component of the electric or magnetic field has the same phase. Such a surface is called an *equiphase surface*. When $h = n\lambda_0/4$ the equiphase surfaces are spheroids with the ends of the antenna as foci. The components E_ρ and E_z along the radial and axial cylindrical coordinates ρ and z must be replaced by the components E_e and E_h, which are tangent, respectively, to a prolate spheroid and the normal hyperboloid as shown in Fig. 8.5–4. The cylindrical component B_θ in the direction $\hat{\boldsymbol{\theta}}$ around the z axis becomes the spheroidal component B_Φ around the same axis. The prolate spheroidal coordinate system k_h, k_e, and Φ is defined as follows:

$$k_e = \frac{R_{2h} + R_{1h}}{2h}; \qquad k_h = \frac{R_{2h} - R_{1h}}{2h}; \qquad \Phi = \tan^{-1}\frac{y}{z} \qquad (8.5-27)$$

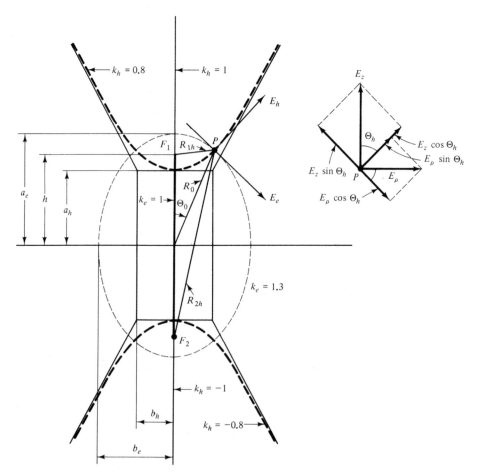

Figure 8.5–4 Spheroidal coordinates k_e and k_h.

where R_{1h} and R_{2h} [defined in (8.5–5a,b)] are the distances from the point $P(k_e,$ k_h, Φ) to the ends of the dipole, at the foci of the spheroids. The length of the antenna, and hence the distance between the foci, is $2h$. The major and minor axes of a typical spheroid are a_e and b_e; the corresponding distances for the orthogonal hyperboloids are a_h and b_h. The coordinates k_e and k_h are related to these axes as follows:

$$k_e = \frac{a_e}{h}; \qquad k_h = \frac{a_h}{h} \tag{8.5–28}$$

The distances are shown in Fig. 8.5–4 for a spheroid with $k_e = 1.3$ and a pair of hyperbolas with $k_h = \pm 0.8$. The axis of the antenna from $z = -h$ to $z = h$ is defined by $k_e = 1$ and the rest of the z axis is defined by $k_h = 1$ and $k_h = -1$. The equatorial plane corresponds to $k_h = 0$. Thus k_e extends from the degenerate straight-line spheroid $k_e = 1$ at the antenna to spheres of infinite radius at $k_e = \infty$. The associated family of hyperboloids of two sheets extends from the equatorial plane where $k_h = 0$ to the degenerate straight-line hyperboloids at $k_h = \pm 1$. The cylindrical coordinates (ρ, z) and the spheroidal coordinates (k_e, k_h) are related as follows:

$$\rho = h[(k_e^2 - 1)(1 - k_h^2)]^{1/2}; \qquad z = hk_ek_h \tag{8.5–29a}$$

The angle Θ_h between the directions of E_z along the z axis and E_h tangent to a hyperboloid is obtained from the equation of the spheroid:

$$\frac{\rho^2}{b_e^2} + \frac{z^2}{a_e^2} = 1 \tag{8.5–29b}$$

It is given by

$$\frac{\partial z}{\partial \rho} = -\tan\Theta_h = \frac{-a_e^2\rho}{b_e^2 z} \tag{8.5–30}$$

Since

$$a_e^2 = b_e^2 + h^2 \tag{8.5–31}$$

it follows that

$$\tan\Theta_h = \frac{k_e(1 - k_h^2)^{1/2}}{k_h(k_e^2 - 1)^{1/2}} \tag{8.5–32a}$$

$$\sin\Theta_h = \frac{k_e(1 - k_h^2)^{1/2}}{(k_e^2 - k_h^2)^{1/2}} \tag{8.5–32b}$$

and

$$\cos\Theta_h = \frac{k_h(k_e^2 - 1)^{1/2}}{(k_e^2 - k_h^2)^{1/2}} \tag{8.5–32c}$$

where Θ_h is the angle between the z axis and the normal to the spheroid at the point P.

The components of the electric field in spheroidal coordinates are

$$E_h = E_z \cos \Theta_h + E_\rho \sin \Theta_h$$

$$= \frac{E_z k_h (k_e^2 - 1)^{1/2} + E_\rho k_e (1 - k_h^2)^{1/2}}{(k_e^2 - k_h^2)^{1/2}} \qquad (8.5\text{--}33a)$$

$$E_e = E_\rho \cos \Theta_h - E_z \sin \Theta_h$$

$$= \frac{E_\rho k_h (k_e^2 - 1)^{1/2} - E_z k_e (1 - k_h^2)^{1/2}}{(k_e^2 - k_h^2)^{1/2}} \qquad (8.5\text{--}33b)$$

The following relations are useful in expressing the complete electromagnetic field in spheroidal coordinates:

$$R_{1h} = a_e - a_h = h(k_e - k_h); \qquad R_{2h} = a_e + a_h = h(k_e + k_h) \qquad (8.5\text{--}34a)$$

$$R_0 \equiv r = (z^2 + \rho^2)^{1/2} = (a_e^2 + a_h^2 - h^2)^{1/2} = h(k_e^2 + k_h^2 - 1)^{1/2} \qquad (8.5\text{--}34b)$$

$$\frac{z - h}{R_{1h}} = \frac{k_e k_h - 1}{k_e - k_h} = \frac{(k_e k_h - 1)(k_e + k_h)}{k_e^2 - k_h^2} \qquad (8.5\text{--}35a)$$

$$\frac{z + h}{R_{2h}} = \frac{k_e k_h + 1}{k_e + k_h} = \frac{(k_e k_h + 1)(k_e - k_h)}{k_e^2 - k_h^2} \qquad (8.5\text{--}35b)$$

$$\frac{z}{r} = \frac{k_e k_h}{(k_e^2 + k_h^2 - 1)^{1/2}} \qquad (8.5\text{--}35c)$$

$$\rho = r \sin \Theta_0 = h[(k_e^2 - 1)(1 - k_h^2)]^{1/2} \qquad (8.5\text{--}35d)$$

Equations (8.5–34a,b) and (8.5–35a–d) are used in (8.5–25a–c) and (8.5–26a–c) to express the complete electromagnetic field of a resonant antenna in spheroidal coordinates:

$$B_\Phi(t) = \frac{-\mu_0 I_0}{2\pi h} \frac{\cos (n\pi k_h/2) \sin (\omega t - n\pi k_e/2)}{[(k_e^2 - 1)(1 - k_h^2)]^{1/2}} \qquad (8.5\text{--}36)$$

$$E_e(t) = \frac{-\zeta_0 I_0}{2\pi h} \frac{\cos (n\pi k_h/2) \sin (\omega t - n\pi k_e/2)}{[(k_e^2 - k_h^2)(1 - k_h^2)]^{1/2}} \qquad (8.5\text{--}37)$$

$$E_h(t) = \frac{\zeta_0 I_0}{2\pi h} \frac{\sin (n\pi k_h/2) \cos (\omega t - n\pi k_e/2)}{[(k_e^2 - k_h^2)(k_e^2 - 1)]^{1/2}} \qquad (8.5\text{--}38)$$

From (8.5–36) to (8.5–38), the time-dependent values of the three field components may be calculated at any point in space with coordinates (k_e, k_h, Φ). Since there is rotational symmetry about the z axis and symmetry with respect to the equatorial plane $z = 0$, it is sufficient to consider the field in one quadrant. The spheroids, k_e = constant, and the orthogonal hyperboloids, k_h = constant, are shown in Fig. 8.5–5 in the range $1 \le k_e \le 2.8$ and $0 \le k_h \le 1$. The upper half of

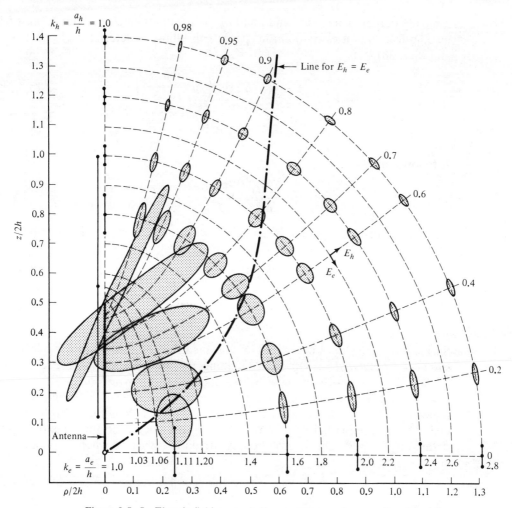

Figure 8.5–5 Electric field near a half-wave antenna of zero radius; $h = \lambda/4$.

the antenna extends along the z axis from $z = 0$ to $z = h$ and coincides with the degenerate spheroid $k_e = 1$ since it is infinitely thin.

The amplitudes and phases of the components of the field may be obtained from (8.5–36) to (8.5–38). The equiphase surfaces are all spheroids with the ends of the antenna as foci. On each spheroid drawn around these two foci, the complete electromagnetic field consists of three mutually perpendicular components, two of the electric field and one of the magnetic field. The component E_e is tangent to an equiphase spheroid. The component E_h is perpendicular to this spheroid and tangent to an orthogonal hyperboloid. The magnetic field B_Φ is tangent to circles around the axis through the antenna and hence tangent to the same spheroid as E_e but perpendicular to it. E_e and B_Φ are always in phase. The component E_h lags E_e at the same point in space by a quarter period in time. Since E_e and E_h are mutually perpendicular, the end point of the resultant electric vector **E** must trace an elliptical path at every point in space. This ellipse is determined from (8.5–37) and (8.5–38) to be

$$E_e^2(t) = E_e^2 \sin^2\left(\omega t - \frac{n\pi k_e}{2}\right) \qquad (8.5\text{–}39a)$$

$$E_h^2(t) = E_h^2 \cos^2\left(\omega t - \frac{n\pi k_e}{2}\right) \qquad (8.5\text{–}39b)$$

It follows that

$$\frac{E_e^2(t)}{E_e^2} + \frac{E_h^2(t)}{E_h^2} = 1 \qquad (8.5\text{–}40)$$

This represents the equation of an ellipse with semiprincipal axes given by

$$E_e = \frac{\zeta_0 I_0}{2\pi h} \frac{\cos\,(n\pi k_h/2)}{[(k_e^2 - k_h^2)(1 - k_h^2)]^{1/2}} \qquad (8.5\text{–}41a)$$

$$E_h = \frac{\zeta_0 I_0}{2\pi h} \frac{\sin\,(n\pi k_h/2)}{[(k_e^2 - k_h^2)(k_e^2 - 1)]^{1/2}} \qquad (8.5\text{–}41b)$$

Such ellipses are shown in Fig. 8.5–5. It is seen that the electric vector is in general *elliptically polarized* in a plane containing the antenna; the magnetic vector is linearly polarized at right angles to this plane.

The ellipse traced by the end of the electric vector degenerates into a straight line when either E_e or E_h vanishes. The former occurs along the z axis beyond the ends of the antenna defined by $k_h = 1$, the latter occurs in the equatorial plane defined by $k_h = 0$. The electric field is circularly polarized when $E_e = E_h$ or

$$\tan \frac{\pi k_h}{2} = \left(\frac{k_e^2 - 1}{1 - k_h^2}\right)^{1/2} \qquad (8.5\text{–}42)$$

The line of circular polarization is shown as a dashed-line in Fig. 8.5–5.

The amplitudes of the magnetic field and the two components of the electric field as a function of the distance from the antenna ($k_e = 1$) are determined by the size of the spheroid on which the field is evaluated. The amplitude factors for B_Φ, E_e, and E_h are, respectively,

$$[(k_e^2 - 1)(1 - k_h^2)]^{-1/2}; \qquad [(k_e^2 - k_h^2)(1 - k_h^2)]^{-1/2}; \qquad [(k_e^2 - k_h^2)(k_e^2 - 1)]^{-1/2}$$

The coordinate k_h defines points along a given spheroid. Since $k_h \leq 1$, B_Φ and E_e decrease with distance essentially as $(k_e^2 - 1)^{-1/2}$ or as $1/k_e$ when $k_e^2 \gg 1$. E_h decreases as $(k_e^2 - 1)^{-1}$ or as $1/k_e^2$ when $k_e^2 \gg 1$. Thus B_Φ and E_e approach inverse first-power laws, and E_h approaches an inverse-square law with increasing distance.

At all points on a given spheroid defined by $k_e =$ constant, the phase ($\omega t - k_e \pi/2$) varies periodically in time, but at every instant its value is the same for all points on the spheroid. A surface that is to remain in the same phase at all times is defined by

$$\omega t - k a_e = \omega t - \frac{k_e \pi}{2} = \text{constant} \tag{8.5-43}$$

Since the time is not constant, this expression is true only if the spheroidal coordinate k_e or the semimajor axis of the spheroids $a_e = k_e h$ increases in time; that is, the spheroidal equiphase surface or wave front expands in time. The time rate of increase of the semimajor axis is obtained from (8.5–43). The axial phase velocity v_{pa} is

$$v_{pa} = \frac{d a_e}{dt} = \frac{\omega}{k_0} = c \doteq 3 \times 10^8 \text{ m/s} \tag{8.5-44}$$

The spheroid of constant phase expands in such a way that its semimajor axis a_e increases in length with the constant velocity c. In all other directions, the outward velocity is not constant since the eccentricity $1/k_e$ of the spheroid decreases as k_e increases from 1 to infinity. The outward velocity in the equatorial plane is given by the time rate of change of the semiminor axis $b_e = (a_e^2 - h^2)^{1/2}$ with $h = \lambda_0/4$. The phase velocity v_{pb} of the intersection of a spheroid with the equatorial plane is

$$v_{pb} = \frac{d b_e}{dt} = \frac{a_e}{b_e} c = c \left[1 + \left(\frac{\lambda_0}{4 b_e} \right)^2 \right]^{1/2} \tag{8.5-45}$$

Very near the antenna, $v_{pb} \gg c$; as the distance from the antenna is increased, b_e approaches a_e, so that v_{pb} approaches c.

The group velocity v_g is defined by

$$v_g = \frac{d\omega}{d k_0} = v_p - \lambda_0 \frac{d v_p}{d \lambda_0} \tag{8.5-46}$$

$$v_{ga} = v_{pa} = c \tag{8.5-47a}$$

$$v_{gb} = \frac{b_e}{a_e} c = c \left[1 + \left(\frac{\lambda_0}{4 b_e} \right)^2 \right]^{-1/2} \tag{8.5-47b}$$

and

$$v_{pb} v_{gb} = c^2 \tag{8.5-47c}$$

The time-dependent electric field near the antenna at successive intervals of a quarter period is shown in Fig. 8.5–6.

It is important to note that even the electric field very close to an antenna cannot be described in terms of an instantaneous electrostatic field. The field near an antenna is calculated in terms of the charge distributions along the antenna at earlier instants. This cannot be taken into account in the electrostatic field.

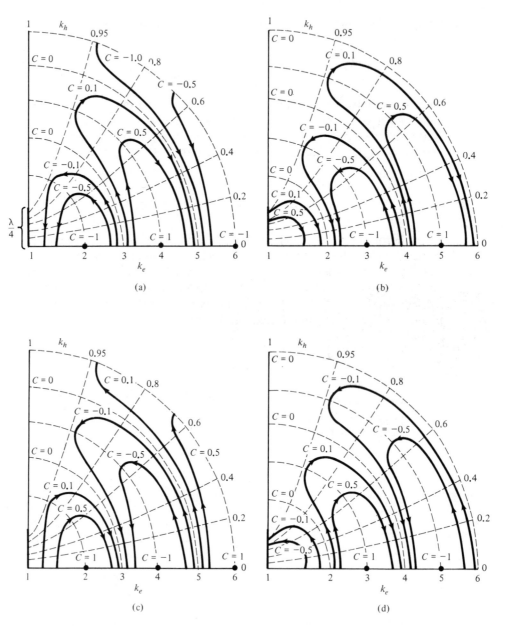

Figure 8.5–6 Instantaneous electric lines near half-wave dipole at successive intervals: (a) $\omega t = 0$, i is maximum up, $q = 0$; (b) $\omega t = \pi/2$, $i = 0$, q is maximum at top; (c) $\omega t = \pi$, i is maximum down, $q = 0$; (d) $\omega t = 3\pi/2$, $i = 0$, $-q$ is maximum at top.

8.6 *THE RADIATION FIELD OF A HALF-WAVE ANTENNA*

At great distances from the half-wave dipole, the spheroidal surfaces of constant phase become spherical, and the orthogonal hyperboloids become radial lines. When

$$R_0^2 \equiv r^2 \gg h^2 \qquad (8.6\text{--}1)$$

it follows that

$$a_e = \frac{R_{2h} + R_{1h}}{2} \doteq r \qquad (8.6\text{--}2a)$$

$$a_h = \frac{R_{2h} - R_{1h}}{2}$$

$$\doteq \tfrac{1}{2} r \left[\left(1 + \frac{2zh}{r^2} \right)^{1/2} - \left(1 - \frac{2zh}{r^2} \right)^{1/2} \right]$$

$$= \frac{hz}{r} \qquad (8.6\text{--}2b)$$

In terms of the spherical coordinates r, Θ, and Φ,

$$a_h \doteq h \cos \Theta; \qquad (1 - k_h^2)^{1/2} = (1 - \cos^2 \Theta)^{1/2} = \sin \Theta \qquad (8.6\text{--}3)$$

With these approximations (8.5–36) to (8.5–38) become

$$B_\Phi^r = \frac{-\mu_0 I_0}{2\pi r} \frac{\cos \left[(\pi/2) \cos \Theta \right]}{\sin \Theta} \sin (\omega t - k_0 r) \qquad (8.6\text{--}4)$$

$$E_\Theta^r = \frac{-\zeta_0 I_0}{2\pi r} \frac{\cos \left[(\pi/2) \cos \Theta \right]}{\sin \Theta} \sin (\omega t - k_0 r) \qquad (8.6\text{--}5)$$

$$E_r^r = \frac{\zeta_0 h I_0}{2\pi r^2} \sin \left[(\pi/2) \cos \Theta \right] \cos (\omega t - k_0 r) \qquad (8.6\text{--}6)$$

where the superscript r denotes the radiation or far zone. The function $F(\Theta) = \cos \left[(\pi/2) \cos \Theta \right] / \sin \Theta$ is shown in Fig. 8.6–1.

The time-dependent Poynting vector $\mathbf{S}(t) = \mu_0^{-1} \mathbf{E}(t) \times \mathbf{B}(t)$ of the antenna in air has the two components

$$S_h(t) = \mu_0^{-1} E_e(t) B_\Phi(t) \qquad (8.6\text{--}7a)$$

$$S_e(t) = -\mu_0^{-1} E_h(t) B_\Phi(t) \qquad (8.6\text{--}7b)$$

The component $S_h(t)$ is directed perpendicularly outward on the surface of any spheroid k_e = constant. The component $S_e(t)$ is directed tangent to the spheroidal

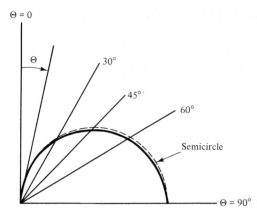

Figure 8.6–1 Vertical field factor $F(\Theta)$ = $\cos\left[(\pi/2)\cos\Theta\right]/\sin\Theta$ (polar plot).

surface.

$$S_h(t) = \frac{\zeta_0 I_0^2}{4\pi^2 h^2}\frac{\cos^2(\pi k_h/2)}{(1 - k_h^2)[(k_e^2 - 1)(k_e^2 - k_h^2)]^{1/2}}\sin^2\left(\omega t - \frac{\pi k_e}{2}\right) \qquad (8.6\text{–}8a)$$

$$S_e(t) = \frac{\zeta_0 I_0^2}{16\pi^2 h^2}\frac{\sin^2(\pi k_h/2)}{(k_e^2 - 1)[(1 - k_h^2)(k_e^2 - k_h^2)]^{1/2}}\sin^2\left(\omega t - \frac{\pi k_e}{2}\right) \qquad (8.6\text{–}8b)$$

The time-dependent power transferred across any spheroidal surface k_e = constant is given by

$$T_e(t) = \int_\Sigma S_h(\mathbf{r}, t)\, d\Sigma \qquad (8.6\text{–}9)$$

where the element of surface on the spheroid is

$$d\Sigma = h^2[(k_e^2 - k_h^2)(k_e^2 - 1)]^{1/2}\, dk_h\, d\Phi \qquad (8.6\text{–}10)$$

It can be shown that

$$T_e(t) = \frac{\zeta_0 I_0^2}{4\pi}\sin^2\left(\omega t - \frac{\pi k_e}{2}\right)\text{Cin }2\pi \qquad (8.6\text{–}11)$$

where

$$\text{Cin }2\pi = \int_0^{2\pi}\frac{1 - \cos x}{x}\, dx = 2.438 \qquad (8.6\text{–}12)$$

The time-average value is independent of k_e. It is

$$\langle T_e(t)\rangle = \frac{\zeta_0 I_0^2}{8\pi}\text{Cin }2\pi = \tfrac{1}{2}I_0^2 R_0^e \qquad (8.6\text{–}13)$$

with R_0^e = 30 Cin 2π = 73.1 Ω.

8.7 *THE RADIATION FIELD OF AN ANTENNA USING THE THREE-TERM CURRENT APPROXIMATION*

The electric field in the radiation zone of an antenna with the distribution of current $I_z(z)$ is shown in Chapter 5 to be

$$\mathbf{E}^r = \hat{\mathbf{\Theta}} E_\Theta^r; \qquad E_\Theta^r = \frac{j\omega\mu}{4\pi} \frac{e^{-jk_0 R_0}}{R_0} \int_{-h}^{h} I_z(z') e^{jk_0 z' \cos\Theta} \sin\Theta \, dz' \qquad (8.7\text{–}1)$$

The three-term approximation for the current is given in (8.3–19) as

$$I_z(z) = \frac{j2\pi V_0^e}{\zeta_0 \Psi_{dR} \cos k_0 h} \left[\sin k_0(h - |z|) + T_U(\cos k_0 z - \cos k_0 h) \right.$$

$$\left. + T_D\left(\cos \frac{k_0 z}{2} - \cos \frac{k_0 h}{2} \right) \right] \qquad (8.7\text{–}2)$$

and the field maintained by this current distribution is obtained when (8.7–2) is substituted into (8.7–1). The result may be expressed as follows:

$$E_\Theta^r = \frac{-V_0^e}{\Psi_{dR}} \frac{e^{-jk_0 R_0}}{R_0} f(\Theta, k_0 h) \qquad (8.7\text{–}3a)$$

where

$$f(\Theta, k_0 h) = [F_m(\Theta, k_0 h) + T_U G_m(\Theta, k_0 h)$$

$$+ T_D D_m(\Theta, k_0 h)] \sec k_0 h \qquad (8.7\text{–}3b)$$

The field functions are

$$F_m(\Theta, k_0 h) = \frac{k_0}{2} \int_{-h}^{h} \sin k_0(h - |z'|) e^{jk_0 z' \cos\Theta} \sin\Theta \, dz'$$

$$= \frac{\cos(k_0 h \cos\Theta) - \cos k_0 h}{\sin\Theta} \qquad (8.7\text{–}4)$$

$$G_m(\Theta, k_0 h) = \frac{k_0}{2} \int_{-h}^{h} (\cos k_0 z' - \cos k_0 h) e^{jk_0 z' \cos\Theta} \sin\Theta \, dz'$$

$$= \frac{\sin k_0 h \cos(k_0 h \cos\Theta) \cos\Theta - \cos k_0 h \sin(k_0 h \cos\Theta)}{\sin\Theta \cos\Theta} \qquad (8.7\text{–}5)$$

$$D_m(\Theta, k_0 h) = \frac{k_0}{2} \int_{-h}^{h} \left(\cos \frac{k_0 z'}{2} - \cos \frac{k_0 h}{2} \right) e^{jk_0 z' \cos\Theta} \sin\Theta \, dz'$$

$$= \left[\frac{2\cos(k_0 h \cos\Theta)\sin(k_0 h/2) - 4\sin(k_0 h \cos\Theta)\cos(k_0 h/2)\cos\Theta}{1 - 4\cos^2\Theta} \right.$$

$$\left. - \frac{\sin(k_0 h \cos\Theta)\cos(k_0 h/2)}{\cos\Theta} \sin\Theta \right] \qquad (8.7\text{–}6)$$

When k_0h is at and near $\pi/2$, the alternative expression for the current is

$$
I_z(z) = \frac{-j2\pi V_0^e}{\zeta_0 \Psi_{dR}} \left[(\sin k_0|z| - \sin k_0h) + T_U' (\cos k_0z - \cos k_0h) \right.
$$

$$
\left. - T_D' \left(\cos \frac{k_0z}{2} - \cos \frac{k_0h}{2} \right) \right] \qquad (8.7-7)
$$

and the far field is given by

$$
E_\Theta^r = \frac{V_0^e}{\Psi_{dR}} \frac{e^{-jk_0R_0}}{R_0} f'(\Theta, k_0h) \qquad (8.7-8a)
$$

where

$$
f'(\Theta, k_0h) = H_m(\Theta, k_0h) + T_U'G_m(\Theta, k_0h) - T_D'D_m(\Theta, k_0h) \qquad (8.7-8b)
$$

The new field function is

$$
H_m(\Theta, k_0h) = \frac{k_0}{2} \int_{-h}^{h} (\sin k_0|z'| - \sin k_0h)e^{jk_0z'\cos\Theta} \sin \Theta \, dz'
$$

$$
= \frac{[1 - \cos k_0h \cos (k_0h \cos \Theta)]\cos \Theta - \sin k_0h \sin (k_0h \cos \Theta)}{\sin \Theta \cos \Theta} \qquad (8.7-9)
$$

$G_m(\Theta, k_0h)$ and $D_m(\Theta, k_0h)$ are as in (8.7–5) and (8.7–6).

In the formulas in (8.7–3a) and (8.7–8a) the field is referred to the driving voltage V_0^e. It can also be referred to the current $I_z(0)$ at the driving point with the substitutions of $I_z(0)/Y_0$ for V_0^e. Y_0 is the driving-point admittance given by (8.3–20). The field in (8.7–3a) is then expressed as follows:

$$
E_\Theta^r = \frac{j\zeta_0 I_z(0)}{2\pi} \frac{e^{-jk_0R_0}}{R_0} f_I(\Theta, k_0h) \qquad (8.7-10)
$$

where

$$
f_I(\Theta, k_0h) = \frac{F_m(\Theta, k_0h) + T_U G_m(\Theta, k_0h) + T_D D_m(\Theta, k_0h)}{\sin k_0h + T_U(1 - \cos k_0h) + T_D[1 - \cos (k_0h/2)]} \qquad (8.7-11)
$$

When k_0h is at or near $\pi/2$, the alternative form of the field becomes

$$
E_\Theta^r = \frac{j\zeta_0 I_z(0)}{2\pi} \frac{e^{-jk_0R_0}}{R_0} f_I'(\Theta, k_0h) \qquad (8.7-12)
$$

where

$$
f_I'(\Theta, k_0h) = - \frac{H_m(\Theta, k_0h) + T_U'G_m(\Theta, k_0h) - T_D'D_m(\Theta, k_0h)}{\sin k_0h - T_U'(1 - \cos k_0h) + T_D'[1 - \cos (k_0h/2)]}
$$

$$
(8.7-13)
$$

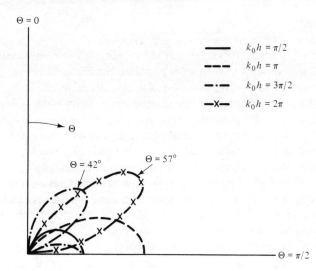

$\Theta = 0$

——— $k_0 h = \pi/2$
- - - $k_0 h = \pi$
-·- $k_0 h = 3\pi/2$
-X- $k_0 h = 2\pi$

Θ

$\Theta = 42°$ $\Theta = 57°$

$\Theta = \pi/2$

Figure 8.7–1 Vertical field factor
$F_m(\Theta, k_0 h) =$
$$\dfrac{\cos\,(k_0 h\,\cos\,\Theta)\,-\,\cos\,k_0 h}{\sin\,\Theta}$$

For a sufficiently thin antenna near resonance a usually adequate approximation of the far field is obtained with the sinusoidal term alone in the current distribution. In this case the electric field in the far zone is given by

$$E_\Theta^r = \frac{j\zeta_0 I_z(0)}{2\pi}\,\frac{e^{-jk_0 R_0}}{R_0}\,\frac{F_m(\Theta, k_0 h)}{\sin k_0 h} \tag{8.7-14}$$

with

$$F_m(\Theta, k_0 h) = \frac{\cos\,(k_0 h\,\cos\,\Theta)\,-\,\cos\,k_0 h}{\sin\,\Theta} \tag{8.7-15}$$

When $k_0 h = \pi/2$, (8.7–14) reduces to (8.6–5). The function $F_m(\Theta, k_0 h)$ is called the *vertical field factor* and is shown graphically in Fig. 8.7–1 for different values of $k_0 h$.

8.8 ANTENNA ARRAYS

The combination of antennas into arrays with a variety of possible special properties constitutes a major concern of antenna engineering. As an application of electromagnetic theory it is adequate to treat the basic aspects of coupled radiating structures, such as distributions of current, self- and mutual impedances, and directional properties in terms of the array of two elements. A thorough understanding of these provides a foundation for arrays of many elements. These can be treated approximately for some applications by assuming rather than deriving the relevant distributions of current. The more accurate approach is appropriate in an application of electromagnetic principles.

Two Coupled Antennas

The electromagnetic field of a single dipole described in Sec. 8.5 is rotationally symmetric and hence nondirectional in the equatorial plane. Such a field is useful, for example, in omnidirectional radio transmission. On the other hand, it is an inefficient field for point-to-point communication, for which a fairly narrow unidirectional beam would be better. Transmitting and receiving systems which have directional properties in the equatorial plane consist of arrays of two or more dipole antennas. The electromagnetic field maintained at a distant point by the array is a superposition of the fields due to the currents in the individual elements. With appropriate relative locations and excitations, the resultant field may be increased in certain directions, decreased in others. In this manner, directive systems can be obtained for various applications. If the elements of an array are close enough, there is necessarily an interaction among the two or more currents. The current in any one element in the array is then a function not only of the length, radius, and driving voltage of that element itself, but also of the relative locations, distributions, directions, amplitudes, and phases of the currents in all the other elements of the array.

The more important properties of arrays can be learned from a study of the very simple array consisting of two parallel nonstaggered elements shown in Fig. 8.8–1. The two antennas are parallel to the z axis and are assumed to be identical. The currents in the two elements are I_{1z} and I_{2z}. The currents at the driving point are I_{10} and I_{20}. V_{10} and V_{20} are the driving voltages, which can be expressed in terms of the currents at the driving points as follows:

$$V_{10} = I_{10}Z_{s1} + I_{20}Z_{12} \tag{8.8–1}$$

$$V_{20} = I_{10}Z_{21} + I_{20}Z_{s2} \tag{8.8–2}$$

Z_{s1} is the input self-impedance of antenna 1 in the presence of antenna 2. It is to

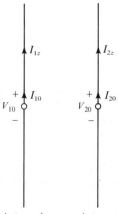

Antenna 1   Antenna 2   **Figure 8.8–1** Two coupled antennas.

be noted that Z_{s1} is different from the input impedance of antenna 1 when it is isolated. Z_{s2} is the input self-impedance of antenna 2 in the presence of antenna 1. Z_{12} is the input mutual impedance of antenna 1 with respect to antenna 2, Z_{21} is the input mutual impedance of antenna 2 with respect to antenna 1. All the self- and mutual impedances are referred to the currents at the driving points, I_{10} and I_{20}. In homogeneous isotropic media, the reciprocal theorem can be applied to show that in general

$$Z_{12} = Z_{21} \tag{8.8-3}$$

Since (8.8–1) and (8.8–2) have the same form as the equations for two coupled circuits, the principle of superposition can be used to determine separately the complex currents due to each applied voltage or to any two pairs of applied voltages. The results are then combined algebraically.

In some practical applications, one antenna is driven and the other is a parasitic reflector or director ($V_{20} = 0$). In this case (8.8–1) and (8.8–2) reduce to

$$V_{10} = I_{10}Z_{s1} + I_{20}Z_{12} \tag{8.8-4}$$

$$0 = I_{10}Z_{21} + I_{20}Z_{s2} \tag{8.8-5}$$

In order to obtain the desired phase and amplitude for the current in the parasitic element, it is possible to connect an impedance Z_L (which may be a tuning reactance $Z_L = jX_L$) at the driving point of the second antenna. In this case, $V_{20} = -I_{20}Z_L$ and (8.8–2) becomes

$$0 = I_{10}Z_{21} + I_{20}(Z_{s2} + Z_L) \tag{8.8-6}$$

When the antennas are extremely thin and are close to a half-wavelength long, it may be an adequate approximation to retain only the sinusoidal term in the distribution of current in the two antennas and to approximate Z_{s1} and Z_{s2} by the self-impedance of the antenna when isolated. The current distribution in general can be expressed in the same form as for the isolated antenna but with different coefficients for the several terms. Furthermore, a study of the simple two-element array provides an insight into the technique of analyzing arrays of more than two elements, such as curtain arrays and circular arrays.

8.9 CURRENTS AND ADMITTANCES FOR THE TWO-ELEMENT ARRAY

The two-element array can be analyzed accurately in terms of two special cases: (1) the zero phase sequence (or symmetrical case), in which the two elements are driven by equal voltages in phase, and (2) the first phase sequence (or antisymmetrical case), in which the two driving voltages are equal but 180° out of phase. These phase sequences correspond respectively to the two-element broadside array and the two-element bidirectional end-fire array, in each of which the distribution of currents along the two elements are the same. By superposition, the general case of arbitrary driving voltages may be obtained. Also included in the general formulation are the array with one element parasitic and center-loaded with an arbitrary lumped impedance and the two-element unidirectional end-fire array or couplet.

Figure 8.9–1 shows an array of two identical parallel antennas of length $2h$, radius a, and spacing b. The integral equation for the currents in the two phase sequences can be derived in the same way as for the isolated antenna. It is

$$\int_{-h}^{h} I_z^{(m)}(z')K_d^{(m)}(z, z')\, dz'$$

$$= \frac{j4\pi}{\zeta_0 \cos k_0 h}\left[\tfrac{1}{2}V^{(m)}\sin k_0(h - |z|) + U^{(m)}(\cos k_0 z - \cos k_0 h)\right] \qquad (8.9\text{–}1)$$

where

$$U^{(m)} = \frac{-j\zeta_0}{4\pi}\int_{-h}^{h} I_z(z')K^{(m)}(h, z')\, dz' \qquad (8.9\text{–}2)$$

$$K^{(m)}(z, z') = \frac{e^{-jk_0 R_{11}}}{R_{11}} \pm \frac{e^{-jk_0 R_{12}}}{R_{12}} \qquad (8.9\text{–}3)$$

$$K_d^{(m)}(z, z') = K^{(m)}(z, z') - K^{(m)}(h, z') \qquad (8.9\text{–}4)$$

$m = 0$ and 1 for the two phase sequences and in (8.9–3) the upper sign is for $m = 0$ and the lower is for $m = 1$.

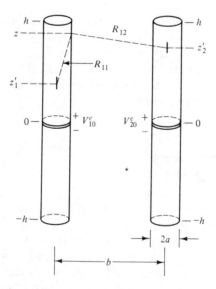

Figure 8.9–1 Two identical parallel antennas.

In most practical applications, the two elements are far enough apart so that $k_0 b \geq 1$ or $b \geq \lambda_0/2\pi$. The current in each of the two coupled elements is then well approximated by the following three-term formula (using the same method as for the isolated antenna):

$$I_z^{(m)}(z) = I_V^{(m)}\left[\sin k_0(h - |z|) + T_U^{(m)}(\cos k_0 z - \cos k_0 h)\right.$$
$$\left. + T_D^{(m)}\left(\cos \frac{k_0 z}{2} - \cos \frac{k_0 h}{2}\right)\right] \qquad (8.9\text{--}5)$$

The functions $I_V^{(m)}$, $T_U^{(m)}$, and $T_D^{(m)}$ are defined as

$$I_V^{(m)} = \frac{j2\pi V_0^{(m)}}{\zeta_0 \Psi_{dR} \cos k_0 h} \qquad (8.9\text{--}6)$$

$$T_U^{(m)} = \frac{\Psi_{dD}^{(m)}[\Psi_V^{(m)}(h) - \Psi_{d\Sigma R}^{(m)} \cos k_0 h] - j\Psi_D^{(m)}(h)\Psi_{dI}^{(m)}}{Q^{(m)}} \qquad (8.9\text{--}7)$$

$$T_D^{(m)} = \frac{-j\{\Psi_{dI}^{(m)}[\Psi_{dUR}^{(m)} \cos k_0 h - \Psi_U^{(m)}(h)] + \Psi_{dUI}^{(m)}[\Psi_V^{(m)}(h) - \Psi_{d\Sigma R}^{(m)} \cos k_0 h]\}}{Q^{(m)}} \qquad (8.9\text{--}8)$$

$$Q^{(m)} = \Psi_{dD}^{(m)}[\Psi_{dUR}^{(m)} \cos k_0 h - \Psi_U^{(m)}(h)] + j\Psi_D^{(m)}(h)\Psi_{dUI}^{(m)} \qquad (8.9\text{--}9)$$

The Ψ functions are defined as follows:

$$\Psi_{dR} = \begin{cases} \Psi_{dR}(0); & k_0 h \leq \dfrac{\pi}{2} \\[2mm] \Psi_{dR}\left(h - \dfrac{\lambda}{4}\right) & k_0 h > \dfrac{\pi}{2} \end{cases} \qquad (8.9\text{--}10)$$

$$\Psi_{dR}(z) = \csc k_0(h - |z|) \int_{-h}^{h} \sin k_0(h - |z'|)[K_R^{(m)}(z,z') - K_R^{(m)}(h,z')] \, dz' \qquad (8.9\text{--}11)$$

where $K^{(m)}(z, z') = K_R^{(m)}(z, z') + jK_I^{(m)}(z, z')$

$$\Psi_{d\Sigma R}^{(m)} = (-1)^{(m)}(1 - \cos k_0 h)^{-1} \int_{-h}^{h} \sin k_0(h - |z'|)$$
$$\times \left(\frac{\cos k_0 R_{12}}{R_{12}} - \frac{\cos k_0 R_{12h}}{R_{12h}}\right) dz' \qquad (8.9\text{--}12)$$

$$\Psi_{dI}^{(m)} = \left(1 - \cos \frac{k_0 h}{2}\right)^{-1} \int_{-h}^{h} \sin k_0(h - |z'|)K_{dI}^{(m)}(0, z') \, dz' \qquad (8.9\text{--}13)$$

$$\Psi_{dUR}^{(m)} = (1 - \cos k_0 h)^{-1} \int_{-h}^{h} (\cos k_0 z' - \cos k_0 h)K_{dR}^{(m)}(0, z') \, dz' \qquad (8.9\text{--}14)$$

$$\Psi_{dUI}^{(m)} = \left(1 - \cos\frac{k_0 h}{2}\right)^{-1} \int_{-h}^{h} (\cos k_0 z' - \cos k_0 h) K_{dI}^{(m)}(0, z')\, dz' \qquad (8.9\text{--}15)$$

$$\Psi_{dD}^{(m)} = \left(1 - \cos\frac{k_0 h}{2}\right)^{-1} \int_{-h}^{h} \left(\cos\frac{k_0 z'}{2} - \cos\frac{k_0 h}{2}\right) K_{d}^{(m)}(0, z')\, dz' \qquad (8.9\text{--}16)$$

$$\Psi_{V}^{(m)}(h) = \int_{-h}^{h} \sin k_0(h - |z'|) K^{(m)}(h, z')\, dz' \qquad (8.9\text{--}17)$$

$$\Psi_{U}^{(m)}(h) = \int_{-h}^{h} (\cos k_0 z' - \cos k_0 h) K^{(m)}(h, z')\, dz' \qquad (8.9\text{--}18)$$

$$\Psi_{D}^{(m)}(h) = \int_{-h}^{h} \left(\cos\frac{k_0 z'}{2} - \cos\frac{k_0 h}{2}\right) K^{(m)}(h, z')\, dz' \qquad (8.9\text{--}19)$$

When $k_0 h$ is at or near $\pi/2$, the current is given by

$$I_z^{(m)}(z) = \frac{-j2\pi V_0^{(m)}}{\zeta_0 \Psi_{dR}} \left[(\sin k_0 |z| - \sin k_0 h) \right.$$

$$\left. + T_U'^{(m)} (\cos k_0 z - \cos k_0 h) - T_D'^{(m)}\left(\cos\frac{k_0 z}{2} - \cos\frac{k_0 h}{2}\right) \right] \qquad (8.9\text{--}20)$$

with

$$T_U'^{(m)} = -\frac{T_U^{(m)} + \sin k_0 h}{\cos k_0 h} \qquad (8.9\text{--}21a)$$

$$T_D'^{(m)} = \frac{T_D^{(m)}}{\cos k_0 h} \qquad (8.9\text{--}21b)$$

The normalized currents in a two-element array with the two-term approximation are shown in Fig. 8.9–2. $I = I'' + jI'$, where I'' is in phase and I' is in phase

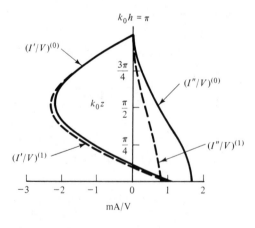

Figure 8.9–2 Zero- and first-phase-sequence currents on two-element array. $\Omega = 2 \ln(2h/a) = 10$, $k_0 b = 1.5$.

quadrature with V_0. The simpler two-term approximation is obtained by the substitution of $(\cos k_0 z - \cos k_0 h)$ for $[\cos (k_0 z/2) - \cos (k_0 h/2)]$.

The currents $I_{1z}(z)$ and $I_{2z}(z)$ in the antennas when they are driven by arbitrary voltages V_{10} and V_{20} are given by

$$I_{1z}(z) = I_z^{(0)}(z) + I_z^{(1)}(z) = V_{10}v(z) + V_{20}w(z) \qquad (8.9\text{-}22a)$$

$$I_{2z}(z) = I_z^{(0)}(z) - I_z^{(1)}(z) = V_{10}w(z) + V_{20}v(z) \qquad (8.9\text{-}22b)$$

where

$$v(z) = \frac{j2\pi}{\zeta_0 \Psi_{dR} \cos k_0 h} \left[\sin k_0(h - |z|) + \frac{1}{2}(T_U^{(0)} + T_U^{(1)})(\cos k_0 z - \cos k_0 h) \right.$$

$$\left. + \frac{1}{2}(T_D^{(0)} + T_D^{(1)}) \left(\cos \frac{k_0 z}{2} - \cos \frac{k_0 h}{2} \right) \right] \qquad (8.9\text{-}23)$$

$$w(z) = \frac{j2\pi}{\zeta_0 \Psi_{dR} \cos k_0 h} \left[\frac{1}{2}(T_U^{(0)} - T_U^{(1)})(\cos k_0 z - \cos k_0 h) \right.$$

$$\left. + \frac{1}{2}(T_D^{(0)} - T_D^{(1)}) \left(\cos \frac{k_0 z}{2} - \cos \frac{k_0 h}{2} \right) \right] \qquad (8.9\text{-}24)$$

When $k_0 h$ is at or near $\pi/2$,

$$v(z) = \frac{-j2\pi}{\zeta_0 \Psi_{dR}} \left[(\sin k_0 |z| - \sin k_0 h) \right.$$

$$+ \frac{1}{2}(T_U'^{(0)} + T_U'^{(1)})(\cos k_0 z - \cos k_0 h)$$

$$\left. - \frac{1}{2}(T_D'^{(0)} + T_D'^{(1)}) \left(\cos \frac{k_0 z}{2} - \cos \frac{k_0 h}{2} \right) \right] \qquad (8.9\text{-}25a)$$

$$w(z) = \frac{-j2\pi}{\zeta_0 \Psi_{dR}} \left[\frac{1}{2}(T_U'^{(0)} - T_U'^{(1)})(\cos k_0 z - \cos k_0 h) \right.$$

$$\left. - \frac{1}{2}(T_D'^{(0)} - T_D'^{(1)}) \left(\cos \frac{k_0 z}{2} - \cos \frac{k_0 h}{2} \right) \right] \qquad (8.9\text{-}25b)$$

The voltages at the driving point are expressed in terms of the currents and impedances in (8.8–1) and (8.8–2). The currents at the driving points can be expressed in terms of the voltages and admittances:

$$I_{10} = I_{1z}(0) = V_{10}Y_{s1} + V_{20}Y_{12} \qquad (8.9\text{-}26a)$$

$$I_{20} = I_{2z}(0) = V_{10}Y_{21} + V_{20}Y_{s2} \qquad (8.9\text{-}26b)$$

where Y_{s1}, Y_{s2} are the self-admittances and Y_{12}, Y_{21} are the mutual admittances.

For two identical antennas,

$$Y_{s1} = Y_{s2} = v(0) \tag{8.9–27}$$

$$Y_{12} = Y_{21} = w(0) \tag{8.9–28}$$

The relations between the admittances and impedances are readily determined. The self- and mutual impedances for a two-element array as a function of the distance b between elements with $k_0 h = \pi$ are shown in Fig. 8.9–3.

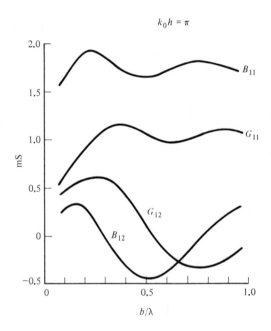

Figure 8.9–3 Self- and mutual admittances of two-element array.

When antenna 2 is parasitic and center-loaded by an arbitrary impedance Z_L, the driving voltage V_{20} is replaced by $-I_{20}Z_L = -I_{20}/Y_L$ and (8.9–26b) becomes

$$I_{20} = V_{10}\left(\frac{Y_{21}}{1 + Y_{s2}Z_L}\right) \tag{8.9–29}$$

It follows from (8.9–22a,b) that

$$I_{1z}(z) = V_{10}\left[v(z) - \left(\frac{Y_{21}}{Y_L + Y_{s2}}\right)w(z)\right] \tag{8.9–30a}$$

$$I_{2z}(z) = V_{10}\left[w(z) - \left(\frac{Y_{21}}{Y_L + Y_{s2}}\right)v(z)\right] \tag{8.9–30b}$$

The driving-point admittance of antenna 1 is

$$Y_{1\,\text{in}} = \frac{I_{10}}{V_{10}} = Y_{s1} - \frac{Y_{21}Y_{12}}{Y_L + Y_{s2}} \qquad (8.9\text{–}31)$$

The driving-point admittances when $Z_L = 0$ and $Z_L = \infty$ can be derived easily. When the parasitic element is tuned to resonance, $Y_L = jB_L$ and $B_L = -B_{s2}$. (Note that $Y_{s2} = G_{s2} + jB_{s2}$.)

The method of phase sequences that has been described for the two-element array can be extended directly to arrays of many elements uniformly spaced around a circle.

8.10 THE FIELD OF A TWO-ELEMENT ARRAY

The radiation field of an array is the vector sum of the fields maintained by the currents in the individual elements. For the two-element array, with the origin of the spherical coordinates R, Θ, Φ midway between the centers of the two elements (Fig. 8.10–1), the fields due to the currents are

$$E^r_{\Theta_1} = -\frac{1}{\Psi_{dR}}\frac{e^{-jk_0R_1}}{R}\left[V_{10}f(\Theta, k_0h) + V_{20}g(\Theta, k_0h)\right] \qquad (8.10\text{–}1)$$

$$E^r_{\Theta_2} = -\frac{1}{\Psi_{dR}}\frac{e^{-jk_0R_2}}{R}\left[V_{10}g(\Theta, k_0h) + V_{20}f(\Theta, k_0h)\right] \qquad (8.10\text{–}2)$$

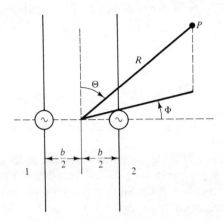

Figure 8.10–1 Two-element array in spherical coordinate system.

where

$$R_1 = R + \frac{b}{2} \cos \Phi \sin \Theta \qquad (8.10\text{–}3\text{a})$$

$$R_2 = R - \frac{b}{2} \cos \Phi \sin \Theta \qquad (8.10\text{–}3\text{b})$$

$$\begin{aligned} f(\Theta, k_0 h) = [F_m(\Theta, k_0 h) + \tfrac{1}{2}(T_U^{(0)} + T_U^{(1)}) G_m(\Theta, k_0 h) \\ + \tfrac{1}{2}(T_D^{(0)} + T_D^{(1)}) D_m(\Theta, k_0 h)] \sec k_0 h \end{aligned} \qquad (8.10\text{–}4)$$

$$\begin{aligned} g(\Theta, k_0 h) = [\tfrac{1}{2}(T_U^{(0)} - T_U^{(1)}) G_m(\Theta, k_0 h) \\ + \tfrac{1}{2}(T_D^{(0)} - T_D^{(1)}) D_m(\Theta, k_0 h)] \sec k_0 h \end{aligned} \qquad (8.10\text{–}5)$$

The field functions F_m, G_m and D_m are defined in (8.7–4) to (8.7–6). When $k_0 h$ is near $\pi/2$, the fields for the currents (8.9–25a,b) are

$$E_{\Theta_1}^r = \frac{1}{\Psi_{dR}} \frac{e^{-jk_0 R_1}}{R} [V_{10} f'(\Theta, k_0 h) + V_{20} g'(\Theta, k_0 h)] \qquad (8.10\text{–}6)$$

$$E_{\Theta_2}^r = \frac{1}{\Psi_{dR}} \frac{e^{-jk_0 R_2}}{R} [V_{10} g'(\Theta, k_0 h) + V_{20} f'(\Theta, k_0 h)] \qquad (8.10\text{–}7)$$

$$\begin{aligned} f'(\Theta, k_0 h) = H_m(\Theta, k_0 h) + \tfrac{1}{2}(T_U'^{(0)} + T_U'^{(1)}) G_m(\Theta, k_0 h) \\ - \tfrac{1}{2}(T_D'^{(0)} + T_D'^{(1)}) D_m(\Theta, k_0 h) \end{aligned} \qquad (8.10\text{–}8)$$

$$\begin{aligned} g'(\Theta, k_0 h) = \tfrac{1}{2}(T_U'^{(0)} - T_U'^{(1)}) G_m(\Theta, k_0 h) \\ - \tfrac{1}{2}(T_D'^{(0)} - T_D'^{(1)}) D_m(\Theta, k_0 h) \end{aligned} \qquad (8.10\text{–}9)$$

$H_m(\Theta)$ is defined in (8.7–9).

The radiation field of an arbitrarily driven two-element array is

$$\begin{aligned} E_\Theta^r &= E_{\Theta_1}^r + E_{\Theta_2}^r \\ &= -\frac{1}{\Psi_{dR}} \frac{e^{-jk_0 R}}{R} \{ [V_{10} f(\Theta, k_0 h) + V_{20} g(\Theta, k_0 h)] e^{-j(k_0 b/2) \cos \Phi \sin \Theta} \\ &\quad + [V_{10} g(\Theta, k_0 h) + V_{20} f(\Theta, k_0 h)] e^{j(k_0 b/2) \cos \Phi \sin \Theta} \} \end{aligned} \qquad (8.10\text{–}10)$$

When $k_0 h$ is near $\pi/2$, $f(\Theta, k_0 h)$ and $g(\Theta, k_0 h)$ are replaced by $-f'(\Theta, k_0 h)$ and $-g'(\Theta, k_0 h)$.

The radiation field of most dipole arrays in which the individual elements are electrically thin and a half-wavelength long or less ($k_0 h \leq \pi/2$) is often quite well approximated by the field due to the leading sinusoidal term alone. This approximation greatly simplifies the calculation. Specifically, the current in each element is approximated by

$$I_z(z) \doteq I_z(0) \frac{\sin k_0(h - |z|)}{\sin k_0 h} \tag{8.10-11}$$

where

$$I_z(0) = V_0 Y_0 = \frac{V_0}{Z_0} \tag{8.10-12}$$

and $Y_0 = 1/Z_0$ is the accurately determined driving-point admittance.

The field in the far zone referred to the driving-point current $I_z(0)$ is

$$E_\Theta^r = \frac{j\zeta_0 I_z(0)}{2\pi} \frac{e^{-jk_0 R}}{R} F_0(\Theta, k_0 h) \tag{8.10-13}$$

where

$$F_0(\Theta, k_0 h) = \frac{F_m(\Theta, k_0 h)}{\sin k_0 h} = \frac{\cos(k_0 h \cos \Theta) - \cos k_0 h}{\sin k_0 h \sin \Theta} \tag{8.10-14}$$

as shown in (8.7–14) and (8.7–15). The far field of an array of two elements is

$$E_\Theta^r = \frac{j\zeta_0}{2\pi} \frac{e^{-jk_0 R}}{R} (I_{10} e^{-j(k_0 b/2)\,\cos\Phi\sin\Theta}$$

$$+ I_{20} e^{j(k_0 b/2)\,\cos\Phi\sin\Theta}) F_0(\Theta, k_0 h) \tag{8.10-15}$$

If $I_{20} = CI_{10}e^{-j\delta}$, where C is an arbitrary amplitude factor and δ is an arbitrary phase angle,

$$E_\Theta^r = \frac{j\zeta_0}{2\pi} \frac{e^{-jk_0 R}}{R} I_{10} (e^{-j(k_0 b/2)\cos\Phi\sin\Theta}$$

$$+ Ce^{j(k_0 b/2)\cos\Phi\sin\Theta - j\delta}) F_0(\Theta, k_0 h) \tag{8.10-16}$$

If the currents in the two antennas have the same amplitude, $C = 1$ and

$$E_\Theta^r = \frac{j\zeta_0}{2\pi} \frac{e^{-jk_0 R}}{R} I_{10} F_0(\Theta, k_0 h) A(\Theta, \Phi) e^{-j\delta/2} \tag{8.10-17}$$

where

$$A(\Theta, \Phi) = 2 \cos \left[\tfrac{1}{2}(\delta - k_0 b \sin \Theta \cos \Phi)\right] \tag{8.10-18}$$

is the array factor.

Thus the vertical characteristic $F_0(\Theta, k_0 h)$ of a single antenna is multiplied by the array factor $A(\Theta, \Phi)$ to give the far field of the two-element array. The horizontal field factor is of particular practical interest. It is obtained from $F_0(\Theta, k_0 h)A(\Theta, \Phi)$ with $\Theta = \pi/2$ [i.e., $F_0(\pi/2, k_0 h)A(\pi/2, \Phi)$]. The vertical field factor for the particular value of Φ for which the horizontal factor is a maximum is important and is given by $F_0(\Theta, k_0 h)A(\Theta, \Phi_m)$.

Some special cases may be derived from the general formulas. When the currents in the two elements are equal in amplitude and in phase ($\delta = 0$) and the elements are spaced a half-wavelength apart ($k_0 b = \pi$), the structure is a *broadside array*. When the elements are a half-wavelength long,

$$F_0\left(\Theta, \frac{\pi}{2}\right)A(\Theta, \Phi) = \frac{\cos\left(\pi/2 \cos\Theta\right)}{\sin\Theta}\, 2\cos\left(\frac{\pi}{2}\sin\Theta\cos\Phi\right) \qquad (8.10\text{–}19)$$

The horizontal factor with $\Theta = \pi/2$ is

$$F_0\left(\frac{\pi}{2}, \frac{\pi}{2}\right)A\left(\frac{\pi}{2}, \Phi\right) = 2\cos\left(\frac{\pi}{2}\cos\Phi\right) \qquad (8.10\text{–}20)$$

and is illustrated in Fig. 8.10–2. It has maxima at $\Phi_m = \pm\pi/2$ and its maximum value is

$$F_0\left(\frac{\pi}{2}, \frac{\pi}{2}\right)A\left(\frac{\pi}{2}, \frac{\pi}{2}\right) = 2 \qquad (8.10\text{–}21)$$

The vertical field factor in the direction of maximum is

$$F_0\left(\Theta, \frac{\pi}{2}\right)A\left(\Theta, \frac{\pi}{2}\right) = 2F_0\left(\Theta, \frac{\pi}{2}\right) = \frac{2\cos\left(\dfrac{\pi}{2}\cos\Theta\right)}{\sin\Theta} \qquad (8.10\text{–}22)$$

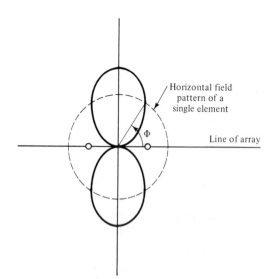

Figure 8.10–2 Horizontal field pattern of two-element broadside array with half-wavelength spacing.

For a broadside array, the vertical factor in the direction Φ_m has the same form as that for a single antenna.

If the currents are equal in amplitude but 180° out of phase ($\delta = \pi$), and $k_0 b$ is maintained at π, the two elements constitute a *bilateral end-fire array*. For $h = \lambda/4$,

$$F_0\left(\Theta, \frac{\pi}{2}\right) A(\Theta, \Phi) = \frac{2 \cos\left(\dfrac{\pi}{2} \cos \Theta\right)}{\sin \Theta} \sin\left(\frac{\pi}{2} \sin \Theta \cos \Phi\right) \qquad (8.10\text{-}23)$$

The horizontal field factor with $\Theta = \pi/2$ is

$$F_0\left(\frac{\pi}{2}, \frac{\pi}{2}\right) A\left(\frac{\pi}{2}, \Phi\right) = 2 \sin\left(\frac{\pi}{2} \cos \Phi\right) \qquad (8.10\text{-}24)$$

This function is shown graphically in Fig. 8.10–3. Its maxima are rotated by an angle $\Phi = 90°$ from the broadside pattern, the phases in the antennas are reversed and the maxima are at $\Phi = 0, \pi$. Again, the vertical field factor in the direction of the maximum horizontal field is the same as for a single element.

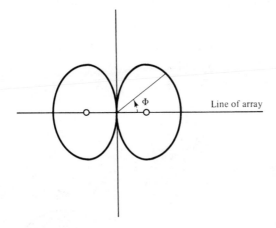

Line of array

Figure 8.10–3 Horizontal field pattern of two-element bilateral end-fire array with half-wavelength spacing.

When the elements are spaced a quarter-wavelength apart ($k_0 b = \pi/2$) and the amplitudes of the currents are equal but 90° out of phase ($\delta = \pi/2$), the structure is an *end-fire array*. For $h = \lambda/4$,

$$F_0\left(\Theta, \frac{\pi}{2}\right) A(\Theta, \Phi)$$

$$= \frac{\cos\left(\dfrac{\pi}{2} \cos \Theta\right)}{\sin \Theta} \left\{ 2 \cos\left[\frac{\pi}{4}(1 - \sin \Theta \cos \Phi)\right] \right\} \qquad (8.10\text{-}25)$$

When $\Theta = \pi/2$, the horizontal pattern is given by

$$F_0\left(\frac{\pi}{2}, \frac{\pi}{2}\right) A\left(\frac{\pi}{2}, \Phi\right) = 2 \cos\left[\frac{\pi}{4}(1 - \cos \Phi)\right] \qquad (8.10\text{–}26)$$

It is shown in Fig. 8.10–4. This is a unidirectional pattern with a single maximum in the direction $\Phi = 0$ with a null in the $\Phi = \pi$ direction. The vertical pattern when $\Phi = 0$ is given by

$$F_0\left(\Theta, \frac{\pi}{2}\right) A\left(\Theta, \frac{\pi}{2}\right) = \frac{\cos\left(\dfrac{\pi}{2} \cos \Theta\right)}{\sin \Theta} \, 2 \cos\left[\frac{\pi}{4}(1 - \sin \Theta)\right] \qquad (8.10\text{–}27)$$

and is shown graphically in Fig. 8.10–5. It differs only slightly from the vertical pattern of a single antenna when $\Phi = 0$ but is quite different in the direction $\Phi = \pi$ where its magnitude is negligible.

The directional properties of the broadside and end-fire arrays can be increased greatly with a larger number of elements. If only the leading sinusoidal term is used for the current distribution and there is a progressive phase difference of δ between the adjacent N elements of the array, the array factor is

$$A(\Theta, \Phi) = \frac{\sin\left[(N/2)(\delta - k_0 b \sin \Theta \cos \Phi)\right]}{N \sin\left[\frac{1}{2}(\delta - k_0 b \sin \Theta \cos \Phi)\right]} \qquad (8.10\text{–}28\text{a})$$

$$= \frac{\sin N\chi}{N \sin \chi} \qquad (8.10\text{–}28\text{b})$$

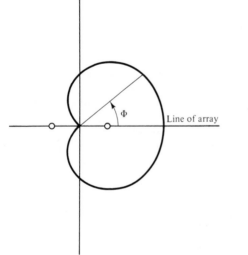

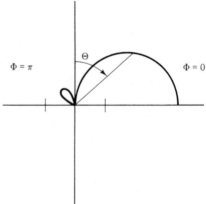

Figure 8.10–4   Horizontal field pattern of two-element end-fire array with quarter-wavelength spacing.

Figure 8.10–5   Vertical field pattern of two-element end-fire array with quarter-wavelength spacing.

with $\chi = \frac{1}{2}(\delta - k_0 b \sin \Theta \cos \Phi)$. Principal maxima of magnitude N occur in directions for which the numerator and denominator of (8.10–28a) vanish. Zeros occur in directions for which only the numerator vanishes. Typical broadside array and end-fire array patterns are shown in Figs. 8.10–6 and 8.10–7. This theory of arrays may be applied to arrays with elements that are arranged collinearly, in a circle or in a curtain.

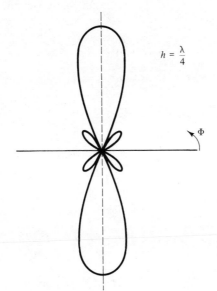

$h = \dfrac{\lambda}{4}$

Φ

Figure 8.10–6 Field pattern of four-element broadside array in $\Theta = \pi/2$ plane with $\lambda/2$ element spacing.

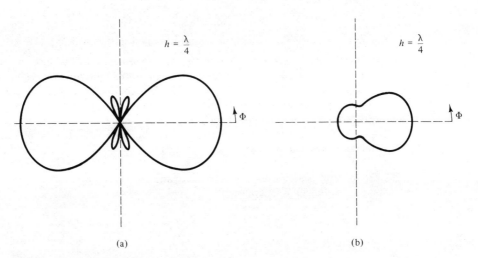

$h = \dfrac{\lambda}{4}$

Φ

$h = \dfrac{\lambda}{4}$

Φ

(a)

(b)

Figure 8.10–7 Field patterns: (a) four-element bilateral end-fire array in $\Theta = \pi/2$ plane with $\lambda/2$ element spacing; (b) four-element end-fire array in $\Theta = \pi/2$ plane with $\lambda/4$ element spacing.

PROBLEMS

1. For an antenna with $a/\lambda = 7.022 \times 10^{-3}$ and $k_0 h = \pi/2$, the parameters Ψ_{dR}, T'_U, and T'_D are given by

$$\Psi_{dR} = 6.218; \qquad T'_U = 3.085 + j3.581; \qquad T'_D = 1.061 + j0.025$$

with $\Omega = 2 \ln (2h/a) = 8.54$. Calculate the current $I_z(z)/V_0^e$ (amperes/volt), the admittance Y_0, and the impedance Z_0 of the antenna. Plot the current distributions $I''_z(z)/V_0^e$, $I'_z(z)/V_0^e$ where

$$\frac{I_z(z)}{V_0^e} = \frac{I''_z(z)}{V_0^e} + j\frac{I'_z(z)}{V_0^e}$$

2. For an antenna with $a/\lambda = 7.022 \times 10^{-3}$ and $k_0 h = \pi$,

$$\Psi_{dR} = 5.737$$

$$T_U = -0.117 + j0.114$$

$$T_D = -0.106 + j0.108$$

$$\Omega = 2 \ln \frac{2h}{a} = 9.92$$

Repeat Problem 1.

3. Repeat Problem 1 using the two-term theory for the current.

$$T' = 2.65 + j3.79$$

Compare the admittance with the result obtained in Problem 1.

4. Repeat Problem 2 using the two-term theory for the current.

$$T = -0.172 + j0.175$$

Compare the admittance Y_0 with the result obtained in Problem 2.

5. A special type of antenna consists of two identical center-driven cylindrical dipoles arranged to be mutually perpendicular and driven from a four-wire line with conductors located in a square as shown in Fig. P8–5. The four conductors of the transmission line extend out a quarter-wavelength (where they end in a short circuit) in order to provide support for the antennas. At the generator end of the line a suitable network provides voltages V_1 (for the vertical line 1) and V_2 (for the horizontal line 2) such that $V_2 = jV_1$. Between the generator and the antennas the four wires are enclosed in a circular metal shield.

   (a) The integral equation for the current in antenna 1 when isolated and center-driven by a delta-function generator is

$$A_1(z) = \frac{\mu_0}{4\pi} \int_{-h}^{h} I_1(z') \frac{e^{-jk_0 R_1}}{R_1} dz'$$

$$= \frac{-j4\pi}{\zeta_0} (C_1 \cos k_0 z + \tfrac{1}{2}V_1^e \sin k_0 |z|)$$

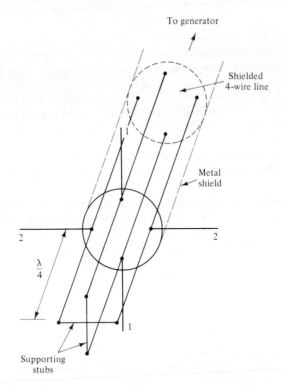

To generator

Shielded
4-wire line

Metal
shield

2   2

$\dfrac{\lambda}{4}$

Supporting
stubs   **Figure P8–5**

What changes in this equation must be made due to the presence of antenna 2? Since the solution of this equation is well known, it would be convenient to adapt it to determine the currents in the crossed dipoles with suitable corrections. Discuss this possibility.

(b) Describe the electromagnetic field (magnitude and direction of electric and magnetic vectors) of the crossed dipoles if each is a half-wavelength long and quite thin:

(1) At distant points, including especially along the extended axis of the transmission line.

(2) In the plane of the antennas both very near the antennas and at distant points.

6. (a) Two identical, parallel, and nonstaggered antennas are separated by a distance $b = 0.25\lambda$. The following data apply to the two phase sequences in the two-term form: $h/\lambda = 0.25$; $a/\lambda = 7.022 \times 10^{-3}$; $\Psi_{dR}^{(m)} = 6.218$ for $m = 0, 1$; $T'^{(0)} = 0.899 + j3.121$; $T'^{(1)} = 4.895 + j2.043$.

Determine the currents, admittances, and impedances of the two driven elements and the far-zone field when the driving voltages are $V_{10} = V$, $V_{20} = jV$ with $V = 1$ V. Sketch curves of the currents and field pattern in the equatorial plane. The lumped correction $B_c = 0.72 \times 10^{-3}$ S is to be added to the self-susceptances B_{s1} and B_{s2}.

(b) Repeat part (a) under identical circumstances except that $V_{20} = 0$.

7. (a) Four identical parallel antennas have their centers at the corners of a square with side equal to $\lambda/4$. The following data apply to the four phase sequences in the two-term form: $h/\lambda = 0.5$; $a/\lambda = 7.022 \times 10^{-3}$; $\Psi_{dR}^{(m)} = 5.737$ for $m = 0, 1, 2, 3$; $T^{(0)} = -0.361 + j0.422$; $T^{(1)} = T^{(3)} = -0.218 + j0.108$; $T^{(2)} = -0.157 + j0.022$.

Determine the currents, admittances, and impedances of each driven element and the far-zone electromagnetic field of the array in the equatorial plane when the driving voltages of the elements when numbered consecutively around the square are $V_{10} = V_{20} = V$, $V_{30} = V_{40} = jV$, where $V = 1$ V. Sketch curves of the currents and field pattern in the equatorial plane. The lumped correction to be added to all self-susceptances is $B_c = 0.72 \times 10^{-3}$ S. Note that this affects the current only very near the driving point.

(b) Repeat part (a) under identical circumstances except that the currents instead of the voltages are specified as follows: $I_1(0) = I_2(0) = I$; $I_3(0) = I_4(0) = jI$, where $I = 10$ mA.

8. Two identical, parallel antennas are connected by two-wire lines to lumped impedances and generators as shown in Fig. P8–8. Determine the impedance seen by V_1^e in each of the following cases. Assume gap and end effects, as well as ohmic losses in antennas, lines, and connecting wires all to be negligible. For the antenna $\Omega = 2 \ln (2h/a) = 15$; $f = 150$ MHz; $R_c = 400$ Ω for the line.

(a) $k_0 h = \pi/2$; $b = 0.2\lambda_0$; $V_2^e = 0$; $s_1 = \lambda_0/4$; $s_2 = \lambda_0/4$; $Z_1 = 0$; $Z_2 = \infty$
(b) $k_0 h = \pi/2$; $b = 0.4\lambda_0$; $V_2^e = V_1^e$; $s_1 = \lambda_0/2$; $s_2 = \lambda_0/2$; $Z_1 = -jX_{s1}$; $Z_2 = -jX_{s2}$
(c) $k_0 h = 3.4$; $b = 0.1\lambda_0$; $V_2^e = 0$; $s_1 = \lambda_0/2$; $s_2 = \lambda_0/2$; $Z_1 = 0$; $Z_2 = 0$
(d) $k_0 h = 3.4$; $b = 0.1\lambda_0$; $V_2^e = 0$; $s_1 = \lambda_0/2$; $s_2 = \lambda_0/4$; $Z_1 = 0$; $Z_2 = 0$
(e) $k_0 h = 3.4$; $b = 0.5\lambda_0$; $V_2^e = -V_1^e$; $s_1 = \lambda_0/2$; $s_2 = \lambda_0/2$; $Z_1 = 0$; $Z_2 = 0$

9. An important directional array consists of two parallel antennas as shown in Fig. P8-8 with $h = \lambda_0/4$, $b = 0.1\lambda_0$, $V_2^e = 0$, s_2 adjustable, $Z_2 = 0$. The current in antenna 2 is to lag that in antenna 1 by 0.4 period, so that the radiation fields due to the two currents are in opposite phase at distant points in the direction toward the driven antenna from the parasite (called *director*). Determine Z_{CD} looking into the line s_2, the relative magnitudes of the currents at A and C, and the impedance Z_{AB}. Neglect gap effects; assume that $\Omega = 15$ for the antenna, $R_c = 400$ Ω for the line. Neglect ohmic losses.

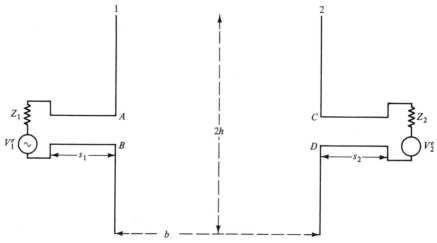

Figure P8–8

10. A modification of the directional array of Problem 9 involves the following; $h = \lambda_0/4$, $b = 0.25\lambda_0$, $V_2^e = 0$, s_2 adjustable, $Z_2 = 0$. The current in antenna 2 is to lead that in antenna 1 by 0.25 period, so that the radiation fields of the two will be in phase at distant points in the direction from the parasite (called *reflector*) to the driven antenna. Determine Z_{CD} looking into the line s_2, the relative magnitudes of the currents at A and C, and the impedance Z_{AB}. Assume that $\Omega = 15$, $R_c = 400\ \Omega$.

11. Determine the effect on the impedance Z_{AB} in Problems 9 and 10 if antenna 1 is a very closely spaced folded dipole instead of a simple dipole. Assume that the significant interaction between the antennas involves only the symmetrical currents in the folded dipole.

12. Determine the driving-point impedance at the terminals of a delta-function generator in Fig. P8–12. Assume that $\Omega = 15$ for the antenna; $R_c = 440\ \Omega$, $\alpha = 2.26 \times 10^{-3}$ nepers/m for the line sections; $f = 150$ MHz. In all cases b is small compared with λ_0; it is exaggerated in the figures. Sketch approximate distributions of "antenna" and "line" currents in Fig. P8–12a.

13. Determine the approximate distribution and magnitudes of antenna and line currents in Fig. P8–12b.

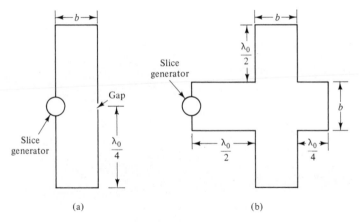

(a)   (b)

Figure P8–12

14. Determine the impedance at the driving terminals AB of the arrays shown in Fig. P8–14. Use the data of Problem 12 for the transmission lines. Assume that $\Omega = 10$ for the antennas. Show the distribution of "zero-order" current on antennas and lines. Neglect gap effects.

15. The two-term approximation for the currents in two parallel antennas is obtained from the three-term formula by the substitution of $(\cos k_0 z - \cos k_0 h)$ for $[\cos (k_0 z/2) - \cos (k_0 h/2)]$. Find the phase sequence currents and admittances with this approximation.

16. Draw the horizontal field pattern of a broadside array of five elements spaced a half-wavelength apart assuming that the currents in all the elements are in phase.

17. Draw the horizontal field pattern of an end-fire array of five elements spaced a quarter-wavelength apart assuming appropriately lagging currents to the right.

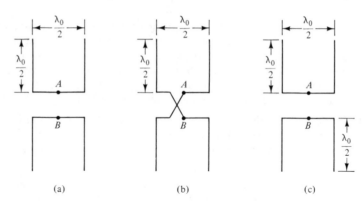

(a)   (b)   (c)

Figure P8–14

18. Derive an expression for the far field of an array of two parallel antennas with the same lengths. Antenna 1 is the driven antenna, antenna 2 is the parasite at a horizontal distance b. Draw the horizontal field pattern when the elements are a half-wavelength long and spaced a half-wavelength apart.

9

Electromagnetic Field Formulation of Electric Circuit Theory; the Loop Antenna

An important and instructive application of electromagnetic field theory is to electric circuits consisting of conductors that may form closed loops and quasi-closed loops with gaps in the conducting path in the form of capacitors with parallel plates.

Formulas for the inductance, resistance, and capacitance can be derived in general terms and then specialized to satisfy the conditions implied in conventional electric circuit theory: uniformly distributed currents, charge concentrations only on the adjacent surfaces of capacitors, and negligible radiation.

9.1 GENERAL FORMULATION OF CIRCUIT THEORY

The circuits to be analyzed are constructed of conductors of circular cross section with radius a such that, with $k_0 = \omega/c$,

$$k_0 a \ll 1 \qquad (9.1\text{--}1)$$

The parallel-plate capacitors to be considered have circular plates of radius b separated by a small distance w. They satisfy the following inequalities:

$$w <<< b; \qquad k_0 w <<< 1 \qquad (9.1\text{--}2)$$

For simplicity, the generator is idealized to consist of a section of conductor of negligible length between the ends of which a charge-separating field $\mathbf{E}^e$ is maintained by forces that are independent of the distribution of current and charge in the network. The conductors forming the circuits have any moderately smooth and continuous configuration. Just as for a straight conductor, the transverse distribution of current and charge is assumed to be independent of the distribution in the axial direction. This means that the internal impedance per unit length z^i of the infinitely long conductor is a good approximation for each unit length of the wires provided that they are far apart compared with their radius a. That is, $a^2 \ll d^2$, where d is the nearest distance between centers of conductors in a circuit.

The conductors in the circuits to be studied may turn and twist in any way. Let an arbitrarily chosen direction along the central axis of the conductor be designated by a unit vector $\hat{s}$. This vector changes its direction from point to point along the conductor if this is not straight. The component of the electric field in the direction of $\hat{s}$ at each point along the surface is

$$\hat{s} \cdot \mathbf{E} = E_s \tag{9.1-3}$$

The total axial current at any cross section is

$$I_s = \int_S J_s \, dS \tag{9.1-4}$$

The current is related to the axial tangential electric field at the surface according to

$$E_s = z^i I_s \tag{9.1-5a}$$

where z^i is the internal impedance per unit length of the conductor as shown in Chapter 3. This relation is an integrated form of

$$E_s = \frac{J_s}{\sigma} \tag{9.1-5b}$$

where J_s is the volume density of conduction current and σ is the conductivity. In regions in which a charge-separating agency is active, in particular in the idealized generator, an axially tangential impressed field E_s^e is maintained between the boundaries of the region at A and B. In this very short path of the circuit,

$$J_s = \sigma(E_s + E_s^e) \tag{9.1-6}$$

or

$$E_s = \frac{J_s}{\sigma} - E_s^e = z^i I_s - E_s^e \tag{9.1-7}$$

z^i is the effective internal impedance per unit length of the charge-separating region, which is assumed to have the conducting properties of a simple conductor, and I_s is the total current through it. If there is a capacitor in the circuit that satisfies (9.1-2), the axial electric field at its outer edge is given by an expression exactly like (9.1-5a) with I_s the total axial current entering or leaving the adjacent surface layers and z^i the internal impedance per unit thickness of the dielectric in the capacitor. The plates themselves are assumed to be perfectly conducting.

The electric field at all points in space outside the conductors and capacitors is given by

$$\mathbf{E} = -\nabla\phi - j\omega\mathbf{A} = \frac{-j\omega}{k_0^2}\left(\nabla\nabla \cdot \mathbf{A} + k_0^2\mathbf{A}\right) \tag{9.1-8}$$

The component of this field axially tangent to the surface of the conductors or the surface of the capacitor at its edge is

$$E_s = \frac{-\partial\phi}{\partial s} - j\omega A_s = \frac{-j\omega}{k_0^2}\left[\frac{\partial}{\partial s}\nabla \cdot \mathbf{A} + k_0^2 A_s\right] \tag{9.1-9}$$

Since the tangential component of the electric field at a boundary is continuous, E_s in (9.1–5a) and E_s in (9.1–9) must be equal at the surface of a conductor or capacitor. That is,

$$(E_s)_{\text{inner}} = (E_s)_{\text{outer}} \tag{9.1-10}$$

The substitution of (9.1–7) and (9.1–9) in (9.1–10) results in

$$-E_s^e + z^i I_s = \frac{-\partial\phi}{\partial s} - j\omega A_s = \frac{-j\omega}{k_0^2}\left[\frac{\partial}{\partial s}(\nabla \cdot \mathbf{A}) + k_0^2 A_s\right] \tag{9.1-11}$$

Along those parts of the circuit between terminals A and B where the charge-separating agency is assumed to be localized, E_s^e differs from zero; elsewhere it is zero.

It is assumed that $\mathbf{E}^e$ may be derived from a scalar potential ϕ^e. Thus

$$\mathbf{E}^e = \nabla\phi^e; \qquad E_s^e = \frac{\partial\phi^e}{\partial s} \tag{9.1-12}$$

The positive gradient is used since $\mathbf{E}^e$ points from lower to higher potentials. This is equivalent to assuming that a generator may be described mathematically by a discontinuity in scalar potential.

Equation (9.1–11) is integrated along the conductors of a circuit from s_1 to s_2 representing the two ends of the circuit. If an idealized generator is assumed to exist between the points A and B somewhere along the conductor,

$$\int_A^B d\phi^e = \int_{s_1}^{s_2} d\phi + j\omega\int_{s_1}^{s_2} A_s\, ds + \int_{s_1}^{s_2} z^i I_s\, ds \tag{9.1-13}$$

Let the driving potential difference or emf maintained by the generator between the terminals A and B be defined by

$$V_{BA}^e \equiv \phi_B^e - \phi_A^e = \int_A^B d\phi^e \tag{9.1-14}$$

which reduces to

$$V_{BA}^e = \int_{s_1}^{s_2} d\phi + j\omega \int_{s_1}^{s_2} A_s \, ds + \int_{s_1}^{s_2} z^i I_s \, ds \qquad (9.1\text{–}15)$$

This is the fundamental equation for an open circuit containing a point generator. It is also a statement of Kirchhoff's voltage law. If the circuit is closed so that s_1 and s_2 coincide, the first integral in (9.1–15) vanishes and

$$V_{BA}^e = \oint z^i I_s \, ds + j\omega \oint \mathbf{A} \cdot d\mathbf{s} \qquad (9.1\text{–}16)$$

This relation must be satisfied by the current in a closed circuit. However, it is not an integral equation for the current.

9.2 TWO COUPLED CIRCUITS

The general equation (9.1–1) may now be extended to determine the general equations for two coupled circuits shown in Fig. 9.2–1. The primary circuit consists of a coil and a capacitor; the secondary consists of a coil and a resistance. The vector potential in the primary circuit is due to the currents in the primary as well as in the secondary. Equation (9.1–16) now becomes

$$V_{BA}^e = V_{10}^e = \oint_{s_1} z_1^i I_{1s} \, ds_1 + j\omega \oint_{s_1} (\mathbf{A}_{11} \cdot d\mathbf{s}_1 + \mathbf{A}_{12} \cdot d\mathbf{s}_2) \qquad (9.2\text{–}1a)$$

$\mathbf{A}_{11}$ is the vector potential at the surface of the primary circuit due to the current I_1 in circuit 1; $\mathbf{A}_{12}$ is the vector potential at the surface of the primary circuit due

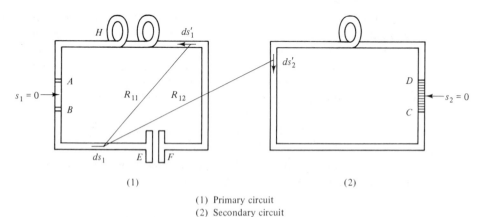

(1)   (2)

(1) Primary circuit
(2) Secondary circuit

Figure 9.2–1   Two coupled circuits, potential difference between $AB = V_{10}^e$; CD is a load resistance.

to the current I_2 in circuit 2. The corresponding equation for the secondary circuit is

$$0 = \oint_{s_2} z_2^i I_{2s}\, ds_2 + j\omega \oint_{s_2} (\mathbf{A}_{22} \cdot d\mathbf{s}_2 + \mathbf{A}_{21} \cdot d\mathbf{s}_2) \qquad (9.2\text{--}1b)$$

It has been shown (Chapter 3) that the contribution to the vector potential at any point in space due to the current I_s in an element ds' of a conductor is

$$d\mathbf{A} = \frac{\mu_0}{4\pi} \frac{I_s'}{R} e^{-jk_0 R}\, d\mathbf{s}' \qquad (9.2\text{--}2)$$

Hence, for the coupled circuits shown in Fig. 9.2–1,

$$\mathbf{A}_{11} = \frac{\mu_0}{4\pi} \oint_{s_1} \frac{I_{1s}'}{R_{11}} e^{-jk_0 R_{11}}\, d\mathbf{s}_1' \qquad (9.2\text{--}3a)$$

The element ds_1' is at the axis of the circular conductor. Equation (9.2–3a) is valid for any conductor and is a good approximation even on the surface except within distances of sharp bends comparable with the radius of the wire. Corresponding to (9.2–3a),

$$\mathbf{A}_{12} = \frac{\mu_0}{4\pi} \oint_{s_2} \frac{I_{2s}'}{R_{12}} e^{-jk_0 R_{12}}\, d\mathbf{s}_2' \qquad (9.2\text{--}3b)$$

$$\mathbf{A}_{22} = \frac{\mu_0}{4\pi} \oint_{s_2} \frac{I_{2s}'}{R_{22}} e^{-jk_0 R_{22}}\, d\mathbf{s}_2' \qquad (9.2\text{--}3c)$$

$$\mathbf{A}_{21} = \frac{\mu_0}{4\pi} \oint_{s_1} \frac{I_{1s}'}{R_{21}} e^{-jk_0 R_{21}}\, d\mathbf{s}_1' \qquad (9.2\text{--}3d)$$

If (9.2–3a–d) are substituted in (9.2–1a,b), two simultaneous equations are obtained in which the distributions of current are the unknowns and they depend on the configuration of the circuit. The general equations become

$$V_{10}^e = \oint_{s_1} z_1^i I_{1s}\, ds_1 + j\frac{\mu_0 \omega}{4\pi} \oint_{s_1} \left(d\mathbf{s}_1 \cdot \oint_{s_1} \frac{I_{1s}'}{R_{11}} e^{-jk_0 R_{11}}\, d\mathbf{s}_1' \right)$$

$$+ j\frac{\mu_0 \omega}{4\pi} \oint_{s_1} \left(d\mathbf{s}_1 \cdot \oint_{s_2} \frac{I_{2s}'}{R_{12}} e^{-jk_0 R_{12}}\, d\mathbf{s}_2' \right) \qquad (9.2\text{--}4a)$$

$$0 = \oint_{s_2} z_2^i I_{2s}\, ds_2 + j\frac{\mu_0 \omega}{4\pi} \oint_{s_2} \left(d\mathbf{s}_2 \cdot \oint_{s_2} \frac{I_{2s}'}{R_{22}} e^{-jk_0 R_{22}}\, d\mathbf{s}_2' \right)$$

$$+ j\frac{\mu_0 \omega}{4\pi} \oint_{s_2} \left(d\mathbf{s}_2 \cdot \oint_{s_1} \frac{I_{1s}'}{R_{21}} e^{-jk_0 R_{21}}\, d\mathbf{s}_1' \right) \qquad (9.2\text{--}4b)$$

In general, it is not possible to solve these equations to obtain the distributions of current.

9.3 DEFINITIONS OF IMPEDANCES, COEFFICIENT OF COUPLING, AND INDUCED VOLTAGE

Equations (9.2–4a,b) can be reduced to standard form by introducing dimensionless distribution functions $f_1(s_1)$ and $f_2(s_2)$ for the currents. Reference currents are chosen at the center of the generator ($s_1 = 0$) in the primary circuit and at the center of the load ($s_2 = 0$) in the secondary circuit. The complex amplitudes of the currents at these points are, respectively, I_{10} and I_{20}. The currents at other points in the circuit are given by

$$I_{1s} = I_{10}f_1(s_1) \qquad (9.3\text{–}1a)$$

$$I_{2s} = I_{20}f_2(s_2) \qquad (9.3\text{–}1b)$$

where $f_1(s_1)$ and $f_2(s_2)$ are dimensionless complex distribution functions.

Using (9.3–1a,b) in (9.2–4a,b) the following standard forms are obtained:

$$V_{10}^e = I_{10}Z_{11} + I_{20}Z_{12} \qquad (9.3\text{–}2a)$$

$$0 = I_{10}Z_{21} + I_{20}Z_{22} \qquad (9.3\text{–}2b)$$

with

$$Z_{11} = Z_1^i + Z_1^e \qquad (9.3\text{–}3a)$$

$$Z_{22} = Z_2^i + Z_2^e \qquad (9.3\text{–}3b)$$

and

$$Z_1^i \equiv \oint_{s_1} z_1^i f_1(s_1)\, ds_1 \qquad (9.3\text{–}4a)$$

$$Z_2^i \equiv \oint_{s_2} z_2^i f_2(s_2)\, ds_2 \qquad (9.3\text{–}4b)$$

$$Z_1^e = \frac{j\omega}{I_{10}} \oint_{s_1} (\mathbf{A}_{11} \cdot d\mathbf{s}_1) = \frac{j\omega\mu_0}{4\pi} \oint_{s_1} \left[d\mathbf{s}_1 \cdot \oint_{s_1} \frac{f_1(s_1')}{R_{11}} e^{-jk_0 R_{11}} \, ds_1' \right] \qquad (9.3\text{–}5a)$$

$$Z_2^e = \frac{j\omega}{I_{20}} \oint_{s_2} (\mathbf{A}_{22} \cdot d\mathbf{s}_2) = \frac{j\omega\mu_0}{4\pi} \oint_{s_2} \left[d\mathbf{s}_2 \cdot \oint_{s_2} \frac{f_2(s_2')}{R_{22}} e^{-jk_0 R_{22}} \, ds_2' \right] \qquad (9.3\text{–}5b)$$

$$Z_{12} = \frac{j\omega}{I_{20}} \oint_{s_1} (\mathbf{A}_{12} \cdot d\mathbf{s}_1) = \frac{j\omega\mu_0}{4\pi} \oint_{s_1} \left[d\mathbf{s}_1 \cdot \oint_{s_2} \frac{f_2(s_2')}{R_{12}} e^{-jk_0 R_{12}} \, ds_2' \right] \qquad (9.3\text{–}6a)$$

$$Z_{21} = \frac{j\omega}{I_{10}} \oint_{s_2} (\mathbf{A}_{21} \cdot d\mathbf{s}_2) = \frac{j\omega\mu_0}{4\pi} \oint_{s_2} \left[d\mathbf{s}_2 \cdot \oint_{s_1} \frac{f_1(s_1')}{R_{21}} e^{-jk_0 R_{21}} \, ds_1' \right] \qquad (9.3\text{–}6b)$$

The complex coefficient Z_{11} is the self-impedance of the primary referred to I_{10}; Z_{22} is the self-impedance of the secondary referred to I_{20}; Z_{12} is the mutual impedance of the primary referred to I_{10}; Z_{21} is the mutual impedance of the secondary referred to I_{20}. Each impedance is changed if another reference current is selected. Therefore, the impedances are functions not only of the geometrical configuration of the circuit (which determines the distribution of current) but also of the arbitrarily selected reference current. The part Z^i of the self-impedance is the internal impedance; the part Z^e of the self-impedance, the external self-imped-ance. The former depends primarily on the internal impedance per unit length of the contour of the conductor, the latter entirely on the interaction of currents in various parts of the same circuit. Both self- and mutual impedances are complex. The real part is the self- or mutual resistance, the imaginary part, the self- or mutual reactance. All the impedances are measured in ohms.

The simultaneous equations (9.3–2a,b) are readily solved for I_{10} and I_{20}. Thus

$$I_{10} = \frac{V_{10}^e Z_{22}}{Z_{11}Z_{22} - Z_{12}Z_{21}} = \frac{V_{10}^e}{Z_{11} - Z_{12}Z_{21}/Z_{22}} \tag{9.3–7a}$$

$$I_{20} = \frac{-V_{10}^e Z_{21}}{Z_{11}Z_{22} - Z_{12}Z_{21}} \tag{9.3–7b}$$

The driving-point impedance is defined to be the ratio of the driving potential difference V_{10}^e across the terminals AB to the current entering one and leaving the other terminal. That is,

$$Z_{AB} = \frac{V_{10}^e}{I_A} = \frac{V_{10}^e}{I_B} \tag{9.3–8}$$

This definition implies that

$$I_A = I_B \tag{9.3–9}$$

Assuming (9.3–9) to be true, (9.3–8) with (9.3–7a) gives

$$Z_{AB} = Z_{11} - \frac{Z_{12}Z_{21}}{Z_{22}} = Z_{11}\left(1 - \frac{Z_{12}Z_{21}}{Z_{11}Z_{22}}\right) \tag{9.3–10}$$

A complex coefficient of coupling between the circuits is defined to be

$$k_{12}^2 = \frac{Z_{12}Z_{21}}{Z_{11}Z_{22}} \tag{9.3–11}$$

Loose coupling exists when

$$k_{12}^2 \ll 1 \tag{9.3–12}$$

so that

$$Z_{AB} \doteq Z_{11} \tag{9.3–13}$$

It is to be noted that Z_{11} in (9.3–13) and Z_{11} in (9.3–10) are not in general the same. This follows from the fact that $f_1(s_1)$ in (9.3–4a) and $f_1(s_1')$ in (9.3–5a) are altered if the relative orientation or the distance separating the two circuits is changed. The self-impedance of the primary circuit is not only a function of the primary but also of the secondary insofar as this contributes to the distribution of current in the primary.

This fact may be clarified by considering the following two cases. First, let the secondary be removed completely so that $\mathbf{A}_{12}$ and Z_{12} vanish. Regardless of the choice of reference points, the second term on the right in (9.3–2a) vanishes, leaving

$$V_{10}^e = I_{10}Z_{11} \qquad (9.3\text{–}14)$$

with Z_{11} defined by (9.3–5a) in which the distribution of the function $f_1(s_1')$ is determined entirely by interactions between currents in the several parts of the primary circuit alone. Now let the secondary be brought close to the primary, but let the circuit be broken at $s_2 = 0$ so that I_{20} vanishes. The second term in (9.3–2a) is again zero, but this time because I_{20} vanishes, not Z_{12}. Since a zero current at $s_2 = 0$ does not in general mean that the current I_{20} must be zero or even small elsewhere in the circuit, it is clear that the distribution function $f_1(s_1')$ is determined not only by interactions between currents in different parts of the primary, but by the interactions of currents in the primary with currents in the secondary. It cannot, therefore, be the same as with the secondary absent, so that Z_{11} is different in the two cases. Therefore, the self-impedance of a circuit as defined in (9.3–5a) is not simply a constant of the geometry of a circuit but a function of the proximity and distribution of currents in coupled circuits.

The following term in (9.3–2a), that is,

$$V_{12}^i \equiv -I_{20}Z_{12} \qquad (9.3\text{–}15)$$

is the voltage induced in the primary by the current in the secondary. The corresponding term in (9.3–2b),

$$V_{21}^i \equiv -I_{10}Z_{21} \qquad (9.3\text{–}16)$$

is the voltage induced in the secondary by the current in the primary. With this notation, (9.3–2a,b) become

$$V_{10}^e + V_{12}^i = I_{10}Z_{11} \qquad (9.3\text{–}17a)$$

$$V_{21}^i = I_{20}Z_{22} \qquad (9.3\text{–}17b)$$

In the case of loose coupling as defined in (9.3–12),

$$V_{10}^e \gg V_{12}^i \qquad (9.3\text{–}18)$$

so that

$$V_{10}^e = I_{10}Z_{11}; \qquad V_{21}^i = I_{20}Z_{22} \qquad (9.3\text{–}19)$$

The foregoing analysis for two circuits with one driven is readily extended to n circuits with any number driven. There are n simultaneous equations such as

(9.3–4a,b) and (9.3–5a,b) with n integrals in each in addition to one involving z^i. If there is no generator, the left side is zero. If distribution functions and reference currents are introduced for the currents, the equations for the n circuits corresponding to (9.3–2a,b) are

$$V_{10}^e = \sum_{j=1}^{n} I_{j0} Z_{1j}$$

$$V_{20}^e = \sum_{j=1}^{n} I_{j0} Z_{2j}$$

$$\vdots$$

$$V_{n0}^e = \sum_{j=1}^{n} I_{j0} Z_{nj}$$

(9.3–20)

The coefficients Z_{kj} in which $k = j$ are the self-impedances defined by expressions similar to (9.3–3) with (9.3–4) and (9.3–5); the coefficients Z_{kj} with $k \neq j$ are mutual impedances defined by expressions like (9.3–6a,b).

The Rayleigh–Carson reciprocal theorem may be used to show that

$$Z_{ij} = Z_{ji}$$

(9.3–21)

It follows that irrespective of circuit configuration or choice of reference currents (which must remain unchanged once they have been chosen), the mutual impedance in circuit i due to circuit j is the same as the mutual impedance in circuit j due to circuit i. In particular, for circuits i and j,

$$\oint_{s_i} \left[d\mathbf{s}_i \cdot \oint_{s_j} \frac{f_j(s_j')}{R_{ij}} e^{-jk_0 R_{ij}} \, d\mathbf{s}_j' \right] = \oint_{s_j} \left[d\mathbf{s}_j \cdot \oint_{s_i} \frac{f_i(s_i')}{R_{ji}} e^{-jk_0 R_{ji}} \, d\mathbf{s}_i' \right]$$

(9.3–22)

9.4 NEAR-ZONE ELECTRIC CIRCUITS

The general formulation of the equations for electric circuits is complicated primarily because the distribution of current is a function of the configuration. If the circuits are confined to the near zone and are so restricted that all the distribution functions such as $f_1(s_1)$ may be set equal to unity, the problem is simplified considerably. The conditions as formulated for two circuits are as follows:

$$f_1(s_1) \doteq f_2(s_2) \doteq 1$$

(9.4–1)

$$k_0 R_{11} \ll 1; \qquad k_0 R_{22} \ll 1; \qquad k_0 R_{12} \ll 1; \qquad k_0 R_{21} \ll 1$$

(9.4–2)

If these conditions are satisfied, it is possible to write

$$f_i(s_i) e^{-jk_0 R_{ij}} \doteq 1; \qquad i = 1, 2; \qquad j = 1, 2$$

(9.4–3)

in each integral in (9.3–3a,b) to (9.3–6a,b). These integrals become

$$Z_1^i = \oint_{s_1} z_1^i \, ds_1; \qquad Z_2^i = \oint_{s_2} z_2^i \, ds_2 \tag{9.4-4}$$

$$Z_1^e = \frac{j\omega\mu_0}{4\pi} \oint_{s_1} \oint_{s_1} \frac{ds_1 \cdot ds_1'}{R_{11}} = jX_1^e; \qquad R_1^e = 0 \tag{9.4-5}$$

$$Z_2^e = \frac{j\omega\mu_0}{4\pi} \oint_{s_2} \oint_{s_2} \frac{ds_2 \cdot ds_2'}{R_{22}} = jX_2^e; \qquad R_2^e = 0 \tag{9.4-6}$$

$$Z_{12} = \frac{j\omega\mu_0}{4\pi} \oint_{s_1} \oint_{s_2} \frac{ds_1 \cdot ds_2'}{R_{12}} = jX_{12}; \qquad R_{12} = 0 \tag{9.4-7}$$

$$Z_{21} = \frac{j\omega\mu_0}{4\pi} \oint_{s_1} \oint_{s_2} \frac{ds_2 \cdot ds_1'}{R_{21}} = jX_{21}; \qquad R_{21} = 0 \tag{9.4-8}$$

All the integrals in (9.4–5) to (9.4–8) are real, so that for two coupled circuits with no parts in common to both, Z_i^e and Z_{ij} are pure imaginaries. Furthermore, the integrals themselves are independent of the frequency and are functions of the geometry of the circuit alone. It follows that

$$X_1^e = \omega L_1^e; \qquad L_1^e = \frac{\mu_0}{4\pi} \oint_{s_1} \oint_{s_1} \frac{ds_1 \cdot ds_1'}{R_{11}} = \frac{1}{I_1} \oint_{s_1} \mathbf{A}_{11} \cdot ds_1 \tag{9.4-9}$$

$$X_2^e = \omega L_2^e; \qquad L_2^e = \frac{\mu_0}{4\pi} \oint_{s_2} \oint_{s_2} \frac{ds_2 \cdot ds_2'}{R_{22}} = \frac{1}{I_2} \oint_{s_2} \mathbf{A}_{22} \cdot ds_2 \tag{9.4-10}$$

$$X_{12} = \omega L_{12}; \qquad L_{12} = \frac{\mu_0}{4\pi} \oint_{s_1} \oint_{s_2} \frac{ds_1 \cdot ds_2'}{R_{12}} = \frac{1}{I_2} \oint_{s_1} \mathbf{A}_{12} \cdot ds_1 \tag{9.4-11}$$

$$X_{21} = \omega L_{21}; \qquad L_{21} = \frac{\mu_0}{4\pi} \oint_{s_1} \oint_{s_2} \frac{ds_2 \cdot ds_1'}{R_{21}} = \frac{1}{I_1} \oint_{s_2} \mathbf{A}_{21} \cdot ds_2 \tag{9.4-12}$$

The primed elements ds' are along the axis, the unprimed elements ds along the surface of the circuit. L_1^e and L_2^e are the external self-inductances of the primary and secondary circuits, respectively. L_{12} and L_{21} are the mutual inductances of the primary circuit due to the current in the secondary circuit and in the secondary circuit due to the current in the primary, respectively. It is seen from (9.4–11) and (9.4–12) that $L_{12} = L_{21}$ in near-zone circuits. The integrals of (9.4–11) and (9.4–12) are known as the *Neumann formulas*.

The formulas (9.4–4) to (9.4–8) are simpler than (9.3–4a,b) to (9.3–6a,b). The integrals of (9.4–4) to (9.4–8) are functions only of the geometry and of the internal impedance per unit length of the circuit. They are independent of distribution functions and reference points for current. It is possible therefore to define the impedance of coils, capacitors, and resistive elements in a manner that is independent of the configuration of the circuit connecting them.

9.5 QUASI-NEAR-ZONE CIRCUITS

The special case of near-zone or conventional circuits defined by (9.4–1) and (9.4–2) is a good approximation for electric circuit theory since this is based on the assumption that the conditions in (9.4–1) and (9.4–2) are very well satisfied. However, as the frequency increases, it is evident that the simple special theory is inadequate. The differences between the special theory and the general theory may be investigated by examining the border-line case between them. The distribution function $f_1(s_1)$ and the phase factor $e^{-jk_0R_{11}}$ are expanded in power series and one term beyond the leading term (which is unity) is retained.

$$e^{-jk_0R_{11}} \doteq 1 - \frac{k_0^2 R_{11}^2}{2!} - jk_0R_{11}\left(1 - \frac{k_0^2 R_{11}^2}{3!}\right) \tag{9.5-1}$$

Since the exact form of the distribution function $f_1(s_1)$ cannot be determined from (9.3–4a) an approximate method is used which depends on the fact that the principal contribution to significant higher-order terms in the factor $f_1(s_1)e^{-jk_0R_{11}}$ is from the exponential factor rather than from $f_1(s_1)$. Hence the approximation

$$f_1(s_1) \doteq 1 \tag{9.5-2}$$

may be retained in the evaluation of Z_1^e.

If (9.5–1) and (9.5–2) are substituted in (9.3–5a), the inner integral becomes

$$\oint_{s_1} \frac{f_1(s_1')}{R_{11}} e^{-jk_0R_{11}} ds_1' \doteq \oint_{s_1} \frac{1}{R_{11}}\left[\left(1 - \frac{k_0^2 R_{11}^2}{2}\right) - jk_0R_{11}\left(1 - \frac{k_0^2 R_{11}^2}{6}\right)\right] ds_1' \tag{9.5-3}$$

$$\oint_{s_1}\oint_{s_1} \frac{f_1(s_1)}{R_{11}} e^{-jk_0R_{11}}(ds_1 \cdot ds_1') \doteq \oint_{s_1}\oint_{s_1}\left(\frac{1}{R_{11}} + j\frac{k_0^3 R_{11}^2}{6}\right)(ds_1 \cdot ds_1') \tag{9.5-4}$$

where only the leading terms have been retained. Equations (9.5–2) and (9.5–4) are good approximations whenever the following inequality is true:

$$k_0^2 R_{11}^2 \ll 1 \tag{9.5-5}$$

This is the condition for the quasi-near-zone circuits.

In the evaluation of Z_1^e, (9.5–2) is a reasonable approximation for all closed loops of any shape that satisfy (9.5–5) for all values of R_{11}. If (9.5–4) is substituted in (9.3–5a), real and imaginary terms are obtained in the form $Z_1^e = R_1^e + jX_1^e$. The imaginary term is simply (9.4–9):

$$X_1^e = \omega L_1^e = \frac{\mu_0\omega}{4\pi} \oint_{s_1}\oint_{s_1} \frac{ds_1 \cdot ds_1'}{R_{11}} \tag{9.5-6}$$

With $\mu_0\omega = k_0\zeta_0$, the real term is

$$R_1^e = \frac{-\zeta_0 k_0^4}{(4\pi)6} \oint_{s_1}\oint_{s_1} R_{11}^2(ds_1 \cdot ds_1') \tag{9.5-7}$$

When $\zeta_0 = 120\pi$ ohms is substituted in (9.5–7), it becomes

$$R_1^e = -5k_0^4 \oint_{s_1} \oint_{s_1} R_{11}^2 (ds_1 \cdot ds_1') \qquad \text{ohms} \qquad (9.5\text{–}8)$$

The general formula (9.5–7) may be evaluated for a plane loop of any shape. In rectangular coordinates,

$$R_{11}^2 = (x - x')^2 + (y - y')^2 = x^2 + x'^2 + y^2 + y'^2 - 2(xx' + yy') \qquad (9.5\text{–}9\text{a})$$

$$ds_1 \cdot ds_1' = dx\, dx' + dy\, dy' \qquad (9.5\text{–}9\text{b})$$

If (9.5–9a,b) are substituted in (9.5–8), all the integrals are around a closed path. Integrals of the types $\oint dx$, $\oint x\, dx$, and $\oint x^2\, dx$ vanish, so that

$$\oint \oint (x^2 + x'^2 + y^2 + y'^2)(dx\, dx' + dy\, dy') = 0 \qquad (9.5\text{–}10)$$

and

$$\oint \oint (xx'\, dx\, dx' + yy'\, dy\, dy') = 0 \qquad (9.5\text{–}11)$$

The remaining integrals are

$$\oint \oint (xx'\, dy\, dy' + yy'\, dx\, dx') = \oint x\, dy \oint x'\, dy' + \oint y\, dx \oint y'\, dx' \qquad (9.5\text{–}12)$$

The integrals that involve primed coordinates give

$$S_1' = \oint x'\, dy' = \oint y'\, dx' \qquad (9.5\text{–}13)$$

where S_1' is the area enclosed by a contour along the axis of the conductor. Similarly,

$$S_1 = \oint x\, dy = \oint y\, dx \qquad (9.5\text{–}14)$$

where S_1 is the area enclosed by a contour along the surface of the conductor. Since conductors of small radius have been assumed in this analysis, a satisfactory approximation is

$$S_1 \doteq S_1' \qquad (9.5\text{–}15)$$

and

$$\oint_{s_1} \oint_{s_1} R_{11}^2\, ds_1 \cdot ds_1' = -4S_1^2 \qquad (9.5\text{–}16)$$

The substitution of (9.5–16) in (9.5–8) results in

$$R_1^e = 20k_0^4 S_1^2 \qquad \text{ohms} \qquad (9.5\text{–}17)$$

This formula may be used as an approximation for the external or radiation resistance of any quasi-near-zone circuit for which the condition (9.5–5) is satisfied.

It is clear from (9.5–17) that every ac circuit that encloses a finite area has a nonvanishing external or radiation resistance. At low frequencies, $k_0 = 2\pi f/c$ is so small that R_1^e is insignificant. At high frequencies, k_0 may be so large that (9.5–5) is not satisfied and (9.5–17) is not a good approximation. Simple circuits which satisfy the condition in (9.5–5) and in which the external or radiation resistance is important are loop antennas. The electrically small loop antenna is often the source of the electromagnetic field in waveguides. Used as a receiving antenna, it is a probe for the detection of electromagnetic fields. Such loop antennas are usually rectangular, square, or circular in shape. For a rectangle of length s and width b,

$$R_1^e = 20k_0^4 s^2 b^2 \qquad \text{ohms} \tag{9.5–18}$$

For a square of side s,

$$R_1^e = 20(k_0 s)^4 \qquad \text{ohms} \tag{9.5–19}$$

For a circle of mean radius b,

$$R_1^e = 20\pi^2 (k_0 b)^4 \qquad \text{ohms} \tag{9.5–20}$$

The self-inductances of these simple circuits can be calculated from (9.5–6). For a rectangle of length s and width b with wire of radius a,

$$L_1^e = \frac{\mu_0}{\pi} \left[b \ln \frac{2bs}{a(b+D)} + s \ln \frac{2bs}{a(s+D)} + 2(a+D-b-s) \right] \tag{9.5–21}$$

where D is the diagonal of the rectangle given by

$$D = (s^2 + b^2)^{1/2} \tag{9.5–22}$$

For a square of side s, the substitution of $s = b$ in (9.5–21) gives

$$L_1^e = \frac{2\mu_0 s}{\pi} \left(\ln \frac{s}{a} + \frac{a}{s} - 0.774 \right) \tag{9.5–23}$$

In most cases, a/s is negligible, so that

$$L_1^e = \frac{2\mu_0 s}{\pi} \left(\ln \frac{s}{a} - 0.77 \right) \tag{9.5–24}$$

If $s_t = 4s$ is the total length of wire,

$$L_1^e = \frac{\mu_0 s_t}{2\pi} \left(\ln \frac{s_t}{a} - 2.16 \right) \tag{9.5–25}$$

For a circular loop of radius b (wire radius a) satisfying the condition $k_0^2 b^2 \ll 1$,

$$L_1^e \doteq \mu_0 b \left(\ln \frac{8b}{a} - 2 \right) \tag{9.5–26}$$

In terms of the total length of wire $s_t = 2\pi b$,

$$L_1^e = \frac{\mu_0 s_t}{2\pi}\left(\ln\frac{4s_t}{\pi a} - 2\right) = \frac{\mu_0 s_t}{2\pi}\left(\ln\frac{s_t}{a} - 1.76\right) \tag{9.5-27}$$

A comparison of (9.5–27) with (9.5–25) shows that the external self-inductance of a loop of wire of length s_t is very nearly the same when the loop is a circle as when it is a square if s_t/a is large. Equation (9.5–27) is therefore a good approximation for loops that do not differ greatly in shape from a circle or square if s_t is the total length of wire.

9.6 MUTUAL IMPEDANCE IN QUASI-NEAR-ZONE CIRCUITS

If each of two coupled circuits is quasi-near-zone so that it is correct to use (9.5–6) and (9.5–7) with subscripts 1 and 2 in

$$Z_1^e = R_1^e + jX_1^e; \qquad Z_2^e = R_2^e + jX_2^e \tag{9.6-1}$$

the self-impedance of each circuit is unaffected by the proximity of the other circuit and its evaluation has been discussed in Sec. 9.5. The mutual impedances are obtained from the general integrals (9.3–6a,b) with the substitution $f(s) = 1$ in each:

$$Z_{12} = \frac{j\omega\mu_0}{4\pi}\oint_{s_1}\left(d\mathbf{s}_1 \cdot \oint_{s_2}\frac{e^{-jk_0R_{12}}}{R_{12}}d\mathbf{s}_2'\right) \tag{9.6-2a}$$

$$Z_{21} = \frac{j\omega\mu_0}{4\pi}\oint_{s_2}\left(d\mathbf{s}_2 \cdot \oint_{s_1}\frac{e^{-jk_0R_{21}}}{R_{21}}d\mathbf{s}_1'\right) \tag{9.6-2b}$$

The real and imaginary parts give expressions for the mutual resistances and reactances.

$$R_{12} = \frac{\zeta_0 k_0}{2\pi}\oint_{s_1}\left(d\mathbf{s}_1 \cdot \oint_{s_2}\frac{\sin k_0R_{12}}{R_{12}}d\mathbf{s}_2'\right) \tag{9.6-3a}$$

$$X_{12} = \frac{\mu_0\omega}{4\pi}\oint_{s_1}\left(d\mathbf{s}_1 \cdot \oint_{s_2}\frac{\cos k_0R_{12}}{R_{12}}d\mathbf{s}_2'\right) \tag{9.6-3b}$$

Expressions for R_{21} and X_{21} are obtained from (9.6–3a,b) by interchanging the subscripts 1 and 2. It is to be noted that whereas the size of each circuit is restricted to the near zone, the distance between the circuits is completely unrestricted. If the circuits are close together so that each is in the near zone with respect to the other, that is,

$$k_0R_{12} \ll 1; \qquad k_0R_{21} \ll 1 \tag{9.6-4}$$

then

$$\sin k_0 R_{12} \doteq k_0 R_{12} - \frac{(k_0 R_{12})^3}{3!} + \cdots \tag{9.6-5a}$$

$$\cos k_0 R_{12} \doteq 1 - \frac{(k_0 R_{12})^2}{2!} + \cdots \tag{9.6-5b}$$

Since $\oint_{s_1} \oint_{s_2} (d\mathbf{s}_2' \cdot d\mathbf{s}_1) = 0$, the leading term in (9.6–3a) is obtained from the second term in (9.6–5a). Then

$$R_{12} \doteq -\frac{\zeta_0 k_0^4}{24\pi} \oint_{s_1} \left(d\mathbf{s}_1 \cdot \oint_{s_2} R_{12}^2 \, d\mathbf{s}_2' \right)$$

$$= -5k_0^4 \oint_{s_1} \oint_{s_2} R_{12}^2 (d\mathbf{s}_1 \cdot d\mathbf{s}_2') \tag{9.6-6a}$$

$$X_{12} \doteq \omega L_{12}$$

$$= \frac{\mu_0 \omega}{4\pi} \oint_{s_1} \oint_{s_2} \frac{d\mathbf{s}_1 \cdot d\mathbf{s}_2'}{R_{12}} \tag{9.6-6b}$$

The mutual resistance term is very small at low frequencies. At radio frequencies, the approximations of (9.6–4) are usually not valid and (9.3–6a,b) have to be used.

9.7 INTERNAL IMPEDANCE OF A CIRCULAR PARALLEL-PLATE CAPACITOR

A useful circuit element in alternating current networks is the parallel-plate capacitor. In its simplest form it consists of two parallel circular metal plates (region 1) each of radius b and thickness d and separated by a dielectric (region 2) with very small thickness w ($w \ll b$, $w \ll \lambda$). At the center of its outer side each plate is connected to a long cylindrical conductor or wire of radius a as shown in Fig. 9.7–1. The z axis is the axis of symmetry; the origin of the cylindrical coordinates ρ, θ, z is at the center of the dielectric disk between the metal plates. The dielectric is a simple medium characterized by the complex permittivity $\tilde{\epsilon}_2 = \epsilon_{e2} - j\sigma_{e2}/\omega$; it is a good dielectric with $(\sigma_{e2}/\omega\epsilon_{e2})^2 \ll 1$. The losses in the highly conducting metal plates are negligible so they can be treated as if perfect conductors in the interior of which all fields vanish.

Since rotational symmetry obtains and the properties of interest are related to the principal current I_z, the current and field distributions in the dielectric can be analyzed in terms of the z component of the vector potential. The appropriate differential equation is (3.9–6). However, since the thickness w of the dielectric

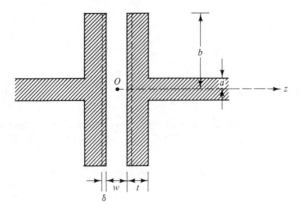

Figure 9.7–1 Circular parallel plate capacitor in series with a long conductor.

is a very small fraction of a wavelength ($w \ll \lambda$), A_z is essentially constant in z between the metal plates so that $\partial^2 A_z/\partial z^2 \sim 0$ and the relevant equation reduces to

$$\frac{1}{\rho}\frac{\partial}{\partial\rho}\left(\rho\,\frac{\partial A_{2z}}{\partial\rho}\right) + k_2^2 A_{2z} = 0; \qquad -\frac{w}{2} \le z \le \frac{w}{2} \qquad (9.7-1)$$

The solution is

$$A_{2z}(\rho) = D_1 J_0(k_2\rho) = A_{2z}(b)\,\frac{J_0(k_2\rho)}{J_0(k_2 b)}; \qquad \rho \le b \qquad (9.7-2)$$

Here

$$k_2 = \beta_2 - j\alpha_2 \quad \text{with } \beta_2 \sim \omega\sqrt{\mu_0\epsilon_{e2}}; \qquad \alpha_2 \sim \frac{\sigma_{e2}}{2}\sqrt{\frac{\mu_0}{\epsilon_{e2}}} \qquad (9.7-3)$$

Also,

$$\frac{\alpha_2^2}{\beta_2^2} = \left(\frac{\sigma_{e2}}{2\omega\epsilon_{e2}}\right)^2 \ll 1 \qquad (9.7-4)$$

The electric and magnetic fields in the dielectric are given by (3.9–22a,b) with $\kappa_2 = k_2$. That is,

$$E_{2z}(\rho) = -j\omega A_{2z}(\rho); \qquad B_{2\theta}(\rho) = -\frac{\partial A_{2z}(\rho)}{\partial\rho} \qquad (9.7-5)$$

It follows that

$$E_{2z}(\rho) = E_{2z}(b)\,\frac{J_0(k_2\rho)}{J_0(k_2 b)}; \qquad B_{2\theta}(\rho) = B_{2\theta}(b)\,\frac{J_1(k_2\rho)}{J_1(k_2 b)} \qquad (9.7-6)$$

The current that enters the lower capacitor plate from the wire is I_z. By

Stokes's theorem with the contour $s = 2\pi a$ and a cap surface $S_1 = \pi a^2$ that is a flat disk across the conductor,

$$\mu_0^{-1} \oint_s B_\theta(a)a \ d\theta = \mu_0^{-1} 2\pi b B_\theta(a) = I_z \tag{9.7-7}$$

With the same contour $s = 2\pi a$ but a cap surface S_2 that encloses the entire metal plate, Stokes's theorem gives

$$\mu_0^{-1} \oint_s B_\theta(a)a \ d\theta = -j\omega\epsilon_0 \int_a^b E_z(\rho)2\pi\rho \ d\rho + j\omega\tilde{\epsilon}_2 \int_0^b E_{2z}(\rho)2\pi\rho \ d\rho \tag{9.7-8}$$

The integral over the edge of the dielectric with area $2\pi bw$ is neglected since it is small and E_ρ on its surface is also small. The first integral in (9.7-8) is over the outside surface of the plate where $E_z(\rho)$ is very small compared to $E_{2z}(\rho)$ on the inside of the plate. Hence this integral can be neglected. It follows that

$$I_z \sim j\omega\tilde{\epsilon}_2 \int_0^b E_{2z}(\rho)2\pi\rho \ d\rho \tag{9.7-9}$$

With (9.7-6) that gives

$$I_z \sim \frac{j\omega\tilde{\epsilon}_2 E_{2z}(b)}{J_0(k_2 b)} \int_0^b J_0(k_2\rho)2\pi\rho \ d\rho = \frac{j\omega\tilde{\epsilon}_2 \cdot 2\pi b E_{2z}(b)}{k_2 J_0(k_2 b)} J_1(k_2 b) \tag{9.7-10}$$

$$E_{2z}(\rho) = \frac{1}{j\omega\tilde{\epsilon}_2} \frac{I_z}{\pi b^2} \frac{k_2 b}{2} \frac{J_0(k_2\rho)}{J_1(k_2 b)} \tag{9.7-11}$$

Similarly,

$$B_{2\theta}(\rho) = \frac{\mu_0 I_z}{2\pi b} \frac{J_1(k_2\rho)}{J_1(k_2 b)} \tag{9.7-12}$$

The surface density of charge on the inside surface of the lower plate is

$$\eta(\rho) \sim -\tilde{\epsilon}_2 E_{2z}(\rho) \tag{9.7-13}$$

That on the upper plate is the negative of that on the lower plate. Evidently,

$$\frac{\eta(\rho)}{\eta(0)} = \frac{E_{2z}(\rho)}{E_{2z}(0)} = J_0(k_2\rho) \tag{9.7-14}$$

This gives the radial distribution of charge on the lower plate. When $|k_2 b| \ll 1$, $J_0(k_2\rho) \sim 1$ and the charge per unit surface is the same at all radii. The internal impedance of the capacitor is $Z^i = wz^i$, where

$$z^i = \frac{E_z(b)}{I_z} = \frac{1}{\pi b^2} \frac{k_2 b}{2j\omega\tilde{\epsilon}_2} \frac{J_0(k_2 b)}{J_1(k_2 b)} \tag{9.7-15}$$

Most capacitors are used at frequencies that are sufficiently low so that $|k_2 b| \ll 1$.

In this case $J_0(k_2b) \sim 1$, $J_1(k_2b) \sim k_2b/2$, so that

$$Z^i = wz^i = \frac{w}{\pi b^2} \frac{1}{j\omega\tilde{\epsilon}_2} = \frac{w}{\pi b^2(\sigma_{2e} + j\omega\epsilon_{2e})} \tag{9.7-16}$$

Let $C_0 \equiv \epsilon_{2e}\pi b^2/w$ be the capacitance, $R_0^i \equiv w/\sigma_{2e}\pi b^2$ the internal series resistance of the capacitor. Then the internal impedance of the capacitor is

$$Z^i = \frac{1}{R_0^i\omega^2C_0^2} - \frac{j}{\omega C_0} \tag{9.7-17}$$

At higher frequencies $Z^i = wz^i$, where z^i is given by (9.7–15).

9.8 IMPEDANCE OF A CIRCULAR CAPACITOR IN A QUASI-NEAR-ZONE CIRCUIT

Figure 9.8–1 shows a circular capacitor symmetrically connected in a rectangular loop of wire with a point generator at AB. The internal impedance Z^i determined in Sec. 9.7 is the impedance between the edges D and E. In order to determine the self-impedance of the entire circuit as observed at AB, it is necessary to evaluate (9.2–1a) around the circuit

$$V_{10}^e = \oint_{s_1} z^i I_{1s} \, ds_1 + j\omega \oint_{s_1} (\mathbf{A}_{11} \cdot d\mathbf{s}_1) \tag{9.8-1}$$

The contour integration for the first term in (9.8–1) can be carried out around a closed contour along the surface of the wire forming the rectangle from B to C, then radially outward along the outer surface of the capacitor plate and around its edge to D, straight across from D to E along the edge, and radially inward to F and finally back to A along the rectangle. From A to C and from F back to A, the circuit is like any other rectangle. For a uniform wire with an internal impedance z_w^i and length s_w, the result is

$$I_1Z_{1w}^i = I_1z_w^i s_w \tag{9.8-2}$$

The contribution by the radial integral from C to D is always negligible for highly

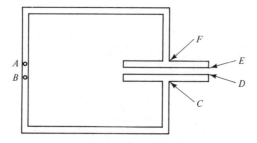

Figure 9.8–1   Circuit with capacitor.

conducting plates. It follows that the entire internal impedance of the capacitor is given by

$$Z_c^i = Z^i \tag{9.8-3}$$

where Z^i is defined in (9.7–17).

In order to determine the contribution to the external impedance of the rectangle in Fig. 9.8–1 by the capacitor, it is necessary to examine the second contour integral in (9.8–1). If the rectangular loop is sufficiently large so that to a good approximation rotational symmetry exists on the outer surfaces of the capacitor plates, the only contributions to the vector potential directed radially along the plates are due to radial currents in the plates themselves. The contributions to the second line integral in (9.8–1) due to integrations radially along the plates and axially across their combined edges are negligible compared with the integration around the rest of the circuit. Therefore, the external impedance of the capacitor may be neglected and an excellent approximation of the entire impedance is the internal impedance Z^i between the edges of the wire surfaces of the plates as determined in Sec. 9.7. The impedance of a capacitor is independent of the rest of the quasi-near-zone circuit provided that connecting wires are perpendicular to the plates.

9.9 MUTUAL IMPEDANCE OF COAXIAL RINGS OF WIRE

The calculation of the impedance of loosely wound single-layer coils depends on the previous determination of the self-impedance of a single circular ring of wire and the mutual impedance of two such rings arranged coaxially. Since the self-impedance can be obtained directly from the integral for mutual impedance by a simple specialization of the parameters, it is convenient to solve first the problem of mutual impedance of two coaxial rings of different size.

Consider two coaxial circular wire rings of radii b_1 and b_2. The radii b_1 and b_2 as well as the axial separation h of the rings are sufficiently small so that all elements of both rings are in the near zone with respect to one another. That is,

$$k_0 R_{12} \ll 1; \qquad k_0 R_{11} \ll 1; \qquad k_0 R_{22} \ll 1 \tag{9.9-1}$$

As shown in Fig. 9.9–1, R_{12} is the distance between an element ds_1 on the inner edge of one ring and an element ds_2' on the axis of the wire of the other ring; R_{11} is the distance between an element ds_1 on the inner edge and an element ds_1' on the axis of the wire of one and the same ring. R_{22} is defined like R_{11} but for the second ring. Subject to (9.9–1), each ring has a current that is sensibly uniform in amplitude around the ring. Therefore, the distribution functions $f_1(s_1)$ and $f_2(s_2)$ for the currents I_{1s} and I_{2s} may be set equal to unity, and the subscript s may be omitted. Let the positive direction for current in each ring be taken in the direction of increasing θ. Each ring may be assumed driven by a slice generator at $\theta = 0$.

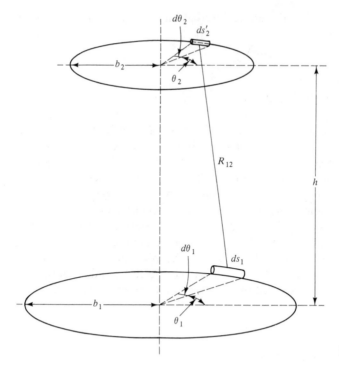

Figure 9.9–1 Coaxial rings of wire.

The quasi-near-zone formula for the mutual impedance $Z_{12} = R_{12} + jX_{12}$ of ring 1 due to the current I_2 in ring 2 is obtained from (9.6–6a,b). It is

$$Z_{12} = R_{12} + jX_{12} = R_{12} + j\omega L_{12} \tag{9.9–2}$$

with

$$L_{12} = \frac{\mu_0}{4\pi} \oint_{s1} \oint_{s2} \frac{d\mathbf{s}_1 \cdot d\mathbf{s}_2'}{R_{12}} \tag{9.9–3}$$

$$R_{12} = -\frac{k_0^4 \zeta_0}{24\pi} \oint_{s1} \oint_{s2} R_{12}^2 \, (d\mathbf{s}_1 \cdot d\mathbf{s}_2') \tag{9.9–4}$$

The cylindrical coordinates (ρ, θ, z) of the two elements are

$$d\mathbf{s}_1 \, (b_1, \theta_1, 0); \qquad d\mathbf{s}_2' \, (b_2, \theta_2, h) \tag{9.9–5}$$

The radii a_1 and a_2 of the wires are assumed to be small compared with b_1, b_2, and h. That is,

$$a_1^2 \ll b_1^2; \qquad a_2^2 \ll b_2^2; \qquad a_{1,2}^2 \ll h^2 \tag{9.9–6}$$

The distance between the elements $d\mathbf{s}_1$ and $d\mathbf{s}_2'$ is most easily obtained by determining first the distance r_{12} between $d\mathbf{s}_1$ and the projections of $d\mathbf{s}_2'$ onto the plane of ring 1. This plane is shown in Fig. 9.9–2, from which

$$r_{12} \doteq [b_1^2 + b_2^2 - 2b_1b_2 \cos(\theta_1 - \theta_2)]^{1/2} \tag{9.9-7}$$

The distance R_{12} between ds_1 and ds_2' follows directly.

$$R_{12} = (h^2 + r_{12}^2)^{1/2} = [h^2 + b_1^2 + b_2^2 - 2b_1b_2 \cos(\theta_1 - \theta_2)]^{1/2} \tag{9.9-8}$$

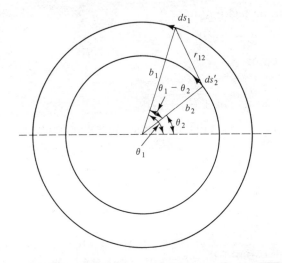

Figure 9.9–2 Construction for determining r_{12} for coaxial rings.

Since the elements ds_1 and ds_2' are tangent to the circles and hence perpendicular to the radii drawn from the common axis of the circles to the elements, the angle between ds_1 and ds_2' is the same as the angle between these radii. Hence

$$d\mathbf{s}_1 \cdot d\mathbf{s}_2' = ds_1 \, ds_2' \cos(\theta_1 - \theta_2)$$

Also,

$$ds_1 = b_1 \, d\theta_1; \qquad ds_2' = b_2 \, d\theta_2 \tag{9.9-9}$$

so that (9.9–3) and (9.9–4) become

$$L_{12} = \frac{\mu_0}{4\pi} \int_0^{2\pi} b_1 \, d\theta_1 \int_0^{2\pi} \frac{b_2 \cos(\theta_1 - \theta_2) \, d\theta_2}{[h^2 + b_1^2 + b_2^2 - 2b_1b_2 \cos(\theta_1 - \theta_2)]^{1/2}} \tag{9.9-10}$$

$$R_{12} = \frac{-\zeta_0 k_0^4}{24\pi} \int_0^{2\pi} b_1 \, d\theta_1 \int_0^{2\pi} [h^2 + b_1^2 + b_2^2 - 2b_1b_2 \cos(\theta_1 - \theta_2)]b_2 \cos(\theta_1 - \theta_2) \, d\theta_2 \tag{9.9-11}$$

The evaluation of (9.9–10) may be carried out as follows. Let $\phi = \theta_2 - \theta_1$; $d\phi = d\theta_2$. Since the integration is completely around the ring, the starting point is of no consequence and the limits are unchanged. With $(1 + \cos \phi) = 2 \cos^2(\phi/2)$, (9.9–10) becomes

$$L_{12} = \frac{\mu_0 b_1 b_2}{4\pi} \int_0^{2\pi} d\theta_1 \int_0^{2\pi} \frac{(2\cos^2\frac{1}{2}\phi - 1) \, d\phi}{[h^2 + (b_1 + b_2)^2 - 4b_1b_2 \cos^2\frac{1}{2}\phi]^{1/2}} \tag{9.9-12}$$

Let the following change in variable and in limits be made in the second integral in (9.9–12), which will be denoted by A: $\phi = \pi - 2\psi$; $d\phi = -2\,d\psi$. At $\phi = 0$, $\psi = \pi/2$; at $\phi = 2\pi$, $\psi = -\pi/2$. Thus, with $\cos\frac{1}{2}\phi = \cos\left(\frac{1}{2}\pi - \psi\right) = \sin\psi$,

$$A = 2\int_{-\pi/2}^{\pi/2} \frac{(2\sin^2\psi - 1)d\psi}{[h^2 + (b_1 + b_2)^2 - 4b_1 b_2 \sin^2\psi]^{1/2}} \tag{9.9–13}$$

Here the denominator can be factored into

$$[h^2 + (b_1 + b_2)^2]^{1/2}(1 - k^2\sin^2\psi)^{1/2} \tag{9.9–14}$$

with the parameter k^2 given by

$$k^2 \equiv \frac{4b_1 b_2}{h^2 + (b_1 + b_2)^2} \tag{9.9–15}$$

Using (9.9–14) in (9.9–13) and noting that the integrand is independent of a change in the sign of ψ,

$$A = \frac{4}{[h^2 + (b_1 + b_2)^2]^{1/2}}\left[\int_0^{\pi/2} \frac{2\sin^2\psi\,d\psi}{(1 - k^2\sin^2\psi)^{1/2}} - \int_0^{\pi/2} \frac{d\psi}{(1 - k^2\sin^2\psi)^{1/2}}\right] \tag{9.9–16}$$

The integrand of the first integral in (9.9–16) is equivalent to

$$\frac{2}{k^2}\left[\frac{1}{(1 - k^2\sin^2\psi)^{1/2}} - (1 - k^2\sin^2\psi)^{1/2}\right] \tag{9.9–17}$$

as can be verified directly. Hence (9.9–16) is equivalent to

$$A = \frac{4}{[h^2 + (b_1 + b_2)^2]^{1/2}}\left[-\frac{2}{k^2}\int_0^{\pi/2}(1 - k^2\sin^2\psi)^{1/2}\,d\psi\right.$$

$$\left. + \left(\frac{2}{k^2} - 1\right)\int_0^{\pi/2}\frac{d\psi}{(1 - k^2\sin^2\psi)^{1/2}}\right] \tag{9.9–18}$$

The two integrals in (9.9–18) are of a standard type known as *complete elliptic integrals of the first and second kinds*. They are denoted by

$$K\left(\frac{\pi}{2}, k\right) = \int_0^{\pi/2}\frac{d\psi}{(1 - k^2\sin^2\psi)^{1/2}} \tag{9.9–19}$$

$$E\left(\frac{\pi}{2}, k\right) = \int_0^{\pi/2}(1 - k^2\sin^2\psi)^{1/2}\,d\psi \tag{9.9–20}$$

The parameter k is often represented in the form $k = \sin\alpha$, so that

$$K\left(\frac{\pi}{2}, \alpha\right) = \int_0^{\pi/2}\frac{d\psi}{(1 - \sin^2\alpha\sin^2\psi)^{1/2}} \tag{9.9–21}$$

$$E\left(\frac{\pi}{2}, \alpha\right) = \int_0^{\pi/2}(1 - \sin^2\alpha\sin^2\psi)^{1/2}\,d\psi \tag{9.9–22}$$

For every value of the parameter k or α, each integral defines a definite number. These numbers are extensively listed in standard tables. With the notations (9.9–19) and (9.9–20) in (9.9–16), this may be substituted for the second integral in (9.9–12). The integration with respect to θ_1 can be performed directly to give 2π.

$$L_{12} = \frac{\mu_0 2b_1 b_2}{[h^2 + (b_1 + b_2)^2]^{1/2}} \left[-\frac{2}{k^2} E\left(\frac{\pi}{2}, k\right) + \left(\frac{2}{k^2} - 1\right) K\left(\frac{\pi}{2}, k\right) \right] \quad (9.9\text{–}23)$$

With (9.9–15) this reduces to

$$L_{12} = \mu_0 (b_1 b_2)^{1/2} \left[-\frac{2}{k} E\left(\frac{\pi}{2}, k\right) + \left(\frac{2}{k} - k\right) K\left(\frac{\pi}{2}, k\right) \right] \quad (9.9\text{–}24)$$

This is the final formula for the mutual inductance of two circular rings entirely within the near zone with respect to each other. When the rings have the same radius, $b_1 = b_2 = b$ and $k = 2b/[h^2 + (2b)^2]^{1/2}$.

The mutual resistance R_{12} is quickly determined from (9.9–11). With the change in variable $\phi = \theta_1 - \theta_2$,

$$R_{12} = -\frac{\zeta_0 k_0^4}{24\pi} \int_0^{2\pi} b_1 \, d\theta_1 \int_0^{2\pi} [(h^2 + b_1^2 + b_2^2) b_2 \cos\phi - 2b_1 b_2^2 \cos^2\phi] \, d\phi \quad (9.9\text{–}25)$$

This is readily integrated to give

$$R_{12} = \frac{\zeta_0}{6\pi} k_0^4 S_1 S_2 \quad (9.9\text{–}26)$$

where $S_1 = \pi b_1^2$ and $S_2 = \pi b_2^2$ are the areas enclosed by the two rings. With $\zeta_0 \doteq 120\pi$ ohms,

$$R_{12}^e = 20 k_0^4 S_1 S_2 \quad (9.9\text{–}27)$$

The mutual resistance of two coaxial rings with uniform currents depends on the area of each ring but not on their axial separation.

9.10 SELF-IMPEDANCE OF A CIRCULAR RING

The external self-impedance of a single ring of wire of radius b measured to the inner edge and driven by a point generator at $\theta = 0$ is readily obtained with (9.9–24) and (9.9–27). It is

$$Z_1^e = R_1^e + jX_1^e = R_1^e + j\omega L_1^e \quad (9.10\text{–}1)$$

with

$$L_1^e = \frac{\mu_0}{4\pi} \oint_{s_1} \oint_{s_1} \frac{d\mathbf{s}_1 \cdot d\mathbf{s}_1'}{R_{11}} \quad (9.10\text{–}2)$$

$$R_1^e = -\frac{\zeta_0 k_0^4}{24\pi} \oint_{s_1} \oint_{s_1} R_{11}^2 (d\mathbf{s}_1 \cdot d\mathbf{s}_1') \quad (9.10\text{–}3)$$

These expressions are the same as (9.9–3) and (9.9–4) except that the primed element is in the center of the wire in ring 1 with radius $(b_1 + a_1)$ instead of in ring 2. The unprimed element is on the inner edge of the ring, that is, at the radius b_1. Accordingly, (9.10–2) and (9.10–3) are like (9.9–3) and (9.9–4) with $(b_1 + a_1)$ written for b_2 and $h = 0$. It follows from (9.9–24) that

$$L_1^e = \mu_0[b_1(b_1 + a_1)]^{1/2}\left[-\frac{2}{k} E\left(\frac{\pi}{2}, k\right) + \left(\frac{2}{k} - k\right)K\left(\frac{\pi}{2}, k\right)\right] \qquad (9.10\text{–}4)$$

with

$$k = \frac{2[b_1(b_1 + a_1)]^{1/2}}{2b_1 + a_1} = \left[1 - \left(\frac{a_1}{2b_1 + a_1}\right)^2\right]^{1/2} \qquad (9.10\text{–}5)$$

The quantity

$$k' = (1 - k^2)^{1/2} = \frac{a_1}{2b_1 + a_1} \qquad (9.10\text{–}6)$$

is small because the following inequality has been imposed: $a_1^2 \ll b_1^2$. The elliptic integrals can be expanded in series in powers of the small quantity k'. Subject to $a_1^2 \ll b_1^2$, which is equivalent to $(k')^2 \ll 1$, it is sufficient to retain the leading terms. These are

$$K \doteq \ln\frac{4}{k'} = \ln\left(\frac{8b_1}{a_1} - 4\right) \qquad (9.10\text{–}7)$$

$$E \doteq 1 \qquad (9.10\text{–}8)$$

With these values in (9.10–4) and with $b_1 \gg a_1$,

$$L_1^e \doteq \mu_0 b_1\left(\ln\frac{8b_1}{a_1} - 2\right) \qquad (9.10\text{–}9)$$

in agreement with (9.5–26). This may be written in terms of the total length of wire in the ring, $s_t = 2\pi b_1$:

$$L_1^e = \frac{\mu_0 s_t}{2\pi}\left(\ln\frac{4s_t}{\pi a_1} - 2\right) = \frac{\mu_0 s_t}{2\pi}\left(\ln\frac{s_t}{a_1} - 1.76\right) \qquad (9.10\text{–}10)$$

The external or radiation self-resistance R_1^e is obtained directly from (9.9–27) with the substitution of $(b_1 + a_1)$ for b_2. In most cases, a_1 is sufficiently small so that $\pi(b_1 + a_1)^2 \doteq \pi b_1^2 = S_1$. Then

$$R_1^e = 20k_0^4 S_1^2 \qquad (9.10\text{–}11)$$

It is interesting to note that for loops which are confined to the near zone, the self-inductance depends primarily on the *total length of wire* in a way that is roughly independent of the shape of the loop, whereas the radiation resistance depends entirely on the *area*.

The total self-impedance of a circular loop is

$$Z_{11} = Z_1^i + Z_1^e = (R_1^i + R_1^e) + j(X_1^i + \omega L_1^e) \qquad (9.10\text{–}12)$$

R_1^i and X_1^i are given by

$$R_1^i = r_1^i s_t; \qquad X_1^i = x_1^i s_t \qquad (9.10\text{–}13)$$

where r_1^i and x_1^i are the internal resistance and reactance per unit length of a circular conductor with rotationally symmetrical current. For a loop that is large compared with the radius of the wire, the distribution of current in a cross section departs only in a negligible degree from rotational symmetry. Formulas for r_1^i and x_1^i are in Sec. 3.9.

As a numerical example, consider a circular loop of copper wire with circumference 0.245 m and radius of wire 4×10^{-4} m at a frequency of 159×10^6 Hz. This satisfies the condition for the near zone, $k_0 R_{11} \ll 1$, since $k_0 R_{11} \doteq 0.26$ with R_{11} the diameter. With (9.10–10) and (9.10–11),

$$L_1^e = \frac{\mu_0 s_t}{2\pi}\left(\ln\frac{s_t}{a} - 1.76\right) = 0.228 \times 10^{-6} \text{ H}$$

$$X_1^e = \omega L_1^e = 228 \text{ } \Omega$$

$$R_1^e = 20 k_0^4 S_1^2 = 0.056 \text{ } \Omega$$

At the given high frequency and with $\sigma = 5.8 \times 10^7$ S/m, $\mu = \mu_0$, and

$$r_1^i = x_1^i = \frac{1}{2\pi a}\sqrt{\frac{\mu\omega}{2\sigma}}$$

it follows that

$$R_1^i = X_1^i = 0.245 \times 1.31 = 0.321 \text{ } \Omega$$

Hence

$$Z_{11} = (R_1^i + R_1^e) + j(X_1^i + X_1^e) = 0.38 + j228 \text{ } \Omega$$

The corresponding values for the square made of the same wire are

$$L_1^e = \frac{\mu_0 s_t}{2\pi}\left(\ln\frac{s_t}{a} - 2.16\right) = 0.209 \times 10^{-6} \text{ H}$$

$$X_1^e = \omega L_1^e = 209 \text{ } \Omega$$

$$R_1^e = 20(k_0 s)^4 = 0.034 \text{ } \Omega$$

$$R_1^i = X_1^i = 0.321 \text{ } \Omega$$

$$Z_{11} = 0.35 + j209 \text{ } \Omega$$

The difference between $R_1^i + R_1^e$ for the circle and the square is due to the greater external or radiation resistance of the circle; the difference in the values of $X_1^i + X_1^e$ is due to the greater external self-inductance of the circle.

9.11 IMPEDANCE OF A HELICAL COIL

The circuit of Fig. 9.11–1 consists of a point generator between A and B; of a helical coil of wire of total length s_c, radius of turns b, and pitch of turns p between C and D; and of a rectangle of connecting wire. The dimensions of the rectangle are s_1 and s_2. All the wire is the same in size and material; its radius is a. The circuit as a whole satisfies the condition for the quasi-near zone so that the distribution function is unity and the current uniform. The impedance of the circuit is

$$Z_{11} = (R_1^i + R_1^e) + j(X_1^i + X_1^e) \qquad (9.11–1)$$

The internal resistance and reactance R_1^i and X_1^i may be computed from the formulas for r_1^i and x_1^i (Sec. 3.9) for a straight wire of length s_t if the turns of the helix are far enough apart so that the distribution of current in the interior of the conductor is approximately rotationally symmetrical. If the turns are close together, more intricate formulas must be derived for r_1^i and x_1^i taking into account the departure from rotational symmetry. For simplicity, it will be assumed below that the coil is loosely wound whenever numerical values of r_1^i and x_1^i are required. The internal impedance of the circuit of total length of wire s_t is

$$Z_1^i = R_1^i + jX_1^i; \qquad R_1^i = r_1^i s_t; \qquad X_1^i = x_1^i s_t \qquad (9.11–2)$$

The contribution due to the coil is

$$Z_{1c}^i = R_{1c}^i + jX_{1c}^i; \qquad R_{1c}^i = r_1^i s_c; \qquad X_{1c}^i = x_1^i s_c \qquad (9.11–3)$$

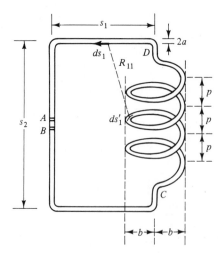

Figure 9.11–1 Circuit including a helical coil. A point generator is between A and B.

The external reactance and resistance in (9.11–1) are defined by (9.5–6) and (9.5–7):

$$X_1^e = \omega L_1^e; \qquad L_1^e = \frac{\mu_0}{4\pi} \oint_{s_1}\oint_{s_1} \frac{d\mathbf{s}_1 \cdot d\mathbf{s}_1'}{R_{11}} \tag{9.11-4}$$

$$R_1^e = \frac{-\zeta_0 k_0^4}{24\pi} \oint_{s_1}\oint_{s_1} R_{11}^2 \, d\mathbf{s}_1 \cdot d\mathbf{s}_1' \tag{9.11-5}$$

As usual, $d\mathbf{s}_1'$ is an element along the axis of the wire; $d\mathbf{s}_1$ is an element along the surface of the wire at its inner edge; R_{11} is the distance between $d\mathbf{s}_1$ and $d\mathbf{s}_1'$. Because the wire forming the coil is not in the same plane as the rest of the circuit, the exact evaluation of (9.11–4) and (9.11–5) is difficult. The problem is very much simplified if certain approximations are made. These depend on the fact that the uniform current in the helix may be considered to be equivalent approximately to a current directed along the axis of the coil from C to D (Fig. 9.11–1) and a series of equal currents in identical circular rings spaced at axial distances p. That is, in evaluating the *external* impedance, the circuit of Fig. 9.11–1 is replaced by that of Fig. 9.11–2. It is assumed that the gaps in the axial conductor between C and D are very small, as are those in the circular rings. The two radial conductors extending from the axis to the circumference of each ring are very close together, almost parallel, and have equal and opposite currents. Since they are either at right angles to other conductors or are symmetrically placed with respect to conductors carrying equal and opposite currents, all integrations along them cancel. That is, the currents in these pairs of conductors contribute nothing to the external impedance. In the arrangement in Fig. 9.11–2 contributions to the contour integral when one of the elements $d\mathbf{s}_1$ and $d\mathbf{s}_1'$ is on a circle of wire and the other is not are *zero* because either the elements are mutually perpendicular so that $d\mathbf{s}_1 \cdot d\mathbf{s}_1'$ vanishes or another pair of elements exists with an equal contribution of opposite sign. The

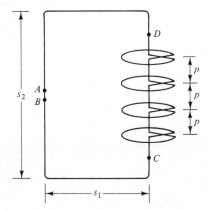

Figure 9.11–2 Circuit approximately equivalent to that of Fig. 9.11–1. For simplicity, the cross section of the wire is not shown.

integral for the complete circuit thus separates into two independent parts in the quasi-near-zone case. The first part is the rectangle of wire; the second part consists of the circular loops of wire. The contour integrals for the circuit thus reduce to components that have already been evaluated, namely, a rectangle and coaxial circular rings. It is necessary merely to add together all the contributions. These include the external self-impedance of the rectangle Z_r^e, the external self-impedance of n identical coaxial rings, and the sum of all mutual impedances between ring i and rings $j = 1$ to $n - 1$, with i taking on all values from 1 to n. The external impedance of the rectangle is Z_r^e, the external self-impedance of ring number i is Z_{ii}^e, the mutual impedance between ring i and ring j is Z_{ij}^e. Assuming the radius a of the wire to be sufficiently small, these impedances are given below:

$$Z_r^e = R_r^e + jX_r^e = R_r^e + j\omega L_r^e \tag{9.11-6}$$

with

$$R_r^e = 20k_0^4 s_1^2 s_2^2 \tag{9.11-7}$$

$$L_r^e = \frac{\mu_0}{\pi} \left\{ s_1 \ln \frac{2s_1 s_2}{a[s_1 + (s_1^2 + s_2^2)^{1/2}]} + s_2 \ln \frac{2s_1 s_2}{a[s_2 + (s_1^2 + s_2^2)^{1/2}]} \right.$$

$$\left. + 2[a + (s_1^2 + s_2^2)^{1/2} - s_1 - s_2] \right\} \tag{9.11-8}$$

as given in (9.5–18) and (9.5–21) with appropriate changes in symbols.

$$Z_{ii}^e = R_{ii}^e + jX_{ii}^e = R_{ii}^e + j\omega L_{ii}^e \tag{9.11-9}$$

with

$$R_{ii}^e = 20\pi^2(k_0 b)^4 \tag{9.11-10}$$

$$L_{ii}^e = \mu_0 b \left(\ln \frac{8b}{a} - 2 \right) \tag{9.11-11}$$

as given in (9.10–11) and (9.10–9).

$$Z_{ij}^e = R_{ij}^e + jX_{ij}^e = R_{ij}^e + j\omega L_{ij}^e \tag{9.11-12}$$

where, using (9.9–27) and (9.9–24) with $b_1 \sim b_2 \sim b$,

$$R_{ij}^e = 20\pi^2(k_0 b)^4 \tag{9.11-13}$$

$$L_{ij}^e = \mu_0 b \left[-\frac{2}{k_{ij}} E\left(\frac{\pi}{2}, k_{ij}\right) + \left(\frac{2}{k_{ij}} - k_{ij}\right) K\left(\frac{\pi}{2}, k_{ij}\right) \right] \tag{9.11-14}$$

and

$$k_{ij} = \frac{2b}{(h_{ij}^2 + 4b^2)^{1/2}} = \frac{1}{[1 + (h_{ij}/2b)^2]^{1/2}} \tag{9.11-15}$$

The axial distance between coil i and coil j is denoted by h_{ij}. Since the pitch is uniform,

$$h_{ij} = p|i - j| \qquad (9.11-16)$$

Let the ratio of pitch to diameter be denoted by

$$\delta \equiv \frac{p}{2b} \qquad (9.11-17)$$

With (9.11–16) and (9.11–17), (9.11–15) becomes

$$k_{ij} = \frac{1}{[1 + \delta^2(i - j)^2]^{1/2}}; \qquad i \neq j \qquad (9.11-18)$$

The external impedance of the circuit is given by

$$Z_{11}^e = \left(R_r^e + \sum_{j=1}^{n} \sum_{i=1}^{n} R_{ij}^e \right) + j\omega \left(L_r^e + \sum_{j=1}^{n} \sum_{i=1}^{n} L_{ij}^e \right) \qquad (9.11-19)$$

The evaluation of the double sum of the L_{ij}^e is straightforward but tedious if there are many turns. If the pitch and diameter are given, the elliptic integrals for successive values of k_{ij} as defined in (9.11–18) may be obtained from tables or curves, and L_{ij} evaluated for $i \neq j$ using (9.11–14). The number of turns may be halved if it is recalled that according to the reciprocal theorem, $L_{ij} = L_{ji}$. That is,

$$\sum_{j=1}^{n} \sum_{i=1}^{n} L_{ij} = \sum_{j=1}^{n} L_{jj} + 2\left(\sum_{j=2}^{n} L_{1j} + \sum_{j=3}^{n} L_{2j} + \cdots + L_{(n-1)n} \right) \qquad (9.11-20)$$

If the pitch is large compared with the radius of the wire so that

$$p^2 \gg a^2 \qquad (9.11-21)$$

the contributions to the double sum by all but the first two turns on each side of every coil are negligible. The condition (9.11–20) is also the one required to permit calculation of the internal impedance using formulas that imply rotational symmetry in the interior of the wire. The determination of the double sum of the R_{ij}^e is simple because $R_{ii}^e = R_{ij}^e = 20\pi^2(k_0 b)^4$. Hence

$$R_{11}^e = R_r^e + n^2 R_{ii}^e = 20k_0^4(s_1^2 s_2^2 + n^2\pi^2 b^4) \qquad (9.11-22)$$

As an example, consider a coil of five turns with radius equal to pitch

$$b = p \qquad (9.11-23)$$

so that

$$\delta = \frac{p}{2b} = \frac{1}{2} \qquad (9.11-24)$$

The radius of the wire satisfies (9.11–21). The possible values of $(i = j)$, $i \neq j$

include 1, 2, 3, 4, so that L_{ij} for the corresponding four values of k_{ij} are required. These values are tabulated below:

$i = j$	k_{ij}	$\dfrac{2}{k_{ij}}$	$\dfrac{2}{k_{ij}} - k_{ij}$	E	K	L_{ij}
1	0.895	2.24	1.34	1.18	2.25	$0.37\mu_0 b$
2	0.707	2.83	2.12	1.35	1.85	$0.12\mu_0 b$
3	0.555	3.60	3.04	1.44	1.72	$0.04\mu_0 b$
4	0.447	4.46	4.01	1.48	1.66	$0.02\mu_0 b$

$$\sum_{j=2}^{5} L_{1j} = \mu_0 b(0.37 + 0.12 + 0.04 + 0.02) \qquad = 0.55\mu_0 b \qquad (9.11{-}25\text{a})$$

$$\sum_{j=3}^{5} L_{2j} = \mu_0 b(0.37 + 0.12 + 0.04) \qquad = 0.53\mu_0 b \qquad (9.11{-}25\text{b})$$

$$\sum_{j=4}^{5} L_{3j} = \mu_0 b(0.37 + 0.12) \qquad = 0.49\mu_0 b \qquad (9.11{-}25\text{c})$$

$$L_{45} = \qquad = 0.37\mu_0 b \qquad (9.11{-}25\text{d})$$

$$2\left(\sum_{j=2}^{5} L_{1j} + \sum_{j=3}^{5} L_{2j} + \sum_{j=4}^{5} L_{3j} + L_{45}\right) = 2(1.95)\mu_0 b = 3.90\mu_0 b \qquad (9.11{-}26)$$

The external inductance of the entire circuit including the five-turn coil is

$$L_{11}^e = L_r^e + \mu_0 b\left[5\left(\ln\frac{8b}{a} - 2\right) + 3.90\right]$$

$$= L_r^e + \mu_0 b\left(5\ln\frac{8b}{a} - 6.11\right) \qquad (9.11{-}27)$$

To continue the numerical data, let the length of wire in the coil be $s_c = 0.245$ m so that the length of each turn is 0.049 m. Since this length is $[p^2 + (2\pi b)^2]^{1/2}$ and $p = b$, it follows that

$$b(1 + 4\pi^2)^{1/2} = 0.049 \text{ m}$$

or

$$p = b = 0.77 \text{ cm} \qquad (9.11{-}28)$$

Let the radius of the wire be

$$a = 0.04 \text{ cm} \qquad (9.11{-}29)$$

Let the rectangle be a square of side 3.85 cm so that the coil replaces one of the four sides. Hence, with (9.5–24),

$$L_r^e = \frac{0.077 \times 4\pi}{\pi \times 10^7} \left(\ln \frac{3.85}{0.04} - 0.77 \right) = 0.114 \times 10^{-6} \text{ H} \qquad (9.11\text{–}30)$$

The use of (9.11–27) leads to

$$L_{11}^e = L_r^e + \frac{0.77 \times 10^{-2} \times 4\pi}{10^7} \left(5 \ln \frac{6.16}{0.04} - 6.11 \right) \text{ H} \qquad (9.11\text{–}31)$$

$$L_{11}^e = 0.295 \times 10^{-6} \text{ H} \qquad (9.11\text{–}32)$$

It is interesting to compare this value with a square with the same total length of wire (i.e., $s_t = s_c + 3 \times 3.85 = 36.05$ cm). The external self-inductance of the square is

$$L_{11}^e = \frac{0.1802 \times 4\pi}{\pi \times 10^7} \left(\ln \frac{9.01}{0.04} - 0.77 \right) = 0.336 \times 10^{-6} \text{ H} \qquad (9.11\text{–}33)$$

The small rectangle with one side replaced by a loosely wound coil of five turns is seen to have a somewhat smaller external self-inductance than the larger square with the same total length of wire. The difference is of order of magnitude of 10 percent.

If the rectangle is very long compared with its width and both are great compared with the radius of the wire, it is possible to write

$$a^2 \ll s_2^2 \ll s_1^2 \qquad (9.11\text{–}34)$$

The self-inductance of the straight piece of wire forming one of the short sides of a long rectangle is independent of the rest of the rectangle. If the helical coil between C and D in Fig. 9.11–1 or its approximate equivalent in Fig. 9.11–2 replaces one of the short sides of a long rectangle, the external inductance of the coil between C and D is also independent of the rest of the rectangle. The axial wire through the center is equivalent to the side of the rectangle, and the circular rings are symmetrically placed as discussed above. Hence the independent self-inductance of a helical coil of length $CD = s_2$ is

$$L_h^e = L_{s_2}^e + \sum_{j=1}^{n} \sum_{i=1}^{n} L_{ij}^e \qquad (9.11\text{–}35)$$

The value of $L_{s_2}^e$ is

$$L_{s_2}^e = \frac{\mu_0 s_2}{2\pi} \left(\ln \frac{2s_2}{a} - 1 \right) \qquad (9.11\text{–}36)$$

For the five-turn helix analyzed above,

$$L_h^e = \frac{\mu_0 s_2}{2\pi} \left(\ln \frac{2s_2}{a} - 1 \right) + \mu_0 b \left(5 \ln \frac{8b}{a} - 6.11 \right) \qquad (9.11\text{–}37)$$

If the rectangle is very long with shorter sides $s_2 = 3.85$ cm and the coil is dimensioned as above, the following inductances are computed:

$$L_{s_2}^e = \frac{0.0385 \times 4\pi}{2\pi \times 10^7} \left(\ln \frac{7.7}{0.04} - 1 \right) = 0.0405 \times 10^{-6} \text{ H} \tag{9.11-38}$$

The contribution to the inductance by the circular rings is the same as before, so that the external self-inductance of the loosely wound helical coil is

$$L_h^e = (0.0405 + 0.185) \times 10^{-6} = 0.226 \times 10^{-6} \text{ H} \tag{9.11-39}$$

The value may be compared with

$$L_1^e = 0.228 \times 10^{-6} \text{ H} \tag{9.11-40}$$

for a circular loop with the same total length of wire and with

$$L_1^e = 0.209 \times 10^{-6} \text{ H} \tag{9.11-41}$$

for a square with the same length of wire.

9.12 INDUCTANCE OF A LONG, CLOSELY WOUND COIL

If a helical coil of radius b and axial length h_c has a single layer of n closely wound turns, the double sum $\sum_{j=1}^n \sum_{i=1}^n L_{ij}$ in (9.11–35) may be evaluated approximately by a double integral. The integral is obtained by replacing the unit of summation 1 by $(n/h_c) \, dz$ in forming the integral. The limits in the sum, 1 and n, become 0 and h_c in the integral. Thus the following approximation is made:

$$L_c^e = \sum_{j=1}^n \sum_{i=1}^n L_{ij} \sim \left(\frac{n}{h_c} \right)^2 \int_0^{h_c} \int_0^{h_c} L_{z_1} L_{z_2} \, dz_1 \, dz_2 \tag{9.12-1}$$

This approximation improves as the coil becomes more like a uniform sheet of current around a cylinder of the same diameter as the coil. In (9.12—1),

$$L_{z_1 z_2} = \mu_0 b \left[-\frac{2}{k_z} E\left(\frac{\pi}{2}, k_z \right) + \left(\frac{2}{k_z} - k_z \right) K\left(\frac{\pi}{2}, k_z \right) \right] \tag{9.12-2}$$

with

$$k_z \equiv \frac{2b}{[(z_2 - z_1)^2 + 4b^2]^{1/2}} \tag{9.12-3}$$

The axial distance between the two elements dz_1 and dz_2 is $(z_2 - z_1)$. The evaluation of the integrals in (9.12–1) is tedious but can be carried out in closed form to give

$$L_c^e = \mu_0 n^2 b \frac{2}{3} \left[\frac{K(\pi/2, k) + (\tan^2 \alpha - 1)E(\pi/2, k)}{\sin \alpha} - \tan^2 \alpha \right] \tag{9.12-4}$$

where

$$k = \frac{2b}{(h_c^2 + 4b^2)^{1/2}} = \sin \alpha \qquad (9.12\text{-}5)$$

and

$$\tan \alpha = \frac{2b}{h_c} \qquad (9.12\text{-}6)$$

The term L_{s_2} in (9.11–35) contributes a negligible amount so that (9.12–4) is the entire external self-inductance of the coil. The factor

$$\frac{2\pi L_c^e}{\mu_0 n^2 b} = \frac{4\pi}{3} \left[\frac{K(\pi/2,\, k) + (\tan^2 \alpha - 1)E(\pi/2,\, k)}{\sin \alpha} - \tan^2 \alpha \right] \qquad (9.12\text{-}7)$$

is tabulated in standard tables. The internal impedance of a long, closely wound coil is not easily determined because rotational symmetry certainly does not obtain even approximately in the interior of the wires forming the winding and will not be analyzed here.

9.13 THE LOOP ANTENNA: THE INTEGRAL EQUATION FOR THE CIRCULAR TRANSMITTING LOOP

The formulation discussed in this section is for relatively thin loops. The loop antenna consists of a circular ring of perfectly conducting wire immersed in an infinite isotropic homogeneous medium characterized by the real effective relative constitutive parameters μ_r, ϵ_{er}, and σ_{er} and driven by an idealized delta-function generator. Let the center of the loop coincide with the origin of a system of cylindrical coordinates ρ, θ, and z as shown in Fig. 9.13–1. The loop lies in the plane $z = 0$ with the generator at $\theta = 0$. It is assumed that the radius b of the loop is large compared with the radius a of the wire and that the latter is small compared with the wavelength. That is,

$$a^2 \ll b^2; \qquad |ka|^2 \ll 1 \qquad (9.13\text{-}1)$$

where k is the propagation constant

$$k = \beta_e - j\alpha_e \qquad (9.13\text{-}2)$$

The total current in the conductor at θ' is $I(\theta')$ in the direction $ds' = b\, d\theta'$. The integral equation for $I(\theta')$ is obtained from the boundary condition that requires the tangential component of the electric field to vanish at the surface of the perfectly conducting wire except in a narrow region at $\theta = 0$ across which a voltage V_0^e is maintained. Thus, since $\mathbf{E} = -\nabla\phi - j\omega\mathbf{A}$,

$$E_\theta = \frac{-V_0^e\, \delta(\theta)}{b} = -\left(\frac{1}{\rho} \frac{\partial\phi}{\partial\theta} + j\omega A_\theta \right); \qquad \rho = b \qquad (9.13\text{-}3)$$

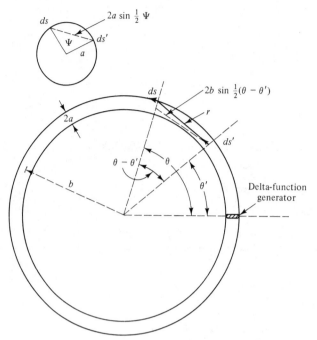

Figure 9.13–1 Circular loop antenna.

The scalar and vector potentials at the element $ds = b\,d\theta$ on the surface of the wire at θ are given by

$$\phi = \frac{1}{4\pi\bar{\epsilon}} \int_{-\pi}^{\pi} q(\theta')W(\theta - \theta')\,d\theta' \tag{9.13–4a}$$

$$A_\theta = \frac{\mu}{4\pi} \int_{-\pi}^{\pi} I(\theta')W(\theta - \theta')\cos(\theta - \theta')\,d\theta' \tag{9.13–4b}$$

where the kernel is

$$W(\theta - \theta') = \frac{b}{2\pi} \int_{-\pi}^{\pi} \frac{e^{-jkr}}{r}\,d\Psi \tag{9.13–4c}$$

with

$$r \doteq \left[4b^2\sin^2\frac{\theta - \theta'}{2} + A^2 \right]^{1/2} \tag{9.13–4d}$$

where

$$A \doteq 2a\sin\frac{\Psi}{2} \tag{9.13–4e}$$

Note that (9.13–4b) with (9.13–4c) assumes that the total current is given by

$$I(\theta) = 2\pi a K_\theta(\theta) \tag{9.13–5}$$

where $K_\theta(\theta)$ is the surface density of current on the perfectly conducting loop. The delta-function generator at $\theta = 0$ is so defined that $\int_{-\pi}^{\pi} bE_\theta \, d\theta = V_0^e$. The equation of continuity for the currents and charges in the loop is

$$\frac{1}{b} \frac{dI(\theta')}{d\theta'} + j\omega q(\theta') = 0 \qquad (9.13\text{-}6)$$

With (9.13–6) and (9.13–4a), it follows that

$$\frac{\partial \phi}{\partial \theta} = -\frac{j}{4\pi \bar\epsilon \omega b} \int_{-\pi}^{\pi} \frac{\partial I(\theta')}{\partial \theta'} \frac{\partial}{\partial \theta'} W(\theta - \theta') \, d\theta' \qquad (9.13\text{-}7)$$

Use has been made of the fact that $(\partial/\partial\theta)W(\theta - \theta') = -(\partial/\partial\theta')W(\theta - \theta')$. This integral may be integrated by parts to give

$$\frac{\partial \phi}{\partial \theta} = \frac{j}{4\pi \bar\epsilon \omega b} \frac{\partial^2}{\partial \theta^2} \int_{-\pi}^{\pi} I(\theta') W(\theta - \theta') \, d\theta' \qquad (9.13\text{-}8)$$

With (9.13–8) and (9.13–4b) substituted in (9.13–3), the following integral equation is obtained:

$$V_0^e \, \delta(\theta) = \frac{j\zeta}{4\pi} \int_{-\pi}^{\pi} K(\theta - \theta')I(\theta') \, d\theta' \qquad (9.13\text{-}9)$$

where the kernel is

$$K(\theta - \theta') = \left[kb\cos(\theta - \theta') + \frac{1}{kb} \frac{\partial^2}{\partial \theta^2} \right] W(\theta - \theta') \qquad (9.13\text{-}10)$$

ζ has been defined earlier as

$$\zeta = \frac{\omega\mu}{k} = \sqrt{\frac{\mu}{\bar\epsilon}} \qquad (9.13\text{-}11)$$

A solution of the integral equation (9.13–9) may be sought with the aid of Fourier-series expansions of the kernel and of the current. First, let the dimensionless quantity $W(\theta - \theta')$ be expanded as follows:

$$W(\theta - \theta') = \sum_{-\infty}^{\infty} K_m e^{-jm(\theta - \theta')} \qquad (9.13\text{-}12)$$

The coefficients K_m can be evaluated if both sides of (9.13–12) are multiplied by $e^{jn\theta}$ and integrated with respect to θ from $-\pi$ to π. Thus

$$\int_{-\pi}^{\pi} W(\theta - \theta')e^{jn\theta} \, d\theta$$

$$= \int_{-\pi}^{\pi} K_n e^{jn\theta'} \, d\theta + \left(\sum_{-\infty}^{n-1} + \sum_{n+1}^{\infty} \right) K_m e^{jm\theta'} \int_{-\pi}^{\pi} e^{j(n-m)\theta'} \, d\theta \qquad (9.13\text{-}13)$$

where the term $m = n$ in the sum has been separated and is written first on the right. Since the last integral in (9.13–13) is zero, it follows that

$$K_n = \frac{1}{2\pi} \int_{-\pi}^{\pi} W(\theta - \theta') e^{jn(\theta - \theta')} \, d\theta = K_{-n} \tag{9.13–14}$$

The next step is to substitute (9.13–12) in (9.13–10) to obtain

$$K(\theta - \theta') = \left\{ \frac{kb}{2} [e^{j(\theta - \theta')} + e^{-j(\theta - \theta')}] \sum_{-\infty}^{\infty} K_m e^{-jm(\theta - \theta')} \right.$$

$$\left. + \frac{1}{kb} \frac{\partial^2}{\partial \theta^2} \sum_{-\infty}^{\infty} K_m e^{-jm(\theta - \theta')} \right\} \tag{9.13–15}$$

Since

$$\sum_{-\infty}^{\infty} K_m e^{-j(m-1)(\theta - \theta')} = \sum_{-\infty}^{\infty} K_{n+1} e^{-jn(\theta - \theta')} \tag{9.13–16a}$$

and

$$\sum_{-\infty}^{\infty} K_m e^{-j(m+1)(\theta - \theta')} = \sum_{-\infty}^{\infty} K_{n-1} e^{-jn(\theta - \theta')} \tag{9.13–16b}$$

it is evident that

$$K(\theta - \theta') = \sum_{-\infty}^{\infty} a_n e^{-jn(\theta - \theta')} \tag{9.13–17}$$

where

$$a_n = \frac{kb}{2} (K_{n+1} + K_{n-1}) - \frac{n^2}{kb} K_n = a_{-n} \tag{9.13–18}$$

It follows that the integral equation (9.13–9) reduces to

$$V_0^e \delta(\theta) = \frac{j\zeta}{4\pi} \sum_{-\infty}^{\infty} a_n \int_{-\pi}^{\pi} e^{-jn(\theta - \theta')} I(\theta') \, d\theta' \tag{9.13–19}$$

The final step in the formal solution of (9.13–19) is to expand the current in a Fourier series as follows:

$$I(\theta) = \sum_{-\infty}^{\infty} I_n e^{-jn\theta} \tag{9.13–20}$$

where the coefficients are given by

$$I_n = \frac{1}{2\pi} \int_{-\pi}^{\pi} I(\theta') e^{jn\theta'} \, d\theta' \tag{9.13–21}$$

A comparison of (9.13–19) and (9.13–21) shows that

$$V_0^e \delta(\theta) = \frac{j\zeta}{2} \sum_{-\infty}^{\infty} a_n I_n e^{-jn\theta} \tag{9.13–22}$$

This is a Fourier series with the coefficients $(j\zeta/2)a_n I_n$. The coefficients are given by

$$\frac{j\zeta a_n I_n}{2} = \frac{1}{2\pi} \int_{-\pi}^{\pi} V_0^e \delta(\theta) e^{jn\theta} \, d\theta = \frac{V_0^e}{2\pi} \tag{9.13–23}$$

if use is made of the properties of the delta function. Thus

$$I_n = \frac{-jV_0^e}{\zeta \pi a_n} \tag{9.13–24}$$

so that

$$I(\theta) = \sum_{-\infty}^{\infty} I_n e^{-jn\theta} = \frac{-jV_0^e}{\zeta \pi} \left(\frac{1}{a_0} + 2 \sum_1^{\infty} \frac{\cos n\theta}{a_n} \right) \tag{9.13–25}$$

The admittance is

$$Y = \frac{I(0)}{V_0^e} = \frac{-j}{\zeta \pi} \left(\frac{1}{a_0} + 2 \sum_1^{\infty} \frac{1}{a_n} \right) \tag{9.13–26}$$

This completes the formal solution of the integral equation (9.13–9) for the current in the circular loop.

The evaluation of (9.13–25) and (9.13–26) depends on the determination of the a_n as defined in (9.13–18). With the notation $\Phi = \theta - \theta'$ and $A = 2a \sin(\Psi/2)$, the coefficients K_n in (9.13–14) may be expressed as follows:

$$K_n = \frac{1}{4\pi^2} \int_{-\pi}^{\pi} d\Psi \int_{-\pi}^{\pi} \frac{e^{jn\Phi} e^{-jkbR(\Phi)}}{R(\Phi)} \, d\Phi \tag{9.13–27a}$$

where

$$bR(\Phi) = r = b \left(4 \sin^2 \frac{\Phi}{2} + \frac{A^2}{b^2} \right)^{1/2} \tag{9.13–27b}$$

If the order of integration is interchanged and the variable of integration is changed from Ψ to A, (9.13–27a) may be expressed in the equivalent form

$$K_n = \frac{2}{\pi} \int_0^{2a} \overline{K_{|n|}}(A) \frac{dA}{\sqrt{4a^2 - A^2}} \tag{9.13–28a}$$

where, for $n \geq 0$,

$$\overline{K_n} = \frac{1}{2\pi} \int_{-\pi}^{\pi} e^{jn\Phi} \frac{e^{-jkbR(\Phi)}}{R(\Phi)} \, d\Phi \tag{9.13–28b}$$

The following difference relation is satisfied:

$$\overline{\Delta_n} = \overline{K_{n+1}} - \overline{K_n} = \frac{j}{\pi} \int_0^{2\pi} \frac{e^{-jkbR(\Phi)}}{R(\Phi)} e^{j(n+1/2)\Phi} \sin \frac{\Phi}{2} d\Phi \qquad (9.13-28c)$$

With the conditions (9.13–1), that is, $|ka|^2 \ll 1$ and $a^2 \ll b^2$ and with

$$A \le 2a \ll \frac{b}{n} \qquad (9.13-29a)$$

it is a satisfactory approximation to set

$$R(\Phi) \doteq 2 \sin \frac{\Phi}{2} \qquad (9.13-29b)$$

so that (9.13–28c) is given by

$$\overline{\Delta_n} = \frac{j}{2\pi} \int_0^{2\pi} e^{-j2kb\sin(\Phi/2) + j(n+1/2)\Phi} d\Phi \qquad (9.13-30a)$$

where terms of the order (a^2/b^2) have been neglected. This integral may be expressed in the form

$$\overline{\Delta_n} = \overline{K_{n+1}} - \overline{K_n} = \Omega_{2n+1}(2kb) + jJ_{2n+1}(2kb) \qquad (9.13-30b)$$

where

$$J_{2n+1}(x) = \frac{1}{\pi} \int_0^\pi \cos[x \sin \Phi - (2n+1)\Phi] d\Phi \qquad (9.13-31)$$

is the Bessel function of order $(2n+1)$ and

$$\Omega_{2n+1}(x) = \frac{1}{\pi} \int_0^\pi \sin[x \sin \Phi - (2n+1)\Phi] d\Phi \qquad (9.13-32)$$

is the Lommel–Weber function of order $(2n+1)$. With the recurrence formula (9.13–30b), all the $\overline{K_n}$ can be determined if $\overline{K_0}$ is known. The evaluation of $\overline{K_0}$ from (9.13–28b) may be formulated as follows:

$$\overline{K_0} = \frac{1}{2\pi} \int_0^{2\pi} \frac{e^{-jkbR(\Phi)} - 1}{R(\Phi)} d\Phi + \frac{1}{2\pi} \int_0^{2\pi} \frac{d\Phi}{R(\Phi)} \qquad (9.13-33)$$

If the approximation (9.13–29b) is introduced in the first integral in (9.13–33), this becomes

$$\frac{1}{2\pi} \int_0^{2\pi} \frac{e^{-jkbR(\Phi)} - 1}{R(\Phi)} d\Phi \doteq \frac{1}{2\pi} \int_0^{2\pi} \frac{e^{-j2kb\sin(\Phi/2)} - 1}{2 \sin(\Phi/2)} d\Phi$$

$$= \int_0^{2kb} dx \left(\frac{-j}{2\pi} \int_0^\pi e^{-jx\sin\Phi} d\Phi \right)$$

$$= -\frac{1}{2} \left[\int_0^{2kb} \Omega_0(x) dx + j \int_0^{2kb} J_0(x) dx \right] \qquad (9.13-34)$$

where $\Omega_0(x) = (1/\pi) \int_0^\pi \sin(x \sin \Phi) \, d\Phi$ and $J_0(x) = (1/\pi) \int_0^\pi \cos(x \sin \Phi) \, d\Phi$. The second integral in (9.13–33) may be expressed as follows:

$$\frac{1}{\pi} \int_0^\pi \frac{d\Phi}{\{[2 \sin(\Phi/2)]^2 + A^2/b^2\}^{1/2}} = \frac{1}{\pi} \int_0^1 \frac{dt}{[(1 - t^2)(t^2 + A^2/4b^2)]^{1/2}}$$

$$= \frac{\mathcal{H}}{\pi} F\left(\mathcal{H}, \frac{\pi}{2}\right) = \frac{\mathcal{H}}{\pi} K(\mathcal{H}) \qquad (9.13\text{--}35)$$

where $\mathcal{H} = (1 + A^2/4b^2)^{-1/2}$ and $K(\mathcal{H})$ is the "complete" elliptic integral of the first kind. With (9.13–1) it is clear that $\mathcal{H} \doteq 1$ and $\mathcal{H}'^2 = 1 - \mathcal{H}^2 \ll 1$. In this case the elliptic integral $K(\mathcal{H})$ can be expanded in a series with the leading term

$$K(\mathcal{H}) = \ln \frac{4}{\mathcal{H}'} \doteq \ln \frac{8b}{A} \qquad (9.13\text{--}36)$$

It follows that

$$\frac{1}{2\pi} \int_0^{2\pi} \frac{d\Phi}{R(\Phi)} \doteq \frac{1}{\pi} \ln \frac{8b}{A} \qquad (9.13\text{--}37)$$

Hence

$$\overline{K_0} \doteq \frac{1}{\pi} \ln \frac{8b}{A} - \frac{1}{2} \left[\int_0^{2kb} \Omega_0(x) \, dx + j \int_0^{2kb} J_0(x) \, dx \right] \qquad (9.13\text{--}38)$$

This formula, in conjunction with (9.13–30b), permits the evaluation of all $\overline{K_n}$. Thus

$$\overline{K_n} = \overline{K_0} + \sum_0^{n-1} \overline{\Delta_n} \qquad (9.13\text{--}39)$$

With (9.13–30a) this formula becomes

$$\overline{K_n} = \overline{K_0} + \frac{j}{\pi} \int_0^\pi e^{-j2kb \sin \Phi} \left(\sum_0^{n-1} e^{j(2m+1)\Phi} \right) d\Phi \qquad (9.13\text{--}40)$$

However, since

$$\sum_0^{n-1} e^{j(2m+1)\Phi} = \frac{e^{j2n\Phi} - 1}{2j \sin \Phi}$$

and since from (9.13–38) and (9.13–34) after a change of variable

$$\overline{K_0} = \frac{1}{\pi} \ln \frac{8b}{A} + \frac{1}{2\pi} \int_0^\pi \frac{e^{-j2kb \sin \Phi} - 1}{\sin \Phi} \, d\Phi \qquad (9.13\text{--}41)$$

it follows that

$$\overline{K_n} = \frac{1}{\pi} \ln \frac{8b}{A} + \frac{1}{2\pi} \int_0^\pi (e^{-j2kb \sin \Phi + j2n\Phi} - 1) \frac{d\Phi}{\sin \Phi} \qquad (9.13\text{--}42)$$

The integral may be rearranged in the form

$$
\overline{K}_n = \frac{1}{\pi} \ln \frac{8b}{A} + \frac{1}{2\pi} \int_0^\pi (e^{-j2kb\sin\Phi} - 1) \frac{e^{j2n\Phi}}{\sin\Phi} \, d\Phi
$$

$$
+ \frac{1}{2\pi} \int_0^\pi (e^{j2n\Phi} - 1) \frac{d\Phi}{\sin\Phi} \qquad (9.13\text{--}43)
$$

If the same procedure used in (9.13–34) is applied, the first integral may be expressed in terms of Bessel functions and Lommel–Weber functions. The last integral yields the harmonic series. Thus

$$
\overline{K}_n = \frac{1}{\pi} \ln \frac{8b}{A} - \frac{2}{\pi} \sum_{m=0}^{n-1} \frac{1}{2m+1} - \frac{1}{2} \int_0^{2kb} [\Omega_{2n}(x) + jJ_{2n}(x)] \, dx \qquad (9.13\text{--}44)
$$

This formula is valid only when (9.13–29a) is satisfied, that is, when $A \ll b/n$.

To obtain a formula that is not restricted by (9.13–29a) the following integral is useful:

$$
N_n = \frac{b}{2\pi} \int_{-\infty}^\infty e^{jn\Phi} \frac{e^{-jk\sqrt{b^2\Phi^2 + A^2}}}{[b^2\Phi^2 + A^2]^{1/2}} \, d\Phi \qquad (9.13\text{--}45)
$$

which can be evaluated to give

$$
N_n = \frac{-j}{2} H_0^{(2)} \left(A \sqrt{k^2 - \frac{n^2}{b^2}} \right) = \frac{1}{\pi} \mathcal{K}_0 \left(A \sqrt{\frac{n^2}{b^2} - k^2} \right)
$$

$$
\doteq -\frac{1}{\pi} \left(\gamma + \ln \frac{1}{2} A \sqrt{\frac{n^2}{b^2} - k^2} \right) \qquad (9.13\text{--}46)
$$

where $\mathcal{K}_0(x)$ is the modified Bessel function of the second kind and order 0. The approximate expression on the right is valid, subject to (9.13–29a). It is readily verified that the difference between N_n and $\overline{K}_n$ is independent of A, at least when n is not too large. It follows that this difference can be determined for all values of A, and specifically when A is small. Note that

$$
\ln \frac{8b}{A} = \frac{1}{2} \ln \left(1 - \frac{k^2 b^2}{n^2} \right) - \ln \frac{1}{2} A \sqrt{\frac{n^2}{b^2} - k^2} + \ln 4n \qquad (9.13\text{--}47a)
$$

$$
\ln \frac{8b}{A} \doteq \mathcal{K}_0 \left(A \sqrt{\frac{n^2}{b^2} - k^2} \right) + \frac{1}{2} \ln \left(1 - \frac{k^2 b^2}{n^2} \right) + \ln 4n + \gamma \qquad (9.13\text{--}47b)
$$

$$
\ln \frac{8b}{A} \doteq \mathcal{K}_0 \left(\frac{nA}{b} \right) + \ln 4n + \gamma \qquad (9.13\text{--}47c)
$$

With (9.13–47c), (9.13–44) becomes

$$\overline{K}_n = \frac{1}{\pi} \mathcal{H}_0 \left(\frac{nA}{b} \right) + \frac{C_n}{\pi} - \frac{1}{2} \int_0^{2kb} [\Omega_{2n}(x) + jJ_{2n}(x)] \, dx \qquad (9.13\text{–}48a)$$

where

$$C_n = \ln 4n + \gamma - 2 \sum_{m=0}^{n-1} \frac{1}{2m + 1} \qquad (9.13\text{–}48b)$$

It is noteworthy that both C_n and the integral on the right in (9.13–48a) are of the order of magnitude of $1/n^2$, whereas $\mathcal{H}_0(nA/b)$ approaches $(\pi b/2nA)e^{-nA/b}$ as $n \to \infty$. Thus (9.13–48a) is not valid when $n \gg b/A$ since $\mathcal{H}_0(nA/b)$ decreases more rapidly than the rest of the terms. Fortunately, this does not cause any difficulties in the final formula.

If (9.13–38) and (9.13–48a) are substituted in (9.13–28a), the final result obtained for unrestricted n is

$$K_0 = \frac{1}{\pi} \ln \frac{8b}{a} - \frac{1}{2} \left[\int_0^{2kb} \Omega_0(x) \, dx + j \int_0^{2kb} J_0(x) \, dx \right] \qquad (9.13\text{–}49a)$$

$$K_{-n} = K_n = \frac{1}{\pi} \left[\mathcal{H}_0 \left(\frac{na}{b} \right) I_0 \left(\frac{na}{b} \right) + C_n \right] - \frac{1}{2} \int_0^{2kb} [\Omega_{2n}(x) + jJ_{2n}(x)] \, dx \qquad (9.13\text{–}49b)$$

where $I_0(na/b)$ is the modified Bessel function of the first kind. Note that the product $\mathcal{H}_0(na/b)I_0(na/b)$ approaches $(\pi b/2na)$ as $n \to \infty$, so that it remains the dominant term. This means that (9.13–49b) is valid for all values of n.

The values of K_n in (9.13–49a,b) can be substituted in (9.13–18) to evaluate a_n. That is,

$$a_n = \frac{kb}{2} (K_{n+1} + K_{n-1}) - \frac{n^2}{kb} K_n \qquad (9.13\text{–}50)$$

For sufficiently large values of n, $K_{n+1} + K_{n-1} \doteq 2K_n$, so that

$$a_n \approx \left(kb - \frac{n^2}{kb} \right) K_n \qquad (9.13\text{–}51)$$

where K_n is given by (9.13–49b).

9.14 *THE ELECTRICALLY SMALL LOOP: CURRENT AND IMPEDANCE*

In many respects the electrically small loop is of greater practical importance than larger loops, owing to its applications in direction finding and in probing to measure and explore magnetic fields. In these uses its properties when electrically small specifically make it uniquely useful; and regardless of the frequency involved, the loop must be kept electrically small if its utility is to be preserved.

The general formulas for the current and the admittance of a circular loop are readily specialized to the electrically small loop since by definition this is small enough to make the first terms in the Fourier series satisfactory approximations. This condition is denoted by

$$|kb| \ll 1 \qquad (9.14\text{--}1)$$

As $|kb|$ approaches zero, a_0 as given in (9.13–18) also becomes vanishingly small; whereas all a_n, $n > 0$, become infinite. It follows that with $|kb|$ sufficiently small the series (9.13–25) and (9.13–26) reduce to the first term. Moreover, since with $n = 0$, (9.13–18) gives

$$a_0 \doteq \frac{kb}{2} (K_1 + K_{-1}) = kbK_1 \qquad (9.14\text{--}2)$$

it follows that

$$I(0) = -\frac{j}{\pi \zeta kbK_1} V_0^e; \qquad Z_0 = j\pi \zeta kbK_1 \qquad (9.14\text{--}3)$$

Thus an important characteristic of the electrically small loop is the constancy of the current around the loop.

The quantity K_1 in (9.14–3) is readily evaluated from (9.13–49b) subject to (9.14–1) and the previously assumed condition (9.13–1), namely, $a^2 \ll b^2$. With the approximate formulas

$$\mathcal{H}_0\left(\frac{a}{b}\right) \doteq -\left(\gamma + \ln \frac{a}{2b}\right); \qquad I_0\left(\frac{a}{b}\right) \doteq 1 \qquad (9.14\text{--}4)$$

$$\Omega_2(x) \doteq -\frac{2x}{3\pi}; \qquad J_2(x) \doteq \frac{x^2}{8} \qquad (9.14\text{--}5)$$

it follows directly that

$$K_1 \doteq \frac{1}{\pi}\left(\ln \frac{8b}{a} - 2 + \frac{2}{3} k^2 b^2\right) - j \frac{k^3 b^3}{6} \qquad (9.14\text{--}6)$$

so that

$$Z_0 = \frac{\pi \zeta}{6} k^4 b^4 + j\omega \mu b \left(\ln \frac{8b}{a} - 2 + \frac{2}{3} k^2 b^2\right) \qquad (9.14\text{--}7a)$$

Alternatively, with $S = \pi b^2$ (the area enclosed by the perfectly conducting loop),

$$Z_0 = \frac{\zeta}{6\pi} k^4 S^2 + j\omega \mu b \left(\ln \frac{8b}{a} - 2 + \frac{2}{3} k^2 b^2\right) \qquad (9.14\text{--}7b)$$

If the loop is not perfectly conducting but has an internal impedance per unit length given by $z^i = r^i + jx^i$, the quantity $Z^i = 2\pi b z^i$ may be added to (9.14–7b).

If the medium in which the loop is immersed is air, $\zeta = \zeta_0 = 120\pi$ ohms, $k = k_0$ is real. It follows that the leading real and imaginary terms are

$$Z_0 = R_0^e + j\omega L_0^e; \qquad R_0^e = 20k_0^4 S^2; \qquad L_0^e = \mu_0 b \left(\ln \frac{8b}{a} - 2 \right) \qquad (9.14\text{--}8)$$

which is in agreement with (9.10–11) and (9.10–9).

If the medium is dissipative so that $k = \beta - j\alpha$ is complex, the leading real and imaginary terms are

$$Z_0 = \zeta_e \left[\frac{\pi}{6} \beta^2 (\beta^2 - 3\alpha^2) b^4 + \frac{4}{3} \alpha \beta^2 b^3 \right] + j\omega\mu b \left(\ln \frac{8b}{a} - 2 \right) \qquad (9.14\text{--}9)$$

where $\zeta_e = \zeta_0 (\mu_r / \epsilon_{er})^{1/2}$. Unless α is very small, the first term in (9.14–9)—which is the radiation term—is negligible, and the resistance is determined by ohmic dissipation in the medium.

9.15 CIRCULAR LOOPS OF MODERATE SIZE, $k_0 b \leq 2.5$

The evaluation of the distribution of current $I(0)$ in a loop that is unrestricted in size requires the summation of the series in (9.13–25), that is,

$$I(\theta) = \frac{-jV_0^e}{\pi\zeta} \left(\frac{1}{a_0} + 2 \sum_{n=1}^{\infty} \frac{\cos n\theta}{a_n} \right) \qquad (9.15\text{--}1)$$

where a_n is given by (9.13–50) or (9.13–51) in terms of K_n. An approximate formula for K_n in (9.13–49b) is

$$K_n \approx \frac{1}{\pi} (\ln n_0 - \ln n) - j \frac{(kb)^{2n+1}}{\Gamma(2n+2)} \qquad (9.15\text{--}2)$$

where

$$n_0 = \frac{2b}{a} e^{-\gamma} \qquad \text{or} \qquad \ln n_0 = \ln \frac{2b}{a} - \gamma \qquad (9.15\text{--}3)$$

This is a good approximation for $kb \leq 2.5$ and $n \geq 5$. Thus

$$a_n \sim \frac{1}{\pi} \left(kb - \frac{n^2}{kb} \right) \left[\ln n_0 - \ln n - j \frac{\pi(kb)^{2n+1}}{\Gamma(2n+2)} \right]; \qquad \begin{cases} n > kb \\ \dfrac{na}{b} < 1 \end{cases} \qquad (9.15\text{--}4)$$

When $|kb| \leq 2.5$, the first five terms of (9.15–1) provide the principal contribution:

$$I(\theta) \doteq \frac{-jV_0^e}{\pi\zeta} \left(\frac{1}{a_0} + 2 \sum_{1}^{4} \frac{\cos n\theta}{a_n} + \Psi(\theta) \right) \qquad (9.15\text{--}5)$$

where

$$\Psi(\theta) = 2 \sum_5^\infty \frac{\cos n\theta}{a_n} \tag{9.15-6}$$

An approximate expression for $I(\theta)$ is

$$I(\theta) \doteq \frac{-jV_0^e}{\zeta\pi} \left\{ \frac{1}{a_0} + 2 \sum_1^4 \frac{\cos n\theta}{a_n} - \frac{2\pi}{\ln(n_0/4.5)} \left[\frac{kb}{4.5} J_1(\theta) \right.\right.$$

$$\left.\left. + \left(\frac{kb}{4.5}\right)^3 J_2(\theta) \right] \right\} \tag{9.15-7}$$

with

$$J_1(\theta) \doteq \int_1^\infty \frac{\ln(n_0/4.5)}{\ln(n_0/4.5) - \ln x} \frac{\cos(4.5\,\theta x)}{x^2} \, dx \tag{9.15-8a}$$

$$J_2(\theta) \doteq \int_1^\infty \frac{\cos(4.5\,\theta x)}{x^4} \, dx \tag{9.15-8b}$$

A suitable parameter for expressing the circumference to radius ratio of the antenna is

$$\Omega = 2 \ln \frac{2\pi b}{a} \tag{9.15-9}$$

Note that

$$\ln \frac{n_0}{4.5} = \frac{\Omega}{2} - 3.226 \tag{9.15-10}$$

Graphs of $J_1(\theta)$ and $J_2(\theta)$ are in Fig. 9.15–1 and of the coefficients $1/a_n$ with $n = 0, 1, 2, 3, 4$ in Fig. 9.15–2 for real values of $k = \beta$ and a range of values of Ω. The distributions of current for βb in the range 0 to 2.5 are shown in Fig. 9.15–3, where the magnitudes and phases referred to $I(0)$ are shown. $\Omega = 2 \ln(2\pi b/a) = 10$ for which $2\pi b/a \doteq 150$.

The admittance is $Y = G + jB = I(0)/V_0^e$. In the general case of dissipative media, the normalized admittance is given by

$$\frac{Y}{\Delta} = \frac{-j(1 - j\alpha/\beta)}{\pi\zeta_0} \left(\frac{1}{a_0} + 2 \sum_1^\infty \frac{1}{a_n} \right) \tag{9.15-11}$$

where a_n is given by (9.13–50), K_0 by (9.13–49a), and K_n by (9.13–49b). The propagation constant k in a_n is complex and is given by

$$k = \beta - j\alpha = \omega \sqrt{\mu\left(\epsilon - \frac{j\sigma}{\omega}\right)} = \omega\sqrt{\mu\epsilon}\,[f(p) - jg(p)] \tag{9.15-12}$$

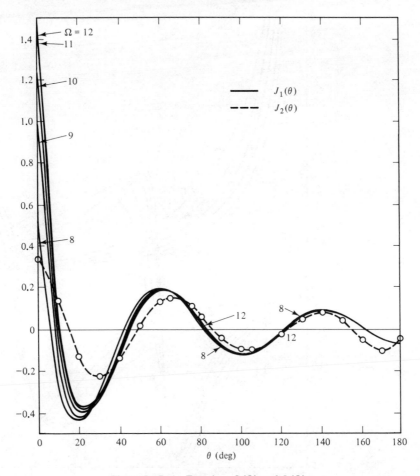

Figure 9.15–1   Functions $J_1(\Phi)$ and $J_2(\Phi)$.

where $p = \sigma/\omega\epsilon$ is the loss tangent and $f(p)$ and $g(p)$ are the functions defined (Chapter 2) by

$$f(p) \pm jg(p) = \sqrt{1 \pm jp} \qquad (9.15\text{–}13)$$

with

$$f(p) = \cosh\left(\frac{1}{2}\sinh^{-1} p\right); \qquad g(p) = \sinh\left(\frac{1}{2}\sinh^{-1} p\right) \qquad (9.15\text{–}14)$$

$f(p)$ and $g(p)$ are tabulated in Appendix II. The normalizing factor Δ in (9.15–11) is defined by

$$\Delta = \sqrt{\frac{\epsilon_r}{\mu_r}} f(p) = \beta \zeta_0/\omega\mu \qquad (9.15\text{–}15)$$

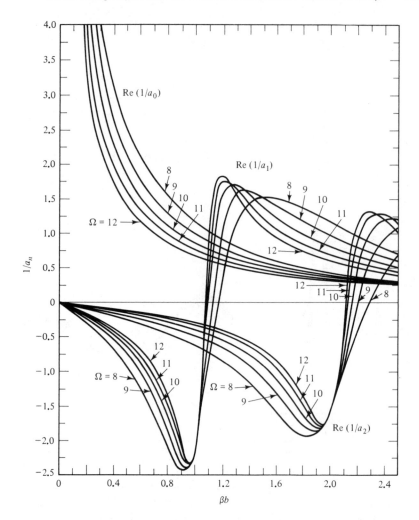

Figure 9.15–2a   Real parts of the functions $1/a_0$, $1/a_1$, and $1/a_2$.

where $\epsilon_r = \epsilon/\epsilon_0$ and $\mu_r = \mu/\mu_0$ are the relative permittivity and permeability of the medium in which the loop is immersed. For air, $\Delta = 1$.

The formula (9.15–1) is approximated by a sum over a finite number of terms. A Fourier series solution with 20 terms is satisfactory for determining the admittance of wire loops which are made of wire that is not too thick ($\Omega \geq 10$) and which are not too large ($\beta b \leq 2.5$) when in air or in an arbitrary dissipative

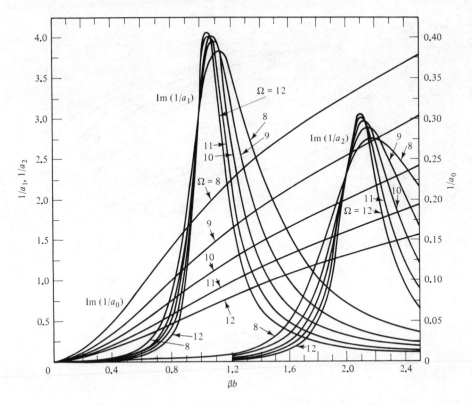

Figure 9.15–2b   Imaginary parts of the functions $1/a_0$, $1/a_1$, $1/a_2$.

medium. The approximation is excellent for the conductance, and somewhat less accurate for the susceptance. Graphs of the normalized admittance $Y/\Delta = G/\Delta + jB/\Delta$ are shown in Fig. 9.15–4. Note that the admittance is insensitive to the size of the loop when α/β approaches unity. This is true particularly of loops near antiresonance that have a high driving-point admittance.

9.16 THE ELECTROMAGNETIC FIELD OF CIRCULAR LOOP ANTENNAS

The electromagnetic field of a circular loop antenna with the distribution of current given by (9.15–1) or, in alternative form, by

$$I(\Phi) = \frac{-jV_0^e}{\pi\zeta} \sum_{-\infty}^{\infty} \frac{e^{jn\Phi}}{a_n} \tag{9.16–1}$$

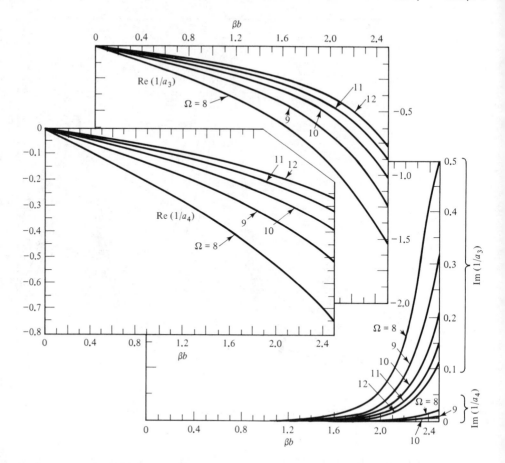

Figure 9.15–2c   Real and imaginary parts of the functions $1/a_3$ and $1/a_4$.

may be determined from the vector potential at an arbitrary point. Consider the loop shown in Fig. 9.16–1, which lies in the xy plane with its center at the origin of the rectangular coordinates x, y, z and the spherical coordinates R_0, Θ, and Φ as shown. The current in the element $b\,d\Phi'$ at Φ' around the loop is $I(\Phi')$ as given in (9.16–1). The element of vector potential dA at the point x, y, z or R_0, Θ, Φ (Fig. 9.16–1) due to this increment of current has the components

$$dA_x = \frac{-\mu}{4\pi} I(\Phi') \frac{e^{-jkR}}{R} \sin \Phi' \, b \, d\Phi' \qquad (9.16\text{–}2)$$

$$dA_y = \frac{\mu}{4\pi} I(\Phi') \frac{e^{-jkR}}{R} \cos \Phi' \, b \, d\Phi' \qquad (9.16\text{–}3)$$

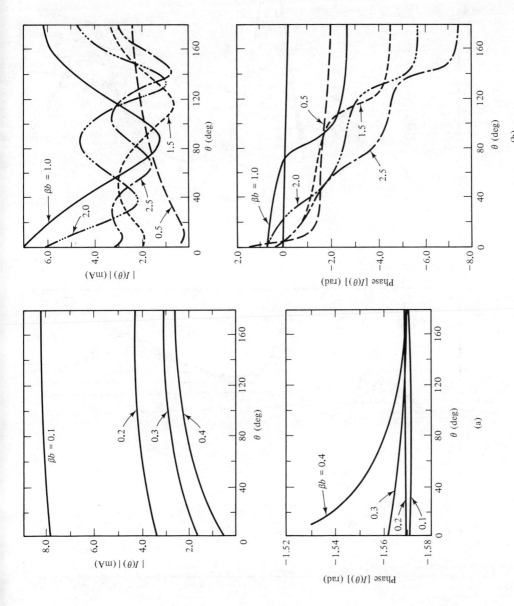

Figure 9.15–3 Magnitude and phase of current distributions: (a) small loop antennas; (b) loop antennas. $\Omega = 2 \ln (2\pi b/a) = 10$.

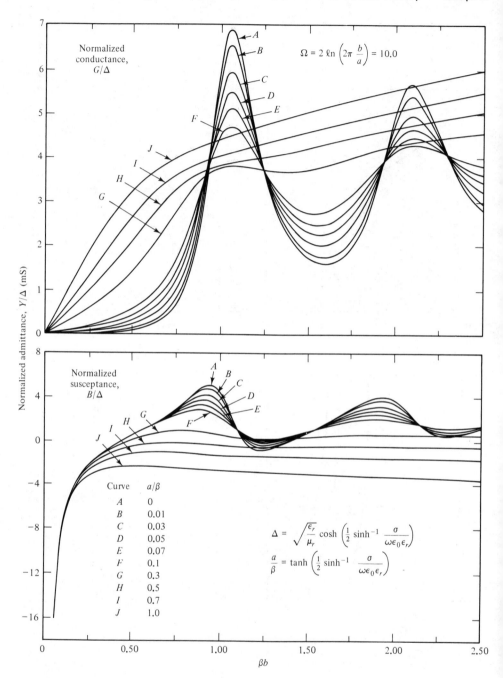

Figure 9.15–4   Normalized admittance of circular loop antenna in a dissipative medium; Wu's theory $\Omega = 10$.

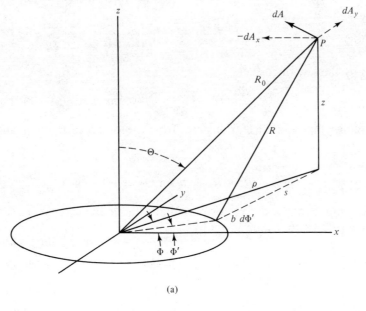

(a)

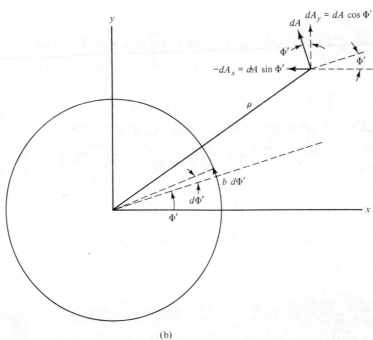

(b)

Figure 9.16–1   Vector potential due to element $I(\Phi')\, b\, d\Phi'$: (a) at arbitrary point; (b) at point in plane of loop.

where

$$R = \sqrt{z^2 + s^2} = \sqrt{(R_0 \cos \Theta)^2 + s^2} \tag{9.16-4a}$$

and by the law of cosines

$$s^2 = (R_0 \sin \Theta)^2 + b^2 - 2bR_0 \sin \Theta \cos (\Phi - \Phi') \tag{9.16-4b}$$

It follows that at sufficient distance from the center of the loop such that

$$b^2 \ll R_0^2 \tag{9.16-5}$$

$$R \doteq R_0 - b \sin \Theta \cos (\Phi - \Phi') \tag{9.16-6}$$

The components of the vector potential (due to all currents in the loop) at points that satisfy (9.16–5) are given by

$$A_x^r = \frac{-b\mu}{4\pi} \frac{e^{-jkR_0}}{R_0} \int_0^{2\pi} I(\Phi')e^{jkb \sin \Theta \cos (\Phi - \Phi')} \sin \Phi' \, d\Phi' \tag{9.16-7a}$$

$$A_y^r = \frac{b\mu}{4\pi} \frac{e^{-jkR_0}}{R_0} \int_0^{2\pi} I(\Phi')e^{jkb \sin \Theta \cos (\Phi - \Phi')} \cos \Phi' \, d\Phi' \tag{9.16-7b}$$

The spherical components of the vector potential are given by

$$A_\Theta^r = (A_x^r \cos \Phi + A_y^r \sin \Phi) \cos \Theta \tag{9.16-8a}$$

$$A_\Phi^r = (-A_x^r \sin \Phi + A_y^r \cos \Phi) \tag{9.16-8b}$$

These are illustrated in Fig. 9.16–2. It may be shown that

$$A_\Theta^r = -2\pi Q \sum_1^n \frac{n}{a_n} \frac{J_n(kb \sin \Theta)}{kb \sin \Theta} \cos \Theta[e^{jn(\Phi + \pi/2)}$$

$$- (-1)^n e^{-jn(\Phi + \pi/2)}] \tag{9.16-9a}$$

$$A_\Phi^r = -j2\pi Q \left\{ \frac{J_0'(kb \sin \Theta)}{a_0} + \sum_1^n \frac{J_n'(kb \sin \Theta)}{a_n} [e^{jn(\Phi + \pi/2)} \right.$$

$$\left. + (-1)^n e^{-jn(\Phi + \pi/2)}] \right\} \tag{9.16-9b}$$

with

$$Q = \frac{-jV_0^e\mu}{4\pi^2\zeta} \frac{be^{-jkR_0}}{R_0} \tag{9.16-9c}$$

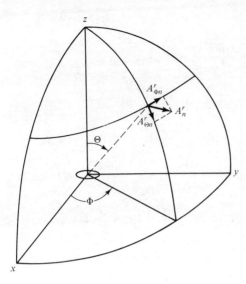

Figure 9.16–2 Spherical components $A^r_{\Theta n}$ and $A^r_{\Phi n}$ of the far-zone vector potential $A^r_n = \hat{x}A^r_{xn} + \hat{y}A^r_{yn}$.

The electromagnetic field is given by

$$\mathbf{E}^r = -j\omega(\hat{\mathbf{\Theta}}A^r_\Theta + \hat{\mathbf{\Phi}}A^r_\Phi) \qquad (9.16\text{–}10\text{a})$$

$$\mathbf{B}^r = -jk(\hat{\mathbf{\Phi}}A^r_\Theta - \hat{\mathbf{\Theta}}A^r_\Phi) \qquad (9.16\text{–}10\text{b})$$

where A^r_Θ and A^r_Φ are in (9.16–9a,b). For loops of moderate size, $|kb| \leq 2.5$, the principal contributions to the current are from the first three terms in the series. For obtaining far-field patterns, the contributions from higher-order terms are usually negligible. The electric-field patterns for the individual modes with $n = 0$, 1, and 2 are readily obtained. Note that

$$J'_0(x) = -J_1(x) \qquad (9.16\text{–}11\text{a})$$

$$J'_1(x) = J_0(x) - \frac{J_1(x)}{x} \qquad (9.16\text{–}11\text{b})$$

$n = 0$:

$$A^r_{\Theta 0} = 0; \qquad A^r_{\Phi 0} = \frac{j2\pi Q}{a_0} J_1(kb \sin \Theta) \qquad (9.16\text{–}12)$$

$$\mathbf{E}^r_0 = \hat{\mathbf{\Phi}}E^r_{\Phi 0}; \qquad \mathbf{B}^r_0 = \hat{\mathbf{\Theta}}B^r_{\Theta 0} \qquad (9.16\text{–}13\text{a})$$

$$E^r_{\Phi 0} = -\frac{\omega}{k} B^r_{\Theta 0} = \frac{2\pi Q\omega}{a_0} J_1(kb \sin \Theta) \qquad (9.16\text{–}13\text{b})$$

$n = 1$:

$$A^r_{\Theta_1} = \frac{4\pi Q}{a_1} \frac{J_1(kb \sin \Theta)}{kb \sin \Theta} \cos \Theta \sin \Phi \tag{9.16-14a}$$

$$A^r_{\Phi_1} = \frac{4\pi Q}{a_1} J'_1 (kb \sin \Theta) \cos \Phi \tag{9.16-14b}$$

$$\mathbf{E}^r_1 = \hat{\mathbf{\Theta}} E^r_{\Theta_1} + \hat{\mathbf{\Phi}} E^r_{\Phi_1}; \qquad \mathbf{B}^r_1 = \hat{\mathbf{\Theta}} B^r_{\Theta_1} + \hat{\mathbf{\Phi}} B^r_{\Phi_1} \tag{9.16-15a}$$

$$E^r_{\Theta_1} = \frac{\omega}{k} B^r_{\Phi_1} = \frac{-j4\pi Q\omega}{a_1} \frac{J_1(kb \sin \Theta)}{kb \sin \Theta} \cos \Theta \sin \Phi \tag{9.16-15b}$$

$$E^r_{\Phi_1} = -\frac{\omega}{k} B^r_{\Theta_1} = \frac{-j4\pi Q\omega}{a_1} J'_1(kb \sin \Theta) \cos \Phi \tag{9.16-15c}$$

$n = 2$:

$$A^r_{\Theta_2} = \frac{j4\pi Q}{a_2} \frac{2J_2(kb \sin \Theta)}{kb \sin \Theta} \cos \Theta \sin 2\Phi \tag{9.16-16a}$$

$$A^r_{\Phi_2} = \frac{j4\pi Q}{a_2} J'_2(kb \sin \Theta) \cos 2\Phi \tag{9.16-16b}$$

$$\mathbf{E}^r_2 = \hat{\mathbf{\Theta}} E^r_{\Theta_2} + \hat{\mathbf{\Phi}} E^r_{\Phi_2}; \qquad \mathbf{B}^r_2 = \hat{\mathbf{\Theta}} B^r_{\Theta_2} + \hat{\mathbf{\Phi}} B^r_{\Phi_2} \tag{9.16-17a}$$

$$E^r_{\Theta_2} = \frac{\omega}{k} B^r_{\Phi_2} = \frac{4\pi Q\omega}{a_2} \frac{2J_2(kb \sin \Theta)}{kb \sin \Theta} \cos \Theta \sin 2\Phi \tag{9.16-17b}$$

$$E^r_{\Phi_2} = \frac{-\omega}{k} B^r_{\Theta_2} = \frac{4\pi Q\omega}{a_2} J'_2(kb \sin \Theta) \cos 2\Phi \tag{9.16-17c}$$

Each of these three modes contributes the principal part of the field in particular ranges of the radius b of the loop. When the loop is electrically small so that $|kb| \ll 1$, the largest mode current is $I_0 \sim 1/a_0$. Therefore, the electromagnetic field consists primarily of E^r_0 and B^r_0 with small contributions from the higher modes. When $|kb| \sim 1$, the largest mode current is $I_1 \sim 1/a_1$ and the electromagnetic field is given by E^r_1 and B^r_1 with relatively small contributions from the other modes. When $|kb| \sim 2$, $I_2 \sim 1/a_2$ is larger than other modes, so E^r_2 and B^r_2 dominate the field.

The distribution of a resonant component of current constitutes a standing wave around the loop which in the case of $I_1(\Phi)$ has maxima at $\Phi = 0$ and π and minima at $\Phi = \pm \pi/2$, where the concentration of charge is the greatest. The charge per unit length has a positive maximum of $\Phi = \pi/2$ and it has a minimum at $\Phi = -\pi/2$.

The electromagnetic field maintained by the resonant current $I_1(\Phi)$ is given by (9.16–15a–c) with $kb = 1$. Thus

$$\mathbf{E}_1^r(\Theta, \Phi) = \hat{\mathbf{\Theta}}E_{\Theta_1}^r(\Theta, \Phi) + \hat{\mathbf{\Phi}}E_{\Phi_1}^r(\Theta, \Phi) \qquad (9.16\text{–}18)$$

where

$$E_{\Theta_1}^r(\Theta, \Phi) = \frac{-j4\pi Q\omega}{a_1}\frac{J_1(\sin\Theta)}{\sin\Theta}\cos\Theta\sin\Phi \qquad (9.16\text{–}19\text{a})$$

$$E_{\Phi_1}^r(\Theta, \Phi) = \frac{-j4\pi Q\omega}{a_1}J_1'(\sin\Theta)\cos\Phi$$

$$= \frac{-j4\pi Q\omega}{a_1}\left[J_0(\sin\theta) - \frac{J_1(\sin\Theta)}{\sin\Theta}\right]\cos\Phi \qquad (9.16\text{–}19\text{b})$$

The field patterns in principal planes are readily obtained. Thus

$$E_{\Theta_1}^r\left(\frac{\pi}{2}, \Phi\right) = 0; \qquad E_{\Phi_1}^r\left(\frac{\pi}{2}, \Phi\right) = \frac{-j4\pi Q\omega}{a_1}0.325\cos\Phi \qquad (9.16\text{–}20)$$

$$E_{\Theta_1}^r(\Theta, 0) = 0; \qquad E_{\Phi_1}^r(\Theta, 0) = \frac{-j4\pi Q\omega}{a_1}\left[J_0(\sin\Theta) - \frac{J_1(\sin\Theta)}{\sin\Theta}\right] \qquad (9.16\text{–}21)$$

$$E_{\Theta_1}^r\left(\Theta, \frac{\pi}{2}\right) = \frac{-j4\pi Q\omega}{a_1}\frac{J_1(\sin\Theta)}{\sin\Theta}\cos\Theta; \qquad E_{\Phi_1}^r\left(\Theta, \frac{\pi}{2}\right) = 0 \qquad (9.16\text{–}22)$$

The maximum field is in the directions $\Theta = 0$ and π. It is

$$E_{\Theta_1}^r(0, \Phi) = \frac{-j2\pi Q\omega}{a_1}\sin\Phi; \qquad E_{\Phi_1}^r(0, \Phi) = \frac{-j2\pi Q\omega}{a_1}\cos\Phi \qquad (9.16\text{–}23)$$

The field $\mathbf{E}_1^r(\Theta, \Phi)$ in the principal planes due to the dipole-mode current $I_1(\Phi)$ is shown in Fig. 9.16–3.

9.17 THE SMALL LOOP: ELECTROMAGNETIC FIELD

The electrically small loop with $|kb| < 1$ has numerous practical applications. When $|kb| < 1$, the electromagnetic field given in (9.16–13a,b) and (9.16–15a,b) may be simplified since the leading term in the expansion of the Bessel functions in powers of $|kb|$ is adequate. Thus, with $J_1(x) \doteq x/2$, $J_1'(x) = J_0(x) - J_1(x)/x \doteq 1/2$,

$$\mathbf{E}_0^r \doteq \hat{\mathbf{\Phi}}\frac{\pi Q\omega}{a_0}kb\sin\Theta \qquad (9.17\text{–}1\text{a})$$

$$\mathbf{E}_1^r \doteq \hat{\mathbf{\Theta}}E_{\Theta_1}^r + \hat{\mathbf{\Phi}}E_{\Phi_1}^r \qquad (9.17\text{–}1\text{b})$$

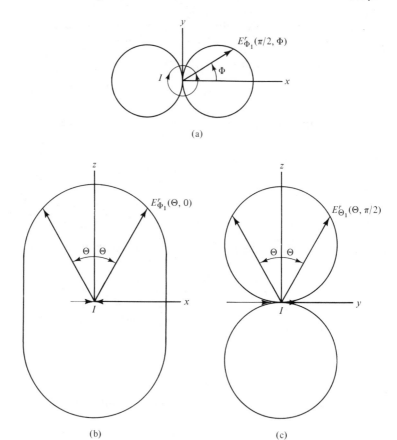

Figure 9.16–3 Dipole-mode field $E_i^r(\Theta,\Phi)$ of a circular loop antenna in air with $kb - \beta b = 1$. The location of the loop and the direction of the dipole-mode currents $I_1(\Phi)$ are indicated. (a) Horizontal pattern of $E_{\Phi 1}(\pi/2, \Phi)$; $E\Theta_1(\pi/2, 0)$ = 0; (b) vertical pattern of $E_{\Phi 1}(\Theta, 0)$; (c) vertical pattern of $E_{\Theta 1}(\Theta, \pi/2)$.

where

$$E_{\Theta_1}^r \doteq \frac{-j2\pi Q\omega}{a_1} \cos \Theta \sin \Phi; \qquad E_{\Phi_1}^r \doteq \frac{-j2\pi Q\omega}{a_1} \cos \Phi \qquad (9.17-1c)$$

The resultant field of the smaller circular loop with $|kb| < 1$ is

$$\mathbf{E}^r = \pi Q\omega \frac{kb}{a_0} [\hat{\boldsymbol{\Phi}} \sin \Theta - j\Gamma_{01}(\hat{\boldsymbol{\Theta}} \cos \Theta \sin \Phi + \hat{\boldsymbol{\Phi}} \cos \Phi)] \qquad (9.17-2a)$$

$$\mathbf{B}^r = -\pi Qk \frac{kb}{a_0} [\hat{\boldsymbol{\Theta}} \sin \Theta + j\Gamma_{01}(\hat{\boldsymbol{\Phi}} \cos \Theta \sin \Phi - \hat{\boldsymbol{\Theta}} \cos \Phi)] \qquad (9.17-2b)$$

where the relative amplitudes of the two modes are given by the coefficient

$$\Gamma_{01} = \frac{2a_0}{kba_1}; \quad |kb| < 1 \tag{9.17-3}$$

It is readily shown with (9.13–18) that when $|kb| \to 0$, $a_0/a_1 \to -(kb)^2$ so that $\Gamma_{01} \to -2kb$. A curve of the ratio Γ_{01} for $k = \beta$, $\alpha = 0$ is shown in Fig. 9.17–1. Clearly, the dipole mode with $n = 1$ contributes significantly to the far field unless $|kb|$ is extremely small.

The currents that maintain the field (9.17–2a,b) are given by

$$I_0(\Phi) + I_1(\Phi) = \frac{-jV_0^e}{\zeta\pi}\left(\frac{1}{a_0} + \frac{2\cos\Phi}{a_1}\right) \tag{9.17-4}$$

where the first term is the circulating mode ($n = 0$) and the second term is the dipole mode ($n = 1$). The ratio $2a_0/a_1$ of the amplitude of the current in the dipole mode to that in the circulating mode is also shown graphically in Fig. 9.17–1. The current in the loop when $\beta b = 0.4$ is shown in a solid line in Fig. 9.17–2 for the entire loop. In the same figure are shown the circulating part of the current $I_0(\Phi)$,

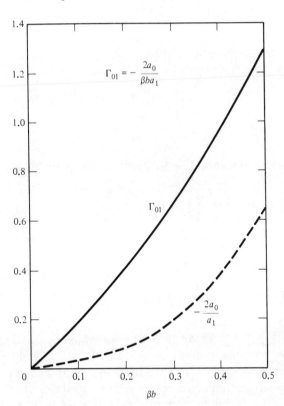

Figure 9.17–1 Ratio of the amplitude of the far field of a circular loop maintained by currents in the dipole ($n = 1$) mode to that maintained by currents in the circulating mode ($n = 0$).

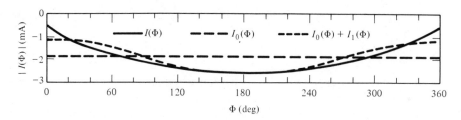

Figure 9.17–2 Current distribution on small loop, $\beta b = 0.4$ and $\Omega = 2 \ln (2\pi b/a) = 10$.

and the algebraic sum of I_0 and the dipole-mode current $I_1(\Phi)$. Only the susceptive part of the current is shown in Fig. 9.17–2 since the conductive part is quite small.

The electric field $\mathbf{E}^r(\Theta, \Phi)$ in the plane $\Theta = \pi/2$ is a circle given by

$$\mathbf{E}^r(\Theta, \Phi) = \hat{\Phi} E_\Phi^r \left(\frac{\pi}{2}, \Phi\right); \quad E_\Phi^r \left(\frac{\pi}{2}, \Phi\right) = E_{\Phi_0}^r \left(\frac{\pi}{2}, \Phi\right) = \pi Q \omega \frac{kb}{a_0}$$

$$E_\Theta^r \left(\frac{\pi}{2}, \Phi\right) = 0 \tag{9.17-5}$$

as shown in Fig. 9.17–3a. In the plane $\Phi = 0, \pi$,

$$\mathbf{E}^r(\Theta, 0) = \hat{\Phi} E_\Phi^r(\Theta, 0); \quad E_\Phi^r(\Theta, 0) = E_{\Phi_0}^r(\Theta, 0) = \pi Q \omega \frac{kb}{a_0} \sin \Theta$$

$$E_\Theta^r(\Theta, 0) = 0 \tag{9.17-6}$$

This is the figure-eight pattern shown in Fig. 9.17–3b. In the plane $\Phi = \pm \pi/2$,

$$\mathbf{E}^r \left(\Theta, \frac{\pi}{2}\right) = \hat{\Phi} E_\Phi^r \left(\Theta, \frac{\pi}{2}\right) + \hat{\Theta} E_\Theta^r \left(\Theta, \frac{\pi}{2}\right) \tag{9.17-7a}$$

$$E_\Phi^r \left(\Theta, \frac{\pi}{2}\right) = E_{\Phi_0}^r \left(\Theta, \frac{\pi}{2}\right) = \pi Q \omega \frac{kb}{a_0} \sin \Theta \tag{9.17-7b}$$

$$E_\Theta^r \left(\Theta, \frac{\pi}{2}\right) = E_{\Theta_1}^r \left(\Theta, \frac{\pi}{2}\right) = -j\pi Q \omega \frac{kb}{a_0} \Gamma_{01} \cos \Theta \tag{9.17-7c}$$

These fields are shown in Fig. 9.17–3c for $kb = 0.5$ for which $\Gamma_{01} = 0.65$. The contribution to the field by the dipole mode ($n = 1$) is the figure-eight pattern with maxima in the broadside direction $\pm z$ of the loop. As the radius of the loop is decreased, the size of this figure eight decreases relative to the sizes of all the other patterns shown which are maintained by the circulating ($n = 0$) mode.

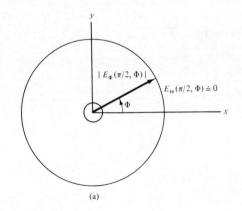

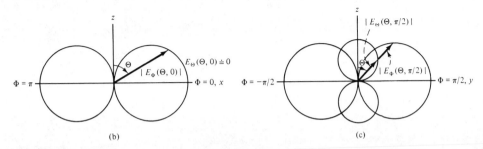

Figure 9.17–3 Field patterns $E(\Theta,\Phi)$ in principal planes of circular loop $kb = 0.5$ and $\Gamma_{01} \doteq 0.65$.

The admittances of the small loop are given by

$$Y = Y_0 + Y_1 = \frac{1}{Z_0} + \frac{1}{Z_1} \tag{9.17–8}$$

where

$$Z_0 = j\zeta\pi a_0 \tag{9.17–9}$$

$$Z_1 = \frac{j\zeta\pi a_1}{2} \tag{9.17–10}$$

The impedances associated with the circulating currents $I_0(\Phi)$ and the dipole-mode currents $I_1(\Phi)$ are in parallel, so that the resultant impedance given by the current $I_0(0) + I_1(0)$ is

$$Z = \frac{Z_0 Z_1}{Z_0 + Z_1} \tag{9.17–11}$$

It is desired to determine the significance of Z_1 on Z for loops with $|kb| < 1$.

The formula for Z_0 is in (9.14–7a); the formula for Z_1 may be derived in a similar manner. They are obtained with

$$a_0 = kbK_1 \doteq \frac{kbD}{\pi} - j\frac{k^4b^4}{6} \tag{9.17–12a}$$

$$a_1 = \frac{kb}{2}(K_2 + K_0) - \frac{K_1}{kb} \doteq \frac{D}{\pi kb} - j\frac{k^2b^2}{3} \tag{9.17–12b}$$

where

$$D = \ln\frac{8b}{a} - 2 + \frac{2}{3}k^2b^2 \tag{9.17–12c}$$

Hence

$$Z_0 = \frac{\pi\zeta}{6}k^4b^4 + j\omega\mu bD \tag{9.17–13a}$$

$$Z_1 = \frac{\pi\zeta}{6}k^2b^2 - j\frac{D}{2\omega\bar{\epsilon}b} \tag{9.17–13b}$$

When the loop is in air, $k = k_0$ and $\zeta = \zeta_0$ so that

$$Z_0 = R_0 + j\omega L_0; \qquad Z_1 = R_1 - \frac{j}{\omega C_1} \tag{9.17–14a}$$

where

$$R_0 = \frac{\pi\zeta_0}{6}k_0^4b^4; \qquad R_1 = \frac{\pi\zeta_0}{6}k_0^2b^2 \tag{9.17–14b}$$

$$L_0 = \mu_0 b\left(\ln\frac{8b}{a} - 2\right); \qquad C_1 = 2\epsilon_0 b\left/\left(\ln\frac{8b}{a} - 2\right)\right. \tag{9.17–14c}$$

The radiation resistance R_1 of the dipole mode ($n = 1$) is two orders of magnitude greater than R_0, the radiation resistance of the circulating mode ($n = 0$). When k_0b is small, R_1 is in series with the high capacitive reactance X_1, whereas R_0 is in series with the low inductive reactance X_0. It follows that I_0 greatly exceeds I_1 when k_0b is sufficiently small and Z_1 has little effect on the radiation field.

PROBLEMS

1. Calculate the impedance of a rectangular loop of sides 1 m by 2 m made of copper wire of diameter 3 mm and driven at a frequency of 5 MHz. If the current in the loop is 5 A, what is the radiated power? Repeat if the same wire forms a circle instead of a rectangle.
2. Calculate the impedance of a circular loop of wire 20 cm in diameter in the presence of a completely closed identical loop. The distance between their centers is 10 cm. The first

loop is driven by a slice generator at 50 MHz. Both loops are made of copper wire of radius 1 mm. What power is radiated if a current of 1 A is in the driven loop?

3. A square loop is constructed of a single turn of heavy copper wire (radius 1 mm; $\sigma = 5.8 \times 10^7$ S/m). The square has sides 1 m long. It is driven at the center of one side by a generator operated at 10 MHz, maintaining a potential difference of 1 kV rms.

   At a distance of 100 km from the driven loop is an identical receiving loop but with a load of 1 Ω connected at the center of one side. The two loops are in the same plane with parallel pairs of sides.

   (a) Determine the impedance seen by the generator in the driven loop and the current in the loop.

   (b) What power and what fraction of the total power are radiated?

   (c) Determine the magnitude of the current in and potential difference across the load in the receiving loop.

4. **(a)** Repeat Problem 3 if the center of each loop is 2.5 m above a perfectly conducting half-space and the planes of the loop are perpendicular to the conducting surface.

   (b) Repeat Problem 3 if the planes of the loops are parallel to the conducting plane.

5. Calculate the inductance of a helical coil of six turns with pitch equal to one-half the radius of the coil which is 5 cm. The radius of the wire is 0.05 cm.

6. Investigate the impedance of an equilateral triangle of wire driven at one apex if the conditions for the quasi-near zone apply.

7. Most commercial television receivers have a loop antenna for UHF reception mounted on the back of the set. Such an antenna is shown in Fig. P9-7. The operating frequencies (voice and picture signals) for two UHF stations are:

   Channel 38:   614–620 MHz

   Channel 56:   722–728 MHz

   (a) Using the center frequencies of 617 and 725 MHz, calculate the electrical size of the above loop ($k_0 b$) for each channel. Also calculate $\Omega = 2 \ln (2\pi b/a)$ for the antenna.

   (b) Using graphs in the literature (e.g., Fig. 9.3.7, King and Harrison, Antennas and Waves) obtain the input impedance of the antenna at the two center frequencies. Use the graph for Ω closest to your calculated value.

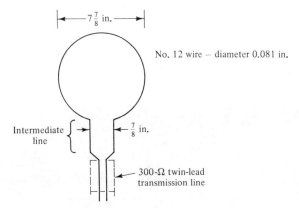

Figure P9-7

(c) If the current distribution on the loop is assumed to be constant, what is the radiation resistance of the loop at the two frequencies? Is this assumption valid?

(d) Calculate the characteristic impedance of the intermediate section of transmission line.

(e) Assuming that the designer of this loop antenna tried to match the real part of the loop impedance to the characteristic impedance of the transmission lines, do you think he used resistance obtained from the constant-current analysis for the loop or the correct value obtained in (b)?

10

Transmission-Line
Theory

Transmission lines are among the most interesting and useful electric circuits. Their primary application is to serve as the connecting link between various more or less distant electrical components, such as a generator and a transmitting antenna, an antenna and a radio or television receiver, or the wiring in a house or factory and the distant alternator. A different application is to electrical measurements. At high frequencies suitably designed transmission lines can be used as high-precision tools for measuring wavelength, conductivity, permittivity, inductance, and capacitance.

To understand the operation of transmission lines, it is necessary to determine the distributions of current and potential difference and the transfer of energy along them with various terminating impedances. This is accomplished by deriving and then solving the basic differential equations. The equations for the current and voltage along a transmission line can be obtained with the help of electric circuit theory or directly from general electromagnetic theory. The electric-circuit theory method is approximate to the extent that it ignores radiation. The transmission line is divided into small elements each of which is represented by an equivalent circuit with lumped constants and this is then analyzed using Kirchhoff's laws. Formulas for the lumped constants—capacitance, conductance, resistance, inductance—must be determined separately. The electromagnetic-theory method proceeds from the fundamental principles and equations for the electromagnetic field. It is, therefore, completely general. It yields not only the underlying differential equations for the voltage and current but also the formulas for the line constants and the conditions that determine their validity.

10.1 THE DERIVATION OF THE DIFFERENTIAL EQUATIONS OF THE TRANSMISSION LINE USING CIRCUIT THEORY

The short section of two-wire line to be analyzed is shown in Fig. 10.1–1a. The two conductors are identical, circular in cross section each with radius a. The distance between centers is b. A short section of coaxial line is shown in Fig. 10.1–1b. The inner conductor has a radius a_1. The concentric outer conductor has an inner radius a_2 and an outer radius a_3. In carrying out the analysis each section of length Δz is treated as equivalent to the circuit shown in Fig. 10.1–2 with fixed values of r, l, c, and g per unit length. The replacement of each length Δz of a transmission line by the circuit of Fig. 10.1–2 implies that all such lengths are

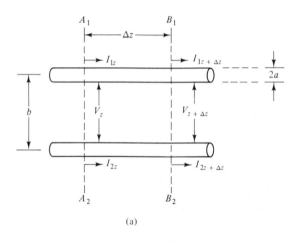

(a)

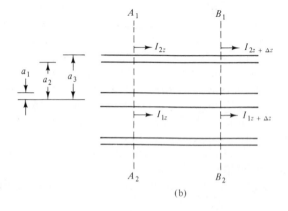

(b)

Figure 10.1–1 Sections of infinite transmission line: (a) two-wire line; (b) coaxial line.

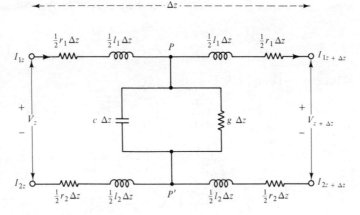

Figure 10.1–2   Equivalent circuit of a section of an infinite transmission line.

exactly alike. This condition is true only for an infinitely long line. Near the ends of a line of finite length the line parameters r, l, c, and g differ from their values far from the ends. Furthermore, the load is usually coupled to the conductors of the line in a short region near their common junctions. If the inductance and capacitance, as well as resistance and leakage conductance per unit length, are treated as constants independent of the location of the element Δz along the line and if the distributed coupling between line and load is ignored, the error introduced becomes a part of the load impedance. It can always be reduced and often made negligible by making the separation of the conductors of the line sufficiently small compared with both the length of the line and the wavelength. The equivalence between the circuits of Figs. 10.1–1 and 10.1–2 is generally true subject to the following restrictions:

$$b \ll \lambda \qquad \text{for a two-wire line}$$

$$a_2 \ll \lambda \qquad \text{for a coaxial line}$$

The circuit of Fig. 10.1–2 may be analyzed as follows: Since Δz is small, the currents and the potential difference at the point $z + \Delta z$ may be expressed in terms of the currents and potential difference at the point z using Maclaurin's expansion:

$$I_{z+\Delta z} = I_z + \left(\frac{dI}{dz}\right)_z \Delta z + \left(\frac{d^2I}{dz^2}\right)_z \frac{(\Delta z)^2}{2} + \cdots \qquad (10.1\text{–}1)$$

$$V_{z+\Delta z} = V_z + \left(\frac{dV}{dz}\right)_z \Delta z + \left(\frac{d^2V}{dz^2}\right)_z \frac{(\Delta z)^2}{2} + \cdots \qquad (10.1\text{–}2)$$

The application of Kirchhoff's voltage law around the rectangle formed by the input and output terminals of the section results in the following equation:

$$\tfrac{1}{2} (I_{1z} + I_{1z+\Delta z}) (r_1 + j\omega l_1) \, \Delta z + V_{z+\Delta z}$$

$$- \tfrac{1}{2} (I_{2z} + I_{2z+\Delta z}) (r_2 + j\omega l_2) \, \Delta z - V_z = 0 \qquad (10.1\text{--}3)$$

With (10.1–1) and (10.1–2) this gives

$$\frac{1}{2} \left[2I_{1z} + \left(\frac{dI_1}{dz}\right)_z \Delta z + \cdots \right](r_1 + j\omega l_1) \, \Delta z$$

$$- \frac{1}{2} \left[2I_{2z} + \left(\frac{dI_2}{dz}\right)_z \Delta z + \cdots \right](r_2 + j\omega l_2) \, \Delta z$$

$$+ \left(\frac{dV}{dz}\right)_z \Delta z + \cdots = 0 \qquad (10.1\text{--}4a)$$

A rearrangement of terms and division by Δz, which is then allowed to approach zero, reduces (10.1–4a) to

$$I_{1z}(r_1 + j\omega l_1) - I_{2z} (r_2 + j\omega l_2) + \left(\frac{dV}{dz}\right)_z = 0 \qquad (10.1\text{--}4b)$$

For a balanced two-wire line that is arranged symmetrically, the following relation is satisfied:

$$I_{2z} = -I_{1z} = -I_z \qquad (10.1\text{--}5)$$

For a coaxial transmission line, (10.1–5) is satisfied for the currents on the inner conductor and on the inner surface of the outer conductor.

    The following notation is convenient for a balanced two-wire line or a coaxial line:

$$r \equiv r_1 + r_2; \qquad l \equiv l_1 + l_2; \qquad z \equiv r + j\omega l \qquad (10.1\text{--}6)$$

r, l, and z are the total resistance, inductance, and impedance per unit length. Note that boldface z is used here to distinguish it from the coordinate z. For the two-wire line, with identical conductors, $r_1 = r_2$ and $l_1 = l_2$.

    The substitution of (10.1–5) and (10.1–6) in (10.1–4b) gives

$$zI_z = -\left(\frac{dV}{dz}\right)_z \qquad (10.1\text{--}7)$$

    Kirchhoff's current law is next applied at the point P noting that the voltage across PP' is $\tfrac{1}{2}(V_z + V_{z+\Delta z})$. Thus

$$I_{1z} = I_{1z+\Delta z} + \tfrac{1}{2} (V_z + V_{z+\Delta z}) (g + j\omega c) \, \Delta z \qquad (10.1\text{--}8)$$

The use of (10.1–1) and (10.1–2) and the division by Δz of the resulting expression yields the following equation in the limit as Δz approaches zero:

$$yV_z = -\left(\frac{dI}{dz}\right)_z \tag{10.1-9}$$

In (10.1–9) y is the total shunt admittance per unit length:

$$y = g + j\omega c \tag{10.1-10}$$

The first-order differential equations (10.1–7) and (10.1–9) are the transmission-line equations. The following second-order differential equations are readily derived from (10.1–7) and (10.1–9):

$$-z\left(\frac{dI}{dz}\right)_z = \left(\frac{d^2V}{dz^2}\right)_z = yzV_z \tag{10.1-11}$$

$$-y\left(\frac{dV}{dz}\right)_z = \left(\frac{d^2I}{dz^2}\right)_z = yzI_z \tag{10.1-12}$$

The complex propagation constant γ and complex wave number k are defined as follows:

$$\gamma^2 \equiv yz = (g + j\omega c)(r + j\omega l) = -k^2 \tag{10.1-13}$$

They can be expressed in terms of their real and imaginary parts. Thus $\gamma = \alpha + j\beta$, $k = \beta - j\alpha$. The law of conservation of electric charge is expressed by the equation of continuity. In complex form it is

$$\frac{dI_z}{dz} + j\omega q_z = 0 \tag{10.1-14}$$

where q_z is the charge per unit length on one conductor. With (10.1–9) it follows that

$$q_z = -\frac{jy}{\omega}V_z = \left(c - j\frac{g}{\omega}\right)V_z \tag{10.1-15}$$

10.2 POTENTIAL FUNCTIONS AS THE FOUNDATION OF TRANSMISSION-LINE THEORY

It was shown in Sec. 3.7 that the scalar and vector potential functions ϕ and $\mathbf{A}$ satisfy the following equations in simple media:

$$\nabla^2\phi + k^2\phi = 0 \tag{10.2-1a}$$

$$\nabla^2\mathbf{A} + k^2\mathbf{A} = 0 \tag{10.2-1b}$$

The solutions of (10.2–1a,b) give the scalar and vector potentials at all points in a homogeneous isotropic medium due to distributions of charge and current in arbitrary configurations of conductors. These are the Helmholtz integrals [(3.8–19) and (3.8–20)]:

$$\phi = \frac{1}{4\pi\epsilon} \int \int_\Sigma \eta' \frac{e^{-jkR}}{R} \, d\sigma' \tag{10.2–2}$$

$$\mathbf{A} = \frac{\mu}{4\pi} \int \int \int_\tau \mathbf{J}' \frac{e^{-jkR}}{R} \, d\tau' \tag{10.2–3}$$

η' is the charge density on the surface element $d\sigma'$ of the conductor, $\mathbf{J}'$ is the volume density of current in the interior volume element $d\tau'$ at $Q'(x', y', z')$. The potentials due to these charge and current distributions are calculated at a point $Q(x, y, z)$ outside the conductors in the medium in which they are embedded. The distance between the source point Q' (x', y', z') locating the element of charge $\eta' \, d\sigma'$ or current $\mathbf{J}' \, d\tau'$ on or in the conductor and the field point $Q(x, y, z)$ where the potential is calculated as

$$R = [(x - x')^2 + (y - y')^2 + (z - z')^2]^{1/2} \tag{10.2–4}$$

The integration in (10.2–2) is carried out over all charged surfaces; that in (10.2–3) over the interior of all current-carrying conductors. It was shown in Chapter 3 that the Helmholtz integrals were derived by taking full account of all the boundary conditions. If the current density is expressed in Cartesian coordinates,

$$\mathbf{J} = \hat{x}J_x + \hat{y}J_y + \hat{z}J_z \tag{10.2–5}$$

The vector potential has three components, given by

$$\mathbf{A} = \hat{x}A_x + \hat{y}A_y + \hat{z}A_z \tag{10.2–6}$$

with

$$A_x = \frac{\mu}{4\pi} \int \int \int_\tau J'_x \frac{e^{-jkR}}{R} \, d\tau' \tag{10.2–7a}$$

$$A_y = \frac{\mu}{4\pi} \int \int \int_\tau J'_y \frac{e^{-jkR}}{R} \, d\tau' \tag{10.2–7b}$$

$$A_z = \frac{\mu}{4\pi} \int \int \int_\tau J'_z \frac{e^{-jkR}}{R} \, d\tau' \tag{10.2–7c}$$

If all the conductors are of sufficiently small cross section, as, for example, a conductor of circular cross section with radius a satisfying the condition

$$|ka| \ll 1 \tag{10.2–8}$$

the total axial current and the total charge per unit length can be defined in each conductor. The potentials calculated from currents and charges in a cylindrical conductor of small cross section along the $\hat{z}$ direction are given in the cylindrical coordinates ρ, θ, z by

$$\phi = \frac{1}{4\pi\epsilon} \int_z q' \frac{e^{-jkR}}{R} \, dz'; \qquad q' = \int_0^{2\pi} \eta' a \, d\theta' \qquad (10.2\text{–}9a)$$

$$\mathbf{A} \doteq \hat{z} A_z; \qquad A_z = \frac{\mu}{4\pi} \int_z I'_z \frac{e^{-jkR}}{R} \, dz'; \qquad I'_z = \int_0^a \int_0^{2\pi} J'_z \rho' \, d\rho' \, d\theta' \qquad (10.2\text{–}9b)$$

Since the conductors are assumed to be electrically thin, J_θ and J_ρ are negligible and the components A_θ and A_ρ (or A_x and A_y) are not required.

It has been shown that with (10.2–1a, b) the scalar and vector potentials are related by the Lorentz condition

$$\nabla \cdot \mathbf{A} + j\frac{k^2}{\omega}\phi = 0$$

so that with $\mathbf{A} = \hat{z}A_z$,

$$\frac{dA_z}{dz} + j\frac{k^2}{\omega}\phi = 0 \qquad (10.2\text{–}10)$$

10.3 THE DERIVATION OF THE TRANSMISSION-LINE EQUATIONS USING ELECTROMAGNETIC FIELD THEORY

In Sec. 10.1 the transmission-line equations derived by network theory apply strictly only to an infinitely long line in which every section is like every other. It is therefore appropriate initially to apply the methods of electromagnetic theory to derive the transmission-line equations for balanced infinitely long two-wire lines.

Figure 10.3–1 shows a uniform two-wire line extending along the z axis of a rectangular coordinate system. The two wires lie in the yz plane with the center of wire 1 at $y = b/2$ and the center of wire 2 at $y = -b/2$. The radius of each wire is a and it satisfies the inequality

$$|ka| \ll 1 \qquad (10.3\text{–}1)$$

For the present it is assumed that the inequality

$$b^2 \gg a^2 \qquad (10.3\text{–}2)$$

is also satisfied so that distributions of current and charge in each conductor may be assumed to be approximately rotationally symmetrical. Let the section of line to the right of the plane $z = s$ or $w = s - z = 0$ be designated the load. The distance along the axis of each conductor to the left of this plane to the axial elements dw' at points Q'_1 and Q'_2 is w'. The total axial current at the point Q'_1 in

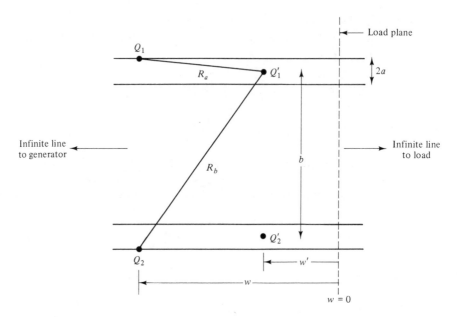

Figure 10.3–1 Section of infinite two-wire line.

conductor 1 is $I_{1z}(w')$ and the charge per unit length is $q_1(w')$. The conditions for a balanced line are assumed to be satisfied:

$$I_{2z}(w) = -I_{1z}(w) = -I_z(w); \qquad q_2(w) = -q_1(w) = -q(w) \qquad (10.3\text{--}3)$$

On the surface of the conductors

$$A_{2z}(w) = -A_{1z}(w); \qquad \phi_2(w) = -\phi_1(w) \qquad (10.3\text{--}4)$$

The current and charge per unit length satisfy the one-dimensional equation of continuity

$$\frac{dI_z(w)}{dw} - j\omega q(w) = 0 \qquad (10.3\text{--}5a)$$

The corresponding equation for the potentials with $\partial/\partial w = -\partial/\partial z$ is given by (10.2–10). It is

$$\frac{\partial A_z(w)}{\partial w} - \frac{jk^2}{\omega}\phi(w) = 0 \qquad (10.3\text{--}5b)$$

The general relation

$$\mathbf{E} + j\omega\mathbf{A} = -\nabla\phi \qquad (10.3\text{--}6a)$$

becomes

$$\frac{\partial\phi}{\partial w} = -\frac{\partial\phi}{\partial z} = E_z + j\omega A_z \qquad (10.3\text{--}6b)$$

for the z component. The following equations are obtained by differentiation and substitution:

$$\frac{\partial^2 \phi}{\partial w^2} + k^2 \phi = \frac{\partial E_z}{\partial w} \tag{10.3-7a}$$

$$\frac{\partial^2 A_z}{\partial w^2} + k^2 A_z = j\frac{k^2}{\omega}E_z \tag{10.3-7b}$$

If the conductors are perfect $E_z = 0$ on their surfaces and ϕ and A_z satisfy the homogeneous equations

$$\frac{\partial^2 \phi}{\partial w^2} + k^2 \phi = 0 \tag{10.3-8a}$$

$$\frac{\partial^2 A_z}{\partial w^2} + k^2 A_z = 0 \tag{10.3-8b}$$

The potential differences between equipotential rings around the surfaces of conductors 1 and 2 at opposite points Q_1 and Q_2 at equal distances w from the plane of the load at $w = 0$ are defined as follows with (10.3–4):

$$V(w) \equiv \phi_1(w) - \phi_2(w) = 2\phi_1(w) \tag{10.3-9}$$

$$W_z(w) \equiv A_{1z}(w) - A_{2z}(w) = 2A_{1z}(w) \tag{10.3-10}$$

The differential equations satisfied by $V(w)$ and $W_z(w)$ are obtained readily as a combination of (10.3–7a,b), (10.3–9), and (10.3–10). They are

$$\frac{\partial^2 V(w)}{\partial w^2} + k^2 V(w) = \frac{\partial}{\partial w}(E_{1z} - E_{2z}) \tag{10.3-11a}$$

$$\frac{\partial^2 W_z(w)}{\partial w^2} + k^2 W_z(w) = j\frac{k^2}{\omega}(E_{1z} - E_{2z}) \tag{10.3-11b}$$

where E_{1z} and E_{2z} are defined on the surfaces of the imperfect conductors.

The potential differences $V(w)$ and $W(w)$ can be evaluated with (10.2–9a,b). Thus

$$V(w) = \frac{1}{2\pi\bar{\epsilon}} \int_{-\infty}^{\infty} q(w')P_L(w,w') \, dw' \tag{10.3-12a}$$

$$W_z(w) = \frac{\mu}{2\pi} \int_{-\infty}^{\infty} I_z(w')P_L(w,w') \, dw' \tag{10.3-12b}$$

where

$$P_L(w,w') = \frac{e^{-jkR_a}}{R_a} - \frac{e^{-jkR_b}}{R_b} \tag{10.3-13}$$

$$R_a = [(w - w')^2 + a^2]^{1/2}; \qquad R_b = [(w - w')^2 + b^2]^{1/2} \tag{10.3-14}$$

Now let the charge per unit length and the current at the point Q' be expanded in Taylor series in terms of the charge per unit length and the current at the point Q. The zero- and first-order terms are

$$q(w') = q(w) + (w' - w) \frac{\partial q(w)}{\partial w} + \cdots \tag{10.3--15a}$$

$$I_z(w') = I_z(w) + (w' - w) \frac{\partial I_z(w)}{\partial w} + \cdots \tag{10.3--15b}$$

With the equation of continuity (10.3--5a), it follows that (10.3--15a,b) become

$$q(w') \doteq q(w) + \frac{1}{j\omega} \frac{\partial^2 I_z(w)}{\partial w^2} (w' - w) \tag{10.3--16a}$$

$$I_z(w') \doteq I_z(w) + j\omega q(w)(w' - w) \tag{10.3--16b}$$

The substitution of (10.3--16a,b) into (10.3--12a,b) gives the general expressions for the potential differences along an infinite line. They are

$$V(w) = \frac{1}{2\pi\bar{\epsilon}} \left[q(w)K_0(w) + \frac{1}{j\omega} \frac{\partial^2 I_z(w)}{\partial w^2} \frac{K_1(w)}{k} \right] \tag{10.3--17a}$$

$$W_z(w) = \frac{\mu}{2\pi} \left[I_z(w)K_0(w) + j\omega q(w) \frac{K_1(w)}{k} \right] \tag{10.3--17b}$$

where

$$K_0(w) \equiv \int_{-\infty}^{\infty} P_L(w,w') \, dw'$$

$$= \int_{-\infty}^{\infty} \left(\frac{1}{R_a} - \frac{1}{R_b} \right) dw' + \int_{-\infty}^{\infty} [F(a) - F(b)] \, dw' \tag{10.3--18}$$

$$\frac{K_1(w)}{k} \equiv \int_{-\infty}^{\infty} (w' - w)P_L(w,w') \, dw'$$

$$= \int_{-\infty}^{\infty} (w' - w)\left(\frac{1}{R_a} - \frac{1}{R_b} \right) dw' + \int_{-\infty}^{\infty} (w' - w)[F(a) - F(b)] \, dw' \tag{10.3--19}$$

with

$$F(a) = \frac{e^{-jkR_a} - 1}{R_a}; \qquad F(b) = \frac{e^{-jkR_b} - 1}{R_b} \tag{10.3--20}$$

It can be shown by direct evaluation that the first integrals in (10.3--18) and (10.3--19) are

$$\int_{-\infty}^{\infty} \left(\frac{1}{R_a} - \frac{1}{R_b} \right) dw' = 2 \ln \frac{b}{a} \tag{10.3--21}$$

$$\int_{-\infty}^{\infty} (w - w')\left(\frac{1}{R_a} - \frac{1}{R_b}\right) dw' = 0 \tag{10.3-22}$$

Subject to the condition

$$|kb|^2 \ll 1 \tag{10.3-23}$$

the second integrals in (10.3–18) and (10.3–19) are negligible. Over that part of the integration for which $(w' - w)^2$ is large compared with b^2 and a^2, R_a and R_b differ negligibly from $|w' - w|$ and from each other. It follows that $F(a) \doteq F(b)$, so that $F(a) - F(b)$ does not contribute to the integrals. The principal contributions to the integrals occur when $|w' - w|$ is small, so that R_a and R_b are of order of magnitude b. When this condition is true,

$$|k^2b^2| \sim |k^2R^2| \ll 1; \qquad e^{-jkR} \doteq 1 - jkR - k^2R^2 + jk^3R^3 \tag{10.3-24}$$

where R may be either R_a or R_b. Hence

$$|F(a) - F(b)| = |k^2(R_b - R_a) - jk^3(R_b^2 - R_a^2)| \tag{10.3-25}$$

The integrand of the first integral is

$$\frac{1}{R_a} - \frac{1}{R_b} = \frac{R_b - R_a}{R_aR_b} \tag{10.3-26}$$

The ratio of the integrand in the first to that in the second integral in both (10.3–18) and (10.3–19) is

$$1 : |k^2R_aR_b - jk^3R_aR_b(R_b + R_a)| \tag{10.3-27}$$

over the range where (10.3–24) is valid. So, with (10.3–24), the second integral is negligible. Therefore, for the infinitely long line, subject to (10.3–23),

$$K_0(w) \doteq K_0 = 2 \ln \frac{b}{a} \tag{10.3-28}$$

$$K_1(w) \doteq 0 \tag{10.3-29}$$

It is to be noted that the radiation from transmission lines is determined from the imaginary part of (10.3–27), which is made negligible by the imposition of (10.3–23).

With (10.3–28) and (10.3–29), the potential differences at a point Q along the transmission line are

$$V(w) \doteq q(w)\frac{j\omega}{y} = \frac{dI_z(w)}{dw}\frac{1}{y} \tag{10.3-30}$$

$$W_z(w) = I_z(w)\, l^e \tag{10.3-31}$$

where

$$y \equiv g + j\omega c \equiv j\omega \frac{2\pi\bar{\epsilon}}{K_0} = j\frac{\omega\pi\bar{\epsilon}}{\ln(b/a)} \tag{10.3-32a}$$

If the transmission line is in a medium which is dissipative, $\bar{\epsilon}$ is a complex quantity given by

$$\bar{\epsilon} = \epsilon_e - j\frac{\sigma_e}{\omega} \tag{10.3-32b}$$

and

$$g = \frac{\pi\sigma_e}{\ln(b/a)}; \qquad c = \frac{\pi\epsilon_e}{\ln(b/a)}; \qquad l^e = \frac{\mu}{\pi}\ln\frac{b}{a} \tag{10.3-32c}$$

These are the leakage conductance, the capacitance and the external inductance per unit length of the infinite two-wire line. Note that in a dissipative medium k is complex

$$k^2 = \omega^2\mu\bar{\epsilon} = -z^e y \tag{10.3-33a}$$

$z^e = j\omega l^e$ is the external impedance per unit length of the parallel line. The assumption of uniform current and charge per unit length is adequate in evaluating the potential differences and constants for an infinite line.

Equation (10.3–30) is one of the two first-order differential equations for the two-wire lines. The other is obtained from (10.3–6b) with (10.3–9) and (10.3–10). Thus

$$\frac{dV(w)}{dw} - j\omega W_z(w) = E_{1z}(w) - E_{2z}(w) \tag{10.3-33b}$$

where $E_{1z}(w)$ and $E_{2z}(w)$ are the axial tangential components of the electric field at the surfaces of the two conductors. The electric field is proportional to the total current:

$$E_{1z}(w) = I_{1z}(w)z_1^i = I_z(w)\frac{z^i}{2} \tag{10.3-34a}$$

$$E_{2z}(w) = I_{2z}(w)z_2^i = -I_z(w)\frac{z^i}{2} \tag{10.3-34b}$$

$I_{2z}(w) = -I_{1z}(w)$ for a balanced line and $z_2^i = z_1^i$ for identical conductors. Hence

$$z^i = z_1^i + z_2^i \tag{10.3-34c}$$

The quantity $z^i = r^i + jx^i$ is the internal impedance per unit length of the two-wire line which was defined in (3.9–50) and is given by

$$z^i = r^i + jx^i = \frac{1+j}{2\pi a}\sqrt{\frac{\mu_c\omega}{2\sigma_c}} \tag{10.3-35}$$

for high frequencies with $a\sqrt{\omega\mu_c\sigma_c} \geq 10$. σ_c is the conductivity of the conductors. $z^i = 2z_1^i$ for a two-wire line. The substitution of (10.3–34a) and (10.3–31) into

(10.3–33b) gives

$$\frac{\partial V(w)}{\partial w} = (z^i + j\omega l^e)I_z(w) \tag{10.3–36a}$$

With (10.3–32a) substituted in (10.3–30),

$$\frac{\partial I_z(w)}{\partial w} = (g + j\omega c)\,V(w) \tag{10.3–36b}$$

(10.3–36a,b) are the one-dimensional transmission-line equations. With $-\partial/\partial z = \partial/\partial w$ and

$$z = z^i + j\omega l^e = r^i + j\omega(l^i + l^e) = r + j\omega l \tag{10.3–37}$$

(where boldface is used for z to avoid confusion with the coordinate z) the equations are

$$\frac{-\partial V(z)}{\partial z} = zI_z(z) \tag{10.3–38}$$

$$\frac{-\partial I_z(z)}{\partial z} = yV(z) \tag{10.3–39}$$

which are (10.1–7) and (10.1–9) derived using network theory. The application of network theory is valid provided that the following conditions are satisfied for a two-wire line.

(1) The two conductors are parallel and identical

(2) $|ka| \ll 1$

(3) $|kb|^2 \ll 1$ (10.3–40)

(4) $b^2 \gg a^2$

(5) The line is infinitely long

10.4 THE COAXIAL LINE

The infinitely long coaxial line consisting of conductor 1 of radius a_1 in a conducting sheath 2 of inner radius a_2 and outer radius a_3 is shown in Fig. 10.4–1. The analysis of the coaxial line may be carried out in a manner similar to that used for the two-wire line. The conditions (10.3–3) for the current and charge are postulated here. Then the potentials at an arbitrary point $Q(\rho, \theta, w)$ in the dielectric medium ($a_1 \leq \rho \leq a_2$) are given by

$$\phi(w) = \frac{1}{4\pi\tilde{\epsilon}} \int_0^{2\pi} \int_{-\infty}^{\infty} q(w')P_\rho(w, w', \theta')\, dw'\, \frac{d\theta'}{2\pi} \tag{10.4–1}$$

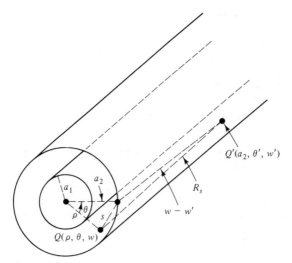

Figure 10.4–1　Coaxial line.

$$A_z(w) = \frac{\mu}{4\pi} \int_0^{2\pi} \int_{-\infty}^{\infty} I_z(w')P_\rho(w, w', \theta')\, dw' \frac{d\theta'}{2\pi} \tag{10.4–2}$$

where

$$P_\rho(w, w', \theta') \equiv \frac{e^{-jkR_\rho}}{R_\rho} - \frac{e^{-jkR_s}}{R_s} \tag{10.4–3}$$

$$R_\rho = [(w - w')^2 + \rho^2]^{1/2} \tag{10.4–4}$$

$$R_s = [(w - w')^2 + s^2]^{1/2}; \qquad s^2 = \rho^2 + a_2^2 - 2a_2\rho \cos\theta' \tag{10.4–5}$$

The charge per unit length and the total current in each conductor are defined as follows in terms of the surface density of charge η and the axial component of the volume density of current J_z.

$$q_1(w) = 2\pi a_1 \eta_1(w) \tag{10.4–6a}$$

$$q_2(w) = 2\pi a_2 \eta_2(w) = -q_1(w) \tag{10.4–6b}$$

$$I_{1z}(w) = 2\pi \int_0^{a_1} J_{1z}(w, \rho)\rho\, d\rho \tag{10.4–7a}$$

$$I_{2z}(w) = 2\pi \int_{a_2}^{a_3} J_{2z}(w, \rho)\rho\, d\rho = -I_{1z}(w) \tag{10.4–7b}$$

In defining s in (10.4–5) it is assumed for simplicity that the entire current $I_{2z}(w)$ is concentrated in a thin layer on the inner surface of the sheath as for a perfect conductor. For a good but imperfect conductor, the field in the sheath is included in the evaluation of the internal impedance. It is to be noted that only R_s in $P_\rho(w, w', \theta')$ involves θ'.

In the evaluation of (10.4–1) and (10.4–2) for the infinite line, $q(w')$ may be replaced by $q(w)$ and $I_z(w')$ by $I_z(w)$, as shown for the two-wire line. This leaves only $P_\rho(w, w', \theta')$ under the sign of integration. The integration with respect to z may be carried out just as for the two-wire line. The condition

$$|ka_2|^2 \ll 1 \tag{10.4–8}$$

is imposed to eliminate higher modes, as discussed in Chapter 12. The steps in the analysis are the same as in (10.3–18) to (10.3–22) and lead to

$$\int_{-\infty}^{\infty} P_\rho(w, w', \theta')\, dw' \doteq 2 \int_0^{\infty} \left(\frac{1}{R_\rho} - \frac{1}{R_s} \right) dw' = 2 \ln \frac{s}{\rho} \tag{10.4–9}$$

With the coaxial line, there is in addition the integration with respect to θ' which does not occur with the two-wire line. Thus

$$\phi(w) = \frac{q(w)}{4\pi\bar{\epsilon}} \int_0^{2\pi} [\ln (\rho^2 + a_2^2 - 2a_2\rho \cos \theta') - 2 \ln \rho] \frac{d\theta'}{2\pi} \tag{10.4–10a}$$

It can be shown with the help of a table of integrals that

$$\phi(w) = \frac{q(w)}{2\pi\bar{\epsilon}} \ln \frac{a_2}{\rho} \tag{10.4–10b}$$

Similarly,

$$A_z(w) = \frac{\mu}{2\pi} I_z(w) \ln \frac{a_2}{\rho} \tag{10.4–11}$$

The potential differences between points $Q_1(a_1, \theta, w)$ and $Q_2(a_2, \theta, w)$ on the surfaces of the two conductors are

$$V(w) = \phi_1(w) - \phi_2(w) = \frac{q(w)}{2\pi\bar{\epsilon}} \ln \frac{a_2}{a_1} \tag{10.4–12}$$

$$W_z(w) = A_{1z}(w) - A_{2z}(w) = \frac{\mu}{2\pi} I_z(w) \ln \frac{a_2}{a_1} \tag{10.4–13}$$

The same results may be obtained even if the term $1/R_s$ is omitted in (10.4–9) and $V(w)$ is evaluated directly. This implies that the potential differences may be determined *entirely* from the charges and currents in the *inner* conductor. The potential differences in (10.4–12) and (10.4–13) may be expressed as in (10.3–30) and (10.3–31), that is,

$$V(w) = \frac{dI_z(w)}{dw} \frac{1}{y} \tag{10.4–14}$$

$$W_z(w) = I_z(w)\, l^e \tag{10.4–15}$$

with

$$y = g + j\omega c; \qquad g = \frac{2\pi\sigma_e}{\ln(a_2/a_1)}; \qquad c = \frac{2\pi\epsilon_e}{\ln(a_2/a_1)} \qquad (10.4\text{--}16a)$$

$$l^e = \frac{\mu}{2\pi} \ln \frac{a_2}{a_1} \qquad (10.4\text{--}16b)$$

The internal impedance per unit length is defined as for the two-wire line,

$$z^i = z_1^i + z_2^i \qquad (10.4\text{--}17)$$

where z_1^i and z_2^i are the internal impedances per unit length of the inner and outer conductors. At high frequencies $(a\sqrt{\omega\sigma_c\mu_c} \geq 10)$

$$z_1^i = \frac{1+j}{2\pi a_1} \sqrt{\frac{\mu_c\omega}{2\sigma_c}}; \qquad z_2^i = \frac{1+j}{2\pi a_2} \sqrt{\frac{\mu_c\omega}{2\sigma_c}} \qquad (10.4\text{--}18)$$

If the same procedure is followed as for the two-wire line, the differential equations obtained are the same as (10.3–38) and (10.3–39) with the line parameters given by (10.4–16a,b) and (10.4–17). The conditions imposed for the validity of these equations are

(1) $|ka_2|^2 \ll 1$

(2) The line is infinitely long

The electric and magnetic fields in the dielectric medium in the coaxial line are obtained from the potential functions. Since there is rotational symmetry and A_z is the only component of the vector potential, the following relations follow from $\mathbf{B} = \nabla \times \mathbf{A}$:

$$B_\rho = 0; \qquad B_\theta = -\frac{\partial A_z}{\partial \rho}; \qquad B_z = 0 \qquad (10.4\text{--}19)$$

Hence, with (10.4–11),

$$\mathbf{B} = \hat{\mathbf{\theta}}B_\theta; \qquad B_\theta = \frac{\mu I_z}{2\pi\rho} \qquad (10.4\text{--}20)$$

The electric field is given by $\mathbf{E} = -\nabla\phi - j\omega\mathbf{A}$. With rotational symmetry and $\mathbf{A} = \hat{\mathbf{z}}A_z$, the only transverse component of the field is E_ρ:

$$E_\rho = -\frac{\partial\phi}{\partial\rho} = \frac{q}{2\pi\tilde{\epsilon}\rho} \qquad (10.4\text{--}21)$$

where ϕ is given by (10.4–10a,b).

10.5 STRIP LINES

There are many applications particularly at the higher frequencies where it is more convenient to have flat strip conductors instead of those with circular cross section discussed in Secs. 10.3 and 10.4. Some possible forms are shown in Fig. 10.5–1. The properties of strip lines do not differ significantly from those of lines with circular cross section. The capacitance per unit length may be determined by electrostatic methods irrespective of the nature of the cross section provided that the width b of each strip and the distance $2h$ between the strips are both small compared with the wavelength.

The capacitance per unit length of a very thin strip (conductivity σ_c, permeability μ_c, and thickness d) in a homogeneous infinite medium (dielectric constant ϵ, permeability μ, and small conductivity σ) over a highly conducting infinite plane surface (conductivity σ_c and permeability μ_c) may be determined by conformal transformations subject to the inequalities $d \ll h \ll b$. It is given by

$$c = \frac{\epsilon b}{h}\frac{f(x)}{x} \tag{10.5–1}$$

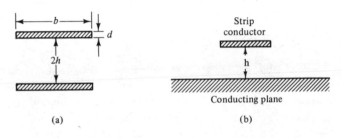

(a)

(b)

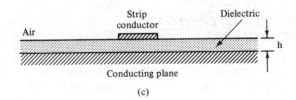

(c)

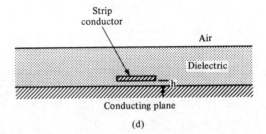

(d)

Figure 10.5–1 Cross sections of strip lines; (a) two-conductor strip line; (b) strip line over conducting plane; (c) strip conductor on dielectric-coated conducting plane (microstrip); (d) strip conductor in dielectric over conducting plane (sandwich line).

where

$$f(x) = 1 + x + \ln(1 + x); \qquad x = \frac{\pi b}{2h} \qquad (10.5\text{--}2)$$

It follows that

$$g = \frac{\sigma b}{h} \frac{f(x)}{x} \qquad (10.5\text{--}3)$$

$$l^e = \frac{\mu \epsilon}{c} = \frac{\mu h}{b} \frac{x}{f(x)} \qquad (10.5\text{--}4)$$

The approximate internal impedance per unit length $z^i = z^i_{\text{strip}} + z^i_{\text{plane}}$ with

$$z^i_{\text{strip}} = \frac{1}{b} \sqrt{\frac{\omega \mu_c}{2\sigma_c}} \frac{x[1 + x + \pi - 2 \ln(\delta/2)]}{f(x)} \qquad (10.5\text{--}5a)$$

$$z^i_{\text{plane}} = \frac{1}{b} \sqrt{\frac{\omega \mu_c}{2\sigma_c}} \frac{x(1 + x)}{f(x)} \qquad (10.5\text{--}5b)$$

where σ_c and μ_c apply to the conductor and

$$\delta = p^2 - 1 + p[p^2 - 1]^{1/2}; \qquad p = 1 + \frac{d}{h} \qquad (10.5\text{--}6)$$

These formulas are good approximations for a plane of finite width provided that it extends on each side a distance equal to the width b of the strip.

10.6 GENERAL SOLUTION OF THE DIFFERENTIAL EQUATIONS FOR AN INFINITE LINE

The first-order differential equations for the scalar potential difference and the current in all the different types of transmission lines analyzed in the previous sections have the form

$$-\frac{\partial V}{\partial z} = z I_z; \qquad -\frac{\partial I_z}{\partial z} = yV \qquad (10.6\text{--}1)$$

where

$$y = g + j\omega c; \qquad z = z^i + j\omega l^e = r^i + j\omega l \qquad (10.6\text{--}2)$$

It is customary to define the total inductance per unit length by

$$l = l^e + l^i; \qquad l^i = \frac{x^i}{\omega} \qquad (10.6\text{--}3)$$

Since x^i is not a linear function of frequency, it follows that l^i is dependent on the frequency. The formulas for l^e, g, c, and x^i depend on the cross section of the conductor. Since the equations for I_z and V are alike in form, it is sufficient to examine one of them.

The differential equation for the voltage is

$$\left(\frac{\partial^2 V}{\partial z^2}\right)_z - \gamma^2 V_z = 0 \tag{10.6-4}$$

where the complex propagation constant γ as defined in Sec. 10.1 is

$$\gamma = \sqrt{zy} = \alpha + j\beta = [(r^i + j\omega l)(g + j\omega c)]^{1/2} \tag{10.6-5}$$

The general solution of (10.6–4) is well known and may be expressed in different forms:

$$V_z = B_1 e^{\gamma z} + B_2 e^{-\gamma z} = C_1 \cosh \gamma z + C_2 \sinh \gamma z$$

$$= D \cosh (\gamma z + \theta) \tag{10.6-6}$$

Alternatively, $w = s - z$ may be substituted in (10.6–6). The B's, C's, D, and θ are complex constants of integration. The expression for the current is most easily obtained from

$$zI_z = -\left(\frac{dV}{dz}\right)_z \tag{10.6-7}$$

Thus, for the exponential form in (10.6–6),

$$zI_z = -\gamma(B_1 e^{\gamma z} - B_2 e^{-\gamma z}) \tag{10.6-8}$$

It is now convenient to introduce Z_c, the characteristic impedance defined by

$$Z_c = \sqrt{\frac{z}{y}} = R_c + jX_c = \left(\frac{r + j\omega l}{g + j\omega c}\right)^{1/2} \tag{10.6-9a}$$

The characteristic admittance is

$$Y_c = \frac{1}{Z_c} = G_c + jB_c \tag{10.6-9b}$$

The current is then given by

$$I_z = Y_c(-B_1 e^{\gamma z} + B_2 e^{-\gamma z}) \tag{10.6-10}$$

Now consider a transmission line that begins at $z = 0$ and continues to $z = s = \infty$. For physical reasons the voltage must vanish at infinity so that $B_1 = 0$. It follows from (10.6–6) that B_2 is the voltage V_0 at $z = 0$. Thus for an infinitely long line ($s = \infty$),

$$V_z = I_z Z_c = V_0 e^{-(\alpha + j\beta)z} \tag{10.6-11}$$

The multiplication throughout by $e^{j\omega t}$ and the selection of the real part as the solution that is consistent with an assumed time dependence of the form

$$v_0(t) = V_0 \cos \omega t = \text{Re}(V_0 e^{j\omega t}) \qquad (10.6\text{--}12a)$$

—which refers the phase to the maximum value of the time-dependent voltage, $v_z(t)$—lead to

$$v_z(t) = V_0 e^{-\alpha z} \cos(\omega t - \beta z) \qquad (10.6\text{--}12b)$$

This solution has a physical interpretation. The voltage v_z is a function of two independent variables, the time t and the distance z along the wire. At a fixed point $z = s$, the voltage varies periodically. The potential is positive on one wire and negative on the other for half of one period. The amplitude increases from zero to a maximum of $V_0 e^{-\alpha z}$ and then decreases to zero sinusoidally. Then the polarity reverses and the voltage decreases to an equal negative extreme and then is again reduced to zero. The phase lag of the voltage at z behind the voltage at $z = 0$ is βz. The cycle repeats. The same variation occurs at every other point z, but the amplitude $V_0 e^{-\alpha z}$ is different and the phase lags that at $z = 0$ by βz. The amplitude decreases exponentially and the phase lag increases linearly with distance from $z = 0$.

If, instead of concentrating on a fixed point along the line, the amplitude all along the line is examined at a given instant, such as $t = 0$, then

$$v_z(0) = V_0 e^{-\alpha z} \cos \beta z \qquad (10.6\text{--}13a)$$

At a quarter-period later, $t = \tau/4$,

$$v_z(\tau/4) = V_0 e^{-\alpha z} \sin \beta z \qquad (10.6\text{--}13b)$$

and at a half-period later, $t = \tau/2$,

$$v_z(\tau/2) = -V_0 e^{-\alpha z} \cos \beta z \qquad (10.6\text{--}13c)$$

These three distributions are shown in Fig. 10.6–1. It appears that as time passes, any given curve such as the one for $t = 0$ moves along the line with its amplitude confined between the limiting curves $V_0 e^{-\alpha z}$ and $-V_0 e^{-\alpha z}$.

This motion may be interpreted by considering specifically the phase of the voltage, that is, $(\omega t - \beta z)$. Points and contours in the distribution of voltage along the semi-infinite line at which the voltages are all in the same phase relative to a complete cycle are defined by $\psi = \omega t - \beta z = $ constant. Since the trigonometric function is multivalued, the current at all points for which the constant differs by $2n\pi$ (n is any integer) is in the same phase as at $z = 0$. The currents at different times and different points along the line which differ in phase by integral multiples of 2π are defined by

$$\omega t - \beta z = \psi_n = \psi_0 - 2n\pi; \qquad n = 0, 1, 2, 3, \ldots \qquad (10.6\text{--}14)$$

where ψ_0 is a constant. The significance of this relation may be seen by first determining the distances z from the input end at which the voltages differ instan-

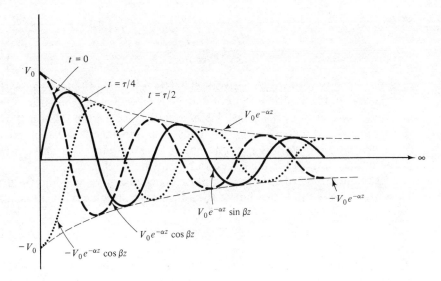

Figure 10.6–1 Instantaneous distribution of voltage along a semi-infinite line at instants differing by a quarter period.

taneously in phase by integral multiples of 2π and then examining what happens to these particular phases with the passage of time. Let an arbitrary instant t_1 be selected. The points characterized by voltages in the phases $\psi_0 - 2n\pi$ are given by

$$z_n = \frac{1}{\beta}(\omega t_1 - \psi_0 + 2n\pi); \qquad n = 0, 1, 2, 3, \ldots \qquad (10.6-15)$$

The distance between two adjacent points differing in phase by 2π is

$$z_{m+1} - z_m = \frac{2\pi}{\beta}; \qquad m \text{ is any integer} \qquad (10.6-16)$$

The distance is the same for all choices of m and it is a fundamental constant of the distribution called the wavelength λ. Hence

$$\lambda = \frac{2\pi}{\beta} \qquad (10.6-17)$$

At any given instant of time, voltages along the semi-infinite line which differ in phase by 2π are separated by distances λ.

The points in the distribution characterized by voltages in a particular phase at a given instant have now been determined. It remains to note how the distance z locating any one such point varies in time. Let both sides of (10.6–14) be differentiated with respect to time. The result is

$$\omega - \beta \frac{dz}{dt} = 0 \qquad (10.6-18a)$$

or

$$v_p \equiv \frac{dz}{dt} = \frac{\omega}{\beta} \qquad (10.6\text{--}18b)$$

v_p is the velocity with which a given phase travels along the line and is termed the *phase velocity*. If the distribution of voltage is viewed along the entire semi-infinite line at a single instant, the phase lag at any distance z from 0 (with respect to the voltage at $z = 0$ at that instant) is βz. Thus β measures the phase angle characteristic of a given semi-infinite line per unit length. It is the *phase constant* per unit length of the infinite line and is measured in radians per meter.

The amplitude of voltage in a particular phase is reduced according to $e^{-\alpha z}$. Thus α measures the logarithm of the ratio of amplitudes $|V_0/V_z|$ per unit length.

$$\alpha = \frac{1}{z} \ln \left| \frac{V_0}{V_z} \right| \qquad (10.6\text{--}19)$$

It is the attenuation constant per unit length of the line. It is measured in nepers per meter.

If the phase constant β is a linear function of the frequency so that

$$\beta = \frac{\omega}{v} \qquad (10.6\text{--}20)$$

where v is a constant independent of frequency, then the phase velocity v_p is the same for all frequencies and equal to v in (10.6–20). Under all other conditions, the phase velocity is different for different frequencies, so that for any complex voltage that is a superposition of components of several frequencies, these components have different phase velocities. In this case *dispersion* is said to occur.

10.7 THE TERMINATED TRANSMISSION LINE

Practical transmission lines are necessarily finite in length extending from a generator with emf V_0^e and internal impedance Z_0 at $z = 0$ to a load impedance Z_s at $z = s$ (Fig. 10.7–1). It follows that the integrals (10.3–21) and (10.4–9) extend only from 0 to s and not from $-\infty$ to $+\infty$ as for an infinitely long line. For the two-wire line,

$$K_0(w) = \int_0^s \left(\frac{1}{R_a} - \frac{1}{R_b} \right) dw'$$

$$\sim 2 \ln \frac{b}{a} - \ln \frac{w + \sqrt{w^2 + b^2}}{w + \sqrt{w^2 + a^2}} - \ln \frac{z + \sqrt{z^2 + b^2}}{z + \sqrt{z^2 + a^2}} \qquad (10.7\text{--}1)$$

where $w = s - z$. It is evident from this expression that at distances from the

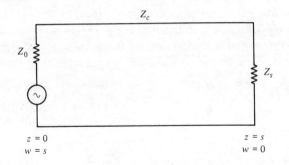

$z = 0$
$w = s$

$z = s$
$w = 0$

Figure 10.7–1 Terminated transmission line.

ends which satisfy the inequalities $z^2 \gg b^2$, $w^2 \gg b^2$, (10.7–1) reduces to $K_0(w) \sim K_0 = 2 \ln (b/a)$, and transmission-line theory as developed for the infinitely long line is valid. Within short distances of the load Z_s this is not true. In particular, at $z = s$ or $w = 0$, $K_0(w) \sim K_0/2 = \ln (b/a)$. This means that the line constants l^e, c, and g as defined in (10.3–32c) are actually functions of w and $l^e(w) \rightarrow l^e/2$, $c(w) \rightarrow 2c$, $g(w) \rightarrow 2g$ as the load is approached. Note, however, that $\gamma(w) = [z(w)y(w)]^{1/2} = \gamma$ even near the load.

The presence of a load impedance at the end $z = s$ of a transmission line introduces complications in addition to the finite length of the line. These are associated with the inductive and capacitive coupling that may exist between the load impedance and the adjacent conductors of the line. These may be positive or negative depending on whether they add to or subtract from the values obtained from the integration along the conductors of the line. If interest is in the determination of the actual impedance of a particular termination, for example, an antenna or a coil, the complete evaluation must be carried out.[*] For most purposes it is adequate and convenient to introduce an apparent terminal impedance which takes full account of the entire effect. It is the impedance that is obtained by direct measurement on the particular transmission line and calculation from formulas that use the constant values of $z(w)$ and $y(w)$ characteristic of the infinite line. In the following analysis of the finite line, all terminal impedances are assumed to be apparent impedances. In many cases the apparent impedance can be obtained from the actual impedance by means of a corrective circuit of lumped capacitances and inductances. When the cross-sectional dimensions of the line are sufficiently small, the apparent impedance differs negligibly from the actual impedance.

The voltage and current on a transmission line were shown in Sec 10.6 to be

$$V_z = B_1 e^{\gamma z} + B_2 e^{-\gamma z} \tag{10.7–2a}$$

$$I_z = \frac{1}{Z_c} (-B_1 e^{\gamma z} + B_2 e^{-\gamma z}) \tag{10.7–2b}$$

[*]This is illustrated, for example, in R. W. P. King, *Transmission-Line Theory*. (New York: Dover Publications, Inc., 1965).

Equation (10.7–2a) for the voltage represents the superposition of voltages in two traveling waves, one traveling from $z = 0$ in the positive z direction and the other traveling toward $z = 0$ in the negative z direction. If these voltages are denoted by V_z^+ and V_z^-,

$$V_z^+ = B_2 e^{-\gamma z} \tag{10.7–3a}$$

$$V_z^- = B_1 e^{+\gamma z} \tag{10.7–3b}$$

and

$$V_z = V_z^+ + V_z^- \tag{10.7–3c}$$

At $z = s$, $V_z = V_s$ the voltage across the load is given by

$$V_s = B_1 e^{\gamma s} + B_2 e^{-\gamma s} \tag{10.7–4a}$$

and the current is given by

$$I_s = \frac{1}{Z_c}(-B_1 e^{\gamma s} + B_2 e^{-\gamma s}) = \frac{V_s}{Z_s} \tag{10.7–4b}$$

(10.7–4a,b) are solved for B_1 and B_2 to obtain:

$$B_1 = \frac{V_s}{2Z_s}(Z_s - Z_c)e^{-\gamma s} \tag{10.7–5a}$$

$$B_2 = \frac{V_s}{2Z_s}(Z_s + Z_c)e^{\gamma s} \tag{10.7–5b}$$

The expressions for the voltage and current (10.7–2a,b) can be written as follows:

$$V_z = \frac{V_s}{2Z_s}[(Z_s - Z_c)e^{-\gamma(s-z)} + (Z_s + Z_c)e^{\gamma(s-z)}] \tag{10.7–6a}$$

$$I_z = \frac{V_s}{2Z_s Z_c}[-(Z_s - Z_c)e^{-\gamma(s-z)} + (Z_s + Z_c)e^{\gamma(s-z)}] \tag{10.7–6b}$$

The voltage reflection coefficient at any point z along the transmission line is given by the ratio of the reflected voltage to the incident voltage at that point. In general, the reflection coefficient (denoted by Γ_z) is a complex quantity. The current reflection coefficient is defined in a similar manner and is the negative of the voltage reflection coefficient. In particular, the voltage reflection coefficients at the terminations Z_s and Z_0 are given by

$$\Gamma_s = \frac{Z_s - Z_c}{Z_s + Z_c}; \qquad \Gamma_0 = \frac{Z_0 - Z_c}{Z_0 + Z_c} \tag{10.7–7}$$

where Z_s is the load impedance at $z = s$ and Z_0 is the impedance of the generator or any impedance in series with the generator at $z = 0$. Equations (10.7–6a,b) can be expressed in terms of the reflection coefficient as follows:

$$V_z = \frac{V_s(Z_s + Z_c)}{2Z_s} [e^{\gamma(s-z)} + \Gamma_s e^{-\gamma(s-z)}] \qquad (10.7\text{--}8a)$$

$$I_z = \frac{V_s(Z_s + Z_c)}{2Z_s Z_c} [e^{\gamma(s-z)} - \Gamma_s e^{-\gamma(s-z)}] \qquad (10.7\text{--}8b)$$

The voltage and current along a transmission line can also be expressed in terms of the generator voltage V_0^e and the reflection coefficient Γ_0. It can be shown that

$$V_z = \frac{V_0^e Z_0}{Z_c + Z_0} \frac{e^{-\gamma z} + \Gamma_s e^{-\gamma(2s-z)}}{1 - \Gamma_0 \Gamma_s e^{-2\gamma s}} \qquad (10.7\text{--}9a)$$

$$I_z = \frac{V_0^e}{Z_c} \frac{e^{-\gamma z} - \Gamma_s e^{-\gamma(2s-z)}}{1 - \Gamma_0 \Gamma_s e^{-2\gamma s}} \qquad (10.7\text{--}9b)$$

When $s \to \infty$ or when $\Gamma_s \to 0$ (matched line), the expressions for voltage and current in (10.7–9a,b) reduce to the form in (10.6–11). These expressions may also be expressed in terms of $w = s - z$ and the instantaneous real solutions are obtained by multiplying by $e^{j\omega t}$ and selecting the real parts. The input impedance at z is given by $Z_{\mathrm{in}z} = V_z/I_z$.

If use is made of the other forms of V_z given in (10.6–6), the following formulas are obtained in terms of $w = s - z$:

$$V_z = \frac{V_0^e Z_c}{D} (Z_s \cosh \gamma w + Z_c \sinh \gamma w) \qquad (10.7\text{--}10a)$$

$$I_z = \frac{V_0^e}{D} (Z_s \sinh \gamma w + Z_c \cosh \gamma w) \qquad (10.7\text{--}10b)$$

where

$$D = (Z_c^2 + Z_0 Z_s) \sinh \gamma s + Z_c(Z_0 + Z_s) \cosh \gamma s \qquad (10.7\text{--}10c)$$

These expressions involve γ and Z_c explicitly together with the terminal impedances Z_0 and Z_s; the reflection coefficients do not appear. The third possible form of the solution is

$$V_z = V_0^e \frac{\sinh \theta_0 \cosh (\gamma w + \theta_s)}{\sinh (\gamma s + \theta_0 + \theta_s)} \qquad (10.7\text{--}11a)$$

$$I_z = \frac{V_0^e}{Z_c} \frac{\sinh \theta_0 \sinh (\gamma w + \theta_s)}{\sinh (\gamma s + \theta_0 + \theta_s)} \qquad (10.7\text{--}11b)$$

where the complex terminal functions are defined by

$$\theta_0 = \rho_0 + j\Phi_0 \equiv \coth^{-1} \frac{Z_0}{Z_c} \qquad (10.7\text{--}12a)$$

$$\theta_s = \rho_s + j\Phi_s \equiv \coth^{-1} \frac{Z_s}{Z_c} \tag{10.7-12b}$$

Also,

$$\rho \equiv \frac{1}{2} \tanh^{-1} \frac{2RR_c}{R^2 + X^2 + R_c^2}; \qquad \Phi \equiv \frac{1}{2} \tan^{-1} \frac{-2XR_c}{R^2 + X^2 - R_c^2} \tag{10.7-13}$$

Note that when $X = 0$, $R < R_c$, $\Phi = \pi/2$ or $3\pi/2$; when $X = 0$, $R > R_c$, $\Phi = 0$ or π; when $X = 0$, $R = R_c$, $\rho = \infty$ (matched line). In the representation of standing-wave distributions of current and voltage the forms (10.7–10a,b) and (10.7–11a,b) are often more convenient than the exponential forms (10.7–9a,b).

10.8 GRAPHICAL REPRESENTATION OF THE NORMALIZED IMPEDANCE OR ADMITTANCE: SMITH CHART

The normalized impedance $z_1 = r_1 + jx_1$ of a transmission line is obtained by dividing the impedance by the characteristic impedance of the transmission line. If it is more convenient to use the admittance, the normalized admittance is obtained by dividing the admittance by the characteristic admittance (or multiplying by the characteristic impedance). Since both the resistance and the reactance can vary from zero to infinity, it is necessary to have a graphical representation that covers the entire range of values of r_1 and x_1. A convenient representation of the resistance and reactance is in the complex plane of the reflection coefficient Γ in polar coordinates. If the normalized impedance is

$$z_1 = \frac{Z}{Z_c} = r_1 + jx_1 \tag{10.8-1}$$

the complex reflection coefficient Γ is

$$\Gamma = |\Gamma|e^{j\psi} = \frac{z_1 - 1}{z_1 + 1} \tag{10.8-2}$$

If z_1 is given by

$$z_1 = \coth \theta = \coth (\rho + j\Phi) \tag{10.8-3}$$

$$\Gamma = e^{-2\theta} \tag{10.8-4}$$

and

$$|\Gamma| = e^{-2\rho}; \qquad \psi = -2\Phi \tag{10.8-5}$$

A circle diagram that represents graphically the transformation from the complex values of $z_1 = r_1 + jx_1$ to the complex values of $\theta = \rho + j\Phi$ is not practical simultaneously for both small and large values of r_1 and x_1. A circle diagram that includes the entire range of r_1 and x_1 from zero to infinity is obtained with the use of a bilinear transformation from the complex z_1 plane to the complex z_1' plane according to

$$z_1' = \frac{2z_1}{z_1 + 1} \tag{10.8-6}$$

The circles of constant ρ have centers in the z_1' plane at $r_1' = 1$, $x_1' = 0$ and radii given by $|e^{-2\theta}| = e^{-2\rho} = |\Gamma|$. Equation (10.8-6) can be solved to derive the equations for the new set of circles:

$$\left(r_1' - 1 - \frac{r_1}{r_1 + 1}\right)^2 + x_1'^2 = \frac{1}{(r_1 + 1)^2} \tag{10.8-7a}$$

$$(r_1' - 2)^2 + \left(x_1' - \frac{1}{x_1}\right)^2 = \frac{1}{x_1^2} \tag{10.8-7b}$$

Equation (10.8-7a) is the equation of a family of circles of constant r_1 in the z_1' plane. They have radii $1/(r_1 + 1)$ with centers at $r_1' = 1 + r_1/(r_1 + 1)$, $x_1' = 0$. Equation (10.8-7b) is the equation of a family of circles of constant x_1 in the z_1' plane with radii $1/|x_1|$ and centers at $r_1' = 2$, $x_1' = 1/x_1$. The two sets of circles are represented graphically on the Smith chart (Fig. 10.8-1). $|\Gamma|$ is the radial distance from the center of the diagram on a scale that has the value 1 for the bounding circle $r_1 = 0$. This follows from the fact that the circles of constant ρ in the z_1' plane are circles of constant $|\Gamma| = e^{-2\rho}$. In the range $\rho = 0$ to ∞, $|\Gamma|$ changes from 1 to 0. The circles of constant Φ are on the chart. Φ ranges from 0 to 180° clockwise around the circle.

The Smith chart is convenient for the graphical solution of various transmission-line problems that involve impedances, admittances, and reflection coefficients. It may be noted that a similar analysis can be carried out for the normalized admittance $y_1 = g_1 + jb_1$. Hence the circles of constant r_1 on the Smith chart also represent circles of constant g_1 and the circles of constant x_1 represent circles of constant b_1.

In general,

$$Z = Z_c \frac{1 + |\Gamma|e^{j\phi}}{1 - |\Gamma|e^{j\phi}} \tag{10.8-8}$$

where $\phi = \psi - 2\beta s$. With the normalized value

$$z_1 = \frac{Z}{R_c} = r_1 + jx_1 \tag{10.8-9}$$

it follows that

$$r_1 + jx_1 = \frac{1 + |\Gamma|e^{j\phi}}{1 - |\Gamma|e^{j\phi}} \tag{10.8-10a}$$

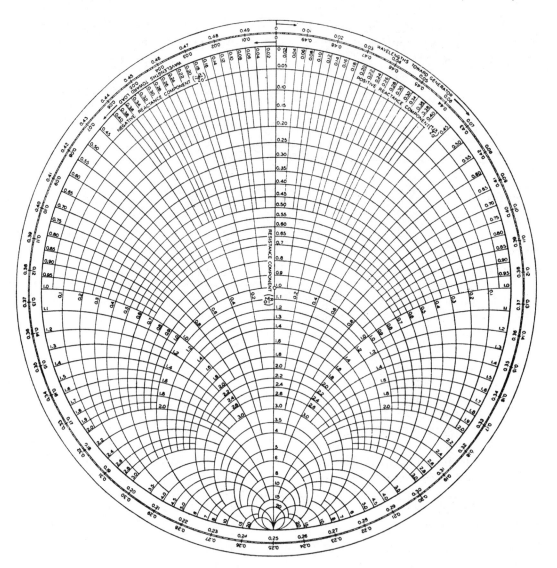

Figure 10.8–1 Smith chart.

and

$$|\Gamma|e^{j\phi} = \frac{r_1 - 1 + jx_1}{r_1 + 1 + jx_1} \qquad (10.8\text{–}10b)$$

Equating magnitudes in (10.8–10b)

$$\left(r_1 - \frac{1 + |\Gamma|^2}{1 - |\Gamma|^2}\right)^2 + x_1^2 = \left(\frac{2|\Gamma|}{1 - |\Gamma|}\right)^2 \qquad (10.8\text{–}11)$$

which is the equation of a circle with its center on the horizontal axis of Fig. 10.8–2 at a distance $(1 + |\Gamma|^2)/(1 - |\Gamma|^2)$ from the origin and with the radius $2|\Gamma|/(1 - |\Gamma|)$. For various values of $|\Gamma|$, the circles surrounding the point $(1, 0)$ in Fig. 10.8–2 are obtained. When the angles are equated,

$$r_1^2 + \left(x_1 - \frac{1}{\tan \phi}\right)^2 = 1 + \frac{1}{\tan^2 \phi} = \frac{1}{\sin^2 \phi} \tag{10.8-12}$$

which is the equation of a circle with its center on the vertical axis at a distance $1/\tan \phi$ below the origin, and with the radius $1/\sin \phi$. For various values of ϕ, the circles passing through the point $(1, 0)$ are obtained. In constructing the chart, $|\Gamma|$ is real and positive, and the circles of constant ϕ are marked with the value of βs corresponding to $-\phi/2$.

The circle diagram is also used to represent the admittance with $g_1 + jb_1$ plotted instead of $r_1 + jx_1$. Hence the diagram may be used to determine both the normalized admittance and the normalized impedance with the same numerical values attached to the coordinate lines marked on the diagram.

10.9 VOLTAGE AND CURRENT DISTRIBUTION ON A TRANSMISSION LINE: RESONANCE, STANDING-WAVE RATIO; TRANSFER OF POWER

The variation of the voltage and current along a transmission line has been expressed in two forms as a function of the distance $w = s - z$ from the load impedance Z_s at $z = s$ in (10.7–10a,b) and (10.7–11a,b). The magnitudes of the voltage and current are most conveniently obtained from the latter. They are

$$|V_z(w)| = \frac{V_0^e S_0}{S_s} [\sinh^2(\alpha w + \rho_s) + \cos^2(\beta w + \Phi_s)]^{1/2} \tag{10.9-1}$$

$$|I_z(w)| = \frac{V_0^e S_0}{S_s R_c} [\sinh^2(\alpha w + \rho_s) + \sin^2(\beta w + \Phi_s)]^{1/2} \tag{10.9-2}$$

where

$$S_0 = [\sinh^2 \rho_0 + \sin^2 \Phi_0]^{1/2} \tag{10.9-3a}$$

$$S_s = [\sinh^2(\alpha s + \rho_0 + \rho_s) + \sin^2(\beta s + \Phi_0 + \Phi_s)] \tag{10.9-3b}$$

When the losses in the line are neglected so that $\alpha \sim 0$ and the termination is a pure reactance so that $\rho_s \sim 0$, the distributions reduce to the following very simple forms:

$$|V_z(w)| \sim \frac{V_0^e S_0}{S_s} |\cos (\beta w + \Phi_s)| \tag{10.9-4a}$$

$$|I_z(w)| \sim \frac{V_0^e S_0}{S_s R_c} |\sin (\beta w + \Phi_s)| \tag{10.9-4b}$$

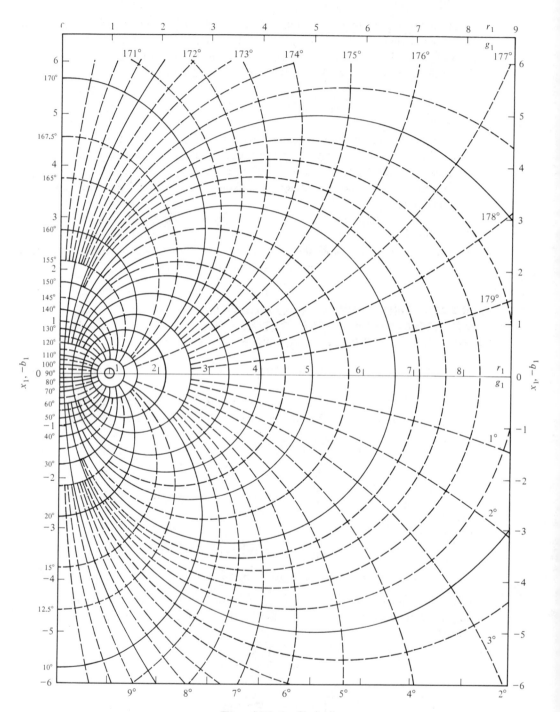

Figure 10.8–2 Circle diagram.

For an ideal short circuit, $\Phi_s = \pi/2$; for an ideal open circuit, $\Phi_s = 0$. Distributions of voltage and current as given in general by (10.9–1) and (10.9–2) can be observed in practice by moving a loosely coupled voltage detector (dipole antenna) or a current detector (loop antenna) along one conductor of the line while all other quantities are kept constant. Typical distribution curves for voltage and current are shown in Fig. 10.9–1. They represent standing waves.

The amplitude of the voltage and current depend on the driving emf V_0^e; on the impedance of, or in series with, the generator as contained in S_0; and on the quantity S_s in (10.9–3b), which contains the attenuation due to the line (αs), due to the generator (ρ_0), and due to the termination (ρ_s), and also the overall effective electrical length βs due to the line, Φ_0 due to the reactance of the generator or an impedance in series with it, and Φ_s due to the reactance of the load. With fixed values of α, ρ_0, and ρ_s, the quantity S_s can be minimized and the amplitudes $|V_z|$ and $|I_z|$ maximized by setting

$$\beta s + \Phi_0 + \Phi_s = n\pi; \qquad n = 1, 2, \ldots \tag{10.9–5}$$

This can be accomplished by tuning the reactances at the load or generator or by adjusting the length s of the transmission line with a section known as a *line stretcher*. When (10.9–5) is satisfied and the amplitudes $|V_z|$ and $|I_z|$ are maximized, the line is said to have been *tuned to resonance*. The condition (10.9–5) is the general condition for resonance of a transmission line.

The ratio of a maximum value of voltage or current in the distribution curve to the adjacent minimum is known as the *standing-wave ratio* (SWR). Hence the voltage standing-wave ratio is

$$S_V = \frac{|V_x(w)|_{max}}{|V_z(w)|_{min}} = \coth\,(\alpha w_{max} + \rho_s) \tag{10.9–6}$$

The current standing-wave ratio is

$$S_I = \frac{|I_z(w)|_{max}}{|I_z(w)|_{min}} = \coth\,(\alpha w_{max} + \rho_s) = S_V \tag{10.9–7}$$

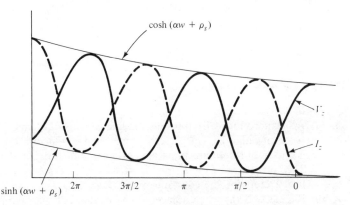

Figure 10.9–1 Current and voltage distributions along a transmission line.

For a low-loss line with $\alpha w \ll \rho_s$, this reduces to

$$S = S_V = S_I = \coth \rho_s \qquad (10.9\text{–}8)$$

The standing-wave ratio is particularly useful in experimental work since it is easily measured by means of a movable current or voltage detector.

With (10.7–11a,b) the time-average power transferred along the line at a distance $w = s - z$ from the load is

$$P_z = \mathrm{Re}\,\tfrac{1}{2}V_z I_z^* = \left| \frac{(V_0^e)^2}{2Z_c \sinh^2(\gamma s + \theta_0 + \theta_s)} \right| \cosh\,[\alpha w + \rho_s$$

$$+ j(\beta w + \Phi_s)] \sinh\,[\alpha w + \rho_s - j(\beta w + \Phi_s)] \qquad (10.9\text{–}9)$$

For a low-loss line with $Z_c \sim R_c$, this expression is readily reduced to

$$P_z = \left| \frac{(V_0^e)^2}{4R_c \sinh^2(\gamma s + \theta_0 + \theta_s)} \right| \sinh\, 2(\alpha w + \rho_s) \qquad (10.9\text{–}10)$$

The power dissipated in the load is obtained with $w = 0$; the power dissipated in the line and load is obtained with $w = s$. The ratio of the power in the load to the total power supplied to the line is

$$W = \frac{P_s}{P_0} = \frac{\sinh 2\rho_s}{\sinh 2(\alpha s + \rho_s)} \qquad (10.9\text{–}11)$$

The ratio of the power dissipated in the line to the total power is $P_L/P_s = 1 - W$. The insertion loss L in the line is defined in terms of the power to the load without the line ($s = 0$) to the power to the load with the line present. It is

$$L\,(dB) = 10 \log_{10} \frac{P_0}{P_s} = 10 \log_{10} \frac{\sinh 2(\alpha s + \rho_s)}{\sinh 2\rho_s} \qquad (10.9\text{–}12)$$

When the termination is matched so that $Z_s = Z_c$, $\rho_s = \infty$ and

$$W = \frac{e^{2\rho_s}}{e^{2(\alpha s + \rho_s)}} = e^{-2\alpha s} \qquad (10.9\text{–}13)$$

PROBLEMS

1. (a) A balanced four-wire transmission line consists of identical and parallel conductors (each of radius a) which are so driven that conductors 1 and 3 on one diagonal of a square of side b are in parallel as are conductors 2 and 4 on the other diagonal. The currents are $I_4 = I_2 = -I_1 = -I_3 = -\tfrac{1}{2}I$, where I is the total current in the line. Determine the constants of the line.
   (b) Repeat part (a) when adjacent pairs (instead of diagonal pairs) of conductors are in parallel.

2. The inner conductor of a horizontal coaxial slotted line is supported along the entire length by a wedge of polystyrene ($\epsilon_r = 2.6$) which occupies a 9° angle. If the wavelength measured along the line is 1.2 m, what would it be if the line were completely air-filled?

3. A transmission line terminated at $z = s$ in its characteristic impedance of 300 Ω is driven at $z = 0$ by a generator with an emf of 100 V and an impedance of $8 + j40$ Ω. The frequency is 100 MHz. The attenuation constant of the line is 0.01 N/m. Determine the time-dependent current and voltage at $z = 10$ m if the instant $t = 0$ is chosen to occur when the emf has a positive maximum in its cycle.

4. The amplitude of the current in a long line terminated in its characteristic impedance is measured at two points 100 m apart. The ratio of the two values is 1.1.
   (a) What is the attenuation constant of the line in nepers per meter?
   (b) What is the ratio of potential differences between the two conductors of the line at two points 20 m apart?

5. Determine the complex reflection coefficient Γ_s in polar form and the terminal functions ρ_s and Φ_s for the following impedances terminating a low loss line with $Z_c \doteq R_c = 300$ Ω.
   (a) $Z_s = 3000 + j0$
   (b) $Z_s = 0 + j3000$
   (c) $Z_s = 3000(1 + j)$
   (d) $Z_s = 0 + j300$
   (e) $Z_s = 200 + j100$
   (f) $Z_s = 0$
   (g) $Z_s = \infty$
   Calculate and verify your answers by using the Smith chart.

6. The capacitance of a capacitor is 15 pF. It is connected as the terminal impedance of a line of characteristic impedance $Z_c \doteq R_c = 400$ Ω. What are the terminal functions ρ and Φ at 150 MHz? Assume the capacitor to be without loss.

7. (a) Determine the input impedance of a section of line of length 20.2 m for which the characteristic impedance $Z_c = 400(1 - j\alpha/\beta)$ and $\alpha = 2 \times 10^{-3}$ N/m at a frequency of 300 MHz. The line is terminated in an impedance $Z_s = 100 - j800$ Ω. First calculate the input impedance and verify your answer by using the Smith chart.
   (b) What would be the input impedance of the line in part (a) if it were lossless?

8. An impedance $Z_s = 20 - j500$ Ω terminates a line with $Z_c = 440(1 - j\alpha/\beta)$, $\alpha = 2.26 \times 10^{-3}$ N/m. The frequency is 150 MHz.
   (a) Determine ρ_s and Φ_s by calculation and by using the Smith chart.
   (b) Determine the shortest distance from the load along the line at which the impedance looking toward the load is a pure resistance. What is this resistance? What are the associated values of ρ_s and Φ_s? What is the standing-wave ratio $S = \coth \rho_s$?
   (c) Repeat part (b) for the next shortest distance for which the impedance is a pure resistance.

9. A low-loss transmission line $[Z_c = 300(1 - j10^{-3}), s = 4.2$ m$]$ connects a generator $(Z_{0a} = 10 + j0$ Ω, $V_0^e = 100$ V, $f = 300$ MHz$)$ to a load $(Z_s = 400 + j0$ $\Omega)$. Determine the following:
   (a) The currents in the generator and the load.
   (b) The current and the voltage on the line halfway between the generator and the load.
   (c) The location and magnitude of each maximum and minimum of current along the line.

(d) The standing-wave ratio.

(e) The efficiency of power transmission to the load.

10. An impedance of $800 - j50 \ \Omega$ is measured on a transmission line for which $Z_c \doteq R_c = 400 \ \Omega$ and $\alpha = 10^{-3}$ N/m. The frequency is 100 MHz.

(a) Where is the voltage minimum nearest the load? What is the standing-wave ratio?

(b) What is the width of the resonance curve obtained by varying the length of the line about the peak occurring at the next to the shortest length of line?

The Insulated Antenna

The insulated antenna is an unusual structure that combines the properties of radiation and transmission in a very interesting and generally useful manner. The integral equation for the device is derived in Chapter 3. The solutions of that equation and some of its many applications are the subject of this chapter.

11.1 THE SOLUTION OF THE INTEGRAL EQUATION FOR THE CURRENT

The integral equation for the current $I(z)$ in a center-driven dipole of length $2h$ and radius a when enclosed in an insulating cylinder of radius b and surrounded by an infinite ambient medium (Fig. 3.12−1) is

$$\int_{-h}^{h} I(z')K(z - z')\, dz' = C \cos k_L z + \frac{iV_0^e}{2\zeta_2} \sin k_L |z| \qquad (11.1\text{-}1)$$

with

$$k_L = \beta_L + i\alpha_L \equiv k_2 \left\{ 1 + H_0^{(1)}(k_4 b) \middle/ \left[k_4 b H_1^{(1)}(k_4 b) \ln \frac{b}{a} \right] \right\}^{1/2} \qquad (11.1\text{-}2)$$

It is assumed that the conductor is perfect and that

$$|k_4^2| \gg |k_2^2|$$

where

$$k_n = \omega\left[\mu\left(\epsilon_n + \frac{i\sigma_n}{\omega}\right)\right]^{1/2}; \qquad n = 2, 4 \qquad (11.1\text{–}3a)$$

The wave impedances of the two media are

$$\zeta_n = \frac{\omega\mu}{k_n} = \left(\frac{\mu}{\epsilon_n + i\sigma_n/\omega}\right)^{1/2}; \qquad n = 2, 4 \qquad (11.1\text{–}3b)$$

The kernel $K(z - z')$ in (11.1–1) is characterized by the peaking property at $z' = z$ illustrated in Fig. 3.12–2, and hence the following approximate relation may be written:

$$I(z)\Psi(z) = \int_{-h}^{h} I(z')K(z - z')\,dz'$$

$$\doteq I(z)\Psi \qquad (11.1\text{–}4)$$

where Ψ is a complex constant that can be evaluated from (11.1–4) as soon as the relative distribution of current $G(z, z') = I(z')/I(z)$ is known.

The substitution of (11.1–4) in (11.1–1) yields

$$I(z) \doteq \frac{1}{\Psi}\left[C \cos k_L z + \frac{iV_0^e}{2\zeta_2}\sin k_L|z|\right] \qquad (11.1\text{–}5)$$

The constant C can be obtained from (11.1–5) since $I(h) = 0$ at $z = h$. Thus

$$C = \frac{-iV_0^e}{2\zeta_2}\tan k_L h \qquad (11.1\text{–}6)$$

Then

$$I(z) = \frac{-iV_0^e}{2\zeta_2\Psi}\frac{\sin k_L(h - |z|)}{\cos k_L h} \qquad (11.1\text{–}7)$$

With (11.1–7), (11.1–4) becomes

$$\Psi(z) \doteq \Psi \sin k_L(h - |z|)$$

$$= \int_0^h \sin k_L(h - z')[K(z - z') + K(z + z')]\,dz' \qquad (11.1\text{–}8)$$

Except in very small ranges near the driving point ($z = 0$) and the ends of the antenna ($|z| = h$), suitable approximate expressions can be used for the integrals to yield

$$\Psi = \left(\frac{k_L}{2\pi k_2}\right)\ln\frac{b}{a} \qquad (11.1\text{–}9)$$

The complete solution for the current in the insulated dipole is, therefore,

$$I(z) = \frac{-iV_0^e}{2Z_c} \frac{\sin k_L(h - |z|)}{\cos k_L h} = I(0) \frac{\sin k_L(h - |z|)}{\sin k_L h}$$ (11.1–10)

where

$$Z_c = \zeta_2 \Psi = \frac{\zeta_2 k_L}{2\pi k_2} \ln \frac{b}{a}$$ (11.1–11)

The associated driving-point admittance is

$$Y_{\text{in}} = \frac{I(0)}{V_0^e} = \frac{-i}{2Z_c} \tan k_L h$$ (11.1–12)

The charge per unit length $q(z)$ on the antenna is obtained from the equation of continuity $dI(z)/dz - i\omega q(z) = 0$:

$$q(z) = \frac{V_0^e \, \bar{\epsilon}_2 \pi}{\ln (b/a)} \frac{\cos k_L(h - z)}{\cos k_L h}; \qquad z \geq 0$$

$$q(-z) = -q(z)$$ (11.1–13)

If the insulator is air, $\zeta_2 = \zeta_0 \doteq 120\pi$ ohms and $k_2 = \omega/c$, with $c = 3 \times 10^8$ m/s.

11.2 THE ELECTROMAGNETIC FIELD OF THE INSULATED ANTENNA

It was shown in Sec. 11.1 that a good approximation of the distribution of current on the insulated antenna is given by (11.1–7) when the ratio $|k_4^2/k_2^2|$ is much greater than 1. The components of the electromagnetic field in the insulation (region 2) with this current are approximately

$$B_{2\phi}(\rho, z) \doteq \frac{\mu_0}{2\pi\rho} I(z) = \frac{\mu_0 I(0)}{2\pi\rho} \frac{\sin k_L(h - |z|)}{\sin k_L h}$$ (11.2–1)

$$E_{2\rho}(\rho, z) \doteq \frac{1}{2\pi\rho\bar{\epsilon}_2} q(z) = \frac{ik_L I(0)}{2\pi\rho\bar{\epsilon}_2\omega} \frac{\cos k_L(h - z)}{\sin k_L h}$$ (11.2–2a)

$$E_{2\rho}(\rho, -z) = -E_{2\rho}(\rho, z)$$ (11.2–2b)

$$E_{2z}(\rho, z) \doteq \int_{\rho'=a}^{\rho} \left[\frac{\partial E_{2\rho}(\rho', z)}{\partial z} - i\omega B_{2\phi}(\rho', z) \right] d\rho'$$

$$= \frac{i\omega\mu_0 I(0) \ln (\rho/a)}{2\pi} \left[\left(\frac{k_L}{k_2} \right)^2 - 1 \right] \frac{\sin k_L(h - |z|)}{\sin k_L h}$$ (11.2–3)

$$E_{2\phi}(\rho, z) \doteq 0; \qquad B_{2\rho}(\rho, z) \doteq 0; \qquad B_{2z}(\rho, z) \doteq 0$$ (11.2–4)

The electromagnetic field at the surface of the insulation ($\rho = b$) in the external medium (region 4) is easily obtained from the boundary conditions

$$B_{4\phi}(b, z) = B_{2\phi}(b, z) \tag{11.2-5}$$

$$E_{4\rho}(b, z) = \frac{\tilde{\epsilon}_2}{\tilde{\epsilon}_4} E_{2\rho}(b, z) \tag{11.2-6}$$

$$E_{4z}(b, z) = E_{2z}(b, z) \tag{11.2-7}$$

It is to be noted that the field is nearly of the transverse electromagnetic type (TEM) in region 2 since the ratio of the longitudinal component to the transverse components of the electric field is very small.

The electric field at any point in the external medium (region 4) can be determined from the electromagnetic field on the surface of the insulator. The evaluation makes use of a general integral.* Subject to the condition $|k_2^2 b^2| \ll 1$, the electric field at all points not too close to the antenna ($\rho \geq 4b$) is given by

$$
E_{4z}(\rho, z) = \frac{i\omega\mu_0 I(0)}{4\pi \sin k_L h} \left\{ \left(1 - \frac{k_L^2}{k_4^2} \right) \int_0^h \sin k_L(h - z')[\psi(z, z') \right.
$$

$$
\left. + \psi(z, -z')] \, dz' + \frac{k_L}{k_4^2} [\psi(z, h) + \psi(z, -h) - 2\psi(z, 0) \cos k_L h] \right\} \tag{11.2-8}
$$

$$
E_{4\rho}(\rho, z) = \frac{-i\omega\mu_0 I(0)}{4\pi \sin k_L h} \frac{k_L \rho}{k_4^2} \int_0^h \cos k_L(h - z') \left[\left(\frac{ik_4}{R_1} - \frac{1}{R_1^2} \right) \psi(z, z') \right.
$$

$$
\left. - \left(\frac{ik_4}{R_2} - \frac{1}{R_2^2} \right) \psi(z, -z') \right] dz' \tag{11.2-9}
$$

where $R_1 = [(z - z')^2 + \rho^2]^{1/2}$, $R_2 = [(z + z')^2 + \rho^2]^{1/2}$, $\psi(z, z') = e^{ik_4 R_1}/R_1$, and $\psi(z, -z') = e^{ik_4 R_2}/R_2$. These expressions are readily integrated numerically to obtain the complete elliptically polarized electric field near the antenna as shown in Fig. 11.2-1.

If the point where the field is to be determined is at a sufficient distance from the antenna, $R_0 \gg h > b$, these formulas can be simplified with the approximations $R_1 \sim R_0 - z' \cos \Theta$, $R_2 \sim R_0 + z' \cos \Theta$ in the exponentials and $R_1 \sim R_0 \sim R_2$ in amplitudes where R_0 and Θ are spherical coordinates. With these substitutions the integrals can be evaluated and the electric field obtained explicitly. In the spherical coordinates R_0, Θ, Φ it is

$$E_R^r(R_0, \Theta) = 0 \tag{11.2-10}$$

$$
E_\Theta^r(R_0, \Theta) = \frac{-i\omega\mu_0 I(0)}{2\pi k_4 R_0} e^{ik_4 R_0} \left[J_0(k_4 b \sin \Theta) \right.
$$

$$
\left. - \frac{H_0^{(1)}(k_4 b)}{H_1^{(1)}(k_4 b) \sin \Theta} J_1(k_4 b \sin \Theta) \right] F_0(\Theta, k_4 h, k_L h) \tag{11.2-11}
$$

---

* R. W. P. King and G. S. Smith, *Antennas in Matter* (Cambridge, Mass.: The MIT Press, 1981), p. 523, eq. (7.12).

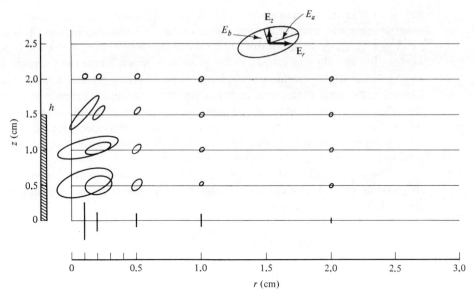

Figure 11.2–1   Electric field near an insulated monopole with $\beta_L h \sim \pi/4$ in a dissipative medium.

$$E_\Phi^r(R_0, \Theta) = 0 \tag{11.2–12}$$

where

$$F_0(\Theta, k_4 h, k_L h) = \frac{[\cos(k_4 h \cos \Theta) - \cos k_L h] \sin \Theta}{[(k_L/k_4) - (k_4/k_L) \cos^2 \Theta] \sin k_L h} \tag{11.2–13}$$

When the insulation is electrically thin in the external medium with $|(k_4 b)^2| \ll 1$, only the first term in the brackets in (11.2–8) contributes significantly to the electric field in the far zone. With $J_0(k_4 b \sin \Theta) \doteq 1$,

$$E_\Theta^r(R_0, \Theta) \doteq \frac{-i\omega\mu_0 I(0)}{2\pi k_4 R_0} e^{ik_4 R_0} F_0(\Theta, k_4 h, k_L h) \tag{11.2–14}$$

11.3 APPLICATIONS OF INSULATED ANTENNAS: EMBEDDED INSULATED ANTENNAS FOR COMMUNICATION AND HEATING

Antennas are located in material media for purposes of communication, telemetry, geophysical exploration, and medical diagnostics. Alternatively, they can be used for heating as in the hyperthermia treatment of tumors or in the extraction of shale oil from the earth. The regions in which the antennas may be embedded for such applications have wide ranges of size, shape, and electrical properties. Some examples are (1) the earth's overburden with its varied crust of sand, soil and stone, fresh and salt water, snow and ice, all bounded above by an interface with air and below by a layer of granitic rock; (2) living organisms such as human beings and animals in air and fish in fresh and salt water; and (3) plasmas such as the ionosphere or as generated in the laboratory. The antennas used in bore holes for geophysical exploration or subsurface heating are physically much larger than similar antennas used at much higher frequencies in living tissues. When the antennas are in perfect dielectrics, a simple scaling in terms of physical size and frequency is possible; on the other hand, when antennas are in media with nonzero conductivity, the principle of similitude requires more complicated relations between large-scale, low-frequency phenomena and similar ones on a smaller scale at a higher frequency. The principle of similitude is discussed in Chapter 6.

The expression (11.1–10) for the current in the insulated dipole center-driven by the emf V_0^e and with open ends may be generalized to include terminations at the ends. A convenient form is

$$I(z) = \frac{-iV_0^e}{2Z_c} \frac{\sin[k_L(h - |z|) + i\theta_h]}{\cos(k_L h + i\theta_h)} \tag{11.3-1}$$

The associated input admittance is

$$Y_{in} = \frac{-i}{2Z_c} \tan(k_L h + i\theta_h) \tag{11.3-2}$$

The complex terminal function is

$$\theta_h = \rho_h - i\Phi_h = \coth^{-1}\frac{Z_h}{Z_c} \tag{11.3-3}$$

[For a base-driven monopole over a conducting plane the factor 2 in the denominators of (11.3–1) and (11.3–2) are to be omitted.] The terminating impedance at each end is $Z_h = R_h - iX_h$. When $Z_h = \infty$ as for an open end, $\theta_h = 0$; when $Z_h = 0$ as for a short-circuited end, $\theta_h = -i\pi/2$.

The formulas (11.3–1) and (11.3–2) for the current in and the admittance of an insulated dipole are like those of conventional transmission-line theory but with more general expressions for the wave number $k_L = (-z_L y_L)^{1/2}$ and for the characteristic impedance $Z_c = (z_L/y_L)^{1/2}$. The impedance per unit length z_L now includes not only the ohmic losses in the ambient medium but also the radiation into it. The former dominate when it is a conductor like seawater, the latter when it is a good dielectric like lake water at high frequencies. y_L is the admittance per unit length.

If the metallic inner conductor is assumed to be perfect and the dielectric is two-layered, the complex wave number and characteristic impedance are

$$k_L = \beta_L + i\alpha_L = k_2 \left[\frac{\ln(c/a)}{\ln(b/a) + n_{23}^2 \ln(c/b)}\right]^{1/2} \left[\frac{\ln(c/a) + F}{\ln(c/a) + n_{24}^2 F}\right]^{1/2} \tag{11.3-4}$$

$$Z_c = \frac{\zeta_2 k_L}{2\pi k_2}\left(\ln\frac{b}{a} + n_{23}^2 \ln\frac{c}{b} + n_{24}^2 F\right) \tag{11.3-5}$$

where a is the radius of the conductor, b the outer radius of an insulator (e.g., air), and c the outer radius of a dielectric tube containing the insulator.

$$F = H_0^{(1)}(k_4 c)/k_4 c H_1^{(1)}(k_4 c) \tag{11.3-6}$$

and

$$\zeta_2 = \frac{\omega\mu_2}{k_2}; \qquad n_{23} = \frac{k_2}{k_3}; \qquad n_{24} = \frac{k_2}{k_4} \tag{11.3-7}$$

When there is only a single insulator with no thin-walled dielectric tube, $c = b$ and

$$k_L = k_2 \left(\ln \frac{b}{a} + F \right)^{1/2} \left(\ln \frac{b}{a} + n_{24}^2 F \right)^{-1/2} \tag{11.3-8}$$

$$Z_c = \frac{\zeta_2}{2\pi} \left(\ln \frac{b}{a} + F \right)^{1/2} \left(\ln \frac{b}{a} + n_{24}^2 F \right)^{1/2} \tag{11.3-9}$$

Since $|n_{24}^2| \ll 1$, b/a may be sufficiently large so that $\ln(b/a) \gg |n_{24}^2|F$. Then $k_L = k_2[1 + F/\ln(b/a)]^{1/2}$, $Z_c = (\zeta_2/2\pi)[1 + F/\ln(b/a)]^{1/2} \ln(b/a)$. These are the same as (11.1-2) and (11.1-11). It follows that k_L/k_2 and Z_c depend only on the dimensionless quantities b/a and $k_4 b$. It is possible to use quite simple scaling.

Radiation from an insulated antenna, unlike that from a bare antenna or open-wire transmission line, is distributed as a load per unit length. When an insulated monopole is terminated in its characteristic impedance, $Z_h = Z_c$, $\theta_h = \infty$, $Y_{in} = 1/Z_c$, and

$$I_z(z) = \frac{V_0^e}{Z_c} \exp(-\alpha_L z) \exp(i\beta_L z) \tag{11.3-10}$$

It follows that the power lost from the monopole per unit length is

$$\frac{P(z) - P(z + \Delta z)}{P(z)\Delta z} = 2\alpha_L \tag{11.3-11}$$

where $P(z) = V(z)I_z^*(z)/2 = [(V_0^e)^2/2R_c]\exp(-\alpha_L z)$. The power lost per wavelength is $2\alpha_L \lambda_L = 4\pi\alpha_L/\beta_L$. The ratio α_L/β_L is a direct measure of the capacity of the insulated antenna to transfer power to the ambient medium by conduction, radiation, or a combination.

Table 11.3-1 gives the ratio α_L/β_L for an air-insulated antenna in lake water ($\epsilon_{er} = 80$, $\sigma_e = 4 \times 10^{-3}$ S/m) and seawater ($\epsilon_{er} = 80$, $\sigma_e = 4$ S/m) at $f = 470$ MHz as a function of b/a.

It is to be noted that for lake water $\alpha_4/\beta_4 = 0.0012$, for seawater $\alpha_4/\beta_4 = 0.639$. Nevertheless, α_L/β_L is larger for radiation into lake water than for conduction into seawater by almost a factor 2. It is significant in the design of insulated antennas that α_L/β_L is quite large when b/a is near 1 and quite small when b/a is much greater

TABLE 11.3–1 Properties of Air-Insulated Antennas in Water (f = 470 MHz)

	α_L/β_L	
b/a	Lake water, $k_4/k_2 = 8.94 + i0.011$	Seawater, $k_4/k_2 = 11.52 + i7.36$
13.0	0.051	0.029
8.9	0.083	0.045
4.0	0.219	0.114
2.6	0.298	0.172

than 1. When b/a is large and α_L/β_L very small, most of the power is transmitted in the *axial* direction along the antenna; this then behaves like a low-loss transmission line. When b/a is near 1 and α_L/β_L is relatively large, the transmission of power is primarily by conduction or radiation in the *radial* direction into the ambient medium and the insulated antenna behaves predominantly as an antenna.

11.4 INSULATED ANTENNAS AND TRANSMISSION LINES FOR BORE HOLES

When an antenna is to be used in a material like the earth or a living organism either for purposes of communication and telemetry or for local heating, its configuration is often constrained by the need to insert it with its feeding transmission line into a hole that is very deep compared to its cross-sectional dimension. Examples are antennas lowered into bore holes in the earth for geophysical exploration, for local heating, or for communication. On a smaller scale are needlelike elements for insertion into living tissue. Usually, the antenna and its transmission line are insulated from the surrounding earth or flesh. The insulation can be the air in the bore hole or any low-loss dielectric with a wave number that is small compared to that of the ambient medium.

As examples consider four different insulated transmission lines and antennas which have interesting and useful properties when inserted in a hole in a general dissipative medium. Table 11.4–1 shows values of k_L and Z_c computed from (11.3–

TABLE 11.4–1 Wave Numbers and Characteristic Impedances[a]

j	a_j (cm)	b/a_j	$k_{Lj} = \beta_{Lj} + i\alpha_{Lj}$ (m^{-1})	α_{Lj}/β_{Lj}	$Z_{cj} = R_{cj} - iX_{cj}$ (Ω)	X_{cj}/R_{cj}
1	0.3175	8.00	12.99 + i0.59	0.045	128.1 + i5.95	0.046
2	0.9525	2.67	13.45 + i1.24	0.092	60.0 + i5.83	0.097
3	1.905	1.33	15.70 + i3.61	0.230	24.1 + i4.91	0.204

[a] $k_2 = 4\pi$ m^{-1}; $k_4 b = 3.33 + i1.77$; $\sigma_{e4} = 4$ S/m; $\epsilon_{er4} = 80$.

8) and (11.3–9) as well as the ratios α_L/β_L and X_c/R_c at $f = 600$ MHz with three different radii for the conductors: $a_1 = 0.3175$ cm, $a_2 = 0.9525$ cm, and $a_3 = 1.905$ cm. The outer radius of the insulation is $b = 2.54$ cm. In the interest of simplicity a thin-walled plastic tube with outer radius c (used when the ambient medium is fluid) is ignored in the calculations. Air is taken to be the insulating material and salt water is the ambient medium.

The four different stuctures have the following characteristics:

Insulated Coaxial Line with Extended Inner Conductor

A simple transmission line and antenna for insertion in a bore hole on any scale is the coaxial line with extended end as shown in Fig. 11.4–1. The shield of a coaxial line with outer radius a_2 extends a distance h_B into an air-filled hole with radius b. The inner conductor projects as the antenna an additional distance h_A. Its radius may be the same as the inner conductor of the coaxial line as in Fig. 11.4–1a, or it may be enlarged to have the radii a_2 or a_3 as in Fig. 11.4–1b and Fig. 11.4–1c. To be determined are conditions and dimensions with which most of the power supplied to the coaxial feed line is transferred to the ambient medium by the currents in the antenna and only a small fraction by the currents on the outer surface of the shield of the coaxial line.

The configurations in Fig. 11.4–1 consist of two different sections of insulated antenna in series with the driving voltage maintained at the open end of the coaxial line. The upper section is end-driven, has the length h_A and the radius a_1, a_2, or a_3 and is terminated in an open end. The current and admittance are given by (11.3–1) and (11.3–2) with $\theta_A = \theta_h = 0$ and with the factor 2 in the denominators of both omitted. The lower section is also end-driven; it consists of the outer surface of the coaxial shield with the length h_B from the upper end to the surface of the ambient medium from where it continues an unspecified distance in a variable direction as a bare conductor in air. The terminating impedance Z_B at the lower end of the insulated monopole of length h_B is indeterminate. If a ground plane is located at the surface of the ambient medium as in Fig. 11.4–1b, $Z_B = 0$ and $\theta_B \doteq -i\pi/2$ in (11.3–1) and (11.3–2). An approximate short circuit is obtained with a metal disk with outer radius b metallically connected to the coaxial shield at any convenient cross section as in Fig. 11.4–1c.

With $\theta_A = 0$ and $\theta_B = -i\pi/2$, the input impedances of the two sections are obtained from (11.3–2), that is, $Z_{in} = 1/Y_{in}$. Thus, with $j = 1, 2,$ or 3,

$$Z_{in\,A} = iZ_{cj} \cot k_{Lj}h_A \qquad (11.4–1a)$$

$$Z_{in\,B} = -iZ_{c2} \tan k_{L2}h_B \qquad (11.4–1b)$$

The impedance terminating the coaxial line is

$$Z = Z_{in\,A} + Z_{in\,B} = R_{in\,A} + R_{in\,B} - i(X_{in\,A} + X_{in\,B}) \qquad (11.4–2)$$

The total power in the load is

$$P_T = |I_{in}|^2 \, (R_{in\,A} + R_{in\,B}) \tag{11.4-3}$$

The fraction of power in the antenna is

$$\frac{P_A}{P_T} = \frac{R_{in\,A}}{R_{in\,A} + R_{in\,B}} \tag{11.4-4}$$

For local heating or radiation to be efficient, (11.4–4) must be as large as possible, preferably near 1. This requires that $R_{in\,A} \gg R_{in\,B}$. Maximum power transfer is achieved when the feeding line and the load are matched. Thus

$$R_{in\,A} + R_{in\,B} \simeq R_{cL}; \qquad X_{in\,A} + X_{in\,B} \simeq 0 \tag{11.4-5}$$

where R_{cL} is the characteristic resistance of the coaxial feed line. In practice, its magnitude is chosen to be between 50 and 100 Ω. In Table 11.4–2 are given the computed input impedances of the monopole antenna with the electrical lengths $\beta_{Lj} h_A = \pi/2$ (near resonance) and $\beta_{Lj} h_A = \pi$ (near antiresonance) for the three radii. Also given is the input impedance of the section of length h_B and conductor radius a_2. The electrical lengths chosen include $\beta_{L2} h_B = 3\pi/2$ for a near maximum of $Z_{in\,B}$, $\beta_{L2} h_B = \pi, 2\pi, 3\pi$, and 4π near successive minima in $Z_{in\,B}$ and $\beta_{L2} h_B = \infty$ for a reflection-free (matched) termination (i.e., $Z_{in\,B} = Z_{c2}$). It is seen that when $\beta_{Lj} h_A = \pi/2$, $Z_{in\,A}$ varies very little with the change in radius and is relatively small. When $\beta_{Lj} h_A = \pi$, $Z_{in\,A}$ varies widely with the radius of the conductor. If $R_{in\,A} \gg R_{in\,B}$ and $(R_{in\,A} + R_{in\,B})$ is to be between 50 and 100 Ω, a satisfactory choice is $\beta_{Lj} h_A = \pi$ with $a = a_3$ and $Z_{in\,A} = 39.0 + i7.9 \; \Omega$ and $\beta_{L2} h_B = \pi$ with $Z_{in\,B} = 16.9 + i1.64 \; \Omega$. The substitution of these values in (11.4–4) results in

$$\frac{P_A}{P_T} = 0.70; \qquad R_{in\,A} + R_{in\,B} = 55.9 \; \Omega \tag{11.4-6}$$

The choice of $\beta_{L2} h_B = \pi$ restricts the length h_B to a small value. This may be unrealistic practically when the short-circuiting termination is a ground plane (Fig. 11.4–1b). The alternative is the metal disk (Fig. 11.4–1c). The value of b/a_2 must

TABLE 11.4–2   Input Impedance of Antenna and Line Sections in Fig. 11.4–1a
($f = 600$ MHz)

$\beta_{Lj} h_A$	h_A (m)	a_j (cm)	$Z_{in\,A}$ (Ω)	$\beta_{L2} h_B$	h_B (m)	a_2 (cm)	$Z_{in\,B}$ (Ω)
$\pi/2$	0.121	0.3175	$9.12 + i0.42$	$3\pi/2$	0.350	0.9525	$146.8 + i14.3$
	0.117	0.9525	$8.70 + i0.84$	π	0.234		$16.9 + i1.64$
	0.100	1.905	$8.34 + i1.70$	2π	0.468		$31.4 + i3.05$
π	0.242	0.3175	$903.3 + i42.0$	3π	0.702		$42.1 + i4.09$
	0.234	0.9525	$212.6 + i20.7$	4π	0.936		$49.3 + i4.78$
	0.200	1.905	$39.0 + i7.9$	∞	∞		$60.0 + i5.83$

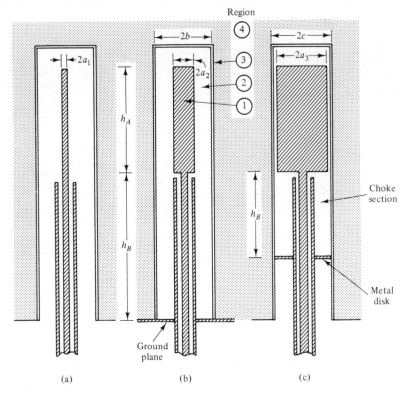

Figure 11.4–1 Insulated coaxial line with extended inner conductor in a semi-infinite ambient medium: (a) simple extension; (b) and (c) enlarged extensions. Region 1, metal; region 2, air; region 3, thin-walled plastic tube; region 4, earth, tissue, water.

be large enough to make α_{L2}/β_{L2} quite small. In this case, $b/a_2 = 2.67$ and $\alpha_{L2}/\beta_{L2} = 0.092$ (Table 11.4–1). A succession of such disks at intervals in electrical length of $\pi/2$ will be even more effective in suppressing currents on the shield of the coaxial line. The disks are also useful centering guides for the coaxial feed line.

Insulated Center-Driven Dipole with Choke Section

The insulated coaxial line with extended inner conductor may be modified as shown on the right in Fig. 11.4–2. The enlarged extension of the inner conductor of the feeding coaxial line with length h_A is now the upper half of an approximately symmetrical center-driven dipole. The lower half consists of a correspondingly enlarged section of the coaxial shield, also with length near h_A and with the same radius a_3 as the upper half. The upper half is a monopole terminated in an open circuit with $Z_A = \infty$ and $\theta_A = 0$, the lower half is terminated in an insulated section which has an inner conductor of radius a_1, which is also the outer radius of the shield of the coaxial line. The terminating impedance of the lower half of the dipole

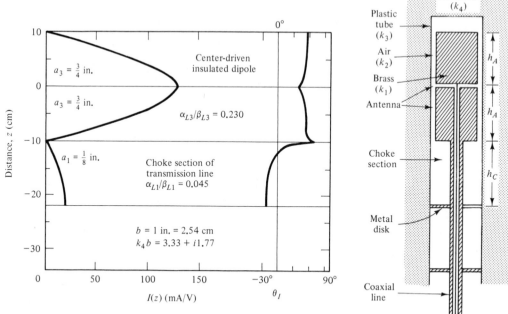

Figure 11.4–2   Current along insulated dipole with choke section.

is the input impedance of the entire section below it. This impedance can be made very large by locating a metal disk at a distance h_C from the lower edge of the dipole such that $\beta_{L1}h_C = \pi/2$. Additional metal disks at intervals of $\pi/2$ in the electrical distance will minimize currents on the shield of the coaxial line. The distribution of current (amplitude and phase) along the insulated dipole antenna and the choke section of transmission line is shown in Fig. 11.4–2. The power ratio may be calculated for half-lengths h_A over a suitable range. It is possible to have up to 99% of the power supplied by the coaxial line transferred to the ambient medium by the antenna.

Insulated Sleeve Dipole

The structure of Fig. 11.4–2 may be modified by having an internal coaxial choke section as shown in Fig. 11.4–3. The central metal structure in Fig. 11.4–3 when isolated in air is the *sleeve dipole*. The generalized transmission-line theory of the insulated antenna applies and the insulated sleeve dipole can be related to the bifurcated coaxial line. The structure in Fig. 11.4–3 consists of two sections of insulated conductor, both of radius a_3, between which the driving voltage is applied by the coaxial line. The upper section of length h_A has an open end so that $\theta_A = 0$ in (11.3–1) and (11.3–2). The lower section of length h_S is terminated in two series-connected sections. One of these is a conventional coaxial line with metal inner and outer conductors with radii a_1 and $a_3 - t$, where t is the thickness of the metal sleeve. The second section is an air-insulated antenna with current on the outside of the feed line (radius a_1) and insulation extending to the ambient medium at radius b. This section has the length h_B to the boundary of the ambient medium where it is effectively terminated in a bare antenna with an indeterminate length and direction in air. If a ground plane is located at the boundary or a metal disk is placed appropriately along its length, this section is terminated at the end of the length h_B in an approximate short circuit with $Z_B \sim 0$ and $\theta_B \sim -i\pi/2$.

Since one of the series sections is a highly conducting coaxial line and the other is an insulated section with b/a_1 large so that α_{L1}/β_{L1} is small, it is to be expected that the equivalent circuit of the bifurcated coaxial line is a good approximation. This is shown in Fig. 11.4–4a, where the reference plane for the phase is at the common junction and the reactive end loads are

$$X_1 = -R_{c1} \cot \beta_{L1}d; \qquad X_2 = \frac{a_3 - a_1}{b - a_1} R_{c1} \cot \beta_{L1}d$$

$$X_3 = \frac{b - a_3}{b - a_1} R_{c1} \cot \beta_{L1}d \qquad (11.4-7)$$

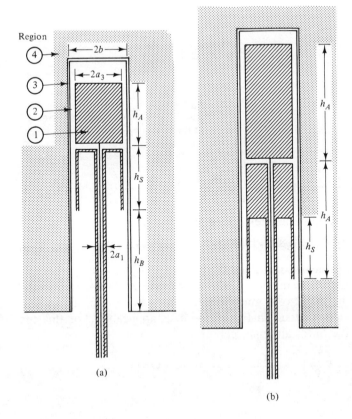

(a)

(b)

Figure 11.4–3 Insulated sleeve dipole; $h_s = \lambda/4$. Region 1, metal; region 2, air; region 3, thin-walled plastic tube; region 4, earth, tissue, water.

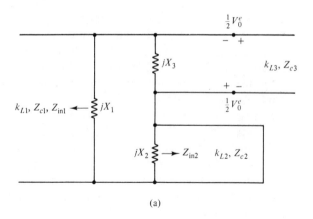

(a)

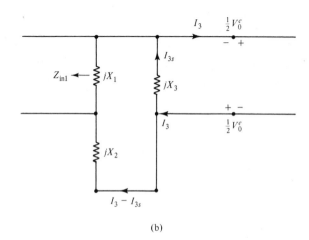

(b)

Figure 11.4–4 (a) Schematic diagram and (b) approximate equivalent circuit of the bifurcated coaxial line.

with d given by

$$\frac{\pi d}{b - a_1} = \frac{b - a_3}{b - a_1} \ln \frac{b - a_1}{b - a_3} + \frac{a_3 - a_1}{b - a_1} \ln \frac{b - a_1}{a_3 - a_1} \qquad (11.4\text{--}8)$$

With the values of the several parameters taken from Table 11.4–1

$$X_1 = 2415 \ \Omega; \qquad X_2 = -1752 \ \Omega; \qquad X_3 = -663 \ \Omega \qquad (11.4\text{--}9)$$

With careful adjustment (at least to within 0.005λ) of h_S to $\beta_S h_S = \pi/2$, where β_S is the wave number of the all-metal sleeve, the impedance Z_{in2} looking into the coaxial-line section can be made very much greater than $|X_2| = 1752 \ \Omega$, so that the approximately equivalent circuit in Fig. 11.4–4b is applicable. For this the impedance terminating the lower antenna section is

$$Z = Z_{in1} \left(\frac{X_2}{X_1} \right)^2 + \frac{iX_2 X_3}{X_1} \sim \frac{iX_2 X_3}{X_1} = i321 \ \Omega \qquad (11.4\text{--}10)$$

where it is assumed that Z_{in1} is kept sufficiently small. The resistive terminal function corresponding to this impedance is $\rho_{hS} = 0.0146$, so that the ratio of the power in the termination to the total power in the lower half of the antenna is

$$\frac{P_B}{P_{hS}} = \frac{\sinh 2\rho_{hS}}{\sinh 2(\rho_{hS} + \alpha_{L3}h_S)} = 0.04 \qquad (11.4\text{--}11)$$

That is, 96% of the power into the lower half of the insulated antenna is transferred by it into the ambient medium. The coaxial choke section is effective in isolating the dipole from the feed line. Note, however, that the sleeve must be adjusted very accurately in length and must not have too small a radius.

With the length of the sleeve fixed at $\beta_S h_S = \pi/2$, it follows that the length of the upper half of the dipole must also have approximately this length for which $\beta_{L3}h_A = 1.96$. At the driving terminals between the halves of the dipole the input impedance differs little from the value $Z = 15.08 + i3.45 \ \Omega$ obtained for the corresponding dipole with choke section. As for this, the input resistance is low for a good match for a typical coaxial feed line. It can be improved by increasing the electrical length $\beta_{L3}h_A$ from $\pi/2$ to π while retaining the electrical length $\beta_S h_S = \pi/2$ for the inside of the sleeve as shown in Fig. 11.4–3b. With $\beta_{L3}h_A = \pi$ and $\beta_S h_S = \pi/2$, the impedance of the dipole differs little from that obtained for the corresponding dipole with choke section (i.e, $Z = 69.0 + i15.75 \ \Omega$), which is readily matched to the coaxial feed line. This increase in length also improves the power ratio to $P_B/P_{hS} = 0.014$.

Series-Connected Insulated Transmission-Line Antenna

The three structures shown in Figs. 11.4–1 to 11.4–3 all include a coaxial transmission line with an outer radius that is small compared to the radius of the insulation. The inner conductor of the coaxial line has to be even smaller. For high-frequency applications such as biological ones, a needlelike outer dimension is required and it may not be possible to have an even thinner inner conductor so that adequate power is radiated into the ambient medium. The series-connected structure shown in Fig. 11.4–5 may be more desirable in such cases. It consists of an end-driven full-wave insulated antenna of length h_3 and radius a_3 (for which $\alpha_{L3}/\beta_{L3} = 0.23$ is quite large) driven, not by a conventional coaxial line, but by a low-loss insulated conductor with the small radius a_1 (for which $\alpha_{L1}/\beta_{L1} = 0.045$ is small). Such a section behaves more like a transmission line than like an antenna. To provide a match from the thin conductor to the thick one, an intermediate quarter-wave matching section is provided to assure a maximum transfer of power to the antenna. The input impedance Z_{in3} of the antenna section with the electrical length $\beta_{L3}h_3$ and an open end is readily determined from (11.3–2) with $Z_{in3} = 1/Y_{in3}$. This is the termination Z_2 for the second section which has $\beta_{L2}h_2 = \pi/2$. With this, Z_{in2} is determined and serves as the termination Z_1 for the first section

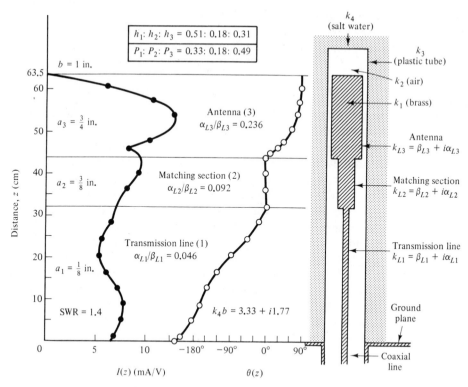

Figure 11.4–5 Current along series-connected insulated antenna.

with the electrical length $\beta_{L1}h_1$. Z_{in1} may again be determined from (11.3–2). Table 11.4–3 lists the several impedances, terminal functions, and lengths as well as the fraction of the total power transferred to the ambient medium by each of the three sections:

$$\frac{P_3}{P_2 + P_3} = \frac{\sinh 2\rho_2}{\sinh 2(\rho_2 + \alpha_{L2}h_2)} \tag{11.4–12}$$

$$\frac{P_2 + P_3}{P_1 + P_2 + P_3} = \frac{\sinh 2\rho_1}{\sinh 2(\rho_1 + \alpha_{L1}h_1)} \tag{11.4–13}$$

TABLE 11.4–3 Factors for Series-Connected Insulated Antenna

j	$Z_j = Z_{in(j+1)}$	Z_{inj} (Ω)	$\theta_j = \rho_j - i\Phi_j$	h_j (cm)	$\beta_{Lj}h_j$	$\alpha_{Lj}h_j$	P_j/P_{tot}
3	∞	$35.1 + i6.8$	0	19.8	3.1086	0.715	0.49
2	$35.1 + i6.8$	$88.4 + i4.5$	$0.674 - i1.484$	11.7	$\pi/2$	0.145	0.18
1	$88.4 + i4.5$	$140.6 - i26.7$	$0.848 - i1.560$	32.0	4.160	0.191	0.33

11.5 *THE ELECTRICALLY SHORT INSULATED ANTENNA AS A PROBE FOR MEASURING ELECTRIC FIELDS IN GENERAL MEDIA*

An interesting and useful property of the insulated antenna is the possibility of constructing it so that its input impedance is very nearly independent of the electrical properties of the medium in which it is embedded. This is important, for example, when it is desired to measure the intensity of the electric field in a region with a permittivity that is not accurately known or that varies from point to point. A probe is evidently needed that has a response to the incident electric field that is independent of the permittivity.

Consider an insulated dipole with inner conductor of length $2h$ and radius a and insulating sleeve of radius b and length $2h_i$, as shown in Fig. 11.5–1. It is center-loaded by an impedance Z_l. The input impedance of the dipole is Z_0. Let the dipole be oriented so the incident electric field E^i is along its axis. The voltage across the load is then given by

$$V_l = 2h_e E^i \frac{Z_l}{Z_0 + Z_l} \qquad (11.5–1)$$

where $2h_e$ is the length known as the *effective length* and $2h_e E^i$ is the open-circuit voltage at the load terminals of the antenna when $Z_l \to \infty$. An electrically short antenna is defined as follows for a bare antenna ($b = a$ in Fig. 11.5–1):

$$|k_1 h| < 1 \qquad (11.5–2)$$

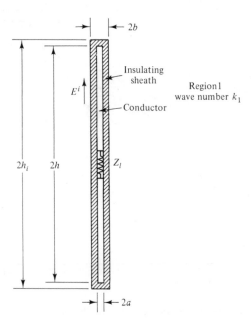

Figure labels: 2b, Insulating sheath, Region 1 wave number k_1, E^i, Conductor, $2h_i$, $2h$, Z_l, $2a$

Figure 11.5–1   Center-loaded insulated dipole.

and for an insulated antenna

$$|k_L h| < 1 \tag{11.5-3}$$

where $k_1 = \beta_1 + i\alpha_1 = \omega[\mu_1(\epsilon_1 + i\sigma_1/\omega)]^{1/2}$ is the wave number of both the medium (region 1) and the current in the bare antenna; $k_L = \beta_L + i\alpha_L$ is the wave number of the current in the insulated antenna as given by (11.1–2). When both (11.5–2) and (11.5–3) are satisfied, the current distributions for bare and insulated antennas are approximately triangular, so that

$$2h_e \sim h \tag{11.5-4}$$

and

$$V_l = hE^i \frac{Z_l}{Z_0 + Z_l} \tag{11.5-5}$$

To make the voltage across the load proportional to the electric field, it is necessary either to make the load impedance Z_l very great compared to the input impedance Z_0 of the antenna ($|Z_l| \gg |Z_0|$), or make Z_0 independent of the constitutive parameters (σ_{1e}, ϵ_{1e}) of the surrounding medium. The first alternative leads to the simple relation $V_l = hE^i$, but involves the difficulty that the input impedance Z_0 of an electrically short antenna is itself very large, so that the condition $|Z_l| \gg |Z_0|$ usually cannot be achieved. This leaves the second alternative, to make Z_0 independent of σ_{1e} and ϵ_{1e}.

Fig. 11.5–2 shows the normalized response $V_n(\epsilon_r) = V_l(\epsilon_r)/V_l(\epsilon_r = 80)$ of

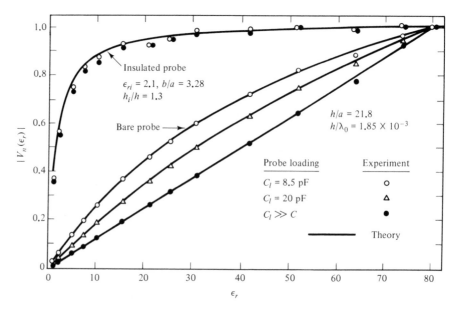

Figure 11.5–2 Normalized electric field response of electrically short, bare, and insulated probes as function of relative dielectric constant ϵ_r of external medium and probe loading C_l.

both bare and insulated electrically short dipoles as a function of the relative permittivity of the ambient medium. The half-length of the conductors is $h/\lambda_0 = 1.85 \times 10^{-3}$, where λ_0 is the wavelength of the source in air. Also $h/a = 21.8$, where a is the radius of the conductor. The insulated antenna has a dielectric coating with radius $b = 3.28a$ and a relative permittivity $\epsilon_{ri} = 2.1$; the length of the insulation is $h_i = 1.3h$. The antennas have capacitive center loads C_l; C is the input capacitance of the antenna.

The normalized response of the bare probe is seen to be a function of the relative permittivity ϵ_{1r} of the ambient medium and of the impedance $Z_l \sim i/\omega C_l$ of the load. On the other hand, the response of the insulated probe is quite uniform for the higher values of ϵ_{1r} with only a 10% variation for $\epsilon_{1r} \geq 12$. This means that the probe could be embedded successively at different locations in a living body or in the earth to measure the local electric field without the need of first determining the permittivity at these locations.

PROBLEMS

1. Determine the field of an air-insulated center-driven dipole antenna immersed in (a) lake water and (b) seawater—at distances R_0 from the center that are large compared to the half-length, h, of the antenna at $f = 500$ MHz and 1 GHz. The radius of the antenna is $a = 0.2$ cm and $b/a = 4.0$. The electrical length of the dipole is $\beta_L h = \pi/2$.

2. Show a diagram of the distribution of current in magnitude and phase along an insulated dipole ($\beta_L h = \pi$) which is air-insulated and immersed in moist earth ($\epsilon_r = 15$, $\sigma = 1.2 \times 10^{-2}$ S/m). What are k_L and Z_c? Determine the electric field at large distances R_0.

12

The Theory
of Waveguides

At high frequencies the transmission of electric power along coaxial or open-wire transmission lines that satisfy the requirement that their cross-sectional dimensions remain electrically small becomes difficult. Thin closely spaced conductors are inadequate to carry either large currents or high voltages so that the power-handling capacity is small. A very useful alternative consists of highly conducting metal pipes with various cross-sectional shapes that are required to be electrically quite large in cross-sectional size. The theory of the transmission of electric power along such pipes is a three-dimensional electromagnetic boundary-value problem that is formulated and analyzed in this chapter.

The analysis can be carried out directly in terms of the electromagnetic vectors and this is often done. An alternative that permits a close parallelism between conventional transmission lines and waveguides is to make use of the Hertz potentials. This route is taken in this chapter.

12.1 THE HERTZ POTENTIALS

The scalar and vector potentials introduced in Chapter 3 are often convenient to formally simplify Maxwell's equations and so facilitate the solution of electromagnetic problems. This arises in part from the fact that the equation for the scalar potential ϕ depends only on the essential density of charge $\bar{\rho}$, the equation for the vector potential only on the essential density of moving charge $\overline{\rho_m \mathbf{v}}$. In many problems the currents that directly generate an electromagnetic field are confined to localized regions in which they are maintained by external sources that are independent of the field they generate. Such localized currents are often confined to a short linear antenna (electric dipole) or a small loop (magnetic dipole) that can be represented by equivalent distributions of the volume densities of polarization and magnetization. A formulation of Maxwell's equations which takes account of this possibility and expresses the electromagnetic field in terms of localized distributions of externally maintained electric and magnetic dipole sources $\mathbf{P}^e$ and $\mathbf{M}^e$ involves the definition of a different set of vector potential functions known as the Hertz potentials.

The first step in developing the new formulation is to separate the externally maintained or intrinsic volume densities of polarization $\mathbf{P}^e$ and magnetization $\mathbf{M}^e$ from the induced volume densities $\mathbf{P} = (\epsilon - \epsilon_0)\mathbf{E}$ and $-\mathbf{M} = (\mu^{-1} - \mu_0^{-1})\mathbf{B}$, which characterize all simply polarizing and magnetizing media. With this separation and the relation $\mathbf{J}_f = \sigma\mathbf{E}$, the essential volume density $\overline{\rho_m \mathbf{v}}$ can be expressed in the following expanded form with the externally maintained dipole sources included explicitly:

$$\overline{\rho_m \mathbf{v}} = \sigma\mathbf{E} - (\mu^{-1} - \mu_0^{-1})\nabla \times \mathbf{B} + \nabla \times \mathbf{M}^e$$
$$+ j\omega(\epsilon - \epsilon_0)\mathbf{E} + j\omega\mathbf{P}^e \quad (12.1\text{--}1)$$

It follows that Maxwell's equation

$$\nabla \times \mathbf{B} = \mu_0(\overline{\rho_m \mathbf{v}} + j\omega\epsilon_0\mathbf{E}) \quad (12.1\text{--}2)$$

becomes

$$\nabla \times \mathbf{B} = \mu(j\omega\bar{\epsilon}\mathbf{E} + \nabla \times \mathbf{M}^e + j\omega\mathbf{P}^e) \quad (12.1\text{--}3)$$

where $j\omega\bar{\epsilon} = \sigma + j\omega\epsilon = \sigma' - j\sigma'' + j\omega(\epsilon' - j\epsilon'') = (\sigma' + \omega\epsilon'') + j\omega(\epsilon' - \sigma''/\omega) = \sigma_e + j\omega\epsilon_e$. When $\nabla \times \mathbf{B}$ in (12.1–3) is used in the definition of the vector potential $\mathbf{A}$, the equation for this becomes

$$\nabla^2\mathbf{A} + \omega^2\mu\bar{\epsilon}\mathbf{A} = -j\omega\mu\mathbf{P}^e - \mu\nabla \times \mathbf{M}^e \quad (12.1\text{--}4)$$

This equation suggests the separation of the vector potential $\mathbf{A}$ into two independent parts of which the one depends on $\mathbf{P}^e$, the other on $\mathbf{M}^e$. This is readily accomplished by setting

$$\mathbf{A} = j\omega\mu\bar{\epsilon}\mathbf{\Pi}_E + \nabla \times \mathbf{\Pi}_M \quad (12.1\text{--}5)$$

When this expression is substituted in (12.1–4), it can be separated into the following independent equations where $k^2 = \omega^2\mu\bar{\epsilon}$:

$$\nabla^2\mathbf{\Pi}_E + k^2\mathbf{\Pi}_E = \frac{-\mathbf{P}^e}{\bar{\epsilon}}; \qquad \nabla^2\mathbf{\Pi}_M + k^2\mathbf{\Pi}_M = -\mathbf{M}^e\mu \quad (12.1\text{--}6)$$

Here the electric Hertz potential or polarization potential $\mathbf{\Pi}_E$ depends only on $\mathbf{P}^e$, the magnetic Hertz potential or magnetization potential $\mathbf{\Pi}_M$ depends only on $\mathbf{M}^e$.

The electromagnetic field can be expressed in terms of the Hertz potentials with the help of the equations that relate it to the scalar and vector potentials. Specifically,

$$\mathbf{B} = \nabla \times \mathbf{A}; \qquad \mathbf{E} = -\nabla\phi - j\omega\mathbf{A} = \frac{-j}{\omega\mu\tilde{\epsilon}} \nabla\nabla \cdot \mathbf{A} - j\omega\mathbf{A} \qquad (12.1\text{-}7)$$

where use has been made of the Lorentz condition $\nabla \cdot \mathbf{A} + j\omega\mu\tilde{\epsilon}\phi = 0$ to define the divergence of the vector potential. With these formulas and (12.1–5) the electromagnetic field is readily expressed in terms of the Hertz potentials. Thus, at all points *outside* the regions in which $\mathbf{P}^e$ and $\mathbf{M}^e$ are defined,

$$\mathbf{B}_E = \frac{jk^2}{\omega}\nabla \times \mathbf{\Pi}_E; \qquad \mathbf{E}_M = -j\omega\nabla \times \mathbf{\Pi}_M \qquad (12.1\text{-}8)$$

$$\mathbf{E}_E = \nabla\nabla\cdot\mathbf{\Pi}_E + k^2\mathbf{\Pi}_E; \qquad \mathbf{B}_M = \nabla \times \nabla \times \mathbf{\Pi}_M = \nabla\nabla\cdot\mathbf{\Pi}_M + k^2\mathbf{\Pi}_M \qquad (12.1\text{-}9)$$

Note that the electromagnetic field is separated into two independent parts. The one, of electric type, has for its sources currents expressed in terms of oscillating electric dipoles; the second, of magnetic type, has for its sources currents around closed loops which are expressed in terms of equivalent oscillating magnetic dipoles. The Hertz potentials can be used where convenient as intermediary functions in the solution of Maxwell's equations subject to the boundary conditions appropriate to particular problems. These include especially the determination of the electromagnetic fields generated by vertical and horizontal electric and magnetic dipoles near the boundary between two different media such as air and ocean, air and earth.

Since the vector wave equations (12.1–6) are difficult to solve, it is advantageous to reduce them to scalar form by noting that P_z^e can be used to represent all $\hat{\mathbf{z}}$ directed currents, M_z^e all currents in the plane perpendicular to $\hat{\mathbf{z}}$. Thus, many problems can be solved with the following source functions and associated Hertz potentials:

$$\mathbf{P}^e = \hat{\mathbf{z}}P_z^e, \quad \mathbf{\Pi}_E = \hat{\mathbf{z}}\Pi_{Ez}; \qquad \mathbf{M}^e = \hat{\mathbf{z}}M_z^e, \quad \mathbf{\Pi}_M = \hat{\mathbf{z}}\Pi_{Mz} \qquad (12.1\text{-}10)$$

This leaves the two scalar equations:

$$\nabla^2\Pi_{Ez} + k^2\Pi_{Ez} = \frac{-P_z^e}{\tilde{\epsilon}}; \qquad \nabla^2\Pi_{Mz} + k^2\Pi_{Mz} = -M_z^e\mu \qquad (12.1\text{-}11)$$

These are the basic equations to be solved for the Hertz potentials. The associated relations for the electromagnetic field are obtained from (12.1–8) and (12.1–9) with $\mathbf{\Pi}_E = \hat{\mathbf{z}}\Pi_{Ez}$ and $\mathbf{\Pi}_M = \hat{\mathbf{z}}\Pi_{Mz}$. They are

$$\mathbf{B}_E = j\omega\mu\tilde{\epsilon}(\nabla \times \hat{\mathbf{z}}\Pi_{Ez}); \qquad \mathbf{E}_M = -j\omega(\nabla \times \hat{\mathbf{z}}\Pi_{Mz}) \qquad (12.1\text{-}12)$$

$$\mathbf{E}_E = \nabla\left(\frac{\partial\Pi_{Ez}}{\partial z}\right) + k^2\hat{\mathbf{z}}\Pi_{Ez}; \qquad \mathbf{B}_M = \nabla\left(\frac{\partial\Pi_{Mz}}{\partial z}\right) + k^2\hat{\mathbf{z}}\Pi_{Mz} \qquad (12.1\text{-}13)$$

If solutions can be obtained for the Hertz potentials in terms of oscillating dipoles that can be represented by localized distributions of the volume densities of polarization and magnetization, the complete electromagnetic field is readily calculated from (12.1–8) and (12.1–9).

12.2 ELECTROMAGNETIC WAVES IN METAL WAVEGUIDES AND GENERALIZED TRANSMISSION LINES: SEPARATION OF AXIAL AND TRANSVERSE PROBLEMS

Conventional transmission lines like the open two-wire line and the coaxial line are useful for the transmission of power from a generator at one end to a load at the other end—which may be very far away. The usefulness of such lines and the theory developed for them are restricted by conditions that require their cross-sectional dimensions to be electrically small. Such a condition is necessary for the open two-wire line in order to make radiation negligible; it is needed for the coaxial line to assure that the solution obtained is complete. As the frequency associated with the currents transmitted along transmission lines is increased, the actual cross-sectional dimensions of the lines must be reduced to keep them electrically small until their power-handling capacities become inadequate. When this occurs, more useful structures to guide electromagnetic waves from a generator to a load must be devised. The most obvious possibility is the coaxial line with a sufficiently large cross-sectional size to handle the power that is to be transmitted. To determine the properties of such a structure it is necessary to reanalyze the coaxial line without the restriction that its transverse dimensions be electrically small. Indeed, it is expedient to formulate the problem even more generally by allowing the cross-sectional shape to be arbitrary before specializing to the circular geometry. Metal-walled tubes in the interior of which electromagnetic fields propagate are known as *waveguides*. They are generally required to have a uniform cross-sectional size and shape along their entire length. That is, the cross section is independent of the axial coordinate, which is taken to be z. The two transverse coordinates can be denoted by u and v until it is desired to use them for circular, ellipsoidal, or rectangular cross sections when they are appropriately specialized.

The Hertz potentials in a slightly modified form are admirably suited for solving for the electromagnetic fields in waveguides in a manner that corresponds formally very closely to the familiar solution for conventional coaxial lines in terms of currents and voltages. For this purpose let the electric Hertz potential—which is generated by an oscillating dipole current density $j\omega P_z^e$—be replaced by the electric-type generalized current function I_E (in amperes) and the magnetic-type Hertz potential—which is generated by an oscillating magnetic dipole density $j\omega M_z^e$—be replaced by the magnetic-type generalized voltage function V_M (in volts). The defining relations are

$$I_E \equiv j\omega\bar{\epsilon}\Pi_{Ez}; \qquad V_M \equiv -j\omega\Pi_{Mz} \qquad (12.2\text{--}1)$$

Let associated voltage and current functions be defined as follows:

$$V_E \equiv \phi = -\nabla \cdot \mathbf{\Pi}_E = -\frac{1}{j\omega\bar{\epsilon}}\frac{\partial I_E}{\partial z}$$

$$I_M \equiv -\mu^{-1}\nabla \cdot \mathbf{\Pi}_M = -\frac{1}{j\omega\mu}\frac{\partial V_M}{\partial z} \tag{12.2-2}$$

With (12.1–11) it follows that

$$\nabla^2 I_E + k^2 I_E = -j\omega P_z^e; \qquad \nabla^2 V_M + k^2 V_M = j\omega\mu M_z^e \tag{12.2-3}$$

And from (12.1–12) and (12.1–13) the associated electromagnetic fields are

$$\mathbf{H}_E = \frac{\mathbf{B}_E}{\mu} = \nabla \times \hat{z}I_E; \qquad \mathbf{E}_M = \nabla \times \hat{z}V_M \tag{12.2-4}$$

$$\mathbf{E}_E = \frac{1}{j\omega\bar{\epsilon}}\left[\nabla\left(\frac{\partial I_E}{\partial z}\right) + k^2\hat{z}I_E\right]$$

$$\mathbf{H}_M = \frac{\mathbf{B}_M}{\mu} = -\frac{1}{j\omega\mu}\left[\nabla\left(\frac{\partial V_M}{\partial z}\right) + k^2\hat{z}V_M\right] \tag{12.2-5}$$

It is convenient to separate the axial and transverse components of the field. Thus with $\nabla = \hat{z}(\partial/\partial z) + \nabla_t$, $\nabla^2 = (\partial^2/\partial z^2) + \nabla_t^2$,

$$H_{Ez} = 0; \qquad\qquad\qquad E_{Mz} = 0 \tag{12.2-6}$$

$$\mathbf{H}_{Et} = \nabla_t \times \hat{z}I_E = -\hat{z} \times \nabla_t I_E; \qquad \mathbf{E}_{Mt} = \nabla_t \times \hat{z}V_M = -\hat{z} \times \nabla_t V_M \tag{12.2-7}$$

$$E_{Ez} = \frac{1}{j\omega\bar{\epsilon}}\left(\frac{\partial^2 I_E}{\partial z^2} + k^2 I_E\right); \qquad H_{Mz} = \frac{1}{j\omega\mu}\left(\frac{\partial^2 V_M}{\partial z^2} + k^2 V_M\right)$$

$$= \frac{-1}{j\omega\epsilon_e}\nabla_t^2 I_E; \qquad\qquad\qquad = \frac{-1}{j\omega\mu}\nabla_t^2 V_M \tag{12.2-8}$$

$$\mathbf{E}_{Et} = \frac{1}{j\omega\bar{\epsilon}}\nabla_t\left(\frac{\partial I_E}{\partial z}\right) = -\nabla_t V_E; \qquad \mathbf{H}_{Mt} = \frac{1}{j\omega\mu}\nabla_t\left(\frac{\partial V_M}{\partial z}\right) = -\nabla_t I_M \tag{12.2-9}$$

The differential equations (12.2–3) can be separated into axial and transverse parts by the method of separation of variables. Specifically, with

$$I_E \equiv I_E(u, v, z) = I_E(z)\Psi_E(u, v); \qquad V_M \equiv V_M(u, v, z) = V_M(z)\Psi_M(u, v) \tag{12.2-10}$$

the equations (12.2–3), for example,

$$\nabla_t^2 I_E(u, v, z) + \frac{\partial^2 I_E(u, v, z)}{\partial z^2} + k^2 I_E(u, v, z) = -j\omega P_z^e \tag{12.2-11}$$

can be expressed in the form

$$-\frac{\nabla_t^2 \Psi_E(u,\, v)}{\Psi_E(u,\, v)} = \frac{\partial^2 I_E(z)/\partial z^2}{I_E(z)} + k^2 \tag{12.2-12}$$

at all points outside the localized regions in which P_z^e differs from zero. Since the left and right sides are independent, they must be equal to a separation constant, κ_E^2. This leads to the following pairs of independent equations when the sequence (12.2–11) and (12.2–12) is repeated for V_M.

$$\nabla_t^2 \Psi_E(u,\, v) + \kappa_E^2 \Psi_E(u,\, v) = 0; \qquad \nabla_t^2 \Psi_M(u,\, v) + \kappa_M^2 \Psi_M(u,\, v) = 0 \tag{12.2-13}$$

$$\frac{\partial^2 I_E(z)}{\partial z^2} - \gamma_E^2 I_E(z) = 0; \qquad \frac{\partial^2 V_M(z)}{\partial z^2} - \gamma_M^2 V_M(z) = 0 \tag{12.2-14}$$

where

$$\gamma_E^2 = \kappa_E^2 - k^2; \qquad \gamma_M^2 = \kappa_M^2 - k^2 \tag{12.2-15}$$

With the transverse and axial problems separated, the next step is to solve the equations for the functions $\Psi(u,\, v)$ which characterize the transverse problems and, in the process, to determine the properties of the separation constants κ^2.

12.3 THE TRANSVERSE PROBLEM AND THE BOUNDARY CONDITIONS ON THE WALLS

The transverse problem consists of the solution of equations of the type

$$\nabla_t^2 \Psi(u,\, v) + \kappa^2 \Psi(u,\, v) = 0 \tag{12.3-1}$$

subject to appropriate boundary conditions on the walls of the waveguide expressed in terms of suitably chosen coordinates u and v. In general, there is a sequence of discrete values of the constant κ which satisfy the equation. These are denoted by κ_a where the single subscript a stands for *two* identifying indices appropriate to the transverse boundaries. The values κ_a are known as *characteristic values* or *eigenvalues* (from the German Eigenwerte) of the transverse equation with its boundary conditions. For perfectly conducting walls the κ_a are real and positive; for highly conducting walls they have very small imaginary parts. The functions $\Psi_a(u,\, v)$ which satisfy the equation are the *characteristic functions* or *eigenfunctions* of the equation with its boundary conditions. In waveguide theory they are known as *mode functions*. Thus the transverse problem requires the solution of (12.3–1) for the eigenvalues and the mode functions.

Since for each transverse mode the constant κ_a has its own value, the longitudinal functions $I_E(z)$ and $V_M(z)$ are different for each κ_a. Thus, for each κ_a there is a solution of the general form (12.2–10). Since the equations (12.2–3) are linear at all points outside the localized regions where P_z^e and M_z^e differ from zero, the general solution is the sum of all modes. That is,

$$I_E(u, v, z) = \sum_a \Psi_{Ea}(u, v) I_{Ea}(z)$$

$$V_M(u, v, z) = \sum_a \Psi_{Ma}(u, v) V_{Ma}(z) \tag{12.3-2}$$

$$V_E(u, v, z) = \sum_a \Psi_{Ea}(u, v) V_{Ea}(z)$$

$$I_M(u, v, z) = \sum_a \Psi_{Ma}(u, v) I_{Ma}(z) \tag{12.3-3}$$

where the functions $I_{Ea}(z)$, $V_{Ea}(z)$, $V_{Ma}(z)$, and $I_{Ma}(z)$ are the longitudinal functions for the individual modes represented by the subscript a which stands for two indices. These functions satisfy the equations

$$\frac{\partial^2 I_{Ea}(z)}{\partial z^2} - \gamma_{Ea}^2 I_{Ea}(z) = 0; \qquad \frac{\partial^2 V_{Ma}(z)}{\partial z^2} - \gamma_{Ma}^2 V_{Ma}(z) = 0 \tag{12.3-4}$$

where

$$\gamma_{Ea}^2 = \kappa_{Ea}^2 - k^2; \qquad \gamma_{Ma}^2 = \kappa_{Ma}^2 - k^2 \tag{12.3-5}$$

Also, with (12.2-2), the functions $V_{Ea}(z)$ and $I_{Ma}(z)$ are defined by

$$V_{Ea}(z) = -\frac{1}{j\omega\tilde{\epsilon}} \frac{\partial I_{Ea}(z)}{\partial z}; \qquad I_{Ma}(z) = -\frac{1}{j\omega\mu} \frac{\partial V_{Ma}(z)}{\partial z} \tag{12.3-6}$$

With (12.3-4) it follows that

$$\frac{\partial V_{Ea}(z)}{\partial z} = -\frac{\gamma_{Ea}^2}{j\omega\tilde{\epsilon}} I_{Ea}(z); \qquad \frac{\partial I_{Ma}(z)}{\partial z} = -\frac{\gamma_{Ma}^2}{j\omega\mu} V_{Ma}(z) \tag{12.3-7}$$

With $k^2 = \omega^2\mu\tilde{\epsilon}$ and the following notation:

$$y_E = j\omega\tilde{\epsilon}; \qquad\qquad z_M = j\omega\mu \tag{12.3-8}$$

$$z_{Ea} = \frac{\gamma_{Ea}^2}{j\omega\tilde{\epsilon}} = \frac{\kappa_{Ea}^2}{j\omega\tilde{\epsilon}} + j\omega\mu; \qquad y_{Ma} = \frac{\gamma_{Ma}^2}{j\omega\mu} = \frac{\kappa_{Ma}^2}{j\omega\mu} + j\omega\tilde{\epsilon} \tag{12.3-9}$$

$$\gamma_{Ea} = (y_E z_{Ea})^{1/2}; \qquad\qquad \gamma_{Ma} = (z_M y_{Ma})^{1/2} \tag{12.3-10}$$

The first-order equations are

$$\frac{\partial V_{Ea}(z)}{\partial z} = -z_{Ea} I_{Ea}(z); \qquad \frac{\partial I_{Ma}(z)}{\partial z} = -y_{Ma} V_{Ma}(z) \tag{12.3-11}$$

$$\frac{\partial I_{Ea}(z)}{\partial z} = -y_E V_{Ea}(z); \qquad \frac{\partial V_{Ma}(z)}{\partial z} = -z_M I_{Ma}(z) \tag{12.3-12}$$

These are the conventional transmission-line equations but with series impedances and shunt admittances per unit length that differ from those for conventional coaxial lines. Although $I_{Ea}(z)$ and $V_{Ma}(z)$ are measured, respectively, in amperes and volts, they are not currents in and potential differences between conductors but potential functions defined in the dielectric medium enclosed by the metal walls.

For perfectly conducting walls κ_E^2 and κ_M^2 are real; also μ, and the effective constants σ_e and ϵ_e are real. Hence, with $k^2 = \omega^2\mu\tilde{\epsilon} = \omega^2\mu\epsilon_e - j\omega\mu\sigma_e$

$$y_E = j\omega\tilde{\epsilon} = g + j\omega c; \qquad\qquad z_M = j\omega\mu = j\omega l \qquad (12.3\text{--}13)$$

$$z_{Ea} = \frac{\kappa_{Ea}^2 - k^2}{j\omega\tilde{\epsilon}} = \frac{1}{g_a' + j\omega c_a'} + j\omega l; \qquad y_{Ma} = \frac{\kappa_{Ma}^2 - k^2}{j\omega\mu} = \frac{1}{j\omega l_a'} + g + j\omega c \qquad (12.3\text{--}14)$$

with $g = \sigma_e$, $c = \epsilon_e$, $l = \mu$, $g_a' = \sigma_e/\kappa_{Ea}^2$, $c_a' = \epsilon_e/\kappa_{Ea}^2$, $l_a' = \mu/\kappa_{Ma}^2$. These constants are characteristic of the equivalent circuits per unit length shown in Fig. 12.3–1a for the electric modes, in Fig. 12.3–1b for the magnetic modes. Note that these circuits both reduce to that of the conventional coaxial line when $g_a' = \infty$, $c_a' = \infty$, $l_a' = \infty$. This is true when $\kappa_E^2 = \kappa_M^2 = 0$.

General solutions of (12.3–4) are

$$I_{Ea}(z) = A_{Ea}\exp(\gamma_{Ea}z) + B_{Ea}\exp(-\gamma_{Ea}z)$$

$$V_{Ma}(z) = A_{Ma}\exp(\gamma_{Ma}z) + B_{Ma}\exp(-\gamma_{Ma}z) \qquad (12.3\text{--}15)$$

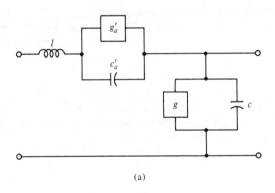

(a)

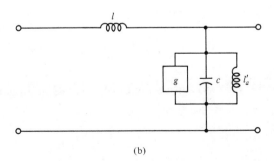

(b)

Figure 12.3–1 Equivalent circuits for waveguides: (a) electric modes; (b) magnetic modes.

$$V_{Ea}(z) = -Z_{Ea}[A_{Ea}\exp(\gamma_{Ea}z) - B_{Ea}\exp(-\gamma_{Ea}z)]$$
$$I_{Ma}(z) = -Y_{Ma}[A_{Ma}\exp(\gamma_{Ma}z) - B_{Ma}\exp(-\gamma_{Ma}z)]$$

(12.3–16)

where the constants A and B are to be evaluated from boundary conditions at $z = 0$, s, and

$$Z_{Ea} \equiv \frac{\gamma_{Ea}}{j\omega\tilde{\epsilon}} = \left(\frac{z_{Ea}}{y_E}\right)^{1/2}; \qquad Y_{Ma} \equiv \frac{\gamma_{Ma}}{j\omega\mu} = \left(\frac{y_{Ma}}{z_M}\right)^{1/2}$$

(12.3–17)

are the *wave impedance* and *wave admittance*, respectively. The electromagnetic field of each mode follows from (12.2–6) to (12.2–9) with (12.3–2) and (12.3–3):

$$E_{Eza} = \frac{\kappa_{Ea}^2}{j\omega\tilde{\epsilon}}I_{Ea}(z)\Psi_{Ea}(u, v); \qquad H_{Mza} = \frac{\kappa_{Ma}^2}{j\omega\mu}V_{Ma}(z)\Psi_{Ma}(u, v)$$

(12.3–18)

$$\mathbf{E}_{Eta} = -V_{Ea}(z)\boldsymbol{\nabla}_t\Psi_{Ea}(u, v); \qquad \mathbf{H}_{Mta} = -I_{Ma}(z)\boldsymbol{\nabla}_t\Psi_{Ma}(u, v)$$

(12.3–19)

$$\mathbf{H}_{Eta} = -I_{Ea}(z)[\hat{\mathbf{z}} \times \boldsymbol{\nabla}_t\Psi_{Ea}(u, v)]; \quad \mathbf{E}_{Mta} = -V_{Ma}(z)[\hat{\mathbf{z}} \times \boldsymbol{\nabla}_t\Psi_{Ma}(u, v)]$$

(12.3–20)

$$H_{Eza} = 0; \qquad\qquad\qquad E_{Mza} = 0$$

(12.3–21)

It is seen that each mode of electric type has a nonvanishing component E_{Eza} and a zero component H_{Eza} in the direction of propagation z. They are called *E modes* or *transverse magnetic* (TM) *modes*. Similarly, each mode of magnetic type has a nonvanishing component H_{Mza} and a zero component E_{Mza} in the direction of propagation. They are called *H modes* or *transverse electric* (TE) *modes*.

If the walls S and any conductors in the waveguide are treated as perfectly conducting, the total electric field must satisfy the boundary condition $\hat{\mathbf{n}} \times \mathbf{E} = 0$, where $\hat{\mathbf{n}}$ is a unit normal perpendicular to S. More generally, the "impedance" boundary condition, $\hat{\mathbf{n}} \times \mathbf{E} = Z_s\mathbf{H}$, where $Z_s = (j\omega\mu/\sigma_w)^{1/2}$ and σ_w is the conductivity of the metal walls, can be used when σ_w is large but not infinite. For a general understanding of the properties of waveguides the simpler condition $\hat{\mathbf{n}} \times \mathbf{E} = 0$ is adequate. This is equivalent to the two conditions

$$(1) \; E_z = 0; \qquad (2) \; \hat{\mathbf{s}} \cdot \mathbf{E}_t = 0$$

(12.3–22)

where $\hat{\mathbf{s}}$ is a unit vector tangent to a bounding surface S along a contour C in a transverse plane (i.e., perpendicular to z). These conditions apply to electric and magnetic modes. For the E modes,

$$(1) \; E_{Ez} = \sum_a E_{Eza} = \sum_a \frac{\kappa_{Ea}^2}{j\omega\tilde{\epsilon}} I_{Ea}(z)\Psi_{Ea}(u, v) = 0 \text{ on } S$$

(12.3–23)

Since this condition must be satisfied for all values of z, it is true in general only when

$$\Psi_{Ea}(u, v) = 0 \text{ on } S$$

(12.3–24)

Since the H modes have no E_z component, the condition $E_{Mz} = 0$ is automatically satisfied. For the E modes,

$$(2)\ \hat{\mathbf{s}} \cdot \mathbf{E}_{Et} = \hat{\mathbf{s}} \cdot \sum_a \mathbf{E}_{Eta} = -\hat{\mathbf{s}} \cdot \sum_a V_{Ea}(z) \nabla_t \Psi_{Ea}(u, v)$$

$$= -\sum_a V_{Ea}(z) \frac{\partial \Psi_{Ea}(u, v)}{\partial s} = 0 \text{ on } C \qquad (12.3\text{–}25)$$

where C is a transverse contour on S. Since this condition must be satisfied for all values of z, it is generally true only when

$$\frac{\partial \Psi_{Ea}(u, v)}{\partial s} = 0 \text{ on } C \text{ for all } a \qquad (12.3\text{–}26)$$

This condition is necessarily satisfied when $\Psi_{Ea}(u, v) = 0$ on C for all a.

For the H modes,

$$(2)\ \hat{\mathbf{s}} \cdot \mathbf{E}_{Mt} = \sum_a \hat{\mathbf{s}} \cdot \mathbf{E}_{Mta} = \sum_a V_{Ma}(z)\hat{\mathbf{s}} \cdot [\hat{\mathbf{z}} \times \nabla_t \Psi_{Ma}(u, v)]$$

$$= \sum_a V_{Ma}(z) \nabla_t \Psi_{Ma}(u, v) \cdot (\hat{\mathbf{s}} \times \hat{\mathbf{z}})$$

$$= \sum_a V_{Ma}(z) \frac{\partial \Psi_{Ma}(u, v)}{\partial n} = 0 \text{ on } C \qquad (12.3\text{–}27)$$

Since this condition must be true for all values of z, it can be generally valid only when

$$\frac{\partial \Psi_{Ma}(u, v)}{\partial n} = 0 \text{ on } C \text{ for all } a \qquad (12.3\text{–}28)$$

It may be concluded that the transverse problem requires the solution of

$$\nabla_t^2 \Psi_{Ea}(u, v) + \kappa_{Ea}^2 \Psi_{Ea}(u, v) = 0; \qquad \nabla_t^2 \Psi_{Ma}(u, v) + \kappa_{Ma}^2 \Psi_{Ma}(u, v) = 0 \qquad (12.3\text{–}29)$$

$$\Psi_{Ea}(u, v) = 0 \text{ on boundaries}; \qquad \frac{\partial \Psi_{Ma}(u, v)}{\partial n} = 0 \text{ on boundaries} \qquad (12.3\text{–}30)$$

The fields are

$$H_{Eza} = 0; \qquad\qquad\qquad E_{Mza} = 0 \qquad (12.3\text{–}31)$$

$$\mathbf{H}_{Eta} = -I_{Ea}(z)[\hat{\mathbf{z}} \times \nabla_t \Psi_{Ea}(u, v)]; \quad \mathbf{E}_{Mta} = -V_{Ma}(z)[\hat{\mathbf{z}} \times \nabla_t \Psi_{Ma}(u, v)] \qquad (12.3\text{–}32)$$

$$E_{Eza} = \frac{\kappa_{Ea}^2}{j\omega\epsilon} I_{Ea}(z)\Psi_{Ea}(u, v); \qquad H_{Mza} = \frac{\kappa_{Ma}^2}{j\omega\mu} V_{Ma}(z)\Psi_{Ma}(u, v) \qquad (12.3\text{–}33)$$

$$\mathbf{E}_{Eta} = -V_{Ea}(z)\nabla_t \Psi_{Ea}(u, v); \qquad \mathbf{H}_{Mta} = -I_{Ma}(z)\nabla_t \Psi_{Ma}(u, v) \qquad (12.3\text{–}34)$$

The expressions for E_{Eza} and H_{Mza} are obtained by combining the expression

$\nabla_t^2 \Psi(u, v)$ with the equation $\nabla_t^2 \Psi(u, v) + \kappa^2 \Psi(u, v) = 0$. Thus

$$\frac{1}{j\omega\tilde{\epsilon}} \nabla_t^2 \Psi_{Ea}(u, v) = -\frac{\kappa_{Ea}^2}{j\omega\tilde{\epsilon}} \Psi_{Ea}(u, v)$$

Note that

$$E_{Eza} = \frac{\kappa_{Ea}^2}{j\omega\tilde{\epsilon}} I_{Ea}(u, v, z) = \kappa_{Ea}^2 \Pi_{Eza}$$

$$H_{Mza} = \frac{\kappa_{Ma}^2}{j\omega\mu} V_{Ma}(u, v, z) = -\frac{\kappa_{Ma}^2}{\mu} \Pi_{Mza}$$

(12.3–35)

That is, E_{Ez} is proportional to $I_{Ea}(u, v, z)$ and Π_{Eza}, H_{Mz} is proportional to $V_{Ma}(u, v, z)$ and Π_{Mza}.

An important special case of the E modes is known as the T *mode* or *transverse electromagnetic mode* (TEM). It arises when $\kappa_E^2 = \kappa_M^2 = 0$, $\nabla_t^2 \Psi_T(u, v) = 0$. The condition $E_{Tz} = \Sigma_a (\kappa_{Ea}^2/j\omega\tilde{\epsilon}) I_{Ea}(z) \Psi_T(u, v) = 0$ is now automatically satisfied at all points. That is, both E_{Tz} and H_{Tz} vanish everywhere. The remaining boundary condition for the T mode is $\partial \Psi_T(u, v)/\partial s = 0$ on C or $\Psi_T(u, v) = \Psi_0 = $ constant on C. For a nontrivial solution the constant Ψ_0 must not be zero. However, the only solution of $\nabla_t^2 \Psi_T(u, v) = 0$ that satisfies $\Psi_T(u, v) = \Psi_0$ on the enclosing boundary C is $\Psi_T(u, v) = \Psi_0$ *everywhere within* C. This means that $\mathbf{E}_T = \mathbf{H}_T = 0$ so that $\mathbf{E}$ and $\mathbf{H}$ vanish identically within C. Hence the T mode cannot exist in a simply connected region enclosed by a single boundary. If the region is doubly (or multiply) connected (as for a coaxial or shielded multiconductor line), the tangential component of $\mathbf{E}$ vanishes on the two (or more) metal boundaries if $\Psi_T(u, v) = \Psi_1$ on surface 1, $\Psi_T(u, v) = \Psi_2$ on surface 2, and so on. In this case $\Psi_T(u, v) = $ constant is not a solution, $\nabla_t \Psi_T(u, v) \neq 0$, and the electromagnetic field does not vanish between the metal surfaces. Thus a T mode (TEM mode) is a possible degenerate case of the E modes. It is readily shown that there is no such degenerate H mode.

It can be proved that when the characteristic functions are not degenerate they satisfy the following orthogonality conditions. The area of integration is a typical cross section A.

$$\int_A \Psi_p(u, v)\Psi_q(u, v) \, dA = 0; \quad p \neq q \qquad (12.3–36)$$

$$\int_A \nabla_t \Psi_p(u, v)\nabla_t \Psi_q(u, v) \, dA = 0; \quad p \neq q \qquad (12.3–37)$$

where Ψ stands for Ψ_E or Ψ_M. The indices p and q correspond to values of the index pairs represented by a. Also valid is the relation

$$\int_A \hat{\mathbf{z}} \cdot [\nabla_t \Psi_{Ep}(u, v) \times \nabla_t \Psi_{Mq}(u, v)] \, dA = 0$$

$$p = 1, 2, \ldots, q, \ldots \qquad (12.3–38)$$

The mode functions are conveniently normalized with the relations

$$\int_A (\nabla_t \Psi_a \cdot \nabla_t \Psi_a) \, dA = 1 \qquad (12.3\text{–}39)$$

$$\kappa_a^2 \int_A \Psi_a^2 \, dA = 1 \qquad (12.3\text{–}40)$$

12.4 THE PROPAGATION CONSTANTS AND WAVE IMPEDANCES AND ADMITTANCES

The properties of a waveguide are most readily visualized when it is terminated in a manner such that with the source at $z = 0$ there are no reflections at the load end $z = s$. This means that all constants A_{Ea} and A_{Ma} are zero so that with $\gamma_{Ea} = \alpha_{Ea} + j\beta_{Ea}$ and $\gamma_{Ma} = \alpha_{Ma} + j\beta_{Ma}$,

$$I_{Ea}(z) = B_{Ea}\exp(-\alpha_{Ea}z)\exp(-j\beta_{Ea}z); \quad V_{Ma}(z) = B_{Ma}\exp(-\alpha_{Ma}z)\exp(-j\beta_{Ma}z) \qquad (12.4\text{–}1)$$

$$V_{Ea}(z) = Z_{Ea}I_{Ea}(z); \qquad\qquad I_{Ma}(z) = Y_{Ma}V_{Ma}(z) \qquad (12.4\text{–}2)$$

These are typical traveling waves progressing in the positive z direction with the phase velocities

$$v_{Ea} = \frac{\omega}{\beta_{Ea}}; \quad v_{Ma} = \frac{\omega}{\beta_{Ma}} \qquad (12.4\text{–}3)$$

and the guide wavelengths

$$\lambda_{Ea} = \frac{2\pi}{\beta_{Ea}}; \quad \lambda_{Ma} = \frac{2\pi}{\beta_{Ma}} \qquad (12.4\text{–}4)$$

The nature of the traveling wave depends greatly on the magnitude of the ratio of attenuation constant to phase constant, viz., α_a/β_a. When this ratio is small, the exponential decrease in amplitude per wavelength of propagation is also small and the guide acts as a low-loss transmission line. On the other hand, when $\alpha_a/\beta_a \geq 1$, the amplitude decreases to a small value in a single wavelength so that propagation over any distance is impossible. In order to study the properties of each mode, it is necessary to separate the real and imaginary parts of the propagation constants. These are defined by

$$\gamma_{Ea} = \alpha_{Ea} + j\beta_{Ea} = (\kappa_{Ea}^2 - k^2)^{1/2} = (z_{Ea}y_E)^{1/2} \qquad (12.4\text{–}5a)$$

$$\gamma_{Ma} = \alpha_{Ma} + j\beta_{Ma} = (\kappa_{Ma}^2 - k^2)^{1/2} = (y_{Ma}z_M)^{1/2} \qquad (12.4\text{–}5b)$$

where, for perfectly conducting walls, κ_a is real. If the dielectric is imperfect

$$k^2 = \omega^2\mu\epsilon_e - j\omega\mu\sigma_e = k_e^2(1 - jp_e) \qquad (12.4\text{–}6)$$

where $p_e = \sigma_e/\omega\epsilon_e \ll 1$ is the loss tangent of the dielectric and $k_e^2 = \omega^2\mu\epsilon_e$. Let

$$\tau_a^2 = \frac{\kappa_a^2}{k_e^2}; \qquad p_a = \frac{p_e}{|\tau_a^2 - 1|} \tag{12.4-7}$$

and note that γ_a must be defined differently in the following three ranges:

(1) $\tau_a^2 > 1$: (stop band)

$$\gamma_a = k_e(\tau_a^2 - 1)^{1/2}(1 + jp_a)^{1/2} \tag{12.4-8}$$

$$= k_e(\tau_a^2 - 1)^{1/2}[f(p_a) + jg(p_a)] \tag{12.4-9}$$

$$\alpha_a = k_e(\tau_a^2 - 1)^{1/2}f(p_a); \qquad \beta_a = k_e(\tau_a^2 - 1)^{1/2}g(p_a) \tag{12.4-10}$$

$$\frac{\alpha_a}{\beta_a} = \frac{f(p_a)}{g(p_a)} > 1 \tag{12.4-11}$$

(2) $\tau_a^2 = 1$: (cutoff)

$$\gamma_a = (j\omega\mu\sigma_e)^{1/2} = (1 + j)\left(\frac{\omega\mu\sigma_e}{2}\right)^{1/2} \tag{12.4-12}$$

$$\alpha_a = \beta_a\left(\frac{\omega\mu\sigma_e}{2}\right)^{1/2} \tag{12.4-13}$$

$$\frac{\alpha_a}{\beta_a} = 1$$

(3) $\tau_a^2 < 1$: (pass band) $\tag{12.4-14}$

$$\gamma_a = jk_e(1 - \tau_a^2)^{1/2}(1 - jp_a)^{1/2} \tag{12.4-15}$$

$$= jk_e(1 - \tau_a^2)^{1/2}[f(p_a) - jg(p_a)] \tag{12.4-16}$$

$$\alpha_a = k_e(1 - \tau_a^2)^{1/2}g(p_a); \qquad \beta_a = k_e(1 - \tau_a^2)^{1/2}f(p_a)$$

$$\frac{\alpha_a}{\beta_a} = \frac{g(p_a)}{f(p_a)} < 1$$

The guide wavelength is

$$\lambda_a = \frac{2\pi}{\beta_a} = \frac{\lambda_e}{(1 - \tau_a^2)^{1/2}}f(p_a) \tag{12.4-17}$$

The functions $f(p)$ and $g(p)$ are defined and tabulated in Appendix II. Note that for a perfect dielectric with $\sigma_e \sim 0$, $p_e = 0$, $p_a = 0$ in all of these expressions. This means that in the range $\tau_a^2 > 1$, $\beta_a = 0$, $\alpha_a/\beta_a = \infty$, so that there is no propagation; at $\tau_a^2 = 1$, $\beta_a = 0$ and again there can be no propagation; at $\tau_a^2 < 1$, $\alpha_a = 0$, $\alpha_a/\beta_a = 0$ and there is propagation with zero attenuation. When the dielectric is perfect, there is a sharp boundary between lossless transmission and no transmission. This is known as the cutoff and is defined by $\tau_a^2 = 1$. Since $\tau_a^2 = \kappa_a^2/k_e^2 = \kappa_a^2/\omega^2\mu\epsilon_e$, it is evident that the waveguide is a high-pass filter. The cutoff frequency and cutoff wavelength are defined as follows:

$$f_{a,\,\text{cutoff}} = \frac{\kappa_a}{2\pi(\mu\epsilon_e)}; \qquad \lambda_{a,\,\text{cutoff}} = \frac{2\pi}{\kappa_a} \qquad (12.4\text{–}18)$$

When the dielectric is perfect and $\tau_a^2 < 1$,

$$\beta_a = k_e(1 - \tau_a^2)^{1/2}; \qquad \lambda_a = \frac{2\pi}{k_e(1 - \tau_a^2)^{1/2}} = \frac{\lambda_e}{(1 - \tau_a^2)^{1/2}} \qquad (12.4\text{–}19)$$

The phase velocity for mode a in the guide is defined by (12.4–3). It is

$$v_a = \frac{v_e}{(1 - \tau_a^2)^{1/2}} \qquad (12.4\text{–}20)$$

where $v_e = 1/(\mu\epsilon_e)^{1/2} = \omega/k_e$. If the dielectric is air, $v_e = c$ and v_a is greater than the velocity of light. The formulas (12.4–3) to (12.4–20) are valid for electric- and magnetic-type modes. Subscripts E or M should precede the subscript a.

In the three ranges of the propagation constant, the wave impedance or admittance have quite different values. These can be calculated from the general definitions

$$Z_{Ea} = \frac{\gamma_{Ea}}{j\omega\epsilon_e} = \left(\frac{z_{Ea}}{y_E}\right)^{1/2}; \qquad Y_{Ma} = \frac{\gamma_{Ma}}{j\omega\mu} = \left(\frac{y_{Ma}}{z_M}\right)^{1/2} \qquad (12.4\text{–}21)$$

with the values of γ_{Ea} and γ_{Ma} obtained from (12.4–8), (12.4–11), and (12.4–14). The principal properties follow from the simple values that are valid when the dielectric is perfect. With $\zeta_e \equiv (\mu/\epsilon_e)^{1/2}$, they are

(1) $\tau_a^2 > 1$: $\quad Z_{Ea} = -j\zeta_e(\tau_{Ea}^2 - 1)^{1/2}; \qquad Y_{Ma} = -j\zeta_e^{-1}(\tau_{Ma}^2 - 1)^{1/2} \qquad (12.4\text{–}22)$

(2) $\tau_a^2 = 1$: $\quad Z_{Ea} = 0; \qquad\qquad\qquad\quad Y_{Ma} = 0 \qquad\qquad\qquad\qquad (12.4\text{–}23)$

(3) $\tau_a^2 < 1$: $\quad Z_{Ea} = \zeta_e(1 - \tau_{Ea}^2)^{1/2}; \qquad Y_{Ma} = \zeta_e^{-1}(1 - \tau_{Ma}^2)^{1/2} \qquad (12.4\text{–}24)$

These formulas show that in the stop band, the wave impedance and admittance are purely reactive, that they go through zero at cutoff and are purely resistive in the pass band.

12.5 DISSIPATION DUE TO IMPERFECTLY CONDUCTING WALLS OF HOLLOW GUIDES*

The attenuation constant α_a defined in (12.4–15) for the pass band takes full account of power losses in an imperfect dielectric in the waveguide. When the real effective conductivity σ_e of the dielectric differs from zero, a leakage conductance g_a' occurs

---

* The generalization to guides with inner conductors is carried out in (12.7–5).

in the impedance per unit length z_{Ea} for electric-type modes and a leakage conductance g in the admittance per unit length y_{Ma} of magnetic-type modes. If the walls of the waveguide are highly but not perfectly conducting (as always in practice with metal walls), the electromagnetic field penetrates the walls. However, when these are highly conducting and sufficiently thick compared with the skin depth $d_s = (2/\omega\mu\sigma_c)^{1/2}$, the field outside the guide is extremely small and can be ignored. When this is true, the contribution to the attenuation by the metal walls can be evaluated separately and added to that due to a good but imperfect dielectric. The procedure is to assume the walls perfectly conducting in evaluating the attenuation due to the dielectric and to assume the dielectric perfect in evaluating the attenuation due to the walls. Since the contribution to the attenuation by a nonzero conductivity of the dielectric has already been determined, it remains to evaluate the contribution by a large but finite conductivity of the metal walls. This is conveniently accomplished by determining the complex power transferred out of a section of the dielectric of length Δz.

Consider a uniform waveguide along the z axis. A source of energy is located in the guide near the end at $z = 0$. Let the section of the guide extending from $z = 0$ to z be enclosed in a surface of integration that is outside the metal walls except at z where it cuts across the guide. Since the electromagnetic field outside the guide is zero, the only contribution to the complex energy transfer function T is from the integration across the guide at z. Thus

$$T_z = \int_{\Sigma_{closed}} (\hat{\mathbf{n}} \cdot \mathbf{S})\, d\Sigma = \int_A S_z\, dA \tag{12.5-1}$$

Similarly, for the section of guide extending from $z = 0$ to $z + \Delta z$,

$$T_{z+\Delta z} = \int_{\Sigma_{closed}} (\hat{\mathbf{n}} \cdot \mathbf{S})\, d\Sigma = \int_A S_{z+\Delta z}\, dA \tag{12.5-2}$$

The difference $T_z - T_{z+\Delta z}$ is the total power dissipated in the length Δz. Since the dielectric is perfect, this must be the power dissipated in the walls of length Δz.

Now let the closed surface be moved inward through the metal walls to enclose only the section of dielectric of length Δz. The net transfer of power across this surface must be zero since it contains no source of energy. Thus, with $\hat{\mathbf{n}}$ an outward normal,

$$\int_{\Sigma_{closed}} \hat{\mathbf{n}} \cdot \mathbf{S}\, d\Sigma = \int_A S_{z+\Delta z}\, dA - \int_A S_z\, dA + \int_\Sigma \hat{\mathbf{n}} \cdot \mathbf{S}\, d\Sigma = 0 \tag{12.5-3}$$

where Σ is the surface of the dielectric in contact with the metal walls. With (12.5-1) and (12.5-2) this formula can be expressed as follows:

$$\frac{dT_z}{dz} = \lim_{\Delta z \to 0} \frac{T_{z+\Delta z} - T_z}{\Delta z} = \lim_{\Delta z \to 0}\left(-\frac{1}{\Delta z}\int_\Sigma \hat{\mathbf{n}} \cdot \mathbf{S}\, d\Sigma\right) = -\oint_s \hat{\mathbf{n}} \cdot \mathbf{S}\, ds \tag{12.5-4}$$

where s is a contour around the perimeter of the dielectric at z. Note that $\Sigma = s\,\Delta z$. Since $\mathbf{S} = \frac{1}{2}\,\mathbf{E} \times \mathbf{H}^*$, $\hat{\mathbf{n}} \cdot \mathbf{S} = \frac{1}{2}\,\hat{\mathbf{n}} \cdot (\mathbf{E} \times \mathbf{H}^*) = \frac{1}{2}\,\mathbf{H}^* \cdot (\hat{\mathbf{n}} \times \mathbf{E})$. However, on the imperfectly conducting walls, $\hat{\mathbf{n}} \times \mathbf{E} = Z_s\mathbf{H}$, where $Z_s = (1 + j)(\omega\mu/2\sigma_c)^{1/2} = (1 + j)(\omega\mu d_s/2)$ is the surface impedance and d_s is the skin depth. It follows that

$$\frac{dT_z}{dz} = -\frac{1}{2}\,Z_s \oint_s \mathbf{H} \cdot \mathbf{H}^*\,ds \tag{12.5-5}$$

For the E modes, $\mathbf{H}$ is entirely transverse and given by (12.3–20). The component in the tangential direction $\hat{\mathbf{s}}$ for each mode is

$$\hat{\mathbf{s}} \cdot \mathbf{H}_{Eta} = -I_{Ea}(z)\hat{\mathbf{s}} \cdot [\hat{\mathbf{z}} \times \nabla_t\Psi_{Ea}(u, v)] = I_{Ea}(z)\hat{\mathbf{n}} \cdot \nabla_t\Psi_{Ea}(u, v)$$

$$= I_{Ea}(z)\,\frac{\partial\Psi_{Ea}(u, v)}{\partial n} \tag{12.5-6}$$

It follows that

$$-\frac{dT_z}{dz} = \frac{1}{2}\,z_{Ea}^i I_{Ea}(z)I_{Ea}^*(z) \tag{12.5-7}$$

where

$$z_{Ea}^i = Z_s \oint \left[\frac{\partial\Psi_{Ea}(u, v)}{\partial n}\right]^2\,ds = \frac{Z_s}{\kappa_{Ea}^2}\,\frac{\oint [\partial\Psi_{Ea}(u, v)/\partial n]^2\,ds}{\displaystyle\int_A \Psi_{Ea}^2(u, v)\,dA} \tag{12.5-8}$$

In the last step use has been made of the normalization condition (12.3–40). The internal series impedance per unit length z_{Ea}^i due to losses in the imperfect conductor must be added to the series impedance for the guide with perfectly conducting walls as given in (12.3–9). Thus

$$z_{Ea} = \frac{\kappa_{Ea}^2}{j\omega\bar{\varepsilon}} + j\omega\mu + z_{Ea}^i \tag{12.5-9}$$

The equivalent circuit in Fig. 12.3–1a is generalized by adding the impedance $z_{Ea}^i = r_{Ea}^i + jx_{Ea}^i$ in series with the inductance l as shown in Fig. 12.5–1a. The generalized propagation constant $\gamma_{Ea} = \sqrt{z_{Ea}y_E}$ and wave impedance $Z_{Ea} = \sqrt{z_{Ea}/y_E}$ are defined with (12.5–9).

The H modes have both transverse and axial components of $\mathbf{H}$ tangent to the walls. Thus for each H mode,

$$\hat{\mathbf{s}} \cdot \mathbf{H}_{Mta} = -I_{Ma}(z)\hat{\mathbf{s}} \cdot \nabla_t\Psi_{Ma}(u, v); \qquad H_{Mza} = \frac{\kappa_{Ma}^2}{j\omega\mu}\,V_{Ma}(z)\Psi_{Ma}(u, v) \tag{12.5-10}$$

It follows that

$$\mathbf{H} \cdot \mathbf{H}^* = I_{Ma}(z)I_{Ma}^*(z)\left[\frac{\partial\Psi_{Ma}(u, v)}{\partial s}\right]^2 + \frac{\kappa_{Ma}^4}{\omega^2\mu^2}\,V_{Ma}(z)V_{Ma}^*(z)\Psi_{Ma}^2(u, v) \tag{12.5-11}$$

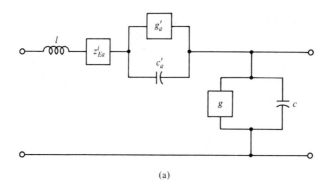

(a)

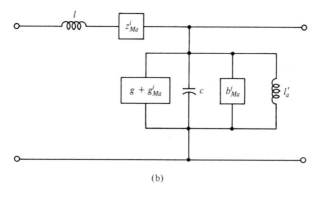

(b)

Figure 12.5–1 Equivalent circuits for waveguides with imperfectly conducting walls: (a) electric modes; (b) magnetic modes.

Hence, with (12.5–5),

$$\frac{dT_z}{dz} = -[\tfrac{1}{2}I_{Ma}(z)I_{Ma}^*(z)z_{Ma}^i + \tfrac{1}{2}V_{Ma}(z)V_{Ma}^*(z)y_{Ma}^i] \tag{12.5–12}$$

where

$$z_{Ma}^i = r_{Ma}^i + jx_{Ma}^i = Z_s \oint \left[\frac{\partial \Psi_{Ma}(u,\,v)}{\partial s}\right]^2 ds$$

$$= \frac{Z_s}{\kappa_{Ma}^2} \frac{\oint [\partial \Psi_{Ma}(u,\,v)/\partial s]^2 \, ds}{\displaystyle\int_A \Psi_{Ma}^2(u,\,v)\, dA} \tag{12.5–13}$$

$$y_{Ma}^i = g_{Ma}^i + jb_{Ma}^i = \frac{Z_s \kappa_{Ma}^4}{\omega^2 \mu^2} \oint \Psi_{Ma}^2(u,\,v)\, ds$$

$$= \frac{Z_s \kappa_{Ma}^2}{\omega^2 \mu^2} \frac{\oint \Psi_{Ma}^2(u,\,v)\, ds}{\displaystyle\int_A \Psi_{Ma}^2(u,\,v)\, dA} \tag{12.5–14}$$

The internal series impedance per unit length, z_{Ma}^i, and the internal shunt admittance per unit length, y_{Ma}^i, due to losses in the imperfect conductor must be added to the corresponding quantities with perfectly conducting walls. Thus, with (12.3–13) and (12.3–14),

$$z_{Ma} = j\omega\mu + z_{Ma}^i = r_{Ma}^i + j(x_{Ma}^i + \omega l) \qquad (12.5\text{–}15)$$

$$y_{Ma} = \frac{1}{j\omega l_a'} + g + j\omega c + y_{Ma}^i = \frac{1}{j\omega l_a'} + (g + g_{Ma}^i) + j(b_{Ma}^i + \omega c) \quad (12.5\text{–}16)$$

The equivalent circuit in Fig. 12.3–1b is generalized by the connection of the impedance z_{Ma}^i in series with the inductance l, the conductance g_{Ma}^i in parallel with the conductance g, and the susceptance b_{Ma}^i in parallel with the capacitance c, as shown in Fig. 12.5–1b. The generalized propagation constant $\gamma_{Ma} = \sqrt{y_{Ma} z_{Ma}}$ and wave admittance $Y_{Ma} = \sqrt{y_{Ma}/z_{Ma}}$ are defined with (12.5–15) and (12.5–16).

12.6 SUMMARY OF WAVEGUIDE MODES: GENERAL TRANSVERSE CROSS SECTION

The electromagnetic field in a tube with imperfectly conducting walls and possible inner conductors is a superposition of the individual fields of an infinite number of E and H modes and of possible T modes when the dielectric is multiply connected because of the presence of inner conductors. Each mode has its own individual properties determined by the formulas and equations shown in Table 12.6–1.

12.7 COAXIAL AND HOLLOW CIRCULAR WAVEGUIDES: THE T MODE

The general theory of uniform waveguides as formulated in the preceding section applies to arbitrary cross sections. Waveguides with circular, rectangular, triangular, and other more complicated transverse shapes are used for a variety of purposes. To make the general theory quantitatively useful, it is necessary to determine for each shape the mode functions $\Psi_a(u, v)$ and the characteristic values κ_a for all propagating modes. The cross-sectional shape most closely related to transmission-line theory is the circle. This shape is of special interest since it is analytically simple to treat first the coaxial waveguide which supports a T mode and becomes a coaxial line when its radius is required to be electrically small and then to allow the radius of the inner conductor to shrink to zero while the radius of the outer tubular conductor is kept constant at an electrically large value. The hollow circular waveguide is obtained in this manner. Actually, it will be convenient to carry out the analysis of the two side by side. Note that in the cylindrical coordinates ρ, θ, z the transverse gradient and Laplacian operators have the forms

$$\nabla_t = \hat{\rho}\,\frac{\partial}{\partial\rho} + \hat{\theta}\,\frac{1}{\rho}\,\frac{\partial}{\partial\theta}; \qquad \nabla_t^2 = \frac{1}{\rho}\,\frac{\partial}{\partial\rho}\,\rho\,\frac{\partial}{\partial\rho} + \frac{1}{\rho^2}\,\frac{\partial^2}{\partial\theta^2}$$

TABLE 12.6-1

	E or TM modes	T or TEM modes	H or TE modes
1. Source functions	$j\omega P_z^e$ (axially directed oscillating electric dipoles or equivalent currents)	$j\omega P_z^e$	$j\omega\mu M_z^e$ (axially directed oscillating magnetic dipoles or currents in closed transverse loops)
2. Potentials	$\begin{aligned} I_E(u, v, z) &= j\omega\bar{\epsilon}\Pi_{Ez}(u, v, z) \\ &= \mu^{-1}A_z(u, v, z) \end{aligned}$ $V_E(u, v, z) = -\dfrac{1}{j\omega\bar{\epsilon}}\dfrac{\partial I_E(u, v, z)}{\partial z}$	$\begin{aligned} I_T(u, v, z) &= j\omega\bar{\epsilon}\Pi_{Ez}(u, v, z) \\ &= \mu^{-1}A_z(u, v, z) \end{aligned}$ $V_T(u, v, z) = -\dfrac{1}{j\omega\bar{\epsilon}}\dfrac{\partial I_T(u, v, z)}{\partial z}$	$V_M(u, v, z) = -j\omega\Pi_{Mz}(u, v, z)$ $I_M(u, v, z) = -\dfrac{1}{j\omega\mu}\dfrac{\partial V_M(u, v, z)}{\partial z}$
3. Modes	$I_E(u, v, z) = \displaystyle\sum_a I_{Ea}(z)\Psi_{Ea}(u, v)$ $V_E(u, v, z) = \displaystyle\sum_a V_{Ea}(z)\Psi_{Ea}(u, v)$	$I_T(u, v, z) = \displaystyle\sum_T I_T(z)\Psi_T(u, v)$ $V_T(u, v, z) = \displaystyle\sum_T V_T(z)\Psi_T(u, v)$	$V_M(u, v, z) = \displaystyle\sum_a V_{Ma}(z)\Psi_{Ma}(u, v)$ $I_M(u, v, z) = \displaystyle\sum_a I_{Ma}(z)\Psi_{Ma}(u, v)$
4. Fields	$E_{Eza} = \dfrac{\kappa_{Ea}^2}{j\omega\bar{\epsilon}}I_{Ea}(z)\Psi_{Ea}(u, v)$ $\mathbf{E}_{Eta} = -V_{Ea}(z)\boldsymbol{\nabla}_t\Psi_{Ea}(u, v)$ $\mathbf{H}_{Eta} = -I_{Ea}(z)[\hat{\mathbf{z}} \times \boldsymbol{\nabla}_t\Psi_{Ea}(u, v)]$ $H_{Eza} = 0$	$E_{Tz} = 0$ $\mathbf{E}_{Tt} = -V_T(z)\boldsymbol{\nabla}_t\Psi_T(u, v)$ $\mathbf{H}_{Tt} = -I_T(z)[\hat{\mathbf{z}} \times \boldsymbol{\nabla}_t\Psi_T(u, v)]$ $H_{Tz} = 0$	$H_{Mza} = \dfrac{\kappa_{Ma}^2}{j\omega\mu}V_{Ma}(z)\Psi_{Ma}(u, v)$ $\mathbf{H}_{Mta} = -I_{Ma}(z)\boldsymbol{\nabla}_t\Psi_{Ma}(u, v)$ $\mathbf{E}_{Mta} = V_{Ma}(z)\boldsymbol{\nabla}_t\Psi_{Ma}(u, v)$ $E_{Mza} = 0$

5. Equations and boundary conditions

$$(\nabla_t^2 + \kappa_{Ea}^2)\Psi_{Ea}(u, v) = 0$$
$$\Psi_{Ea}(u, v) = 0 \text{ on } C$$

$$\left(\frac{\partial^2}{\partial z^2} - \gamma_{Ea}^2\right)I_{Ea}(z) = 0$$

$$\gamma_{Ea} = (z_{Ea}y_E)^{1/2}$$

$$Z_{Ea} = \left(\frac{z_{Ea}}{y_E}\right)^{1/2}$$

$$\nabla_t^2\Psi_T(u, v) = 0$$
$$\Psi_T(u, v) = \Psi_1 \text{ on } C_1$$
$$= \Psi_2 \text{ on } C_2, \text{ etc.}$$

$$\left(\frac{\partial^2}{\partial z^2} - \gamma_T^2\right)I_T(z) = 0$$

$$\gamma_T = (z_T y_T)^{1/2}$$

$$Z_T = \left(\frac{z_T}{y_T}\right)^{1/2}$$

$$(\nabla_t^2 + \kappa_{Ma}^2)\Psi_{Ma}(u, v) = 0$$
$$\frac{\partial\Psi_{Ma}(u, v)}{\partial n} = 0 \text{ on } C$$

$$\left(\frac{\partial^2}{\partial z^2} - \gamma_{Ma}^2\right)V_{Ma}(z) = 0$$

$$\gamma_{Ma} = (y_{Ma}z_{Ma})^{1/2}$$

$$Y_{Ma} = \left(\frac{y_{Ma}}{z_{Ma}}\right)^{1/2}$$

6. First-order axial equations

$$-\frac{\partial V_{Ea}(z)}{\partial z} = z_{Ea}I_{Ea}(z)$$

$$-\frac{\partial I_{Ea}(z)}{\partial z} = y_E V_{Ea}(z)$$

$$-\frac{\partial V_T(z)}{\partial z} = z_T I_T(z)$$

$$-\frac{\partial I_T(z)}{\partial z} = y_T V_T(z)$$

$$-\frac{\partial I_{Ma}(z)}{\partial z} = y_{Ma} V_{Ma}(z)$$

$$-\frac{\partial V_{Ma}(z)}{\partial z} = z_{Ma} I_{Ma}(z)$$

7. Equivalent circuit parameters

$$z_{Ea} = j\omega\mu + \frac{\kappa_{Ea}^2}{j\omega\bar\epsilon} + z_{Ea}^i$$

$$y_E = j\omega\bar\epsilon = \sigma_e + j\omega\epsilon_e$$

$$z_T = j\omega\mu + z_T^i$$

$$y_T = j\omega\bar\epsilon = \sigma_e + j\omega\epsilon_e$$

$$y_{Ma} = j\omega\bar\epsilon + \frac{\kappa_{Ma}^2}{j\omega\mu} + y_{Ma}^i$$

$$z_{Ma} = j\omega\mu + z_{Ma}^i$$

Perfectly conducting walls: $z_{Ea}^i = z_T^i = z_{Ma}^i = 0$; $\quad y_{Ma}^i = 0$

Perfect dielectric: $\bar\epsilon = \epsilon_e - j\sigma_e/\omega = \epsilon_e$; $\quad \sigma_e = 0$

Consider first the T mode or TEM mode in a coaxial guide with a single, central inner conductor with the radius a_1, an outer conductor with the inner radius a_2, as shown in Fig. 12.7–1. The boundary conditions $\Psi_T(\rho, \theta) = \Psi_T(a_1)$ at $\rho = a_1$, $\Psi_T(\rho, \theta) = \Psi_T(a_2)$ at $\rho = a_2$, *for all* θ require that $\partial\Psi_T(\rho, \theta)/\partial\theta = 0$ so that $\Psi_T(\rho, \theta) = \Psi_T(\rho)$ and (12.3–1) reduces to

$$\nabla_t^2 \Psi_T(\rho) = \frac{1}{\rho} \frac{\partial}{\partial\rho} \rho \frac{\partial \Psi_T(\rho)}{\partial\rho} = 0 \qquad (12.7–1)$$

so that

$$\Psi_T(\rho) = C \ln \rho + D = \frac{\Psi_T(a_1) \ln (a_2/\rho)}{\ln (a_2/a_1)} \qquad (12.7–2)$$

In the last step in (12.7–2) the constant $\Psi_T(a_2)$ has been set equal to zero without loss in generality. If the normalization condition (12.3–39) is applied, it follows that

$$\int_{a_1}^{a_2} \left[\frac{\partial \Psi_T(\rho)}{\partial\rho} \right]^2 \cdot 2\pi\rho \, d\rho = \frac{2\pi\Psi_T^2(a_1)}{\ln (a_2/a_1)} = 1 \qquad (12.7–3)$$

When this relation is inserted in (12.7–2), the result is

$$\Psi_T(\rho) = \frac{\ln (a_2/\rho)}{[2\pi \ln (a_2/a_1)]^{1/2}} \qquad (12.7–4)$$

From Table 12.6–1 the circuit parameters are completely defined except z_T^i,

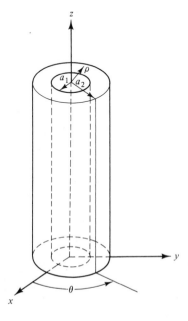

Figure 12.7–1 Coaxial waveguide.

the series impedance due to imperfectly conducting walls. This is evaluated from (12.5–8) in a readily generalized form. If the surface impedance of the inner and outer conductors is the same,

$$z_T^i = Z_s \left\{ \int_0^{2\pi} \left[\frac{\partial \Psi(\rho)}{\partial \rho} \right]_{\rho=a_1}^2 a_1 + \left[\frac{\partial \Psi(\rho)}{\partial \rho} \right]_{\rho=a_2}^2 a_2 \right\} d\theta$$

$$= \frac{Z_s}{\ln (a_2/a_1)} \left(\frac{1}{a_1} + \frac{1}{a_2} \right)$$

$$(12.7\text{–}5)$$

The circuit parameters, propagation constant, and wave impedance for the T mode in the coaxial cylinder are

$$y_T = j\omega\bar{\epsilon} = \sigma_e + j\omega\epsilon_e; \qquad z_T = z_T^i + j\omega\mu \qquad (12.7\text{–}6)$$

$$\gamma_T = \sqrt{z_T y_T} = \sqrt{(\sigma_e + j\omega\epsilon_e)(z_T^i + j\omega\mu)}$$

$$(12.7\text{–}7)$$

$$Z_T = \sqrt{\frac{z_T}{y_T}} = \sqrt{\frac{z_T^i + j\omega\mu}{\sigma_e + j\omega\epsilon_e}}$$

If the walls are treated as perfectly conducting, $z_T^i = 0$, and

$$\gamma_T = \omega\sqrt{-\mu\bar{\epsilon}} = jk; \qquad Z_T = \sqrt{\frac{\mu}{\bar{\epsilon}}} = \zeta \qquad (12.7\text{–}8)$$

The complete solution for the T mode is

$$I_T(\rho, \theta, z) = I_T(z)\Psi_T(\rho) = I_T(z) \frac{\ln (a_2/\rho)}{[2\pi \ln (a_2/a_1)]^{1/2}} \qquad (12.7\text{–}9)$$

$$V_T(\rho, \theta, z) = V_T(z) \frac{\ln (a_2/\rho)}{[2\pi \ln (a_2/a_1)]^{1/2}} \qquad (12.7\text{–}10)$$

The electromagnetic field is

$$E_{T\rho}(\rho, z) = -\frac{\partial \Psi(\rho)}{\partial \rho} V_T(z) = \frac{V_T(z)}{[2\pi \ln (a_2/a_1)]^{1/2}} \frac{1}{\rho}; \qquad E_{T\theta}(\rho, z) = 0$$

$$E_{Tz}(\rho, z) = 0 \qquad (12.7\text{–}11)$$

$$H_{T\rho}(\rho, z) = 0; \qquad H_{T\theta} = -\frac{\partial \Psi_T(\rho)}{\partial \rho} I_T(z) = \frac{I_T(z)}{[2\pi \ln (a_2/a_1)]^{1/2}} \frac{1}{\rho}$$

$$H_{Tz}(\rho, z) = 0 \qquad (12.7\text{–}12)$$

The boundary conditions on the normal ρ component of the electric field and the tangential θ component of the magnetic field can be used to obtain the surface densities of charge and current on the highly (perfectly) conducting inner and outer tubes. Thus

$$E_\rho(a_1, z) = \frac{\eta_{1f}(z)}{\bar{\epsilon}}; \qquad E_\rho(a_2, z) = \frac{-\eta_{2f}(z)}{\bar{\epsilon}} \qquad (12.7\text{–}13)$$

$$H_\theta(a_1, z) = K_{1z}(z); \qquad H_\theta(a_2, z) = -K_{2z}(z) \qquad (12.7\text{–}14)$$

The potential functions $I_T(\rho, z)$ and $V_T(\rho, z)$, although dimensionally in amperes and volts, are not actually currents in the sense of moving charges in the conducting walls or voltages in the sense of potential differences between the conductors. The physically meaningful current and voltage are readily introduced. Thus the actual current $I_1(z)$ in the central conductor and the voltage $V(z)$ between them are defined as follows:

$$I_1(z) = 2\pi a_1 H_\theta(a_1, z) = \sqrt{\frac{2\pi}{\ln (a_2/a_1)}} I_T(z) = \frac{I_T(z)}{\Psi_T(a_1)} \qquad (12.7\text{–}15)$$

$$V(z) = \int_{a_1}^{a_2} E_\rho(\rho, z)\, d\rho = \sqrt{\frac{\ln (a_2/a_1)}{2\pi}}\, V_T(z) = V_T(z)\Psi_T(a_1) \qquad (12.7\text{–}16)$$

The first-order equations are

$$-\frac{\partial I_T(z)}{\partial z} = -\frac{\partial I_1(z)}{\partial z}\sqrt{\frac{\ln (a_2/a_1)}{2\pi}}$$

$$= y_T V_T(z) = y_T \sqrt{\frac{2\pi}{\ln (a_2/a_1)}}\, V(z) \qquad (12.7\text{–}17)$$

$$-\frac{\partial V_T(z)}{\partial z} = -\frac{\partial V(z)}{\partial z}\sqrt{\frac{2\pi}{\ln (a_2/a_1)}}$$

$$= z_T I_T(z) = z_T \sqrt{\frac{\ln (a_2/a_1)}{2\pi}}\, I_1(z) \qquad (12.7\text{–}18)$$

It follows that

$$\frac{-\partial I_1(z)}{\partial z} = yV(z); \qquad \frac{-\partial V(z)}{\partial z} = zI_1(z) \qquad (12.7\text{–}19)$$

where

$$z = z_T \frac{\ln (a_2/a_1)}{2\pi} = \frac{Z_s}{2\pi}\left(\frac{1}{a_1} + \frac{1}{a_2}\right) + j\frac{\omega\mu}{2\pi}\ln\frac{a_2}{a_1} \qquad (12.7\text{–}20)$$

$$y = y_T \frac{2\pi}{\ln (a_2/a_1)} = \frac{2\pi\sigma_e}{\ln (a_2/a_1)} + j\omega\frac{2\pi\epsilon_e}{\ln (a_2/a_1)} \qquad (12.7\text{–}21)$$

(Note that boldface $\mathbf{z}$ is used here for the impedance per unit length to distinguish it from the coordinate z.) These are the equations and parameters of the coaxial line. They are seen to apply without restriction as to the cross-sectional size of the coaxial cylinders. However, they apply only to the T mode, and this is the complete solution only when all higher modes are nonpropagating (i.e., beyond cutoff in the stop band). This is true only when the radius of the outer tube is electrically small, that is, when $|ka| \ll 1$.

The propagation constant γ_T and wave impedance Z_T are unchanged when $I_1(z)$ and $V(z)$ are used instead of the potentials $I_T(z)$ and $V_T(z)$. However, the wave impedance is not the characteristic impedance Z_c of the coaxial line. Note that

$$\frac{V(z)}{I_1(z)} = \frac{V_T(z)}{I_T(z)} \frac{\ln\,(a_2/a_1)}{2\pi}$$

For a *traveling wave on a matched line*, $V_T(z)/I_T(z) = \zeta_e$ and $V(z)/I_1(z) = Z_c$. It follows that $Z_c = (\zeta_e/2\pi) \ln (a_2/a_1)$. Evidently the impedance $Z_{\mathrm{in}T}(z) = V_T(z)/I_T(z)$ looking into an *arbitrarily terminated section* of coaxial waveguide is different from the impedance $Z_{\mathrm{in}}(z) = V(z)/I_1(z)$ defined for the same structure as a coaxial line. However, the *normalized* impedances,

$$z_{\mathrm{in}T}(z) = \frac{Z_{\mathrm{in}T}(z)}{Z_T} = \frac{Z_{\mathrm{in}}(z)}{Z_c} = z_{\mathrm{in}}(z)$$

are the same and it is they that occur in the reflection coefficient and the terminal functions.

The electric and magnetic fields in the T mode are readily expressed in terms of the actual currents and charges per unit length on the inner conductor if use is made of (12.7–13) and (12.7–14). Since the charge per unit length for the rotationally symmetric T mode is simply $q_1(z) = 2\pi a_1 \eta_{1f}(z)$ and the total current $I_1(z) = 2\pi a_1 K_{1z}(z)$, it follows that $E_\rho(a_1, z) = q_1(z)/2\pi a_1\bar{\epsilon}$, $H_\theta(a, z) = I_1(z)/2\pi a_1$, so that

$$E_\rho(\rho, z) = \frac{q_1(z)}{2\pi\bar{\epsilon}\rho}; \qquad H_\theta(\rho, z) = \frac{I_1(z)}{2\pi\rho} \qquad (12.7\text{–}22)$$

These fields are shown schematically in Fig. 12.7–2 for a lossless guide that is infinitely long so that $E_\rho(\rho, z)$ and $H_\theta(\rho, z)$ are in phase, with $E_\rho(\rho, z)/H_\rho(\rho, z) = V_T(z)/I_T(z) = Z_T = \zeta_e = (\mu/\epsilon_e)^{1/2}$. Each component is a traveling wave, $E_\rho(\rho, z) = E_\rho(\rho, 0)e^{-\gamma z} = E_\rho(\rho, 0)e^{-jkz}$ since for the lossless guide $\gamma = (zy)^{1/2} = (-\omega^2\mu\epsilon_e)^{1/2} = jk$. Since $k = \omega/v_e = 2\pi/\lambda_T$, it follows that the instantaneous field at any given cross section z travels to the right with the phase velocity $v_e = (\mu\epsilon_e)^{-1/2}$. The cross-sectional distributions are periodic in the distance $\lambda_T = 2\pi/k$, the wavelength of the T mode in the guide. Note that in a lossless guide $E_\rho(\rho, z)$ and $H_\theta(\rho, z)$ are in phase when the distribution is that of a traveling wave, in phase quadrature when the distribution is that of a standing wave. In the latter the locations of maximum

$E_\rho(\rho, z)$ and maximum $H_\theta(\rho, z)$ are displaced by $\lambda_T/4$. They remain fixed in location instead of traveling with the phase velocity v_e.

12.8 COAXIAL AND HOLLOW CIRCULAR WAVEGUIDES: HIGHER MODES

When the cross section of the coaxial line is unrestricted in size, the electromagnetic field is a superposition of the T mode and other possible modes. The characteristic equation for the mode functions $\Psi(\rho, \theta)$ of the electric or magnetic type is

$$\left(\frac{1}{\rho}\frac{\partial}{\partial \rho}\rho\frac{\partial}{\partial \rho} + \frac{1}{\rho^2}\frac{\partial^2}{\partial \theta^2} + \kappa^2\right)\Psi(\rho, \theta) = 0 \qquad (12.8–1)$$

with the boundary conditions

$$\Psi_E(a_1, \theta) = \Psi_E(a_2, \theta) = 0; \qquad\qquad \text{for } E \text{ modes} \quad (12.8–2a)$$

$$\left[\frac{\partial \Psi_M(\rho, \theta)}{\partial \rho}\right]_{\rho = a_1} = \left[\frac{\partial \Psi_M(\rho, \theta)}{\partial \rho}\right]_{\rho = a_2} = 0; \qquad \text{for } H \text{ modes} \quad (12.8–2b)$$

Equation (12.8–2a) can be solved by the method of separation of variables. With

$$\Psi(\rho, \theta) = R(\rho)\Theta(\theta) \qquad (12.8–3)$$

and the separation constant m^2, (12.8–1) becomes

$$\left(\frac{d^2}{d\theta^2} + m^2\right)\Theta(\theta) = 0 \qquad (12.8–4)$$

$$\left[\frac{d^2}{d\rho^2} + \frac{1}{\rho}\frac{d}{d\rho} + \kappa^2\left(1 - \frac{m^2}{\kappa^2\rho^2}\right)\right]R(\rho) = 0 \qquad (12.8–5)$$

These equations have the solutions:

$$\Theta(\theta) = \cos{(m\theta + D)} \qquad (12.8–6)$$

$$Z(\rho) = AJ_m(\kappa\rho) + BN_m(\kappa\rho) \qquad (12.8–7)$$

where $J_m(\kappa\rho)$ and $N_m(\kappa\rho)$ are Bessel functions of the first and second kinds. They are shown graphically for real arguments in Fig. 12.8–1. Since the choice of D merely locates the orientation of the zero line of the θ coordinate, it is usually convenient to set $D = 0$ without loss of generality. It follows that

$$\Psi(\rho, \theta) = [AJ_m(\kappa\rho) + BN_m(\kappa\rho)] \cos{m\theta} \qquad (12.8–8)$$

where A and B are constants. Since it is necessary that $\cos{m\theta} = \cos{(m\theta + 2m\pi)}$, it follows that $m = 0, 1, 2, \ldots$

In order to determine A, B, and κ it is necessary to impose the boundary

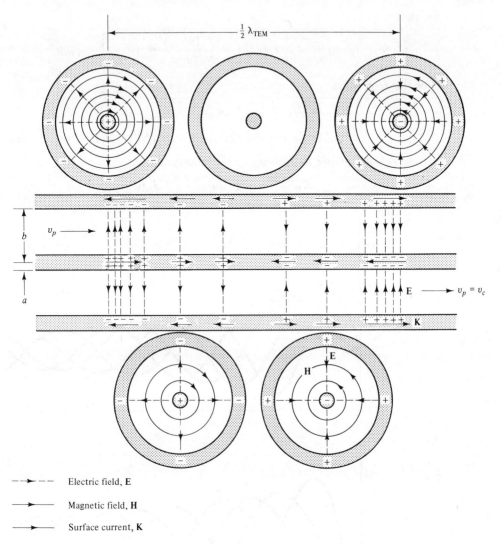

—→ — — —   Electric field, **E**

———————→   Magnetic field, **H**

———————→   Surface current, **K**

Figure 12.7–2 TEM mode in infinitely long (or equivalent) coaxial line with sufficiently small cross section ($b \ll \lambda_{\text{TEM}}/2\pi$).

conditions. For the E modes:

$$Z_m(\kappa_E a_1) = A_E J_m(\kappa_E a_1) + B_E N_m(\kappa_E a_1) = 0 \qquad (12.8\text{–}9)$$

$$Z_m(\kappa_E a_2) = A_E J_m(\kappa_E a_2) + B_E N_m(\kappa_E a_2) = 0 \qquad (12.8\text{–}10)$$

It follows that

$$\frac{B_E}{A_E} = -\frac{J_m(\kappa_E a_1)}{N_m(\kappa_E a_1)} = -\frac{J_m(\kappa_E a_2)}{N_m(\kappa_E a_2)} \qquad (12.8\text{–}11)$$

so that

$$\Psi_{Emn}(\rho, \theta) = A_E\left[J_m(\kappa_E\rho) - \frac{J_m(\kappa_E a_1)}{N_m(\kappa_E a_1)} N_m(\kappa_E\rho)\right] \cos m\theta \qquad (12.8\text{--}12)$$

For the H modes,

$$Z'_m(\kappa_M a_1) = A_M J'_m(\kappa_M a_1) + B_M N'_m(\kappa_M a_1) = 0 \qquad (12.8\text{--}13)$$

$$Z'_m(\kappa_M a_2) = A_M J'_m(\kappa_M a_2) + B_M N'_m(\kappa_M a_2) = 0 \qquad (12.8\text{--}14)$$

where $J'_m(\kappa a_1) = [dJ_m(\kappa\rho)/d\kappa\rho]_{\rho = a_1}$. It follows that

$$\frac{B_M}{A_M} = -\frac{J'_m(\kappa_M a_1)}{N'_m(\kappa_M a_1)} = -\frac{J'_m(\kappa_M a_2)}{N'_m(\kappa_M a_2)} \qquad (12.8\text{--}15)$$

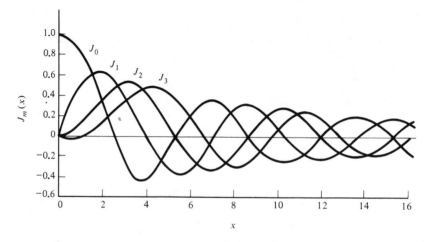

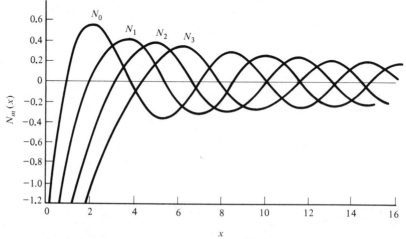

Figure 12.8–1 Bessel functions of the first kind $J_n(k_c\rho)$ and the second kind $N_n(k_c\rho)$.

and

$$\Psi_{Mmn}(\rho, \theta) = A_M\left[J_m(\kappa_M\rho) - \frac{J'_m(\kappa_M a_1)}{N'_m(\kappa_M a_1)} N_m(\kappa_M\rho)\right] \cos m\theta \qquad (12.8{-}16)$$

The constants A_E and A_M can be evaluated with the help of the normalizing condition

$$\kappa_E^2 \int_{a_1}^{a_2} \int_0^{2\pi} \Psi_E^2(\rho, \theta)\rho \, d\theta \, d\rho = 1$$

$$\kappa_M^2 \int_{a_1}^{a_2} \int_0^{2\pi} \Psi_M^2(\rho, \theta)\rho \, d\theta \, d\rho = 1 \qquad (12.8{-}17)$$

Specifically, since

$$\int_0^{2\pi} \cos^2 m\theta \, d\theta = \pi(1 + \delta_{0m}) \qquad (12.8{-}18)$$

where $\delta_{0m} = 1$ when $m = 0$ and $\delta_{0m} = 0$ when $m \neq 0$, it follows that

$$\kappa_E^2 \int_{a_1}^{a_2} Z_m^2(\kappa_E\rho)\rho \, d\rho = -\frac{\kappa_E^2 a_2^2}{2} Z_{m+1}(\kappa_E a_2)Z_{m-1}(\kappa_E a_2)$$

$$+ \frac{\kappa_E^2 a_1^2}{2} Z_{m+1}(\kappa_E a_1)Z_{m-1}(\kappa_E a_1)$$

$$= \frac{1}{\pi(1 + \delta_{0m})} \qquad (12.8{-}19)$$

The terms $Z_m^2(\kappa_E a_2)$ and $Z_m^2(\kappa_E a_1)$ which occur in the integrated expression have been set equal to zero due to (12.8–9) and (12.8–10). However, with (12.8–9), (12.8–10), and the relation $Z_{m-1} + Z_{m+1} = (2m/x)Z_m$, it follows that $Z_{m-1} = -Z_{m+1}$, so that (12.8–19) gives

$$\frac{\kappa_E^2 a_2^2}{2} Z_{m+1}^2(\kappa_E a_2) - \frac{\kappa_E^2 a_1^2}{2} Z_{m+1}^2(\kappa_E a_1) = \frac{1}{\pi(1 + \delta_{0m})} \qquad (12.8{-}20)$$

But

$$Z_{m+1}(\kappa_E a) = \frac{A_E}{N_m(\kappa_E a)} [J_{m+1}(\kappa_E a)N_m(\kappa_E a) - J_m(\kappa_E a)N_{m+1}(\kappa_E a)]$$

$$= \frac{A_E}{N_m(\kappa_E a)} \frac{2}{\pi\kappa_E a} \qquad (12.8{-}21)$$

When (12.8–21) is used in (12.8–20), this gives

$$A_E = \frac{[\pi/2(1 + \delta_{0m})]^{1/2}}{[N_m^{-2}(\kappa_E a_2) - N_m^{-2}(\kappa_E a_1)]^{1/2}} \qquad (12.8{-}22)$$

for the amplitude factor in (12.8–12).

Note that in the limit $a_1 \to 0$, $N_m(\kappa_E a_1) \to \infty$, so that (12.8–12) and (12.8–22) become

$$\Psi_{Emn}(\rho, \theta) = A_E J_m(\kappa_E \rho) \cos m\theta \qquad (12.8\text{–}23)$$

with

$$A_E = \left[\frac{\pi}{2(1 + \delta_{0m})} \right]^{1/2} N_m(\kappa_E a_2) \qquad (12.8\text{–}24a)$$

Alternatively, since in (12.8–11), $J_m(\kappa_E a_2) = 0$ when $a_1 = 0$, it follows from (12.8–21) that $N_m(\kappa_E a_2) = 2/\pi\kappa_E a_2 J_{m+1}(\kappa_E a_2)$ and

$$A_E = \left[\frac{2}{\pi(1 + \delta_{0m})} \right]^{1/2} \frac{1}{\kappa_E a_2 J_{m+1}(\kappa_E a_2)} \qquad (12.8\text{–}24b)$$

This is the amplitude for the E modes in a hollow circular guide for use with (12.8–19).

For the H modes, the normalization is defined by

$$\kappa_M^2 \int_{a_1}^{a_2} Z_m^2(\kappa_M \rho)\rho \, d\rho = \frac{1}{2} (\kappa_M^2 a_2^2 - m^2) Z_m^2(\kappa_M a_2)$$

$$- \frac{1}{2} (\kappa_M^2 a_1^2 - m^2) Z_m^2(\kappa_M a_1) = \frac{1}{\pi(1 + \delta_{0m})} \qquad (12.8\text{–}25)$$

In (12.8–25) use has been made of (12.8–13) and (12.8–14) together with the general formulas for cylinder functions: $Z_m' = -(m/x)Z_m + Z_{m-1} = (m/x)Z_m - Z_{m+1}$ which yield $Z_{m-1} = Z_{m+1} = (m/x)Z_m$ when $Z_m' = 0$. With

$$Z_m(\kappa_M a) = \frac{A_M}{N_m'(\kappa_M a)} [J_m(\kappa_M a)N_m'(\kappa_M a) - N_m(\kappa_M a)J_m'(\kappa_M a)]$$

$$= \frac{A_M}{N_m'(\kappa_M a)} \frac{2}{\pi\kappa_M a} \qquad (12.8\text{–}26)$$

it follows that

$$A_M = \frac{\sqrt{\pi/2(1 + \delta_{0m})}}{\{[1 - m^2/\kappa_M^2 a_2^2][N_m'(\kappa_M a_2)]^{-2} - [1 - m^2/\kappa_M^2 a_1^2][N_m'(\kappa_M a_1)]^{-2}\}^{1/2}}$$

$$(12.8\text{–}27)$$

In the limit, $a_1 \to 0$, $N_m(\kappa_M a_1) \to \infty$, so that (12.8–16) and (12.8–27) reduce to

$$\Psi_{Mmn}(\rho, \theta) = A_M J_m(\kappa_M \rho) \cos m\theta \qquad (12.8\text{–}28)$$

with

$$A_M = \left[\frac{\pi}{2(1 + \delta_{0m})} \right]^{1/2} \frac{N_m'(\kappa_M a_2)}{(1 - m^2/\kappa_M^2 a_2^2)^{1/2}} \qquad (12.8\text{–}29)$$

Alternatively, since with $a_1 \to 0$, (12.8–15) gives $J'_m(\kappa_M a_2) = 0$, it follows that (12.8–26) gives $N'_m(\kappa_M a_2) = 2/\pi\kappa_M a_2 J_m(\kappa_M a_2)$ and

$$A_M = \left[\frac{2}{\pi(1 + \delta_{0m})} \right]^{1/2} \frac{1}{(\kappa_M^2 a_2^2 - m^2)^{1/2} J_m(\kappa_M a_2)} \tag{12.8–30}$$

This is the amplitude for the H modes in a hollow circular guide for use with (12.8–28).

It remains to determine the characteristic values κ_E and κ_M for the E and H modes in coaxial and hollow circular guides. They can be evaluated from the equations obtained from (12.8–9), (12.8–10), (12.8–13), and (12.8–14) when the determinant of the coefficients of A and B is set equal to zero. This determinant must vanish if A and B are to have values other than zero. Thus, for the coaxial guide,

$$J_m(\kappa_E a_1)N_m(\kappa_E a_2) - J_m(\kappa_E a_2)N_m(\kappa_E a_1) = 0 \tag{12.8–31}$$

$$J'_m(\kappa_M a_1)N'_m(\kappa_M a_2) - J'_m(\kappa_M a_2)N'_m(\kappa_M a_1) = 0 \tag{12.8–32}$$

For the hollow circular guide, $a_1 \to 0$, $N_m(\kappa a_1) \to \infty$ and

$$J_m(\kappa_E a_2) = 0; \qquad J'_m(\kappa_M a_2) = 0 \tag{12.8–33}$$

The solution of these equations yields in each case a set of roots in the form $\kappa_E = \kappa_{Emn}$, $\kappa_M = \kappa_{Mmn}$, where $m = 0, 1, 2, \ldots$ is the order of the Bessel function and $n = 1, 2, \ldots$ is the number of the root. With $u = u_{mn} = \kappa_{mn} a_2$ and $c = a_1/a_2$, (12.8–31) and (12.8–32) can be expressed in the form

$$J_m(u)N_m(cu) - J_m(cu)N_m(u) = 0 \tag{12.8–34}$$

$$J'_m(u)N'_m(cu) - J'_m(cu)N'_m(u) = 0 \tag{12.8–35}$$

The roots of these equations are u_{mn}. A few values are in Tables 12.8–1 and 12.8–2 where the quantity $(1 - c)u_{mn}$ or $(1 + c)u_{mn}$ or u_{mn} is tabulated for different values of c, m, and n. The values for $c = a_1/a_2 = 0$ apply to the hollow guide with $a_1 = 0$.

The components of the electromagnetic field for each mode in a coaxial or hollow circular waveguide are

$$(E_{Ez})_{mn} = \frac{\kappa_{Emn}^2}{j\omega\bar{\epsilon}} \Psi_{Emn}(\rho, \theta) I_{Emn}(z)$$

$$\tag{12.8–36}$$

$$(H_{Mz})_{mn} = \frac{\kappa_{Mmn}^2}{j\omega\mu} \Psi_{Mmn}(\rho, \theta) V_{Mmn}(z)$$

$$(E_{E\rho})_{mn} = -\frac{\partial \Psi_{Emn}(\rho, \theta)}{\partial \rho} V_{Emn}(z)$$

$$\tag{12.8–37}$$

$$(H_{M\rho})_{mn} = -\frac{\partial \Psi_{Mmn}(\rho, \theta)}{\partial \rho} I_{Mmn}(z)$$

TABLE 12.8–1   Roots for Electric-Type Modes

$$(u_{mn} = \kappa_{Emn}a_2, \; c = a_1/a_2)$$

	mn					
	01	11	21	02	12	22

(A) Roots of $J_m(u)N_m(cu) - J_m(cu)N_m(u) = 0$:

c	$(1 - c)u_{mn}$					
1.0	π	π	π	2π	2π	2π
0.833	3.140	3.142	3.161	6.282	6.285	6.293
0.667	3.135	3.146	3.237	6.280	6.293	6.382
0.500	3.123	3.161	3.4	6.273	6.312	6.43
0.333	3.096	3.271		6.258	6.357	
0.25	3.073	3.336		6.243	6.403	

(B) Roots of $J_m(u) = 0$:

c	u_{mn}					
0	2.405	3.832	5.136	5.52	7.016	8.417

TABLE 12.8–2   Roots for Magnetic-Type Modes

$$(u_{mn} = \kappa_{Mmn}a_2, \; c = a_1/a_2)$$

	mn					
	11	21	01	12	22	02

(A) Roots of $J'_m(u)N'_m(cu) - J'_m(cu)N'_m(u) = 0$:

c	$(1 + c)u_{mn}$		$(1 - c)u_{mn}$			
1.0	2.000	4.000	π	π	π	π
0.833	2.002	4.006	3.145	3.151	3.167	3.193
0.667	2.013	4.020	3.161	3.188	3.27	3.40
0.500	2.031	4.023	3.197	3.282	3.5	
0.333	2.056	3.908	3.271	3.516		
0.250	2.055	3.760	3.336	3.753		

(B) Roots of $J'_m(u) = 0$:

c	u_{mn}					
0	1.841	3.054	3.832	5.331	6.706	7.016

$$(E_{E\theta})_{mn} = -\frac{1}{\rho}\frac{\partial\Psi_{Emn}(\rho,\theta)}{\partial\theta}V_{Emn}(z)$$

$$(H_{M\theta})_{mn} = -\frac{1}{\rho}\frac{\partial\Psi_{Mmn}(\rho,\theta)}{\partial\theta}I_{Mmn}(z) \tag{12.8–38}$$

$$(H_{Ez})_{mn} = 0; \qquad (E_{Mz})_{mn} = 0 \tag{12.8–39}$$

$$(H_{E\rho})_{mn} = \frac{1}{\rho}\frac{\partial\Psi_{Emn}(\rho,\theta)}{\partial\theta}I_{Emn}(z)$$

$$(E_{M\rho})_{mn} = -\frac{1}{\rho}\frac{\partial\Psi_{Mmn}(\rho,\theta)}{\partial\theta}V_{Mmn}(z) \tag{12.8–40}$$

$$(H_{E\theta})_{mn} = -\frac{\partial\Psi_{Emn}(\rho,\theta)}{\partial\rho}I_{Emn}(z)$$

$$(E_{M\theta})_{mn} = \frac{\partial\Psi_{Mmn}(\rho,\theta)}{\partial\rho}V_{Mmn}(z) \tag{12.8–41}$$

where $\Psi_{Emn}(\rho,\theta)$ is given by (12.8–12) or (12.8–23) and $\Psi_{Mmn}(\rho,\theta)$ is given by (12.8–16) or (12.8–28). Also,

$$\frac{\partial\Psi_{Emn}(\rho,\theta)}{\partial\rho} = \kappa_{Emn}A_{Emn}\left[J'_m(\kappa_{Emn}\rho) - \frac{J_m(\kappa_{Emn}a_1)}{N_m(\kappa_{Emn}a_1)}N'_m(\kappa_{Emn}\rho)\right]\cos m\theta \tag{12.8–42}$$

$$\frac{\partial\Psi_{Emn}(\rho,\theta)}{\partial\theta} = -mA_{Emn}\left[J_m(\kappa_{Emn}\rho) - \frac{J_m(\kappa_{Emn}a_1)}{N_m(\kappa_{Emn}a_1)}N_m(\kappa_{Emn}\rho)\right]\sin m\theta \tag{12.8–43}$$

$$\frac{\partial\Psi_{Mmn}(\rho,\theta)}{\partial\rho} = \kappa_{Mmn}A_{Mmn}\left[J'_m(\kappa_{Mmn}\rho) - \frac{J'_m(\kappa_{Mmn}a_1)}{N'_m(\kappa_{Mmn}a_1)}N'_m(\kappa_{Mmn}\rho)\right]\cos m\theta \tag{12.8–44}$$

$$\frac{\partial\Psi_{Mmn}(\rho,\theta)}{\partial\theta} = -mA_{Mmn}\left[J_m(\kappa_{Mmn}\rho) - \frac{J'_m(\kappa_{Mmn}a_1)}{N'_m(\kappa_{Mmn}a_1)}N_m(\kappa_{Mmn}\rho)\right]\sin m\theta \tag{12.8–45}$$

For the hollow guide with $a_1 = 0$, the second term in each of the brackets is zero.

The propagation constants are

$$\gamma_{Emn} = (\kappa_{Emn} - k^2)^{1/2}; \qquad \gamma_{Mmn} = (\kappa_{Mmn}^2 - k^2)^{1/2} \tag{12.8–46}$$

and the pass band is defined by

$$\tau_{Emn} = \frac{\kappa_{Emn}}{k_e} < 1; \qquad \tau_{Mmn} = \frac{\kappa_{Mmn}}{k_e} < 1 \tag{12.8–47}$$

where $k_e = \omega(\mu\epsilon_e)^{1/2} = 2\pi/\lambda_e$. The cutoff wavelengths are

$$\lambda_{e\text{ cutoff}} = \frac{2\pi}{\kappa_{Emn}}; \qquad \lambda_{e\text{ cutoff}} = \frac{2\pi}{\kappa_{Mmn}} \tag{12.8–48}$$

Figures 12.8–2 to 12.8–7 show the fields schematically of some higher modes in coaxial and hollow circular waveguides.

The surface densities of charge and current on the walls are obtained directly from the boundary conditions. Thus, for each mode, but with subscripts omitted:

$$\eta(a_2, \theta) = -\bar{\epsilon}E_\rho(a_2, \theta); \qquad \eta(a_1, \theta) = \bar{\epsilon}E_\rho(a_1, \theta) \qquad (12.8\text{–}49)$$

$$K_z(a_2, \theta) = -H_\theta(a_2, \theta); \qquad K_z(a_1, \theta) = H_\theta(a_1, \theta) \qquad (12.8\text{–}50)$$

$$K_\theta(a_2, \theta) = H_z(a_2, \theta); \qquad K_\theta(a_1, \theta) = -H_z(a_1, \theta) \qquad (12.8\text{–}51)$$

Note that for all E or TM modes, $H_z = 0$, so that $K_{E\theta} = 0$; that is, there are only axially directed currents. For the H modes there are, in general, both axial and transverse currents. However, for the H_{0n} modes $\sin m\theta = 0$, so that $\partial/\partial\theta = 0$. This means that $(H_{M\theta})_{0n} = 0$, so that $(K_{Mz})_{0n} = 0$. Hence all currents are transverse. The surface currents are shown schematically in Figs. 12.8–8 and 12.8–9.

12.9 THE HOLLOW RECTANGULAR WAVEGUIDE

In the absence of an inner conductor, the rectangular waveguide with cross section defined by $0 \leq x \leq a$, $0 \leq y \leq b$ with $b \geq a$ cannot support a T mode. The characteristic equation for the E and H modes is

$$\left(\frac{\partial^2}{\partial x^2} + \frac{\partial^2}{\partial y^2} + \kappa^2\right)\Psi(x, y) = 0 \qquad (12.9\text{–}1)$$

with the boundary conditions

$$\Psi_E(0, y) = \Psi_E(a, y) = \Psi_E(x, 0) = \Psi_E(x, b) = 0; \qquad \text{for } E \text{ modes} \qquad (12.9\text{–}2)$$

$$\left[\frac{\partial\Psi_M(x, y)}{\partial x}\right]_{x=0,a} = \left[\frac{\partial\Psi_M(x, y)}{\partial y}\right]_{y=0,b} = 0; \qquad \text{for } H \text{ modes} \qquad (12.9\text{–}3)$$

With

$$\Psi(x, y) = X(x)Y(y) \qquad (12.9\text{–}4)$$

(12.9–1) is readily separated into

$$\left(\frac{\partial^2}{\partial x^2} + \kappa_x^2\right)X(x) = 0; \qquad \left(\frac{\partial^2}{\partial y^2} + \kappa_y^2\right)Y(y) = 0 \qquad (12.9\text{–}5a)$$

with

$$\kappa^2 = \kappa_x^2 + \kappa_y^2 \qquad (12.9\text{–}5b)$$

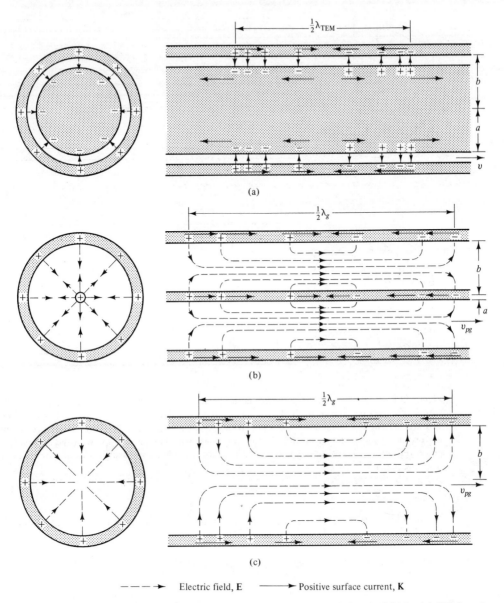

$----\blacktriangleright$ Electric field, **E** $\longrightarrow$ Positive surface current, **K**

Figure 12.8–2 Distributions of surface charge, surface current, and electric field for (a) TEM mode in a coaxial pipe with b unrestricted, $(b - a) \ll \lambda/2\pi$; (b) $TM_{0.1}$ mode in coaxial pipe with $b > \lambda_{TEM}/2.61$, $a \ll b$; (c) $TM_{0.1}$ mode in hollow cylinder with $b > \lambda_{TEM}/2.61$. All pipes are infinitely long or the equivalent.

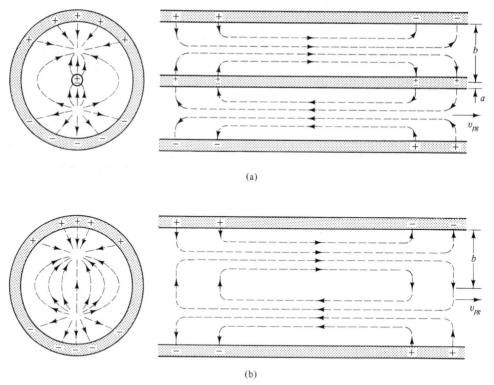

(a)

(b)

Figure 12.8–3 Surface charge and electric-field distributions for (a) $TM_{1,1}$ mode in a coaxial line with $b > \lambda_{TEM}/1.63$, $a \ll b$; (b) $TM_{1,1}$ mode in a hollow cylinder with $b > \lambda_{TEM}/1.63$. All surface currents are parallel to the axis; the line and the pipe are assumed to be infinitely long (or equivalent) so that cross-sectional surfaces of constant phase travel along the pipe with the appropriate phase velocity v_{pg}. Lines that appear to end in space actually go vertically down into or come up from the paper.

Nontrivial solutions of these equations that satisfy the boundary conditions (12.9–2) or (12.9–3) are

$$\Psi_E(x, y) = C_E \sin \frac{m\pi x}{a} \sin \frac{n\pi y}{b} \qquad (12.9\text{–}6)$$

$$\Psi_M(x, y) = C_M \cos \frac{m\pi x}{a} \cos \frac{n\pi y}{b} \qquad (12.9\text{–}7)$$

where $\kappa^2 = (m\pi/a)^2 + (n\pi/b)^2$. Note that the lowest nonzero E mode has $m = 1, n = 1$; the lowest possible H mode with $a > b$ has $m = 1, n = 0$. The arbitrary constants C_E and C_M can be determined with the normalization condition

$$\kappa^2 \int_0^a \int_0^b \Psi^2(x, y) \, dx \, dy = 1 \qquad (12.9\text{–}8)$$

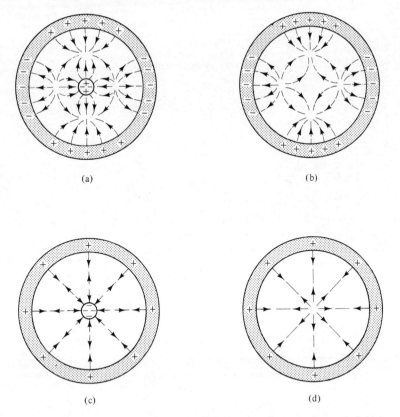

Figure 12.8–4 Cross-sectional distribution of electric charge and electric field for (a) coaxial $TM_{2,1}$ with $b > \lambda_{TEM}/1.2$; (b) cylinder $TM_{2,1}$ with $b > \lambda_{TEM}/1.2$; (c) coaxial $TM_{0,2}$ with $b < \lambda_{TEM}1.14$; (d) cylinder $TM_{0,2}$ with $b < \lambda_{TEM}/1.14$. Note that the $TM_{0,2}$ modes are rotationally symmetrical—all $TM_{0,m}$ modes are. Electric lines that appear to end in space actually bend down into or up from the paper.

The results are

$$C_E \equiv C_{Emn} = \frac{2}{\kappa_E (ab)^{1/2}} = \frac{2}{\pi[(m^2 b/a) + (n^2 a/b)]^{1/2}} \qquad (12.9\text{–}9)$$

$$C_M = C_{Mmn} = \frac{2}{\kappa_M (ab)^{1/2}} = \frac{2}{\pi[(m^2 b/a) + (n^2 a/b)]^{1/2}} \qquad (12.9\text{–}10)$$

The components of the electromagnetic field are

$$(E_{Ex})_{mn} = - \frac{\partial \Psi_{Emn}(x, y)}{\partial x} V_{Emn}(z) \qquad (12.9\text{–}11)$$

$$(H_{Mx})_{mn} = - \frac{\partial \Psi_{Mmn}(x, y)}{\partial x} I_{Mmn}(z)$$

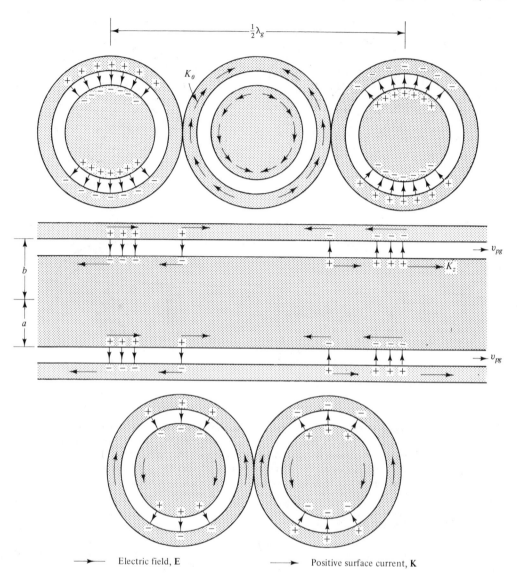

Electric field, **E**   Positive surface current, **K**

Figure 12.8–5 TE$_{1,1}$ mode in a coaxial pipe subject to $(b - a) \ll \lambda_{\text{TEM}}/2\pi$, $b > \lambda_{\text{TEM}}/2\pi$. The circles represent cross sections in the pipe directly above or below. $\lambda_g > \lambda_{\text{TEM}}$; $v_{pg} > v_c$.

$$(E_{Ey})_{mn} = -\frac{\partial \Psi_{Emn}(x, y)}{\partial y} V_{Emn}(z) \tag{12.9–12}$$

$$(H_{My})_{mn} = -\frac{\partial \Psi_{Mmn}(x, y)}{\partial y} I_{Mmn}(z)$$

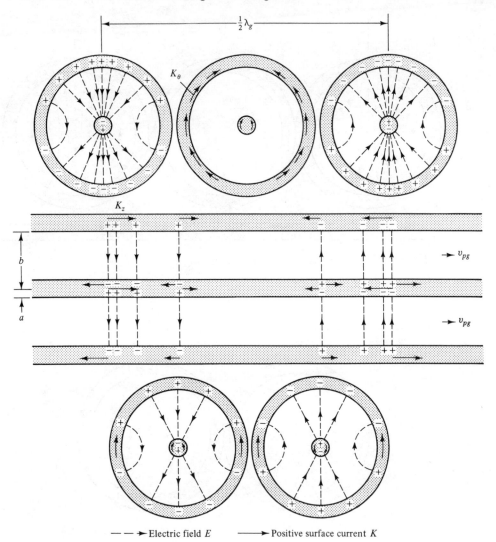

Figure 12.8–6 $\text{TE}_{1,1}$ mode in a coaxial pipe subject to $a \ll b$, $b > \lambda_{\text{TEM}}/3.41$. $\lambda_g > \lambda_{\text{TEM}}$, $v_{pg} > v_c$.

$$(E_{Ez})_{mn} = \frac{\kappa_{Emn}^2}{j\omega\bar{\epsilon}} \Psi_{Emn}(x, y)I_{Emn}(z)$$

$$(H_{Mz})_{mn} = \frac{\kappa_{Mmn}^2}{j\omega\mu} \Psi_{Mmn}(x, y)V_{Mmn}(z) \qquad (12.9\text{–}13)$$

$$(H_{Ex})_{mn} = \frac{\partial\Psi_{Emn}(x, y)}{\partial y} I_{Emn}(z)$$

$$(E_{Mx})_{mn} = -\frac{\partial\Psi_{Mmn}(x, y)}{\partial y} V_{Mmn}(z) \qquad (12.9\text{–}14)$$

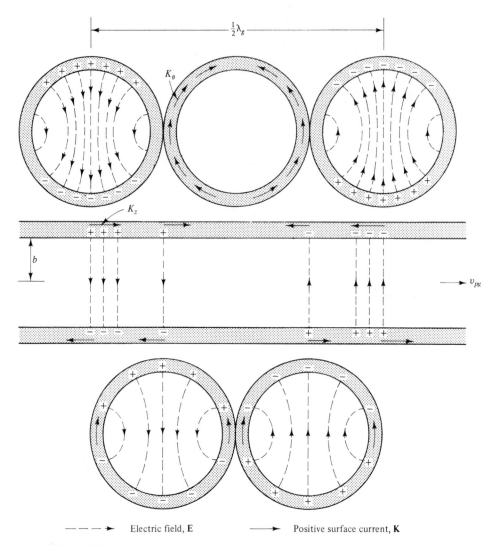

Figure 12.8–7 $TE_{1,1}$ mode in a cylindrical pipe subject to $b > \lambda_{TEM}/3.41$. $\lambda_g > \lambda_{TEM}$, $v_{pg} > v_c$.

$$(H_{Ey})_{mn} = - \frac{\partial \Psi_{Emn}(x, y)}{\partial x} I_{Emn}(z)$$

$$(E_{My})_{mn} = \frac{\partial \Psi_{Mmn}(x, y)}{\partial x} V_{Mmn}(z)$$

$$(12.9\text{–}15)$$

$$(H_{Ez})_{mn} = 0; \qquad (E_{Mz})_{mn} = 0 \qquad (12.9\text{–}16)$$

Note that

$$\frac{\partial \Psi_{Emn}(x, y)}{\partial x} = C_{Emn} \frac{m\pi}{a} \cos \frac{m\pi x}{a} \sin \frac{n\pi y}{b} \qquad (12.9\text{--}17)$$

$$\frac{\partial \Psi_{Emn}(x, y)}{\partial y} = C_{Emn} \frac{n\pi}{b} \sin \frac{m\pi x}{a} \cos \frac{n\pi y}{b} \qquad (12.9\text{--}18)$$

$$\frac{\partial \Psi_{Mmn}(x, y)}{\partial x} = -C_{Mmn} \frac{m\pi}{a} \sin \frac{m\pi x}{a} \cos \frac{n\pi y}{b} \qquad (12.9\text{--}19)$$

$$\frac{\partial \Psi_{Mmn}(x, y)}{\partial y} = -C_{Mmn} \frac{n\pi}{b} \cos \frac{m\pi x}{a} \sin \frac{n\pi y}{b} \qquad (12.9\text{--}20)$$

The dominant and single most important mode in the rectangular waveguide

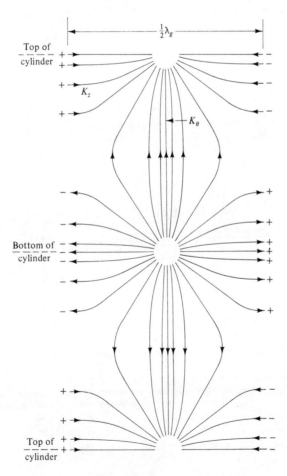

Figure 12.8–8 Surface current of the TE$_{1,1}$ mode on the inside of the metal cylinder of Fig. 12.8–7. The cylinder is represented as cut along the top center and laid flat.

is the H_{10} mode. Its components are

$$(H_{Mx})_{10} = C_{M10} \frac{\pi}{a} \sin \frac{\pi x}{a} I_{M10}(z); \qquad (E_{Mx})_{10} = 0 \qquad (12.9\text{-}21)$$

$$(H_{My})_{10} = 0; \qquad (E_{My})_{10} = -C_{M10} \frac{\pi}{a} \sin \frac{\pi x}{a} V_{M10}(z) \quad (12.9\text{-}22)$$

$$(H_{Mz})_{10} = \frac{-\pi^2}{j\omega a^2 \mu} C_{M10} \cos \frac{\pi x}{a} V_{M10}(z); \quad (E_{Mz})_{10} = 0 \qquad (12.9\text{-}23)$$

where

$$C_{M10} = \frac{2}{\pi} \left(\frac{a}{b} \right)^{1/2} \quad \text{and} \quad \kappa_{M10} = \frac{\pi}{a} \qquad (12.9\text{-}24)$$

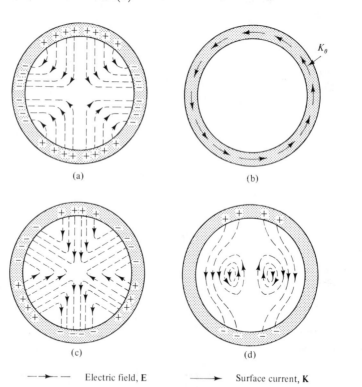

(a)   (b)

(c)   (d)

— ▸ — Electric field, **E**   ——▸ Surface current, **K**

Figure 12.8–9 Higher TE modes in an axially nonresonant cylinder: (a) $TE_{2,1}$ $(b > \lambda_{TEM}/2.06)$ at instant of maximum charge density; (b) $TE_{0,1}$ $(b > \lambda_{TEM}/1.64)$ at instant of maximum surface current density (charge density always zero), **E** in the dielectric is also circular but with maximum a quarter period later; (c) $TE_{3,1}$ $(b > \lambda_{TEM}/1.50)$ at instant of maximum charge density; (d) $TE_{1,2}$ $(b > \lambda_{TEM}/1.18)$ at instant of maximum charge density.

The cutoff wavelength is $\lambda_{e\ \text{cutoff}} = 2a$, so that the pass band requires that $\lambda_e <$ $2a$. The fields are shown schematically in Figs. 12.9–1 and 12.9–2.

12.10 METHODS OF DRIVING AND LOADING COAXIAL AND HOLLOW WAVEGUIDES

The distributions of current on the interior walls of and the associated electric and magnetic fields in coaxial and hollow metal waveguides differ for each mode. To excite a particular mode, the method of driving must be appropriately designed. Similarly, to receive power transmitted along a waveguide, a load must be properly connected. Some of the simpler connecting circuits are considered in this section. The underlying principle is found in the source functions for the E and H modes. The former are excited by electric dipoles, the latter by magnetic dipoles or currents circulating in loops. Current-carrying wires located along the required E lines, current-carrying loops around the required H lines are suitable approximate realizations of the theoretical source functions.

In conventional coaxial (and open-wire) transmission lines, the two terminals of a symmetrical generator always can be connected or coupled directly to the two conductors (or groups of conductors) of the line because these carry equal and opposite currents and charges. Most generators designed for use with coaxial or hollow waveguides have a coaxial line output, although the dimensions of the line may be such that the TM_{01} mode is excited as well as the TEM mode. In either case, the currents on the inner surface of the outer conductor and the outer surface of the inner conductor are axially directed and rotationally symmetrical, so that total currents can be defined easily for both the TEM mode (in which they are equal and opposite) and the TM_{01} mode (in which they are codirectional, with the inner current smaller). The transfer from one coaxial line to another is made easily if both lines use the TEM mode or the TM_{01} mode, or if one uses the TEM and the other the TM_{01} mode. The simplest connection for the TEM mode is shown in Fig. 12.10–1a. The coaxial line from the generator is terminated in its characteristic impedance by the use of the double-stub tuner. The currents are shown by arrows.

It is a simple matter to transfer from a TEM or TM_{01} mode in a coaxial waveguide to a mode such as the cylindrical TM_{01} or the rectangular TM_{11} mode in a hollow waveguide. These are characterized by exclusively axial surface currents that obviously can be excited using the arrangement of Fig. 12.10–1b, where the inner conductor of the coaxial line is extended a short distance into the guide. The arrangement is now similar to that of a monopole antenna erected perpendicular to a conducting plane of infinite extent. In this case, the infinite plane is folded over into an infinitely long guide. A standing-wave distribution of oscillating current and charge is maintained on the projecting inner conductor or antenna, and a rotationally symmetric traveling-wave distribution is maintained on the inner

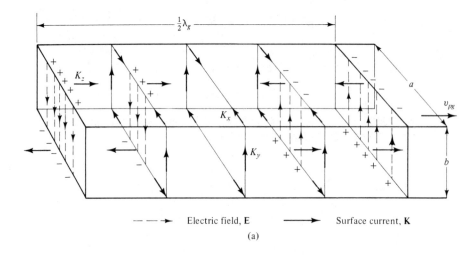

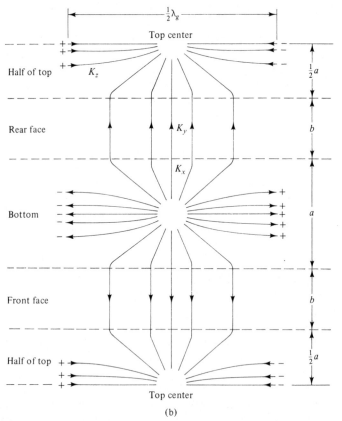

Figure 12.9–1 (a) TE$_{1,0}$ mode in a rectangular pipe with $a > \frac{1}{2}\lambda_{\text{TEM}}$, $b < \frac{1}{2}\lambda_{\text{TEM}}$; (b) surface current of TE$_{1,0}$ mode on the inside of a section of an axially nonresonant rectangular pipe. The pipe of part (a) is represented as cut lengthwise along the center of the top and laid flat. Note that arrows indicate the direction of positive surface current; all charge concentrations move to the right.

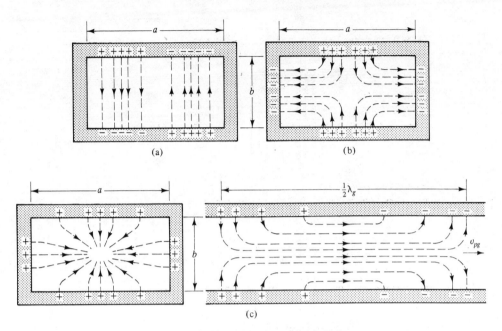

Figure 12.9–2 Higher modes in a rectangular pipe: (a) $TE_{2,0}$ mode, $b > \lambda_{TEM}$; (b) $TE_{1,1}$ mode, $2ab/\sqrt{a^2 + b^2} > \lambda_{TEM}$; (c) $TM_{1,1}$ mode $2ab/\sqrt{a^2 + b^2} > \lambda_{TEM}$.

surface of the waveguide. The electromagnetic field of an antenna, however, decreases rapidly outward along the axis. Hence the field and the current and charge distributions on the waveguide even a short distance beyond the end of the antenna are not affected *directly* by the periodically varying charge in the antenna. The distribution of current and charge is that of the cylindrical TM_{01} mode or rectangular TM_{11} mode.

To excite the TE modes, in particular the dominant TE_{11} mode in a circular guide or the TE_{10} mode in a rectangular guide, a standing-wave pattern must be excited in the transverse plane. This is shown in Fig. 12.10–2. The inner conductor of the resonant end of a coaxial line is arranged as an antenna to extend across the center of and perpendicular to the axis of the pipe. The maximum current may be fixed at the center of the antenna by means of the adjustable stub at the top. A standing-wave distribution is excited in the transverse plane of Fig. 12.10–2a. By suitably adjusting the large movable piston in the guide, the electromagnetic field due to currents up and down on the inner face of the piston and the field due to currents in the antenna itself may be adjusted to be essentially in phase at points to the right of the antenna. The resulting distributions of current and the electromagnetic field at a distance of a wavelength or more from the antenna are characteristic of the TE_{11} mode alone if the size of the guide is appropriately chosen. In the near zone of the antenna, the pattern is a complicated superposition of many modes. The corresponding arrangement for a rectangular waveguide is shown in Fig. 12.10–2c. A double-stub tuner (or a movable single-stub tuner) must be

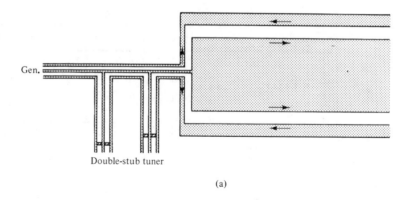

Double-stub tuner

(a)

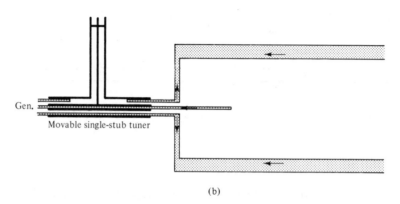

Movable single-stub tuner

(b)

Figure 12.10–1 Methods of driving (a) the TEM mode in a coaxial pipe with b unrestricted and $(b - a) \ll \lambda_{\text{TEM}}/2$; (b) the $TM_{0,1}$ mode in a hollow cylinder with $b > \lambda_{\text{TEM}}/2.61$ using a conventional coaxial line. Arrows show the location of maximum currents at the appropriate instant.

provided at the end of the coaxial line if this is to be terminated to be nonresonant. If the line is short and resonant, a telescoping section of coaxial line may be used instead of the tuner.*

The electromagnetic fields and the associated currents on the walls of waveguides as given in Secs. 12.8 and 12.9 involve the voltage function $V(z)$ and the current function $I(z)$ which are solutions of the transmission-line equations discussed in Chapter 10. They can be expressed in the exponential form (10.7–9a,b) or the forms (10.7–10a,b) and (10.7–11a,b) that make use of hyperbolic functions of complex argument. All of these solutions explicitly or implicitly involve the terminating impedances Z_0 at $z = 0$ and Z_s at $z = s$ as well as the characteristic

---

* A detailed discussion of the excitation of waveguides and the field distribution is given in R. E. Collin, *Field Theory of Guided Waves* (New York: McGraw-Hill Book Company, 1960).

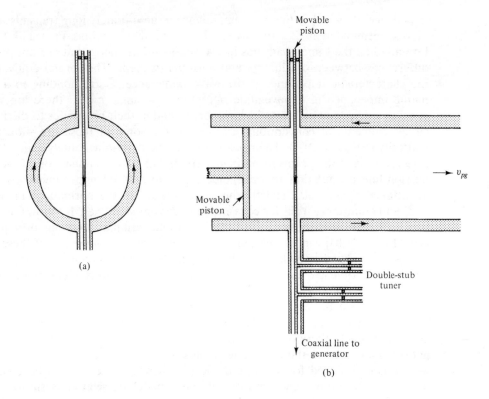

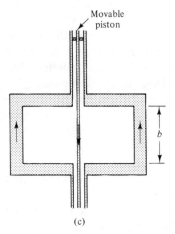

Figure 12.10–2 Method of driving (a) and (b) TE$_{1,1}$ mode in a hollow cylinder; (c) and (b) TE$_{1,0}$ mode in a rectangular pipe. Arrows denote maximum current.

impedance Z_c, which is the input impedance of an infinitely long transmission line or one terminated in Z_c. In waveguide theory the functions $V(z)$ and $I(z)$ are formally like the same quantities in transmission-line theory but are not potential differences between conductors and currents in them. The formal equivalent of the characteristic impedance is the wave impedance. Corresponding to a terminating impedance in a waveguide operated at a single mode, there are various generalized "circuit elements" which correspond in their properties to the familiar circuit elements of conventional circuit theory. Some of these are shown schematically in Fig. 12.10–3. Their properties can be determined analytically, but this is beyond the scope of this introductory treatment. In the application of the transmission-line formulation to waveguides, the normalized impedance Z_a/Z_{Ea} and admittance Y_a/Y_{Ma} play exactly the same role as do the normalized impedance Z/Z_c and admittance Y/Y_c for conventional transmission lines. It follows that complex reflection coefficients Γ [see (10.7–7)] and terminal functions $\theta = \rho + j\Phi$ [see (10.7–12a,b)] can be defined for waveguide terminations. All of these quantities—normalized impedance, reflection coefficient, terminal function—can be determined for each termination and each mode by direct measurements on a waveguide. Examples are shown in Fig. 12.10–4.

In general there are two types of loads for hollow waveguides. The first consists of structures specifically designed for a particular mode in a given waveguide and which, therefore, have no input terminals in the sense used in ordinary networks confined to the near zone. Loads of the second type are conventional.

Loads designed for a single mode in a waveguide can often be connected directly to the output end with or without a matching section as shown in Fig. 12.10–5. Important examples are electromagnetic horns such as shown in Fig. 12.10–6. The lines of flow of sheets of current on the inner surfaces of a horn of rectangular cross section driven in the TE_{10} mode are sketched in Fig. 12.10–7 at an instant when the current is at a maximum and the charge density zero in the standing-wave distribution. The radiated electromagnetic field at distant points is due primarily to these sheets of current.

In addition to electromagnetic horns, the load for a particular mode may be another waveguide of different size or cross-sectional shape. An example is shown in Fig. 12.10–8, where a rectangular waveguide is connected to a circular one that permits the use of a rotatable joint and this in turn is connected to a rectangular guide terminated in a horn.

To drive conventional two-terminal antennas or other loads from a waveguide, it is usually necessary to make use of a closely spaced coaxial line or a two-wire line to connect directly to the load. This is readily done in the manner already described in conjunction with Fig. 12.10–1 for driving a hollow waveguide from a conventional coaxial line. The reciprocal theorem can be invoked to verify that the load and generator can be interchanged and the equality of transmission and reception maintained.

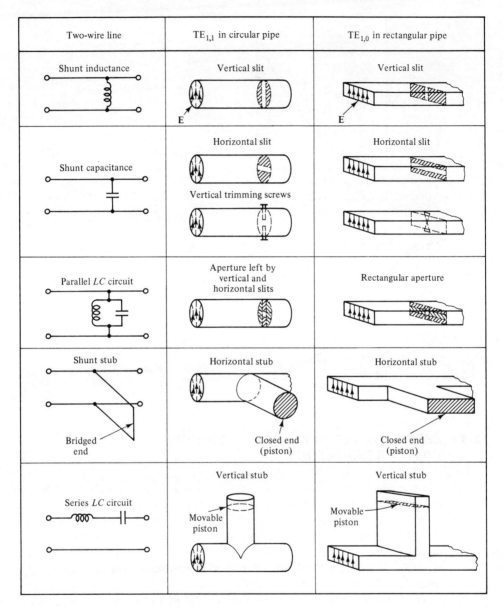

Figure 12.10–3   Roughly comparable circuit elements.

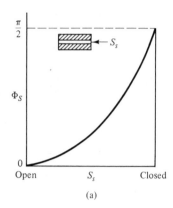

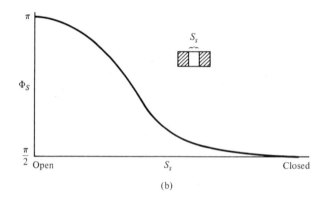

(a)

(b)

Figure 12.10–4 Phase functions for (a) horizontal and (b) vertical slits in a rectangular transmission circuit using the $TE_{1,0}$ mode.

12.11 RESONANT SECTIONS OF WAVEGUIDE; CAVITY RESONATORS

If a waveguide of the coaxial or the hollow type is closed at both ends by a metal piston, the distributions of oscillating current and charge may be tuned to resonance in the manner described in Sec. 10.9. Depending on the method of driving and the size of the guide, these distributions may be of the TM or the TE type. When the metal piston at one end is moved, the currents and charges and consequently the electric and magnetic fields are maximized in amplitude. The resulting axial standing wave has the same transverse distribution characteristic of the given mode as a traveling wave. However, the distributions of maximum charge (and of maximum electric field) are fixed at distances of $\lambda_g/2$ instead of traveling with a phase velocity along the guide. The maxima of current and magnetic field occur a quarter period after the maxima of charge and electric field at fixed positions displaced by $\lambda_g/4$ from these.

The electromagnetic field for such resonant circuits is obtained from the

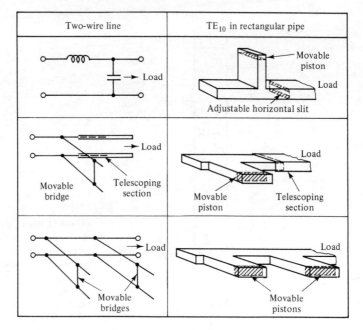

Figure 12.10–5 Roughly comparable matching sections.

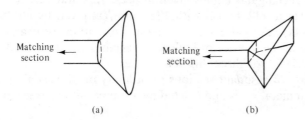

(a)   (b)

Figure 12.10–6 (a) Horn for $TE_{1,1}$ mode in a circular pipe; (b) horn for $TE_{1,0}$ mode in a rectangular pipe.

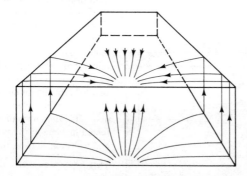

Figure 12.10–7 Lines of flow of surface current in an electromagnetic horn.

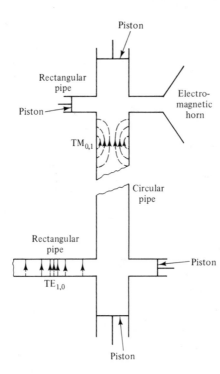

Piston

Rectangular pipe

Piston

Electro-magnetic horn

$TM_{0,1}$

Circular pipe

Rectangular pipe

Piston

$TE_{1,0}$

Piston

Figure 12.10–8 Cross section showing the transfer from the $TE_{1,0}$ mode in a rectangular pipe to the rotationally symmetrical $TM_{0,1}$ mode in a circular cylinder and back to a rectangular pipe. Pistons for matching are also shown. Dashed lines show the approximate distribution of electric field for a short distance in each pipe at a particular instant.

solutions for circular and rectangular guides given in Secs. 12.8 and 12.9 [(12.8–36) to (12.8–41); (12.9–11) to (12.9–16)] with $V(z)$ and $I(z)$ given by (10.9–1) and (10.9–2) with (10.9–4a). Typical axially resonant distributions of charge on the walls and electric field in the interior of hollow waveguides are shown in Figs. 12.11–1, 12.11–2, and 12.11–3 for three different modes.

The essential property of resonant sections of ordinary or of coaxial guides is that they are enclosed in metal walls and excited by a source within. It has been

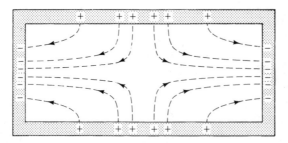

Figure 12.11–1 Charge and electric field distributions at the instant when these are maximum in the standing-wave $TM_{0,1,1}$ mode in a closed circular cylinder of inner radius b and length s. Rotational symmetry about the axis obtains. A quarter period later, charge and electric field are everywhere zero; sheets of maximum current are from $+$ to $-$ along the walls; a circular magnetic field obtains with oppositely directed maxima at the two circular end walls and zero at the center

$$TM\lambda_{0,1,1} = \frac{2}{\sqrt{1/s^2 + 0.58/b^2}}.$$

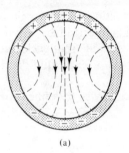

(a)

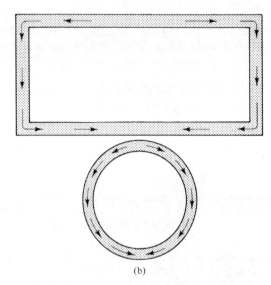

(b)

Figure 12.11–2   (a) Distribution of charge and electric field at the instant of maximum in the standing-wave $TE_{1.1.1}$ mode in a closed circular cylinder of inner radius b and length s. The distribution is the same for each cross section, but the density of the charge and the intensity of the electric field are largest at the center and diminish cosinusoidally to zero at the end walls; (b) distribution of surface current a quarter period after (a). There is a standing-wave pattern in the transverse and in the axial planes. Current is from top center to bottom center around side and end walls. The distributions are not rotationally symmetrical. $TE\lambda_{1.1.1} = \dfrac{2}{\sqrt{1/s^2 + 0.34/b^2}}$.

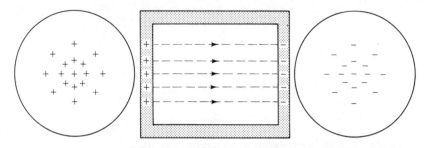

Figure 12.11–3   Distribution of charge and electric field at the instant of maximum in the standing-wave $TM_{0.1.0}$ mode in a closed circular cylinder of inner radius b. A quarter period later radial currents are on the inner surface of each end face and uniform currents parallel to the axis are on the cylindrical surface. The magnetic field is circular and the same in magnitude and direction in every cross section. $TM\lambda_{0.1.0} = 2.61b$.

shown that every closed, highly conducting shell can be excited in an infinite sequence of resonant modes, each characterized by a particular standing-wave distribution of surface current. The resonant frequencies are determined by the size and shape of the cavity. The method of excitation determines the mode (TE or TM type) that exists in the cavity.

The quality factor Q of a resonant cavity is defined for each resonant mode by

$$Q = \frac{f_r}{\Delta f} \qquad (12.11-1)$$

where f_r is the resonant frequency and Δf is the bandwidth between the frequencies $f_r \pm \Delta f/2$ for which the amplitude of the surface current, the surface charge and the electric and magnetic fields is reduced to 0.707 of its value at the resonant frequency f_r. The quality factor, in general, may be written as

$$Q = 2\pi \frac{\text{maximum energy stored}}{\text{energy dissipated per cycle}} \qquad (12.11-2)$$

PROBLEMS

1. A hollow cylindrical waveguide is to be used to transmit power at 2 GHz from a generator with coaxial output to an antenna.
   (a) Calculate the inside diameter of the cylindrical waveguide in order to pass the TE_{11} mode but not the TM_{01} mode.
   (b) Determine the ratio of the radius of the inner conductor to the outer conductor of a coaxial waveguide which is to pass only the TE_{11} mode.

2. A 4-GHz oscillator is to be used to drive a radiating horn antenna using a short coaxial line and a long hollow waveguide. What should be the cross-sectional dimensions of the rectangular waveguide?

3. A hollow rectangular waveguide is to be used for transmission at 3 GHz in the TE_{10} mode.
   (a) What are the cross-sectional dimensions a and b for the waveguide?
   (b) Show a sketch of a cross section where at a given instant the electric field distribution and the charge distribution are maximum. What are the currents and where do they exist?
   (c) Sketch the charge, current, and field distributions a quarter-period later.
   (d) Under what conditions will a rectangular waveguide be nonresonant in the axial direction yet resonant in each transverse plane?

4. A hollow cylindrical waveguide with circular cross section is to be used to transmit a pure 1-GHz signal from a generator to an antenna. The waveguide diameter is 22 cm.
   (a) What mode or modes can be excited at this frequency?
   (b) A standing-wave detector measures the distance between two adjacent resonant peaks. How far apart are they?

5. Compute the following quantities for a hollow rectangular waveguide with sides $a = 6.35$ cm, $b = 3.15$ cm when filled with air. The wavelength of the generator is 9.725 cm. The TE_{10} mode is excited.
   (a) The cutoff frequency and cutoff wavelength.
   (b) The wavelength λ_g in the guide.
   (c) The phase velocity.
   (d) The wave impedance.

6. Repeat Problem 5 when the waveguide is filled with polystyrene.

7. Determine the field in a cylindrical cavity excited in the
   (a) TM_{011} mode
   (b) TE_{111} mode
   when the cavity has an inner radius 0.5 cm and length 1 cm. Sketch the fields and the currents.

8. Repeat Problem 7 for a rectangular cavity excited in the TE_{101} mode with cross-sectional dimensions 0.5 and 0.25 and length 1 cm. $a > b$.

13

Waves Along Dielectric Rods; Optical Fiber Transmission

The invention of the laser made generators of coherent and polarized electromagnetic waves at optical frequencies ($f \sim 3 \times 10^{14}$ Hz, $\lambda \sim 1$ µm) possible. For the transmission of information at such high frequencies conventional coaxial lines and metal-pipe waveguides are impractical. A more appropriate waveguide is the dielectric rod in the form of a hair-thin optical fiber drawn from very pure glass. The properties of such a rod differ significantly from those of the metallic waveguide in that for single-mode operation the electromagnetic field extends outside the dielectric core instead of being contained in it as with metallically bounded guides. As a consequence, the field is more complicated.

In this introductory study no attempt is made to provide a comprehensive exposition of optical-fiber technology. Rather, attention is directed to the principal characteristics of the dielectric rod as a waveguide in general and then to the special conditions appropriate to optical frequencies.

13.1 THE FIELD OF THE DIELECTRIC-ROD WAVEGUIDE

The structure of interest is shown in Fig. 13.1–1. It consists of a uniform circular dielectric rod of radius a (region 1) characterized by the wave number $k_1 = \beta_1 + i\alpha_1$ embedded in an infinite medium (region 2) with the wave number $k_2 = \beta_2 + i\alpha_2$. It is assumed that both media are nearly perfect dielectrics, so that $\beta_1 \gg \alpha_1$, $k_1 \sim \beta_1$; $\beta_2 \gg \alpha_2$, $k_2 \sim \beta_2$. The analytical formulation parallels that for the circular

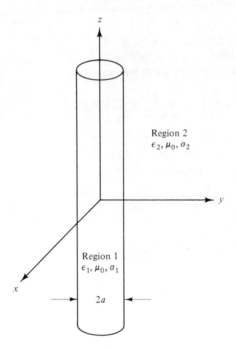

Figure 13.1–1   Dielectric-rod waveguide.

metal waveguide up to the imposition of quite different boundary conditions. Specifically, with the z axis along the axis of the dielectric rod, the electromagnetic field in general includes components of both electric and magnetic types as given by (12.2–4) to (12.2–9). They are defined in terms of the modified Hertz potentials,

$$I_E \equiv j\omega\epsilon\Pi_{Ez}; \qquad V_M \equiv -j\omega\Pi_{Mz} \tag{13.1–1}$$

and as components of the vector potential

$$\mathbf{A} = \hat{\mathbf{z}}j\omega\mu\epsilon\Pi_{Ez} + \nabla \times \hat{\mathbf{z}}\Pi_{Mz} \tag{13.1–2}$$

where, in cylindrical coordinates, $\nabla = \hat{\boldsymbol{\rho}}(\partial/\partial\rho) + \hat{\boldsymbol{\theta}}(\partial/\rho\partial\theta) + \hat{\mathbf{z}}(\partial/\partial z)$, so that $\nabla \times \hat{\mathbf{z}} = -\hat{\boldsymbol{\theta}}(\partial/\partial\rho) + \hat{\boldsymbol{\rho}}(\partial/\rho\partial\theta)$. The general solutions for $I_E(\rho, \theta, z)$ and $V_M(\rho, \theta, z)$ are sums of modes as given by

$$I_E(\rho, \theta, z) = \sum_a \Psi_{Ea}(\rho, \theta)I_{Ea}(z)$$
$$V_M(\rho, \theta, z) = \sum_a \Psi_{Ma}(\rho, \theta)V_{Ma}(z) \tag{13.1–3}$$

where the single subscript a stands for two identifying subscripts. The general solutions for $I_{Ea}(z)$ and $V_{Ma}(z)$ and the associated functions $V_{Ea}(z)$ and $I_{Ma}(z)$ for waves traveling in the positive z direction are

$$I_{Ea}(z) = I_{Ea} \exp(-\gamma_{Ea}z); \qquad V_{Ma}(z) = V_{Ma} \exp(-\gamma_{Ma}z) \tag{13.1–4a}$$

$$V_{Ea}(z) = Z_{Ea}I_{Ea}(z); \qquad I_{Ma}(z) = Y_{Ma}V_{Ma}(z) \tag{13.1–4b}$$

For simplicity, $I_{Ea}(0) \equiv I_{Ea}$, $V_{Ma}(0) \equiv V_{Ma}$. In (13.1–4a,b) the propagation constants are $\gamma_{Ea} = (\kappa_{Ea}^2 - k^2)^{1/2}$, $\gamma_{Ma} = (\kappa_{Ma}^2 - k^2)^{1/2}$; the electric wave impedance is $Z_{Ea} = \gamma_{Ea}/j\omega\epsilon$; the magnetic wave admittance is $Y_{Ma} = \gamma_{Ma}/j\omega\mu$.

The mode functions $\Psi_a(\rho, \theta)$ satisfy the equation

$$\left[\frac{1}{\rho}\frac{\partial}{\partial\rho}\left(\rho\frac{\partial}{\partial\rho}\right) + \frac{1}{\rho^2}\frac{\partial^2}{\partial\theta^2} + \kappa_a^2\right]\Psi_a(\rho, \theta) = 0 \tag{13.1–5}$$

subject to appropriate boundary conditions at $\rho = a$.

The electromagnetic field of each mode is given by (12.3–18) to (12.3–21). In cylindrical coordinates the components are

$$E_{E\rho a} = -V_{Ea}(z)\frac{\partial\Psi_{Ea}(\rho, \theta)}{\partial\rho}; \qquad H_{M\rho a} = -I_{Ma}(z)\frac{\partial\Psi_{Ma}(\rho, \theta)}{\partial\rho} \tag{13.1–6}$$

$$E_{E\theta a} = -V_{Ea}(z)\frac{1}{\rho}\frac{\partial\Psi_{Ea}(\rho, \theta)}{\partial\theta}; \qquad H_{M\theta a} = -I_{Ma}(z)\frac{1}{\rho}\frac{\partial\Psi_{Ma}(\rho, \theta)}{\partial\theta} \tag{13.1–7}$$

$$E_{Eza} = \frac{\kappa_{Ea}^2}{j\omega\epsilon}I_{Ea}(z)\Psi_{Ea}(\rho, \theta); \qquad H_{Mza} = \frac{\kappa_{Ma}^2}{j\omega\mu}V_{Ma}(z)\Psi_{Ma}(\rho, \theta) \tag{13.1–8}$$

$$H_{E\rho a} = I_{Ea}(z)\frac{1}{\rho}\frac{\partial\Psi_{Ea}(\rho, \theta)}{\partial\theta}; \qquad E_{M\rho a} = -V_{Ma}(z)\frac{1}{\rho}\frac{\partial\Psi_{Ma}(\rho, \theta)}{\partial\theta} \tag{13.1–9}$$

$$H_{E\theta a} = -I_{Ea}(z)\frac{\partial\Psi_{Ea}(\rho, \theta)}{\partial\rho}; \qquad E_{M\theta a} = V_{Ma}(z)\frac{\partial\Psi_{Ma}(\rho, \theta)}{\partial\rho} \tag{13.1–10}$$

$$H_{Eza} = 0; \qquad E_{Mza} = 0 \tag{13.1–11}$$

Solutions of (13.1–5) appropriate to the finite cylindrical region 1 ($\rho \leq a$) and the infinite region 2 ($\rho \geq a$) are

$$\rho \leq a: \quad \Psi_{Ea}(\rho, \theta) = J_m(\kappa_{Ea}\rho)\cos(m\theta + D_{Ea})$$

$$\Psi_{Ma}(\rho, \theta) = J_m(\kappa_{Ma}\rho)\cos(m\theta + D_{Ma}) \tag{13.1–12}$$

$$\rho \geq a: \quad \Psi_{Ea}(\rho, \theta) = H_m(\kappa_{Ea}\rho)\cos(m\theta + D_{Ea})$$

$$\Psi_{Ma}(\rho, \theta) = H_m(\kappa_{Ma}\rho)\cos(m\theta + D_{Ma}) \tag{13.1–13}$$

Here the subscript a stands for mn. Note that the superscript (1) or (2) has been omitted from the Hankel functions. It remains to be selected. The choice of the origins for $\theta = \theta_E$ and $\theta = \theta_M$ depends on the orientation and polarization of the source. In the analysis of metal waveguides, the E and H modes are independent and can be excited separately, so that it was convenient to set $D_{Ea} = D_{Ma} = 0$. With the dielectric rod waveguide the E and H modes are independent only for the rotationally symmetric modes. With $m > 0$ they together form the hybrid HE_{mn} and EH_{mn} modes. For these θ_E and θ_M are not independent and in order to satisfy the boundary conditions it is necessary to impose an appropriate relation between D_{Ea} and D_{Ma}.

When the values of (13.1–12) and (13.1–13) are substituted in (13.1–6) to (13.1–11) with (13.1–4a,b), the results for region 1 ($\rho \leq a$) are

$$E_{E\rho 1} = -V_{Ea1}(z)\, \kappa_{Ea1} J'_m(\kappa_{Ea1}\rho) \cos(m\theta + D_E)$$

$$H_{M\rho 1} = -I_{Ma1}(z)\kappa_{Ma1} J'_m(\kappa_{Ma1}\rho) \cos(m\theta + D_M) \qquad (13.1\text{–}14)$$

$$E_{E\theta 1} = V_{Ea1}(z)\frac{m}{\rho} J_m(\kappa_{Ea1}\rho) \sin(m\theta + D_E)$$

$$H_{M\theta 1} = I_{Ma1}(z)\frac{m}{\rho} J_m(\kappa_{Ma1}\rho) \sin(m\theta + D_M) \qquad (13.1\text{–}15)$$

$$E_{Ez1} = \frac{\kappa_{Ea1}^2}{j\omega\epsilon_1} I_{Ea1}(z) J_m(\kappa_{Ea1}\rho) \cos(m\theta + D_E)$$

$$H_{Mz1} = \frac{\kappa_{Ma1}^2}{j\omega\mu_1} V_{Ma1}(z) J_m(\kappa_{Ma1}\rho) \cos(m\theta + D_M) \qquad (13.1\text{–}16)$$

$$H_{E\rho 1} = -I_{Ea1}(z)\frac{m}{\rho} J_m(\kappa_{Ea1}\rho) \sin(m\theta + D_E)$$

$$E_{M\rho 1} = V_{Ma1}(z)\frac{m}{\rho} J_m(\kappa_{Ma1}\rho) \sin(m\theta + D_M) \qquad (13.1\text{–}17)$$

$$H_{E\theta 1} = -I_{Ea1}(z)\kappa_{Ea1} J'_m(\kappa_{Ea1}\rho) \cos(m\theta + D_E) \qquad (13.1\text{–}18)$$

$$E_{M\theta 1} = V_{Ma1}(z)\kappa_{Ma1} J'_m(\kappa_{Ma1}\rho) \cos(m\theta + D_M)$$

$$H_{Ez1} = 0; \qquad E_{Mz1} = 0 \qquad (13.1\text{–}19)$$

(The subscript a has been omitted from E, H and D for simplicity.) Note that $V_{Ea}(z) = (\gamma_{Ea}/j\omega\epsilon)I_{Ea}(z)$ and $I_{Ma}(z) = (\gamma_{Ma}/j\omega\mu)V_{Ma}(z)$.

The formulas for region 2 ($\rho \geq a$) are like the above, with H written for J and the subscript 1 replaced by 2.

13.2 APPLICATION OF THE BOUNDARY CONDITIONS: CHARACTERISTIC EQUATION

The boundary conditions at $\rho = a$ between regions 1 and 2 require the tangential components of both **E** and **H** to be continuous. In addition, it is assumed that both regions are nonmagnetic so that $\mu_1 = \mu_2 = \mu_0$. In the dielectric rod waveguide the conducting walls of the hollow metal pipe waveguide with their distributions of surface current and charge are absent and as a consequence the E and H modes can exist independently only for the rotationally symmetric modes with $m = 0$ when E_{Ez}, $E_{E\rho}$, and $H_{E\theta}$ are independent of H_{Mz}, $H_{M\rho}$, and $E_{M\theta}$. For all other modes the boundary conditions cannot be satisfied independently for the E modes and H modes since both must be present. Thus the general boundary conditions are

$$E_{Ez1}(a, \theta, z) = E_{Ez2}(a, \theta, z)$$

$$E_{E\theta1}(a, \theta, z) + E_{M\theta1}(a, \theta, z) = E_{E\theta2}(a, \theta, z) + E_{M\theta2}(a, \theta, z) \tag{13.2–1}$$

$$H_{Mz1}(a, \theta, z) = H_{Mz2}(a, \theta, z)$$

$$H_{M\theta1}(a, \theta, z) + H_{E\theta1}(a, \theta, z) = H_{M\theta2}(a, \theta, z) + H_{E\theta2}(a, \theta, z) \tag{13.2–2}$$

All the boundary equations contain a factor $\exp(-\gamma_{Ea}z)$ or $\exp(-\gamma_{Ma}z)$. For a single wave to propagate along the rod, the equations cannot be satisfied unless all the γ_a's are equal for each propagating mode. For the hybrid modes that involve both E and H modes, this means that

$$\gamma_{Ea} = \gamma_{Ma} = \gamma_a \tag{13.2–3a}$$

Since $\gamma_a^2 = \kappa_a^2 - k^2$, it follows that

$$\kappa_{Ea} = \kappa_{Ma} \equiv \kappa_a = \sqrt{\gamma_a^2 + k^2} \tag{13.2–3b}$$

for each region with appropriate subscript 1 or 2 on κ_a and k. With $k^2 = \omega^2\mu\epsilon$ and with the choice of

$$D_E = D_{Ea} = -\frac{\pi}{2}; \qquad D_M = D_{Ma} = 0 \tag{13.2–4}$$

the boundary conditions give

$$(E_{z1} = E_{z2}); \qquad I_{E1}\left(\frac{\kappa_1}{k_1}\right)^2 J_m(\kappa_1 a) = I_{E2}\left(\frac{\kappa_2}{k_2}\right)^2 H_m(\kappa_2 a) \tag{13.2–5}$$

$$(H_{z1} = H_{z2}); \qquad V_{M1}\kappa_1^2 J_m(\kappa_1 a) = V_{M2}\kappa_2^2 H_m(\kappa_2 a) \tag{13.2–6}$$

$$(E_{\theta1} = E_{\theta2}); \qquad -I_{E1}\frac{m}{a}\frac{\gamma}{j\omega\epsilon_1}J_m(\kappa_1 a) + V_{M1}\kappa_1 J'_m(\kappa_1 a)$$

$$= -I_{E2}\frac{m}{a}\frac{\gamma}{j\omega\epsilon_2}H_m(\kappa_2 a) + V_{M2}\kappa_2 H'_m(\kappa_2 a) \tag{13.2–7}$$

$$(H_{\theta1} = H_{\theta2}); \qquad -I_{E1}\kappa_1 J'_m(\kappa_1 a) + V_{M1}\frac{m}{a}\frac{\gamma}{j\omega\mu}J_m(\kappa_1 a)$$

$$= -I_{E2}\kappa_2 H'_m(\kappa_2 a) + V_{M2}\frac{m}{a}\frac{\gamma}{j\omega\mu}H_m(\kappa_2 a) \tag{13.2–8}$$

Note that the subscript a has been omitted for greater simplicity in writing. From (13.2–5) and (13.2–6) and with $k_2^2/k_1^2 = \epsilon_2/\epsilon_1$.

$$\frac{I_{E2}}{I_{E1}} = \frac{\epsilon_2}{\epsilon_1} \left(\frac{\kappa_1}{\kappa_2}\right)^2 \frac{J_m(\kappa_1 a)}{H_m(\kappa_2 a)} \tag{13.2-9}$$

$$\frac{V_{M2}}{V_{M1}} = \left(\frac{\kappa_1}{\kappa_2}\right)^2 \frac{J_m(\kappa_1 a)}{H_m(\kappa_2 a)} \tag{13.2-10}$$

These relations can be used to eliminate I_{E2} and V_{M2} from (13.2–7) and (13.2–8), which then become

$$I_{E1}\left[\frac{m}{a}\frac{\gamma}{j\omega\epsilon_1}J_m(\kappa_1 a) - \left(\frac{\kappa_1}{\kappa_2}\right)^2 \frac{\epsilon_2}{\epsilon_1}\frac{J_m(\kappa_1 a)}{H_m(\kappa_2 a)}\frac{m}{a}\frac{\gamma}{j\omega\epsilon_2}H_m(\kappa_2 a)\right]$$

$$= V_{M1}\left[\kappa_1 J_m'(\kappa_1 a) - \left(\frac{\kappa_1}{\kappa_2}\right)^2 \frac{J_m(\kappa_1 a)}{H_m(\kappa_2 a)}\kappa_2 H_m'(\kappa_2 a)\right] \tag{13.2-11}$$

$$I_{E1}\left[\kappa_1 J_m'(\kappa_1 a) - \left(\frac{\kappa_1}{\kappa_2}\right)^2 \frac{\epsilon_2}{\epsilon_1}\frac{J_m(\kappa_1 a)}{H_m(\kappa_2 a)}\kappa_2 H_m'(\kappa_2 a)\right]$$

$$= V_{M1}\left[\frac{m}{a}\frac{\gamma}{j\omega\mu}J_m(\kappa_1 a) - \left(\frac{\kappa_1}{\kappa_2}\right)^2 \frac{J_m(\kappa_1 a)}{H_m(\kappa_2 a)}\frac{m}{a}\frac{\gamma}{j\omega\mu}H_m(\kappa_2 a)\right] \tag{13.2-12}$$

It follows from (13.2–11) that

$$\frac{I_{E1}}{V_{M1}} = -\frac{j\omega\epsilon_1 a}{m\gamma}\frac{\kappa_1^2\kappa_2^2}{\kappa_1^2 - \kappa_2^2}\left[\frac{1}{\kappa_1}\frac{J_m'(\kappa_1 a)}{J_m(\kappa_1 a)} - \frac{1}{\kappa_2}\frac{H_m'(\kappa_2 a)}{H_m(\kappa_2 a)}\right] \tag{13.2-13a}$$

Similarly from (13.2–12),

$$\frac{I_{E1}}{V_{M1}} = -\frac{m\gamma\epsilon_1}{j\omega\mu a}\frac{\kappa_1^2 - \kappa_2^2}{\kappa_1^2\kappa_2^2}\left[\frac{\epsilon_1}{\kappa_1}\frac{J_m'(\kappa_1 a)}{J_m(\kappa_1 a)} - \frac{\epsilon_2}{\kappa_2}\frac{H_m'(\kappa_2 a)}{H_m(\kappa_2 a)}\right]^{-1} \tag{13.2-13b}$$

If these two formulas for I_{E1}/V_{M1} are equated, the following general characteristic equation is obtained for electromagnetic waves along an infinitely long dielectric cylinder. Note that $k_2^2 = \omega^2\mu_2\epsilon_2$, $k_1^2 = \omega^2\mu_1\epsilon_1$.

$$\left[\frac{1}{\kappa_1}\frac{J_m'(\kappa_1 a)}{J_m(\kappa_1 a)} - \frac{1}{\kappa_2}\frac{H_m'(\kappa_2 a)}{H_m(\kappa_2 a)}\right]\left[\frac{\epsilon_1/\epsilon_2}{\kappa_1}\frac{J_m'(\kappa_1 a)}{J_m(\kappa_1 a)} - \frac{1}{\kappa_2}\frac{H_m'(\kappa_2 a)}{H_m(\kappa_2 a)}\right]$$

$$= -\frac{m^2\gamma^2}{a^2 k_2^2}\left(\frac{\kappa_1^2 - \kappa_2^2}{\kappa_1^2\kappa_2^2}\right)^2 \tag{13.2-14}$$

Since

$$\gamma^2 = \kappa_1^2 - k_1^2 = \kappa_2^2 - k_2^2 \quad \text{and} \quad \kappa_1^2 - \kappa_2^2 = k_1^2 - k_2^2 \tag{13.2-15}$$

the right side of (13.2–14) can be rearranged as follows:

$$\frac{-m^2(\kappa_2^2 - k_2^2)}{a^2 k_2^2} \frac{(k_1^2 - k_2^2)(\kappa_1^2 - \kappa_2^2)}{\kappa_1^4 \kappa_2^4}$$

$$= \frac{m^2}{a^2} \frac{[\kappa_1^2 - (\epsilon_1/\epsilon_2)\kappa_2^2](\kappa_1^2 - \kappa_2^2)}{\kappa_1^4 \kappa_2^4} \tag{13.2–16}$$

With the shorthand notation, $x_1 = \kappa_1 a$, $x_2 = \kappa_2 a$ (13.2–14) becomes

$$\left[\frac{1}{x_1}\frac{J_m'(x_1)}{J_m(x_1)} - \frac{1}{x_2}\frac{H_m'(x_2)}{H_m(x_2)}\right]\left[\frac{\epsilon_1/\epsilon_2}{x_1}\frac{J_m'(x_1)}{J_m(x_1)} - \frac{1}{x_2}\frac{H_m'(x_2)}{H_m(x_2)}\right]$$

$$= \frac{m^2(x_1^2 - x_2^2)(x_1^2 - x_2^2\epsilon_1/\epsilon_2)}{x_1^4 x_2^4} \tag{13.2–17}$$

With (13.2–15) it follows that

$$x_1^2 - x_2^2 = a^2(k_1^2 - k_2^2) \tag{13.2–18}$$

where k_1^2 and k_2^2 are known. Finally, (13.2–17) and (13.2–18) must be solved simultaneously for x_1 and x_2 to determine κ_1 and κ_2 for each value of m and from them the propagation constant γ. An alternative way of writing the right side of (13.2–18) is

$$x_1^2 - x_2^2 = \left(\frac{\pi d}{\lambda_2}\right)^2\left(\frac{\epsilon_1}{\epsilon_2} - 1\right) \tag{13.2–19}$$

where $d = 2a$ is the diameter of the dielectric rod and λ_2 is the wavelength for plane waves in the outer region 2.

To have unattenuated propagation along the dielectric rod, it is necessary that $\gamma = j\beta$, where β is real. For no radiation radially into region 2, $\rho > a$, the following must be true:

$$H_m(\kappa_2\rho) \equiv H_m^{(1)}(\kappa_2\rho) \quad \text{with } \kappa_2 = \kappa_{2R} + j\kappa_{2I}, \ \kappa_{2I} \gg \kappa_{2R} \tag{13.2–20}$$

In this case, for large arguments,

$$H_m^{(1)}(\kappa_2\rho) \to \frac{2}{\sqrt{\pi\kappa_2\rho}} \exp\left[j\left(\kappa_2\rho - \frac{2m+1}{4}\pi\right)\right]$$

$$= \frac{2}{\sqrt{\pi\kappa_2\rho}} \exp\left(-\kappa_{2I}\rho\right) \exp\left[j\left(\kappa_{2R}\rho - \frac{2m+1}{4}\pi\right)\right] \tag{13.2–21}$$

This function decays rapidly in the radial direction. Note that with $\kappa_2 \sim j\kappa_{2I}$, $x_2^2 = \kappa_2^2 a^2 \sim -\kappa_{2I}^2 a^2$ in (13.2–18).

Since

$$\gamma^2 = -\beta^2 = \kappa_1^2 - k_1^2 = \kappa_2^2 - k_2^2 \sim -\kappa_{2I}^2 - k_2^2 \tag{13.2–22}$$

it follows that for each mode (subscript a omitted)

$$\kappa_1^2 = k_1^2 - \beta^2; \qquad \kappa_{2I}^2 = \beta^2 - k_2^2 \qquad (13.2\text{--}23)$$

Since $k_1 = \beta_1 - j\alpha_1 \sim \beta_1$ is essentially real with the guiding rod a nearly perfect dielectric, and since the phase velocity $v = \omega/\beta$ of the guided wave is greater than the velocity $v_1 = \omega/\beta_1$ in an unbounded region 1, it follows that κ_1^2 is positive and therefore κ_1 is real. Thus

$$\kappa_1 = \sqrt{\beta_1^2 - \beta^2} \qquad (13.2\text{--}24)$$

On the other hand, since region 2 is also a nearly perfect dielectric with $k_2 = \beta_2 - j\alpha_2 \sim \beta_2$, it follows that

$$\kappa_{2I} = \sqrt{\beta^2 - \beta_2^2} \qquad (13.2\text{--}25)$$

The velocity $v = \omega/\beta$ of the guided wave is smaller than the velocity $v_2 = \omega/\beta_2$ characteristic of an infinite region 2, larger than $v_1 = \omega/\beta_1$ characteristic of region 1. Thus

$$v_1 < v < v_2; \qquad \beta_1 > \beta > \beta_2 \qquad (13.2\text{--}26)$$

From the relations

$$\beta^2 = \beta_2^2 - \kappa_2^2 = \beta_1^2 - \kappa_1^2 \qquad (13.2\text{--}27)$$

it follows that

$$\left(\frac{\beta}{\beta_2}\right)^2 = \left(\frac{\lambda_2}{\lambda}\right)^2 = \left[1 - x_2^2\left(\frac{\lambda_2}{\pi d}\right)^2\right] = \left[\frac{\epsilon_1}{\epsilon_2} - x_1^2\left(\frac{\lambda_2}{\pi d}\right)^2\right] \qquad (13.2\text{--}28)$$

where β is the wave number and λ the wavelength of the guided wave along the dielectric rod.

13.3 THE CIRCULAR E_{01} AND H_{01} MODES

When $m = 0$ the characteristic equation (13.2–17) reduces to two independent equations, respectively, for the circular magnetic (H_{0n}) and circular electric (E_{0n}) modes. They are

$$(H_{0n}); \qquad x_1 \frac{J_0(x_1)}{J_0'(x_1)} = x_2 \frac{H_0(x_2)}{H_0'(x_2)} \qquad (13.3\text{--}1)$$

$$(E_{0n}); \qquad x_1 \frac{J_0(x_1)}{J_0'(x_1)} = \frac{\epsilon_1 x_2}{\epsilon_2} \frac{H_0(x_2)}{H_0'(x_2)} \qquad (13.3\text{--}2)$$

The relation (13.2–18) continues to apply. From the solution of these equations with (13.2–18) and their use in (13.2–25), the wave number β and wavelength λ of the guided wave have been evaluated for H_{01} and E_{01} modes. Graphs of $\lambda/\lambda_2 = \beta_2/\beta$ as a function of d/λ_2 with ϵ_1/ϵ_2 as the parameter are shown in Fig. 13.3–

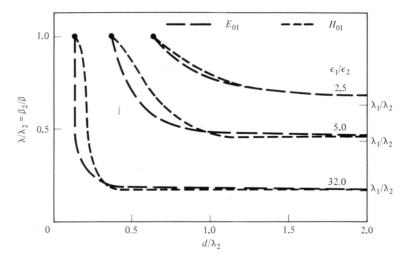

Figure 13.3–1 Relative wavelengths λ/λ_2 of the guided H_{01} and E_{01} waves as a function of d/λ_2. (Data of D. G. Kiely.)

1. These are quite similar for the two circular modes. Both are characterized by a cutoff as d/λ_2 is reduced. This occurs when λ increases to λ_2 (or β decreases to β_2) and propagation along the dielectric rod without radiation becomes impossible. On the other hand, as d/λ_2 is increased, λ decreases asymptotically toward $\lambda_1 = \lambda_2/\sqrt{\epsilon_1/\epsilon_2}$ (or β increases toward $\beta_1 = \beta_2\sqrt{\epsilon_1/\epsilon_2}$), where λ_1 is the wavelength and β_1 the propagation constant of an unbounded region 1.

The electromagnetic fields for the E_{01} and H_{01} modes are given directly by (13.1–14) to (13.1–19) with $m = 0$ and the use of (13.2–9) and (13.2–10). Note that the E_{01} mode has only the components E_z, E_ρ, and H_θ; the H_{01} mode only the components H_z, H_ρ, and E_θ. When d/λ_2 is small, $\lambda/\lambda_2 = \beta_2/\beta$ is close to unity and most of the field propagates in region 2 outside the dielectric rod (region 1). As d/λ_2 is increased and λ/λ_2 decreases toward λ_1/λ_2, the field is transferred more and more into the rod with less and less in the outside region 2. If the small dielectric losses in very good but imperfect dielectrics are included, the total energy loss due to time lag in polarization response is significantly greater when the field is confined primarily in the dielectric rod than when it propagates along the rod but primarily in the much larger volume of the outside region 2. Note that the field decreases exponentially in the radial direction in region 2 so that it is confined to a relatively small transverse area in both cases, but the cross-sectional area of the rod is much smaller than the cross-sectional area of the cylindrical region outside the rod in which the field travels. This means that for efficient transmission the dielectric rod should have as small a radius as possible without actually reaching cutoff. Since the E_{01} and H_{01} modes are difficult to excite and maintain and because they do have a cutoff, the higher HE_{11} mode which is more easily excited and maintained and which has no cutoff is usually preferred.

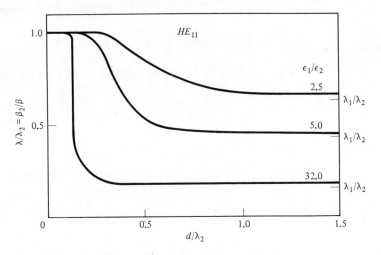

Figure 13.4–1   Relative wavelengths λ/λ_2 of the guided HE_{11} wave as a function of d/λ_2. (Data of D. G. Kiely.)

13.4 THE HYBRID HE_{11} MODE

The solution of the general equations (13.2–17) and (13.2–18) for the phase constant β of the lowest propagating mode with $m = 1$ leads to the graphs in Fig. 13.4–1 for the HE_{11} mode. These show λ/λ_2 or β_2/β as a function of the normalized diameter $d = 2a$ of the central dielectric rod (region 1) with ϵ_1/ϵ_2 as the parameter. The curves resemble those of the E_{01} and H_{01} modes shown in Fig. 13.3–1 with one significant difference: there is no cutoff as d/λ_2 is reduced. When d/λ_2 reaches a value that produces cutoff for the E_{01} and H_{01} modes, nonradiating propagation continues in the HE_{11} mode with $\lambda/\lambda_2 = \beta_2/\beta \sim 1$.

The electromagnetic field of the HE_{11} mode is quite complicated since it involves all six components. They are obtained by combining the E and H modes with $m = 1$ from (13.1–14) to (13.1–19) with (13.1–4), (13.1–5), (13.2–3), (13.2–4), (13.2–9), and (13.2–10); also $D_M = 0$, $D_E = -\pi/2$. The superscript (1) is omitted from the Hankel function for convenience. For region 1 with $\rho \le a$, the components of the field are

$$E_{\rho 1} = \left[V_{M1} \frac{1}{\rho} J_1(\kappa_1 \rho) - I_{E1} \frac{\beta \kappa_1}{\omega \epsilon_1} J_1'(\kappa_1 \rho) \right] \sin \theta\, e^{-j\beta z} \qquad (13.4\text{–}1)$$

$$E_{\theta 1} = \left[V_{M1} \kappa_1 J_1'(\kappa_1 \rho) - I_{E1} \frac{\beta}{\omega \epsilon_1} \frac{1}{\rho} J_1(\kappa_1 \rho) \right] \cos \theta\, e^{-j\beta z} \qquad (13.4\text{–}2)$$

$$E_{z1} = \left[I_{E1} \frac{\kappa_1^2}{j\omega \epsilon_1} J_1(\kappa_1 \rho) \right] \sin \theta\, e^{-j\beta z} \qquad (13.4\text{–}3)$$

$$H_{\rho 1} = \left[-V_{M1}\frac{\beta\kappa_1}{\omega\mu}J_1'(\kappa_1\rho) + I_{E1}\frac{1}{\rho}J_1(\kappa_1\rho) \right] \cos\theta\, e^{-j\beta z} \qquad (13.4\text{--}4)$$

$$H_{\theta 1} = \left[V_{M1}\frac{\beta}{\omega\mu}\frac{1}{\rho}J_1(\kappa_1\rho) - I_{E1}\kappa_1 J_1'(\kappa_1\rho) \right] \sin\theta\, e^{-j\beta z} \qquad (13.4\text{--}5)$$

$$H_{z1} = V_{M1}\frac{\kappa_1^2}{j\omega\mu}J_1(\kappa_1\rho) \cos\theta\, e^{-j\beta z} \qquad (13.4\text{--}6)$$

with

$$\frac{\beta I_{E1}}{\omega\epsilon_1 V_{M1}} = -\frac{a\kappa_1^2\kappa_2^2}{\kappa_1^2 - \kappa_2^2}\left[\frac{1}{\kappa_1}\frac{J_1'(\kappa_1 a)}{J_1(\kappa_1 a)} - \frac{1}{\kappa_2}\frac{H_1'(\kappa_2 a)}{H_1(\kappa_2 a)} \right] \qquad (13.4\text{--}7)$$

For region 2 with $\rho \geq a$,

$$E_{\rho 2} = \left[V_{M2}\frac{1}{\rho}H_1(\kappa_2\rho) - I_{E2}\frac{\beta\kappa_2}{\omega\epsilon_2}H_1'(\kappa_2\rho) \right] \sin\theta\, e^{-j\beta z} \qquad (13.4\text{--}8)$$

$$E_{\theta 2} = \left[V_{M2}\kappa_2 H_1'(\kappa_2\rho) - I_{E2}\frac{\beta}{\omega\epsilon_2}\frac{1}{\rho}H_1(\kappa_2\rho) \right] \cos\theta\, e^{-j\beta z} \qquad (13.4\text{--}9)$$

$$E_{z2} = \left[I_{E2}\frac{\kappa_2^2}{j\omega\epsilon_2}H_1(\kappa_2\rho) \right] \sin\theta\, e^{-j\beta z} \qquad (13.4\text{--}10)$$

$$H_{\rho 2} = \left[-V_{M2}\frac{\beta\kappa_2}{\omega\mu}H_1'(\kappa_2\rho) + I_{E2}\frac{1}{\rho}H_1(\kappa_2\rho) \right] \cos\theta\, e^{-j\beta z} \qquad (13.4\text{--}11)$$

$$H_{\theta 2} = \left[V_{M2}\frac{\beta}{\omega\mu}\frac{1}{\rho}H_1(\kappa_2\rho) - I_{E2}\kappa_2 H_1'(\kappa_2\rho) \right] \sin\theta\, e^{-j\beta z} \qquad (13.4\text{--}12)$$

$$H_{z2} = V_{M2}\frac{\kappa_2^2}{j\omega\mu}H_1(\kappa_2\rho) \cos\theta\, e^{-j\beta z} \qquad (13.4\text{--}13)$$

with

$$\frac{I_{E2}}{\epsilon_2 V_{M2}} = \frac{I_{E1}}{\epsilon_1 V_{M1}} \qquad (13.4\text{--}14)$$

where I_{E1}/V_{M1} is given by (13.4–7). Also, from (13.2–10),

$$V_{M2} = V_{M1}\left(\frac{\kappa_1}{\kappa_2}\right)^2\frac{J_1(\kappa_1 a)}{H_1(\kappa_2 a)} \qquad (13.4\text{--}15)$$

The superscript (1) has been omitted from the Hankel functions.

Graphs of the electric and magnetic field lines for the HE_{11} mode are shown in Fig. 13.4–2 in the transverse (vertical) plane in (a) and the longitudinal (horizontal) plane in (b). The magnitudes of the electric fields $|E_\rho(\theta = \pi/2)|$ and $|E_\theta(\theta = 0)|$ are shown, respectively, in Fig. 13.4–3a and b. At $\theta = \pi/2$ the electric field

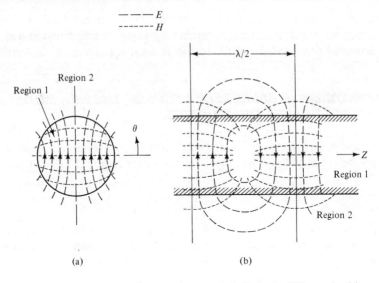

Figure 13.4–2 Lines of electric and magnetic fields in the HE_{11} mode: (a) cross section of dielectric rod; (b) longitudinal section. (Data of D. G. Kiely.)

is perpendicular to the boundary between regions 1 and 2 and, in accordance with the boundary conditions, is discontinuous. At $\theta = 0$ the electric field is tangent to the boundary across which it is continuous. The field is seen to have a maximum at the center of the rod and to decrease radially outward. The field is largely in the central rod (region 1) when λ/λ_2 is significantly smaller than 1, which occurs when d/λ_2 is relatively large (see Fig. 13.4–1). The field is primarily in the outer region 2 when λ/λ_2 is near 1, which occurs when d/λ_2 is relatively small (see Fig. 13.4–1).

Since the HE_{11} mode has no cutoff, the radius of the central rod can be kept

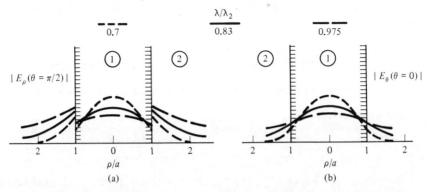

Figure 13.4–3 Magnitude of the electric field at (a) $\theta = \pi/2$; (b) $\theta = 0$, HE_{11} mode. (Data of D. G. Kiely.)

small and most of the field located in the outer region 2. In this way dielectric losses are reduced and single-mode long-distance transmission with low attenuation achieved provided the outer region is a sufficiently good dielectric.

13.5 APPROXIMATE REPRESENTATION OF THE HE₁₁ MODE; APPLICATION TO FIBER OPTICS

When the permittivity ϵ_1 of the dielectric rod (region 1, $\rho \le a$) is only slightly greater than the permittivity ϵ_2 of the outer region 2 ($\rho > a$), so that

$$\delta \equiv 1 - \frac{\epsilon_2}{\epsilon_1} \ll 1 \tag{13.5-1}$$

the factor $\kappa_1^2 - \kappa_2^2 = k_1^2 - k_2^2 = \omega^2\mu(\epsilon_1 - \epsilon_2)$—which occurs in (13.4–7)—is quite small. (In optical waveguides δ may be as small as 0.01.) Furthermore, with $\kappa_1 \sim \kappa_2$, the bracketed term in (13.4–7) is readily evaluated and found not to be small. It follows that

$$\frac{\beta I_{E1}}{\omega\epsilon_1} \gg V_{M1}; \qquad \frac{\beta I_{E2}}{\omega\epsilon_2} \gg V_{M2}; \qquad I_{E1} \gg \frac{\beta V_{M1}}{\omega\mu}; \qquad I_{E2} \gg \frac{\beta V_{M2}}{\omega\mu} \tag{13.5-2}$$

This means that the HE_{11} mode is predominantly an E mode with the electromagnetic field approximated by (13.4–1) to (13.4–13) with the terms multiplied by V_{M1} and V_{M2} omitted. A further examination of the components of the field and of Fig. 13.4–2a indicates that $E_y = E_\rho \sin \theta + E_\theta \cos \theta$ and $H_x = H_\rho \cos \theta - H_\theta \sin \theta$ are substantially greater, respectively, than E_x, E_z and H_y, H_z. Accordingly, the field is quite well approximated by

$$E_{1y} = -I_{E1}\frac{\beta\kappa_1}{2\omega\epsilon_1}F_1(\rho, \theta)e^{-j\beta z}; \qquad H_{1x} = I_{E1}\frac{\kappa_1}{2}F_1(\rho, \theta)e^{-j\beta z} \tag{13.5-3}$$

$$E_{2y} = -I_{E2}\frac{-j\beta\kappa_2}{2\omega\epsilon_2}F_2(\rho, \theta)e^{-j\beta z}; \qquad H_{2x} = I_{E2}\frac{-j\kappa_2}{2}F_2(\rho, \theta)e^{-j\beta z} \tag{13.5-4}$$

where

$$F_1(\rho, \theta) = 2[J_1'(\kappa_1\rho) \sin^2 \theta + (\kappa_1\rho)^{-1}J_1(\kappa_1\rho) \cos^2 \theta] \tag{13.5-5}$$

$$F_2(\rho, \theta) = 2j[H_1'(\kappa_2\rho) \sin^2 \theta + (\kappa_2\rho)^{-1}H_1(\kappa_2\rho) \cos^2 \theta] \tag{13.5-6}$$

and, from (13.2–9),

$$I_{E2} = I_{E1}\frac{\epsilon_2}{\epsilon_1}\left(\frac{\kappa_1}{\kappa_2}\right)^2\frac{J_1(\kappa_1 a)}{H_1(\kappa_2 a)} \tag{13.5-7}$$

As before, the superscript (1) has been omitted from the Hankel functions. With the relation $J_1'(x) = -(1/x)J_1(x) + J_0(x)$ and a similar one for the Hankel function, and with the trigonometric formulas $\cos 2\theta = \cos^2 \theta - \sin^2 \theta = 1 - 2 \sin^2 \theta$, the following alternatives to (13.5–5) and (13.5–6) are obtained:

$$F_1(\rho, \theta) = \left\{ J_0(\kappa_1\rho) + \left[\frac{2}{\kappa_1\rho} J_1(\kappa_1\rho) - J_0(\kappa_1\rho) \right] \cos 2\theta \right\} \qquad (13.5\text{–}8)$$

$$F_2(\rho, \theta) = j \left\{ H_0^{(1)}(\kappa_2\rho) + \left[\frac{2}{\kappa_2\rho} H_1^{(1)}(\kappa_2\rho) - H_0^{(1)}(\kappa_2\rho) \right] \cos 2\theta \right\} \qquad (13.5\text{–}9)$$

It is now readily verified that in ranges of interest the difference terms which are multiplied by $\cos 2\theta$ are quite small compared to the leading rotationally symmetric terms $J_0(\kappa_1\rho)$ and $H_0^{(1)}(\kappa_2\rho)$. This means that the radial dependence of E_y and H_x is almost independent of the direction of polarization, so that

$$F_1(\rho, \theta) \sim F_1(\rho) = J_0(\kappa_1\rho); \qquad \rho \le a \qquad (13.5\text{–}10)$$

$$F_2(\rho, \theta) \sim F_2(\rho) = jH_0^{(1)}(\kappa_2\rho); \qquad \rho > a \qquad (13.5\text{–}11)$$

Note that for perfect dielectrics κ_1 is real, $\kappa_2 \sim j\kappa_{2l}$ is a pure imaginary, and $jH_0^{(1)}(\kappa_2\rho) = jH_0^{(1)}(j\kappa_{2l}\rho)$ is real.

Because the approximations (13.5–10) and (13.5–11) are rotationally symmetric, the boundary conditions on E_y at $\rho = a$ cannot be correctly satisfied at all angles θ. However, since with (13.5–1) ϵ_1 is only slightly greater than ϵ_2, this is of no consequence and approximate continuity of E_y can be assumed with

$$I_{E2} \sim I_{E1} \frac{\epsilon_2\kappa_1 J_0(\kappa_1 a)}{\epsilon_1\kappa_2 H_0^{(1)}(\kappa_2 a)} \qquad (13.5\text{–}12)$$

With (13.5–12), (13.5–4) becomes

$$E_{2y} \sim -I_{E1} \frac{\beta\kappa_1}{2\omega\epsilon_1} \frac{J_0(\kappa_1 a)}{jH_0^{(1)}(\kappa_2 a)} F_2(\rho)e^{-j\beta z}; \qquad \rho > a \qquad (13.5\text{–}13)$$

The radial dependence represented by the function

$$F(\rho) = \begin{cases} J_0(\kappa_1\rho); & \rho \le a \\[2ex] \dfrac{J_0(\kappa_1 a)}{H_0^{(1)}(\kappa_2 a)} H_0^{(1)}(\kappa_2\rho); & \rho > a \end{cases} \qquad (13.5\text{–}14)$$

is quite closely approximated by the Gaussian distribution for all radial distances, that is,

$$F(\rho) \sim \exp\left[-\frac{1}{2} \left(\frac{\rho}{\rho_0} \right)^2 \right] \qquad (13.5\text{–}15)$$

where ρ_0, called the *spot size*, is defined by

$$\rho_0^2 = \frac{a^2}{\ln\left[a^2(k_1^2 - k_2^2) \right]} \qquad (13.5\text{–}16)$$

Note that $k_1^2 - k_2^2 = k_0^2 (\epsilon_{1r} - \epsilon_{2r})$ where $k_0 = \omega/c$ is the wave number for free space. The wave number β for the guided wave is given by the following formula. Note that β is not linear in the frequency.

$$(\beta a)^2 = (k_1 a)^2 - \ln [a^2(k_1^2 - k_2^2)] - 1 \tag{13.5-17}$$

so that

$$(\kappa_1 a)^2 = (k_1^2 - \beta^2)a^2 = \ln [a^2(k_1^2 - k_2^2)] - 1 \tag{13.5-18}$$

The spot size is equal to the cross section of the rod when

$$\rho_0 = a; \quad \ln[a^2(k_1^2 - k_2^2)] = 1 \quad \text{or} \quad a\sqrt{k_1^2 - k_2^2} = e^{0.5} = 1.65 \tag{13.5-19}$$

This is sufficiently below the cutoff for the higher modes, which is given by $a \sqrt{k_1^2 - k_2^2} \leq 2.41$.

The total power transferred along the dielectric guide is readily evaluated from the axially directed component of the complex Poynting vector, that is,

$$S_z = \hat{\mathbf{z}} \cdot (\tfrac{1}{2}\mathbf{E} \times \mathbf{H}^*) = -\tfrac{1}{2}E_y H_x^* \tag{13.5-20}$$

With (13.5–3), (13.5–4), and (13.5–8),

$$S_z(\rho) = \frac{1}{2} I_{E1}^2 \frac{\beta}{\omega\epsilon_1} \left(\frac{\kappa_1}{2}\right)^2 F^2(\rho) = I_{E1}^2 \frac{\beta\kappa_1^2}{8\omega\epsilon_1} \exp\left(\frac{-\rho^2}{\rho_0^2}\right) \tag{13.5-21}$$

The total power transferred along the guide is

$$P = \int_0^\infty 2\pi S_z(\rho)\rho \, d\rho = I_{E1}^2 \frac{\pi\beta}{8\omega\epsilon_1} \kappa_1^2\rho_0^2 \tag{13.5-22}$$

The fraction of the total power within the radius a of the central core [i.e., in region 1 $(\rho \leq a)$] is

$$\frac{P_1}{P_{\text{total}}} = 1 - \exp\left(\frac{-a^2}{\rho_0^2}\right) = 1 - \frac{1}{\ln [a^2(k_1^2 - k_2^2)]} \tag{13.5-23}$$

The last step involves (13.5–16).

When $\rho_0 = a$ with $a \sqrt{k_1^2 - k_2^2} = 1.65$, $P_1/P_{\text{total}} = 0.63$, $P_2/P_{\text{total}} = 0.37$. If the frequency is increased to the cutoff value for higher modes, $a \sqrt{k_1^2 - k_2^2} = 2.41$, $\rho_0 = 0.75a$, and $P_1/P_{\text{total}} = 0.83$, $P_2/P_{\text{total}} = 0.17$. If the frequency is decreased by a factor of 0.7 so that $a \sqrt{k_1^2 - k_2^2} = 1.155$, $\rho_0 = 1.86a$, $P_1/P_{\text{total}} = 0.25$, and $P_2/P_{\text{total}} = 0.75$. Thus single-mode operation is possible with most of the field in the central rod (region 1) or in the outside region 2, depending on the magnitude of the quantity $a \sqrt{k_1^2 - k_2^2} = ak_0 \sqrt{\epsilon_{1r} - \epsilon_{2r}}$.

Single-mode optical-fiber transmission is carried out at wavelengths ranging from $\lambda_0 \sim 0.85$ μm to $\lambda_0 \sim 1.55$ μm. The corresponding frequencies are $f = c/\lambda_0 = 3.5 \times 10^{14}$ to 1.9×10^{14} Hz. For a wavelength of $\lambda_0 \sim 1$ μm, $f = 3 \times 10^{14}$ Hz and $k_0 = 2\pi/\lambda_0 = 2\pi \times 10^6$ m^{-1}. If $\delta = 1 - (\epsilon_2/\epsilon_1)$ is as small as $\delta = 0.01$ and $\epsilon_{1r} \sim 5$, it follows that $k_1 = 4.5\pi \times 10^6$ m^{-1}, $ak_1\sqrt{\delta} = a \sqrt{k_1^2 - k_2^2} = 4.5a\pi \times 10^5$. For $ak_1 \sqrt{\delta} = 1.65$, $a = 1.1$ μm—a very small radius.

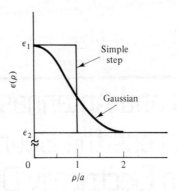

Figure 13.5–1   Cross-sectional profiles of the permittivity of an optical fiber.

An actual optical-fiber waveguide is not simply a dielectric fiber in air. It consists of an inner glass fiber (region 1) surrounded by a glass cladding (region 2) that has a permittivity ϵ_2 slightly lower than ϵ_1 of the central core and is sufficiently thick so that the radially exponentially attenuated field decays to a negligible value at its surface. Modifications include a variety of cross-sectional profiles of the permittivity that introduce somewhat different properties. Instead of the simple step from ϵ_1 to ϵ_2 at $\rho = a$, the permittivity may be a more complicated function of the radius. The simple step and the continuous and theoretically very convenient Gaussian profile are shown in Fig. 13.5–1. Other possibilities include parabolic, triangular, and variously segmented profiles. The attenuation due to dielectric losses and dispersion are different for different profiles. Thus the simple step profile can be selected with the particular value $a\sqrt{k_1^2 - k_2^2} = 2.718$—which is only slightly above the cutoff for higher modes—at which waveguide dispersion (due to the nonlinear dependence of the wave number β on the frequency) is zero.

If the radius of the central rod is increased to many times the cutoff value for higher modes, the field is contained almost entirely in the central core. Single-mode transmission is replaced by multimode transmission. Since each propagating mode has a different phase velocity, a field that is coherent and linearly polarized near the source becomes incoherent and unpolarized as it progresses. Attenuation is typically higher in multimode than in single-mode transmission.

14

Waves and Antennas Near and Across the Boundary Between Electrically Different Half-Spaces

The propagation of electromagnetic waves in the presence of a boundary between two quite different materials like air and earth (salt and fresh water, soil, sand, ice, etc.) has a long and interesting history in classical electromagnetism. It has application in all radio communication between points near the surface of the earth and this is true whether the transmitting and receiving antennas are both in the air (broadcast radio), both in the earth (communication between tunnels, mine shafts, submarines), or one is in each medium (communication with submarines). It underlies geophysical exploration by means of electromagnetic waves generated and received by antennas on the surface of the earth, or on the floor of the ocean.

In this introductory study, attention will be directed to two related aspects that differ in the nature of the source with significant consequences. The first topic is a generalization of the discussion at the end of Chapter 2 of the transmission of electromagnetic waves into the earth from normal to arbitrary angles of incidence; the second topic is the electromagnetic field generated by a vertical electric dipole on the horizontal boundary between air and a general dissipative half-space.

14.1 PLANE WAVE INCIDENT ON A PLANE BOUNDARY: PLANE-WAVE REFLECTION AND TRANSMISSION COEFFICIENTS

A plane electromagnetic wave traveling in the direction of the vector wave number $\mathbf{k} = \hat{\mathbf{k}}k$ [where in general $k = (k_x^2 + k_y^2 + k_z^2)^{1/2}$] is represented by $e^{i\mathbf{k}\cdot\mathbf{r}}$, where $\mathbf{r}$

518

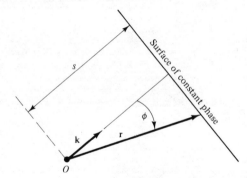

Figure 14.1–1 Wave front or surface of constant phase in plane wave.

is the vector drawn from the arbitrarily located origin to a point on the wave front— a plane surface of constant phase as shown in Fig. 14.1–1. It is clear that $\mathbf{k} \cdot \mathbf{r} = kr \cos \phi = ks = $ constant is a plane perpendicular to $\mathbf{k}$.

Consider plane electromagnetic waves traveling in region 2 ($z < 0$; k_2, $\mu_2 = \mu_0$, ϵ_2, σ_2) in the direction $\hat{\mathbf{k}}_2^i$ as shown in Fig. 14.1–2. The waves are characterized by $\partial/\partial x = 0$, $k_{2x} = 0$, for all components of the field. When this is true, Maxwell's equations in Cartesian coordinates can be separated into two mutually independent parts. These are the field of magnetic type, including E_x, B_y, and B_z given by

$$E_x^i = E_{0x}^i e^{i\mathbf{k}_2^i \cdot \mathbf{r}} = E_{0x}^i e^{i(k_{2y}y + k_{2z}z)} \tag{14.1–1}$$

$$B_y^i = \frac{i}{\omega} \frac{\partial E_x^i}{\partial z}; \qquad B_z^i = \frac{i}{\omega} \frac{\partial E_x^i}{\partial y} \tag{14.1–2}$$

and the field of electric type including B_x, E_y, and E_z given by

$$B_x^i = B_{0x}^i e^{i\mathbf{k}_2^i \cdot \mathbf{r}} = B_{0x}^i e^{i(k_{2y}y + k_{2z}z)} \tag{14.1–3}$$

$$E_y^i = \frac{i\omega}{k_2^2} \frac{\partial B_x^i}{\partial z}; \qquad E_z^i = \frac{-i\omega}{k_2^2} \frac{\partial B_x^i}{\partial y} \tag{14.1–4}$$

Note that

$$k_{2y} = k_2 \sin \Theta^i; \qquad k_{2z} = k_2 \cos \Theta^i \tag{14.1–5}$$

The *plane of incidence* for these waves is the plane that contains the vector $\mathbf{k}_2^i$ and its projection onto the boundary plane $z = 0$. In Fig. 14.1–2 the plane of incidence is the yz plane. For the waves of magnetic (electric) type, the magnetic (electric) field is in the plane of incidence, the electric (magnetic) field perpendicular to it and parallel to the boundary plane. Note that the field of magnetic (electric) type is related to the radiation field of a vertical magnetic (electric) dipole on the boundary surface.

In order to satisfy the boundary conditions

$$\hat{\mathbf{n}}_1 \times \mathbf{E}_1 + \hat{\mathbf{n}}_2 \times \mathbf{E}_2 = 0 \quad \text{or} \qquad E_{1x} = E_{2x} \text{ at } z = 0 \tag{14.1–6a}$$

$$\mu_1^{-1}(\hat{\mathbf{n}}_1 \times \mathbf{B}_1) + \mu_2^{-1}(\hat{\mathbf{n}}_2 \times \mathbf{B}_2) = 0 \quad \text{or} \quad \mu_1^{-1}B_{1y} = \mu_2^{-1}B_{2y} \text{ at } z = 0 \tag{14.1–6b}$$

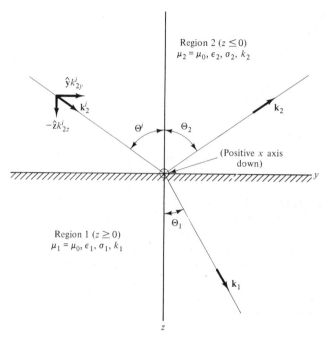

Figure 14.1–2 Plane wave incident on the plane boundary $z = 0$ at angle Θ^i.

$$\bar{\epsilon}_1 \hat{n}_1 \cdot \mathbf{E}_1 + \bar{\epsilon}_2 \hat{n}_2 \cdot \mathbf{E}_2 = 0 \quad \text{or} \quad \bar{\epsilon}_1 E_{1z} = \bar{\epsilon}_2 E_{2z} \text{ at } z = 0 \qquad (14.1\text{–}6\text{c})$$

$$\hat{n}_1 \cdot \mathbf{B}_1 + \hat{n}_2 \cdot \mathbf{B}_2 = 0 \quad \text{or} \quad B_{1z} = B_{2z} \text{ at } z = 0 \qquad (14.1\text{–}6\text{d})$$

it is necessary to assume a solution of the form

$$\mathbf{E}_2 = \mathbf{E}^i + \mathbf{E}^s; \qquad \mathbf{B}_2 = \mathbf{B}^i + \mathbf{B}^s \qquad (14.1\text{–}7)$$

where $\mathbf{E}^s$ and $\mathbf{B}^s$ are the reflected or scattered field. The field $\mathbf{E}_1$, $\mathbf{B}_1$ in region 1 is the transmitted field. The field of magnetic type can be expressed in terms of reflection and transmission coefficients as follows:

$$E_{2x} = E_{0x}^i \left[e^{i(k_{2y}y + k_{2z}z)} + f_{mr} e^{i(k_{2y}y - k_{2z}z)} \right] \qquad (14.1\text{–}8\text{a})$$

$$B_{2y} = \frac{-k_{2z}}{\omega} E_{0x}^i \left[e^{i(k_{2y}y - k_{2z}z)} - f_{mr} e^{i(k_{2y}y - k_{2z}z)} \right] \qquad (14.1\text{–}8\text{b})$$

$$B_{2z} = \frac{-k_{2y}}{\omega} E_{0x}^i \left[e^{i(k_{2y}y + k_{2z}z)} + f_{mr} e^{i(k_{2y}y - k_{2z}z)} \right] \qquad (14.1\text{–}8\text{c})$$

$$E_{1x} = f_{mt} E_{0x}^i e^{i(k_{1y}y + k_{1z}z)} \qquad (14.1\text{–}9\text{a})$$

$$B_{1y} = \frac{-k_{1z}}{\omega} f_{mt} E_{0x}^i e^{i(k_{1y}y + k_{1z}z)} \qquad (14.1\text{–}9\text{b})$$

$$B_{1z} = \frac{-k_{1y}}{\omega} f_{mt} E_{0x}^i e^{i(k_{1y}y + k_{1z}z)} \qquad (14.1\text{–}9\text{c})$$

The corresponding expressions for the field of electric type are

$$B_{2x} = B_{0x}^i[e^{i(k_{2y}y + k_{2z}z)} + f_{er}e^{i(k_{2y}y - k_{2z}z)}] \tag{14.1–10a}$$

$$E_{2y} = \frac{-\omega k_{2z}}{k_2^2}B_{0x}^i[e^{i(k_{2y}y + k_{2z}z)} - f_{er}e^{i(k_{2y}y - k_{2z}z)}] \tag{14.1–10b}$$

$$E_{2z} = \frac{\omega k_{2y}}{k_2^2}B_{0x}^i[e^{i(k_{2y}y + k_{2z}z)} + f_{er}e^{i(k_{2y}y - k_{2z}z)}] \tag{14.1–10c}$$

$$B_{1x} = \frac{\mu_1}{\mu_2}f_{et}B_{0x}^i e^{i(k_{1y}y + k_{1z}z)} \tag{14.1–11a}$$

$$E_{1y} = \frac{-\omega\mu_1 k_{1z}}{\mu_2 k_1^2}f_{et}B_{0x}^i e^{i(k_{1y}y + k_{1z}z)} \tag{14.1–11b}$$

$$E_{1z} = \frac{\omega\mu_1 k_{1y}}{\mu_2 k_1^2}f_{et}B_{0x}^i e^{i(k_{1y}y - k_{1z}z)} \tag{14.1–11c}$$

When these components are substituted in the boundary conditions (14.1–6a–d) at $z = 0$, they require that

$$e^{ik_{2y}^i y} = e^{ik_{2y}y} = e^{ik_{1y}y} \quad \text{for all values of } y \tag{14.1–12}$$

so that with (14.1–5),

$$k_2 \sin \Theta^i = k_2 \sin \Theta_2 = k_1 \sin \Theta_1 \tag{14.1–13}$$

It follows that

$$\Theta^i = \Theta_2 \tag{14.1–14a}$$

and the complex index of refraction is

$$N_{12} \equiv \frac{k_1}{k_2} = \left(\frac{\mu_1\bar{\epsilon}_1}{\mu_2\bar{\epsilon}_2}\right)^{1/2} = \frac{\sin \Theta_2}{\sin \Theta_1} \tag{14.1–14b}$$

This is a generalized form of Snell's law. Note that when region 2 is air with $\bar{\epsilon}_2 = \epsilon_0$, $\mu_2 = \mu_0$, it follows that k_2 and Θ_2 are real. On the other hand, when region 1 is a general simple medium with $\bar{\epsilon}_1$ and $k_1 = \beta_1 + i\alpha_1$ complex, the angle Θ_1 must also be complex.

In addition to (14.1–12), the substitution of (14.1–8a–c) and (14.1–9a–c) into the boundary conditions (14.1–6a–d) yields

$$1 + f_{mr} = f_{mt} \tag{14.1–15}$$

$$\mu_2^{-1}k_{2z}(1 - f_{mr}) = \mu_1^{-1}k_{1z}f_{mt}$$

or $$\tag{14.1–16}$$

$$(1 - f_{mr}) \cos \Theta_2 = \frac{\zeta_2}{\zeta_1}f_{mt} \cos \Theta_1$$

where the wave impedances are

$$\zeta_1 = \left(\frac{\mu_1}{\tilde{\epsilon}_1}\right)^{1/2}; \qquad \zeta_2 = \left(\frac{\mu_2}{\tilde{\epsilon}_2}\right)^{1/2} \tag{14.1-17}$$

When (14.1–15) and (14.1–16) are solved for the reflection coefficient f_{mr} and the transmission coefficient f_{mt}, these turn out to be

$$f_{mr} = \frac{Y_{m2} - Y_{m1}}{Y_{m2} + Y_{m1}}; \qquad f_{mt} = \frac{2Y_{m2}}{Y_{m2} + Y_{m1}} \tag{14.1-18}$$

where

$$Y_{m2} = \zeta_2^{-1} \cos \Theta_2; \qquad Y_{m1} = \zeta_1^{-1} \cos \Theta_1 \tag{14.1-19}$$

are generalized admittances in siemens.

The corresponding expressions for the waves of electric type are

$$f_{er} = \frac{Z_{e2} - Z_{e1}}{Z_{e2} + Z_{e1}}; \qquad f_{et} = \frac{2Z_{e2}}{Z_{e2} + Z_{e1}} \tag{14.1-20}$$

where

$$Z_{e2} = \zeta_2 \cos \Theta_2; \qquad Z_{e1} = \zeta_1 \cos \Theta_1 \tag{14.1-21}$$

are generalized impedances in ohms. Note that when region 2 is air, ζ_2 and Θ_2 are real so that Y_{m2} and Z_{e2} are real. Note also that (14.1–18) and (14.1–20) are formally like the reflection and transmission coefficients for sections of waveguide or transmission line. In order to study the properties of the reflection coefficients, let

$$f_{mr} = |f_{mr}| e^{i\psi_{mr}}; \qquad f_{er} = |f_{er}| e^{i\psi_{er}} \tag{14.1-22}$$

14.2 REFLECTION COEFFICIENTS FOR PERFECT NONCONDUCTORS

When both regions 1 and 2 are perfect nonconductors with region 2 air, $\sigma_1 = \sigma_2 = 0$ and $\tilde{\epsilon}_1 = \epsilon_1 = \epsilon_{1r}\epsilon_0$, $\tilde{\epsilon}_2 = \epsilon_0$, where ϵ_1 is real. Under these conditions the magnitudes and angles of the two reflection coefficients are, respectively, very different from each other. They are shown graphically in Fig. 14.2–1 for $\epsilon_{1r} = 4$, 10, and 80. The coefficient of magnetic type (for an incident plane wave with its magnetic field in the plane of incidence and the electric field perpendicular to it and parallel to the boundary) has the constant angle $\psi_{mr} = 180°$ for all values of Θ_2 and ϵ_{1r}, and a magnitude that has a flat minimum at $\Theta_2 = 0°$ and increases continuously and smoothly to $|f_{mr}| = 1$ at $\Theta_2 = 90°$. Thus, when $\mu_1 = \mu_2$, f_{mr} is always a negative quantity. The coefficient of electric type (for an incident plane wave with its electric field in the plane of incidence and the magnetic field perpendicular to it and parallel to the boundary) has a flat maximum at $\Theta_2 = 0°$,

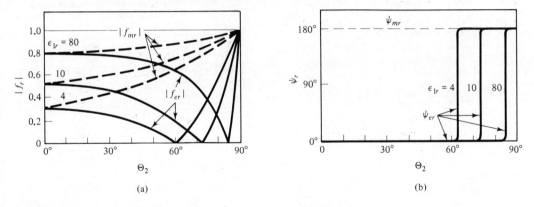

Figure 14.2–1 Complex reflection coefficients for an incident plane wave; $\mu_1 = \mu_2 = \mu_0$.

decreases to a null between $\Theta_2 = 50°$ and $90°$, and then increases steeply to 1. At the null the angle ψ_{er} jumps by $180°$.

The angle of incidence $\Theta^i = \Theta_2$ at which the reflection coefficient vanishes and its angle changes by $180°$ is known as the *Brewster angle* (denoted here by Θ_{2B}). Respectively for incident waves of the magnetic or electric types, the reflection coefficient vanishes when:

Magnetic type: $Y_{m1} = Y_{m2}$

$$\Theta_{2B} = \cos^{-1}\left[\frac{(\mu_1\epsilon_1/\mu_2\epsilon_2) - 1}{(\mu_1/\mu_2)^2 - 1}\right]^{1/2} \tag{14.2–1}$$

When $\epsilon_1 = \epsilon_2$,

$$\Theta_{2B} = \cos^{-1}\left(\frac{\mu_1}{\mu_2} + 1\right)^{-1/2} = \tan^{-1}\left(\frac{\mu_1}{\mu_2}\right)^{1/2} \tag{14.2–2}$$

Electric type: $Z_{e1} = Z_{e2}$

$$\Theta_{2B} = \cos^{-1}\left[\frac{(\mu_1\epsilon_1/\mu_2\epsilon_2) - 1}{(\epsilon_1/\epsilon_2)^2 - 1}\right]^{1/2} \tag{14.2–3}$$

When $\mu_1 = \mu_2$,

$$\Theta_{2B} = \cos^{-1}\left(\frac{\epsilon_1}{\epsilon_2} + 1\right)^{-1/2} = \tan^{-1}\left(\frac{\epsilon_1}{\epsilon_2}\right)^{1/2} \tag{14.2–4}$$

With $\epsilon_2 = \epsilon_0$ for air, $\Theta_{2B} = 54°.7, 63°.4, 72°.5, 83°.7$, respectively, for $\epsilon_{1r} = 2, 4, 10,$ and 81. No Brewster angle exists for waves of magnetic type when $\mu_1 = \mu_2$; for waves of electric type when $\epsilon_1 = \epsilon_2$. At the Brewster angle the two half-spaces are perfectly matched at $z = 0$ and the electromagnetic waves proceed from the one to the other *without reflection*.

When $\mu_1 = \mu_2$ and the angle of incidence in region 2 is the Brewster angle for a field of electric type, propagation in region 1 is at an angle Θ_1 which satisfies Snell's law:

$$\sin \Theta_1 = \frac{k_2}{k_1} \sin \Theta_2 = \frac{k_2}{k_1} \sin \Theta_{2B} = \frac{k_2}{k_1} (1 - \cos^2 \Theta_{2B})^{1/2} \qquad (14.2\text{-}5)$$

With (14.2-4),

$$\sin \Theta_1 = \frac{k_2}{k_1} \left[1 - \frac{1}{(\epsilon_1/\epsilon_2) + 1} \right]^{1/2}$$

$$= \frac{k_2}{k_1} \left[\frac{\epsilon_1/\epsilon_2}{(\epsilon_1/\epsilon_2) + 1} \right]^{1/2} = \cos \Theta_{2B} \qquad (14.2\text{-}6)$$

Hence

$$\Theta_1 = \frac{\pi}{2} - \Theta_{2B} \qquad (14.2\text{-}7)$$

This means that $\mathbf{k}_1$ is perpendicular to $\mathbf{k}_2$. Since $\mathbf{E}_1$ and $\mathbf{P}_1 = (\epsilon_1 - \epsilon_0)\mathbf{E}_1$ are perpendicular to $\mathbf{k}_1$, they must be parallel to $\mathbf{k}_2$. Thus the oscillating dipoles that determine the volume density of polarization $\mathbf{P}$ in region 1 have their axes—along which they do not radiate—parallel to $\mathbf{k}_2$. Thus they generate no field in this direction, so that there is no reflected field.

For waves of magnetic type, when $\epsilon_1 = \epsilon_2$, $\mathbf{k}_1$ is also perpendicular to $\mathbf{k}_2$; $\mathbf{B}_1$ and $-\mathbf{M}_1 = (\mu_1^{-1} - \mu_0^{-1})\mathbf{B}_1$ are perpendicular to $\mathbf{k}_1$ and parallel to $\mathbf{k}_2$. The oscillating magnetic dipoles (or equivalent loops) have their axes—along which they do not radiate—parallel to $\mathbf{k}_2$. It follows that they generate no field in this direction, so that there is no reflected field.

Since f_{er} and f_{mr} differ except at normal and grazing incidence, an incident field that is arbitrarily polarized (circular, elliptical, unpolarized) is reflected as a linearly polarized field when the angle of incidence is the Brewster angle.

14.3 TOTAL INTERNAL REFLECTION: SURFACE WAVES

In the preceding discussion of the reflection coefficients for perfect dielectrics, it was assumed that the incident wave is in air with $\epsilon_2 = \epsilon_0$, the refracted wave in a dielectric with $\epsilon_1 = \epsilon_{1r}\epsilon_0$. In this case Snell's law is

$$\sin \Theta_1 = \frac{k_2}{k_1} \sin \Theta_2 \qquad (14.3\text{-}1)$$

where $\Theta_2 = \Theta^i$ and $k_2/k_1 = 1/\epsilon_{1r}^{1/2} < 1$. It follows that for every real angle Θ_2 there is a smaller real angle Θ_1. The refracted wave is bent toward the normal.

If the incident wave originates in a dielectric with $\epsilon_2 = \epsilon_{2r}\epsilon_0$ and the refracted wave is in air with $\epsilon_1 = \epsilon_0$, the relation (14.3-1) is still valid but now $k_2/k_1 = \epsilon_{2r}^{1/2} > 1$, so that there can be real angles Θ_1 only for $(k_2/k_1) \sin \Theta_2 \leq 1$.

At the critical angle $(k_2/k_1) \sin \Theta_{2c} = 1$, $\Theta_1 = \pi/2$. To understand this new situation, note that

$$k_{1y} = k_1 \sin \Theta_1 = k_2 \sin \Theta_2 = k_{2y} \qquad (14.3\text{--}2)$$

$$k_{1z} = k_1 \cos \Theta_1 = k_1(1 - \sin^2 \Theta_1)^{1/2} = (k_1^2 - k_2^2 \sin^2 \Theta_2)^{1/2}; \qquad k_2 \sin \Theta_2 \leq k_1$$

$$= i(k_2^2 \sin^2 \Theta_2 - k_1^2)^{1/2}; \qquad k_2 \sin \Theta_2 > k_1 \qquad (14.3\text{--}3)$$

Let

$$\alpha_1 \equiv (k_2^2 \sin^2 \Theta_2 - k_1^2)^{1/2}; \qquad k_{1z} = i\alpha_1; \qquad k_2 \sin \Theta_2 > k_1 \qquad (14.3\text{--}4)$$

Also,

$$Z_{e1} = \zeta_1 \cos \Theta_1 = i\zeta_1 k_1^{-1}\alpha_1 = \frac{i\alpha_1}{\omega\epsilon_1} \qquad (14.3\text{--}5)$$

$$Y_{m1} = \zeta_1^{-1} \cos \Theta_1 = i\zeta_1^{-1}k_1^{-1}\alpha_1 = \frac{i\alpha_1}{\omega\mu_1} \qquad (14.3\text{--}6)$$

The reflection and transmission coefficients are

$$f_{mr} = \frac{Y_{m2} - i\alpha_1/\omega\mu_1}{Y_{m2} + i\alpha_1/\omega\mu_1}; \qquad f_{er} = \frac{Z_{e2} - i\alpha_1/\omega\epsilon_1}{Z_{e2} + i\alpha_1/\omega\epsilon_1} \qquad (14.3\text{--}7)$$

$$f_{mt} = \frac{2Y_{m2}}{Y_{m2} + i\alpha_1/\omega\mu_1}; \qquad f_{et} = \frac{2Z_{e2}}{Z_{e2} + i\alpha_1/\omega\epsilon_1} \qquad (14.3\text{--}8)$$

The electromagnetic fields for the magnetic type are like (14.1–8a–c) with (14.3–7) in region 2. In region 1 with (14.3–2) they are

$$E_{1x} = f_{mt} E_{0x}^i e^{-\alpha_1 z} e^{ik_{2y}y}; \qquad k_{2y} > k_1 \qquad (14.3\text{--}9a)$$

$$B_{1y} = \frac{-i\alpha_1}{\omega} f_{mt} E_{0x}^i e^{-\alpha_1 z} e^{ik_{2y}y}; \qquad k_{2y} > k_1 \qquad (14.3\text{--}9b)$$

$$B_{1z} = \frac{-k_{1y}}{\omega} f_{mt} E_{0x}^i e^{-\alpha_1 z} e^{ik_{2y}y}; \qquad k_{2y} > k_1 \qquad (14.3\text{--}9c)$$

where f_{mt} is given by (14.3–8). Similarly, for the electric type the field is like (14.1–10a–c) in region 2 with f_{er} as given in (14.3–7). In region 1 the field is

$$B_{1x} = \frac{\mu_1}{\mu_2} f_{et} B_{0x}^i e^{-\alpha_1 z} e^{ik_{2y}y}; \qquad k_{2y} > k_1 \qquad (14.3\text{--}10a)$$

$$E_{1y} = \frac{-i\omega\mu_1\alpha_1}{\mu_2 k_1^2} f_{et} B_{0x}^i e^{-\alpha_1 z} e^{ik_{2y}y}; \qquad k_{2y} > k_1 \qquad (14.3\text{--}10b)$$

$$E_{1z} = \frac{\omega\mu_1 k_{1y}}{\mu_2 k_1^2} f_{et} B_{0x}^i e^{-\alpha_1 z} e^{ik_{2y}y}; \qquad k_{2y} > k_1 \qquad (14.3\text{--}10c)$$

where f_{et} is given in (14.3–8).

At angles of incidence $\Theta^i = \Theta_2$ for which $k_{2y} = k_2 \sin \Theta_2 < k_1$, the fields in both regions obey Snell's law for real wave numbers and real angles. Since $k_2 > k_1$, $\Theta_1 > \Theta_2$, so that the direction $\hat{\mathbf{k}}_1$ of the refracted waves is bent away from the normal to the surface. At angles of incidence $\Theta^i = \Theta_2$ for which $k_2 \sin \Theta_2 > k_1$, there is no transmitted wave into region 1. The field is totally reflected. However, although there are no electromagnetic waves propagating into region 1, there are electromagnetic waves traveling in region 1 very close to and along the surface. The components of the field in region 1 all have the form $\exp(-\alpha_1 z) \exp(ik_{2y} y)$. When multiplied by $\exp(-i\omega t)$, the real part is

$$e^{-\alpha_1 z} \cos(\omega t - k_{2y} y) \tag{14.3-11}$$

This is a wave traveling in the y direction with the wave number $k_{2y} = k_2 \sin \Theta_2$. It exists only very close to the surface since it is exponentially attenuated with the attenuation constant α_1 defined in (14.3-4). At the critical angle, $\alpha_1 = 0$; but when k_2 is large compared to k_1, α_1 increases rapidly as the angle of incidence is made larger than the critical angle. A wave that travels along a boundary surface and is exponentially attenuated in the direction perpendicular to the surface is known as a *surface wave*. It accompanies the incident and reflected fields in region 2 by providing a field that satisfies the boundary conditions across the surface and is of no significance even a short distance from the surface. Note that the exponential attenuation in region 1 is not associated with losses in either region—both are assumed to be perfect dielectrics such as glass and air or distilled water and air.

14.4 PLANE WAVES INCIDENT AT AN ARBITRARY ANGLE ON A DISSIPATIVE HALF-SPACE

The electromagnetic field in a dissipative half-space when generated by a plane wave normally incident from the air is discussed in detail in Chapter 2. When the angle of incidence is not zero, the field in the dissipative half-space is much more complicated, as indicated in Sec. 14.1. The general expressions for the fields in both regions are given by (14.1-8a-c) and (14.1-9a-c) for the magnetic type and by (14.1-10a-c) and (14.1-11a-c) for the electric type. Snell's law in the generalized form given in (14.1-14a,b) applies with

$$k_1 = \beta_1 + i\alpha_1 = \omega(\mu_0 \bar{\epsilon}_1)^{1/2} = \omega\left[\mu_0\left(\epsilon_{1e} + \frac{i\sigma_{1e}}{\omega}\right)\right]^{1/2} \tag{14.4-1}$$

$$k_2 = k_0 = \omega(\mu_0 \epsilon_0)^{1/2} \tag{14.4-2}$$

The complex index of refraction is

$$\frac{k_1}{k_2} = N_{12} = \left(\epsilon_{1er} + \frac{i\sigma_{1e}}{\omega\epsilon_0}\right)^{1/2} \tag{14.4-3}$$

where it has been assumed that $\mu_1 = \mu_0$.

With (14.4–1) and (14.4–2) and $\Theta_1 = \Theta_{1r} + i\Theta_{1i}$, Snell's law has the form

$$\beta_1 \sin \Theta_{1r} \cosh \Theta_{1i} - \alpha_1 \cos \Theta_{1r} \sinh \Theta_{1i} = k_2 \sin \Theta_2 \qquad (14.4\text{–}4a)$$

$$\alpha_1 \sin \Theta_{1r} \cosh \Theta_{1i} + \beta_1 \cos \Theta_{1r} \sinh \Theta_{1i} = 0 \qquad (14.4\text{–}4b)$$

Thus, when region 1 is dissipative, Snell's law consists of two equations. Of these, the second can be solved for Θ_{1i} in the form

$$\tanh \Theta_{1i} = -\frac{\alpha_1}{\beta_1} \tan \Theta_{1r} \qquad (14.4\text{–}5)$$

With (14.4–5), (14.4–4a) becomes

$$\beta_1 \sin \Theta_{1r} \left\{ \frac{1 + \alpha_1^2/\beta_1^2}{[1 + (\alpha_1^2/\beta_1^2) \tan^2 \Theta_{1r}]^{1/2}} \right\} = k_2 \sin \Theta_2 \qquad (14.4\text{–}6)$$

which is Snell's law in terms of real angles for a plane wave entering a dissipative half-space at the angle Θ_{1r}. When $(\alpha_1/\beta_1)^2 \ll 1$, this reduces to the simple form for perfect dielectrics. Since Snell's law requires that $k_{1y} = k_{2y} = k_2 \sin \Theta_2$, propagation in both regions in the direction parallel to the interface is the same as for perfect dielectrics. Propagation normal to the interface is governed by $k_{2z} = k_2 \cos \Theta_2$ in region 2 and in region 1 by

$$k_{1z} = \beta_{1z} + i\alpha_{1z} = k_1 \cos \Theta_1 = k_1(1 - \sin^2 \Theta_1)^{1/2}$$
$$= (k_1^2 - k_2^2 \sin^2 \Theta_2)^{1/2} \qquad (14.4\text{–}7)$$

With $k_1 = \beta_1 + i\alpha_1$ and

$$p_1 = 2\alpha_1\beta_1(\beta_1^2 - \alpha_1^2 - k_2^2 \sin^2 \Theta_2)^{-1/2} \qquad (14.4\text{–}8)$$

$$k_{1z} = (\beta_1^2 - \alpha_1^2 - k_2^2 \sin^2 \Theta_2)^{1/2}(1 + ip_1)^{1/2} \qquad (14.4\text{–}9)$$

It follows that

$$\beta_{1z} = (\beta_1^2 - \alpha_1^2 - k_2^2 \sin^2 \Theta_2)^{1/2} f(p_1) \qquad (14.4\text{–}10)$$

$$\alpha_{1z} = (\beta_1^2 - \alpha_1^2 - k_2^2 \sin^2 \Theta_2)^{1/2} g(p_1) \qquad (14.4\text{–}11)$$

where $f(p_1)$ and $g(p_1)$ are given in Appendix II.

With (14.4–7) to (14.4–11) the general expressions for the fields of electric type in the two regions have the following forms:

$$B_{2x} = B_{0x}^i e^{ik_2 y \sin \Theta_2}(e^{ik_2 z \cos \Theta_2} + f_{er} e^{-ik_2 z \cos \Theta_2}) \qquad (14.4\text{–}12a)$$

$$E_{2y} = \frac{-B_{0x}^i \omega \cos \Theta_2}{k_2} e^{ik_2 y \sin \Theta_2}(e^{ik_2 z \cos \Theta_2} - f_{er} e^{-ik_2 z \cos \Theta_2}) \qquad (14.4\text{–}12b)$$

$$E_{2z} = \frac{B_{0x}^i \omega \sin \Theta_2}{k_2} e^{ik_2 y \sin \Theta_2}(e^{ik_2 z \cos \Theta_2} + f_{er} e^{-ik_2 z \cos \Theta_2}) \qquad (14.4\text{–}12c)$$

$$B_{1x} = \frac{B_{0x}^i \mu_1 f_{et}}{\mu_2} e^{-\alpha_{1z}z} e^{i(k_2 y \sin \Theta_2 + \beta_{1z}z)} \tag{14.4-13a}$$

$$E_{1y} = \frac{-B_{0x}^i \omega \mu_1 (\beta_{1z} + i\alpha_{1z}) f_{et}}{\mu_2 (\beta_1 + i\alpha_1)^2} e^{-\alpha_{1z}z} e^{i(k_2 y \sin \Theta_2 + \beta_{1z}z)} \tag{14.4-13b}$$

$$E_{1z} = \frac{B_{0x}^i \omega \mu_1 k_2 \sin \Theta_2}{\mu_2 (\beta_1 + i\alpha_1)^2} f_{et} e^{-\alpha_{1z}z} e^{i(k_2 y \sin \Theta_2 + \beta_{1z}z)} \tag{14.4-13c}$$

where

$$f_{er} = \frac{Z_{e2} - Z_{e1}}{Z_{e2} + Z_{e1}}; \qquad f_{et} = \frac{2Z_{e2}}{Z_{e2} + Z_{e1}} \tag{14.4-14}$$

Here

$$Z_{e2} = \zeta_2 \cos \Theta_2; \qquad Z_{e1} = \zeta_1 \cos \Theta_1 = \frac{k_{1z}}{\omega \epsilon_1} = \frac{\beta_{1z} + i\alpha_{1z}}{\omega \epsilon_1} \tag{14.4-15}$$

The field in region 2 includes a direct and reflected wave substantially like that for a perfect dielectric. It consists of a standing wave in the z direction perpendicular to the boundary and a traveling wave along the boundary. The field in region 1 when this is dissipative differs from that in a perfect dielectric primarily and significantly in the appearance of the exponential attenuation $\exp(-\alpha_{1z}z)$ in the direction of z. This governs the amplitude in its dependence on z. The attenuation constant α_{1z} as given by (14.4–11) is a function of the angle of incidence $\Theta^i = \Theta_2$, but for any and every angle of incidence, the amplitude of the wave entering region 1 decays in the direction of increasing z. Thus the amplitude fronts are horizontal planes for all angles of incidence, whereas the phase fronts proceed in a direction determined by Θ_{1r}. These are the characteristics of an *inhomogeneous* wave. The normal to the phase fronts has the angle Θ_{1r} as determined by Snell's law, the normal to the amplitude fronts always has the angle $\pi/2$. Very similar results are obtained for the waves of magnetic type. The two types coincide at normal incidence.

The behavior of the plane-wave reflection coefficients $f_{mr} = |f_{mr}|e^{i\psi_{mr}}$ and $f_{er} = |f_{er}|e^{i\psi_{er}}$ when region 1 is dissipative with different values of $N_{12}^2 = k_1^2/k_2^2$ is shown in Figs. 14.4–1 and 14.4–2. The deep minima in $|f_{er}|$ and phase changes by nearly 180° in ψ_{er} at the pseudo-Brewster's angle are clearly evident.

14.5 FIELD OF A VERTICAL ELECTRIC DIPOLE OVER A DISSIPATIVE HALF-SPACE

The radiation field of a vertical electric dipole at a height d in air over a conducting earth is well approximated by the superposition of the direct field and the reflected field. At sufficient distances this is given by the plane-wave reflection coefficient.

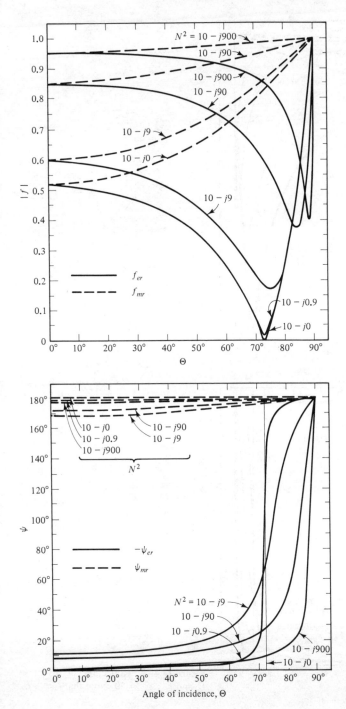

Figure 14.4–1 Reflection coefficients f_{er} and f_{mr}; $\epsilon_{1r} = 10$. Note that $j = -i$.

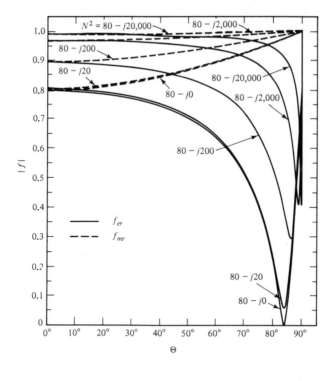

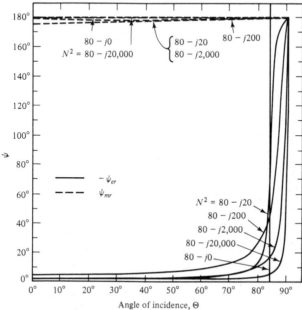

Figure 14.4–2 Reflection coefficients f_{er} and f_{emi}; $\epsilon_{1r} = 80$. Note that $j = -i$.

The complete rotationally symmetric electric field, in polar coordinates R, Θ, Φ, for an electric dipole with effective length $2h_e$ and moment $2I_zh_e$ is

$$E_{2\Theta} = \frac{-i\omega\mu_0 I_z h_e}{2\pi} \frac{e^{ik_2R}}{R} (e^{-ik_2d\cos\Theta} + f_{er}e^{ik_2d\cos\Theta}) \sin\Theta \qquad (14.5\text{-}1)$$

Note that, in agreement with the earlier sections, the positive z axis (the origin for the coordinate Θ) is vertically downward into the dissipative region 1. The dipole is at $z = -d$. The term in parentheses in (14.5-1) is the array factor $A_e(\Theta)$.

The plane-wave reflection coefficient is that defined by (14.4-14). In terms of the complex index of refraction, $N = k_1/k_2 = (\epsilon_{1er} + i\sigma_{1e}/\omega\epsilon_0)^{1/2}$, and angle of incidence $\Theta^i = \Theta_2 = \Theta$, it is

$$f_{er} = \frac{N^2 \cos\Theta_2 - (N^2 - \sin^2\Theta_2)^{1/2}}{N^2 \cos\Theta_2 + (N^2 - \sin^2\Theta_2)^{1/2}} \qquad (14.5\text{-}2)$$

Note that $|f_{er}|$ as shown in Figs. 14.4-1 and 14.4-2 has values near 1 when $\Theta_2 = 0$ and $\pi/2$ and has a deep minimum between $\Theta_2 = 0$ and $\pi/2$ at the pseudo-Brewster angle. The field factor, $A_e(\Theta) \sin\Theta = (1 + f_{er}) \sin\Theta$, for a dipole on the earth $(d = 0)$ for (14.5-1) is shown graphically in Fig. 14.5-1 for a wide range of values of N^2 corresponding to earth and water. It is seen that along the surface of the earth where $\Theta_2 = \pi/2$, the field vanishes except for $N^2 = i\infty$, the value for a perfectly conducting earth. The corresponding patterns for a quarter-wave monopole erected on the earth—shown in Fig. 14.5-2—also give a vanishing field along the surface of the earth. This suggests that radio communication along the surface of the earth between vertical transmitting and receiving antennas is impossible. Since very satisfactory communication is actually maintained between such antennas, it is evident that the formula (14.5-1) is not complete. Although it gives the complete radiation field, it does not include a surface wave that exists in the air along the boundary.

14.6 GENERAL INTEGRALS FOR THE ELECTROMAGNETIC FIELD OF A VERTICAL ELECTRIC DIPOLE ALONG THE BOUNDARY BETWEEN AIR AND A DISSIPATIVE HALF-SPACE

The determination of the electromagnetic field generated by a vertical electric dipole on the horizontal boundary between air and earth is a classical problem that dates back to the early days of radio communication. It is perhaps most commonly formulated with the help of the Hertz potentials. However, it is an instructive exercise to proceed directly from Maxwell's equations and the integrals of the field so obtained can be evaluated directly with certain approximations. To provide a broad insight into different aspects of the electromagnetic surface waves generated by a vertical electric dipole, this will initially be located at a depth d in the dissipative half-space near its boundary with air and the complete field at all points in the

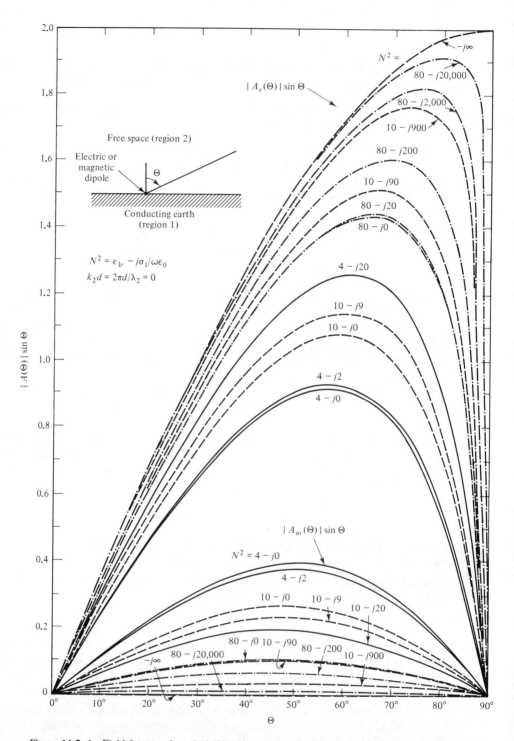

Figure 14.5–1 Field factors of vertical electric and magnetic dipoles on conducting earth; $d = 0$. Note that $j = -i$.

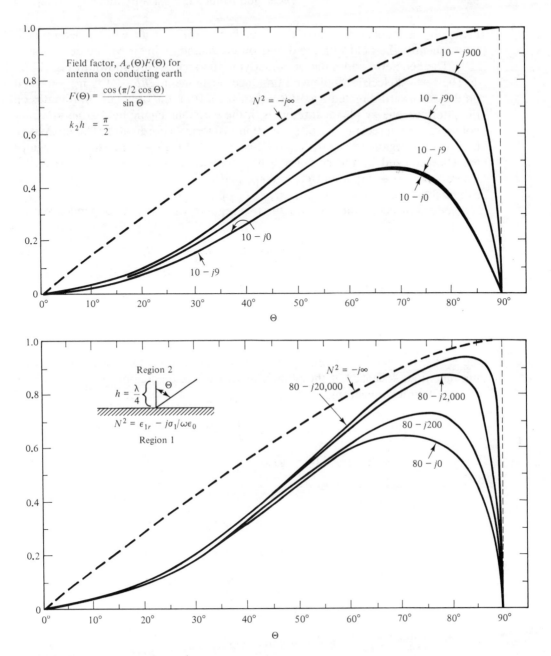

Figure 14.5–2 Vertical field pattern of λ/4 monopole on a conducting earth or water. Note that $j = -i$.

dissipative half-space and near the boundary surface in air will be determined. Subsequently, the field with the dipole on the boundary in air will be determined.

The geometry and notation underlying the analysis are shown in Fig. 14.6–1. The vertical electric dipole with unit electric moment ($2hI_z = 1$ A·m) is located on the downward-directed z axis at a distance d from the origin of coordinates on the interface, the xy plane. Interest is in the electromagnetic field at an arbitrary point (x, y, z) in rectangular or (ρ, θ, z) in cylindrical coordinates. The dissipative half-space is region 1 ($z > 0$); region 2 is air ($z < 0$). In general, the two regions are characterized by the complex wave numbers $k_1 = \beta_1 + i\alpha_1 = \omega(\mu_1\bar{\epsilon}_1)^{1/2}$, where $\bar{\epsilon}_1 = \epsilon_1 + i\sigma_1/\omega$ and $\epsilon_1 = \epsilon_0\epsilon_{1r}$, and $k_2 = \beta_2 + i\alpha_2 = \omega(\mu_2\bar{\epsilon}_2)^{1/2}$. It is assumed that $\mu_1 = \mu_2 = \mu_0$; for air, $\bar{\epsilon}_2 = \epsilon_0$, $\alpha_2 = 0$.

Maxwell's curl equations for the two regions with the time dependence $e^{-i\omega t}$ and $j = 1, 2$ are

$$\nabla \times \mathbf{E}_j = i\omega\mathbf{B}_j = i\omega(\hat{x}B_{jx} + \hat{y}B_{jy}); \qquad B_{jz} = 0 \qquad (14.6\text{–}1a)$$

$$\nabla \times \mathbf{B}_j = -\frac{ik_j^2}{\omega}\mathbf{E}_j + \mu_0\hat{z}J_z \qquad (14.6\text{–}1b)$$

where

$$J_z = \delta(x)\delta(y)\delta(z - d) \qquad (14.6\text{–}2)$$

and $\delta(x)$ is the delta function. J_z as defined in (14.6–2) is the current density of unit amplitude localized at the point $z = d$. Let these equations be transformed with the Fourier transform

$$\mathbf{E}(x, y, z) = (2\pi)^{-2}\int_{-\infty}^{\infty} d\xi \int_{-\infty}^{\infty} d\eta\, e^{i(\xi x + \eta y)}\overline{\mathbf{E}}(\xi, \eta, z) \qquad (14.6\text{–}3a)$$

$$\overline{J}_z(\xi, \eta, z) = \delta(z - d) \qquad (14.6\text{–}3b)$$

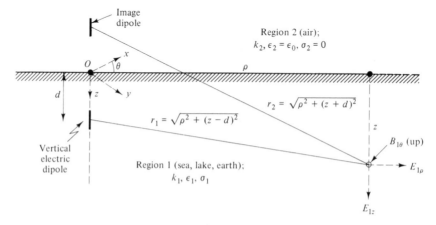

Figure 14.6–1 Vertical dipole at depth d in region 1; electromagnetic field calculated at (ρ, z).

With (14.6–3a,b), equations (14.6–1a,b) become

$$i\eta\overline{E}_{jz} - \frac{\partial\overline{E}_{jy}}{\partial z} = i\omega\overline{B}_{jx}; \qquad \frac{\partial\overline{E}_{jx}}{\partial z} - i\xi\overline{E}_{jz} = i\omega\overline{B}_{jy}$$

$$i\xi\overline{E}_{jy} - i\eta\overline{E}_{jx} = 0 \tag{14.6-4}$$

$$-\frac{\partial\overline{B}_{jy}}{\partial z} = -\frac{ik_j^2}{\omega}\overline{E}_{jx}; \qquad \frac{\partial\overline{B}_{jx}}{\partial z} = -\frac{ik_j^2}{\omega}\overline{E}_{jy}$$

$$i\xi\overline{B}_{jy} - i\eta\overline{B}_{jx} = -\frac{ik_j^2}{\omega}\overline{E}_{jz} + \mu_0\delta(z - d) \tag{14.6-5}$$

It follows that $\overline{E}_{jy} = (\eta/\xi)\overline{E}_{jx}$, $\overline{B}_{jy} = -(\xi/\eta)\overline{B}_{jx}$. Suitable combinations of (14.6–4) and (14.6–5) yield the following ordinary differential equations:

$$\left(\frac{d^2}{dz^2} + \gamma_j^2\right)\overline{B}_{jx} = -i\eta\mu_0\delta(z - d) \tag{14.4-6}$$

with

$$\gamma_j = (k_j^2 - \xi^2 - \eta^2)^{1/2}; \qquad j = 1, 2 \tag{14.6-7}$$

From solutions of (14.6–6) for $\overline{B}_{jx}$, the other components of the field can be determined from

$$\overline{B}_{jy} = -\frac{\xi}{\eta}\overline{B}_{jx} \tag{14.6-8}$$

$$\overline{E}_{jx} = \frac{-i\omega}{k_j^2}\frac{\partial\overline{B}_{jy}}{\partial z} = \frac{i\omega}{k_j^2}\frac{\xi}{\eta}\frac{\partial\overline{B}_{jx}}{\partial z} \tag{14.6-9}$$

$$\overline{E}_{jy} = \frac{i\omega}{k_j^2}\frac{\partial\overline{B}_{jx}}{\partial z} \tag{14.6-10}$$

$$\overline{E}_{jz} = \frac{\omega}{\eta k_j^2}\left(\frac{d^2}{dz^2} + k_j^2\right)\overline{B}_{jx} \tag{14.6-11}$$

The solutions of (14.6–6) that do not become infinite at $z = \pm\infty$ are

$$\overline{B}_{1x}(z) = C_1 e^{i\gamma_1 z} - \frac{\eta\mu_0}{2\gamma_1} e^{i\gamma_1|z - d|}; \qquad z \geq 0 \tag{14.6-12a}$$

$$\overline{B}_{2x}(z) = C_2 e^{-i\gamma_2 z}; \qquad\qquad\qquad z \leq 0 \tag{14.6-12b}$$

The boundary condition $\overline{B}_{1x}(0) = \overline{B}_{2x}(0)$ leads to $C_1 = C_2 + (\eta\mu_0/2\gamma_1)\exp(i\gamma_1 d)$, so that

$$\overline{B}_{1x}(z) = C_2 e^{i\gamma_1 z} + \frac{\eta\mu_0}{2\gamma_1}\left(e^{i\gamma_1(z + d)} - e^{i\gamma_1|z - d|}\right) \tag{14.6-13}$$

With (14.6–9), (14.6–12b), and (14.6–13),

$$\overline{E}_{1x}(z) = \frac{-\omega}{k_1^2}\left[\frac{\gamma_1\xi}{\eta}\,C_2 e^{i\gamma_1 z} + \frac{\xi\mu_0}{2}\left(e^{i\gamma_1(z+d)} \mp e^{i\gamma_1|z-d|}\right)\right]; \qquad \begin{cases} z > d \\ 0 \le z \le d \end{cases} \qquad (14.6\text{–}14a)$$

$$\overline{E}_{2x}(z) = \frac{\omega}{k_2^2}\frac{\gamma_2\xi}{\eta}\,C_2 e^{-i\gamma_2 z}; \qquad\qquad\qquad\qquad z \le 0 \qquad (14.6\text{–}14b)$$

The boundary condition $\overline{E}_{1x}(0) = \overline{E}_{2x}(0)$ leads to

$$C_2 = \frac{-\mu_0\eta k_2^2 e^{i\gamma_1 d}}{k_1^2\gamma_2 + k_2^2\gamma_1} \qquad (14.6\text{–}15)$$

When this value of C_2 is substituted in (14.6–14a, b), (14.6–12b), and (14.6–13), the final expressions for $\overline{E}_{1x}(z)$, $\overline{E}_{2x}(z)$, and $\overline{B}_{2x}(z)$ are obtained. The corresponding solutions for $\overline{E}_{1y}(z)$, $\overline{E}_{2y}(z)$, $\overline{B}_{1y}(z)$, and $\overline{B}_{2y}(z)$ follow from $\overline{E}_{jy} = (\eta/\xi)\overline{E}_{jx}$, $\overline{B}_{jy} = -(\xi/\eta)\overline{B}_{jx}$. Finally, $\overline{E}_{1z}(z)$ and $\overline{E}_{2z}(z)$ are obtained with (14.6–11). The results are as follows:

Region 1, $z > 0$:

$$\overline{B}_{1x}(\xi, \eta, z) = -\mu_0\eta\left[\frac{e^{i\gamma_1|z-d|}}{2\gamma_1} - \frac{e^{i\gamma_1(z+d)}}{2\gamma_1} + \frac{k_2^2 e^{i\gamma_1(z+d)}}{k_1^2\gamma_2 + k_2^2\gamma_1}\right] \qquad (14.6\text{–}16)$$

$$\overline{B}_{1y}(\xi, \eta, z) = -\frac{\xi}{\eta}\overline{B}_{1x}(\xi, \eta, z) \qquad (14.6\text{–}17)$$

$$\overline{E}_{1x}(\xi, \eta, z) = \frac{-\omega\mu_0\xi}{k_1^2}\left[\mp\frac{e^{i\gamma_1|z-d|}}{2} + \frac{e^{i\gamma_1(z+d)}}{2}\right.$$
$$\left. - \frac{k_2^2\gamma_1 e^{i\gamma_1(z+d)}}{k_1^2\gamma_2 + k_2^2\gamma_1}\right]; \qquad \begin{cases} z > d, \\ 0 \le z \le d \end{cases} \qquad (14.6\text{–}18)$$

$$\overline{E}_{1y}(\xi, \eta, z) = \frac{\eta}{\xi}\overline{E}_{1x}(\xi, \eta, z) \qquad (14.6\text{–}19)$$

$$\overline{E}_{1z}(\xi, \eta, z) = \frac{-\omega\mu_0(\xi^2 + \eta^2)}{k_1^2}\left[\frac{e^{i\gamma_1|z-d|}}{2\gamma_1} - \frac{e^{i\gamma_1(z+d)}}{2\gamma_1} + \frac{k_2^2 e^{i\gamma_1(z+d)}}{k_1^2\gamma_2 + k_2^2\gamma_1}\right] \qquad (14.6\text{–}20)$$

Region 2, $z < 0$:

$$\overline{B}_{2x}(\xi, \eta, z) = -\mu_0\eta\frac{k_2^2 e^{i\gamma_1 d}e^{-i\gamma_2 z}}{k_1^2\gamma_2 + k_2^2\gamma_1} \qquad (14.6\text{–}21)$$

$$\overline{B}_{2y}(\xi, \eta, z) = -\frac{\xi}{\eta}\overline{B}_{2x}(\xi, \eta, z) \qquad (14.6\text{–}22)$$

$$\overline{E}_{2x}(\xi, \eta, z) = -\omega\mu_0\xi \frac{\gamma_2 e^{i\gamma_1 d} e^{-i\gamma_2 z}}{k_1^2\gamma_2 + k_2^2\gamma_1} \tag{14.6-23}$$

$$\overline{E}_{2y}(\xi, \eta, z) = \frac{\eta}{\xi} \overline{E}_{2x}(\xi, \eta, z) \tag{14.6-24}$$

$$\overline{E}_{2z}(\xi, \eta, z) = -\omega\mu_0(\xi^2 + \eta^2) \frac{e^{i\gamma_1 d} e^{-i\gamma_2 z}}{k_1^2\gamma_2 + k_2^2\gamma_1} \tag{14.6-25}$$

The substitution of (14.6–16) to (14.6–25) into (14.6–3a) yields the general integrals for the components of the electric and magnetic fields in Cartesian coordinates. These are readily converted to cylindrical coordinates with the relations: $x = \rho \cos \theta$, $y = \rho \sin \theta$, $\xi = \lambda \cos \theta'$, $\eta = \lambda \sin \theta'$, $\rho = (x^2 + y^2)^{1/2}$, $\lambda = (\xi^2 + \eta^2)^{1/2}$, $d\xi \, d\eta = \lambda \, d\theta' \, d\lambda$, $\xi x + \eta y = \lambda\rho \cos (\theta - \theta')$, $E_\rho = E_x \cos \theta + E_y \sin \theta$, $B_\theta = -B_x \sin \theta + B_y \cos \theta$. If use is made of the integral representations of the Bessel functions, that is,

$$J_0(\lambda\rho) = \frac{1}{2\pi} \int_0^{2\pi} e^{i\lambda\rho \cos(\theta - \theta')} \, d\theta'$$

$$J_1(\lambda\rho) = \frac{-i}{2\pi} \int_0^{2\pi} e^{i\lambda\rho \cos(\theta - \theta')} \cos (\theta - \theta') \, d\theta'$$

it follows that $B_{jz}(\rho, z) = 0$, $B_{j\rho}(\rho, z) = 0$, $E_{j\theta}(\rho, z) = 0$, $j = 1, 2$, and

$$B_{1\theta}(\rho, z) = \frac{i\mu_0}{2\pi} \int_0^\infty \left[\frac{e^{i\gamma_1|z - d|}}{2\gamma_1} - \frac{e^{i\gamma_1(z + d)}}{2\gamma_1} \right.$$

$$\left. + \frac{k_2^2 e^{i\gamma_1(z + d)}}{k_1^2\gamma_2 + k_2^2\gamma_1} \right] J_1(\lambda\rho)\lambda^2 \, d\lambda; \qquad z > 0 \quad (14.6\text{--}26)$$

$$E_{1\rho}(\rho, z) = \frac{i\omega\mu_0}{2\pi k_1^2} \int_0^\infty \left[\pm \frac{e^{i\gamma_1|z - d|}}{2} - \frac{e^{i\gamma_1(z + d)}}{2} \right.$$

$$\left. + \frac{k_2^2\gamma_1 e^{i\gamma_1(z + d)}}{k_1^2\gamma_2 + k_2^2\gamma_1} \right] J_1(\lambda\rho)\lambda^2 \, d\lambda; \qquad \begin{cases} z > d, \\ 0 \le z \le d \end{cases} \quad (14.6\text{--}27)$$

$$E_{1z}(\rho, z) = \frac{-\omega\mu_0}{2\pi k_1^2} \int_0^\infty \left[\frac{e^{i\gamma_1|z - d|}}{2\gamma_1} - \frac{e^{i\gamma_1(z + d)}}{2\gamma_1} \right.$$

$$\left. + \frac{k_2^2 e^{i\gamma_1(z + d)}}{k_1^2\gamma_2 + k_2^2\gamma_1} \right] J_0(\lambda\rho)\lambda^3 \, d\lambda; \qquad z > 0 \quad (14.6\text{--}28)$$

$$B_{2\theta}(\rho, z) = \frac{i\mu_0}{2\pi} \int_0^\infty \frac{k_2^2 e^{i\gamma_1 d} e^{-i\gamma_2 z}}{k_1^2\gamma_2 + k_2^2\gamma_1} J_1(\lambda\rho)\lambda^2 \, d\lambda; \qquad z < 0 \quad (14.6\text{--}29)$$

$$E_{2\rho}(\rho, z) = \frac{-i\omega\mu_0}{2\pi} \int_0^\infty \frac{\gamma_2 e^{i\gamma_1 d} e^{-i\gamma_2 z}}{k_1^2\gamma_2 + k_2^2\gamma_1} J_1(\lambda\rho)\lambda^2 \, d\lambda; \qquad z < 0 \qquad (14.6\text{--}30)$$

$$E_{2z}(\rho, z) = \frac{-\omega\mu_0}{2\pi} \int_0^\infty \frac{e^{i\gamma_1 d} e^{-i\gamma_2 z}}{k_1^2\gamma_2 + k_2^2\gamma_1} J_0(\lambda\rho)\lambda^3 \, d\lambda; \qquad z < 0 \qquad (14.6\text{--}31)$$

These are the general integrals for the electromagnetic field of a vertical electric dipole at $z = d$ in the dissipative half-space.

14.7 EVALUATION OF THE FIELD OF A VERTICAL DIPOLE NEAR A PLANE BOUNDARY BETWEEN AIR AND A DISSIPATIVE HALF-SPACE

It is possible to evaluate the integrals in (14.6–26) to (14.6–28) for the field in region 1—the dissipative half-space where $z > 0$—subject to the following conditions:

$$|k_1| \geq 3|k_2|; \qquad \rho \geq 5z; \qquad \rho \geq 5d \qquad (14.7\text{--}1)$$

Note that the condition $|k_1^2| = k_0^2|\epsilon_{1r} + i\sigma_1/\omega\epsilon_0| \geq 9\,|k_2^2|$ does not require region 1 to be conducting. It may be lake water or earth as well as seawater, and the range of frequencies is not otherwise restricted.

  The first step in the evaluation of (14.6–26) to (14.6–28) is to note that in each formula the first term is the appropriate component of the direct field $(B_{1\theta}^d, E_{1\rho}^d, E_{1z}^d)$ of a vertical dipole at $z = d$ in an infinite, rather than semi-infinite region 1. This follows directly since when $k_2 = k_1$, the third term in each formula cancels the second term. Furthermore, the second term in each formula is the field $(B_{1\theta}^i, E_{1\rho}^i, E_{1z}^i)$ of an image dipole at $z = -d$ but, because of the $-$ sign, with its electric moment reversed. The remaining third term includes the surface-wave or lateral-wave field $(B_{1\theta}^L, E_{1\rho}^L, E_{1z}^L)$ and a correction to the reflected field. Thus, with (5.9–3) to (5.9–5) and with $r_1 = [(z - d)^2 + \rho^2]^{1/2}$, $r_2 = [(z + d)^2 + \rho^2]^{1/2}$,

$$B_{1\theta}^d(\rho, z) = \frac{i\mu_0}{4\pi} \int_0^\infty \frac{e^{i\gamma_1|z - d|}}{\gamma_1} J_1(\lambda\rho)\lambda^2 \, d\lambda$$

$$= \frac{-\mu_0}{4\pi} e^{ik_1 r_1} \left(\frac{ik_1}{r_1} - \frac{1}{r_1^2}\right)\frac{\rho}{r_1} \qquad (14.7\text{--}2)$$

$$B_{1\theta}^i(\rho, z) = \frac{-i\mu_0}{4\pi} \int_0^\infty \frac{e^{i\gamma_1(z + d)}}{\gamma_1} J_1(\lambda\rho)\lambda^2 \, d\lambda$$

$$= \frac{\mu_0}{4\pi} e^{ik_1 r_2} \left(\frac{ik_1}{r_2} - \frac{1}{r_2^2}\right)\frac{\rho}{r_2} \qquad (14.7\text{--}3)$$

$$E_{1\rho}^d(\rho, z) = \frac{i\omega\mu_0}{4\pi k_1^2} \int_0^\infty \pm \, e^{i\gamma_1|z-d|} J_1(\lambda\rho)\lambda^2 \, d\lambda$$

$$= \frac{i\omega\mu_0}{4\pi k_1^2} e^{ik_1 r_1} \left(\frac{k_1^2}{r_1} + \frac{3ik_1}{r_1^2} - \frac{3}{r_1^3} \right) \frac{\rho}{r_1} \frac{z-d}{r_1} \tag{14.7-4}$$

$$E_{1\rho}^i(\rho, z) = \frac{-i\omega\mu_0}{4\pi k_1^2} \int_0^\infty e^{i\gamma_1(z+d)} J_1(\lambda\rho)\lambda^2 \, d\lambda$$

$$= \frac{-i\omega\mu_0}{4\pi k_1^2} e^{ik_1 r_2} \left(\frac{k_1^2}{r_2} + \frac{3ik_1}{r_2^2} - \frac{3}{r_2^3} \right) \frac{\rho}{r_2} \frac{z+d}{r_2} \tag{14.7-5}$$

$$E_{1z}^d(\rho, z) = \frac{-\omega\mu_0}{4\pi k_1^2} \int_0^\infty \frac{e^{i\gamma_1|z-d|}}{\gamma_1} J_0(\lambda\rho)\lambda^3 \, d\lambda$$

$$= \frac{-\omega\mu_0}{4\pi k_1^2} e^{ik_1 r_1} \left[\left(\frac{ik_1^2}{r_1} - \frac{k_1}{r_1^2} - \frac{i}{r_1^3} \right) - \left(\frac{z-d}{r_1} \right)^2 \left(\frac{ik_1^2}{r_1} - \frac{3k_1}{r_1^2} - \frac{3i}{r_1^3} \right) \right] \tag{14.7-6}$$

$$E_{1z}^i(\rho, z) = \frac{\omega\mu_0}{4\pi k_1^2} \int_0^\infty \frac{e^{i\gamma_1(z+d)}}{\gamma_1} J_0(\lambda\rho)\lambda^3 \, d\lambda$$

$$= \frac{\omega\mu_0}{4\pi k_1^2} e^{ik_1 r_2} \left[\left(\frac{ik_1^2}{r_2} - \frac{k_1}{r_2^2} - \frac{i}{r_2^3} \right) - \left(\frac{z+d}{r_2} \right)^2 \left(\frac{ik_1^2}{r_2} - \frac{3k_1}{r_2^2} - \frac{3i}{r_2^3} \right) \right] \tag{14.7-7}$$

With the conditions (14.7-1), $r_1 \sim r_2 \sim \rho$ in amplitudes; $\rho/r_1 \sim \rho/r_2 \sim 1$ and terms in $(z-d)^2/r_1^2$ and $(z+d)^2/r_2^2$ are negligible. It follows that

$$B_{1\theta}^d(\rho, z) + B_{1\theta}^i(\rho, z) \sim \frac{-\mu_0}{4\pi} (e^{ik_1 r_1} - e^{ik_1 r_2}) \left(\frac{ik_1}{\rho} - \frac{1}{\rho^2} \right) \tag{14.7-8}$$

$$E_{1\rho}^d(\rho, z) + E_{1\rho}^i(\rho, z)$$

$$\sim \frac{i\omega\mu_0}{4\pi k_1^2} \left(\frac{z-d}{\rho} e^{ik_1 r_1} - \frac{z+d}{\rho} e^{ik_1 r_2} \right) \left(\frac{k_1^2}{\rho} + \frac{3ik_1}{\rho^2} - \frac{3}{\rho^3} \right) \tag{14.7-9}$$

$$E_{1z}^d(\rho, z) + E_{1z}^i(\rho, z) \sim \frac{-\omega\mu_0}{4\pi k_1^2} (e^{ik_1 r_1} - e^{ik_1 r_2}) \left(\frac{ik_1^2}{\rho} - \frac{k_1}{\rho^2} - \frac{i}{\rho^3} \right) \tag{14.7-10}$$

It is now necessary to evaluate the remaining terms given by the third integral in each of the formulas (14.6–26) to (14.6–28). They include the surface-wave or lateral-wave terms $[B_{1\theta}^L(\rho, z)$, and $E_{1\rho}^L(\rho, z)$, and $E_{1z}^L(\rho, z)]$ and a correction term $[E_{1\rho}^c(\rho, z)]$ of the reflected field. Thus

$$B_{1\theta}^L(\rho, z) = \frac{i\mu_0}{2\pi} F_{1\theta}(\rho, z+d)$$

$$\tag{14.7-11}$$

$$F_{1\theta}(\rho, z+d) = k_2^2 \int_0^\infty \frac{e^{i\gamma_1(z+d)}}{k_1^2\gamma_2 + k_2^2\gamma_1} J_1(\lambda\rho)\lambda^2 \, d\lambda$$

$$E^L_{1\rho}(\rho, z) + E^c_{1\rho}(\rho, z) = \frac{i\omega\mu_0}{2\pi k_1^2} F_{1\rho}(\rho, z + d)$$

$$F_{1\rho}(\rho, z + d) = k_2^2 \int_0^\infty \frac{\gamma_1 e^{i\gamma_1(z+d)}}{k_1^2\gamma_2 + k_2^2\gamma_1} J_1(\lambda\rho)\lambda^2 \, d\lambda \qquad (14.7\text{--}12)$$

$$E^L_{1z}(\rho, z) = \frac{-\omega\mu_0}{2\pi k_1^2} F_{1z}(\rho, z + d)$$

$$F_{1z}(\rho, z + d) = k_2^2 \int_0^\infty \frac{e^{i\gamma_1(z+d)}}{k_1^2\gamma_2 + k_2^2\gamma_1} J_0(\lambda\rho)\lambda^3 \, d\lambda \qquad (14.7\text{--}13)$$

In (14.7–12) $E^c_{1\rho}(\rho, z)$ is a correction of the reflected field.

These integrals can be evaluated subject to (14.7–1) and the following approximation:

$$F_{1m}(\rho, z + d) \sim F_{1m}(\rho, 0)e^{ik_1(z+d)}; \qquad m = \theta, \rho, z \qquad (14.7\text{--}14)$$

This approximation has been verified by a direct comparison of the results obtained with (14.7–14) and by the numerical evaluation of the integrals. It can be justified physically by noting that (14.7–14) implies that the surface wave originates in the air above and propagates vertically downward in region 1. This is true for the amplitude fronts but only approximately for the phase fronts which continue downward at small angles from the normal that depend on the angle of incidence. This difference is unimportant at small depths and significant at large depths only in very low loss materials. The expressions obtained for the field with (14.7–14) and verified numerically support this physical picture. Note that the approximation (14.7–14) is not required when the dipole is on the surface ($d = 0$) and the field is observed on the surface ($z = 0$).

The three integrals $F_{1\theta}(\rho, 0)$, $F_{1\rho}(\rho, 0)$, and $F_{1z}(\rho, 0)$ as given by (14.7–11) to (14.7–13) have been evaluated* subject to (14.7–1). The results are

$$B^L_{1\theta}(\rho, z) = \frac{-\mu_0 k_2^2}{2\pi k_1^2} e^{ik_1(z+d)} e^{ik_2\rho} f(k_2\rho, k_1) \qquad (14.7\text{--}15)$$

$$E^L_{1\rho}(\rho, z) = \frac{\omega}{k_1} B^L_{1\theta}(\rho, z); \qquad E^c_{1\rho}(\rho, z) = \frac{i\omega\mu_0 k_2^2}{2\pi k_1^3} e^{ik_1(z+d)} \frac{e^{ik_1\rho}}{\rho^2} \qquad (14.7\text{--}16)$$

$$E^L_{1z}(\rho, z) = \frac{\omega\mu_0 k_2^3}{2\pi k_1^4} e^{ik_1(z+d)} e^{ik_2\rho} g(k_2\rho, k_1) \qquad (14.7\text{--}17)$$

where

$$f(k_2\rho, k_1) = \frac{ik_2}{\rho} - \frac{1}{\rho^2} - \frac{k_2^3}{k_1}\left(\frac{\pi}{k_2\rho}\right)^{1/2} e^{-ik_2^3\rho/2k_1^2} \mathscr{F}(k_2\rho, k_1) \qquad (14.7\text{--}18a)$$

---

* R. W. P. King, "New Formulas for the Electromagnetic Field of a Vertical Electric Dipole in a Dielectric or Conducting Half-Space Near Its Horizontal Interface," *J. Appl. Phys.*, *53*, 8476–8482 (1982); Erratum, *56*, 3366 (1984).

$$g(k_2\rho, k_1) = f(k_2\rho, k_1) - \frac{i}{k_2\rho^3} \qquad (14.7\text{--}18b)$$

with

$$\mathcal{F}(k_2\rho, k_1) = \frac{1}{2}(1 + i) - C_2\left(\frac{k_2^3\rho}{2k_1^2}\right) - iS_2\left(\frac{k_2^3\rho}{2k_1^2}\right) \qquad (14.7\text{--}19a)$$

Here

$$C_2(u) + iS_2(u) = \int_0^u \frac{e^{it}}{(2\pi t)^{1/2}} dt \qquad (14.7\text{--}19b)$$

is the Fresnel integral. Note that when $k_2\rho \geq 3$, the term $i/k_2\rho^3$ in (14.7–18b) is negligible compared with $f(k_2\rho, k_1)$ so that $g(k_2\rho, k_1) \sim f(k_2\rho, k_1)$. Note also that when $k_2\rho > |k_1^2/k_2^2|$,

$$f(k_2\rho, k_1) \sim g(k_2\rho, k_1) \sim \frac{-k_1^2}{k_2^2\rho^2} \qquad (14.7\text{--}20)$$

since $(k_2^3/k_1)(\pi/k_2\rho)^{1/2} \exp(-ik_2^3\rho/2k_1^2) \, \mathcal{F}(k_2\rho, k_1) \sim (ik_2/\rho) + (k_1^2/k_2^2\rho^2)$.

The complete field at depth z in region 1 due to a vertical dipole at depth d also in region 1 is the sum of the direct and reflected fields given in (14.7–8) to (14.7–10) and the surface-wave and correction terms in (14.7–15) to (14.7–17). However, for all practically available materials that form the surface of the earth, the attenuation constant α_1 is sufficiently large so that all terms with the factor $\exp(ik_1\rho) = \exp(-\alpha_1\rho)\exp(i\beta_1\rho)$ decay rapidly with radial distance. Only close to the source is the magnitude of the direct and reflected fields comparable with that of the surface-wave field. Here their superposition can produce an interference pattern. This is observable in lake water with low conductivity and in earth and salt water at very high frequencies. At all moderate and large distances from the source, the surface-wave field is alone significant in region 1 and along its boundary with region 2.

The field in air (region 2) along the boundary with region 1 is readily obtained with the boundary conditions:

$$B_{2\theta}(\rho, 0) = B_{1\theta}(\rho, 0); \qquad E_{2\rho}(\rho, 0) = E_{1\rho}(\rho, 0)$$

$$E_{2z}(\rho, 0) = \frac{k_1^2}{k_2^2} E_{2z}(\rho, 0) \qquad (14.7\text{--}21)$$

When the vertical dipole is in region 2 on the boundary ($d = 0$) instead of in region 1 ($d \geq 0$), the electromagnetic field in region 1 is given by (14.6–29) to (14.6–31) with subscripts 1 and 2 interchanged, $-z$ replaced by z, $2hI_z = 1$ replaced by $2hI_z = -1$, and $d = 0$; that is, by

$$B_{1\theta}(\rho, z) = \frac{i\mu_0 k_1^2}{2\pi} \int_0^\infty \frac{e^{i\gamma_1 z}}{k_1^2\gamma_2 + k_2^2\gamma_1} J_1(\lambda\rho)\lambda^2 \, d\lambda; \qquad z > 0 \quad (14.7\text{--}22)$$

$$E_{1\rho}(\rho, z) = \frac{i\omega\mu_0}{2\pi} \int_0^\infty \frac{\gamma_1 e^{i\gamma_1 z}}{k_1^2 \gamma_2 + k_2^2 \gamma_1} J_1(\lambda\rho)\lambda^2 \, d\lambda; \qquad z > 0 \quad (14.7\text{-}23)$$

$$E_{1z}(\rho, z) = \frac{-\omega\mu_0}{2\pi} \int_0^\infty \frac{e^{i\gamma_1 z}}{k_1^2 \gamma_2 + k_2^2 \gamma_1} J_0(\lambda\rho)\lambda^3 \, d\lambda; \qquad z > 0 \quad (14.7\text{-}24)$$

Since the integrals in these formulas are the same as those in (14.7–11) to (14.7–13) when $d = 0$, it follows with (14.7–15) to (14.7–17) that

$$B_{1\theta}(\rho, z) = \frac{-\mu_0}{2\pi} e^{ik_{1z}z} e^{ik_2\rho} f(k_2\rho, k_1) \qquad (14.7\text{-}25)$$

$$E_{1\rho}(\rho, z) = \frac{-\omega\mu_0}{2\pi k_1} e^{ik_{1z}z} \left[e^{ik_2\rho} f(k_2\rho, k_1) - \frac{ie^{ik_1\rho}}{\rho^2} \right] \qquad (14.7\text{-}26)$$

$$E_{1z}(\rho, z) = \frac{\omega\mu_0 k_2}{2\pi k_1^2} e^{ik_{1z}z} e^{ik_2\rho} g(k_2\rho, k_1) \qquad (14.7\text{-}27)$$

This is the field at depth z in region 1 when the vertical dipole is located in region 2 on the boundary surface, $d = 0$. It is greater than the corresponding field when the same dipole is in region 1 at $d = 0$ by the large factor k_1^2/k_2^2.

The field along the surface $z = 0$ in region 2 (air) is obtained from the field at $z = 0$ in region 1 with the boundary conditions. The results are

$$B_{2\theta}(\rho, 0) = \frac{-\mu_0}{2\pi} e^{ik_2\rho} f(k_2\rho, k_1) \qquad (14.7\text{-}28)$$

$$E_{2\rho}(\rho, 0) = \frac{-\omega\mu_0}{2\pi k_1} \left[e^{ik_2\rho} f(k_2\rho, k_1) - \frac{ie^{ik_1\rho}}{\rho^2} \right] \qquad (14.7\text{-}29)$$

$$E_{2z}(\rho, 0) = \frac{\omega\mu_0}{2\pi k_2} e^{ik_2\rho} g(k_2\rho, k_1) \qquad (14.7\text{-}30)$$

where $f(k_2\rho, k_1)$ and $g(k_2\rho, k_1)$ are defined in (14.7–18a,b). This is the field in air along the boundary surface ($z = 0$) when the vertical dipole is also in air on the boundary ($d = 0$).

14.8 PROPERTIES OF THE FIELD OF A VERTICAL DIPOLE NEAR A PLANE BOUNDARY BETWEEN AIR AND A DISSIPATIVE HALF-SPACE

In conjunction with radio transmission along the surface of the earth, the electromagnetic field of a vertical electric dipole in the air on the earth is of primary interest. The electric field of such a dipole is given by (14.5–1) supplemented with (14.7–29) and (14.7–30) and simplified for distances for which $k_2\rho \gg 1$ and $\exp(ik_1\rho) = \exp(-\alpha_1\rho)\exp(i\beta_1\rho)$ is negligibly small. Since E_Θ given by (14.5–1) vanishes at $\Theta = \pi/2$ or $z = 0$, the complete field in air along the boundary is the surface-wave field. In the air it is

$$\mathbf{E}_2(\rho, 0) \sim \frac{-\omega\mu_0}{2\pi k_1} f(k_2\rho, k_1) e^{ik_2\rho} \left[\hat{\boldsymbol{\rho}} - \hat{\mathbf{z}}\left(\frac{k_1}{k_2}\right) \right]; \qquad z = 0 \text{ in region 2} \qquad (14.8\text{-}1)$$

In the earth,

$$\mathbf{E}_1(\rho, z) \sim \frac{-\omega\mu_0}{2\pi k_1} f(k_2\rho, k_1) e^{ik_2\rho} e^{ik_{1z}} \left[\hat{\boldsymbol{\rho}} - \hat{\mathbf{z}}\left(\frac{k_2}{k_1}\right) \right]; z \geq 0 \text{ in region 1} \qquad (14.8\text{-}2)$$

The time-dependent electric vectors are

$$\mathbf{E}_2(\rho, 0; t) = \text{Re} \left\{ K(\rho) e^{-i\theta\kappa(\rho,0)} \left[\hat{\boldsymbol{\rho}} e^{-i\omega t} - \hat{\mathbf{z}} \left| \frac{k_1}{k_2} \right| e^{-i(\omega t - \theta_1)} \right] \right\} \qquad (14.8\text{-}3)$$

$$\mathbf{E}_1(\rho, z; t) = \text{Re} \left\{ K(\rho) e^{-i\theta\kappa(\rho,z)} \left[\hat{\boldsymbol{\rho}} e^{-i\omega t} - \hat{\mathbf{z}} \left| \frac{k_2}{k_1} \right| e^{-i(\omega t + \theta_1)} \right] \right\} \qquad (14.8\text{-}4)$$

where

$$\theta_1 = \tan^{-1} \frac{\alpha_1}{\beta_1}; \qquad K(\rho) e^{-i\theta\kappa(\rho,z)} = \frac{-\omega\mu_0}{2\pi k_1} f(k_2\rho, k_1) e^{ik_2\rho} e^{ik_{1z}} \qquad (14.8\text{-}5)$$

These equations can be reduced to the following forms:

$$\left| \frac{k_2}{k_1} \right|^2 E_{2z}^2(\rho, 0; t) - 2 \left| \frac{k_2}{k_1} \right| E_{2z}(\rho, 0; t) E_{2\rho}(\rho, 0; t) \cos \theta_1 \qquad (14.8\text{-}6)$$

$$+ E_{2\rho}^2(\rho, 0; t) - \sin^2 \theta_1 = 0$$

$$\left| \frac{k_1}{k_2} \right|^2 E_{1z}^2(\rho, z; t) - 2 \left| \frac{k_1}{k_2} \right| E_{1z}(\rho, z; t) E_{1\rho}(\rho, z; t) \cos \theta_1 \qquad (14.8\text{-}7)$$

$$+ E_{1\rho}^2(\rho, z; t) - \sin^2 \theta_1 = 0$$

which are the equations of ellipses with their major axes rotated with respect to the z axis by the angles

$$\psi_1 = \frac{1}{2} \tan^{-1} \frac{2|k_2/k_1| \cos \theta_1}{1 - |k_1^2/k_2^2|} = \frac{1}{2} \tan^{-1} \frac{2k_2\beta_1}{|k_1|^2} \sim \frac{k_2\beta_1}{|k_1|^2} \qquad (14.8\text{-}8)$$

$$\psi_2 = \frac{\pi}{2} - \frac{1}{2} \tan^{-1} \frac{2k_2\beta_1}{|k_1|^2} \sim \frac{\pi}{2} - \frac{k_2\beta_1}{|k_1|^2} \qquad (14.8\text{-}9)$$

Use has been made of the condition $|k_2/k_1|^2 \leq \frac{1}{9}$.

These equations indicate that the surface wave that travels radially outward from the dipole along the surface $z = 0$ in region 2 (air) consists of a linearly polarized transverse magnetic field $B_{2\theta}(\rho, 0)$ and an elliptically polarized electric field with its major axis inclined slightly from the plane perpendicular to the direction of propagation $\hat{\boldsymbol{\rho}}$. Note that $|E_{2z}/E_{2\rho}| \sim |k_1/k_2|$ is large and ψ_1 is small. The surface wave is not a transverse plane wave but an inhomogeneous wave since it has the component $E_{2\rho}$ in the direction of propagation. If region 1 were a perfect conductor with $\sigma_1 = \infty$ and $k_1 = \infty$, $E_{2\rho}(\rho, 0)$ and the angle of tilt ψ_1 would be zero and the remaining field (characterized by $B_{2\theta}$ and E_{2z}) would proceed as a transverse plane wave.

The field in region 1, the dissipative half-space, consists of a wave that travels vertically downward from the surface. It is composed of a transverse magnetic field $B_{1\theta}(\rho, z)$ and an elliptically polarized electric field. This has its major axis inclined slightly from the plane perpendicular to the direction of propagation, $\hat{\mathbf{z}}$. Note that $|E_{1\rho}/E_{1z}| \sim |k_1/k_2|$ is large and $(\pi/2) - \psi_2$ is small.

The general behavior of $E_{2z}(\rho, 0)$ due to a vertical dipole in air on the surface of seawater is shown as a function of the radial distance ρ in Fig. 14.8–1 in the form $20 \log_{10}|E_{2z}(\rho, 0)|$. It is seen that in the range $0 \le k_2\rho \le 1$, $|E_{2z}(\rho, 0)|$ is proportional to $1/\rho^3$; in the range $1 \le k_2\rho \le 8|k_1^2/k_2^2|$, $|E_{2z}(\rho, 0)|$ is proportional to $1/\rho$; and finally in the range $k_2\rho \ge 8|k_1^2/k_2^2|$ it is proportional to $1/\rho^2$. This general behavior obtains for all media that satisfy the condition $|k_1| \ge 3k_2$. It indicates that in the intermediate range the surface-wave field decreases at the same rate as the far field of an isolated dipole in air, but that in the far field the rate of decrease of the surface wave is much more rapid as $1/\rho^2$. This suggests that for optimum transmission, frequencies should be selected for which the maximum radial distance of interest does not extend beyond the intermediate range. Graphs of $20 \log_{10}|E_{2z}(\rho, 0)|$ for a wide range of frequencies and radial distances are shown in Fig. 14.8–2. It is seen that in the near-field range, unit dipoles at low frequencies have stronger signals, but that as the distance is increased to the intermediate or far field, higher frequencies are superior.

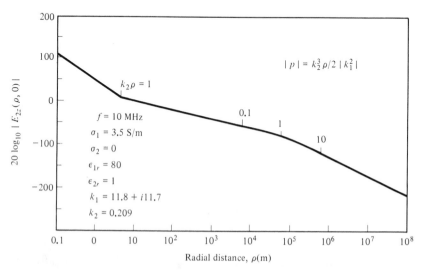

Figure 14.8–1 $E_{2z}(\rho, 0)$ of vertical unit dipole in air on the surface of seawater.

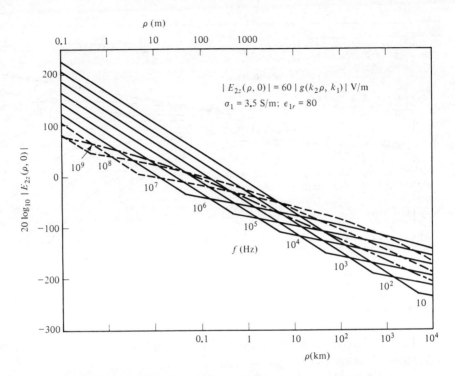

Figure 14.8–2 Magnitude of $E_{2z}(\rho, 0)$ due to a vertical unit electric dipole in air above the surface of seawater.

The surface-wave field generated by a vertical monopole on the shore of the ocean provides a useful means for communicating with submarines at a radial distance ρ and a depth z in the ocean. For antennas with electrical lengths $k_2 h \leq \pi/2$, the distribution of current for calculating the radiated field is adequately approximated by

$$I_z(z) = \frac{V_0^e \sin k_2(h - |z|)}{Z_0 \sin k_2 h} \tag{14.8–10}$$

where Z_0 is the driving-point impedance. For the calculation of the field in the equatorial plane, the antenna can be represented by an equivalent monopole with the uniform current $I_z(0)$ and an effective length h_e defined by

$$I_z(0)h_e = \int_0^h I_z(z)\, dz \tag{14.8–11}$$

The electrical effective length is

$$k_2 h_e = \int_0^h \frac{\sin k_2(h - z)}{\sin k_2 h} k_2\, dz = \tan \frac{k_2 h}{2} \tag{14.8–12}$$

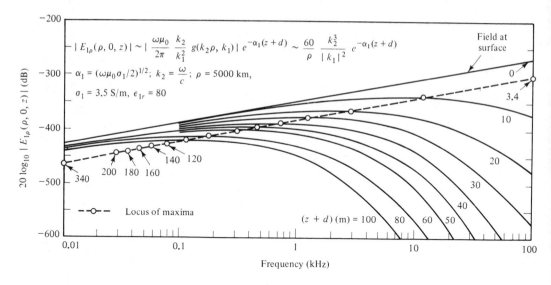

Figure 14.8–3 Magnitude of $E_{1\rho}(\rho, 0, z)$ at depth z and $\rho = 5000$ km due to horizontal electric dipole at depth d in seawater as a function of frequency with $z + d$ as parameter.

For an electrically short antenna with $k_2^2 h^2 \ll 1$, $h_e \sim h/2$. If the antenna is sufficiently top-loaded, $h_e \sim h$.

The radial electric field at $\rho = 5000$ km generated by a horizontal dipole with $I_z(0)h_e = 1$ A·m is shown in Fig. 14.8–3 with the combined depth $(z + d)$ as the parameter. It is seen that for depths of 10 to 20 m, frequencies in the range $f = 10$ to 20 kHz provide a maximum field. For greater depths, lower frequencies are better. For example, at $z + d = 50$ m, the maximum field is generated with $f = 1$ kHz. The field of a vertical dipole on the surface of the sea—of interest here— differs from that shown in Fig. 14.8–3 with $d = 0$ only by the factor $k_1/k_2 = (\sigma_1/\omega\varepsilon_1)^{1/2}$. This raises the level by 80 dB at $f = 0.01$ kHz and by 40 dB atr 100 kHz. The conclusions are the same. It is important to note that physical limitations make it impossible to construct vertical monopoles with a large enough electric moment at $f = 1$ kHz whereas this is not true of horizontal dipoles in the sea.

Historical and Critical Retrospect

In formulating the theoretical foundations of electromagnetism a purely logical approach was chosen in the interest of simplicity and coherence. As a result, the historical development of the science of electricity was ignored completely, and only casual mention was made of those illustrious men of genius whose imagination, perseverance, and skill erected one of the truly great monuments to the human intellect. Although a brief summary is not adequate to do justice to the amazing and fascinating account of intellectual adventure and achievement which is the true history of electrical science, a short outline of its highlights will nevertheless, serve to illuminate the following important lessons which may be learned from the story of electromagnetism: (1) An entire field of engineering and of technological industry may owe its existence to abstract mathematical theory; and (2) preconception and tradition may be as important as scientific fact in determining the nonmathematical descriptive text of a physical theory.

The history of electromagnetism may be treated conveniently under three headings, which characterize the guiding principles in successive periods of its growth. From the purely mathematical point of view, the combined achievements in all three periods form a single, powerful tool for predicting new measured data in intricate electrophysical and electrotechnical experiments. On the other hand, the physical theories and the philosophical implications that are associated with the mathematical symbolism differ greatly and fundamentally in successive periods. Indeed, it is amazing that different physical pictures or philosophical interpretations could be so irreconcilable and so positive when there is no conflict in the mathematics for which they seek to provide the explanatory text. The basic conflict in the several interpretations of electromagnetic action center about the fundamental physical and psychological differences between *action by contact* and *action at a distance*. Shall it be postulated that two bodies can interact only by direct contact with each other or with a continuous material medium that makes contact with both? Or shall it be assumed that nature may be so constituted that the interaction of two or more bodies requires no contact and may be accomplished when they are widely separated in a matterless void? In other words, do gravitational attraction, and electric and magnetic attraction and repulsion, actually require contact with a physically meaningful material medium between two interaction entities so that no experimentally verifiable theory can be devised without it? Or is it adequate to proceed on the assumption that such a medium does not actually exist and, hence, is not required in a consistent and comprehensive theory? In studying the answer to this question care must be taken to distinguish between arguments based on traditions of thought and evidence based on experimental observation.

The concept of action by contact or through a material medium is familiar to all in the sense of pushing a wheelbarrow or pressing on the pedal of a hydraulic brake. Historically, it is as old as coordinated thought. Aristotle believed in action by contact. Newton in a letter to Bentley considered the very idea that one body could act on another through empty space absurd. Perhaps the concept of action at a distance might have been acceptable to early scientific thinkers if they had been more interested in natural law as revealed in visible nature than in systems of their own design based on their preconceptions. For example, if they had merely taken for granted that the falling of a body is due to an interaction between it and the earth, and that the motion of the planets is possible because of mutual action between them and the sun, action at a distance might have been accepted as readily as action by contact. Actually, for the evolving scientific mind, an explanation for the process of *throwing* a stone was somehow so much nearer at hand than for just *dropping* the stone that action by contact became reasonable, action at a distance puzzling and absurd. Thus, if the stone was not set in motion by *a visible push*, there must be an invisible something to force it to the surface of the earth.

In spite of his own belief in action by contact or through a mechanical medium, it was Newton who was largely responsible for the first widespread acceptance of the concept of action at a distance. He expressed the law of gravitation in terms of a force acting at a distance to attract two masses. This formulation was destined to plant in the minds of generations to come the idea that action at a distance between masses without reference to a medium might be a fundamental property of nature. But for more than half a century action by contact with a medium dominated scientific thought, and attempts were made, notably by the great mathematician Euler in the years 1746–1748, to formulate comprehensive theories of gravitation, of the transmission of light (following earlier work by Huygens), and of the interaction of permanent magnets in which an all-pervading material medium, the ether, played a dominant role. Euler described the ether as a perfectly elastic substance of very small density.

Action at a Distance between Charges and Currents. In view of the widespread and traditional belief in action by contact in a medium in other fields, it is surprising that early work in electrostatics and electromagnetism developed in a way that led to the general acceptance of the conception of *action at a distance between electric charges and currents.* This may be traced directly to Newton's inverse square law of gravitation which served as a model in the formulation of the laws of force in electrostatics and magnetostatics. The inverse square law of electric charges was first stated by Priestly in 1766 using an electrometer. It was discussed by Cavendish in 1771 and finally formulated by Coulomb in 1785 following experiments using a torsion balance. The inverse square law for magnetic poles was first expressed by Michell in 1750. The magnetic effects of currents were investigated especially by Oersted, Biot, Savart, and Faraday in the time from 1820 to 1821. In this period, Laplace formulated a law of action at a distance between elements of current and magnetic poles. In 1823, Ampère conducted his celebrated experimental researches and formulated his fundamental law of force between currents, a law of action at a distance. Ohm's law was announced in 1826; Faraday formulated his law of induction in 1832; Lenz's law followed in 1834. In this period the theory of potential was developed by Gauss and independently by Green. Later work on induction due to moving conductors carrying currents and due to the rise and decay of currents was published by Neumann and Weber in 1845–1847. In 1845, Kirchhoff worked out his well-known laws for closed circuits.

In 1846, Weber began a systematic study of Ampère's fundamental law of instantaneous action at a distance from the point of view of the interaction between discrete charges. He took into account both the velocities and accelerations of the charges. Other important work in the formulation of a fundamental law of electromagnetic action at a distance was done by Grassmann, Riemann, and Clausius. The first suggestion of a *propagation of such action with a finite velocity* appears to be due to Gauss (1845). In 1858, the celebrated mathematician Riemann presented a paper in which he introduced a finite velocity of propagation of electromagnetic effects and showed this velocity to be equal to that of light. In 1867, the Danish physicist L. Lorenz (not to be confused with the Dutch physicist H. A. Lorentz) extended earlier work by Neumann to obtain formulas for *retarded scalar and vector potentials* essentially like those which are the basis of the theory of retarded action at a distance without an intervening medium. Recognizing that light is electromagnetic in nature, Lorenz concluded that a hypothetical, all-pervading medium such as the ether was no more required for the transmission of light than for other electromagnetic phenomena.

The actual formulation of the general laws of electromagnetism in terms of a retarded action at a distance between charges and currents without an accessory medium was thus within reach. It was, however, not to be enunciated for another half-century because the work of Lorenz was completely eclipsed by the independent, brilliant, and comprehensive theory of Maxwell which was destined to entangle all electromagnetic phenomena in the ether hypothesis.

Action in an Elastic Ether.   In 1847, Kelvin exhibited analogies between electric and elastic phenomena and made a first attempt to treat electrical experiments by the equations of the theory of elasticity. This work, together with Faraday's conceptions of lines and tubes of electric and magnetic force, inspired Maxwell to begin in 1864 the investigation of electrical phenomena from the point of view of conditions existing in a hypothetical, continuous, all-pervading mechanical medium, rather than in terms of interactions of discrete charges and currents. As a first step, he showed that the magnetic **B** vector could be represented formally by the velocity of flow in an incompressible fluid. He ultimately devised a set of field equations governing all electromagnetic phenomena. They were so general that their solutions included Coulomb's law, Ampère's formula, the laws of induction, and the propagation of electromagnetic effects with the velocity of light. Maxwell, like Lorenz, identified light as an electromagnetic phenomenon, but he concluded that all electromagnetic effects must be propagated in the optical ether. Thus while Lorenz, proceeding from the laws of action between charges and currents and apparently completely unaware of the work three years earlier by Maxwell, wished to dispense with the optical ether as unnecessary, Maxwell's *initial assumption* that all electromagnetic effects resided in a mechanical medium necessarily led him to the opposite conclusion. Instead of abandoning the optical ether as unnecessary, Maxwell elevated it to the most fundamental role in electromagnetic theory. Indeed, electromagnetism following Maxwell became a mechanical theory of the state of an elastic medium.

After Maxwell's intricate analytical formulation had been simplified by Heaviside and Hertz, the field equations became the core of electromagnetic theory. Although the equations themselves in no way required the Maxwellian interpretation in terms of the ether, this was universally accepted, and the great success of the equations in predicting and coordinating diverse electromagnetic effects was assumed to substantiate the validity not only of the equations, but of the mechanical models in terms of which they had been derived. Faraday's lines and tubes of force became essential properties of the ether along with strains and displacements. Energy was pictured as stored in the strained medium, much as if this were filled with stretched rubber bands, and in 1884 Poynting defined the vector bearing his name which was assumed to govern the flow of energy distributed in the medium. In 1887, Hertz demonstrated experimentally the existence of what were interpreted to be electromagnetic waves in the ether. In a series of brilliant experiments he measured the velocity of propagation and showed their properties of reflection, refraction, and polarization. *Thus the first radio transmission was achieved in order to confirm the mathematical predictions of the field equations of Maxwell.*

From the point of view of mechanically minded physicists, the entire Maxwellian theory together with all its assumptions and interpretations was ideal, it fitted so perfectly into their preconceptions. Every triumph of the field equations became new proof of the correctness not only of the equations, but more significantly of the physical pictures devised by Faraday and Maxwell. In time, especially in the views of practical scientists, the pictures rather than the intricate equations became identified with the theory. The lines of force, the displacements, and the stored energy became increasingly real in the minds of students as texts and teachers used them and embellished them. The mathematical predictions of Maxwell's equations thus became completely enveloped in an ether the existence of which had never been proved either real or necessary and the strange properties of which were dogmatically inculcated into the minds of students as truth itself.

Retarded Action at a Distance between Charges and Currents.   In 1895, the Dutch physicist H. A. Lorentz took a great step in coordinating the early restricted theories of action between charges and currents with Maxwell's general theory of the state of the ether. Lorentz conceived matter to contain electric charges (electrons) that act on each other in various ways to produce all electromagnetic (including optical) effects. Instead of acting on one another at a distance, Lorentz assumed the charges and their motions to establish in Maxwell's ether precisely those conditions of strain and displacement required by the Maxwellian interpretation of the field equations. In Lorentz's theory these conditions constitute the so-called *electromagnetic field*, and it is the field that characterizes and is propagated by the ether. From its primary Maxwellian role as the ultimate seat of all electromagnetic and optical phenomena the ether is reduced to play only a secondary part in Lorentz's theory, where it is a mere means of transporting electrical effects from one charge to another. Evidently, the ether was no longer indispensable, it was needed only to satisfy the traditional preconception that action at a distance is a priori inadmissible, that natural law necessarily requires action by contact. Clearly, a mere willingness to admit the possibility that electric charges can exert forces directly on one another even when separated in vacuum would make the ether completely unnecessary in the electron theory. Thus in Lorentz's theory the significance of the ether is more pyschological than physical.

In formulating his theory of the ether, Maxwell must have assumed that incontestable experimental evidence for the existence of the ether would be forthcoming in due course. But this has not been the case. Experiments skillfully devised to verify the existence of the ether, notably the ether-drift experiments, have without exception failed. Every experimental inquiry has given the same answer: All natural phenomena proceed exactly as if there were no ether. Furthermore, as a result of the theory of relativity, the ether hypothesis is faced with insurmountable logical difficulties., For example, the constancy of the electromagnetic velocity with respect to moving observers demands that each observer have his own ether that moves with him. Unless and until conclusive evidence is provided that a material, all-pervading ether exists in a true physical sense, it must be concluded that *there is no ether*. Therefore, Maxwell's entire theory of the electromagnetic ether must necessarily become a part of the history and no longer a part of the practice of physical science. The Maxwell field equations, on the other hand, will continue to be the mathematical heart of macroscopic electromagnetic theory.

In spite of the negative answers of ether-drift experiments, and the general acceptance of the theory of relativity, the Maxwellian ether with all its implications is still made to play an apparently significant part in technical work. Those who continue to believe in the ether argue that nature may be so constituted as to conceal forever its existence. Since anything may be assumed to exist on the same basis, this is wishful, not scientific, thinking. Even engineers and physicists who are willing to agree that the ether is nonexistent and unnecessary, nevertheless continue to do all their electromagnetic thinking and teaching in terms of Maxwellian pictures. Such adherence to traditions in the face of contrary scientific evidence is carefully avoided in this book. Since the action of charge on charge requires no medium, none is introduced and all explanations are made without it. The fundamental law of macroscopic electromagnetism as expressed in the field and force equations is interpreted as a retarded action at a distance. The universal constants appearing in this law are assumed to be defined operationally in terms of experiments devised to measure them. They are, of course, not assigned a *localized* physical significance as properties of a medium. The electromagnetic field and the fields of the potential functions serve merely as intermediate steps in a mathematical calculation of action between statistical distributions of charge and current.

Appendix

1

Differential Operators; Vector Formulas and Identities

A.I–1   *DIFFERENTIAL OPERATORS*

The equations for potentials involve the differential operators

$$\nabla^2 \phi \equiv \nabla \cdot \nabla \phi \tag{A.I–1}$$

$$\nabla^2 \mathbf{A} \equiv \nabla \nabla \cdot \mathbf{A} - \nabla \times \nabla \times \mathbf{A} \tag{A.I–2}$$

and hence

$$\nabla \phi; \quad \nabla \cdot \mathbf{A}; \quad \nabla \times \mathbf{A} \tag{A.I–3}$$

In order to express (A.I–1) and (A.I–2) in systems of orthogonal coordinates, it is necessary to obtain expressions for the operators ∇, ∇ ; $\nabla \times$ in general orthogonal curvilinear coordinates U, V, W and then specialize these to each system.

Let the curvilinear coordinate axes of U, V, W be drawn at a point A shown in Fig. A.I–1. At the point of intersection, the tangents to these axes are mutually perpendicular. In advancing along the positive U axis a distance ds_U, from A to D, the coordinate U changes by an amount dU. Similarly, traversing a distance ds_V in passing from A to B along the positive V axis, V changes by dV. In moving a distance ds_W from A to E along the W axis, W changes by dW. It is important to note that U, V, W do not necessarily have the dimension of a simple length as is true in the Cartesian system. However, dU, dV, dW are always functionally

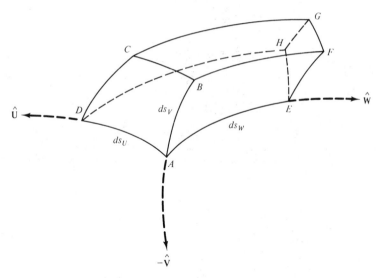

Figure A.I–1 Orthogonal curvilinear coordinates.

related, respectively, to ds_U, ds_V, ds_W. That is, for example,

$$ds_U = f(dU); \quad ds_V = g(dV); \quad ds_W = h(dW) \qquad \text{(A.I–4)}$$

By expanding the functions f, g, h in Maclaurin series about the origin at A,

$$ds_U = f(0) + f'(0)\, dU + f''(0)\frac{dU^2}{2!} + \cdots$$

$$ds_V = g(0) + g'(0)\, dV + g''(0)\frac{dV^2}{2!} + \cdots \qquad \text{(A.I–5)}$$

$$ds_W = h(0) + h'(0)\, dW + h''(0)\frac{dW^2}{2!} + \cdots$$

Here $f(0), f'(0), g(0)$, and so on, are in general functions of all three variables U, V, W. In order to evaluate these functions it is to be noted that

$$ds_U = 0 \quad \text{when } dU = 0 \quad \text{so that } f(0) = 0$$

$$ds_V = 0 \quad \text{when } dV = 0 \quad \text{so that } g(0) = 0 \qquad \text{(A.I–6)}$$

$$ds_W = 0 \quad \text{when } dW = 0 \quad \text{so that } h(0) = 0$$

If dU, dV, dW are sufficiently small, infinitesimals of higher order than the first may be neglected. In this case

$$ds_U = e_1\, dU \quad \text{with } e_1 = f'(0)$$

$$ds_V = e_2\, dV \quad \text{with } e_2 = g'(0) \qquad \text{(A.I–7)}$$

$$ds_W = e_3\, dW \quad \text{with } e_3 = h'(0)$$

The determination of e_1, e_2, e_3 is not difficult in Cartesian, cylindrical, and spherical coordinates. It is accomplished by inspection from appropriately specialized figures like Fig. A.I–1 by determining the factor by which each coordinate must be multiplied to give a length. The calculation in other cases is illustrated later, where the e factors are evaluated for rotationally symmetrical confocal coordinates.

The evaluation of $\nabla\phi$ in general curvilinear coordinates using e_1, e_2, e_3 is simple. Let $\hat{U}$, $\hat{V}$, $\hat{W}$ be unit vectors along the orthogonal curvilinear axes. Then

$$\nabla\phi = (\hat{U}\cdot\nabla\phi)\hat{U} + (\hat{V}\cdot\nabla\phi)\hat{V} + (\hat{W}\cdot\nabla\phi)\hat{W} \tag{A.I–8}$$

$$\nabla\phi = \frac{\partial\phi}{\partial s_U}\hat{U} + \frac{\partial\phi}{\partial s_V}\hat{V} + \frac{\partial\phi}{\partial s_W}\hat{W} \tag{A.I–9}$$

$$\nabla\phi = \frac{\partial\phi}{e_1\partial U}\hat{U} + \frac{\partial\phi}{e_2\partial V}\hat{V} + \frac{\partial\phi}{e_3\partial W}\hat{W} \tag{A.I–10}$$

The evaluation of $\nabla\cdot\mathbf{A}$ is accomplished by direct calculation from the definition of the divergence. This is,

$$\nabla\cdot\mathbf{A} = \lim_{d\tau\to 0}\frac{\int(\hat{n}\cdot\mathbf{A})\,d\sigma}{d\tau} \tag{A.I–11}$$

Let $d\tau$ be an element of volume in orthogonal curvilinear coordinates:

$$d\tau = ds_U\,ds_V\,ds_W = e_1 e_2 e_3\,dU\,dV\,dW \tag{A.I–12}$$

The surface elements are of the form

$$d\sigma_{UV} = ds_U\,ds_V = e_1 e_2\,dU\,dV \tag{A.I–13}$$

and similarly on the other surfaces. Consider the two opposite faces $ABCD$ and $EFGH$. On $ABCD$,

$$\int(\hat{n}\cdot\mathbf{A})\,d\sigma_{ABCD} = -A_W\,d\sigma_{UV} \tag{A.I–14}$$

The positive normal across $ABCD$ is taken along the positive axis of W. The value on $EFGH$ is obtained by expanding in a power series in ds_W and retaining only first-order terms.

$$\int(\hat{n}\cdot\mathbf{A})\,d\sigma_{EFGH} = A_W\,d\sigma_{UV} + \frac{\partial}{\partial s_W}(A_W\,d\sigma_{UV})\,ds_W \tag{A.I–15}$$

The value of $\int(\hat{n}\cdot\mathbf{A})\,d\sigma$ for $ABCD$ and $EFGH$ is the sum of (A.I–14) and (A.I–15). It is

$$\int(\hat{n}\cdot\mathbf{A})_{UV}\,d\sigma = \frac{\partial}{\partial s_W}(A_W\,d\sigma_{UV})\,ds_W = \frac{\partial}{\partial W}(A_W e_1 e_2\,dU\,dV)\,dW \tag{A.I–16a}$$

By cyclic permutation,

$$\int (\hat{\mathbf{n}} \cdot \mathbf{A})_{VW} \, d\sigma = \frac{\partial}{\partial U} (A_U e_2 e_3 \, dV \, dW) \, dU \qquad \text{(A.I–16b)}$$

$$\int (\hat{\mathbf{n}} \cdot \mathbf{A})_{WU} \, d\sigma = \frac{\partial}{\partial V} (A_V e_3 e_1 \, dW \, dU) \, dV \qquad \text{(A.I–16c)}$$

The use of (A.I.–16a–c) in (A.I–11) leads to

$$\nabla \cdot \mathbf{A} \equiv \lim_{d\tau \to 0} \frac{1}{e_1 e_2 e_3 \, dU \, dV \, dW} \left[\frac{\partial}{\partial U} (A_U e_2 e_3 \, dV \, dW) \, dU \right.$$

$$\left. + \frac{\partial}{\partial V} (A_V e_3 e_1 \, dW \, dU) \, dV + \frac{\partial}{\partial W} (A_W e_1 e_2 \, dU \, dV) \, dW \right] \qquad \text{(A.I–17)}$$

Since U, V, and W are independent of one another, cancellation leads to

$$\nabla \cdot \mathbf{A} \equiv \frac{1}{e_1 e_2 e_3} \left[\frac{\partial}{\partial U} (e_2 e_3 A_U) + \frac{\partial}{\partial V}(e_3 e_1 A_V) + \frac{\partial}{\partial W} (e_1 e_2 A_W) \right] \qquad \text{(A.I–18)}$$

The evaluation of $\nabla \times \mathbf{A}$ is accomplished in a similar manner from the alternative definition

$$\hat{\mathbf{N}} \cdot (\nabla \times \mathbf{A}) = \lim_{d\sigma \to 0} \frac{\oint (\mathbf{A} \cdot d\mathbf{s})}{d\sigma} \qquad \text{(A.I–19)}$$

where $\hat{\mathbf{N}}$ is normal to the surface of the element $d\sigma$ around which $d\mathbf{s}$ is integrated. Evidently,

$$\hat{\mathbf{W}} \cdot (\nabla \times \mathbf{A}) = \lim_{d\sigma \to 0} \frac{\oint (\mathbf{A} \cdot d\mathbf{s})}{d\sigma} \qquad \text{(A.I–20)}$$

with $\hat{\mathbf{W}}$ normal to the surface $ABCD$ (Fig. A.I–1) around which $d\mathbf{s}$ is integrated. The area of this element of surface is

$$d\sigma_{UV} = ds_U \, ds_V = e_1 e_2 \, dU \, dV \qquad \text{(A.I–21)}$$

$$\int_{AB} (\mathbf{A} \cdot d\mathbf{s}) = A_V \, ds_V \qquad \text{(A.I–22)}$$

$$\int_{AD} (\mathbf{A} \cdot d\mathbf{s}) = A_U \, ds_U \qquad \text{(A.I–23)}$$

$$\int_{BC} (\mathbf{A} \cdot d\mathbf{s}) = A_U \, ds_U + \frac{\partial}{\partial s_V}(A_U \, ds_U) \, ds_V \qquad \text{(A.I–24)}$$

$$\int_{DC} (\mathbf{A} \cdot d\mathbf{s}) = A_V \, ds_V + \frac{\partial}{\partial s_U}(A_V \, ds_V) \, ds_U \qquad \text{(A.I–25)}$$

In order that the right-hand-screw relation be satisfied, the positive direction of integration must be $DCBA$. The combination of (A.I–22) to (A.I–25) algebraically leaves only

$$\oint_{DCBA} (\mathbf{A} \cdot d\mathbf{s}) = \frac{\partial}{\partial s_U}(A_V \, ds_V) \, ds_U - \frac{\partial}{\partial s_V}(A_U \, ds_U) \, ds_V \qquad \text{(A.I–26)}$$

The appropriate substitution for ds_U and ds_V combined with the fact that U and V are independent lead to

$$\oint_{DCBA} (\mathbf{A} \cdot d\mathbf{s}) = \left[\frac{\partial}{\partial U}(e_2 A_V) - \frac{\partial}{\partial V}(e_1 A_U) \right] dV \, dU \qquad \text{(A.I–27)}$$

The substitution of (A.I–27) in (A.I–20) and appropriate cancellation result in

$$\hat{\mathbf{W}} \cdot (\nabla \times \mathbf{A}) = \frac{1}{e_1 e_2} \left[\frac{\partial}{\partial U}(e_2 A_V) - \frac{\partial}{\partial V}(e_1 A_U) \right] \qquad \text{(A.I–28a)}$$

$$\hat{\mathbf{U}} \cdot (\nabla \times \mathbf{A}) = \frac{1}{e_2 e_3} \left[\frac{\partial}{\partial V}(e_3 A_W) - \frac{\partial}{\partial W}(e_2 A_V) \right] \qquad \text{(A.I–28b)}$$

$$\hat{\mathbf{V}} \cdot (\nabla \times \mathbf{A}) = \frac{1}{e_3 e_1} \left[\frac{\partial}{\partial W}(e_1 A_U) - \frac{\partial}{\partial U}(e_3 A_W) \right] \qquad \text{(A.I–28c)}$$

The general curvilinear formula for $\nabla^2 \phi$ is obtained by forming $(\nabla \cdot \nabla \phi)$ with (A.I–10) and (A.I–18). It is

$$\nabla^2 \phi = \frac{1}{e_1 e_2 e_3} \left[\frac{\partial}{\partial U}\left(\frac{e_2 e_3}{e_1} \frac{\partial \phi}{\partial U} \right) + \frac{\partial}{\partial V}\left(\frac{e_3 e_1}{e_2} \frac{\partial \phi}{\partial V} \right) + \frac{\partial}{\partial W}\left(\frac{e_1 e_2}{e_3} \frac{\partial \phi}{\partial W} \right) \right] \qquad \text{(A.I–29)}$$

A formula for $(\nabla \times \nabla \times \mathbf{A})$ may be obtained from the substitution of the expressions for $\hat{\mathbf{U}} \cdot (\nabla \times \mathbf{A})$, $\hat{\mathbf{V}} \cdot (\nabla \times \mathbf{A})$, $\hat{\mathbf{W}} \cdot (\nabla \times \mathbf{A})$, respectively, for A_U, A_V, A_W, in (A.I–28). Since no simplification is obtained in this way, it is usually more convenient to evaluate $(\nabla \times \nabla \times \mathbf{A})$ directly as required.

In order to determine an expression for the quantity $\nabla^2 \mathbf{A}$ [where $\mathbf{A}$ is a vector to be carefully distinguished from the scalar ϕ in (A.I–29)], which appears in the important Helmholtz equation for the vector potential, it is necessary to proceed from its definition

$$\nabla^2 \mathbf{A} \equiv \nabla \nabla \cdot \mathbf{A} - \nabla \times \nabla \times \mathbf{A} \qquad \text{(A.I–30)}$$

The formula for $\nabla^2 \mathbf{A}$ in general orthogonal coordinates is excessively complicated. In fact, in all systems of coordinates *except only rectangular ones*, the expressions for the components of $\nabla^2 \mathbf{A}$ involve so many terms as to be practically useless. In Cartesian coordinates, on the other hand, its form is very simple and readily calculated. Thus, with

$$\mathbf{A} = \hat{\mathbf{x}} A_x + \hat{\mathbf{y}} A_y + \hat{\mathbf{z}} A_z \qquad \text{(A.I–31)}$$

and

$$\nabla^2 \mathbf{A} = \hat{\mathbf{x}}[\hat{\mathbf{x}} \cdot (\nabla^2 \mathbf{A})] + \hat{\mathbf{y}}[\hat{\mathbf{y}} \cdot (\nabla^2 \mathbf{A})] + \hat{\mathbf{z}}[\hat{\mathbf{z}} \cdot (\nabla^2 \mathbf{A})] \qquad \text{(A.I–32)}$$

the x component of $\nabla^2 \mathbf{A}$ must equal the x component of the expression to the right in (A.I–30). That is,

$$\hat{\mathbf{x}} \cdot (\nabla^2 \mathbf{A}) = \frac{\partial}{\partial x}(\nabla \cdot \mathbf{A}) - [\nabla \times (\nabla \times \mathbf{A})]_x \qquad \text{(A.I–33a)}$$

The use of the formulas for divergence and curl already derived gives

$$\hat{\mathbf{x}} \cdot (\nabla^2 \mathbf{A}) = \frac{\partial}{\partial x}\left(\frac{\partial A_x}{\partial x} + \frac{\partial A_y}{\partial y} + \frac{\partial A_z}{\partial z}\right) - \frac{\partial}{\partial y}(\nabla \times \mathbf{A})_z + \frac{\partial}{\partial z}(\nabla \times \mathbf{A})_y \qquad \text{(A.I–33b)}$$

This is equivalent to

$$\hat{\mathbf{x}} \cdot (\nabla^2 \mathbf{A}) = \frac{\partial^2 A_x}{\partial x^2} + \frac{\partial^2 A_y}{\partial x \partial y} + \frac{\partial^2 A_z}{\partial x \partial z}$$
$$- \frac{\partial}{\partial y}\left(\frac{\partial A_y}{\partial x} - \frac{\partial A_x}{\partial y}\right) + \frac{\partial}{\partial z}\left(\frac{\partial A_x}{\partial z} - \frac{\partial A_z}{\partial x}\right) \qquad \text{(A.I–33c)}$$

$$\hat{\mathbf{x}} \cdot (\nabla^2 \mathbf{A}) = \frac{\partial^2 A_x}{\partial x^2} + \frac{\partial^2 A_x}{\partial y^2} + \frac{\partial^2 A_x}{\partial z^2} = \nabla^2 A_x \qquad \text{(A.I–34a)}$$

Similarly,

$$\hat{\mathbf{y}} \cdot (\nabla^2 \mathbf{A}) = \frac{\partial^2 A_y}{\partial x^2} + \frac{\partial^2 A_y}{\partial y^2} + \frac{\partial^2 A_y}{\partial z^2} = \nabla^2 A_y \qquad \text{(A.I–34b)}$$

$$\hat{\mathbf{z}} \cdot (\nabla^2 \mathbf{A}) = \frac{\partial^2 A_z}{\partial x^2} + \frac{\partial^2 A_z}{\partial y^2} + \frac{\partial^2 A_z}{\partial z^2} = \nabla^2 A_z \qquad \text{(A.I–34c)}$$

It thus appears that in Cartesian coordinates each rectangular component of the vector $\mathbf{A}$ in $\nabla^2 \mathbf{A}$ satisfies precisely the same equation as the scalar ϕ. It is especially important to remember that this is *not* true in cylindrical or spherical coordinates.

With the substitution of the values of e_1, e_2, e_3 appropriate for each coordinate system as listed with the formulas for $\nabla \phi$, $\nabla \cdot \mathbf{A}$, $\nabla \times \mathbf{A}$, and $\nabla^2 \phi$, the results, given in Table A.I–1 are obtained for Cartesian, cylindrical, and spherical coordinates. $\nabla^2 \mathbf{A}$ is listed only in Cartesian coordinates as explained above.

Where it is not possible to determine the e factors by inspection, it is possible to proceed from the definitions (A.I–7). Thus

$$ds_U^2 = dx_U^2 + dy_U^2 + dz_U^2 = e_1^2\, dU^2 \qquad \text{(A.I–35)}$$

Then

$$e_1^2 = \left(\frac{ds_U}{dU}\right)^2 = \left(\frac{\partial x}{\partial U}\right)^2 + \left(\frac{\partial y}{\partial U}\right)^2 + \left(\frac{\partial z}{\partial U}\right)^2 \qquad \text{(A.I–36a)}$$

TABLE A.I–1 Differential Operators

	Cartesian (x, y, z)	Cylindrical (ρ, θ, z)	Spherical[a] (Θ, Φ, R)
e_1	1	1	R
e_2	1	ρ	$R\sin\Theta$
e_3	1	1	1

$\nabla\phi$

Cartesian:
$$\hat{x}\frac{\partial\phi}{\partial x} + \hat{y}\frac{\partial\phi}{\partial y} + \hat{z}\frac{\partial\phi}{\partial z}$$

Cylindrical:
$$\hat{\rho}\frac{\partial\phi}{\partial \rho} + \hat{\theta}\frac{1}{\rho}\frac{\partial\phi}{\partial \theta} + \hat{z}\frac{\partial\phi}{\partial z}$$

Spherical:
$$\hat{\Theta}\frac{1}{R}\frac{\partial\phi}{\partial \Theta} + \hat{\Phi}\frac{1}{R\sin\Theta}\frac{\partial\phi}{\partial \Phi} + \hat{R}\frac{\partial\phi}{\partial R}$$

$\nabla^2\phi$

Cartesian:
$$\frac{\partial^2\phi}{\partial x^2} + \frac{\partial^2\phi}{\partial y^2} + \frac{\partial^2\phi}{\partial z^2}$$

Cylindrical:
$$\frac{1}{\rho}\frac{\partial}{\partial \rho}\left(\rho\frac{\partial\phi}{\partial \rho}\right) + \frac{1}{\rho^2}\frac{\partial^2\phi}{\partial \theta^2} + \frac{\partial^2\phi}{\partial z^2}$$

Spherical:
$$\frac{1}{R^2\sin\Theta}\frac{\partial}{\partial \Theta}\left(\sin\Theta\frac{\partial\phi}{\partial \Theta}\right) + \frac{1}{R^2\sin^2\Theta}\frac{\partial^2\phi}{\partial \Phi^2} + \frac{1}{R^2}\frac{\partial}{\partial R}\left(R^2\frac{\partial\phi}{\partial R}\right)$$

$\nabla\cdot A$

Cartesian:
$$\frac{\partial A_x}{\partial x} + \frac{\partial A_y}{\partial y} + \frac{\partial A_z}{\partial z}$$

Cylindrical:
$$\frac{1}{\rho}\frac{\partial}{\partial \rho}(\rho A_\rho) + \frac{1}{\rho}\frac{\partial A_\theta}{\partial \theta} + \frac{\partial A_z}{\partial z}$$

Spherical:
$$\frac{1}{R\sin\Theta}\frac{\partial}{\partial \Theta}(A_\Theta\sin\Theta) + \frac{1}{R\sin\Theta}\frac{\partial A_\Phi}{\partial \Phi} + \frac{1}{R^2}\frac{\partial}{\partial R}(R^2 A_R)$$

$\nabla\times A$

Cartesian:
$$(\nabla\times A)_x = \frac{\partial A_z}{\partial y} - \frac{\partial A_y}{\partial z}$$
$$(\nabla\times A)_y = \frac{\partial A_x}{\partial z} - \frac{\partial A_z}{\partial x}$$
$$(\nabla\times A)_z = \frac{\partial A_y}{\partial x} - \frac{\partial A_x}{\partial y}$$

Cylindrical:
$$(\nabla\times A)_\rho = \frac{1}{\rho}\frac{\partial A_z}{\partial \theta} - \frac{\partial A_\theta}{\partial z}$$
$$(\nabla\times A)_\theta = \frac{\partial A_\rho}{\partial z} - \frac{\partial A_z}{\partial \rho}$$
$$(\nabla\times A)_z = \frac{1}{\rho}\left[\frac{\partial}{\partial \rho}(\rho A_\theta) - \frac{\partial A_\rho}{\partial \theta}\right]$$

Spherical:
$$(\nabla\times A)_\Theta = \frac{1}{R}\left[\frac{1}{\sin\Theta}\frac{\partial A_R}{\partial \Phi} - \frac{\partial}{\partial R}(RA_\Phi)\right]$$
$$(\nabla\times A)_\Phi = \frac{1}{R}\left[\frac{\partial}{\partial R}(RA_\Theta) - \frac{\partial A_R}{\partial \Theta}\right]$$
$$(\nabla\times A)_R = \frac{1}{R\sin\Theta}\left[\frac{\partial}{\partial \Theta}(A_\Phi\sin\Theta) - \frac{\partial A_\Theta}{\partial \Phi}\right]$$

$\nabla^2 A$

Cartesian:
$$\hat{x}\,\nabla^2 A_x + \hat{y}\,\nabla^2 A_y + \hat{z}\,\nabla^2 A_z$$
$$\nabla^2 A_x = \frac{\partial^2 A_x}{\partial x^2} + \frac{\partial^2 A_x}{\partial y^2} + \frac{\partial^2 A_x}{\partial z^2}$$
$$\nabla^2 A_y = \frac{\partial^2 A_y}{\partial x^2} + \frac{\partial^2 A_y}{\partial y^2} + \frac{\partial^2 A_y}{\partial z^2}$$
$$\nabla^2 A_z = \frac{\partial^2 A_z}{\partial x^2} + \frac{\partial^2 A_z}{\partial y^2} + \frac{\partial^2 A_z}{\partial z^2}$$

[a] Note that the spherical coordinates are written in the order Θ, Φ, R, *not* R, Θ, Φ.

since $dx_U/dU = \partial x/\partial U$, and so on. Similarly,

$$e_2^2 = \left(\frac{ds_V}{dV}\right)^2 = \left(\frac{\partial x}{\partial V}\right)^2 + \left(\frac{\partial y}{\partial V}\right)^2 + \left(\frac{\partial z}{\partial V}\right)^2 \qquad \text{(A.I–36b)}$$

$$e_3^2 = \left(\frac{ds_W}{W}\right)^2 = \left(\frac{\partial x}{\partial W}\right)^2 + \left(\frac{\partial y}{\partial W}\right)^2 + \left(\frac{\partial z}{\partial W}\right)^2 \qquad \text{(A.I–36c)}$$

The calculation of e_1, e_2, e_3 in rotationally symmetrical confocal coordinates follows. The equation of an ellipsoid in Cartesian coordinates is

$$\frac{x^2}{a^2} + \frac{y^2}{b^2} + \frac{z^2}{c^2} = 1 \qquad \text{(A.I–37)}$$

An hyperboloid of one sheet is defined by

$$\frac{x^2}{a^2} + \frac{y^2}{b^2} - \frac{z^2}{c^2} = 1 \qquad \text{(A.I–38)}$$

An hyperboloid of two sheets is defined by

$$\frac{x^2}{a^2} - \frac{y^2}{b^2} - \frac{z^2}{c^2} = 1 \qquad \text{(A.I–39)}$$

Here a, b, c are the semimajor axes; $a \geq b \geq c$. If rotational symmetry prevails, there are two possibilities: the prolate (stretched) ellipsoid with its related orthogonal hyperboloid defined by $b = c$, and the oblate (flattened) ellipsoid with its related orthogonal hyperboloid defined by $a = b$.

For the prolate surfaces with $a > b$ and $r^2 = y^2 + z^2$,

$$\frac{x^2}{a^2} + \frac{r^2}{b^2} = 1 \qquad \text{(A.I–40)}$$

$$\frac{x^2}{a^2} - \frac{r^2}{b^2} = 1 \qquad \text{(A.I–41)}$$

The hyperboloid has two sheets.

For the oblate surfaces with $a > c$ and $r^2 = x^2 + y^2$,

$$\frac{r^2}{a^2} + \frac{z^2}{c^2} = 1 \qquad \text{(A.I–42)}$$

$$\frac{r^2}{a^2} - \frac{z^2}{c^2} = 1 \qquad \text{(A.I–43)}$$

For the prolate case the eccentricity e is defined by

$$e^2 = a^2 - b^2 \text{ (ellipse)} \qquad \text{or} \qquad b^2 = a^2 - e^2 \qquad \text{(A.I–44)}$$

$$e^2 = a^2 + b^2 \text{ (hyperbola)} \qquad \text{or} \qquad b^2 = e^2 - a^2 \qquad \text{(A.I–45)}$$

Accordingly, the equations for the prolate case are

$$\frac{x^2}{a^2} + \frac{r^2}{a^2 - e^2} = 1 \quad \text{(ellipsoid)} \tag{A.I–46}$$

$$\frac{x^2}{a^2} - \frac{r^2}{e^2 - a^2} = 1 \quad \text{(hyperboloid of two sheets)} \tag{A.I–47}$$

The eccentricity for the oblate case is given by (A.I–44) and (A.I–45) with c written for b. Hence

$$\frac{r^2}{a^2} + \frac{z^2}{a^2 - e^2} = 1 \quad \text{(ellipsoid)} \tag{A.I–48}$$

$$\frac{r^2}{a^2} - \frac{z^2}{e^2 - a^2} = 1 \quad \text{(hyperboloid of one sheet)} \tag{A.I–49}$$

The oblate case follows directly from the prolate one with an interchange of coordinates, that is, with r substituted for x and z for r.

A family of confocal ellipsoids of revolution is defined by

$$\frac{x^2}{u^2} + \frac{r^2}{u^2 - e^2} = 1; \quad u > e \tag{A.I–50}$$

The associated orthogonal hyperboloids are

$$\frac{x^2}{v^2} - \frac{r^2}{e^2 - v^2} = 1; \quad -e < v < e \tag{A.I–51}$$

The variables u, v are ellipsoidal and hyperboloidal coordinates. The parameters e_1, e_2, e_3 may be calculated from

$$e_1 = \sqrt{\left(\frac{\partial x}{\partial u}\right)^2 + \left(\frac{\partial y}{\partial u}\right)^2 + \left(\frac{\partial z}{\partial u}\right)^2} = \sqrt{\left(\frac{\partial x}{\partial u}\right)^2 + \left(\frac{\partial r}{\partial u}\right)^2} \tag{A.I–52a}$$

With $r^2 = y^2 + z^2$,

$$e_2 = \sqrt{\left(\frac{\partial x}{\partial v}\right)^2 + \left(\frac{\partial y}{\partial v}\right)^2 + \left(\frac{\partial z}{\partial v}\right)^2} = \sqrt{\left(\frac{\partial x}{\partial v}\right)^2 + \left(\frac{\partial r}{\partial v}\right)^2} \tag{A.I–52b}$$

$$e_3 = \sqrt{\left(\frac{\partial x}{\partial w}\right)^2 + \left(\frac{\partial y}{\partial w}\right)^2 + \left(\frac{\partial z}{\partial w}\right)^2} = \sqrt{\left(\frac{\partial x}{\partial \theta}\right)^2 + \left(\frac{\partial r}{\partial \theta}\right)^2} \tag{A.I–52c}$$

The coordinate w is identically the angle of rotation θ about the axis of symmetry

x. It is related to y and z by

$$y = r \cos \theta; \qquad \frac{\partial y}{\partial \theta} = -r \sin \theta$$

$$z = r \sin \theta; \qquad \frac{\partial z}{\partial \theta} = r \cos \theta \tag{A.I-53}$$

$$e_3 = \sqrt{\left(\frac{\partial y}{\partial \theta}\right)^2 + \left(\frac{\partial z}{\partial \theta}\right)^2} = r; \qquad \frac{\partial x}{\partial \theta} = 0 \tag{A.I-54}$$

The solution of (A.I–50) and (A.I–51) for r and x gives

$$x^2 = u^2 - \frac{u^2 r^2}{u^2 - e^2} = v^2 + \frac{v^2 r^2}{e^2 - v^2} \tag{A.I-55a}$$

$$r^2 \left(\frac{v^2}{e^2 - v^2} + \frac{u^2}{u^2 - e^2}\right) = u^2 - v^2 \tag{A.I-55b}$$

$$r^2 \left[\frac{e^2(u^2 - v^2)}{(e^2 - v^2)(u^2 - e^2)}\right] = u^2 - v^2 \tag{A.I-55c}$$

$$r^2 = \frac{(e^2 - v^2)(u^2 - e^2)}{e^2} \tag{A.I-55d}$$

$$x^2 = u^2 \left(1 - \frac{e^2 - v^2}{e^2}\right) = \frac{u^2 v^2}{e^2} \tag{A.I-55e}$$

$$r = \frac{\sqrt{(e^2 - v^2)(u^2 - e^2)}}{e}; \qquad x = \frac{uv}{e} \tag{A.I-56}$$

Hence

$$\frac{\partial x}{\partial u} = \frac{v}{e}; \qquad \frac{\partial x}{\partial v} = \frac{u}{e} \tag{A.I-57}$$

$$\frac{\partial r}{\partial u} = \frac{u}{e}\sqrt{\frac{e^2 - v^2}{u^2 - e^2}}; \qquad \frac{\partial r}{\partial v} = -\frac{v}{e}\sqrt{\frac{u^2 - e^2}{e^2 - v^2}} \tag{A.I-58}$$

$$e_1 = \sqrt{\left(\frac{v}{e}\right)^2 + \frac{u^2}{e^2}\frac{e^2 - v^2}{u^2 - e^2}} = \sqrt{\frac{u^2 - v^2}{u^2 - e^2}} \tag{A.I-59a}$$

$$e_2 = \sqrt{\left(\frac{u}{e}\right)^2 + \frac{v^2}{e^2}\frac{u^2 - e^2}{e^2 - v^2}} = \sqrt{\frac{u^2 - v^2}{e^2 - v^2}} \tag{A.I-59b}$$

$$e_3 = r = \frac{\sqrt{(e^2 - v^2)(u^2 - e^2)}}{e}$$ (A.I–59c)

It follows that

$$\nabla\phi = \sqrt{\frac{u^2 - e^2}{u^2 - v^2}} \frac{\partial\phi}{\partial u} \hat{\mathbf{u}} + \sqrt{\frac{e^2 - v^2}{u^2 - v^2}} \frac{\partial\phi}{\partial v} \hat{\mathbf{v}}$$

$$+ \frac{e}{\sqrt{(e^2 - v^2)(u^2 - e^2)}} \frac{\partial\phi}{\partial\theta} \hat{\boldsymbol{\theta}}$$ (A.I–60)

$$\nabla \cdot \mathbf{A} = \frac{e}{u^2 - v^2} \left\{ \frac{\partial}{\partial u} \left[\frac{\sqrt{(u^2 - v^2)(u^2 - e^2)}}{e} A_u \right] \right.$$

$$+ \frac{\partial}{\partial v} \left[\frac{\sqrt{(u^2 - v^2)(e^2 - v^2)}}{e} A_v \right]$$

$$\left. + \frac{\partial}{\partial\theta} \left[\frac{u^2 - v^2}{\sqrt{(u^2 - e^2)(e^2 - v^2)}} A_\theta \right] \right\}$$ (A.I–61)

$$\hat{\mathbf{u}} \cdot (\nabla \times \mathbf{A}) = \frac{e}{\sqrt{(u^2 - v^2)(u^2 - e^2)}} \left\{ \frac{\partial}{\partial v} \left[\frac{\sqrt{(e^2 - v^2)(u^2 - e^2)}}{e} A_\theta \right] \right.$$

$$\left. - \frac{\partial}{\partial\theta} \left[\sqrt{\frac{u^2 - v^2}{e^2 - v^2}} A_v \right] \right\}$$ (A.I–62a)

$$\hat{\mathbf{v}} \cdot (\nabla \times \mathbf{A}) = \frac{e}{\sqrt{(u^2 - v^2)(e^2 - v^2)}} \left\{ \frac{\partial}{\partial\theta} \left[\sqrt{\frac{u^2 - v^2}{u^2 - e^2}} A_u \right] \right.$$

$$\left. - \frac{\partial}{\partial u} \left[\frac{\sqrt{(e^2 - v^2)(u^2 - e^2)}}{e} A_\theta \right] \right\}$$ (A.I–62b)

$$\hat{\boldsymbol{\theta}} \cdot (\nabla \times \mathbf{A}) = \frac{\sqrt{(u^2 - e^2)(e^2 - v^2)}}{u^2 - v^2} \left[\frac{\partial}{\partial u} \left(\sqrt{\frac{u^2 - v^2}{e^2 - v^2}} A_v \right) \right.$$

$$\left. - \frac{\partial}{\partial v} \left(\sqrt{\frac{u^2 - v^2}{u^2 - e^2}} A_u \right) \right]$$ (A.I–62c)

$$\nabla^2\phi = \frac{e}{u^2 - v^2} \left\{ \frac{\partial}{\partial u} \left(\frac{u^2 - e^2}{e} \frac{\partial\phi}{\partial u} \right) \right.$$

$$+ \frac{\partial}{\partial v} \left(\frac{e^2 - v^2}{e} \frac{\partial\phi}{\partial v} \right)$$

$$\left. + \frac{\partial}{\partial\theta} \left[\frac{e(u^2 - v^2)}{(u^2 - e^2)(e^2 - v^2)} \frac{\partial\phi}{\partial\theta} \right] \right\}$$ (A.I–63)

A.I–2 VECTOR FORMULAS AND IDENTITIES WITH CARTESIAN EQUIVALENTS

Scalar product:

$$\mathbf{A} \cdot \mathbf{B} = \mathbf{B} \cdot \mathbf{A} = AB \cos (A, B) = A_x B_x + A_y B_y + A_z B_z$$

$$\mathbf{A} \cdot (\mathbf{B} + \mathbf{C}) = \mathbf{A} \cdot \mathbf{B} + \mathbf{A} \cdot \mathbf{C}$$

Vector product:

$$\mathbf{A} \times \mathbf{B} = -\mathbf{B} \times \mathbf{A} = \mathbf{C} = \begin{vmatrix} \hat{\mathbf{x}} & \hat{\mathbf{y}} & \hat{\mathbf{z}} \\ A_x & A_y & A_z \\ B_x & B_y & B_z \end{vmatrix}$$

$\mathbf{C}$ is an axial vector perpendicular to the plane containing $\mathbf{A}$ and $\mathbf{B}$ and pointing in the direction of advance of a right-hand screw when $\mathbf{A}$ is turned into $\mathbf{B}$ through the shortest arc.

$$C = AB \sin (A, B)$$

$$\mathbf{A} \times (\mathbf{B} + \mathbf{C}) = \mathbf{A} \times \mathbf{B} + \mathbf{A} \times \mathbf{C}$$

$$\mathbf{A} \times \mathbf{A} = 0$$

Double products:

$$\mathbf{A} \cdot (\mathbf{B} \times \mathbf{C}) = \mathbf{B} \cdot (\mathbf{C} \times \mathbf{A}) = \mathbf{C} \cdot (\mathbf{A} \times \mathbf{B})$$

$$\mathbf{A} \cdot (\mathbf{A} \times \mathbf{B}) = 0$$

$$\mathbf{A} \times (\mathbf{B} \times \mathbf{C}) = \mathbf{B}(\mathbf{A} \cdot \mathbf{C}) - \mathbf{C}(\mathbf{A} \cdot \mathbf{B}) \qquad \text{(polar vector)}$$

Differential operators:

$$\boldsymbol{\nabla}\phi = \lim_{\Delta\tau \to 0} \frac{\displaystyle\int_\Sigma \hat{\mathbf{n}}\phi \, d\sigma}{\Delta\tau} = \hat{\mathbf{x}} \frac{\partial\phi}{\partial x} + \hat{\mathbf{y}} \frac{\partial\phi}{\partial y} + \hat{\mathbf{z}} \frac{\partial\phi}{\partial z}$$

is the rate of change $\partial\phi/ds$ in the direction of its maximum;

$$\boldsymbol{\nabla} \cdot \mathbf{A} = \lim_{\Delta\tau \to 0} \frac{\displaystyle\int_\Sigma (\hat{\mathbf{n}} \cdot \mathbf{A}) \, d\sigma}{\Delta\tau} = \frac{\partial A_x}{\partial x} + \frac{\partial A_y}{\partial y} + \frac{\partial A_z}{\partial z}$$

$$\boldsymbol{\nabla} \times \mathbf{A} = \lim_{\Delta\tau \to 0} \frac{\displaystyle\int_\Sigma (\hat{\mathbf{n}} \times \mathbf{A}) \, d\sigma}{\Delta\tau} = \begin{vmatrix} \hat{\mathbf{x}} & \hat{\mathbf{y}} & \hat{\mathbf{z}} \\ \dfrac{\partial}{\partial x} & \dfrac{\partial}{\partial y} & \dfrac{\partial}{\partial z} \\ A_x & A_y & A_z \end{vmatrix}$$

$$= \hat{\mathbf{x}}(\boldsymbol{\nabla} \times \mathbf{A})_x + \hat{\mathbf{y}}(\boldsymbol{\nabla} \times \mathbf{A})_y + \hat{\mathbf{z}}(\boldsymbol{\nabla} \times \mathbf{A})_z$$

$$\nabla^2\phi = \nabla \cdot (\nabla\phi) = \frac{\partial^2\phi}{\partial x^2} + \frac{\partial^2\phi}{\partial y^2} + \frac{\partial^2\phi}{\partial z^2}$$

$$\nabla^2\mathbf{A} = \nabla(\nabla \cdot \mathbf{A}) - \nabla \times (\nabla \times \mathbf{A}) = \hat{\mathbf{x}}\nabla^2 A_x + \hat{\mathbf{y}}\nabla^2 A_y + \hat{\mathbf{z}}\nabla^2 A_z$$

Double operations:

$$\nabla(\nabla \cdot \mathbf{A}) = \hat{\mathbf{x}}\frac{\partial}{\partial x}(\nabla \cdot \mathbf{A}) + \hat{\mathbf{y}}\frac{\partial}{\partial y}(\nabla \cdot \mathbf{A}) + \hat{\mathbf{z}}\frac{\partial}{\partial z}(\nabla \cdot \mathbf{A})$$

$$\nabla \cdot (\nabla\phi) = \nabla^2\phi = \frac{\partial^2\phi}{\partial x^2} + \frac{\partial^2\phi}{\partial y^2} + \frac{\partial^2\phi}{\partial z^2}$$

$$\nabla \cdot (\nabla \times \mathbf{A}) = 0$$

$$\nabla \times (\nabla\phi) = 0$$

$$\nabla \times (\nabla \times \mathbf{A}) = \begin{vmatrix} \hat{\mathbf{x}} & \hat{\mathbf{y}} & \hat{\mathbf{z}} \\ \dfrac{\partial}{\partial x} & \dfrac{\partial}{\partial y} & \dfrac{\partial}{\partial z} \\ (\nabla \times \mathbf{A})_x & (\nabla \times \mathbf{A})_y & (\nabla \times \mathbf{A})_z \end{vmatrix}$$

$\mathbf{A} \cdot (\nabla\phi) \equiv (\mathbf{A} \cdot \nabla)\phi = A_x\dfrac{\partial\phi}{\partial x} + A_y\dfrac{\partial\phi}{\partial y} + A_z\dfrac{\partial\phi}{\partial z}$ is the rate of change $d\phi/ds$ in the direction of $\mathbf{A}$ multiplied by $\mathbf{A}$

$\hat{\mathbf{a}} \cdot (\nabla\phi) \equiv (\hat{\mathbf{a}} \cdot \nabla)\phi$ is the rate of change $d\phi/ds$ in the direction of the unit vector $\hat{\mathbf{a}}$

$(\mathbf{A} \cdot \nabla)\mathbf{B} = \hat{\mathbf{x}}(\mathbf{A} \cdot \nabla B_x) + \hat{\mathbf{y}}(\mathbf{A} \cdot \nabla B_y) + \hat{\mathbf{z}}(\mathbf{A} \cdot \nabla B_z)$ is the rate of change $d\mathbf{B}/ds$ in the direction of $\mathbf{A}$ multiplied by $\mathbf{A}$

$(\hat{\mathbf{a}} \cdot \nabla)\mathbf{B} = \hat{\mathbf{x}}(\hat{\mathbf{a}} \cdot \nabla B_x) + \hat{\mathbf{y}}(\hat{\mathbf{a}} \cdot \nabla B_y) + \hat{\mathbf{z}}(\hat{\mathbf{a}} \cdot \nabla B_z)$ is the rate of change $d\mathbf{B}/ds$ in the direction of the unit vector $\hat{\mathbf{a}}$; it is the directional derivative of $\mathbf{B}$ in the direction $\hat{\mathbf{a}}$

Operations on products:

$$\nabla(\phi\psi) = \phi(\nabla\psi) + \psi(\nabla\phi)$$

$$\nabla \cdot (\phi\mathbf{A}) = \phi(\nabla \cdot \mathbf{A}) + \nabla\phi \cdot \mathbf{A}$$

$$\nabla \cdot (\mathbf{A} \times \mathbf{B}) = \mathbf{B} \cdot (\nabla \times \mathbf{A}) - \mathbf{A} \cdot (\nabla \times \mathbf{B})$$

$$\nabla \times (\phi\mathbf{A}) = \phi(\nabla \times \mathbf{A}) + \nabla\phi \times \mathbf{A}$$

$$\nabla \times (\mathbf{A} \times \mathbf{B}) = \mathbf{A}(\nabla \cdot \mathbf{B}) - \mathbf{B}(\nabla \cdot \mathbf{A}) + (\mathbf{B} \cdot \nabla)\mathbf{A} - (\mathbf{A} \cdot \nabla)\mathbf{B}$$

Appendix

Tables of Functions f(h), g(h)

$$f(h) \equiv + \sqrt{\tfrac{1}{2}(\sqrt{1 + h^2} + 1)} = \cosh(\tfrac{1}{2} \sinh^{-1} h)$$

$$g(h) \equiv + \sqrt{\tfrac{1}{2}(\sqrt{1 + h^2} - 1)} = \sinh(\tfrac{1}{2} \sinh^{-1} h)$$

in the expression

$$\sqrt{1 \pm jh} = f(h) \pm jg(h)$$

as computed by G. W. Pierce.*

$$h^2 \ll 1; \quad f(h) \sim 1; \quad g(h) \sim \frac{h}{2}$$

$$h \gg 1; \quad f(h) \sim g(h) \sim \sqrt{\frac{h}{2}}$$

*Proc. Am. Acad. Arts Sci., 57, 175 (1922).

APPENDIX II

h	$f(h)$	$g(h)$	h	$f(h)$	$g(h)$	h	$f(h)$	$g(h)$	h	$f(h)$	$g(h)$
.000	1.000	.0000	.072	1.001	.0360	.144	1.003	.0718	.216	1.006	.1074
.002	1.000	.0010	.074	1.001	.0370	.146	1.003	.0728	.218	1.006	.1084
.004	1.000	.0020	.076	1.001	.0380	.148	1.003	.0738	.220	1.006	.1094
.006	1.000	.0030	.078	1.001	.0390	.150	1.003	.0748	.222	1.006	.1104
.008	1.000	.0040	.080	1.001	.0400	.152	1.003	.0758	.224	1.006	.1114
.010	1.000	.0050	.082	1.001	.0410	.154	1.003	.0768	.226	1.006	.1124
.012	1.000	.0060	.084	1.001	.0420	.156	1.003	.0778	.228	1.006	.1133
.014	1.000	.0070	.086	1.001	.0430	.158	1.003	.0788	.230	1.007	.1143
.016	1.000	.0080	.088	1.001	.0440	.160	1.003	.0798	.232	1.007	.1153
.018	1.000	.0090	.090	1.001	.0450	.162	1.003	.0807	.234	1.007	.1163
.020	1.000	.0100	.092	1.001	.0460	.164	1.003	.0817	.236	1.007	.1172
.022	1.000	.0110	.094	1.001	.0470	.166	1.003	.0827	.238	1.007	.1182
.024	1.000	.0120	.096	1.001	.0480	.168	1.004	.0837	.240	1.007	.1192
.026	1.000	.0130	.098	1.001	.0490	.170	1.004	.0847	.242	1.007	.1202
.028	1.000	.0140	.100	1.001	.0499	.172	1.004	.0857	.244	1.007	.1211
.030	1.000	.0150	.102	1.001	.0509	.174	1.004	.0867	.246	1.007	.1221
.032	1.000	.0160	.104	1.001	.0519	.176	1.004	.0877	.248	1.007	.1230
.034	1.000	.0170	.106	1.001	.0529	.178	1.004	.0887	.250	1.008	.1240
.036	1.000	.0180	.108	1.001	.0539	.180	1.004	.0896	.252	1.008	.1250
.038	1.000	.0190	.110	1.002	.0549	.182	1.004	.0906	.254	1.008	.1260
.040	1.000	.0200	.112	1.002	.0559	.184	1.004	.0916	.256	1.008	.1270
.042	1.000	.0210	.114	1.002	.0569	.186	1.004	.0926	.258	1.008	.1280
.044	1.000	.0220	.116	1.002	.0579	.188	1.004	.0936	.260	1.008	.1289
.046	1.000	.0230	.118	1.002	.0589	.190	1.004	.0946	.262	1.008	.1299
.048	1.000	.0240	.120	1.002	.0599	.192	1.005	.0956	.264	1.008	.1309
.050	1.000	.0250	.122	1.002	.0609	.194	1.005	.0966	.266	1.009	.1319
.052	1.000	.0260	.124	1.002	.0619	.196	1.005	.0975	.268	1.009	.1329
.054	1.000	.0270	.126	1.002	.0629	.198	1.005	.0985	.270	1.009	.1338
.056	1.000	.0280	.128	1.002	.0639	.200	1.005	.0995	.272	1.009	.1348
.058	1.000	.0290	.130	1.002	.0649	.202	1.005	.1005	.274	1.009	.1358
.060	1.000	.0300	.132	1.002	.0659	.204	1.005	.1015	.276	1.009	.1368
.062	1.000	.0310	.134	1.002	.0669	.206	1.005	.1025	.278	1.009	.1377
.064	1.001	.0320	.136	1.002	.0679	.208	1.005	.1035	.280	1.010	.1387
.066	1.001	.0330	.138	1.002	.0689	.210	1.005	.1044	.282	1.010	.1397
.068	1.001	.0340	.140	1.002	.0698	.212	1.006	.1054	.284	1.010	.1406
.070	1.001	.0350	.142	1.003	.0708	.214	1.006	.1064	.286	1.010	.1416

APPENDIX II

h	$f(h)$	$g(h)$	h	$f(h)$	$g(h)$	h	$f(h)$	$g(h)$	h	$f(h)$	$g(h)$
.286	1.010	.1416	.358	1.015	.1762	.430	1.022	.2104	.502	1.029	.2438
.288	1.010	.1426	.360	1.016	.1772	.432	1.022	.2114	.504	1.029	.2447
.290	1.010	.1435	.362	1.016	.1782	.434	1.022	.2123	.506	1.030	.2456
.292	1.010	.1445	.364	1.016	.1792	.436	1.022	.2132	.508	1.030	.2465
.294	1.011	.1455	.366	1.016	.1801	.438	1.023	.2141	.510	1.030	.2475
.296	1.011	.1464	.368	1.016	.1811	.440	1.023	.2150	.512	1.030	.2484
.298	1.011	.1474	.370	1.016	.1820	.442	1.023	.2159	.514	1.031	.2493
.300	1.011	.1484	.372	1.017	.1829	.444	1.023	.2170	.516	1.031	.2502
.302	1.011	.1494	.374	1.017	.1839	.446	1.023	.2179	.518	1.031	.2512
.304	1.011	.1503	.376	1.017	.1848	.448	1.024	.2188	.520	1.031	.2520
.306	1.011	.1513	.378	1.017	.1858	.450	1.024	.2197	.522	1.031	.2529
.308	1.011	.1522	.380	1.017	.1867	.452	1.024	.2207	.524	1.032	.2538
.310	1.012	.1532	.382	1.017	.1877	.454	1.024	.2216	.526	1.032	.2549
.312	1.012	.1541	.384	1.018	.1886	.456	1.024	.2225	.528	1.032	.2558
.314	1.012	.1551	.386	1.018	.1896	.458	1.025	.2234	.530	1.032	.2566
.316	1.012	.1561	.388	1.018	.1905	.460	1.025	.2244	.532	1.033	.2575
.318	1.012	.1570	.390	1.018	.1915	.462	1.025	.2253	.534	1.033	.2584
.320	1.012	.1580	.392	1.018	.1924	.464	1.025	.2262	.536	1.033	.2593
.322	1.013	.1590	.394	1.018	.1934	.466	1.025	.2271	.538	1.033	.2602
.324	1.013	.1600	.396	1.019	.1943	.468	1.026	.2281	.540	1.034	.2612
.326	1.013	.1609	.398	1.019	.1953	.470	1.026	.2290	.542	1.034	.2621
.328	1.013	.1619	.400	1.019	.1962	.472	1.026	.2300	.544	1.034	.2630
.330	1.013	.1629	.402	1.019	.1972	.474	1.026	.2309	.546	1.034	.2639
.332	1.013	.1638	.404	1.019	.1981	.476	1.026	.2318	.548	1.034	.2648
.334	1.013	.1648	.406	1.020	.1991	.478	1.027	.2327	.550	1.035	.2658
.336	1.014	.1658	.408	1.020	.2001	.480	1.027	.2337	.552	1.035	.2667
.338	1.014	.1667	.410	1.020	.2010	.482	1.027	.2346	.554	1.035	.2676
.340	1.014	.1677	.412	1.020	.2020	.484	1.027	.2355	.556	1.035	.2685
.342	1.014	.1686	.414	1.020	.2029	.486	1.028	.2364	.558	1.036	.2694
.344	1.014	.1696	.416	1.020	.2038	.488	1.028	.2374	.560	1.036	.2703
.346	1.014	.1705	.418	1.021	.2048	.490	1.028	.2383	.562	1.036	.2712
.348	1.015	.1715	.420	1.021	.2057	.492	1.028	.2392	.564	1.036	.2721
.350	1.015	.1724	.422	1.021	.2067	.494	1.028	.2401	.566	1.036	.2730
.352	1.015	.1734	.424	1.021	.2076	.496	1.029	.2411	.568	1.037	.2739
.354	1.015	.1743	.426	1.022	.2085	.498	1.029	.2420	.570	1.037	.2748
.356	1.015	.1753	.428	1.022	.2095	.500	1.029	.2429	.572	1.037	.2757

h	$f(h)$	$g(h)$	h	$f(h)$	$g(h)$	h	$f(h)$	$g(h)$	h	$f(h)$	$g(h)$
.572	1.037	.2757	1.04	1.106	.4705	1.76	1.230	.715	2.48	1.356	.915
.574	1.038	.2766	1.06	1.109	.4782	1.78	1.233	.722	2.50	1.359	.920
.576	1.038	.2775	1.08	1.113	.4857	1.80	1.237	.728	2.52	1.362	.925
.578	1.038	.2784	1.10	1.116	.4933	1.82	1.241	.733	2.54	1.366	.930
.580	1.038	.2793	1.12	1.119	.5007	1.84	1.244	.740	2.56	1.369	.935
.582	1.039	.2802	1.14	1.122	.5082	1.86	1.248	.745	2.58	1.373	.939
.584	1.039	.2811	1.16	1.126	.5155	1.88	1.251	.751	2.60	1.376	.945
.586	1.039	.2820	1.18	1.129	.5229	1.90	1.255	.757	2.62	1.379	.950
.588	1.039	.2829	1.20	1.132	.530	1.92	1.258	.763	2.64	1.383	.955
.590	1.040	.2838	1.22	1.135	.538	1.94	1.262	.769	2.66	1.386	.960
.592	1.040	.2847	1.24	1.139	.544	1.96	1.265	.775	2.68	1.390	.964
.594	1.040	.2856	1.26	1.142	.552	1.98	1.269	.780	2.70	1.393	.969
.596	1.040	.2865	1.28	1.146	.559	2.00	1.272	.786	2.72	1.396	.974
.598	1.041	.2874	1.30	1.149	.565	2.02	1.275	.792	2.74	1.400	.979
.60	1.041	.2882	1.32	1.153	.572	2.04	1.279	.798	2.76	1.403	.984
.62	1.044	.2972	1.34	1.156	.580	2.06	1.283	.803	2.78	1.407	.989
.64	1.046	.3059	1.36	1.160	.586	2.08	1.286	.809	2.80	1.410	.993
.66	1.049	.3148	1.38	1.163	.593	2.10	1.290	.814	2.82	1.413	.998
.68	1.051	.3235	1.40	1.167	.600	2.12	1.293	.820	2.84	1.416	1.003
.70	1.054	.3321	1.42	1.170	.607	2.14	1.297	.825	2.86	1.420	1.007
.72	1.057	.3408	1.44	1.174	.613	2.16	1.300	.831	2.88	1.423	1.012
.74	1.060	.3493	1.46	1.178	.620	2.18	1.304	.836	2.90	1.426	1.017
.76	1.062	.3578	1.48	1.181	.627	2.20	1.307	.842	2.92	1.429	1.022
.78	1.065	.3662	1.50	1.184	.634	2.22	1.310	.847	2.94	1.433	1.026
.80	1.068	.3745	1.52	1.188	.640	2.24	1.314	.852	2.96	1.436	1.031
.82	1.071	.3828	1.54	1.191	.647	2.26	1.317	.858	2.98	1.440	1.035
.84	1.074	.3911	1.56	1.195	.653	2.28	1.321	.863	3.00	1.443	1.040
.86	1.077	.3993	1.58	1.198	.659	2.30	1.324	.869	3.02	1.446	1.044
.88	1.080	.4074	1.60	1.202	.666	2.32	1.328	.874	3.04	1.449	1.049
.90	1.083	.4155	1.62	1.205	.672	2.34	1.331	.879	3.06	1.453	1.053
.92	1.086	.4236	1.64	1.209	.678	2.36	1.335	.884	3.08	1.456	1.058
.94	1.089	.4316	1.66	1.212	.685	2.38	1.338	.889	3.10	1.459	1.062
.96	1.093	.4394	1.68	1.216	.691	2.40	1.342	.894	3.12	1.462	1.067
.98	1.096	.4473	1.70	1.219	.697	2.42	1.345	.900	3.14	1.465	1.072
1.0	1.099	.4552	1.72	1.223	.703	2.44	1.349	.904	3.16	1.469	1.076
1.02	1.102	.4628	1.74	1.226	.710	2.46	1.352	.910	3.18	1.472	1.080

APPENDIX II

h	$f(h)$	$g(h)$	h	$f(h)$	$g(h)$	h	$f(h)$	$g(h)$	h	$f(h)$	$g(h)$
3.18	1.472	1.080	3.90	1.585	1.230	4.62	1.692	1.365	5.34	1.794	1.490
3.20	1.475	1.085	3.92	1.588	1.234	4.64	1.695	1.369	5.36	1.797	1.493
3.22	1.478	1.089	3.94	1.591	1.238	4.66	1.698	1.372	5.38	1.800	1.495
3.24	1.482	1.093	3.96	1.595	1.241	4.68	1.701	1.376	5.40	1.802	1.497
3.26	1.485	1.098	3.98	1.598	1.245	4.70	1.704	1.379	5.42	1.804	1.502
3.28	1.489	1.102	4.00	1.601	1.249	4.72	1.706	1.383	5.44	1.807	1.506
3.30	1.492	1.106	4.02	1.604	1.253	4.74	1.709	1.387	5.46	1.810	1.509
3.32	1.495	1.110	4.04	1.607	1.257	4.76	1.712	1.390	5.48	1.813	1.513
3.34	1.498	1.115	4.06	1.610	1.261	4.78	1.715	1.394	5.50	1.815	1.515
3.36	1.502	1.119	4.08	1.613	1.265	4.80	1.718	1.397	5.52	1.818	1.518
3.38	1.505	1.123	4.10	1.616	1.269	4.82	1.721	1.400	5.54	1.821	1.522
3.40	1.508	1.127	4.12	1.619	1.273	4.84	1.723	1.403	5.56	1.824	1.526
3.42	1.511	1.132	4.14	1.622	1.276	4.86	1.726	1.407	5.58	1.827	1.529
3.44	1.514	1.136	4.16	1.625	1.280	4.88	1.729	1.411	5.60	1.829	1.531
3.46	1.517	1.140	4.18	1.628	1.284	4.90	1.732	1.415	5.62	1.831	1.535
3.48	1.520	1.145	4.20	1.631	1.288	4.92	1.735	1.418	5.64	1.834	1.537
3.50	1.523	1.149	4.22	1.634	1.291	4.94	1.738	1.422	5.66	1.837	1.540
3.52	1.526	1.153	4.24	1.637	1.295	4.96	1.741	1.425	5.68	1.840	1.544
3.54	1.529	1.158	4.26	1.640	1.299	4.98	1.743	1.429	5.70	1.842	1.548
3.56	1.533	1.161	4.28	1.642	1.303	5.00	1.746	1.432	5.72	1.844	1.550
3.58	1.536	1.165	4.30	1.645	1.307	5.02	1.749	1.435	5.74	1.847	1.553
3.60	1.539	1.170	4.32	1.648	1.310	5.04	1.752	1.439	5.76	1.850	1.557
3.62	1.542	1.174	4.34	1.651	1.314	5.06	1.755	1.442	5.78	1.853	1.560
3.64	1.545	1.178	4.36	1.654	1.318	5.08	1.758	1.445	5.80	1.855	1.562
3.66	1.549	1.181	4.38	1.657	1.322	5.10	1.760	1.449	5.82	1.857	1.566
3.68	1.552	1.186	4.40	1.660	1.325	5.12	1.763	1.452	5.84	1.860	1.568
3.70	1.555	1.190	4.42	1.663	1.328	5.14	1.766	1.455	5.86	1.863	1.571
3.72	1.558	1.194	4.44	1.666	1.332	5.16	1.769	1.459	5.88	1.866	1.575
3.74	1.561	1.198	4.46	1.669	1.336	5.18	1.772	1.463	5.90	1.868	1.579
3.76	1.564	1.202	4.48	1.672	1.340	5.20	1.774	1.465	5.92	1.871	1.581
3.78	1.567	1.206	4.50	1.675	1.343	5.22	1.777	1.468	5.94	1.874	1.584
3.80	1.570	1.210	4.52	1.678	1.346	5.24	1.780	1.473	5.96	1.877	1.588
3.82	1.573	1.214	4.54	1.681	1.350	5.26	1.783	1.477	5.98	1.880	1.592
3.84	1.576	1.218	4.56	1.684	1.354	5.28	1.786	1.480	6.00	1.882	1.594
3.86	1.579	1.222	4.58	1.686	1.358	5.30	1.788	1.482	6.02	1.885	1.598
3.88	1.582	1.226	4.60	1.689	1.362	5.32	1.791	1.486	6.04	1.888	1.600

h	$f(h)$	$g(h)$	h	$f(h)$	$g(h)$	h	$f(h)$	$g(h)$	h	$f(h)$	$g(h)$
6.04	1.888	1.600	6.76	1.979	1.708	7.48	2.066	1.808	8.20	2.151	1.904
6.06	1.890	1.604	6.78	1.981	1.710	7.50	2.069	1.811	8.22	2.153	1.908
6.08	1.892	1.606	6.80	1.984	1.713	7.52	2.071	1.815	8.24	2.156	1.911
6.10	1.895	1.609	6.82	1.986	1.716	7.54	2.074	1.816	8.26	2.158	1.912
6.12	1.898	1.613	6.84	1.989	1.719	7.56	2.076	1.819	8.28	2.160	1.915
6.14	1.900	1.616	6.86	1.991	1.722	7.58	2.078	1.822	8.30	2.163	1.918
6.16	1.903	1.619	6.88	1.994	1.725	7.60	2.081	1.825	8.32	2.165	1.920
6.18	1.905	1.621	6.90	1.996	1.727	7.62	2.083	1.828	8.34	2.167	1.923
6.20	1.908	1.625	6.92	1.999	1.730	7.64	2.086	1.830	8.36	2.170	1.926
6.22	1.911	1.628	6.94	2.001	1.733	7.66	2.088	1.833	8.38	2.172	1.928
6.24	1.913	1.630	6.96	2.004	1.736	7.68	2.090	1.836	8.40	2.174	1.930
6.26	1.916	1.633	6.98	2.006	1.739	7.70	2.093	1.838	8.42	2.176	1.933
6.28	1.918	1.636	7.00	2.009	1.743	7.72	2.095	1.841	8.44	2.179	1.935
6.30	1.921	1.639	7.02	2.011	1.745	7.74	2.097	1.844	8.46	2.181	1.938
6.32	1.924	1.643	7.04	2.014	1.748	7.76	2.100	1.847	8.48	2.183	1.941
6.34	1.926	1.645	7.06	2.016	1.751	7.78	2.102	1.850	8.50	2.186	1.944
6.36	1.929	1.649	7.08	2.019	1.754	7.80	2.104	1.852	8.52	2.188	1.946
6.38	1.931	1.652	7.10	2.021	1.756	7.82	2.108	1.854	8.54	2.190	1.949
6.40	1.934	1.655	7.12	2.023	1.758	7.84	2.109	1.857	8.56	2.193	1.952
6.42	1.936	1.657	7.14	2.026	1.761	7.86	2.112	1.860	8.58	2.195	1.954
6.44	1.939	1.661	7.16	2.028	1.764	7.88	2.114	1.863	8.60	2.197	1.956
6.46	1.941	1.664	7.18	2.030	1.767	7.90	2.117	1.866	8.62	2.199	1.959
6.48	1.944	1.667	7.20	2.033	1.770	7.92	2.119	1.868	8.64	2.202	1.962
6.50	1.946	1.669	7.22	2.035	1.772	7.94	2.122	1.871	8.66	2.204	1.965
6.52	1.949	1.672	7.24	2.038	1.775	7.96	2.124	1.873	8.68	2.207	1.967
6.54	1.951	1.675	7.26	2.040	1.778	7.98	2.126	1.876	8.70	2.209	1.970
6.56	1.954	1.678	7.28	2.042	1.781	8.00	2.128	1.878	8.72	2.211	1.972
6.58	1.956	1.682	7.30	2.045	1.784	8.02	2.130	1.881	8.74	2.214	1.975
6.60	1.959	1.684	7.32	2.047	1.786	8.04	2.133	1.884	8.76	2.216	1.977
6.62	1.962	1.687	7.34	2.050	1.789	8.06	2.135	1.886	8.78	2.218	1.980
6.64	1.964	1.690	7.36	2.052	1.792	8.08	2.138	1.889	8.80	2.220	1.982
6.66	1.966	1.692	7.38	2.054	1.795	8.10	2.140	1.892	8.82	2.222	1.984
6.68	1.969	1.696	7.40	2.057	1.798	8.12	2.142	1.894	8.84	2.225	1.987
6.70	1.971	1.698	7.42	2.059	1.800	8.14	2.144	1.897	8.86	2.227	1.989
6.72	1.974	1.702	7.44	2.062	1.803	8.16	2.147	1.900	8.88	2.229	1.992
6.74	1.976	1.705	7.46	2.064	1.805	8.18	2.149	1.902	8.90	2.231	1.994

h	$f(h)$	$g(h)$	h	$f(h)$	$g(h)$	h	$f(h)$	$g(h)$	h	$f(h)$	$g(h)$
8.90	2.231	1.994	9.7	2.319	2.092	13.3	2.678	2.484	16.9	2.995	2.823
8.92	2.233	1.997	9.8	2.329	2.104	13.4	2.687	2.494	17.0	3.003	2.832
8.94	2.236	2.000	9.9	2.340	2.116	13.5	2.696	2.503	17.1	3.011	2.841
8.96	2.238	2.002	10.0	2.351	2.128	13.6	2.705	2.513	17.2	3.019	2.850
8.98	2.240	2.005	10.1	2.362	2.140	13.7	2.714	2.523	17.3	3.027	2.857
9.00	2.242	2.007	10.2	2.372	2.151	13.8	2.723	2.533	17.4	3.035	2.866
9.02	2.244	2.009	10.3	2.382	2.162	13.9	2.732	2.543	17.5	3.044	2.876
9.04	2.247	2.012	10.4	2.392	2.173	14.0	2.741	2.553	17.6	3.052	2.883
9.06	2.249	2.014	10.5	2.403	2.185	14.1	2.750	2.563	17.7	3.060	2.892
9.08	2.251	2.016	10.6	2.413	2.196	14.2	2.759	2.573	17.8	3.068	2.901
9.10	2.253	2.019	10.7	2.423	2.207	14.3	2.768	2.582	17.9	3.076	2.910
9.12	2.255	2.021	10.8	2.433	2.218	14.4	2.777	2.592	18.0	3.084	2.919
9.14	2.258	2.024	10.9	2.443	2.230	14.5	2.787	2.602	18.1	3.093	2.926
9.16	2.260	2.026	11.0	2.454	2.242	14.6	2.796	2.612	18.2	3.101	2.935
9.18	2.262	2.029	11.1	2.465	2.253	14.7	2.805	2.621	18.3	3.109	2.943
9.20	2.264	2.031	11.2	2.475	2.264	14.8	2.814	2.631	18.4	3.117	2.952
9.22	2.266	2.034	11.3	2.484	2.274	14.9	2.823	2.640	18.6	3.133	2.969
9.24	2.269	2.036	11.4	2.494	2.285	15.0	2.832	2.649	18.8	3.148	2.986
9.26	2.271	2.039	11.5	2.504	2.296	15.1	2.841	2.659	19.0	3.164	3.002
9.28	2.273	2.042	11.6	2.514	2.307	15.2	2.850	2.669	19.2	3.180	3.019
9.30	2.275	2.044	11.7	2.524	2.318	15.3	2.859	2.678	19.4	3.196	3.035
9.32	2.277	2.046	11.8	2.534	2.329	15.4	2.867	2.686	19.6	3.211	3.052
9.34	2.280	2.049	11.9	2.544	2.340	15.5	2.876	2.697	19.8	3.227	3.068
9.36	2.282	2.051	12.0	2.554	2.350	15.6	2.885	2.706	20.0	3.243	3.085
9.38	2.284	2.053	12.1	2.563	2.360	15.7	2.892	2.715	20.2	3.258	3.101
9.40	2.286	2.056	12.2	2.573	2.370	15.8	2.901	2.724	20.4	3.273	3.117
9.42	2.288	2.058	12.3	2.582	2.381	15.9	2.911	2.733	20.6	3.288	3.132
9.44	2.291	2.061	12.4	2.592	2.392	16.0	2.919	2.742	20.8	3.303	3.148
9.46	2.293	2.063	12.5	2.601	2.401	16.1	2.928	2.751	21.0	3.318	3.164
9.48	2.295	2.065	12.6	2.610	2.411	16.2	2.936	2.760	21.2	3.333	3.179
9.50	2.297	2.067	12.7	2.620	2.423	16.3	2.945	2.769	21.4	3.348	3.195
9.52	2.299	2.069	12.8	2.630	2.432	16.4	2.953	2.778	21.6	3.362	3.210
9.54	2.302	2.072	12.9	2.640	2.443	16.5	2.961	2.787	21.8	3.377	3.226
9.56	2.304	2.075	13.0	2.650	2.454	16.6	2.970	2.796	22.0	3.392	3.241
9.58	2.306	2.077	13.1	2.660	2.465	16.7	2.977	2.805	22.2	3.408	3.256
9.60	2.308	2.080	13.2	2.669	2.475	16.8	2.987	2.814	22.4	3.421	3.272

h	$f(h)$	$g(h)$	h	$f(h)$	$g(h)$	h	$f(h)$	$g(h)$	h	$f(h)$	$g(h)$
22.4	3.421	3.272	29.6	3.912	3.782	36.8	4.348	4.232	72	6.042	5.959
22.6	3.436	3.287	29.8	3.924	3.795	37	4.359	4.243	73	6.084	6.001
22.8	3.450	3.303	30.0	3.937	3.808	38	4.417	4.302	74	6.123	6.041
23.0	3.465	3.318	30.2	3.950	3.821	39	4.473	4.360	75	6.166	6.084
23.2	3.480	3.333	30.4	3.963	3.835	40	4.528	4.416	76	6.205	6.124
23.4	3.494	3.348	30.6	3.976	3.848	41	4.584	4.473	77	6.245	6.165
23.6	3.509	3.363	30.8	3.989	3.862	42	4.638	4.528	78	6.285	6.205
23.8	3.523	3.378	31.0	4.002	3.875	43	4.692	4.584	79	6.326	6.246
24.0	3.538	3.393	31.2	4.014	3.888	44	4.743	4.636	80	6.364	6.284
24.2	3.551	3.407	31.4	4.026	3.900	45	4.797	4.691	81	6.403	6.326
24.4	3.565	3.423	31.6	4.039	3.913	46	4.848	4.744	82	6.443	6.364
24.6	3.579	3.437	31.8	4.051	3.925	47	4.899	4.795	83	6.481	6.403
24.8	3.593	3.452	32.0	4.063	3.938	48	4.950	4.848	84	6.519	6.442
25.0	3.607	3.467	32.2	4.075	3.951	49	5.000	4.900	85	6.558	6.481
25.2	3.621	3.481	32.4	4.087	3.963	50	5.050	4.950	86	6.597	6.519
25.4	3.635	3.495	32.6	4.100	3.976	51	5.099	5.000	87	6.633	6.557
25.6	3.647	3.509	32.8	4.112	3.988	52	5.147	5.049	88	6.671	6.596
25.8	3.661	3.524	33.0	4.124	4.001	53	5.198	5.099	89	6.708	6.633
26.0	3.676	3.538	33.2	4.136	4.014	54	5.244	5.148	90	6.746	6.671
26.2	3.689	3.552	33.4	4.148	4.026	55	5.293	5.198	91	6.782	6.708
26.4	3.702	3.565	33.6	4.161	4.039	56	5.339	5.244	92	6.819	6.746
26.6	3.716	3.579	33.8	4.173	4.051	57	5.387	5.292	93	6.857	6.783
26.8	3.729	3.592	34.0	4.185	4.064	58	5.433	5.340	94	6.892	6.819
27.0	3.742	3.606	34.2	4.197	4.076	59	5.478	5.386	95	6.928	6.856
27.2	3.756	3.620	34.4	4.209	4.088	60	5.524	5.433	96	6.964	6.892
27.4	3.769	3.634	34.6	4.220	4.100	61	5.568	5.477	97	7.001	6.929
27.6	3.783	3.648	34.8	4.232	4.112	62	5.612	5.522	98	7.035	6.964
27.8	3.796	3.662	35.0	4.244	4.124	63	5.656	5.567	99	7.071	7.000
28.0	3.810	3.676	35.2	4.256	4.136	64	5.701	5.613	100	7.105	7.035
28.2	3.823	3.689	35.4	4.267	4.148	65	5.746	5.658	101	7.141	7.070
28.4	3.836	3.703	35.6	4.279	4.160	66	5.789	5.701	102	7.176	7.106
28.6	3.848	3.716	35.8	4.290	4.172	67	5.832	5.745	103	7.212	7.142
28.8	3.861	3.730	36.0	4.302	4.184	68	5.875	5.788	104	7.247	7.178
29.0	3.874	3.743	36.2	4.314	4.196	69	5.916	5.830	105	7.279	7.211
29.2	3.887	3.756	36.4	4.325	4.208	70	5.959	5.875	106	7.315	7.246
29.4	3.899	3.769	36.6	4.336	4.219	71	6.000	5.917	107	7.349	7.280

APPENDIX II

h	$f(h)$	$g(h)$	h	$f(h)$	$g(h)$	h	$f(h)$	$g(h)$	h	$f(h)$	$g(h)$
107	7.349	7.280	143	8.485	8.426	179	9.487	9.434	215	10.392	10.344
108	7.383	7.315	144	8.515	8.456	180	9.513	9.460	216	10.416	10.368
109	7.416	7.349	145	8.544	8.485	181	9.539	9.487	217	10.440	10.392
110	7.450	7.383	146	8.573	8.515	182	9.566	9.513	218	10.464	10.416
111	7.483	7.416	147	8.602	8.544	183	9.592	9.539	219	10.488	10.440
112	7.517	7.450	148	8.631	8.573	184	9.618	9.566	220	10.512	10.464
113	7.550	7.483	149	8.660	8.602	185	9.644	9.592	221	10.536	10.488
114	7.583	7.517	150	8.689	8.631	186	9.670	9.618	222	10.559	10.512
115	7.616	7.550	151	8.718	8.660	187	9.695	9.644	223	10.583	10.536
116	7.649	7.583	152	8.747	8.689	188	9.721	9.670	224	10.607	10.559
117	7.681	7.616	153	8.775	8.718	189	9.747	9.695	225	10.630	10.583
118	7.714	7.649	154	8.803	8.747	190	9.772	9.721	226	10.654	10.607
119	7.746	7.681	155	8.832	8.775	191	9.798	9.747	227	10.677	10.630
120	7.778	7.714	156	8.860	8.803	192	9.823	9.772	228	10.700	10.654
121	7.810	7.746	157	8.888	8.832	193	9.849	9.798	229	10.724	10.677
122	7.842	7.778	158	8.916	8.860	194	9.874	9.823	230	10.747	10.700
123	7.874	7.810	159	8.944	8.888	195	9.900	9.849	231	10.770	10.724
124	7.906	7.842	160	8.972	8.916	196	9.925	9.874	232	10.794	10.747
125	7.937	7.874	161	9.000	8.944	197	9.950	9.900	233	10.817	10.770
126	7.969	7.906	162	9.028	8.972	198	9.975	9.925	234	10.840	10.794
127	8.000	7.937	163	9.055	9.000	199	10.000	9.950	235	10.863	10.817
128	8.031	7.969	164	9.083	9.028	200	10.025	9.975	236	10.886	10.840
129	8.062	8.000	165	9.110	9.055	201	10.050	10.000	237	10.909	10.863
130	8.093	8.031	166	9.138	9.083	202	10.075	10.025	238	10.932	10.886
131	8.124	8.062	167	9.165	9.110	203	10.100	10.050	239	10.954	10.909
132	8.155	8.093	168	9.192	9.138	204	10.124	10.075	240	10.977	10.932
133	8.185	8.124	169	9.220	9.165	205	10.149	10.100	241	11.000	10.954
134	8.216	8.155	170	9.247	9.192	206	10.174	10.124	242	11.023	10.977
135	8.246	8.185	171	9.274	9.220	207	10.198	10.149	243	11.045	11.000
136	8.277	8.216	172	9.301	9.247	208	10.223	10.174	244	11.068	11.023
137	8.307	8.246	173	9.327	9.274	209	10.247	10.198	245	11.091	11.045
138	8.337	8.277	174	9.354	9.301	210	10.271	10.223	246	11.113	11.068
139	8.367	8.307	175	9.381	9.327	211	10.296	10.247	247	11.136	11.091
140	8.397	8.337	176	9.407	9.354	212	10.320	10.271	248	11.158	11.113
141	8.426	8.367	177	9.434	9.381	213	10.344	10.296	249	11.180	11.136
142	8.456	8.397	178	9.460	9.407	214	10.368	10.320	250	11.203	11.158

Appendix

Tables of Bessel Functions*

TABLE 1. $J_0(y)$

y	0	0.1	0.2	0.3	0.4	0.5	0.6	0.7	0.8	0.9
0	1.0000	0.9975	0.9900	0.9776	0.9604	0.9385	0.9120	0.8812	0.8463	0.8075
1	0.7652	0.7196	0.6711	0.6201	0.5669	0.5118	0.4554	0.3980	0.3400	0.2818
2	0.2239	0.1666	0.1104	0.0555	0.0025	−0.0484	−0.0968	−0.1424	−0.1850	−0.2243
3	−0.2601	−0.2921	−0.3202	−0.3443	−0.3643	−0.3801	−0.3918	−0.3992	−0.4026	−0.4018
4	−0.3971	−0.3887	−0.3766	−0.3610	−0.3423	−0.3205	−0.2961	−0.2693	−0.2404	−0.2097
5	−0.1776	−0.1443	−0.1103	−0.0758	−0.0412	−0.0068	0.0270	0.0599	0.0917	0.1220
6	0.1506	0.1773	0.2017	0.2238	0.2433	0.2601	0.2740	0.2851	0.2931	0.2981
7	0.3001	0.2991	0.2951	0.2882	0.2786	0.2663	0.2516	0.2346	0.2154	0.1944
8	0.1717	0.1475	0.1222	0.0960	0.0692	0.0419	0.0146	−0.0125	−0.0392	−0.0653
9	−0.0903	−0.1142	−0.1367	−0.1577	−0.1768	−0.1939	−0.2090	−0.2218	−0.2323	−0.2403
10	−0.2459	−0.2490	−0.2496	−0.2477	−0.2434	−0.2366	−0.2276	−0.2164	−0.2032	−0.1881
11	−0.1712	−0.1528	−0.1330	−0.1121	−0.0902	−0.0677	−0.0446	−0.0213	0.0020	0.0250
12	0.0477	0.0697	0.0908	0.1108	0.1296	0.1469	0.1626	0.1766	0.1887	0.1988
13	0.2069	0.2129	0.2167	0.2183	0.2177	0.2150	0.2101	0.2032	0.1943	0.1836
14	0.1711	0.1570	0.1414	0.1245	0.1065	0.0875	0.0679	0.0476	0.0271	0.0064
15	−0.0142	−0.0346	−0.0544	−0.0736	−0.0919	−0.1092	−0.1253	−0.1401	−0.1533	−0.1650

For $y \geq 16$, an approximate formula is

$$J_0(y) \doteq \sqrt{\frac{2}{\pi y}} \left[\sin\left(y + \frac{\pi}{4}\right) + \frac{1}{8y} \sin\left(y - \frac{\pi}{4}\right) \right]$$

* These tables are taken from N. W. McLachlan, *Bessel Functions for Engineers* (Oxford: Clarendon Press, 1934), by permission of the author and the publisher.

TABLE 2. $J_1(y)$

y	0	0.1	0.2	0.3	0.4	0.5	0.6	0.7	0.8	0.9
0	0.0000	0.0499	0.0995	0.1483	0.1960	0.2423	0.2867	0.3290	0.3688	0.4059
1	0.4401	0.4709	0.4983	0.5220	0.5419	0.5579	0.5699	0.5778	0.5815	0.5812
2	0.5767	0.5683	0.5560	0.5399	0.5202	0.4971	0.4708	0.4416	0.4097	0.3754
3	0.3391	0.3009	0.2613	0.2207	0.1792	0.1374	0.0955	0.0538	0.0128	−0.0272
4	−0.0660	−0.1033	−0.1386	−0.1719	−0.2028	−0.2311	−0.2566	−0.2791	−0.2985	−0.3147
5	−0.3276	−0.3371	−0.3432	−0.3460	−0.3453	−0.3414	−0.3343	−0.3241	−0.3110	−0.2951
6	−0.2767	−0.2559	−0.2329	−0.2081	−0.1816	−0.1538	−0.1250	−0.0953	−0.0652	−0.0349
7	−0.0047	0.0252	0.0543	0.0826	0.1096	0.1352	0.1592	0.1813	0.2014	0.2192
8	0.2346	0.2476	0.2580	0.2657	0.2708	0.2731	0.2728	0.2697	0.2641	0.2559
9	0.2453	0.2324	0.2174	0.2004	0.1816	0.1613	0.1395	0.1166	0.0928	0.0684
10	0.0435	0.0184	−0.0066	−0.0313	−0.0555	−0.0789	−0.1012	−0.1224	−0.1422	−0.1603
11	−0.1768	−0.1913	−0.2039	−0.2143	−0.2225	−0.2284	−0.2320	−0.2333	−0.2323	−0.2290
12	−0.2234	−0.2157	−0.2060	−0.1943	−0.1807	−0.1655	−0.1487	−0.1307	−0.1114	−0.0912
13	−0.0703	−0.0489	−0.0271	−0.0052	0.0166	0.0380	0.0590	0.0791	0.0984	0.1165
14	0.1334	0.1488	0.1626	0.1747	0.1850	0.1934	0.1999	0.2043	0.2066	0.2069
15	0.2051	0.2013	0.1955	0.1879	0.1784	0.1672	0.1544	0.1402	0.1247	0.1080

For $y \geq 16$, an approximate formula is

$$J_1(y) \doteq \sqrt{\frac{2}{\pi y}} \left[\sin\left(y - \frac{\pi}{4}\right) + \frac{3}{8y} \sin\left(y + \frac{\pi}{4}\right) \right]$$

TABLE 3. $N_0(y)$

y	0	0.1	0.2	0.3	0.4	0.5	0.6	0.7	0.8	0.9
0	−∞	−1.534	−1.081	−0.8073	−0.6060	−0.4445	−0.3085	−0.1907	−0.0868	0.0056
1	0.0883	0.1622	0.2281	0.2865	0.3379	0.3824	0.4204	0.4520	0.4774	0.4968
2	0.5104	0.5183	0.5208	0.5181	0.5104	0.4981	0.4813	0.4605	0.4359	0.4079
3	0.3769	0.3431	0.3071	0.2691	0.2296	0.1890	0.1477	0.1061	0.0645	0.0234
4	−0.0169	−0.0561	−0.0938	−0.1296	−0.1633	−0.1947	−0.2235	−0.2494	−0.2723	−0.2921
5	−0.3085	−0.3216	−0.3313	−0.3374	−0.3402	−0.3395	−0.3354	−0.3282	−0.3177	−0.3044
6	−0.2882	−0.2694	−0.2483	−0.2251	−0.1999	−0.1732	−0.1452	−0.1162	−0.0864	−0.0563
7	−0.0259	0.0042	0.0339	0.0628	0.0907	0.1173	0.1424	0.1658	0.1872	0.2065
8	0.2235	0.2381	0.2501	0.2595	0.2662	0.2702	0.2715	0.2700	0.2659	0.2592
9	0.2499	0.2383	0.2245	0.2086	0.1907	0.1712	0.1502	0.1279	0.1045	0.0804
10	0.0557	0.0307	0.0056	−0.0193	−0.0437	−0.0675	−0.0904	−0.1122	−0.1326	−0.1516
11	−0.1688	−0.1843	−0.1977	−0.2091	−0.2183	−0.2252	−0.2299	−0.2322	−0.2322	−0.2298
12	−0.2252	−0.2184	−0.2095	−0.1986	−0.1858	−0.1712	−0.1551	−0.1375	−0.1187	−0.0989
13	−0.0782	−0.0569	−0.0352	−0.0134	0.0085	0.0301	0.0512	0.0717	0.0913	0.1099
14	0.1272	0.1431	0.1575	0.1703	0.1812	0.1903	0.1974	0.2025	0.2056	0.2065
15	0.2055	0.2023	0.1972	0.1902	0.1813	0.1706	0.1584	0.1446	0.1295	0.1132

For $y \geq 16$, an approximate formula is

$$N_0(y) \doteq \sqrt{\frac{2}{\pi y}} \left[\sin\left(y - \frac{\pi}{4}\right) - \frac{1}{8y} \sin\left(y + \frac{\pi}{4}\right) \right]$$

TABLE 4. $N_1(y)$

y	0	0.1	0.2	0.3	0.4	0.5	0.6	0.7	0.8	0.9
0	$-\infty$	-6.459	-3.324	-2.293	-1.781	-1.471	-1.260	-1.103	-0.9781	-0.8731
1	-0.7812	-0.6981	-0.6211	-0.5485	-0.4791	-0.4123	-0.3476	-0.2847	-0.2237	-0.1644
2	-0.1070	-0.0517	0.0015	0.0523	0.1005	0.1459	0.1884	0.2276	0.2635	0.2959
3	0.3247	0.3496	0.3707	0.3879	0.4010	0.4102	0.4154	0.4167	0.4141	0.4078
4	0.3979	0.3846	0.3680	0.3484	0.3260	0.3010	0.2737	0.2445	0.2136	0.1812
5	0.1479	0.1137	0.0792	0.0445	0.0101	-0.0238	-0.0568	-0.0887	-0.1192	-0.1481
6	-0.1750	-0.1998	-0.2223	-0.2422	-0.2596	-0.2741	-0.2857	-0.2945	-0.3002	-0.3029
7	-0.3027	-0.2995	-0.2934	-0.2846	-0.2731	-0.2591	-0.2428	-0.2243	-0.2039	-0.1817
8	-0.1581	-0.1331	-0.1072	-0.0806	-0.0535	-0.0262	0.0011	0.0280	0.0544	0.0799
9	0.1043	0.1275	0.1491	0.1691	0.1871	0.2032	0.2171	0.2287	0.2379	0.2447
10	0.2490	0.2508	0.2502	0.2471	0.2416	0.2337	0.2236	0.2114	0.1973	0.1813
11	0.1637	0.1446	0.1243	0.1029	0.0807	0.0579	0.0348	0.0114	-0.0118	-0.0347
12	-0.0571	-0.0787	-0.0994	-0.1189	-0.1371	-0.1538	-0.1689	-0.1821	-0.1935	-0.2028
13	-0.2101	-0.2152	-0.2182	-0.2190	-0.2176	-0.2140	-0.2084	-0.2007	-0.1912	-0.1798
14	-0.1666	-0.1520	-0.1359	-0.1186	-0.1003	-0.0810	-0.0612	-0.0408	-0.0202	0.0005
15	0.0211	0.0413	0.0609	0.0799	0.0979	0.1148	0.1305	0.1447	0.1575	0.1686

For $y \geq 16$, an approximate formula is

$$N_1(y) \doteq \sqrt{\frac{2}{\pi y}} \left[\sin\left(y - \frac{3\pi}{4}\right) + \frac{3}{8y} \sin\left(y - \frac{\pi}{4}\right) \right]$$

TABLE 5. $J_0(yj^{-1/2}) = J_0(yj^{3/2}) = M_0(y)e^{j\theta_0(y)}$

y	$M_0(y)$	$\log_{10}\sqrt{y}M_0(y)$	$\theta_0(y)$	y	$M_0(y)$	$\log_{10}\sqrt{y}M_0(y)$	$\theta_0(y)$
0.00	1.000	...	0.00°	2.0	1.229	0.2401	52.29°
0.05	1.000	$\bar{1}$.3995	0.04	2.1	1.274	0.2663	56.74
0.10	1.000	$\bar{1}$.5000	0.14	2.2	1.325	0.2933	61.22
0.15	1.000	$\bar{1}$.5880	0.32	2.3	1.381	0.3210	65.71
0.20	1.000	$\bar{1}$.6505	0.57	2.4	1.443	0.3493	70.19
0.25	1.000	$\bar{1}$.6990	0.90	2.5	1.511	0.3783	74.65
0.30	1.000	$\bar{1}$.7386	1.29	2.6	1.586	0.4077	79.09
0.35	1.000	$\bar{1}$.7721	1.75	2.7	1.666	0.4375	83.50
0.40	1.000	$\bar{1}$.8012	2.29	2.8	1.754	0.4676	87.87
0.45	1.001	$\bar{1}$.8269	2.90	2.9	1.849	0.4980	92.21
0.50	1.001	$\bar{1}$.8499	3.58	3.0	1.950	0.5286	96.52
0.55	1.001	$\bar{1}$.8708	4.33	3.1	2.059	0.5594	100.79
0.60	1.002	$\bar{1}$.8900	5.15	3.2	2.176	0.5902	105.03
0.65	1.003	$\bar{1}$.9077	6.04	3.3	2.301	0.6212	109.25
0.70	1.004	$\bar{1}$.9242	7.01	3.4	2.434	0.6521	113.43
0.75	1.005	$\bar{1}$.9397	8.04	3.5	2.576	0.6830	117.60
0.80	1.006	$\bar{1}$.9543	9.14	3.6	2.728	0.7140	121.75
0.85	1.008	$\bar{1}$.9682	10.31	3.7	2.889	0.7449	125.87
0.90	1.010	$\bar{1}$.9815	11.55	3.8	3.061	0.7758	129.99
0.95	1.013	$\bar{1}$.9943	12.86	3.9	3.244	0.8067	134.10
1.00	1.016	0.0067	14.23	4.0	3.439	0.8375	138.19
1.05	1.019	0.0187	15.66	4.5	4.618	0.9910	158.59
1.10	1.023	0.0304	17.16	5.0	6.231	1.1441	178.93
1.15	1.027	0.0419	18.72	5.5	8.447	1.2969	199.28
1.20	1.032	0.0533	20.34	6.0	11.50	1.4498	219.62
1.25	1.038	0.0645	22.02	7.0	21.55	1.7560	260.29
1.30	1.044	0.0756	23.75	8.0	40.82	2.0624	300.92
1.35	1.051	0.0867	25.54	9.0	77.96	2.3690	341.52
1.40	1.059	0.0978	27.37	10.0	149.8	2.6756	382.10
1.45	1.067	0.1089	29.26	11.0	289.5	2.9824	422.66
1.50	1.077	0.1201	31.19	12.0	561.8	3.2892	463.22
1.55	1.087	0.1314	33.16	14.0	2,137	3.9029	544.32
1.60	1.098	0.1428	35.17	16.0	8,217	4.5168	625.40
1.65	1.111	0.1544	37.22	18.0	$3{,}185_1$	5.1307	706.46
1.70	1.124	0.1661	39.30	20.0	$1{,}242_2$	5.7447	787.52
1.75	1.139	0.1779	41.41	25.0	$3{,}809_3$	7.2798	990.15
1.80	1.154	0.1900	43.54	30.0	$1{,}192_5$	8.8150	1,192.75
1.85	1.171	0.2022	45.70	35.0	$3{,}786_6$	10.3502	1,395.35
1.90	1.189	0.2146	47.88	40.0	$1{,}215_8$	11.8856	1,597.94
1.95	1.208	0.2273	50.08	45.0	$3{,}929_9$	13.4209	1,800.53

$(1{,}215_8 \text{ represents } 1{,}215 \times 10^8.)$

For $y > 45$, approximate formulas are

$$\log_{10} M_0(y) = 0.3071y + \frac{0.0384}{y} - 0.3991 - \frac{1}{2}\log_{10}y$$

$$\theta_0(y) = 40.°514y - \frac{5.°06}{y} - 22.°5$$

If interpolation is required when $y \geq 2.7$, the function $\log_{10}[\sqrt{y}M_0(y)]$ should be used and $M_0(y)$ computed from it.

TABLE 6. $J_1(yj^{-1/2})e^{j180°} = J_1(yj^{3/2}) = M_1(y)e^{j\theta_1(y)}$

y	$M_1(y)$	$\log_{10}\sqrt{y}M_1(y)$	$\theta_1(y)$	y	$M_1(y)$	$\log_{10}\sqrt{y}M_1(y)$	$\theta_1(y)$
0.00	0.0000	. . .	135.00°	2.25	1.199	0.2548	170.50°
0.05	0.0250	$\bar{3}$.7474	135.02	2.30	1.232	0.2715	172.03
0.10	0.0500	$\bar{2}$.1990	135.07	2.35	1.266	0.2881	173.58
0.15	0.0750	$\bar{2}$.4631	135.16	2.40	1.301	0.3045	175.16
0.20	0.1000	$\bar{2}$.6505	135.29	2.45	1.337	0.3207	176.76
0.25	0.1250	$\bar{2}$.7959	135.45	2.50	1.374	0.3368	178.39
0.30	0.1500	$\bar{2}$.9147	135.64	2.55	1.411	0.3529	180.03
0.35	0.1750	$\bar{1}$.0151	135.88	2.60	1.450	0.3688	181.70
0.40	0.2000	$\bar{1}$.1021	136.15	2.65	1.489	0.3846	183.39
0.45	0.2250	$\bar{1}$.1788	136.45	2.70	1.530	0.4004	185.10
0.50	0.2500	$\bar{1}$.2475	136.79	2.80	1.615	0.4317	188.57
0.55	0.2751	$\bar{1}$.3096	137.17	2.90	1.705	0.4628	192.11
0.60	0.3001	$\bar{1}$.3663	137.58	3.00	1.800	0.4938	195.71
0.65	0.3252	$\bar{1}$.4185	138.03	3.10	1.901	0.5247	199.37
0.70	0.3502	$\bar{1}$.4669	138.51	3.20	2.009	0.5555	203.03
0.75	0.3753	$\bar{1}$.5119	139.03	3.30	2.124	0.5863	206.83
0.80	0.4004	$\bar{1}$.5541	139.58	3.40	2.246	0.6171	210.62
0.85	0.4256	$\bar{1}$.5937	140.17	3.50	2.376	0.6479	214.44
0.90	0.4508	$\bar{1}$.6311	140.80	3.60	2.515	0.6788	218.30
0.95	0.4760	$\bar{1}$.6665	141.46	3.70	2.664	0.7096	222.17
1.00	0.5013	$\bar{1}$.7001	142.16	3.80	2.823	0.7405	226.07
1.05	0.5267	$\bar{1}$.7321	142.89	4.00	3.173	0.8025	233.90
1.10	0.5521	$\bar{1}$.7627	143.66	4.25	3.681	0.8801	243.77
1.15	0.5776	$\bar{1}$.7920	144.46	4.50	4.278	0.9579	253.67
1.20	0.6032	$\bar{1}$.8201	145.29	5.00	5.809	1.1136	273.55
1.25	0.6290	$\bar{1}$.8471	146.17	5.5	7.925	1.2692	293.48
1.30	0.6548	$\bar{1}$.8731	147.07	6.0	10.85	1.4245	313.45
1.35	0.6808	$\bar{1}$.8982	148.02	6.5	14.90	1.5795	333.46
1.40	0.7070	$\bar{1}$.9225	148.99	7.0	20.50	1.7343	353.51
1.45	0.7333	$\bar{1}$.9460	150.00	7.5	28.27	1.8889	373.59
1.50	0.7598	$\bar{1}$.9688	151.04	8.0	39.07	2.0434	393.69
1.55	0.7866	$\bar{1}$.9909	152.12	9.0	74.97	2.3520	433.96
1.60	0.8136	0.0125	153.23	10.0	144.7	2.6604	474.28
1.65	0.8408	0.0335	154.38	11.0	280.4	2.9685	514.63
1.70	0.8684	0.0539	155.55	12.0	545.6	3.2765	555.02
1.75	0.8962	0.0739	156.76	14.0	2,084	3.8920	635.84
1.80	0.9244	0.0935	158.00	16.0	8,038	4.5072	716.72
1.85	0.9530	0.1127	159.27	18.0	$3{,}123_1$	5.1222	797.63
1.90	0.9819	0.1315	160.57	20.0	$1{,}220_2$	5.7370	878.57
1.95	1.011	0.1499	161.90	25.0	$3{,}755_3$	7.2736	1,080.98
2.00	1.041	0.1680	163.27	30.0	$1{,}178_5$	8.8099	1,283.45
2.05	1.072	0.1859	164.66	35.0	$3{,}748_6$	10.3459	1,485.94
2.10	1.102	0.2035	166.08	40.0	$1{,}204_8$	11.8817	1,688.46
2.15	1.134	0.2208	167.53	45.0	$3{,}899_9$	13.4175	1,890.98
2.20	1.166	0.2379	169.00	50.0	$1{,}270_{11}$	14.9532	2,093.52

$(1{,}178_5$ represents $1{,}718 \times 10^5)$

For $y > 50$, approximate formulas are

$$\log_{10} M_1(y) \doteq 0.3071y - \frac{0.1152}{y} - 0.3991 - \frac{1}{2}\log_{10} y$$

$$\theta_1(y) \doteq 40.°514y + \frac{15.°19}{y} + 67.°5$$

If interpolation is required when $y \geq 3.3$, the function $\log_{10}[\sqrt{y}\,M_1(y)]$ should be used and $M_1(y)$ computed from it.

Appendix

IV

Material Constants

Constants of Conductors and Semiconductors

Materials in order of decreasing conductivity	Conductivity, σ_e at 20°C constant (S/m)	Relative dielectric constant, ϵ_{er} at 20°C
Silver	6.14×10^7	1
Copper, annealed	5.80×10^7	1
Copper, hard-drawn	5.65×10^7	1
Aluminum	3.54×10^7	1
Tungsten	1.81×10^7	1
Zinc	1.74×10^7	1
Brass (30% Zn)	$1.2-1.5 \times 10^7$	1
Nickel	1.28×10^7	1
Iron, pure	1.00×10^7	1
Steel	$0.5-1 \times 10^7$	1
Tin	0.87×10^7	1
Manganin (84% Cu, 12% Mn, 4% Ni)	0.23×10^7	1
Constantin (60% Cu, 40% Ni)	0.20×10^7	1
Nichrome	0.10×10^7	1
Salt water	$3-5$	80
Wet earth	$10^{-2}-10^{-3}$	5-15
Lake water	$10^{-2}-10^{-3}$	80
Distilled water	2×10^{-4}	81
Dry earth	$10^{-4}-10^{-5}$	2-6

Constants of Dielectrics at Low Frequencies

Material in order of decreasing conductivity	Conductivity, σ_e at 20°C (S/m)	Relative dielectric constant, ϵ_{er} at 20°C
Slate	10^{-6}	6.6–7.4
Bakelite	10^{-8}–10^{-10}	5.5
Wood, paraffined	10^{-8}–10^{-11}	2–7
Mica	10^{-11}–10^{-15}	2.1
Glass	10^{-12}	6–8
Shellac	10^{-14}	2.7–3.7
Petroleum	10^{-14}	2.0–3.2
Paraffin	10^{-14}–10^{-16}	1.9–2.3
Rubber, hard	10^{-14}–10^{-16}	2.0–3.2

Constants of Dielectrics at High Frequencies

Material	Relative dielectric constant, ϵ_{er}	Loss tangent, $\sigma_e/\omega\epsilon_e = \epsilon''/\epsilon'$	Frequency (MHz)
Bakelite (sheet)	3.57	0.080	3000
Glass	3.8–8.7	0.001–0.01	1200
Lucite	$\Big\{$ 2.58	0.0090	1200
	2.56	0.0087	3000
Mycalex, red	5.91	0.0030	1200
Mycalex, white	5.74	0.0033	1200
Paraffin	2.17	0.00019	1200
Polyethylene	2.26	0.00031–0.0023	1200
Polystyrene	2.45	0.00028–0.00090	1200
Rubber, hard, black	2.69	0.00059	3000
Rubber, soft, black	3.15	0.0058	1200

Source: C. R. Englund, *Bell Syst. Tech. J.*, *23*, 125 (January 1944).

Dielectric Materials[a]

Material	T (°C)		1×10^2	1×10^3	1×10^4	1×10^5	1×10^6	1×10^7	1×10^8	3×10^8	3×10^9	1×10^{10}	2.5×10^{10}
								Frequency (Hz)					
Ice	−12	ϵ_{cr}									3.2	3.17	
		$\sigma_e/\omega\epsilon_e$									9	7	
Sodium chloride	85	ϵ_{cr}	5.90	5.90	5.90	5.90	5.90	5.90					5.90
		$\sigma_e/\omega\epsilon_e$	<1	<1	<1	<2	<2	<2					<5
		ϵ_{cr}	6.35	6.11	6.00	5.98	5.98	5.98					5.97
		$\sigma_e/\omega\epsilon_e$	170	240	70	6	<2	<2					<3.9
Phosphate glass (with 2% iron oxide)	25	ϵ_{cr}	5.25	5.25	5.25	5.25	5.25	5.25	5.24	5.23	5.17	5.00	4.93
		$\sigma_e/\omega\epsilon_e$	22	18	16	15	14	16	20	25	46	42	34
Sandy soil Dry	25	ϵ_{cr}	3.42	2.91	2.75	2.65	2.59	2.55		2.55	2.55	2.53	
		$\sigma_e/\omega\epsilon_e$	0.196	0.08	0.034	0.020	0.017	0.016		0.0100	0.0062	0.0036	
2–18% moisture	25	ϵ_{cr}	3.23	2.72	2.50	2.50	2.50	2.50		2.50	2.50	2.50	
		$\sigma_e/\omega\epsilon_e$	0.64	0.13	0.056	0.030	0.025	0.025		0.026	0.03	0.065	
Clay soil Dry	25	ϵ_{cr}	4.73	3.94	3.27	2.79	2.57	2.44		2.38	2.27	2.16	
		$\sigma_e/\omega\epsilon_e$	0.12	0.12	0.12	0.10	0.065	0.04		0.020	0.015	0.013	
20.09% moisture	25	ϵ_{cr}					21.6			20	11.3		
		$\sigma_e/\omega\epsilon_e$		7800	1000		1.7			0.52	0.25		
Laminated fiberglass	24	ϵ_{cr}	14.2	9.8	7.2	5.9	5.3	5.0	4.8	4.54	4.40	4.37	
		$\sigma_e/\omega\epsilon_e$	2500	2600	1600	880	460	340	260	240	290	360	
Teflon	22	ϵ_{cr}	2.1	2.1	2.1	2.1	2.1	2.1	2.1	2.1	2.1	2.08	2.08
		$\sigma_e/\omega\epsilon_e$	<5	<3	<3	<3	<2	<2	<2	1.5	1.5	3.7	6
	100	ϵ_{cr}	2.04	2.04	2.04	2.04	2.04	2.04				2.04	
		$\sigma_e/\omega\epsilon_e$	10	4	2	<3	<2	<2				5.1	

Material	T (°C)	Quantity											
Polystyrene	25	ϵ_{er}	2.56	2.56	2.56	2.56	2.56	2.56	2.55	2.55	2.55	2.54	2.54
		$\sigma_e/\omega\epsilon_e$	<0.5	<0.5	<0.5	0.5	0.7	<2	<1	3.5	3.3	4.3	12
	80	ϵ_{er}	2.54	2.54	2.54	2.54	2.54	2.54	2.54	2.54	2.54	2.53	
		$\sigma_e/\omega\epsilon_e$	9	2	<1	<2	<2	<2	<3	2.7	4.5	5.3	
Styrofoam	25	ϵ_{er}	1.03	1.03	1.03	1.03	1.03	1.03			1.03	1.03	
		$\sigma_e/\omega\epsilon_e$	<2	<1	<1	<1	<2	<2			1		
Water, conductivity	25	ϵ_{er}				78.2	78.2	78.2	78	77.5	76.7	55	34
		$\sigma_e/\omega\epsilon_e$				4000	400	46	50	160	1570	5400	2650
	85	ϵ_{er}				58	58	58	58	57	56.5	54	
		$\sigma_e/\omega\epsilon_e$				12400	1240	125	30	73	547	2600	
Aqueous sodium chloride 0.1 molal solution	25	ϵ_{er}				78.2				76	75.5	54	
		$\sigma_e/\omega\epsilon_e$				24×10^6				7800	2400	5600	
0.7 molal solution	25	ϵ_{er}				78.2						50	
		$\sigma_e/\omega\epsilon_e$				130×10^6						6600	
Methyl alcohol	25	ϵ_{er}					31	31.0	31.0	30.9	23.9	8.9	
		$\sigma_e/\omega\epsilon_e$					2000	260	380	800	6400	8100	
Ethyl alcohol	25	ϵ_{er}					24.5	24.1	23.7	22.3	6.5	1.7	
		$\sigma_e/\omega\epsilon_e$					900	330	620	2700	2500	680	
Vaseline	25	ϵ_{er}	2.16	2.16	2.16	2.16	2.16	2.16	2.16		2.16	2.16	
		$\sigma_e/\omega\epsilon_e$	3	2	<2	<1	<1	<3	<4		6.6	10	
	80	ϵ_{er}	2.10	2.10	2.10	2.10	2.10				2.10	2.10	
		$\sigma_e/\omega\epsilon_e$	16	3.6	0.9	<1	<1				9.2	22	

[a] Values of $\sigma_e/\omega\epsilon_e$ are multiplied by 10^4.

Source: A. Von Hippel, *Dielectric Materials and Applications* (Cambridge, Mass.: The MIT Press, 1954).

Appendix

V

Dimensions and Units

A.V-1 *THEORETICAL ANALOGUES AND EXPERIMENTAL MEASUREMENTS*

The primary purpose of the mathematical model of electromagnetism is to calculate theoretical analogues of directly observable measurements—these to be determined by means of suitably arranged experiments. Such an aim evidently can be achieved only if the equations constituting the mathematical structure have been so devised, and the symbols appearing in them have been so defined, that each equation or set of equations may be reduced to a relation between direct theoretical analogues of actual measurements or pointer readings. This does not mean that each, or even a single one, of the symbols appearing in a given equation has a direct experimental analogue. In fact, this usually is not the case. It does mean, however, that every symbol in every equation forming a part of the mathematical model must somewhere and somehow be functionally related to other symbols that do have such analogues. For example, the electric and magnetic vectors **E** and **B** have no experimental analogues as written in the field equations. But through the force equation they are functionally related to deflection variables of the mathematical-mechanical model. And these may be chosen to have direct experimental analogues.

Every measurement expresses how many times an arbitrary scale unit is contained in an observed scale interval that constitutes the measurement. The existence of basic scale units or standards is thus essential to all quantitative determinations. The establishment of standards may be described in terms of two interrelated parts. The first of these involves the selection of a minimum number of "fundamental dimensions" from the symbols of the mathematical model. Such dimensions may themselves have direct experimental analogues, or they may be functionally related to quantities that have such analogues. The second part is concerned with the experimental determination of the pointer readings in terms of which the unit dimension is to be defined. Both the selection of fundamental dimensions and the experimental technique used to specify the units offer a wide variety of possibilities.* In every case, however, the basic criteria should be those of simplicity, convenience, and accuracy. In particular, the functional relationship between a proposed fundamental dimension and a quantity for which direct experimental analogues can be provided should be a simple one. Where possible, it is desirable to choose dimensions which themselves have direct experimental analogues. Furthermore, the experimental technique involved in obtaining the pointer readings in terms of which the unit is to be specified must be simple and easily and accurately reproduced.

It is possible to distinguish among three different methods of establishing fundamental units for a given choice of fundamental dimension.†

1. An arbitrary standard is constructed once and for all, and permanent constancy and invariance are *assumed* for it. Examples of this method are the setting aside of carefully selected pieces of matter such as the standard meter and the standard kilogram.

2. Proportionality constants appearing in equations in the mathematical model that involve the quantity chosen as a fundamental dimension are given arbitrary numerical values. The unit mass might be defined in terms of Newton's law of universal gravitation,

$$f = \frac{km_1 m_2}{r^2}$$

By arbitrarily setting the constant k equal to 1, the definition of unit mass would be that mass which when placed a unit distance from a like mass attracts this with a unit of force. As another example, the unit of temperature might be defined in a similar way in terms of the general gas formula, by first assigning an arbitrary numerical value to the gas constant.

---

*Ernst Weber, "A Proposal to Abolish the Absolute Electrical Units Systems," *Trans. AIEE*, *51* (1932).

†G. Mie, *Handbuch der Experimentalphysik*, Vol. XI/1 (1932), p. 648.

3. Fundamental dimensions that are simply related to direct theoretical analogues of accurate pointer readings are assigned numerical values.

   a. The fundamental dimension of length is theoretically closely related to the wavelength of a red cadmium line, which, in turn, has a direct spectroscopic pointer-reading analogue. The unit of length, the meter, is defined by assigning the numerical value

   $$\lambda_{cad, red} = \frac{1}{1,553,164.13} \text{ meter}$$

   to the wavelength of the cadmium line.

   b. Mechanical force is itself a direct theoretical analogue of the equilibrium pointer reading of a balance. A unit of force may be defined by assigning the value 0.999972 gram to the weight of a cubic centimeter of water at 4°C.

   c. The dimension of temperature is theoretically related to mechanical or electrical changes in various devices, such as the expansion of a gas when heated. The unit of temperature may be defined by assigning the value 100.00 degrees to the temperature interval between the melting and boiling points of water under standard barometric pressure as observed for the calibration of any convenient device.

A brief consideration of these three methods suggests the following criticism. The first method is simple and experimentally convenient. But its accuracy is hardly adequate for high-precision measurements, since it is based on the hope rather than on the knowledge that the sample chosen remains invariant in time. The only way to verify that, in spite of all precautions, it does not change is to provide a dependable way for measuring it from time to time. Such a method, as for example the spectroscopic measurement of the standard meter, then supersedes the standard sample. It is evident also that only a very limited number of dimensions can be assigned even a reasonably permanent unit in this way. The second method is mathematically simple, but it is more unfortunate from the experimental point of view. It actually does not specify any series of experimental operations for determining the pointer-reading analogues of the dimension chosen, it merely imagines an experiment. This is as likely as not inconvenient, inaccurate, and even impossible. The third method is the most satisfactory and the most common. By proper choice of dimension and a careful selection of an experiment which combines a high degree of precision with convenience, a unit may be specified accurately in a way that is both experimentally and theoretically convenient.

If fundamental dimensions are assigned units according to the third method, then the proportionality constant in important relations such as Newton's law of gravitation and the general gas formula are determined by the variables in the relation. By a suitably arranged experiment in which all these variables may be assigned numerical values by substituting pointer readings for theoretical analogues, the constant may be computed as a number with dimensions. This is done, for example, both in Newton's law and in the general gas formula.

A.V-2   THE DIMENSIONAL FORMULATION OF THE ELECTROMATHEMATICAL MODEL

In the mathematical description of the atomic model in terms of continuous functions, six densities were constructed to characterize the average condition of charge and of moving charge. In addition to using the dimensions of length and time, these were defined in terms of one specifically electrical dimension introduced to stand for the property of matter called charge. Thus the mathematical analogue of the atomic model deals with the four dimensions symbolically represented by L, M, T, and Q. The unit names introduced for these dimensions are for length, the meter; for mass, the kilogram; for time, the second; and for charge, the coulomb.

In order to examine the dimensional character of the field and potential equations, these may be written in terms of the quantities involved insofar as these can be assigned. Quantities to which dimensions can be assigned in these equations are

$$\bar{\rho} \approx \frac{Q}{L^3}; \qquad \overline{\rho_m \mathbf{v}} \approx \frac{Q}{L^2 T} \tag{A.V-1}$$

From their definitions, it is easily verified that the vector operators which occur have the following dimensional equivalents:

$$\nabla \approx \frac{1}{L}; \qquad \nabla^2 \approx \frac{1}{L^2} \tag{A.V-2}$$

Since the field equations and the potential equations are analytically equivalent, it is evidently not necessary to examine the dimensional characteristics of both. The relation between the field vectors and the potentials is defined by

$$\mathbf{E} = -\nabla \phi - \dot{\mathbf{A}}$$
$$\mathbf{B} = \nabla \times \mathbf{A} \tag{A.V-3}$$

Consequently, their dimensional interdependence is simply

$$E \approx \frac{\phi}{L}$$

$$B \approx \frac{A}{L} \tag{A.V-4}$$

Because of their symmetrical form, the potential equations (3.4–19) and (3.4–20) are more convenient than the field equations. The three terms in each lead to the following quasi-dimensional forms:

$$\frac{\phi}{L^2} \approx \frac{\phi \epsilon_0 \mu_0}{T^2} \approx \frac{Q/L^3}{\epsilon_0} \tag{A.V-5a}$$

$$\frac{A}{L^2} \approx \frac{A \epsilon_0 \mu_0}{T^2} \approx \frac{Q}{TL^2} \mu_0 \tag{A.V-5b}$$

The potential equation of continuity (3.4–17) has the dimensional form

$$\frac{A}{L} \approx \frac{\phi \epsilon_0 \mu_0}{T} \tag{A.V-5c}$$

In (A.V-5a–c), ϕ, A, ϵ_0, and μ_0 as yet cannot be written dimensionally.

From these relations it is apparent, as was already evident from the potential equations themselves, that the product $(\epsilon_0 \mu_0)$ has the dimensions

$$\epsilon_0 \mu_0 \approx \frac{T^2}{L^2} \tag{A.V-6}$$

This is the square of a reciprocal velocity that has been assigned the symbol c. Thus

$$c^2 = \frac{1}{\mu_0 \epsilon_0} \approx \frac{L^2}{T^2} \tag{A.V-7}$$

The substitution of this dimensional equivalent for $(\epsilon_0 \mu_0)$ in (A.V-5a–c) with a rearrangement of terms, gives

$$\epsilon_0 \approx \frac{1}{\phi} \frac{Q}{L}$$

$$\mu_0 \approx \frac{AT}{Q} \approx \frac{\phi T^2}{QL} \tag{A.V-8}$$

$$A \approx \phi \frac{T}{L}$$

This is a set of three relations that involve four quantities as yet without dimensional equivalents. Since three relations are not sufficient to define four quantities, it is convenient for the present to assign an independent dimension to the scalar potential ϕ. Let its dimensional symbol be V, and its unit name the volt. All purely electrical quantities can now be expressed in terms of combinations of the four dimensions L, T, Q, V and the corresponding units the meter, the second, the coulomb, and the volt. In most electrical problems, the use of the fundamental dimensions L, T, and Q and the auxiliary dimension V leads to simple and useful dimensional forms.

Proceeding to a consideration of the force equation (4.1–5), it follows from (A.V-4) that

$$E \approx \frac{V}{L}; \qquad B \approx \frac{VT}{L^2} \tag{A.V-9}$$

Accordingly, the electromagnetic force F has the following dimensions and units:

$$F \approx \frac{QV}{L} \qquad \frac{\text{coulomb-volts}}{\text{meter}} \tag{A.V-10}$$

The mechanical equivalent of electricity A_e is defined by the force equation. Dimensionally, it must be*

$$A_e = -\frac{F_m}{F} \approx \frac{MLT^{-2}}{QVL^{-1}} \qquad \frac{\text{newtons}}{\text{coulomb-volts per meter}} \tag{A.V-11}$$

or

$$A_e \approx \frac{ML^2T^{-2}}{QV} \qquad \frac{\text{mechanical joules}}{\text{coulomb-volts}} \tag{A.V-12}$$

Here the numerator has the dimension of mechanical energy, the denominator that of electromagnetic energy.

If in the relations (A.V-11) and (A.V-12), V is treated as a fifth independent dimension, A_e is a dimensional constant. Its numerical value is the ratio between mechanical and electrical energy functions; it must be determined experimentally and treated as a fundamental constant.** On the other hand, if V is used as an auxiliary dimension, it may be expressed in terms of the four fundamental dimensions L, T, M, and Q. This is readily done (following method 2 described in Sec. A.V-1) with the assignment of a dimensionless numerical value of unity to the constant A_e.† In this case the fundamental dimensions of V are given by

$$V \approx \frac{ML^2}{QT^2} \tag{A.V-13}$$

and the unit of V, the volt, may be defined directly or indirectly in terms of the kilogram, meter, second, and coulomb.

---

*1 newton = 1 mechanical joule/meter = force to accelerate 1 kilogram 1 meter/second/second.

**This is the system adopted by Mie, "Elektrodynamik," *Handbuch der Experimentalphysik*," Vol. XI/1, (1932).

†This is done in the rationalized practical system (SI units) used in this text.

Unit Nomenclature

Electromagnetic unit		Mechnical unit
coulomb/second	is equivalent to	ampere
coulomb-volt		joule[a]
coulomb-volt/second ⎫		
ampere-volt ⎬		watt[a]
joule/second ⎭		
coulomb/volt		farad
volt-second/coulomb ⎫		
volt/ampere ⎬		ohm
coulomb/volt-second ⎫		
ampere/volt ⎬		siemens
1/ohm ⎭		
volt-second/square meter		weber/m² or tesla
volt-second/coulomb/second ⎫		
volt-second/ampere		
weber/ampere ⎬		henry
tesla-square meter/ampere ⎭		

[a] Note that the unit names joule and watt may be taken to be purely electrical in the same way as the calorie and calorie/second are purely thermal. The relation of the electrical joule to the mechanical joule, or erg $\times$ 10⁷, is then contained in the mechanical equivalent of electricity A_e. If, as in the practical system (SI units) used in this text, A_e is set equal to 1, mechanical and electrical joules are identical.

A.V-3 ELECTROMAGNETIC UNITS AND CONSTANTS

The dimensional formulation of the electromagnetic model outlined in the preceding two sections involves the selection of a fundamental electrical dimension Q for electric charge and an auxiliary electrical dimension V for the scalar potential. The practical unit of charge, the coulomb, and the practical unit of potential, the volt, must be specified in terms of one of the three methods outlined in Sec. A.V-1. If both Q and V are treated as independent dimensions (Mie), the coulomb and the volt may be defined directly in terms of pointer-reading analogues of quantities related to Q and V in specially devised and easily and accurately reproduced experiments. The most convenient method from the experimental point of view involves the definition of the coulomb (e.g., in terms of the electrolysis of silver) and the definition of the volt-second per coulomb or ohm in terms of a standard resistance (such as a suitable column of mercury). The units so defined are the international coulomb and the international volt. With Q and V thus defined, further experiments must be performed to determine numerical values for the fundamental constants μ_0 and ϵ_0 (and for the mechanical equivalent of electricity A_e if this is not set equal to 1).

In the practical system of units used in this text (SI units), the coulomb but not the volt is treated as an independent unit. In this system, the dimensionless value of unity is assigned to the mechanical equivalent of electricity A_e, so that any experiment performed to determine A_e may be used to define the product QV in mechanical joules (which are here identical with electrical joules). Similarly, the experiment defining VT/Q in terms of the standard ohm may be used to define the ratio V/Q. With the coulomb-volt and volt per coulomb uniquely specified, two equations in two unknowns are available to define the coulomb and the volt in terms of the ohm and the mechanical joule and, hence, in terms of the meter, kilogram, ohm, and second. The fundamental constants μ_0 and ϵ_0 are determined experimentally, and from them the characteristic velocity $c = 1/\sqrt{\mu_0\epsilon_0}$ and the characteristic resistance $\zeta_0 = \sqrt{\mu_0/\epsilon_0}$ may be calculated. Or, c and one of μ_0 and ϵ_0 may be measured experimentally.

Based on the definition of the international coulomb and volt, Mie* has calculated the following as the best numerical values in the rationalized system:

$$\epsilon_0 = 0.8859 \times 10^{-11} \text{ farad/meter}$$

$$\mu_0 = 1.25598 \times 10^{-6} \text{ henry/meter}$$

$$c = \frac{1}{\sqrt{\mu_0\epsilon_0}} = 2.9979 \times 10^8 \text{ meters/second} \doteq 3 \times 10^8 \text{ meters/second}$$

$$A_e = 1.00043 \text{ mechanical joules/electrical joule}$$

$$\frac{A_q}{A_e} = 4.1842 \text{ electrical joules/calorie}$$

In the rationalized practical system, slightly different values obtain because A_e is by definition unity so that the product QV is slightly changed. The values are

$$\epsilon_0 = \frac{10^7}{4\pi c^2} = 0.8854 \times 10^{-11} \text{ farad/meter}$$

$$\mu_0 = 4\pi \times 10^{-7} = 1.257 \times 10^{-6} \text{ henry/meter}$$

$$c = \frac{1}{\sqrt{\mu_0\epsilon_0}} = 2.9979 \times 10^8 \text{ meters/second} \doteq 3 \times 10^8 \text{ meters/second}$$

$$\zeta_0 = \sqrt{\frac{\mu_0}{\epsilon_0}} = 376.7 \text{ ohms}$$

$$A_e = 1$$

$$\frac{A_q}{A_e} = A_q = 4.186 \text{ joules/calorie}$$

*"Electrodynamik," *Handbuch der Experimentalphysik,* Vol. XI/1 (1932), p. 484.

Table of Dimensions and Units

Electric quantities[a]

Symbol	Quantity	Dimensions	Units
ρ	volume density of charge	$\dfrac{Q}{L^3}$	$\dfrac{\text{coulomb}}{\text{meter}^3}$
$\bar{\rho}$	essential volume characteristic of charge		
η	surface density of charge	$\dfrac{Q}{L^2}$	$\dfrac{\text{coulomb}}{\text{meter}^2}$
$\bar{\eta}$	essential surface characteristic of charge		
$\mathbf{P}$	volume density of polarization	$\dfrac{Q}{L^2}$	$\dfrac{\text{coulomb}}{\text{meter}^2}$
$\mathbf{D}$	auxiliary electric vector		
ϕ	scalar potential	V	volt
$\mathbf{\Pi}_e$	polarization potential	VL	volt-meter
$\mathbf{E}$	electric vector	$\dfrac{V}{L}$	$\dfrac{\text{volt}}{\text{meter}}$
U_e	electric energy function	QV	joule
ϵ_0	universal electric constant	$\dfrac{Q}{VL}$	$\dfrac{\text{farad}}{\text{meter}}$
χ_e	electric susceptibility	numeric	
ϵ_{er}	relative dielectric constant		

Magnetic quantities

Symbol	Quantity	Dimensions	Units
$\mathbf{J}$	volume density of moving charge	$\dfrac{Q}{L^2T}$	$\dfrac{\text{ampere}}{\text{meter}^2}$
$\rho_m \mathbf{v}$	essential volume characteristic of moving charge		
$\mathbf{K}$	surface density of moving charge	$\dfrac{Q}{LT}$	$\dfrac{\text{ampere}}{\text{meter}}$
$\eta_m \mathbf{v}$	essential surface characteristic of moving charge		
$\mathbf{M}$	volume density of magnetization	$\dfrac{Q}{LT}$	$\dfrac{\text{ampere}}{\text{meter}}$
$\mathbf{H}$	auxiliary magnetic vector		
$\mathbf{A}$	vector potential	$\dfrac{VT}{L}$	$\dfrac{\text{volt-second}}{\text{meter}}$
$\mathbf{\Pi}_m$	magnetization potential	VT	volt-second
$\mathbf{B}$	magnetic vector	$\dfrac{VT}{L^2}$	$\dfrac{\text{volt-second}}{\text{meter}^2}$
U_m	magnetic energy function	QV	joule
μ_0	universal magnetic constant	$\dfrac{VT^2}{QL}$	$\dfrac{\text{henry}}{\text{meter}}$
χ_m	magnetic susceptibility	numeric	
μ_r	relative permeability		

Electromagnetic quantities

$\mathbf{F}$	electromagnetic force function	$\left\{\begin{array}{c}\dfrac{QV}{L}\\[4pt]\dfrac{ML}{T^2}\end{array}\right.$	$\dfrac{\text{joule}}{\text{meter}}$ newton when $A_e = 1$ as in this book
$\mathbf{T}$	electromagnetic torque	$\dfrac{ML^2}{T^2}$	newton-meter when $A_e = 1$
σ	conductivity	$\dfrac{Q}{VTL}$	$\dfrac{\text{siemens}}{\text{meter}}$
T	electromagnetic energy transfer function	$\dfrac{QV}{T}$	watt
$\mathbf{S}$	Poynting vector	$\dfrac{QV}{L^2T}$	$\dfrac{\text{watt}}{\text{meter}^2}$
c	characteristic velocity of propagation	$\dfrac{L}{T}$	$\dfrac{\text{meter}}{\text{second}}$
ζ_0	characteristic resistance	$\dfrac{VT}{Q}$	ohm
A_e	mechanical equivalent of electricity	$\dfrac{ML^2T^{-2}}{QV}$	$\dfrac{\text{mechanical joule}}{\text{electrical joule}} = 1$ in the practical system used in this book (SI units)

[a] The dimensions of an electric quantity are changed to those of the analogous magnetic quantity by multiplying Q by L/T and V by T/L. The dimensions of a magnetic quantity are changed to those of the analogous electric quantity by multiplying Q by T/L and V by L/T.

Some writers using the practical system feel that if the ohm is established as a fundamental unit, resistance should be introduced with a symbol R in all dimensional formulas instead of Q. However, resistance is not a fundamental concept as is electric charge, and its use together with M, L, and T leads to intricate dimensional formulas such as

$$Q \approx M^{1/2}LR^{-1/2}T^{-1/2}; \qquad V \approx M^{1/2}LR^{1/2}T^{-3/2}$$

Since dimensional analysis in no way requires that the dimensions of experimentally convenient fundamental units be used in preference to theoretically more fundamental quantities, it is in any case desirable to retain Q rather than R as a fourth fundamental dimension and V as an often convenient auxiliary dimension. Fractional powers do not appear in the simple dimensional formulas using Q or Q and V. The accompanying table of dimensions is written in terms of Q and V. Whenever required, the relation (A.V-13) that expresses V in terms of M, L, T, and Q may be introduced.

The practical system of units (SI units) described above and used throughout this text is not the only system in common use. Although other systems may have certain advantages in exclusively theoretical work, especially for those who are accustomed to them, only the practical system has the *unique and compelling characteristic that it alone is entirely adequate for both theoretical and experimental work*, so that confusing conversion tables are not required. *One system is always simpler than two* when, as in applied physics and engineering, theory and experiment must work in close cooperation. It is the responsibility of writers using systems other than the practical to convert their results to the practical system or at least to provide adequate conversion tables.

Appendix

Evaluation of the Field of a Dipole Antenna With a Sinusoidal Current Distribution

The leading term in the distribution of current in a very thin antenna is given by

$$I_z(z) = I_z(0) \frac{\sin k_0(h - |z|)}{\sin k_0 h} \tag{A.VI-1}$$

where $I_z(0)$ is the input current. The associated vector potential is

$$\mathbf{A} = \hat{z} A_z = \hat{z} \frac{\mu_0}{4\pi} \int_{-h}^{h} I_z(z') \frac{e^{-jk_0 R}}{R} dz' \tag{A.VI-2}$$

With (A.VI-1) this becomes

$$A_z = \frac{\mu_0 I_z(0)}{4\pi} [\sin k_0 h \, C_{k\rho}(h, z) - \cos k_0 h \, S_{k\rho}(h, z)] \tag{A.VI-3}$$

where

$$
\begin{aligned}
C_{k\rho}(h, z) &= \int_{-h}^{h} \cos k_0 z' \frac{e^{-jk_0 R}}{R} dz' \\
&= \int_{0}^{h} \cos k_0 z' \left(\frac{e^{-jk_0 R_1}}{R_1} + \frac{e^{-jk_0 R_2}}{R_2} \right) dz' \tag{A.VI-4}
\end{aligned}
$$

$$
\begin{aligned}
S_{k\rho}(h, z) &= \int_{-h}^{h} \sin k_0 |z'| \frac{e^{-jk_0 R}}{R} dz' \\
&= \int_{0}^{h} \sin k_0 z' \left(\frac{e^{-jk_0 R_1}}{R_1} + \frac{e^{-jk_0 R_2}}{R_2} \right) dz' \tag{A.VI-5}
\end{aligned}
$$

In these integrals

$$R = R_1 = [(z - z')^2 + \rho^2]^{1/2}; \qquad R_2 = [(z + z')^2 + \rho^2]^{1/2} \qquad \text{(A.VI–6)}$$

where ρ is the radial coordinate measured from the z axis.

The magnetic field is evaluated from

$$\mathbf{B(r)} = \nabla \times \mathbf{A(r)} \qquad \text{(A.VI–7a)}$$

When $\mathbf{A(r)} = \hat{z}A_z(\mathbf{r})$,

$$\mathbf{B(r)} = \hat{\boldsymbol{\theta}}B_\theta(\mathbf{r}) = -\hat{\boldsymbol{\theta}}\frac{\partial A_z(\mathbf{r})}{\partial \rho} \qquad \text{(A.VI–7b)}$$

With (A.VI–3),

$$B_\theta(\mathbf{r}) = \frac{-\mu_0 I(0)}{4\pi \sin k_0 h}\left[\sin k_0 h \frac{\partial C_{k\rho}(h,\, z)}{\partial \rho} - \cos k_0 h \frac{\partial S_{k\rho}(h,\, z)}{\partial \rho}\right] \qquad \text{(A.VI–8)}$$

The differentiation of the functions $C_{k\rho}(h,\, z)$ and $S_{k\rho}(h,\, z)$ as defined in (A.VI–4) and (A.VI–5) may be carried out as follows. The exponential integral

$$\text{Ei}(v) = \int_\infty^{-v} \frac{e^{-u}}{u}\, du \qquad \text{(A.VI–9)}$$

can be differentiated with respect to a parameter ρ contained in v as follows:

$$\begin{aligned}\frac{\partial}{\partial \rho}\,\text{Ei}(v) &= \frac{\partial}{\partial \rho}\int_\infty^{-v} \frac{e^{-u}}{u}\, du \\ &= \left(\frac{\partial}{\partial v}\int_\infty^{-v} \frac{e^{-u}}{u}\, du\right)\frac{dv}{d\rho}\end{aligned} \qquad \text{(A.VI–10)}$$

Since u is an independent variable, the differentiation of the definite integral reduces to a single term obtained by substituting the upper limit in the integrand. Thus,

$$\frac{\partial}{\partial \rho}\,\text{Ei}(v) = \frac{e^v}{v}\frac{dv}{d\rho} \qquad \text{(A.VI–11)}$$

The functions $C_{k\rho}(h,\, z)$ and $S_{k\rho}(h,\, z)$ may be expressed in the form

$$C_{k\rho}(h,\, z) = \frac{1}{2}\,(I_1 + I_2 + I_3 + I_4) \qquad \text{(A.VI–12)}$$

$$S_{k\rho}(h,\, z) = \frac{-j}{2}\,(I_1 - I_2 + I_3 - I_4) \qquad \text{(A.VI–13)}$$

where

$$\begin{aligned}I_1 &= e^{jk_0 z}\int_0^h e^{jk_0(z'-z)}\frac{e^{-jk_0 R_1}}{R_1}\, dz' \\ &= e^{jU_0}\int_{-U_0}^{U_1} e^{jU}\frac{e^{-jW}}{W}\, dU\end{aligned} \qquad \text{(A.VI–14a)}$$

$$I_2 = e^{-jk_0 z} \int_0^h e^{-jk_0(z'-z)} \frac{e^{-jk_0 R_1}}{R_1} \, dz'$$

$$= e^{-jU_0} \int_{-U_0}^{U_1} e^{-jU} \frac{e^{-jW}}{W} \, dU \qquad \text{(A.VI–14b)}$$

$$I_3 = e^{-jk_0 z} \int_0^h e^{jk_0(z'+z)} \frac{e^{-jk_0 R_2}}{R_2} \, dz'$$

$$= e^{-jU_0} \int_{U_0}^{U_2} e^{jU} \frac{e^{-jW}}{W} \, dU \qquad \text{(A.VI–14c)}$$

$$I_4 = e^{jk_0 z} \int_0^h e^{-jk_0(z'+z)} \frac{e^{-jk_0 R_2}}{R_2} \, dz'$$

$$= e^{jU_0} \int_{U_0}^{U_2} e^{-jU} \frac{e^{-jW}}{W} \, dU \qquad \text{(A.VI–14d)}$$

where

$$W = (U^2 + A^2)^{1/2}; \qquad A = k_0 \rho \qquad \text{(A.VI–15a)}$$

$$U = k_0 u = \begin{cases} k_0(z' - z); & \text{in } I_1 \text{ and } I_2 \\ k_0(z' + z); & \text{in } I_3 \text{ and } I_4 \end{cases} \qquad \text{(A.VI–15b)}$$

$$U_0 = k_0 z; \qquad U_1 = k_0 u_1 = k_0(h - z)$$

$$U_2 = k_0 u_2 = k_0(h + z) \qquad \text{(A.VI–15c)}$$

The following expression is obtained:

$$\frac{\partial C_{k\rho}(h, z)}{\partial \rho} = \frac{1}{2} e^{jk_0 z} \left(- \frac{e^{-jk_0(R_{2h} + u_2)}}{R_{2h} + u_2} \frac{\rho}{R_{2h}} \right.$$

$$+ \frac{e^{-jk_0(R_{1h} - u_1)}}{R_{1h} - u_1} \frac{\rho}{R_{1h}} \left. \right)$$

$$+ \frac{1}{2} e^{-jk_0 z} \left(- \frac{e^{-jk_0(R_{1h} + u_1)}}{R_{1h} + u_1} \frac{\rho}{R_{1h}} \right.$$

$$+ \frac{e^{-jk_0(R_{2h} - u_2)}}{R_{2h} - u_2} \frac{\rho}{R_{2h}} \left. \right) \qquad \text{(A.VI–16)}$$

Combining terms gives

$$\frac{\partial C_{k\rho}(h, z)}{\partial \rho} = \frac{\rho}{2} \left[- \frac{e^{-jk_0(R_{2h} + h)}}{R_{2h}(R_{2h} + u_2)} + \frac{e^{-jk_0(R_{1h} - h)}}{R_{1h}(R_{1h} - u_1)} \right.$$

$$\left. - \frac{e^{-jk_0(R_{1h} + h)}}{R_{1h}(R_{1h} + u_1)} + \frac{e^{-jk_0(R_{2h} - h)}}{R_{2h}(R_{2h} - u_2)} \right] \qquad \text{(A.VI–17)}$$

$$R_{1h} = (u_1^2 + \rho^2)^{1/2}; \qquad R_{2h} = (u_2^2 + \rho^2)^{1/2}$$

Since

$$(R_{2h} + u_2)(R_{2h} - u_2) = R_{2h}^2 - u_2^2$$

$$= u_2^2 + \rho^2 - u_2^2 = \rho^2 \qquad \text{(A.VI–18a)}$$

and similarly

$$(R_{1h} + u_1)(R_{1h} - u_1) = \rho^2 \qquad \text{(A.VI–18b)}$$

it is possible by reducing to a common denominator, to obtain

$$\frac{\partial C_{k\rho}(h, z)}{\partial \rho} = \frac{1}{2\rho} \left\{ \frac{e^{-jk_0 R_{2h}}}{R_{2h}} [(R_{2h} + u_2)e^{jk_0 h} - (R_{2h} - u_2)e^{-jk_0 h}] \right.$$

$$\left. + \frac{e^{-jk_0 R_{1h}}}{R_{1h}} [(R_{1h} + u_1)e^{jk_0 h} - (R_{1h} - u_1)e^{-jk_0 h}] \right\}$$

$$= \frac{e^{-jk_0 R_{2h}}}{\rho} \left(\frac{u_2 \cos k_0 h}{R_{2h}} + j \sin k_0 h \right)$$

$$+ \frac{e^{-jk_0 R_{1h}}}{\rho} \left(\frac{u_1 \cos k_0 h}{R_{1h}} + j \sin k_0 h \right)$$

$$= \frac{\cos k_0 h}{\rho} \left(\frac{u_2 e^{-jk_0 R_{2h}}}{R_{2h}} + \frac{u_1 e^{-jk_0 R_{1h}}}{R_{1h}} \right)$$

$$+ j \frac{\sin k_0 h}{\rho} (e^{-jk_0 R_{2h}} + e^{-jk_0 R_{1h}}) \qquad \text{(A.VI–19)}$$

Similarly,

$$\frac{\partial S_{k\rho}(h, z)}{\partial \rho} = \frac{j}{2} e^{jk_0 z} \left[- \frac{e^{-jk_0(R_{2h} + u_2)}}{R_{2h} + u_2} \frac{\rho}{R_{2h}} \right.$$

$$\left. - \frac{e^{-jk_0(R_{1h} - u_1)}}{R_{1h} - u_1} \frac{\rho}{R_{1h}} + \frac{2e^{-jk_0(r + z)}}{r + z} \frac{\rho}{r} \right]$$

$$+ \frac{j}{2} e^{-jk_0 z} \left[- \frac{e^{-jk_0(R_{1h} + u_1)}}{R_{1h} + u_1} \frac{\rho}{R_{1h}} \right.$$

$$\left. - \frac{e^{-jk_0(R_{2h} - u_2)}}{R_{2h} - u_2} \frac{\rho}{R_{2h}} + \frac{2e^{-jk_0(r - z)}}{r - z} \frac{\rho}{r} \right] \qquad \text{(A.VI–20)}$$

Collecting terms gives

$$
\frac{\partial S_{k\rho}(h, z)}{\partial \rho} = \frac{j}{2\rho} \left\{ \frac{e^{-jk_0 R_{2h}}}{R_{2h}} \left[(R_{2h} + u_2) e^{jk_0 h} \right. \right.
$$
$$
\left. + (R_{2h} - u_2) e^{-jk_0 h} \right]
$$
$$
- \frac{e^{-jk_0 R_{1h}}}{R_{1h}} \left[(R_{1h} + u_1) e^{jk_0 h} \right.
$$
$$
\left. \left. + (R_{1h} - u_1) e^{-jk_0 h} \right] + 4 e^{-jk_0 r} \right\}
$$
$$
= -j \frac{\cos k_0 h}{\rho} \left(e^{-jk_0 R_{2h}} + e^{-jk_0 R_{1h}} \right)
$$
$$
+ \frac{\sin k_0 h}{\rho} \left(\frac{u_2 e^{-jk_0 R_{2h}}}{R_{2h}} + \frac{u_1 e^{-jk_0 R_{1h}}}{R_{1h}} \right.
$$
$$
\left. + j \frac{2}{\rho} e^{-jk_0 r} \right) \tag{A.VI–21}
$$

The combination of (A.VI–19) and (A.VI–21) to form (A.VI–8) leads to the following simple result:

$$
B_\theta(\mathbf{r}) = \frac{j \mu_0 I_z(0)}{4\pi\rho} \left(e^{-jk_0 R_{1h}} + e^{-jk_0 R_{2h}} - 2 \cos k_0 h \, e^{-jk_0 r} \right) \tag{A.VI–22}
$$

where

$$
R_{2h} = [(h + z)^2 + \rho^2]^{1/2}; \qquad R_{1h} = [(h - z)^2 + \rho^2]^{1/2}
$$
$$
r = (z^2 + \rho^2)^{1/2} \tag{A.VI–23}
$$

The electric field associated with the magnetic field in (A.VI–22) is most easily obtained from the relation $\nabla \times \mathbf{B}(\mathbf{r}) = j\omega\mu_0\epsilon_0\mathbf{E}(\mathbf{r})$. Since $\mathbf{B}(\mathbf{r}) = \hat{\theta} B_\theta(\mathbf{r})$, it follows that the components of the electric field are

$$
E_\rho(\mathbf{r}) = -\frac{j\omega}{k_0^2} [\nabla \times \mathbf{B}(\mathbf{r})]_\rho \tag{A.VI–24}
$$

$$
E_\theta(\mathbf{r}) = -\frac{j\omega}{k_0^2} [\nabla \times \mathbf{B}(\mathbf{r})]_\theta = 0 \tag{A.VI–25}
$$

$$
E_z(\mathbf{r}) = -\frac{j\omega}{k_0^2} [\nabla \times \mathbf{B}(\mathbf{r})]_z \tag{A.VI–26}
$$

$$
[\nabla \times \mathbf{B}(\mathbf{r})]_\rho = -\frac{1}{\rho} \frac{\partial}{\partial z} (\rho B_\theta); \qquad [\nabla \times \mathbf{B}(\mathbf{r})]_z = \frac{1}{\rho} \frac{\partial}{\partial \rho} (\rho B_\theta) \tag{A.VI–27}
$$

From (A.VI–22),

$$(\rho B_\theta) = \frac{j\mu_0 I_z(0)}{4\pi} (e^{-jk_0R_{1h}} + e^{-jk_0R_{2h}} - 2\cos k_0h\, e^{-jk_0r}) \tag{A.VI–28}$$

It may be noted that

$$\frac{\partial R_{2h}}{\partial z} = \frac{z+h}{R_{2h}}; \qquad \frac{\partial R_{1h}}{\partial z} = \frac{z-h}{R_{1h}}; \qquad \frac{\partial r}{\partial z} = \frac{z}{r} \tag{A.VI–29}$$

$$\frac{\partial R_{2h}}{\partial \rho} = \frac{\rho}{R_{2h}}; \qquad \frac{\partial R_{1h}}{\partial \rho} = \frac{\rho}{R_{1h}}; \qquad \frac{\partial r}{\partial \rho} = \frac{\rho}{r} \tag{A.VI–30}$$

With (A.VI–29) and (A.VI–30),

$$\frac{\partial}{\partial z}(\rho B_\theta) = \frac{\mu_0 k_0 I_z(0)}{4\pi} \left(\frac{z+h}{R_{2h}} e^{-jk_0R_{2h}} + \frac{z-h}{R_{1h}} e^{-jk_0R_{1h}} \right.$$
$$\left. - \frac{2z}{r}\cos k_0h\, e^{-jk_0r} \right) \tag{A.VI–31}$$

$$\frac{\partial}{\partial \rho}(\rho B_\theta) = \frac{\mu_0 k_0 I_z(0)}{4\pi} \left(\frac{e^{-jk_0R_{2h}}}{R_{2h}} + \frac{e^{-jk_0R_{1h}}}{R_{1h}} \right.$$
$$\left. - \frac{2}{r}\cos k_0h\, e^{-jk_0r} \right) \tag{A.VI–32}$$

It follows that

$$E_\rho(\mathbf{r}) = \frac{j\omega\mu_0 I_z(0)}{4\pi k_0 \rho} \left(\frac{z-h}{R_{1h}} e^{-jk_0R_{1h}} + \frac{z+h}{R_{2h}} e^{-jk_0R_{2h}} \right. \tag{A.VI–33}$$
$$\left. - \frac{2z}{r}\cos k_0h\, e^{-jk_0r} \right)$$

$$E_\theta(\mathbf{r}) = 0 \tag{A.VI–34}$$

$$E_z(\mathbf{r}) = \frac{-j\omega\mu_0 I_z(0)}{4\pi k_0} \left(\frac{e^{-jk_0R_{1h}}}{R_{1h}} + \frac{e^{-jk_0R_{2h}}}{R_{2h}} \right. \tag{A.VI–35}$$
$$\left. - \frac{2}{r}\cos k_0h\, e^{-jk_0r} \right)$$

Bibliography

A selected list of books is given for supplementary reading and for more elementary as well as more advanced study.

ELECTROMAGNETIC THEORY

ABRAHAM, M., AND R. BECKER, *The Classical Theory of Electricity and Magnetism.* Glasgow: Blackie & Son Ltd., 1932.

FRANK, N. H., *Introduction to Electricity and Optics.* New York: McGraw-Hill Book Company, 1950.

JACKSON, J. D., *Classical Electrodynamics.* New York: John Wiley & Sons, Inc., 1975.

JEANS, J. H., *The Mathematical Theory of Electricity and Magnetism.* Cambridge: Cambridge University Press, 1933.

JORDAN, E. C., AND K. BALMAIN, *Electromagnetic Waves and Radiating Systems.* Englewood Cliffs, N.J.: Prentice-Hall, Inc., 1950.

KING, R. W. P., AND C. W. HARRISON, JR., *Antennas and Waves.* Cambridge, Mass.: The MIT Press, 1969.

MASON, M., AND W. WEAVER, *The Electromagnetic Field.* New York: Dover Publications, Inc., 1929.

MAXWELL, J. C., *Electricity and Magnetism.* New York: Dover Publications, Inc., 1891.

PANOFSKY, W. K. H., AND M. PHILLIPS, *Classical Electricity and Magnetism.* Reading, Mass.: Addison-Wesley Publishing Company, Inc., 1955.

PLONSEY, R., AND R. E. COLLIN, *Principles and Applications of Electromagnetic Fields*. New York: McGraw-Hill Book Company, 1961.

SCHELKUNOFF, S., *Electromagnetic Waves*. New York: Van Nostrand Reinhold Company, Inc., 1943.

SOMMERFELD, A., *Electrodynamics*. New York: Academic Press, Inc., 1952.

STRATTON, J. A., *Electromagnetic Theory*. New York: McGraw-Hill Book Company, 1941.

TRANSMISSION LINES, WAVEGUIDES, SCATTERING AND ANTENNAS

COLLIN, R. E., *Foundations for Microwave Engineering*. New York: McGraw-Hill Book Company, 1966.

FLUGGE, S., ed., *Handbuch der Physik*, Vol. XVI. Berlin: Springer-Verlag, 1958.

GANDHI, O. P., *Principles of Microwave Circuits*. Elmsford, N.Y.: Pergamon Press, Inc., 1983.

HAUS, H. A., *Waves and Fields in Optoelectronics*. Englewood Cliffs, N.J.: Prentice-Hall, Inc., 1984.

KING, R. W. P., *Theory of Linear Antennas*. Cambridge, Mass.: Harvard University Press, 1956.

KING, R. W. P., *Transmission-Line Theory*. New York: Dover Publications, Inc., 1965.

KING, R. W. P., R. B. MACK, AND S. S. SANDLER, *Arrays of Cylindrical Dipoles*. New York: Cambridge University Press, 1968.

KING, R. W. P., H. R. MIMNO, AND A. H. WING, *Transmission Lines, Antennas and Waveguides*. New York: Dover Publications, Inc., 1965.

KING, R. W. P., AND G. S. SMITH, *Antennas in Matter*. Cambridge, Mass.: The MIT Press, 1981.

KING, R. W. P., AND T. T. WU, *The Scattering and Diffraction of Waves*. Cambridge, Mass.: Harvard University Press, 1959.

LIAO, S. Y., *Microwave Devices and Circuits*. Englewood Cliffs, N.J.: Prentice-Hall, Inc., 1985.

ENGINEERING MATHEMATICS

ARFKEN, G., *Mathematical Methods of Physicists*. New York: Academic Press, Inc., 1970.

SOKOLNIKOFF, I. S., AND R. M. REDHEFFER, *Mathematics of Physics and Modern Engineering*. New York: McGraw-Hill Book Company, 1958.

WYLIE, C. R., JR., *Advanced Engineering Mathematics*. New York: McGraw-Hill Book Company, 1975.

VECTOR ANALYSIS

HINCHEY, F. A., *Vectors and Tensors for Engineers and Scientists*. New York: Halstead Press, 1976.

LASS, H., *Vector and Tensor Analysis*. New York: McGraw-Hill Book Company, 1950.

SHERCLIFF, J. A., *Vector Fields*. New York: Cambridge University Press, 1977.

BESSEL FUNCTIONS

McLACHLAN, N. W., *Bessel Functions for Engineers*. New York: Oxford University Press, Inc., 1934.

DIFFERENTIAL AND INTEGRAL EQUATIONS

FRANK, P., AND R. VON MISES, *Die Differentialgleichungen der Physik*. Vol. I, New York: Mary S. Rosenberg, 1973.

INCE, E. L., *Ordinary Differential Equations*. New York: Dover Publications, Inc., 1953.

LOVITT, W. V., *Linear Integral Equations*. New York: Dover Publications, Inc., 1950.

FOURIER TRANSFORMS AND GREEN'S FUNCTIONS

ARSAC, J., *Fourier Transforms and Theory of Distributions*. Englewood Cliffs, N.J.: Prentice-Hall, Inc., 1966.

SNEDDON, I. N., *The Use of Integral Transforms*. New York: McGraw-Hill Book Company, 1972.

STAKGOLD, I., *Green's Functions and Boundary Value Problems*. New York: Wiley-Interscience, 1979.

TABLES

ABRAMOWITZ, M., AND I. A. STEGUN, *Handbook of Mathematics Functions*. New York: Dover Publications, Inc., 1965.

JAHNKE, E., AND F. EMDE, *Tables of Functions*. New York: Dover Publications, Inc., 1945.

KING, R. W. P., *Tables of Antenna Characteristics*. New York: Plenum Press, 1971.

PIERCE, B. O., *Tables of Integrals*. Boston: Ginn and Co., 1956.

Index
of Principal Symbols

Symbol	Name	Units	Page where defined
A	Vector potential	$\dfrac{\text{Volt-second}}{\text{meter}}$	153
a	Radius of conductor	meter	74
a_e	Semi-major axis of ellipse	meter	302
a_h	Semi-minor axis of ellipse	meter	302
B	Magnetic vector (Magnetic flux density)	$\dfrac{\text{volt-second}}{\text{square meter}}, \dfrac{\text{weber}}{\text{square meter}}$	87
b	Radius of insulator	meter	167
C	Typical vector		153
C_o	Capacitance of capacitor	farad	350
C	Capacitance per unit length of transmission line	$\dfrac{\text{farad}}{\text{meter}}$	399
c	Velocity of light	$\dfrac{\text{meter}}{\text{second}}$	96
D	Auxiliary electric vector (Displacement)	$\dfrac{\text{coulomb}}{\text{square meter}}$	95
D	Density of mass	$\dfrac{\text{kilogram}}{\text{meter}}$	6
D	Directivity		240
d_c	Mean free path	meter	8

d_s	Skin depth	meter	129
$\mathbf{E}$	Electric vector	$\dfrac{\text{volt}}{\text{meter}}$	87
$\mathbf{E}_e$	Impressed electric field	$\dfrac{\text{volt}}{\text{meter}}$	125
E	Elliptic integral of second kind		354
e	Electric charge	coulomb	2
$\mathbf{F}$	Electromagnetic force	newton	193
$\mathbf{F}_M$	Mechanical force	newton	193
F	Fresnel integral		269
f	Frequency	hertz	
f	Function		127
f_{er}, f_{mn}	Reflection coefficients		525
f_{et}, f_{mt}	Transmission coefficients		525
G	Gain of antenna		240
g	Leakage conductance per unit length of transmission line, of waveguide	$\dfrac{\text{siemens}}{\text{m}}$	399, 455
g	Function		127
$\mathbf{H}$	Auxiliary magnetic vector (magnetic field)	$\dfrac{\text{ampere}}{\text{meter}}$	95
h	Half-length of dipole antenna	meter	287
I	Total current in conductor	ampere	76
I_E	Modified electric Hertz potential	ampere	451
I_M	Current function in waveguides	ampere	452
$\mathbf{J}$	Volume density of current	$\dfrac{\text{ampere}}{\text{square meter}}$	36
$\mathbf{K}$	Surface density of current	$\dfrac{\text{ampere}}{\text{meter}}$	38
K	Elliptic integral of first kind		354
K_p	Space radiation function		239
$\mathscr{K}$	Kernel in integral equation		185
$\mathbf{k}$	Surface density of polarization	$\dfrac{\text{coulomb}}{\text{meter}}$	32
k	Wave number in simple medium	$\dfrac{1}{\text{meter}}$	127
k_o	Wave number in free space	$\dfrac{1}{\text{meter}}$	126
k_e, k_k	Spheroidal coordinates	dimensionless	303
k_L	Wave number of insulated antenna	$\dfrac{1}{\text{meter}}$	434
L_1	Self inductance of circuit 1	henry	342
L_{12}	Mutual inductance of circuits 1 and 2	henry	342
L	Dimensional symbol of length		1

Symbol	Name	Units	Page where defined
$\varepsilon_r', \varepsilon_r''$	Real and imaginary parts of $\varepsilon_r = \varepsilon_r' - j\varepsilon_r''$		122
ζ	Characteristic impedance	ohm	
ζ_c	Real effective characteristic impedance of simple medium	ohm	128
ζ_o	Characteristic impedance of free space	ohm	96
η	Surface density of charge	$\dfrac{\text{coulomb}}{\text{square meter}}$	12
η	Coordinate in Fourier transformed space		534
$\bar{\eta}$	Essential surface density of charge	$\dfrac{\text{coulomb}}{\text{square meter}}$	29
$\overline{\eta_{m-}^v}$	Essential surface density of moving charge	$\dfrac{\text{ampere}}{\text{meter}}$	74
θ	Spherical coordinate	radian, degree	
$\hat{\theta}$	Unit vector		
θ	Cylindrical coordinate	radian, degree	
$\hat{\theta}$	Unit vector		
λ	Wavelength in simple medium	meter	
λ_a	Guide wavelength in waveguide	meter	460
λ_E, λ_M	Wavelengths of electric and magnetic types in waveguide	meter	459
λ_o	Wavelength in free space (air)	meter	126
μ	Absolute permeability of simple medium	$\dfrac{\text{henry}}{\text{meter}}$	109
μ_r	Relative permeability of simple medium		109
μ_o	Absolute permittivity of free space	$\dfrac{\text{henry}}{\text{meter}}$	89
μ_r', μ_r''	Real and imaginary parts of $\mu_r = \mu_r' - j\mu_r''$		122
ξ	Coordinate in Fourier transformed space		534
π_E, π_M	Hertz potentials, electric and magnetic	volt seconds; volt meter	449
ρ	Volume density of charge	$\dfrac{\text{coulomb}}{\text{cubic meter}}$	6
ρ	Cylindrical coordinate	meter	
ρ_f	Volume density of free charge	$\dfrac{\text{coulomb}}{\text{meter}}$	
ρ_o	Spot size in optical fiber	square meters	514
$\overline{\rho_{m-}^v}$	Essential volume density of moving charge	$\dfrac{\text{ampere}}{\text{square meter}}$	74

Symbol	Name	Units	Page where defined
Σ, σ	Surface	square meter	23
σ	Conductivity	$\dfrac{\text{siemens}}{\text{meter}}$	109
σ_e	Real effective conductivity of simple medium	$\dfrac{\text{siemens}}{\text{meter}}$	123
σ', σ''	Real and imaginary parts of $\sigma = \sigma' - j\sigma''$	$\dfrac{\text{siemens}}{\text{meter}}$	122
τ	Volume	cubic meter	7
τ_a	Cut off parameter in waveguide		460
Φ	Spherical coordinate	radian, degree	
$\hat{\Phi}$	Unit vector		
ϕ	Scalar potential	volt	153
χ_s	Extinction coefficient		127
χ_e, χ_m	Electric, magnetic susceptibility		109
ψ	Expansion function for antenna		290
ψ_E, ψ_M	Mode function of electric and magnetic types in waveguides		454
ψ_T	Mode function of T-mode in waveguide		468
ω	Angular velocity	$\dfrac{\text{radian}}{\text{second}}$	118

Index